Astronomical Tables of the Sun, Moon, and Planets

Astronomical Tables of the Sun, Moon, and Planets

Jean Meeus
Vereniging voor Sterrenkunde Belgium

Foreword by
Robert C. Victor
Planet Columnist for **Sky and Telescope** Magazine
Astronomer, Abrams Planetarium, Michigan State University

Published by:

Willmann-Bell, Inc.
P. O. Box 3125
Richmond, Virginia 23235 ☎ (804)
United States of America 320-7016

Serving Astronomers Worldwide

Since 1973

Library of Congress Cataloging in Publication Data

Meeus, Jean.
  Astronomical tables of the sun, moon, and planets.

  Bibliography: p.
  Includes index.
  1. Ephemerides. I. Title.
QB12.M44 1983   528   83-5762
ISBN 0-943396-02-6

BT 6420-85   5/1/85

Printed in the United States of America
10 9 8 7 6 5 4 3 2 1

Foreword

This book will interest all who thirst for knowledge about past, present, and future celestial events. In compiling these listings of thousands of separate astronomical phenomena, Jean Meeus has created a valuable reference for all who love the sky.

Meeus' clear style of explanation liberally sprinkled with examples quickly becomes apparent in his introductory note on time reckoning and the two different time scales used throughout this work. In the first chapter Meeus lists planetary phenomena, concentrating on the 30-year interval 1976-2005. Instead of merely listing, in almanac fashion, all events in simple chronological order, the author provides, in this one chapter, over 60 separate tables and columns of events, conveniently arranged into categories. The listing of mutual planetary conjunctions alone, for example, includes 21 separate tables.

In the section on planetary phenomena, we can find such data as the ranges in angular distance from the sun for Mercury and Venus at their greatest elongations; the maximum angular distance Venus can appear to wander from the ecliptic (the farthest possible for any naked-eye planet); a list of transits of Earth across the solar disk as seen from other planets (observers on Mars, Jupiter, and Uranus would all witness such an event in 1984); and the closest approach to Earth of all the planets Venus through Pluto during the 20th Century.

Meeus points out several periodicities between planetary events, and many more can be discovered by examining his tables. Long-time skywatchers gradually become aware of celestial rhythms as events unfold year by year; but now, with such data conveniently arranged before us, our consciousness of cycles is quickened. Mercury's elongations, for example, occur about 3 days later every 13 years. Venus-sun alignments repeat about 2½ days earlier every 8 years. Scanning the tables for Mars, the reader will note that oppositions recur at 25- to 27-month intervals, and that very close ones recur at intervals of 15 to 17 years. Scanning the columns for magnitude, distance, and apparent size of Mars, the reader will note the ebb and flow connecting the distant oppositions (in 1963, 1980, 1995) with the close ones (1971, 1988, 2003). Elsewhere in the book, a longer listing of Mars oppositions shows that the approach in the year 2003 will be the closest so far in the Christian era, and that there won't be a closer approach of Mars until the year 2287.

In each trip around the sun, Saturn displays two maxima and two minima in its brilliance at opposition. The maxima result from the "open" aspect of the rings as the planet nears the aphelion and perihelion points of its 30-year orbit; the minima correspond to times of edge-on views of the rings. A separate table lists passages of sun and Earth through Saturn's ring plane; telescopic viewers can expect the next edgewise presentation of the rings in 1995-1996.

Dates of northernmost and southernmost journeys in our sky are given for most of the planets, as well as dates of celestial equator crossings for the five outer ones. Most of the 1980's will be a poor time for northern observers to study the four gas giants: Meeus' tables show Jupiter reaching a southernmost point in 1984, Neptune reaching one in 1986, and Saturn and Uranus riding low in 1989. Even Mars will pass farther south in 1986 (nearly 29° south of the equator) than at any time since 1907. But recovery will follow quickly. At its very close opposition of 1988 Mars will be well placed near the equator, while Jupiter will cross that boundary in 1987 and reach its northernmost position in 1990.

Providing separate lists for each planetary pairup, while giving the angular distance from the sun for each event, Meeus tells us when to expect the next similar conjunction, and whether it will be seen in the morning or evening sky. Multiple conjunctions between single pairs of planets are bracketed for emphasis; five triple conjunctions between gas giants are slated for 1981-1993, beginning with the much-publicized "Great Conjunction" of Jupiter-Saturn in 1981 and concluding with a Uranus-Neptune thrice pairing in 1993.

"Quasi-conjunctions" (close approaches between bright planets without a conjunction in right ascension) are omitted in conventional almanacs, but you'll find them here!

Of interest to historical investigators are tables which help reconstruct the appearance of the sky at times in the distant past. Given are such data as: Dates and times of equinoxes and solstices; oppositions of Mars, Jupiter, and Saturn; inferior and superior conjunctions of Venus; illuminated fraction of the moon; celestial coordinates of 48 bright zodiacal stars; and solar conjunctions with 12 bright stars. Continuing at least four centuries into the future, these tables are also useful to those who wish to predict future sky happenings.

Researchers delving into Earth-sun relationships may well employ the tables of sunspot activity dating back over 200 years.

Students of the moon will find tables of the four principal lunar phases and of the sun's selenographic colongitude. The latter enable the user to fix the location of the moon's terminator (day-night boundary) and the angle of solar illumination on any lunar feature for any hour of observation since the invention of the telescope through the year 2399. Tables giving the dates of equinoxes and solstices on Mars will aid in interpreting observations as far back as the telescopic drawings by Huygens and Cassini, through the recent Viking orbiter and lander missions, all the way to the middle of the 21st century.

Sky events for the next several decades are especially well covered. A listing of all occultations of planets and bright stars through the year 2000 is supplied, with data and calculator programs for computation of local circumstances and graze limits. Mathematical procedures are clearly outlined, so microcomputer users can write their own programs. Lunar distances are given for each perigee and apogee until the year 2005. Eclipse buffs will appreciate the listing of solar eclipses through the year 2050. Begin travel plans now! Additional data for the same period enable the reader to obtain the times of the principal stages of all eclipses of the moon.

Data on past and future transits of Mercury and Venus enable the calculation of the four contact times of the planetary and solar disks. By comparing your local sunrise and sunset times, you can determine which lunar eclipses and which planet transits (including those of Venus coming up in 2004 and 2012 can be seen from your area!

Also included is an assortment of easy-to-use tables for calculating such past and future information as the Julian date, the weekday of a given date, and the date of Easter.

Jean Meeus' books and articles on celestial phenomena have been enthusiastically received in many countries by professional astronomers and amateur skywatchers alike. His writings are standard references in the field. Serving interests ranging from serious research to simple sky gazing, this book offers a wide and fascinating selection of data on past, present, and future sky events.

Robert C. Victor
Planet columnist for *Sky and Telescope* magazine
Astronomer, Abrams Planetarium, Michigan State University

TABLE OF CONTENTS

TABLES-Part 7, Other Tables

NOTE ON TIME RECKONING

CALENDAR DATES

For the calendar dates, the Julian Calendar is used till 1582 October 4, while the Gregorian Calendar is used from 1582 October 15 onwards.

The dates are written as is usual in Astronomy, namely in the order year – month – day of month. For instance, 1982 May 16 means the 16th day of May, 1982.

The dates are based either on the Ephemeris Time, or on the Universal Time (see below). In Part 1, Universal Time is used. For instance, on page 1-18 we read that on 1981 October 6 Saturn was in conjunction with the Sun ; this means that this conjunction took place between 0^h00^m and 24^h00^m Universal Time on 1981 October 6.

EPHEMERIS TIME AND UNIVERSAL TIME

The Ephemeris Time (ET) is a uniform time based on the planetary motions. The Universal Time (UT), necessary for civil life, is based on the rotation of the Earth. Universal Time is the same as Greenwich Civil Time.

Because the Earth's rotation is slowing down — and, moreover, with unpredictable irregularities — UT is not a uniform time. Since the astronomers need a uniform time scale for their calculation of accurate ephemerides, they use ET in that case.

The exact value of the difference ΔT = ET − UT at a given epoch can be deduced only from observations. Table A on the next page gives the value of ΔT, in seconds of time, for the *beginning* of some years. The values for 1680 to 1898, due to Spencer Jones, have been taken from André Danjon's *Astronomie Générale* (Paris, 1959), page 121.

For the year 1982, the value ΔT = +53 seconds can be used. For the next years, increase ΔT by +1 second per year, in order to obtain an extrapolated provisional value of ΔT.

For epochs outside the time interval 1680 – 2000, an *approximate* value of ΔT, in *minutes*, can be calculated from

$$\Delta T = +0.41 + 1.2053\,T + 0.4992\,T^2$$

where T is the time in centuries since the year 1900. This leads to Table B. It should be noted

TABLE   A

Subtract the value given by this Table in order to convert ET to UT							
Year	ΔT	*Year*	ΔT	*Year*	ΔT	*Year*	ΔT
1680	-14^s	1853	$+ 3^s$	1914	$+15^s$	1948	$+28^s$
1710	-12	1858	3	1916	17	1950	29
1727	$- 7$	1862	3	1918	19	1952	30
1737	$- 3$	1868	$+ 1$	1920	20	1954	31
1747	$- 1$	1873	$- 5$	1922	21	1956	31
1755	$+ 1$	1878	$- 8$	1924	22	1958	32
1772	6	1883	$- 8$	1926	23	1960	33
1785	8	1888	$- 8$	1928	23	1962	34
1792	8	1895	$- 7$	1930	23	1964	35
1801	6	1898	$- 6$	1932	23	1966	37
1809	5	1900	$- 4$	1934	24	1968	38
1813	5	1902	$- 2$	1936	24	1970	40
1822	4	1904	$+ 1$	1938	24	1972	42
1832	1	1906	4	1940	24	1974	44
1837	0	1908	7	1942	25	1976	46
1843	1	1910	10	1944	26	1978	49
1848	2	1912	12	1946	27	1980	51

TABLE   B

Approximate values of ΔT					
Year	*Minutes*	*Year*	*Minutes*	*Year*	*Minutes*
0	158	1000	30	2075	4
100	140	1100	23	2190	8
200	124	1200	16	2300	13
300	109	1300	11	2400	19
400	95	1400	7	2500	26
500	81	1490	4	2600	33
600	69	1600	1	2700	42
700	58	1700	0	2800	52
800	48	1950	0	2900	62
900	38	2000	1	3000	74

that for the very past and future, the tabulated values of ΔT may be in error by several minutes, as the fluctuations due to the variable rotation of the Earth are unknown for these epochs.

Tabulated times in ET may be converted into Universal Time by *subtracting* from them the quantity ΔT, since we have UT = ET − ΔT. In Part 1 of this book, all times have already been converted from ET to UT.

E x a m p l e 1. — On page 4-10 we find the following times for lunar phases :

Last Quarter	1980 May 7	$20^h51^m26^s$ ET
New Moon	May 14	12 01 01
First Quarter	May 21	19 16 47

In 1980, ΔT was equal to +51 seconds. Consequently, the phases above took place at the following UT instants :

1980 May 7	$20^h50^m35^s$ UT
May 14	12 00 10
May 21	19 15 56

E x a m p l e 2. — On page 3-42, we find that in A.D. 1999 the March equinox will take place on March 21 at $1^h47^m11^s$ ET. According to the rule given above, we expect that in A.D. 1999 the quantity ΔT will be +70 seconds. Consequently, in 1999 the spring equinox will occur on March 21 at $1^h47^m11^s$ − 70^s = $1^h46^m01^s$ UT, which may be rounded to 1^h46^m UT.

For other time zones, this reduces for example to :

March 21	at	2^h46^m	Central European Time,
March 20	at	20 46	Eastern Standard Time,
March 20	at	17 46	Pacific Standard Time.

E x a m p l e 3. — On page 2-3, we find that in A.D. 133 the opposition of Mars took place on January 17 at 0^h19^m ET.

In 133, the quantity ΔT was equal to approximately +135 minutes, or 2 hours and 15 minutes. Consequently, in A.D. 133 Mars' opposition occurred on January 16 at 22^h04^m UT.

1

PLANETARY   PHENOMENA

1976 – 2005

PLANETARY PHENOMENA   1976 - 2005

CONTENTS

INTRODUCTION

General Remarks

Notwithstanding its title, the present part contains some data for years outside the period 1976 - 2005.

Most data in the present part have been deduced by the author from accurate ephemerides calculated by Edwin Goffin of the Public Observatory of Antwerp, Belgium.

All times are expressed in *Universal Time*, or Greenwich Civil Time. All dates, too, are based on the UT, even for years before 1925 when the Greenwich *Mean* Time (*i.e.* measured from Greenwich Mean Noon) was in use in the astronomical ephemerides. Thus, when we write that Neptune reached the ascending node of its orbit on 1920 June 2, this means that the passage through that node took place between 0^h00^m and 24^h00^m UT on 1920 June 2.

Contents

Tables I and II give the dates of all inferior and superior conjunctions of Mercury with the Sun taking place from A.D. 1976 to 2005.

Table III lists, from 1976 to 2005, the greatest elongations of Mercury, that is when the angular distance between Mercury and the center of the Sun's disk is a maximum.

During the period 1976-2005, the extreme values of Mercury's greatest elongations are :

        17°52'   on 1977 September 21, 1990 September 24, and 2003 September 27 ;
        27°50'   on 2000 March 28.

There is a periodicity of 13 years, a periodicity of 33 years, and a much better periodicity of 46 years (which is a combination of those of 13 and 33 years), for instance :

13 years		*33 years*		*46 years*	
1953 Aug 13	18°50'W	1939 Aug 28	18°16'W	1913 Aug 22	18°26'W
1966 Aug 16	18°43'W	1972 Aug 25	18°20'W	1959 Aug 23	18°25'W
1979 Aug 19	18°36'W	2005 Aug 23	18°24'W	2005 Aug 23	18°24'W
1992 Aug 21	18°30'W				
2005 Aug 23	18°24'W				

Table IV, giving the dates of all inferior conjunctions of Venus with the Sun during the period 1960-2024, was first published in the *Journal of the British Astronomical Association*, Vol. 81, page 114 (February 1971).

During the 20th century, the greatest geocentric latitudes of Venus took place at the inferior conjunctions of 1919 September 13 and 1927 September 10, when in both cases it was −8°43′. Venus will reach no greater geocentric latitude (at inferior conjunction) until the second half of the 21st century (for instance +8°50′ at the inferior conjunctions of 2081 March 6 and 2089 March 3).

Venus is nearest to the Earth near the times of its inferior conjunctions. The times and values of these minimum distances (in astronomical units) are given in Table IVa for the years 1960 to 2005.

Table V gives the dates of the superior conjunctions of Venus with the Sun which take place during the period 1960 - 2005.

Table VI gives, from 1976 to 2005, all the greatest elongations of Venus, that is when the angular distance between Venus and the center of the Sun's disk is a maximum. During the 20th century, the greatest and least of Venus' greatest elongations are the following ones :

$$47°19′ \quad \text{on} \quad 1901 \text{ December } 5$$
$$45°23′ \quad \text{on} \quad 1999 \text{ June } 11$$

Tables VII to XII contain information about the oppositions of the planets Mars to Pluto with the Sun which take place during the period 1960 - 2005. In each table the columns give successively :
— the Julian Day, the calendar date and the Universal Time of the geometric opposition in celestial longitude ;
— the common heliocentric longitude of the planet and the Earth at the time of opposition, referred to the mean equinox of date ;
— the geocentric declination of the planet at the time of opposition, referred to the mean equinox of date ;
— the planet's magnitude at the time of opposition (except for Pluto);
— the instant (calendar date and UT) when the planet is nearest to the Earth ;
— the value of this least distance, respectively in astronomical units and in millions of kilometers ;
— the value of the apparent diameter of the planet at the time of least distance to Earth. For Jupiter and Saturn, the equatorial diameter is given. This information is not given for Pluto.

The stellar magnitudes of Mars, Jupiter and Saturn have been obtained from the formulae of G. Müller (1893). Those of Uranus and Neptune are calculated from

$$\text{Uranus :} \quad −6.85 + 5 \log r\Delta$$
$$\text{Neptune :} \quad −7.05 + 5 \log r\Delta$$

where r and Δ are the planet's distances to the Sun and to the Earth, respectively, expressed in astronomical units.

The following values have been used for the diameters at unit distance :

Mars 9".36   Jupiter 196".94   Saturn 166".66

Uranus 68".56   Neptune 73".12

During the period 1960 - 2005, a *transit of the Earth* over the Sun's disk occurs
at the following dates :

for Mars :	1984 May 11	*for Uranus :*	1981 May 19
			1982 May 24
for Jupiter :	1972 Jun 24		1983 May 29
	1977 Dec 23		1984 Jun 1
	1984 Jun 29		1985 Jun 6
	1989 Dec 27		1986 Jun 11
	1996 Jul 4		1987 Jun 16
	2002 Jan 1		1988 Jun 20
for Saturn :	1961 Jul 19	*for Neptune :*	2000 Jul 27
	1990 Jul 14		2001 Jul 30
	2005 Jan 13		2002 Aug 2
			2003 Aug 4
			2004 Aug 6
			2005 Aug 8

During the 20th century, that is the years 1901 to 2000, the least distances of
the planets Venus - Pluto to the Earth are as follows :

Venus	1906 Nov 30	0.26506 *a.u.*	40 *millions of km*
Mars	1924 Aug 22	0.37284	56
Jupiter	1951 Oct 2	3.94871	591
Saturn	1914 Dec 21	8.03033	1201
Uranus	1966 Mar 9	17.29211	2587
Neptune	1901 Dec 22	28.90113	4324
Pluto	1990 May 7	28.67852	4290

During the period 1960 to 2005, the extreme values of the intervals between successive oppositions are as follows :

for Jupiter :   395.44 days (between the oppositions of 2004 and 2005)
                402.84 days (between the oppositions of 1962 and 1963)

for Saturn :   377.16 days (between the oppositions of 1988 and 1989)
               379.17 days (between the oppositions of 1972 and 1973)

During the period 1960 to 2005, the extreme values of the intervals between the
opposition and the least distance to Earth are as follows :

for Jupiter :   +1.5 day (oppositions of 1971, 1983, 1995)
                -1.5 day (oppositions of 1988 and 2000)

for Saturn :   +5 hours (1963 and 1986)
               -5 hours (opposition of 1978)

for Uranus :   +22 hours (1962)
               -23 hours (2002)

Table XIII gives the dates of all conjunctions of the superior planets with the Sun taking place from A.D. 1960 to 2005.

Tables XIV and XV contain the times of the perihelion and aphelion passages of Earth and Mars, respectively, and the corresponding values of the radius vector expressed in astronomical units.

Table XVI contains the dates of the perihelion and aphelion passages of the outer planets, and the corresponding values of the radius vector : Jupiter from 1950 to 2010, Saturn and Uranus from 1900 to 2050, Neptune and Pluto from 1850 to 2100.

Table XVII contains the times of the passages of Mars through the nodes of its orbit, from 1960 to 2005.

Table XVIII contains the dates on which the outer planets reach the nodes of their orbits and their greatest heliocentric latitudes : Jupiter from 1960 to 2005, Saturn and Uranus from 1940 to 2005, Neptune and Pluto from 1900 to 2020. All these data refer to the ecliptic of date.

Table XIX provides some informations about the extreme northern and southern declinations of Venus and Mars. Furthermore, the complete list is given of the dates when the outer planets cross the celestial equator and reach their extreme northern or southern declinations : Jupiter for the years 1960 - 2005, Saturn for 1951 - 2005, Uranus and Neptune for 1920 - 2005, Pluto for 1960 - 2005.
All these data refer to the apparent declinations of the planets, that is the corrections for light-time, aberration and nutation are taken into account.

Table XX gives the complete list of all planetary conjunctions in right ascension during the period 1976 - 2005. Conjunctions with Pluto are not taken in consideration. The conjunctions are given separately for each couple of planets. The columns give the date and time (UT) of the conjunction in right ascension, the difference in declination between the two bodies at that instant, and the elongation or angular distance to the Sun (E = East from the Sun = visible in the evening sky; W = West from the Sun = visible in the morning sky).
The difference in declination is positive (negative) if the first planet passes to the North (South) of the second planet. For instance, on 1989 May 16 at 7^h UT Mercury will be 0°34' North of Venus.
The tabulated elongation is that of the planet which is closest to the Sun.
Multiple conjunctions are indicated with a parenthesis. It appears that *quintuple* conjunctions are possible between Mercury and Venus (as in 2004-2005), and between Mercury and Mars (as in 1978-1979).
The time of conjunction in right ascension is not the same as that of the least apparent separation. The difference can be important in some cases. For example,

Venus passed 1°17′ North of Mars on 1977 May 13, and 1°13′ South of it on 1977
June 3 ; but there was only *one* least separation, namely 0°38′ on 1977 May 18.
Another striking example occurs in August 1991, when there will be a triple conjunction between Mercury and Venus, but only one least distance between the two
planets, namely 2°05′ on August 8.

It should further be noted that a conjunction in right ascension is not the same
as a conjunction in ecliptic longitude. In limiting cases, there may be a conjunction in right ascension but not in longitude, or inversely. For instance, in July
1979 there was *no* conjunction in right ascension between Mercury and Jupiter,
but there were *two* conjunctions in longitude between these two planets, namely
on July 11 and on July 18.

Conjunctions which occur close to the Sun are not observable. An accurate limit
cannot be given, since the visibility depends on the magnitude of the planets and
of the obliquity of the ecliptic on the local horizon near the time of sunrise or
sunset. But a general rule is that a conjunction is certainly not visible if the
two planets are less than 10 degrees from the Sun.

Some interesting facts can be seen by examination of the lists. For example, for
the planets Mercury-Venus the difference of the declinations at the time of conjunction in right ascension can exceed 10 degrees, as for instance on 1999 August 26.

For the planets Mercury-Mars, single conjunctions are much rarer than multiple
ones. During the period 1976-2005 there are only two single conjunctions in right
ascension, namely in 1987 and in 2002, against 9 triple ones and 3 quintuple ones.

For the Mercury-Saturn conjunctions, the dates show a remarkable periodicity.
During 4 or 5 consecutive years, the conjunctions occur at almost the same date of
the year, for example :

   1978 Sep 13   1979 Sep 11   1980 Sep 9   1981 Sep 10

and then there is a jump of approximately fifty days (1982 Nov 1), giving the beginning of a new series (1982-1985), etc. Triple conjunctions occur when one series
ends and another begins.

While triple conjunctions between Venus and Mars are frequent (seven cases during
the period 1976-2005), there is only one triple conjunction during the same period
for each of the couples Venus-Jupiter, Venus-Saturn, Venus-Uranus and Venus-Neptune.

During the period 1976-2005 there is no triple conjunction of Mars with either
Saturn, Uranus or Neptune. The last triple conjunction between Mars and Saturn took
place in 1945-1946, between Mars and Uranus in 1964-1965, between Mars and Neptune
in 1932-1933.

During the period 1976-2005, Mars always passes South of Neptune ($\Delta\delta < 0$).

From 1982 to 2002, all Mars-Jupiter conjunctions will occur East from the Sun,
that is in the evening sky. From 1954 to 2004 all Mars-Saturn conjunctions take
place every second year, and only during the months January to July.

The Mars-Uranus conjunctions, from 1986 to 2001, occur at intervals of 2 years minus one or two weeks, while the elongation from the Sun gradually changes from West to East. Then there is a discontinuity, and in 2003 another series begins. Similar series exist for the conjunctions of Mars with Jupiter, Saturn and Neptune.

In 1981 there was a triple conjunction between Jupiter and Saturn. The previous one between these planets took place in 1940-1941, while the next one will not occur before 2238-2239.

Triple conjunctions between Jupiter and Uranus occurred in 1927-1928, 1954-1955 and 1968-1969 ; between Jupiter and Neptune in 1919-1920 and 1971 ; between Saturn and Uranus in 1896-1897 ; between Saturn and Neptune in 1952-1953 when, moreover, both Saturn and Neptune came in triple conjunction with the star Spica !

As seen from the Earth, the angular distance between Uranus and Neptune will be less than 5° from 1990 December 10 to 1995 April 25, and then again from 1995 June 12 to 1996 January 12.

Those conjunctions in right ascension of the period 1976-2005 for which $|\Delta\delta|$ is less than $0°06'00''$, are given with more details in Table XXa. The times of the conjunction in right ascension and of least separation are given to the nearest minute of time, while the difference of the declinations (at the time of conjunction in right ascension) and the least distance between the centers of the two planets are given to the nearest second of a degree.

During the period 1976-2005, the least geocentric separation between two planets will be 40″ (Mars-Uranus, on 1988 February 22), and there will be no occultation of one planet by another one.

Sometimes two planets approach each other within 5°, although there is no conjunction in right ascension between these bodies. Such approaches-without-conjunction have been called *quasi-conjunctions* by the author [*L'Astronomie*, Vol. 88, 141 (April 1974)]. Table XXI lists all quasi-conjunctions between the naked-eye planets from 1976 to 2005. The instants of the least distances are given, together with the values of the least distances themselves.

Table XXII gives the heliocentric conjunctions between the outer planets, from 1950 to 2005, except for the couples Jupiter-Pluto and Saturn-Pluto. After the date, the common heliocentric longitude, referred to the mean equinox of date, is given.

Table XXIII gives the dates of the passages of the Earth and the Sun through the ring-plane of Saturn from 1840 to 2060.

April 1977

I
INFERIOR CONJUNCTIONS OF MERCURY

1976 Jan 23	1982 Feb 1	1988 Feb 11	1994 Feb 20	2000 Mar 1
1976 May 20	1982 Jun 1	1988 Jun 13	1994 Jun 25	2000 Jul 6
1976 Sep 22	1982 Oct 2	1988 Oct 11	1994 Oct 21	2000 Oct 30
1977 Jan 6	1983 Jan 16	1989 Jan 25	1995 Feb 3	2001 Feb 13
1977 Apr 30	1983 May 12	1989 May 23	1995 Jun 5	2001 Jun 16
1977 Sep 5	1983 Sep 15	1989 Sep 24	1995 Oct 5	2001 Oct 14
1977 Dec 21	1983 Dec 31	1990 Jan 9	1996 Jan 18	2002 Jan 27
1978 Apr 11	1984 Apr 22	1990 May 3	1996 May 15	2002 May 27
1978 Aug 18	1984 Aug 28	1990 Sep 8	1996 Sep 17	2002 Sep 27
1978 Dec 5	1984 Dec 14	1990 Dec 24	1997 Jan 2	2003 Jan 11
1979 Mar 24	1985 Apr 3	1991 Apr 14	1997 Apr 25	2003 May 7 *T*
1979 Jul 31	1985 Aug 10	1991 Aug 21	1997 Aug 31	2003 Sep 11
1979 Nov 20	1985 Nov 28	1991 Dec 8	1997 Dec 17	2003 Dec 27
1980 Mar 6	1986 Mar 16	1992 Mar 26	1998 Apr 6	2004 Apr 17
1980 Jul 11	1986 Jul 23	1992 Aug 2	1998 Aug 13	2004 Aug 23
1980 Nov 3	1986 Nov 13 *T*	1992 Nov 21	1998 Dec 1	2004 Dec 10
1981 Feb 17	1987 Feb 27	1993 Mar 9	1999 Mar 19	2005 Mar 29
1981 Jun 22	1987 Jul 4	1993 Jul 15	1999 Jul 26	2005 Aug 5
1981 Oct 18	1987 Oct 28	1993 Nov 6 *T*	1999 Nov 15 *T*	2005 Nov 24

T means that there is a transit of Mercury over the Sun's disk

II
SUPERIOR CONJUNCTIONS OF MERCURY

1976 Apr 1	1982 Apr 11	1988 Apr 20	1994 Apr 30 *O*	2000 May 9 *O*
1976 Jul 15	1982 Jul 25	1988 Aug 3	1994 Aug 13	2000 Aug 22
1976 Nov 7 *O*	1982 Nov 19	1988 Dec 1	1994 Dec 14	2000 Dec 25
1977 Mar 16	1983 Mar 26	1989 Apr 4	1995 Apr 14	2001 Apr 23
1977 Jun 30	1983 Jul 9	1989 Jul 18	1995 Jul 28	2001 Aug 5
1977 Oct 18	1983 Oct 30	1989 Nov 10 *O*	1995 Nov 23	2001 Dec 4
1978 Feb 27	1984 Mar 8	1990 Mar 19	1996 Mar 28	2002 Apr 7
1978 Jun 14	1984 Jun 23	1990 Jul 2	1996 Jul 11	2002 Jul 21
1978 Sep 30	1984 Oct 10	1990 Oct 22	1996 Nov 1	2002 Nov 14 *O*
1979 Feb 9	1985 Feb 19	1991 Mar 2	1997 Mar 11	2003 Mar 21
1979 May 29	1985 Jun 7	1991 Jun 17	1997 Jun 25	2003 Jul 5
1979 Sep 13	1985 Sep 22	1991 Oct 3	1997 Oct 13	2003 Oct 25
1980 Jan 21	1986 Feb 1	1992 Feb 12	1998 Feb 22	2004 Mar 4
1980 May 13 *O*	1986 May 23	1992 May 31	1998 Jun 10	2004 Jun 18
1980 Aug 26	1986 Sep 5	1992 Sep 15	1998 Sep 25	2004 Oct 5
1980 Dec 31	1987 Jan 12	1993 Jan 23	1999 Feb 4	2005 Feb 14
1981 Apr 27	1987 May 7 *O*	1993 May 16 *O*	1999 May 25	2005 Jun 3
1981 Aug 10	1987 Aug 20	1993 Aug 29	1999 Sep 8	2005 Sep 18
1981 Dec 10	1987 Dec 23	1994 Jan 3	2000 Jan 16	

O means that Mercury is occulted by the Sun's disk

III
GREATEST ELONGATIONS OF MERCURY

Date	Elong.		Date	Elong.		Date	Elong.		Date	Elong.
1976 Jan 7	19°14'E		1983 Aug 19	27°25'E		1991 Mar 27	18°47'E		1998 Nov 11	22°57'E
Feb 16	26 15 W		Oct 1	17 53 W		May 12	26 12 W		Dec 20	21 38 W
Apr 28	20 35 E		Dec 13	20 37 E		Jul 25	27 02 E		1999 Mar 3	18 11 E
Jun 15	23 14 W		1984 Jan 22	24 19 W		Sep 7	18 01 W		Apr 16	27 35 W
Aug 26	27 20 E		Apr 3	19 06 E		Nov 19	22 23 E		Jun 28	25 33 E
Oct 7	17 59 W		May 19	25 35 W		Dec 27	22 14 W		Aug 14	18 48 W
Dec 20	20 10 E		Jul 31	27 16 E		1992 Mar 9	18 17 E		Oct 24	24 17 E
1977 Jan 29	24 56 W		Sep 14	17 54 W		Apr 23	27 17 W		Dec 3	20 23 W
Apr 10	19 28 E		Nov 25	21 50 E		Jul 6	26 05 E		2000 Feb 15	18 09 E
May 27	24 55 W		1985 Jan 3	22 51 W		Aug 21	18 30 W		Mar 28	27 50 W
Aug 8	27 24 E		Mar 17	18 27 E		Oct 31	23 43 E		Jun 9	24 03 E
Sep 21	17 52 W		May 1	26 53 W		Dec 9	20 53 W		Jul 27	19 48 W
Dec 3	21 18 E		Jul 14	26 33 E		1993 Feb 21	18 07 E		Oct 6	25 31 E
1978 Jan 11	23 28 W		Aug 28	18 15 W		Apr 5	27 48 W		Nov 15	19 20 W
Mar 24	18 41 E		Nov 8	23 09 E		Jun 17	24 43 E		2001 Jan 28	18 26 E
May 9	26 23 W		Dec 17	21 26 W		Aug 4	19 20 W		Mar 11	27 28 W
Jul 22	26 55 E		1986 Feb 28	18 09 E		Oct 14	25 00 E		May 22	22 27 E
Sep 4	18 04 W		Apr 13	27 39 W		Nov 22	19 45 W		Jul 9	21 08 W
Nov 16	22 35 E		Jun 25	25 20 E		1994 Feb 4	18 16 E		Sep 18	26 32 E
Dec 24	22 01 W		Aug 11	18 56 W		Mar 19	27 41 W		Oct 29	18 34 W
1979 Mar 8	18 15 E		Oct 21	24 28 E		May 30	23 07 E		2002 Jan 11	19 01 E
Apr 21	27 24 W		Nov 30	20 12 W		Jul 17	20 31 W		Feb 21	26 35 W
Jul 3	25 54 E		1987 Feb 12	18 10 E		Sep 26	26 08 E		May 4	20 58 E
Aug 19	18 36 W		Mar 26	27 49 W		Nov 6	18 52 W		Jun 21	22 44 W
Oct 29	23 55 W		Jun 7	23 49 E		1995 Jan 19	18 44 E		Sep 1	27 13 E
Dec 7	20 42 W		Jul 25	19 58 W		Mar 1	27 01 W		Oct 13	18 04 W
1980 Feb 19	18 07 E		Oct 4	25 41 E		May 12	21 34 E		Dec 26	19 52 E
Apr 2	27 49 W		Nov 13	19 13 W		Jun 29	22 01 W		2003 Feb 4	25 21 W
Jun 14	24 29 E		1988 Jan 26	18 30 E		Sep 9	26 58 E		Apr 16	19 46 E
Aug 1	19 29 W		Mar 8	27 22 W		Oct 20	18 15 W		Jun 3	24 26 W
Oct 11	25 11 E		May 19	22 13 E		1996 Jan 2	19 28 E		Aug 14	27 26 E
Nov 19	19 36 W		Jul 6	21 21 W		Feb 11	25 55 W		Sep 27	17 52 W
1981 Feb 2	18 19 E		Sep 15	26 40 E		Apr 23	20 14 E		Dec 9	20 56 E
Mar 16	27 37 W		Oct 26	18 28 W		Jun 10	23 42 W		2004 Jan 17	23 55 W
May 27	22 53 E		1989 Jan 9	19 07 E		Aug 21	27 24 E		Mar 29	18 53 E
Jul 14	20 44 W		Feb 18	26 25 W		Oct 3	17 55 W		May 14	25 59 W
Sep 23	26 17 E		May 1	20 46 E		Dec 15	20 27 E		Jul 27	27 07 E
Nov 3	18 45 W		Jun 18	22 59 W		1997 Jan 24	24 32 W		Sep 9	17 58 W
1982 Jan 16	18 49 E		Aug 29	27 16 E		Apr 6	19 13 E		Nov 21	22 11 E
Feb 26	26 52 W		Oct 10	18 01 W		May 22	25 22 W		Dec 29	22 27 W
May 9	21 21 E		Dec 23	20 01 E		Aug 4	27 19 E		2005 Mar 12	18 20 E
Jun 26	22 16 W		1990 Feb 1	25 09 W		Sep 16	17 53 W		Apr 26	27 10 W
Sep 6	27 04 E		Apr 13	19 37 E		Nov 28	21 38 E		Jul 9	26 15 E
Oct 17	18 11 W		May 31	24 41 W		1998 Jan 6	23 04 W		Aug 23	18 24 W
Dec 30	19 36 E		Aug 11	27 25 E		Mar 20	18 32 E		Nov 3	23 31 E
1983 Feb 8	25 43 W		Sep 24	17 52 W		May 4	26 44 W		Dec 12	21 05 W
Apr 21	20 04 E		Dec 6	21 07 E		Jul 17	26 41 E			
Jun 8	23 58 W		1991 Jan 14	23 42 W		Aug 31	18 11 W			

E = greatest eastern elongation   *W = greatest western elongation*

IV
INFERIOR CONJUNCTIONS OF VENUS

Besides the instant, the geocentric latitude of Venus at the time of conjunction
is given. The sign + (or —) indicates that Venus is north (or south) of the ecliptic. Because the path of Venus, at the time of inferior conjunction, is nearly
parallel to the ecliptic, the planet's geocentric latitude is approximately equal
to its least distance to the *center* of the Sun. At the inferior conjunctions of
2004 and 2012, there is a transit of Venus over the Sun.

1961 Apr 10	24 h	+7°08'		1993 Apr 1	13 h	+7°52'
1962 Nov 12	20	-4 04		1994 Nov 2	23	-5 24
1964 Jun 19	23	-1 49		1996 Jun 10	16	-0 30
1966 Jan 26	9	+6 55		1998 Jan 16	11	+5 49
1967 Aug 29	22	-8 30		1999 Aug 20	12	-8 07
1969 Apr 8	15	+7 20		2001 Mar 30	4	+8 01
1970 Nov 10	9	-4 25		2002 Oct 31	12	-5 42
1972 Jun 17	15	-1 29		2004 Jun 8	9	-0 11 T
1974 Jan 23	21	+6 40		2006 Jan 13	24	+5 30
1975 Aug 27	13	-8 26		2007 Aug 18	4	-7 59
1977 Apr 6	6	+7 31		2009 Mar 27	19	+8 10
1978 Nov 7	22	-4 45		2010 Oct 29	1	-5 59
1980 Jun 15	7	-1 10		2012 Jun 6	1	+0 09 T
1982 Jan 21	10	+6 24		2014 Jan 11	12	+5 11
1983 Aug 25	5	-8 20		2015 Aug 15	19	-7 51
1985 Apr 3	22	+7 42		2017 Mar 25	10	+8 18
1986 Nov 5	10	-5 05		2018 Oct 26	14	-6 15
1988 Jun 12	24	-0 50		2020 Jun 3	18	+0 29
1990 Jan 18	23	+6 07		2022 Jan 9	1	+4 51
1991 Aug 22	20	-8 14		2023 Aug 13	11	-7 41

IVa
VENUS' LEAST DISTANCES TO EARTH

1961 Apr 10	20 h	0.28371 a.u.		1983 Aug 25	13 h	0.28712 a.u.
1962 Nov 13	12	0.26792		1985 Apr 3	17	0.28260
1964 Jun 19	21	0.28949		1986 Nov 6	6	0.26964
1966 Jan 25	12	0.26881		1988 Jun 12	22	0.28918
1967 Aug 30	8	0.28636		1990 Jan 18	3	0.26734
1969 Apr 8	11	0.28334		1991 Aug 23	4	0.28745
1970 Nov 11	2	0.26842		1993 Apr 1	8	0.28224
1972 Jun 17	14	0.28942		1994 Nov 3	19	0.27023
1974 Jan 23	1	0.26830		1996 Jun 10	14	0.28905
1975 Aug 27	22	0.28675		1998 Jan 15	16	0.26688
1977 Apr 6	2	0.28296		1999 Aug 20	19	0.28783
1978 Nov 8	16	0.26903		2001 Mar 29	22	0.28187
1980 Jun 15	6	0.28932		2002 Nov 1	8	0.27088
1982 Jan 20	14	0.26783		2004 Jun 8	7	0.28888

V
SUPERIOR CONJUNCTIONS OF VENUS

1960 Jun 22	1973 Apr 9	1986 Jan 19	1998 Oct 30
1962 Jan 27	1974 Nov 6	1987 Aug 23	2000 Jun 11 *O*
1963 Aug 30	1976 Jun 18 *O*	1989 Apr 4	2002 Jan 14
1965 Apr 12	1978 Jan 22	1990 Nov 1	2003 Aug 18
1966 Nov 9	1979 Aug 25	1992 Jun 13 *O*	2005 Mar 31
1968 Jun 20	1981 Apr 7	1994 Jan 17	
1970 Jan 24	1982 Nov 4	1995 Aug 20	
1971 Aug 27	1984 Jun 15 *O*	1997 Apr 2	

O means that Venus is occulted by the Sun's disk

VI
GREATEST ELONGATIONS OF VENUS

1977 Jan 24	47°00'E	1987 Jan 15	46°58'W	1997 Nov 6	47°08'E
1977 Jun 15	45 48 W	1988 Apr 3	45 56 E	1998 Mar 27	46 30 W
1978 Aug 29	46 07 E	1988 Aug 22	45 51 W	1999 Jun 11	45 23 E
1979 Jan 18	46 58 W	1989 Nov 8	47 09 E	1999 Oct 30	46 29 W
1980 Apr 5	45 54 E	1990 Mar 30	46 29 W	2001 Jan 17	47 06 E
1980 Aug 24	45 52 W	1991 Jun 13	45 23 E	2001 Jun 8	45 50 W
1981 Nov 11	47 11 E	1991 Nov 2	46 31 W	2002 Aug 22	46 00 E
1982 Apr 1	46 27 W	1993 Jan 19	47 04 E	2003 Jan 11	46 58 W
1983 Jun 16	45 24 E	1993 Jun 10	45 49 W	2004 Mar 29	46 00 E
1983 Nov 4	46 32 W	1994 Aug 24	46 03 E	2004 Aug 17	45 49 W
1985 Jan 22	47 02 E	1995 Jan 13	46 58 W	2005 Nov 3	47 06 E
1985 Jun 12	45 48 W	1996 Apr 1	45 58 E		
1986 Aug 27	46 05 E	1996 Aug 20	45 50 W		

E = greatest eastern elongation   *W = greatest western elongation*

VII
OPPOSITIONS OF MARS

	At opposition						Nearest to Earth				
J.D.	Date	UT	Helioc. long.	δ	Magn.	Date	UT	a.u.	10^6 km	Diam.
		h	° ′	° ′			h			″
2437298.93	1960 Dec 30	10	98 45	+26 49	-1.3	1960 Dec 25	6	0.60682	91	15.42
2438064.99	1963 Feb 4	12	134 57	+20 42	-1.0	1963 Feb 3	3	0.67045	100	13.96
2438829.01	1965 Mar 9	12	168 44	+ 8 08	-1.0	1965 Mar 12	1	0.66848	100	14.00
2439595.97	1967 Apr 15	11	204 49	- 7 43	-1.3	1967 Apr 21	18	0.60121	90	15.57
2440373.16	1969 May 31	16	250 00	-23 57	-2.0	1969 Jun 9	4	0.47954	72	19.52
2441173.78	1971 Aug 10	7	317 00	-22 15	-2.6	1971 Aug 12	3	0.37569	56	24.91
2441980.64	1973 Oct 25	3	31 34	+10 18	-2.3	1973 Oct 17	4	0.43604	65	21.47
2442762.08	1975 Dec 15	14	82 58	+26 03	-1.6	1975 Dec 9	0	0.56549	85	16.55
2443530.50	1978 Jan 22	0	121 37	+24 06	-1.1	1978 Jan 19	3	0.65320	98	14.33
2444294.73	1980 Feb 25	6	155 47	+13 27	-1.0	1980 Feb 26	6	0.67732	101	13.82
2445059.92	1982 Mar 31	10	190 23	- 1 21	-1.2	1982 Apr 5	7	0.63512	95	14.74
2445831.86	1984 May 11	9	230 51	-18 06	-1.7	1984 May 19	11	0.53146	80	17.61
2446621.72	1986 Jul 10	5	287 41	-27 44	-2.4	1986 Jul 16	11	0.40356	60	23.19
2447432.64	1988 Sep 28	3	5 14	- 2 07	-2.5	1988 Sep 22	3	0.39314	59	23.81
2448223.35	1990 Nov 27	20	65 20	+22 38	-1.8	1990 Nov 20	4	0.51692	77	18.11
2448995.44	1993 Jan 7	23	107 40	+26 16	-1.2	1993 Jan 3	14	0.62610	94	14.95
2449760.60	1995 Feb 12	2	142 54	+18 10	-1.0	1995 Feb 11	14	0.67570	101	13.85
2450524.83	1997 Mar 17	8	176 46	+ 4 40	-1.1	1997 Mar 20	17	0.65939	99	14.20
2451293.23	1999 Apr 24	18	214 06	-11 37	-1.5	1999 May 1	17	0.57846	87	16.18
2452074.24	2001 Jun 13	18	262 46	-26 30	-2.1	2001 Jun 21	23	0.45016	67	20.79
2452880.25	2003 Aug 28	18	335 01	-15 49	-2.7	2003 Aug 27	10	0.37271	56	25.11
2453681.83	2005 Nov 7	8	45 01	+15 54	-2.1	2005 Oct 30	3	0.46406	69	20.17

VIII
OPPOSITIONS OF JUPITER

	At opposition						Nearest to Earth				
J.D.	Date	UT	Helioc. long.	δ	Magn.		Date	UT	a.u.	$10^6 km$	Diam.
		h	° '	° '				h			"
2437105.57	1960 Jun 20	2	268 44	-23 07	-2.2		1960 Jun 21	12	4.23894	634	46.46
2437505.93	1961 Jul 25	10	302 14	-20 16	-2.3		1961 Jul 26	6	4.09505	613	48.09
2437908.11	1962 Aug 31	15	337 43	- 9 58	-2.4		1962 Aug 31	13	3.98532	596	49.42
2438310.95	1963 Oct 8	11	14 27	+ 4 11	-2.5		1963 Oct 7	12	3.95245	591	49.83
2438712.89	1964 Nov 13	9	51 03	+16 50	-2.4		1964 Nov 11	23	4.00972	600	49.12
2439112.87	1965 Dec 18	9	86 15	+23 00	-2.3		1965 Dec 17	1	4.13302	618	47.65
2439510.71	1967 Jan 20	5	119 27	+20 47	-2.2		1967 Jan 19	9	4.27604	640	46.06
2439906.95	1968 Feb 20	11	150 53	+12 19	-2.1		1968 Feb 20	9	4.39277	657	44.83
2440302.45	1969 Mar 21	23	181 09	+ 0 59	-2.0		1969 Mar 22	14	4.45092	666	44.25
2440698.13	1970 Apr 21	15	211 06	-10 27	-2.0		1970 Apr 22	21	4.43615	664	44.39
2441094.87	1971 May 23	9	241 35	-19 30	-2.1		1971 May 24	21	4.35176	651	45.26
2441493.40	1972 Jun 24	22	273 26	-23 13	-2.2		1972 Jun 26	6	4.21970	631	46.67
2441894.03	1973 Jul 30	13	307 12	-19 11	-2.4		1973 Jul 31	6	4.07869	610	48.29
2442296.34	1974 Sep 5	20	342 52	- 8 04	-2.5		1974 Sep 5	15	3.97761	595	49.51
2442699.12	1975 Oct 13	15	19 39	+ 6 12	-2.5		1975 Oct 12	13	3.95623	592	49.78
2443100.84	1976 Nov 18	8	56 07	+18 11	-2.4		1976 Nov 16	22	4.02336	602	48.95
2443500.52	1977 Dec 23	0	91 04	+23 11	-2.3		1977 Dec 21	17	4.15108	621	47.44
2443898.13	1979 Jan 24	15	124 02	+19 53	-2.2		1979 Jan 23	22	4.29222	642	45.88
2444294.24	1980 Feb 24	18	155 17	+10 47	-2.1		1980 Feb 24	17	4.40255	659	44.73
2444689.74	1981 Mar 26	6	185 29	- 0 43	-2.0		1981 Mar 26	24	4.45245	666	44.23
2445085.51	1982 Apr 26	0	215 28	-11 58	-2.0		1982 Apr 27	7	4.42999	663	44.46
2445482.43	1983 May 27	22	246 04	-20 26	-2.1		1983 May 29	11	4.34016	649	45.38
2445881.17	1984 Jun 29	16	278 05	-23 08	-2.2		1984 Jun 30	23	4.20568	629	46.83
2446281.98	1985 Aug 4	11	312 02	-18 01	-2.4		1985 Aug 5	3	4.06677	608	48.43
2446684.38	1986 Sep 10	21	347 52	- 6 11	-2.5		1986 Sep 10	13	3.97189	594	49.58
2447087.10	1987 Oct 18	14	24 40	+ 8 05	-2.5		1987 Oct 17	12	3.95915	592	49.74
2447488.62	1988 Nov 23	3	61 03	+19 22	-2.4		1988 Nov 21	16	4.03398	603	48.82
2447888.09	1989 Dec 27	14	95 48	+23 11	-2.3		1989 Dec 26	9	4.16550	623	47.28
2448285.51	1991 Jan 29	0	128 35	+18 51	-2.1		1991 Jan 28	8	4.30610	644	45.74
2448681.52	1992 Feb 29	0	159 40	+ 9 13	-2.0		1992 Feb 29	3	4.41181	660	44.64
2449076.99	1993 Mar 30	12	189 47	- 2 25	-2.0		1993 Mar 31	7	4.45423	666	44.21
2449472.86	1994 Apr 30	9	219 47	-13 25	-2.0		1994 May 1	17	4.42272	662	44.53
2449869.97	1995 Jun 1	11	250 31	-21 14	-2.1		1995 Jun 2	23	4.32470	647	45.54
2450268.98	1996 Jul 4	11	282 46	-22 53	-2.2		1996 Jul 5	18	4.18607	626	47.05
2450670.06	1997 Aug 9	13	317 00	-16 40	-2.4		1997 Aug 10	1	4.04908	606	48.64
2451072.62	1998 Sep 16	3	353 03	- 4 11	-2.5		1998 Sep 15	17	3.96289	593	49.70
2451475.29	1999 Oct 23	19	29 56	+10 00	-2.5		1999 Oct 22	14	3.96288	593	49.70
2451876.59	2000 Nov 28	2	66 09	+20 26	-2.4		2000 Nov 26	15	4.04937	606	48.63
2452275.74	2002 Jan 1	6	100 38	+23 01	-2.3		2001 Dec 31	1	4.18747	626	47.03
2452672.88	2003 Feb 2	9	133 06	+17 43	-2.1		2003 Feb 1	19	4.32715	647	45.51
2453068.70	2004 Mar 4	5	163 58	+ 7 38	-2.0		2004 Mar 4	9	4.42566	662	44.50
2453464.14	2005 Apr 3	15	193 58	- 4 03	-2.0		2005 Apr 4	14	4.45664	667	44.19

IX

OPPOSITIONS OF SATURN

	At opposition					Nearest to Earth				
J.D.	Date	UT	Helioc. long.	δ	Magn.	Date	UT	a.u.	10^6 km	Diam.
		h	∘ ′	∘ ′			h			″
2437122.76	1960 Jul 7	6	285 07	-22 11	+0.3	1960 Jul 7	10	9.03630	1352	18.44
2437499.97	1961 Jul 19	11	296 32	-21 00	+0.3	1961 Jul 19	15	9.00005	1346	18.52
2437877.28	1962 Jul 31	19	308 04	-18 56	+0.4	1962 Jul 31	23	8.94220	1338	18.64
2438254.72	1963 Aug 13	5	319 46	-16 04	+0.5	1963 Aug 13	10	8.86542	1326	18.80
2438632.34	1964 Aug 24	20	331 40	-12 29	+0.6	1964 Aug 25	0	8.77330	1312	19.00
2439010.14	1965 Sep 6	15	343 49	- 8 20	+0.8	1965 Sep 6	18	8.66973	1297	19.22
2439388.16	1966 Sep 19	16	356 16	- 3 45	+0.8	1966 Sep 19	18	8.55922	1280	19.47
2439766.41	1967 Oct 2	22	9 01	+ 1 06	+0.5	1967 Oct 2	23	8.44689	1264	19.73
2440144.88	1968 Oct 15	9	22 05	+ 6 01	+0.3	1968 Oct 15	10	8.33812	1247	19.99
2440523.56	1969 Oct 29	1	35 28	+10 46	+0.1	1969 Oct 29	1	8.23843	1232	20.23
2440902.43	1970 Nov 11	22	49 07	+15 05	-0.1	1970 Nov 11	21	8.15333	1220	20.44
2441281.45	1971 Nov 25	23	63 01	+18 39	-0.2	1971 Nov 25	20	8.08794	1210	20.61
2441660.57	1972 Dec 9	2	77 05	+21 10	-0.3	1972 Dec 8	22	8.04649	1204	20.71
2442039.74	1973 Dec 23	6	91 15	+22 23	-0.3	1973 Dec 23	2	8.03186	1202	20.75
2442418.88	1975 Jan 6	9	105 25	+22 10	-0.2	1975 Jan 6	5	8.04528	1204	20.72
2442797.95	1976 Jan 20	11	119 30	+20 32	-0.1	1976 Jan 20	7	8.08603	1210	20.61
2443176.89	1977 Feb 2	9	133 26	+17 42	+0.0	1977 Feb 2	5	8.15141	1219	20.45
2443555.66	1978 Feb 16	4	147 08	+13 55	+0.3	1978 Feb 15	23	8.23723	1232	20.23
2443934.23	1979 Mar 1	17	160 33	+ 9 30	+0.5	1979 Mar 1	14	8.33814	1247	19.99
2444312.58	1980 Mar 14	2	173 39	+ 4 44	+0.8	1980 Mar 13	24	8.44780	1264	19.73
2444690.69	1981 Mar 27	5	186 26	- 0 07	+0.6	1981 Mar 27	4	8.55942	1280	19.47
2445068.60	1982 Apr 9	2	198 55	- 4 51	+0.5	1982 Apr 9	2	8.66692	1297	19.23
2445446.30	1983 Apr 21	19	211 07	- 9 17	+0.4	1983 Apr 21	20	8.76574	1311	19.01
2445823.84	1984 May 3	8	223 05	-13 17	+0.3	1984 May 3	10	8.85252	1324	18.83
2446201.24	1985 May 15	18	234 50	-16 41	+0.2	1985 May 15	21	8.92436	1335	18.67
2446578.52	1986 May 28	0	246 25	-19 23	+0.2	1986 May 28	5	8.97866	1343	18.56
2446955.71	1987 Jun 9	5	257 52	-21 17	+0.2	1987 Jun 9	9	9.01350	1348	18.49
2447332.88	1988 Jun 20	9	269 15	-22 18	+0.2	1988 Jun 20	13	9.02779	1351	18.46
2447710.04	1989 Jul 2	13	280 37	-22 24	+0.2	1989 Jul 2	17	9.02116	1350	18.47
2448087.23	1990 Jul 14	18	292 01	-21 34	+0.3	1990 Jul 14	22	8.99384	1345	18.53
2448464.51	1991 Jul 27	0	303 30	-19 50	+0.3	1991 Jul 27	4	8.94664	1338	18.63
2448841.90	1992 Aug 7	10	315 08	-17 16	+0.4	1992 Aug 7	13	8.88119	1329	18.77
2449219.45	1993 Aug 19	23	326 57	-13 58	+0.5	1993 Aug 20	2	8.79997	1316	18.94
2449597.19	1994 Sep 1	16	339 00	-10 01	+0.7	1994 Sep 1	19	8.70620	1302	19.14
2449975.13	1995 Sep 14	15	351 20	- 5 36	+0.9	1995 Sep 14	17	8.60386	1287	19.37
2450353.29	1996 Sep 26	19	3 58	- 0 50	+0.7	1996 Sep 26	20	8.49761	1271	19.61
2450731.68	1997 Oct 10	4	16 54	+ 4 05	+0.4	1997 Oct 10	3	8.39239	1255	19.86
2451110.28	1998 Oct 23	19	30 10	+ 8 56	+0.2	1998 Oct 23	17	8.29342	1241	20.10
2451489.07	1999 Nov 6	14	43 42	+13 27	-0.0	1999 Nov 6	12	8.20570	1228	20.31
2451868.02	2000 Nov 19	12	57 30	+17 21	-0.2	2000 Nov 19	11	8.13334	1217	20.49
2452247.08	2001 Dec 3	14	71 29	+20 18	-0.3	2001 Dec 3	12	8.08059	1209	20.62
2452626.22	2002 Dec 17	17	85 36	+22 03	-0.3	2002 Dec 17	14	8.05194	1205	20.70
2453005.37	2003 Dec 31	21	99 46	+22 25	-0.3	2003 Dec 31	17	8.05012	1204	20.70
2453384.45	2005 Jan 13	23	113 53	+21 20	-0.2	2005 Jan 13	19	8.07562	1208	20.64

X
OPPOSITIONS OF URANUS

	At opposition					Nearest to Earth				
J.D.	Date	UT	Helioc. long.	δ	Magn.	Date	UT	a.u.	10^6 km	Diam.
		h	° ′	° ′			h			″
2436973.27	1960 Feb 8	19	139 01	+15 49	5.68	1960 Feb 9	15	17.41469	2605	3.94
2437343.20	1961 Feb 12	17	143 45	+14 20	5.67	1961 Feb 13	14	17.37890	2600	3.95
2437713.14	1962 Feb 17	15	148 30	+12 44	5.66	1962 Feb 18	13	17.34903	2595	3.95
2438083.10	1963 Feb 22	14	153 15	+11 04	5.66	1963 Feb 23	11	17.32536	2592	3.96
2438453.09	1964 Feb 27	14	158 01	+ 9 19	5.65	1964 Feb 28	11	17.30820	2589	3.96
2438823.09	1965 Mar 3	14	162 48	+ 7 31	5.65	1965 Mar 4	11	17.29714	2588	3.96
2439193.12	1966 Mar 8	15	167 35	+ 5 39	5.65	1966 Mar 9	11	17.29211	2587	3.96
2439563.16	1967 Mar 13	16	172 23	+ 3 46	5.65	1967 Mar 14	12	17.29242	2587	3.96
2439933.21	1968 Mar 17	17	177 11	+ 1 51	5.65	1968 Mar 18	13	17.29785	2588	3.96
2440303.28	1969 Mar 22	19	181 58	- 0 04	5.65	1969 Mar 23	14	17.30802	2589	3.96
2440673.35	1970 Mar 27	20	186 45	- 2 00	5.66	1970 Mar 28	15	17.32292	2591	3.96
2441043.41	1971 Apr 1	22	191 32	- 3 54	5.66	1971 Apr 2	15	17.34244	2594	3.95
2441413.49	1972 Apr 5	24	196 17	- 5 46	5.67	1972 Apr 6	17	17.36664	2598	3.95
2441783.55	1973 Apr 11	1	201 01	- 7 36	5.68	1973 Apr 11	16	17.39568	2602	3.94
2442153.62	1974 Apr 16	3	205 45	- 9 23	5.68	1974 Apr 16	17	17.42980	2607	3.93
2442523.66	1975 Apr 21	4	210 26	-11 06	5.69	1975 Apr 21	16	17.46923	2613	3.92
2442893.70	1976 Apr 25	5	215 07	-12 45	5.71	1976 Apr 25	16	17.51389	2620	3.91
2443263.73	1977 Apr 30	5	219 46	-14 18	5.72	1977 Apr 30	14	17.56382	2628	3.90
2443633.74	1978 May 5	6	224 23	-15 46	5.73	1978 May 5	13	17.61827	2636	3.89
2444003.73	1979 May 10	5	228 59	-17 08	5.74	1979 May 10	11	17.67668	2644	3.88
2444373.71	1980 May 14	5	233 34	-18 23	5.76	1980 May 14	9	17.73803	2654	3.87
2444743.67	1981 May 19	4	238 06	-19 31	5.77	1981 May 19	6	17.80179	2663	3.85
2445113.61	1982 May 24	3	242 37	-20 32	5.79	1982 May 24	3	17.86700	2673	3.84
2445483.52	1983 May 29	1	247 07	-21 24	5.81	1983 May 28	24	17.93315	2683	3.82
2445853.41	1984 Jun 1	22	251 34	-22 08	5.82	1984 Jun 1	19	17.99944	2693	3.81
2446223.28	1985 Jun 6	19	255 59	-22 44	5.84	1985 Jun 6	15	18.06582	2703	3.80
2446593.10	1986 Jun 11	14	260 23	-23 11	5.85	1986 Jun 11	8	18.13186	2712	3.78
2446962.90	1987 Jun 16	10	264 44	-23 29	5.87	1987 Jun 16	3	18.19778	2722	3.77
2447332.66	1988 Jun 20	4	269 03	-23 39	5.88	1988 Jun 19	19	18.26365	2732	3.75
2447702.41	1989 Jun 24	22	273 20	-23 40	5.90	1989 Jun 24	12	18.32967	2742	3.74
2448072.10	1990 Jun 29	14	277 36	-23 33	5.91	1990 Jun 29	2	18.39557	2752	3.73
2448441.78	1991 Jul 4	7	281 49	-23 18	5.93	1991 Jul 3	17	18.46100	2762	3.71
2448811.43	1992 Jul 7	22	286 01	-22 54	5.94	1992 Jul 7	6	18.52537	2771	3.70
2449181.06	1993 Jul 12	13	290 12	-22 24	5.96	1993 Jul 11	21	18.58816	2781	3.69
2449550.65	1994 Jul 17	4	294 21	-21 46	5.97	1994 Jul 16	9	18.64859	2790	3.68
2449920.23	1995 Jul 21	17	298 29	-21 01	5.98	1995 Jul 20	23	18.70600	2798	3.67
2450289.77	1996 Jul 25	7	302 36	-20 10	6.00	1996 Jul 24	11	18.75986	2806	3.65
2450659.30	1997 Jul 29	19	306 42	-19 13	6.01	1997 Jul 28	24	18.80951	2814	3.64
2451028.79	1998 Aug 3	7	310 46	-18 11	6.02	1998 Aug 2	10	18.85466	2821	3.64
2451398.27	1999 Aug 7	18	314 49	-17 04	6.03	1999 Aug 6	22	18.89523	2827	3.63
2451767.71	2000 Aug 11	5	318 51	-15 52	6.04	2000 Aug 10	8	18.93165	2832	3.62
2452137.13	2001 Aug 15	15	322 51	-14 37	6.04	2001 Aug 14	18	18.96424	2837	3.62
2452506.53	2002 Aug 20	1	326 51	-13 18	6.05	2002 Aug 19	2	18.99340	2841	3.61
2452875.91	2003 Aug 24	10	330 50	-11 55	6.05	2003 Aug 23	12	19.01912	2845	3.60
2453245.27	2004 Aug 27	18	334 49	-10 30	6.06	2004 Aug 26	20	19.04160	2849	3.60
2453614.62	2005 Sep 1	3	338 47	- 9 02	6.06	2005 Aug 31	5	19.06039	2851	3.60

XI
OPPOSITIONS OF NEPTUNE

	At opposition					Nearest to Earth				
J.D.	*Date*	*UT*	*Helioc. long.*	δ	*Magn.*	*Date*	*UT*	*a.u.*	$10^6\,km$	*Diam.*
		h	° ′	° ′			h			″
2437052.56	1960 Apr 28	1	217 46	-12 22	7.70	1960 Apr 28	22	29.32425	4387	2.49
2437420.05	1961 Apr 30	13	219 57	-13 04	7.70	1961 May 1	10	29.32218	4387	2.49
2437787.53	1962 May 3	1	222 08	-13 44	7.69	1962 May 3	22	29.31934	4386	2.49
2438155.01	1963 May 5	12	224 18	-14 23	7.69	1963 May 6	9	29.31657	4386	2.49
2438522.50	1964 May 6	24	226 28	-15 01	7.69	1964 May 7	20	29.31427	4385	2.49
2438889.99	1965 May 9	12	228 39	-15 38	7.69	1965 May 10	6	29.31296	4385	2.49
2439257.48	1966 May 11	24	230 49	-16 13	7.69	1966 May 12	18	29.31252	4385	2.49
2439624.97	1967 May 14	11	232 59	-16 47	7.69	1967 May 15	4	29.31256	4385	2.49
2439992.47	1968 May 15	23	235 10	-17 20	7.69	1968 May 16	17	29.31259	4385	2.49
2440359.97	1969 May 18	11	237 20	-17 51	7.69	1969 May 19	3	29.31208	4385	2.49
2440727.48	1970 May 20	24	239 31	-18 21	7.69	1970 May 21	16	29.31073	4385	2.49
2441094.98	1971 May 23	12	241 42	-18 49	7.69	1971 May 24	4	29.30802	4384	2.49
2441462.50	1972 May 24	24	243 53	-19 16	7.69	1972 May 25	15	29.30404	4384	2.50
2441830.01	1973 May 27	12	246 04	-19 41	7.69	1973 May 28	4	29.29869	4383	2.50
2442197.52	1974 May 30	0	248 14	-20 04	7.69	1974 May 30	15	29.29252	4382	2.50
2442565.04	1975 Jun 1	13	250 25	-20 25	7.69	1975 Jun 2	3	29.28595	4381	2.50
2442932.55	1976 Jun 3	1	252 36	-20 45	7.69	1976 Jun 3	14	29.27971	4380	2.50
2443300.07	1977 Jun 5	14	254 46	-21 03	7.69	1977 Jun 6	2	29.27441	4379	2.50
2443667.58	1978 Jun 8	2	256 57	-21 19	7.69	1978 Jun 8	13	29.27000	4379	2.50
2444035.10	1979 Jun 10	14	259 07	-21 33	7.69	1979 Jun 11	1	29.26644	4378	2.50
2444402.62	1980 Jun 12	3	261 18	-21 45	7.69	1980 Jun 12	12	29.26313	4378	2.50
2444770.15	1981 Jun 14	16	263 29	-21 55	7.69	1981 Jun 14	24	29.25982	4377	2.50
2445137.68	1982 Jun 17	4	265 40	-22 03	7.69	1982 Jun 17	13	29.25579	4377	2.50
2445505.21	1983 Jun 19	17	267 52	-22 09	7.69	1983 Jun 19	24	29.25075	4376	2.50
2445872.75	1984 Jun 21	6	270 03	-22 14	7.68	1984 Jun 21	13	29.24429	4375	2.50
2446240.29	1985 Jun 23	19	272 15	-22 16	7.68	1985 Jun 24	1	29.23640	4374	2.50
2446607.83	1986 Jun 26	8	274 26	-22 16	7.68	1986 Jun 26	14	29.22723	4372	2.50
2446975.36	1987 Jun 28	21	276 37	-22 14	7.68	1987 Jun 29	1	29.21714	4371	2.50
2447342.90	1988 Jun 30	10	278 49	-22 09	7.68	1988 Jun 30	14	29.20707	4369	2.50
2447710.44	1989 Jul 2	23	281 00	-22 03	7.68	1989 Jul 3	1	29.19743	4368	2.50
2448077.97	1990 Jul 5	11	283 11	-21 55	7.68	1990 Jul 5	13	29.18878	4367	2.51
2448445.51	1991 Jul 8	0	285 22	-21 45	7.68	1991 Jul 8	1	29.18090	4365	2.51
2448813.05	1992 Jul 9	13	287 34	-21 33	7.67	1992 Jul 9	12	29.17374	4364	2.51
2449180.60	1993 Jul 12	2	289 46	-21 18	7.67	1993 Jul 12	2	29.16686	4363	2.51
2449548.14	1994 Jul 14	15	291 58	-21 02	7.67	1994 Jul 14	13	29.15978	4362	2.51
2449915.69	1995 Jul 17	5	294 10	-20 44	7.67	1995 Jul 17	2	29.15207	4361	2.51
2450283.24	1996 Jul 18	18	296 22	-20 24	7.67	1996 Jul 18	14	29.14330	4360	2.51
2450650.80	1997 Jul 21	7	298 34	-20 02	7.67	1997 Jul 21	3	29.13332	4358	2.51
2451018.34	1998 Jul 23	20	300 46	-19 38	7.67	1998 Jul 23	16	29.12188	4357	2.51
2451385.89	1999 Jul 26	9	302 59	-19 13	7.66	1999 Jul 26	4	29.10959	4355	2.51
2451753.45	2000 Jul 27	23	305 11	-18 46	7.66	2000 Jul 27	17	29.09690	4353	2.51
2452120.99	2001 Jul 30	12	307 23	-18 17	7.66	2001 Jul 30	4	29.08475	4351	2.51
2452488.53	2002 Aug 2	1	309 35	-17 47	7.66	2002 Aug 1	17	29.07346	4349	2.52
2452856.07	2003 Aug 4	14	311 47	-17 15	7.66	2003 Aug 4	4	29.06334	4348	2.52
2453223.63	2004 Aug 6	3	314 00	-16 42	7.66	2004 Aug 5	17	29.05443	4346	2.52
2453591.17	2005 Aug 8	16	316 12	-16 07	7.66	2005 Aug 8	5	29.04636	4345	2.52

XII
OPPOSITIONS OF PLUTO

	At opposition				Nearest to Earth			
J.D.	Date	UT	Helioc. long.	δ	Date	UT	a.u.	10^6 km
		h	° ′	° ′		h		
2436989.00	1960 Feb 24	12	154 54	+21 29	1960 Feb 27	10	32.82723	4911
2437356.19	1961 Feb 25	16	156 51	+21 07	1961 Feb 28	14	32.58912	4875
2437723.42	1962 Feb 27	22	158 51	+20 44	1962 Mar 2	20	32.35543	4840
2438090.68	1963 Mar 2	4	160 52	+20 18	1963 Mar 5	0	32.12706	4806
2438457.98	1964 Mar 3	11	162 56	+19 51	1964 Mar 6	8	31.90420	4773
2438825.31	1965 Mar 5	19	165 02	+19 22	1965 Mar 8	14	31.68692	4740
2439192.69	1966 Mar 8	4	167 10	+18 50	1966 Mar 10	23	31.47506	4709
2439560.10	1967 Mar 10	14	169 20	+18 17	1967 Mar 13	8	31.26818	4678
2439927.55	1968 Mar 12	1	171 32	+17 42	1968 Mar 14	19	31.06633	4647
2440295.04	1969 Mar 14	13	173 47	+17 05	1969 Mar 17	6	30.86942	4618
2440662.57	1970 Mar 17	2	176 03	+16 27	1970 Mar 19	17	30.67802	4589
2441030.13	1971 Mar 19	15	178 22	+15 46	1971 Mar 22	6	30.49244	4562
2441397.72	1972 Mar 21	5	180 42	+15 04	1972 Mar 23	18	30.31347	4535
2441765.35	1973 Mar 23	20	183 04	+14 20	1973 Mar 26	9	30.14180	4509
2442133.01	1974 Mar 26	12	185 28	+13 34	1974 Mar 28	22	29.97843	4485
2442500.71	1975 Mar 29	5	187 53	+12 47	1975 Mar 31	14	29.82413	4462
2442868.43	1976 Mar 30	22	190 20	+11 58	1976 Apr 2	4	29.67935	4440
2443236.19	1977 Apr 2	16	192 49	+11 08	1977 Apr 4	21	29.54450	4420
2443603.97	1978 Apr 5	11	195 19	+10 16	1978 Apr 7	13	29.41923	4401
2443971.78	1979 Apr 8	7	197 51	+ 9 22	1979 Apr 10	6	29.30349	4384
2444339.62	1980 Apr 10	3	200 24	+ 8 28	1980 Apr 12	1	29.19676	4368
2444707.48	1981 Apr 12	23	202 58	+ 7 32	1981 Apr 14	19	29.09914	4353
2445075.36	1982 Apr 15	21	205 33	+ 6 35	1982 Apr 17	15	29.01060	4340
2445443.25	1983 Apr 18	18	208 09	+ 5 37	1983 Apr 20	9	28.93133	4328
2445811.17	1984 Apr 20	16	210 45	+ 4 38	1984 Apr 22	5	28.86165	4318
2446179.08	1985 Apr 23	14	213 22	+ 3 39	1985 Apr 24	23	28.80214	4309
2446547.01	1986 Apr 26	12	215 59	+ 2 39	1986 Apr 27	20	28.75354	4301
2446914.93	1987 Apr 29	10	218 35	+ 1 39	1987 Apr 30	14	28.71641	4296
2447282.86	1988 May 1	9	221 12	+ 0 39	1988 May 2	9	28.69149	4292
2447650.80	1989 May 4	7	223 48	- 0 22	1989 May 5	4	28.67883	4290
2448018.72	1990 May 7	5	226 24	- 1 23	1990 May 7	23	28.67852	4290
2448386.64	1991 May 10	3	229 00	- 2 23	1991 May 10	19	28.68986	4292
2448754.55	1992 May 12	1	231 34	- 3 23	1992 May 12	12	28.71252	4295
2449122.46	1993 May 14	23	234 09	- 4 23	1993 May 15	8	28.74594	4300
2449490.35	1994 May 17	20	236 42	- 5 22	1994 May 18	2	28.78973	4307
2449858.23	1995 May 20	17	239 14	- 6 21	1995 May 20	21	28.84351	4315
2450226.08	1996 May 22	14	241 45	- 7 19	1996 May 22	13	28.90709	4324
2450593.92	1997 May 25	10	244 15	- 8 15	1997 May 25	7	28.98061	4335
2450961.73	1998 May 28	5	246 43	- 9 11	1998 May 27	23	29.06417	4348
2451329.52	1999 May 31	0	249 09	-10 05	1999 May 30	15	29.15833	4362
2451697.28	2000 Jun 1	19	251 34	-10 58	2000 Jun 1	6	29.26327	4378
2452065.01	2001 Jun 4	12	253 57	-11 49	2001 Jun 3	20	29.37946	4395
2452432.71	2002 Jun 7	5	256 18	-12 39	2002 Jun 6	10	29.50636	4414
2452800.38	2003 Jun 9	21	258 38	-13 28	2003 Jun 8	22	29.64359	4435
2453168.04	2004 Jun 11	13	260 56	-14 14	2004 Jun 10	12	29.79034	4457
2453535.65	2005 Jun 14	4	263 12	-14 59	2005 Jun 12	23	29.94585	4480

XIII
CONJUNCTIONS OF SUPERIOR PLANETS WITH THE SUN

Mars	Jupiter	Saturn	Uranus	Neptune	Pluto
			1960 Aug 14	1960 Nov 1	1960 Aug 29
1961 Dec 14	1961 Jan 5 O	1961 Jan 11 O	1961 Aug 19	1961 Nov 3	1961 Aug 31
	1962 Feb 8	1962 Jan 22	1962 Aug 24	1962 Nov 6	1962 Sep 2
	1963 Mar 16	1963 Feb 3	1963 Aug 29	1963 Nov 8	1963 Sep 5
1964 Feb 17	1964 Apr 22	1964 Feb 15	1964 Sep 2	1964 Nov 9	1964 Sep 6
	1965 May 30	1965 Feb 26	1965 Sep 8	1965 Nov 12	1965 Sep 9
1966 Apr 29 O	1966 Jul 5 O	1966 Mar 10	1966 Sep 13	1966 Nov 14	1966 Sep 11
	1967 Aug 8	1967 Mar 23	1967 Sep 18	1967 Nov 17	1967 Sep 13
1968 Jun 21	1968 Sep 9	1968 Apr 5	1968 Sep 22	1968 Nov 18	1968 Sep 15
	1969 Oct 9	1969 Apr 18	1969 Sep 27	1969 Nov 20	1969 Sep 18
1970 Aug 2	1970 Nov 9	1970 May 2	1970 Oct 2	1970 Nov 23	1970 Sep 20
	1971 Dec 10	1971 May 17	1971 Oct 7	1971 Nov 25	1971 Sep 23
1972 Sep 7		1972 May 31	1972 Oct 11	1972 Nov 27	1972 Sep 24
	1973 Jan 10 O	1973 Jun 15	1973 Oct 16	1973 Nov 29	1973 Sep 27
1974 Oct 14	1974 Feb 13	1974 Jun 30	1974 Oct 21	1974 Dec 1	1974 Sep 30
	1975 Mar 22	1975 Jul 15 O	1975 Oct 26	1975 Dec 4	1975 Oct 2
1976 Nov 25 O	1976 Apr 27	1976 Jul 29	1976 Oct 30	1976 Dec 5	1976 Oct 4
	1977 Jun 4	1977 Aug 13	1977 Nov 4	1977 Dec 8	1977 Oct 7
	1978 Jul 10 O	1978 Aug 27	1978 Nov 9	1978 Dec 10	1978 Oct 10
1979 Jan 20	1979 Aug 13	1979 Sep 10	1979 Nov 14	1979 Dec 12	1979 Oct 13
	1980 Sep 13	1980 Sep 23	1980 Nov 18 O	1980 Dec 14	1980 Oct 14
1981 Apr 2	1981 Oct 14	1981 Oct 6	1981 Nov 22 O	1981 Dec 16	1981 Oct 17
	1982 Nov 13	1982 Oct 18	1982 Nov 27 O	1982 Dec 19	1982 Oct 20
1983 Jun 3	1983 Dec 14	1983 Oct 31	1983 Dec 2 O	1983 Dec 21	1983 Oct 23
		1984 Nov 11	1984 Dec 5 O	1984 Dec 22	1984 Oct 25
1985 Jul 18	1985 Jan 14 O	1985 Nov 23	1985 Dec 10 O	1985 Dec 25	1985 Oct 28
	1986 Feb 18	1986 Dec 4	1986 Dec 14 O	1986 Dec 27	1986 Oct 31
1987 Aug 25	1987 Mar 27	1987 Dec 16	1987 Dec 19 O	1987 Dec 29	1987 Nov 2
	1988 May 2	1988 Dec 26	1988 Dec 22 O	1988 Dec 31	1988 Nov 4
1989 Sep 29	1989 Jun 9		1989 Dec 27 O		1989 Nov 7
	1990 Jul 15 O	1990 Jan 6	1990 Dec 31	1990 Jan 2	1990 Nov 10
1991 Nov 8 O	1991 Aug 17	1991 Jan 18 O		1991 Jan 5	1991 Nov 13
	1992 Sep 17	1992 Jan 29	1992 Jan 5	1992 Jan 7	1992 Nov 15
1993 Dec 27	1993 Oct 18	1993 Feb 9	1993 Jan 8	1993 Jan 8	1993 Nov 17
	1994 Nov 17	1994 Feb 21	1994 Jan 12	1994 Jan 11	1994 Nov 20
	1995 Dec 18 O	1995 Mar 6	1995 Jan 17	1995 Jan 13	1995 Nov 23
1996 Mar 4		1996 Mar 17	1996 Jan 21	1996 Jan 16	1996 Nov 25
	1997 Jan 19	1997 Mar 30	1997 Jan 24	1997 Jan 17	1997 Nov 27
1998 May 12 O	1998 Feb 23	1998 Apr 13	1998 Jan 28	1998 Jan 19	1998 Nov 30
	1999 Apr 1	1999 Apr 27	1999 Feb 2	1999 Jan 22	1999 Dec 3
2000 Jul 1	2000 May 8	2000 May 10	2000 Feb 6	2000 Jan 24 O	2000 Dec 4
	2001 Jun 14	2001 May 25	2001 Feb 9	2001 Jan 26 O	2001 Dec 7
2002 Aug 10	2002 Jul 20	2002 Jun 9	2002 Feb 13	2002 Jan 28 O	2002 Dec 9
	2003 Aug 22	2003 Jun 24	2003 Feb 17	2003 Jan 30 O	2003 Dec 12
2004 Sep 15	2004 Sep 21	2004 Jul 8 O	2004 Feb 22	2004 Feb 2 O	2004 Dec 13
	2005 Oct 22	2005 Jul 23	2005 Feb 25	2005 Feb 3 O	2005 Dec 16

O means that the planet is occulted by the Sun's disk

XIV

EARTH IN PERIHELION				EARTH IN APHELION			
1960 Jan 4	19^h	0.983254	$a.u.$	1960 Jul 2	22^h	1.016740	$a.u.$
1961 Jan 2	17	208		1961 Jul 5	20	719	
1962 Jan 2	5	298		1962 Jul 4	5	744	
1963 Jan 4	19	272		1963 Jul 4	14	716	
1964 Jan 2	21	264		1964 Jul 5	12	751	
1965 Jan 2	19	0.983319		1965 Jul 3	9	1.016722	
1966 Jan 3	23	269		1966 Jul 5	12	659	
1967 Jan 2	6	287		1967 Jul 5	15	735	
1968 Jan 4	18	294		1968 Jul 2	20	736	
1969 Jan 3	1	248		1969 Jul 5	20	707	
1970 Jan 1	21	0.983308		1970 Jul 4	15	1.016748	
1971 Jan 4	18	265		1971 Jul 4	4	728	
1972 Jan 3	4	215		1972 Jul 5	17	755	
1973 Jan 2	12	276		1973 Jul 3	15	753	
1974 Jan 4	10	262		1974 Jul 5	1	684	
1975 Jan 2	13	0.983288		1975 Jul 6	3	1.016745	
1976 Jan 4	11	319		1976 Jul 3	3	731	
1977 Jan 3	10	262		1977 Jul 5	21	662	
1978 Jan 1	23	310		1978 Jul 5	0	706	
1979 Jan 4	23	298		1979 Jul 3	21	702	
1980 Jan 3	15	0.983254		1980 Jul 5	18	1.016726	
1981 Jan 2	2	318		1981 Jul 3	23	749	
1982 Jan 4	11	294		1982 Jul 4	14	684	
1983 Jan 2	16	263		1983 Jul 6	10	745	
1984 Jan 3	22	278		1984 Jul 3	7	759	
1985 Jan 3	20	0.983221		1985 Jul 5	10	1.016690	
1986 Jan 2	5	278		1986 Jul 5	10	745	
1987 Jan 4	23	300		1987 Jul 4	1	738	
1988 Jan 3	24	251		1988 Jul 5	24	718	
1989 Jan 1	22	307		1989 Jul 4	12	724	
1990 Jan 4	17	0.983302		1990 Jul 4	5	1.016653	
1991 Jan 3	3	280		1991 Jul 6	15	703	
1992 Jan 3	15	323		1992 Jul 3	12	740	
1993 Jan 4	3	282		1993 Jul 4	22	667	
1994 Jan 2	6	301		1994 Jul 5	19	724	
1995 Jan 4	11	0.983302		1995 Jul 4	2	1.016742	
1996 Jan 4	7	222		1996 Jul 5	19	718	
1997 Jan 1	23	267		1997 Jul 4	19	754	
1998 Jan 4	21	299		1998 Jul 4	0	697	
1999 Jan 3	13	280		1999 Jul 6	23	719	
2000 Jan 3	6	0.983320		2000 Jul 3	24	1.016742	
2001 Jan 4	9	285		2001 Jul 4	14	643	
2002 Jan 2	14	289		2002 Jul 6	4	689	
2003 Jan 4	5	320		2003 Jul 4	6	729	
2004 Jan 4	18	264		2004 Jul 5	11	694	
2005 Jan 2	1	296		2005 Jul 5	5	742	

XV

MARS IN PERIHELION					MARS IN APHELION			
1960 May 26	14^h	1.38155 $a.u.$			1961 May 5	4^h	1.66586 $a.u.$
1962 Apr 13	13	1.38137			1963 Mar 23	1	1.66587
1964 Feb 29	16	1.38150			1965 Feb 6	23	1.66595
1966 Jan 16	10	1.38149			1966 Dec 26	3	1.66601
1967 Dec 4	12	1.38123			1968 Nov 12	0	1.66601
1969 Oct 21	16	1.38135			1970 Sep 30	1	1.66603
1971 Sep 8	14	1.38152			1972 Aug 17	4	1.66599
1973 Jul 26	12	1.38139			1974 Jul 5	2	1.66589
1975 Jun 13	16	1.38141			1976 May 21	23	1.66591
1977 Apr 30	11	1.38154			1978 Apr 9	2	1.66603
1979 Mar 18	10	1.38124			1980 Feb 25	1	1.66605
1981 Feb 2	16	1.38125			1982 Jan 12	1	1.66602
1982 Dec 21	15	1.38150			1983 Nov 30	3	1.66592
1984 Nov 7	10	1.38147			1985 Oct 17	1	1.66591
1986 Sep 25	14	1.38133			1987 Sep 3	22	1.66599
1988 Aug 12	13	1.38153			1989 Jul 22	1	1.66604
1990 Jun 30	10	1.38133			1991 Jun 9	2	1.66603
1992 May 17	14	1.38120			1993 Apr 25	23	1.66604
1994 Apr 4	16	1.38141			1995 Mar 14	2	1.66599
1996 Feb 20	11	1.38148			1997 Jan 29	2	1.66593
1998 Jan 7	11	1.38132			1998 Dec 16	22	1.66590
1999 Nov 25	13	1.38149			2000 Nov 2	21	1.66600
2001 Oct 12	8	1.38141			2002 Sep 21	1	1.66614
2003 Aug 30	11	1.38114			2004 Aug 7	23	1.66614
2005 Jul 17	16	1.38129					

XVI
THE OUTER PLANETS IN PERIHELION AND APHELION

Jupiter in	perihelion	1951 Nov 21	4.94841 a.u.
	aphelion	1957 Oct 23	5.45640
	perihelion	1963 Sep 26	4.95133
	aphelion	1969 Sep 3	5.45375
	perihelion	1975 Aug 12	4.95263
	aphelion	1981 Jul 28	5.45352
	perihelion	1987 Jul 10	4.95242
	aphelion	1993 Jun 14	5.45428
	perihelion	1999 May 20	4.95047
	aphelion	2005 Apr 14	5.45652
Saturn in	aphelion	1900 Jul 10	10.06986 a.u.
	perihelion	1915 Feb 20	9.01354
	aphelion	1929 Nov 11	10.04668
	perihelion	1944 Sep 8	9.02885
	aphelion	1959 May 29	10.06641
	perihelion	1974 Jan 8	9.01531
	aphelion	1988 Sep 11	10.04440
	perihelion	2003 Jul 26	9.03090
	aphelion	2018 Apr 17	10.06565
	perihelion	2032 Nov 28	9.01491
	aphelion	2047 Jul 15	10.04615
Uranus in	aphelion	1925 Apr 1	20.09733 a.u.
	perihelion	1966 May 19	18.28478
	aphelion	2009 Feb 27	20.09880
	perihelion	2050 Aug 13	18.28300
Neptune in	perihelion	1876 Sep 2	29.81482 a.u.
	(a)	1881 Dec 19	29.82130
	(b)	1886 Jul 18	29.81738
	aphelion	1959 Jul 21	30.33129
	(c)	1965 Sep 30	30.32219
	(d)	1968 Nov 22	30.32349
	perihelion	2042 Sep 8	29.80587
	(a)	2050 Jan 2	29.81626
	(b)	2050 May 29	29.81626
Pluto in	aphelion	1866 Jul 12	49.28205 a.u.
	perihelion	1989 Sep 12	29.64730

(a) *Maximum of the radius vector, which does not correspond to an aphelion.*

(b) *Second minimum of the radius vector.*

(c) *Minimum of the radius vector, which does not correspond to a perihelion.*

(d) *Second maximum of the radius vector.*

The radius vector of Pluto will be smaller than that of Neptune from
1979 January 21 to 1999 March 14.

XVII

MARS IN ASCENDING NODE			MARS IN DESCENDING NODE		
1960 Sep 25	11^h		1961 Oct 13	2^h	
1962 Aug 13	9		1963 Aug 31	1	
1964 Jun 30	9		1965 Jul 18	0	
1966 May 18	8		1967 Jun 4	24	
1968 Apr 4	6		1969 Apr 21	22	
1970 Feb 20	5		1971 Mar 9	21	
1972 Jan 8	5		1973 Jan 24	21	
1973 Nov 25	4		1974 Dec 12	19	
1975 Oct 13	3		1976 Oct 29	18	
1977 Aug 30	2		1978 Sep 16	18	
1979 Jul 18	0		1980 Aug 3	16	
1981 Jun 3	23		1982 Jun 21	15	
1983 Apr 21	23		1984 May 8	14	
1985 Mar 8	22		1986 Mar 26	13	
1987 Jan 24	20		1988 Feb 11	12	
1988 Dec 11	20		1989 Dec 29	11	
1990 Oct 29	19		1991 Nov 16	10	
1992 Sep 15	17		1993 Oct 3	9	
1994 Aug 3	17		1995 Aug 21	7	
1996 Jun 20	16		1997 Jul 8	6	
1998 May 8	15		1999 May 26	5	
2000 Mar 25	14		2001 Apr 12	5	
2002 Feb 10	13		2003 Feb 28	4	
2003 Dec 29	11		2005 Jan 15	3	
2005 Nov 15	11				

XVIII

ZERO AND EXTREME LATITUDES OF THE OUTER PLANETS

JUPITER

1960 Nov 4	Descending node	
1963 Aug 21	Greatest heliocentric latitude South :	-1°18'22"
1966 May 30	Ascending node	
1969 Jul 18	Greatest heliocentric latitude North :	+1°18'21"
1972 Sep 14	Descending node	
1975 Jul 2	Greatest heliocentric latitude South :	-1°18'20"
1978 Apr 10	Ascending node	
1981 May 28	Greatest heliocentric latitude North :	+1°18'21"
1984 Jul 27	Descending node	
1987 May 15	Greatest heliocentric latitude South :	-1°18'19"
1990 Feb 19	Ascending node	
1993 Apr 7	Greatest heliocentric latitude North :	+1°18'18"
1996 Jun 6	Descending node	
1999 Mar 25	Greatest heliocentric latitude South :	-1°18'17"
2001 Dec 30	Ascending node	
2005 Feb 17	Greatest heliocentric latitude North :	+1°18'13"

SATURN

1946 Feb 28	Ascending node	
1953 Mar 22	Greatest heliocentric latitude North :	+2°29'20"
1961 Apr 4	Descending node	
1968 Nov 22	Greatest heliocentric latitude South :	-2°29'19"
1975 Aug 10	Ascending node	
1982 Aug 27	Greatest heliocentric latitude North :	+2°29'09"
1990 Sep 4	Descending node	
1998 Apr 20	Greatest heliocentric latitude South :	-2°29'07"
2005 Jan 8	Ascending node	

URANUS

1945 Jul 20	Ascending node	
1965 May 25	Greatest heliocentric latitude North :	+0°46'21"
1984 Dec 21	Descending node	

NEPTUNE

1920 Jun 2	Ascending node	
1961 Dec 19	Greatest heliocentric latitude North :	+1°46'23"
2003 Aug 10	Descending node	

PLUTO

1930 Sep 9	Ascending node *(a)*	
1980 Feb 4	Greatest heliocentric latitude North :	+17°08'14"
2018 Oct 15	Descending node *(b)*	

(a) Passage through the node on the ecliptic of 1950.0 on 1930 September 6

(b) Passage through the node on the ecliptic of 1950.0 on 2018 October 21

XIX
ZERO AND EXTREME VALUES OF THE DECLINATIONS OF THE PLANETS

VENUS

Each year Venus reaches its greatest northern declination. Sometimes this extreme value is less than +22°, as in 1977. During the period 1960-2005, Venus' declination reaches a maximum greater than +27° on the following dates :

```
1964 May 6    +27°33'
1972 May 6    +27°37'
1980 May 6    +27°41'
1988 May 6    +27°44'
1996 May 5    +27°47'  (greatest northern declination of Venus
2004 May 5    +27°49'                during the 20th century)
```

During the same period 1960-2005, Venus reaches a maximum southern declination south of δ = −26° on the following dates :

```
1965 Nov 7    -26°45'
1973 Nov 7    -26°48'
1981 Nov 7    -26°52'
1989 Nov 7    -26°57'
1997 Nov 6    -27°01'
2005 Nov 6    -27°05'
```

Venus' southernmost declination during the 20th century was −27°51', on 1906 November 3. On 1874 November 6, Venus reached an extreme southern declination of −28°05'.

MARS

During the period 1900-2005, Mars reaches a maximum northern declination north of +26°, or a maximum southern declination south of −28° on the following dates :

```
1914 Jan 26    +27°11'          1907 Jul 29    -28°54'
1929 Jan  1    +26°47'          1954 Jul 19    -28°22'
1946 Feb  7    +26°35'          1986 Aug  2    -28°43'
1961 Jan 17    +27°13'
1975 Dec 20    +26°04'
1993 Jan 30    +27°01'
```

JUPITER

```
1960 Nov  2    -23°25'53"
1963 Apr 14      0°   S → N
1966 May 11    +23°24'28"
1968 Dec  4      0°   N → S ⎫
1969 Mar  2      0°   S → N ⎬
1969 Aug  2      0°   N → S ⎭
1972 Oct 10    -23°28'51"
1975 Mar 28      0°   S → N
1978 Apr 21    +23°27'06"
1980 Nov 10      0°   N → S ⎫
1981 Apr  9      0°   S → N ⎬
1981 Jul  8      0°   N → S ⎭
1984 Sep 17    -23°29'46"
1987 Mar 13      0°   S → N
1990 Mar 27    +23°29'27"
1992 Oct 22      0°   N → S
1996 Sep 11    -23°23'59"
1999 Feb 24      0°   S → N
2002 Mar 12    +23°27'24"
2004 Oct  6      0°   N → S
```

S → N means that the planet crosses the celestial equator from South to North.

N → S means that the planet crosses the celestial equator from North to South.

Triple crossings are indicated with a parenthesis.

XIX (continued)

SATURN	1951 Sep 25	0° N → S
	1959 Nov 2	-22°46'10"
	1967 Apr 12	0° S → N ⎞
	1967 Nov 18	0° N → S ⎟
	1967 Dec 20	0° S → N ⎠
	1974 May 14	+22°45'05"
	1980 Nov 1	0° N → S ⎞
	1981 Mar 30	0° S → N ⎟
	1981 Jul 29	0° N → S ⎠
	1989 Oct 2	-22°46'51"
	1996 May 25	0° S → N ⎞
	1996 Aug 30	0° N → S ⎟
	1997 Feb 16	0° S → N ⎠
	2004 Apr 2	+22°48'46"

URANUS	1927 May 1	0° S → N ⎞
	1927 Sep 16	0° N → S ⎟
	1928 Feb 21	0° S → N ⎠
	1950 Feb 26	+23°42'51"
	1968 Oct 25	0° N → S ⎞
	1969 Mar 26	0° S → N ⎟
	1969 Aug 11	0° N → S ⎠
	1989 Sep 1	-23°42'27"

NEPTUNE	1943 Nov 1	0° N → S ⎞
	1944 Mar 7	0° S → N ⎟
	1944 Sep 4	0° N → S ⎠
	1986 Oct 29	-22°21'37"

PLUTO	1987 Nov 12	0° N → S ⎞
	1988 Feb 27	0° S → N ⎟
	1988 Sep 5	0° N → S ⎠

XX

PLANETARY CONJUNCTIONS

(a) MERCURY – VENUS

1976 May 25	2^h	-1°48'	7°W
1976 Jul 24	14	+0 27	10°E
1976 Sep 6	4	-5 12	22°E
1977 Mar 27	19	-8 19	12°E
1977 Dec 24	13	+2 47	7°W
1978 Mar 12	22	+1 20	12°E
1978 Mar 28	19	+4 03	16°E
1978 Oct 27	4	+5 08	16°E
1979 Aug 3	13	-5 47	6°W
1979 Sep 22	20	-0 28	8°E
1979 Nov 8	20	-2 05	20°E
1980 Jun 1	18	+0 19	20°E
1981 Feb 22	1	+4 53	10°W
1981 May 3	19	+1 13	7°E
1981 Jun 9	11	-1 42	17°E
1982 Jan 9	14	-5 27	17°E
1982 Oct 5	10	-2 59	7°W
1982 Dec 2	18	-1 21	7°E
1983 Jan 7	10	+2 04	16°E
1983 Aug 6	6	+5 47	24°E
1984 Apr 29	24	+0 41	12°W
1984 Jun 25	3	+0 49	3°E
1984 Aug 16	16	-6 07	17°E
1985 Mar 23	2	-5 17	16°E
1985 Dec 4	4	+1 38	11°W
1986 Feb 6	13	-0 37	5°E
1986 Mar 8	13	+4 56	13°E
1986 Oct 18	14	+4 23	24°E
1987 Jul 12	1	-4 50	12°W
1987 Aug 18	22	+0 30	2°W
1987 Oct 20	1	-3 28	15°E
1988 Jun 8	18	-2 22	7°E
1989 Feb 1	3	+3 56	15°W
1989 Apr 4	14	+0 18	1°E
1989 May 16	7	+0 34	11°E
1990 Feb 4	6	-7 06	25°W
1990 Sep 14	15	-3 21	11°W
1990 Oct 16	5	+0 02	4°W
1990 Dec 18	23	+1 27	12°E

1991 Aug 7	6^h	+2°08'	22°E
1991 Aug 20	19	+3 43	5°E
1991 Aug 29	5	+6 04	12°W
1992 Apr 5	22	+2 20	17°W
1992 May 28	2	+0 30	4°W
1992 Jul 25	15	-6 03	12°E
1993 Apr 16	11	-8 25	23°W
1993 Nov 14	13	+0 45	15°W
1993 Dec 25	8	-0 59	5°W
1994 Feb 15	19	+5 04	7°E
1994 Nov 12	18	+5 26	16°W
1995 Jun 19	7	-3 36	17°W
1995 Jul 20	14	+0 23	9°W
1995 Sep 28	21	-5 01	11°E
1996 Jun 23	12	+1 36	19°W
1997 Jan 12	13	+2 43	19°W
1997 Mar 2	0	-0 51	8°W
1997 Apr 21	18	+3 04	5°E
1998 Jan 26	17	-8 11	17°W
1998 Aug 25	23	-2 45	16°W
1998 Sep 11	0	+0 21	13°W
1998 Nov 28	12	+0 15	7°E
1999 Aug 26	13	+10 10	13°W
2000 Mar 15	0	+2 30	22°W
2000 Apr 28	9	-0 21	12°W
2000 Jul 2	9	-4 57	6°E
2001 Apr 6	21	-10 00	14°W
2002 Jan 26	1	+4 28	3°E
2002 Nov 3	8	+6 34	7°W
2003 May 28	0	-2 21	22°W
2003 Jun 21	2	-0 25	16°W
2003 Sep 7	3	-6 07	5°E
2004 Jun 13	1	+1 21	7°W
2004 Dec 29	5	+1 12	22°W
2005 Jan 14	1	-0 21	19°W
2005 Mar 28	23	+4 49	1°W
2005 Jun 27	20	-0 05	23°E
2005 Jul 7	8	-1 38	26°E

XX (continued)

(b) MERCURY – MARS

1976	Nov 13	10^h	-0°25'	3°E
1977	Jan 12	12	+4 05	14°W
1977	Feb 12	19	-0 07	21°W
1978	Nov 5	7	-1 52	20°E
1978	Nov 29	20	+0 05	13°E
1979	Feb 4	4	-1 03	4°W
1979	Apr 1	22	+2 33	15°W
1979	May 5	8	-2 07	22°W
1981	Jan 23	22	-0 18	15°E
1981	Feb 10	18	+4 00	11°E
1981	Apr 23	11	-0 36	5°W
1983	Apr 9	12	+1 24	14°E
1983	May 8	2	+0 49	7°E
1983	Jul 2	24	-0 12	8°W
1985	Jun 15	15	+0 52	10°E
1985	Aug 15	5	-5 26	8°W
1985	Sep 4	21	-0 01	16°W
1987	Aug 21	7	+0 40	2°E
1989	Aug 5	22	+0 01	18°E
1989	Sep 22	24	-4 26	2°E
1989	Oct 27	3	+1 02	9°W
1991	Oct 14	19	-0 15	8°E
1991	Dec 13	15	+2 56	11°W
1992	Jan 10	20	+0 39	19°W
1993	Oct 6	17	-2 20	23°E
1993	Oct 28	6	-2 29	17°E
1994	Jan 1	10	-0 50	2°W
1994	Feb 27	1	+4 26	14°W
1994	Apr 4	2	-1 28	22°W
1995	Dec 23	9	-1 08	16°E
1996	Jan 13	0	+2 48	12°E
1996	Mar 23	20	-0 54	4°W
1996	May 31	5	-3 42	19°W
1996	Jun 14	14	-3 05	22°W
1998	Mar 11	14	+1 11	15°E
1998	Mar 30	5	+4 06	10°E
1998	Jun 5	11	-0 16	6°W
2000	May 19	15	+1 07	12°E
2000	Jul 7	3	-5 40	2°W
2000	Aug 10	13	-0 05	12°W
2002	Jul 25	12	+0 40	5°E
2004	Jul 10	23	+0 10	22°E
2004	Aug 17	3	-6 14	10°E
2004	Sep 29	12	+0 51	5°W

(c) MERCURY – JUPITER

1976	Apr 12	18^h	+1°57'	11°E
1977	Jun 20	7	+0 09	12°W
1978	Jun 24	8	+1 46	12°E
1979	Aug 30	11	+0 40	13°W
1980	Sep 3	12	+0 19	7°E
1981	Sep 13	19	-2 49	23°E
1981	Oct 20	6	-1 59	4°W
1981	Nov 6	0	+1 12	18°W
1982	Nov 16	9	-0 50	2°W
1983	Nov 26	6	-2 31	14°E
1985	Jan 31	5	-1 18	13°W
1986	Feb 9	14	-1 00	7°E
1987	Apr 19	12	-1 21	17°W
1988	Apr 25	14	+1 21	5°E
1989	Jul 2	17	-0 35	17°W
1990	Jul 7	8	+1 30	6°E
1991	Jul 15	8	-0 05	25°E
1991	Aug 22	22	-5 20	4°W
1991	Sep 10	10	-0 04	18°W
1992	Sep 16	6	+0 29	2°E
1993	Sep 24	12	-2 01	18°E
1994	Nov 28	18	-0 24	9°W
1995	Dec 8	10	-2 08	8°E
1997	Feb 12	14	-1 02	19°W
1998	Feb 22	14	-1 05	1°E
1999	May 1	10	-1 45	22°W
2000	May 8	22	+0 52	0°W
2001	May 16	17	+2 47	21°E
2001	Jun 18	15	-3 39	3°W
2001	Jul 12	22	-1 56	21°W
2002	Jul 20	13	+1 15	0°W
2003	Jul 26	1	+0 23	20°E
2004	Sep 28	21	+0 40	5°W
2005	Oct 6	7	-1 28	13°E

XX (continued)

(d) MERCURY – SATURN

1976 Jul 21	8^h	+1°20′	7°E
1977 Jul 20	1	+0 21	20°E
⎧ 1978 Jul 31	22	-4 43	22°E
1978 Aug 4	6	-5 26	20°E
⎩ 1978 Sep 13	15	+0 04	14°W
1979 Sep 11	23	-0 09	2°W
1980 Sep 9	13	-1 26	12°E
1981 Sep 10	15	-3 34	22°E
1982 Nov 1	6	-0 40	12°W
1983 Oct 31	10	-1 57	1°E
1984 Oct 29	20	-3 14	11°E
⎧ 1985 Oct 30	21	-4 22	21°E
1985 Dec 2	18	+0 17	9°W
⎩ 1985 Dec 16	18	+0 28	21°W
1986 Dec 19	15	-1 19	13°W
1987 Dec 18	23	-2 17	3°W
1988 Dec 16	21	-2 46	9°E
⎧ 1989 Dec 16	22	-2 30	19°E
1990 Jan 10	5	+2 49	3°W
⎩ 1990 Feb 3	15	+0 13	25°W
1991 Feb 5	16	-1 14	16°W
1992 Feb 4	16	-1 28	5°W
1993 Feb 1	22	-0 55	7°E
⎧ 1994 Feb 2	4	+1 21	17°E
1994 Feb 17	16	+5 19	4°E
⎩ 1994 Mar 24	8	-0 15	27°W
1995 Mar 26	0	-0 35	18°W
1996 Mar 23	11	+0 20	5°W
1997 Mar 21	3	+2 06	9°E
1998 May 12	16	-0 49	25°W
1999 May 13	18	+0 40	14°W
2000 May 10	3	+2 16	1°E
2001 May 7	16	+3 40	15°E
2002 Jul 2	11	-0 14	19°W
2003 Jul 1	1	+1 33	5°W
2004 Jun 26	21	+2 06	10°E
2005 Jun 26	6	+1 25	22°E

(e) MERCURY – URANUS

1976 Nov 2	19^h	+0°12′	3°W
1977 Oct 29	1	-0 44	6°E
1978 Oct 24	18	-1 43	15°E
⎧ 1979 Oct 22	0	-2 48	22°E
1979 Nov 25	2	+1 42	10°W
⎩ 1979 Dec 5	0	+2 19	20°W
1980 Dec 3	14	+0 55	15°W
1981 Nov 29	9	-0 08	6°W
1982 Nov 24	17	-1 00	3°E
1983 Nov 20	4	-1 47	11°E
⎧ 1984 Nov 15	14	-2 27	19°E
1984 Dec 23	21	+3 00	17°W
⎩ 1984 Dec 26	8	+2 52	20°W
1985 Dec 29	12	+0 40	18°W
1986 Dec 25	14	-0 24	10°W
1987 Dec 20	20	-1 10	1°W
1988 Dec 14	23	-1 43	8°E
1989 Dec 10	13	-2 02	16°E
⎧ 1990 Dec 10	7	-1 20	20°E
1990 Dec 18	6	+0 37	13°E
⎩ 1991 Jan 23	17	+0 27	22°W
1992 Jan 20	3	-0 38	15°W
1993 Jan 14	8	-1 16	6°W
1994 Jan 9	2	-1 37	4°E
1995 Jan 4	1	-1 39	12°E
⎧ 1996 Jan 1	1	-0 53	19°E
1996 Jan 16	15	+3 17	5°E
⎩ 1996 Feb 16	22	+0 15	25°W
1997 Feb 13	0	-0 54	19°W
1998 Feb 8	5	-1 23	10°W
1999 Feb 2	18	-1 31	1°W
2000 Jan 28	4	-1 19	9°E
⎧ 2001 Jan 22	17	-0 24	17°E
2001 Feb 14	13	+4 38	5°W
⎩ 2001 Mar 10	11	+0 08	27°W
2002 Mar 9	3	-1 13	22°W
2003 Mar 4	13	-1 32	14°W
2004 Feb 26	23	-1 27	5°W
2005 Feb 20	1	-1 00	5°E

XX (continued)

(f) MERCURY – NEPTUNE

1976 Nov 25	15^h	-3°12'	10°E
1977 Nov 20	8	-3 45	17°E
1978 Nov 17	23	-3 59	22°E
1978 Dec 2	12	-1 10	8°E
1978 Dec 31	19	-0 15	21°W
1979 Dec 27	7	-1 23	14°W
1980 Dec 20	14	-2 09	6°W
1981 Dec 14	12	-2 44	2°E
1982 Dec 8	13	-3 13	10°E
1983 Dec 3	7	-3 33	18°E
1984 Dec 2	14	-2 46	20°E
1984 Dec 6	17	-1 49	16°E
1985 Jan 13	6	-0 41	21°W
1986 Jan 8	10	-1 41	14°W
1987 Jan 2	14	-2 19	6°W
1987 Dec 27	10	-2 46	3°E
1988 Dec 20	9	-3 04	11°E
1989 Dec 15	4	-3 07	18°E
1990 Jan 13	20	+2 33	11°W
1990 Jan 27	23	+0 46	25°W
1991 Jan 26	14	-1 04	21°W
1992 Jan 21	11	-1 53	14°W
1993 Jan 14	11	-2 22	5°W
1994 Jan 8	5	-2 40	3°E
1995 Jan 2	2	-2 45	11°E
1995 Dec 28	2	-2 24	19°E
1996 Jan 20	13	+2 57	4°W
1996 Feb 11	15	+0 04	26°W
1997 Feb 7	20	-1 21	21°W
1998 Feb 2	11	-1 59	13°W
1999 Jan 27	8	-2 18	5°W
2000 Jan 20	22	-2 23	4°E
2001 Jan 13	17	-2 13	12°E
2002 Jan 9	5	-1 20	19°E
2002 Jan 26	13	+3 13	2°E
2002 Feb 24	13	-0 30	26°W
2003 Feb 20	24	-1 34	20°W
2004 Feb 15	9	-1 58	13°W
2005 Feb 8	1	-2 04	4°W

(g) VENUS – MARS

1976 Sep 10	22^h	+0°24'	23°E
1977 May 13	18	+1 17	40°W
1977 Jun 3	13	-1 13	45°W
1978 Aug 14	15	-1 11	45°E
1978 Oct 20	8	-6 39	24°E
1979 May 20	6	-1 03	26°W
1981 Apr 4	15	-0 42	1°W
1983 Feb 18	22	-0 32	25°E
1983 Sep 14	19	-8 33	29°W
1983 Oct 28	13	-1 43	46°W
1985 Feb 8	2	+2 41	45°E
1985 Feb 15	20	+3 45	43°E
1985 Oct 4	23	+0 06	26°W
1987 Aug 24	6	+0 15	1°E
1989 Jul 12	12	+0 28	26°E
1991 Jun 23	11	+0 16	45°E
1991 Jul 22	6	-3 36	35°E
1992 Feb 19	22	+0 51	29°W
1994 Jan 6	6	+0 18	3°W
1995 Nov 22	22	-0 11	24°E
1996 Jun 30	4	-4 12	26°W
1996 Sep 4	15	-2 51	45°W
1997 Oct 26	23	-2 07	47°E
1997 Dec 22	12	+1 07	33°E
1998 Aug 5	3	-0 51	23°W
2000 Jun 21	20	-0 18	3°E
2002 May 10	21	+0 18	28°E
2004 Dec 5	7	+1 15	27°W

XX (continued)

<table>
<tr><td colspan="6">(h) VENUS – JUPITER</td></tr>
<tr><td>1976 May 11</td><td>14^h</td><td>-0°10'</td><td>10°W</td></tr>
<tr><td>1977 Jul 30</td><td>6</td><td>-1 33</td><td>41°W</td></tr>
<tr><td>1978 May 29</td><td>2</td><td>+1 36</td><td>31°E</td></tr>
<tr><td>1979 Aug 16</td><td>12</td><td>+0 33</td><td>2°W</td></tr>
<tr><td>1980 Oct 30</td><td>20</td><td>+0 27</td><td>37°W</td></tr>
<tr><td>1981 Aug 28</td><td>1</td><td>-0 53</td><td>36°E</td></tr>
<tr><td>1982 Nov 11</td><td>9</td><td>-0 17</td><td>2°E</td></tr>
<tr><td>1984 Jan 27</td><td>2</td><td>+0 51</td><td>35°W</td></tr>
<tr><td>1984 Nov 24</td><td>21</td><td>-2 03</td><td>40°E</td></tr>
<tr><td>1986 Feb 11</td><td>3</td><td>-0 38</td><td>6°E</td></tr>
<tr><td>1987 May 4</td><td>22</td><td>-0 38</td><td>29°W</td></tr>
<tr><td>1988 Mar 6</td><td>20</td><td>+2 24</td><td>43°E</td></tr>
<tr><td>1989 May 23</td><td>4</td><td>+0 50</td><td>13°E</td></tr>
<tr><td>1990 Aug 12</td><td>23</td><td>+0 03</td><td>21°W</td></tr>
<tr><td>1991 Jun 17</td><td>23</td><td>+1 14</td><td>45°E</td></tr>
<tr><td>1991 Aug 23</td><td>7</td><td>-9 36</td><td>4°W</td></tr>
<tr><td>1991 Oct 17</td><td>3</td><td>-2 28</td><td>45°W</td></tr>
<tr><td>1992 Aug 23</td><td>3</td><td>+0 17</td><td>19°E</td></tr>
<tr><td>1993 Nov 8</td><td>17</td><td>+0 23</td><td>17°W</td></tr>
<tr><td>1995 Jan 14</td><td>9</td><td>+2 49</td><td>46°W</td></tr>
<tr><td>1995 Nov 19</td><td>12</td><td>-1 16</td><td>23°E</td></tr>
<tr><td>1997 Feb 5</td><td>24</td><td>-0 20</td><td>14°W</td></tr>
<tr><td>1998 Apr 23</td><td>2</td><td>+0 18</td><td>45°W</td></tr>
<tr><td>1999 Feb 23</td><td>21</td><td>+0 09</td><td>28°E</td></tr>
<tr><td>2000 May 17</td><td>11</td><td>+0 01</td><td>7°W</td></tr>
<tr><td>2001 Aug 5</td><td>24</td><td>-1 12</td><td>38°W</td></tr>
<tr><td>2002 Jun 3</td><td>18</td><td>+1 39</td><td>34°E</td></tr>
<tr><td>2003 Aug 21</td><td>6</td><td>+0 34</td><td>1°E</td></tr>
<tr><td>2004 Nov 4</td><td>21</td><td>+0 36</td><td>34°W</td></tr>
<tr><td>2005 Sep 2</td><td>12</td><td>-1 22</td><td>39°E</td></tr>
</table>

<table>
<tr><td colspan="6">(i) VENUS – SATURN</td></tr>
<tr><td>1976 Jul 19</td><td>1^h</td><td>+0°47'</td><td>9°E</td></tr>
<tr><td>1977 Sep 18</td><td>13</td><td>-0 24</td><td>30°W</td></tr>
<tr><td>1978 Jul 10</td><td>11</td><td>+0 07</td><td>40°E</td></tr>
<tr><td>1979 Sep 6</td><td>19</td><td>-0 26</td><td>4°E</td></tr>
<tr><td>1980 Nov 3</td><td>22</td><td>-0 33</td><td>36°W</td></tr>
<tr><td>1981 Aug 25</td><td>22</td><td>-1 58</td><td>35°E</td></tr>
<tr><td>1982 Oct 22</td><td>22</td><td>-1 10</td><td>3°W</td></tr>
<tr><td>1983 Dec 17</td><td>11</td><td>+0 09</td><td>42°W</td></tr>
<tr><td>1984 Oct 8</td><td>17</td><td>-2 36</td><td>30°E</td></tr>
<tr><td>1985 Dec 5</td><td>11</td><td>-1 06</td><td>11°W</td></tr>
<tr><td>1987 Jan 24</td><td>20</td><td>+1 47</td><td>46°W</td></tr>
<tr><td>1987 Nov 20</td><td>16</td><td>-2 08</td><td>23°E</td></tr>
<tr><td>1989 Jan 16</td><td>16</td><td>-0 33</td><td>19°W</td></tr>
<tr><td>1989 Nov 15</td><td>19</td><td>-3 54</td><td>47°E</td></tr>
<tr><td>1990 Feb 7</td><td>5</td><td>+7 06</td><td>28°W</td></tr>
<tr><td>1990 Feb 14</td><td>18</td><td>+6 32</td><td>35°W</td></tr>
<tr><td>1991 Jan 1</td><td>15</td><td>-1 11</td><td>15°E</td></tr>
<tr><td>1992 Feb 29</td><td>2</td><td>+0 08</td><td>27°W</td></tr>
<tr><td>1992 Dec 21</td><td>16</td><td>-1 04</td><td>45°E</td></tr>
<tr><td>1994 Feb 14</td><td>3</td><td>-0 02</td><td>7°E</td></tr>
<tr><td>1995 Apr 13</td><td>17</td><td>+0 34</td><td>34°W</td></tr>
<tr><td>1996 Feb 3</td><td>2</td><td>+1 16</td><td>39°E</td></tr>
<tr><td>1997 Mar 31</td><td>21</td><td>+0 57</td><td>1°W</td></tr>
<tr><td>1998 May 29</td><td>2</td><td>+0 16</td><td>39°W</td></tr>
<tr><td>1999 Mar 20</td><td>21</td><td>+2 34</td><td>32°E</td></tr>
<tr><td>2000 May 18</td><td>20</td><td>+1 14</td><td>6°W</td></tr>
<tr><td>2001 Jul 15</td><td>5</td><td>-0 44</td><td>42°W</td></tr>
<tr><td>2002 May 7</td><td>18</td><td>+2 25</td><td>27°E</td></tr>
<tr><td>2003 Jul 8</td><td>8</td><td>+0 49</td><td>11°W</td></tr>
<tr><td>2004 Sep 1</td><td>1</td><td>-1 57</td><td>45°W</td></tr>
<tr><td>2005 Jun 25</td><td>21</td><td>+1 18</td><td>23°E</td></tr>
</table>

XX (continued)

(j) VENUS - URANUS				
1976 Sep 30	22^h	-0°27'	28°E
1977 Nov 20	10	+0 53	15°W
1978 Sep 27	24	-5 59	40°E
1978 Nov 2	20	-6 19	6°E
1978 Dec 24	15	+3 16	43°W
1979 Oct 27	16	-0 13	17°E
1980 Dec 16	2	+1 14	27°W
1981 Oct 7	11	-2 26	44°E
1982 Nov 22	15	+0 02	5°E
1984 Jan 10	13	+1 50	38°W
1984 Oct 29	24	-1 27	35°E
1985 Dec 18	9	+0 16	8°W
1987 Jan 31	17	+3 04	46°W
1987 Nov 24	10	-0 56	24°E
1989 Jan 12	17	+0 33	20°W
1989 Nov 8	2	-3 15	47°E
1990 Dec 19	10	-0 36	12°E
1992 Feb 7	7	+0 54	32°W
1992 Nov 26	11	-1 54	41°E
1994 Jan 13	13	-0 22	1°W
1995 Mar 2	5	+1 30	42°W
1995 Dec 20	13	-1 16	31°E
1997 Feb 7	12	-0 12	13°W
1998 Mar 19	7	+3 20	46°W
1999 Jan 13	19	-0 56	18°E
2000 Mar 4	0	-0 04	25°W
2000 Dec 23	21	-1 19	46°E
2002 Feb 7	10	-0 45	6°E
2003 Mar 28	13	+0 03	37°W
2004 Jan 15	1	-0 56	36°E
2005 Mar 4	4	-0 41	7°W

(k) VENUS - NEPTUNE				
1976 Jan 12	4^h	+0°22'	38°W
1976 Oct 31	6	-2 56	35°E
1977 Dec 17	1	-1 07	9°W
1979 Jan 26	18	+1 54	47°W
1979 Nov 20	5	-2 15	22°E
1981 Jan 5	22	-0 39	22°W
1981 Oct 30	1	-4 32	46°E
1982 Dec 10	1	-1 46	9°E
1984 Jan 25	23	+0 02	35°W
1984 Nov 13	19	-3 07	38°E
1985 Dec 30	4	-1 19	5°W
1987 Feb 11	13	+1 17	45°W
1987 Dec 3	10	-2 22	26°E
1989 Jan 19	4	-0 52	18°W
1989 Nov 15	15	-4 24	47°E
1990 Dec 23	3	-1 51	13°E
1992 Feb 8	15	-0 17	31°W
1992 Nov 27	13	-3 01	41°E
1994 Jan 12	7	-1 24	1°W
1995 Feb 26	10	+0 41	43°W
1995 Dec 16	17	-2 16	30°E
1997 Feb 1	11	-0 59	15°W
1997 Dec 7	20	-2 46	42°E
1998 Jan 9	17	+4 00	10°E
1998 Mar 7	10	+3 46	45°W
1999 Jan 5	9	-1 45	16°E
2000 Feb 22	6	-0 30	28°W
2000 Dec 11	21	-2 32	44°E
2002 Jan 25	13	-1 20	3°E
2003 Mar 12	20	+0 11	40°W
2003 Dec 30	7	-1 53	33°E
2005 Feb 14	19	-0 58	11°W

XX (continued)

(1) MARS – JUPITER

1977 Sep 4	22^h	+0°30′	70°W
1979 Dec 13	17	+1 40	101°W
1980 Mar 2	19	+3 11	170°E
1980 May 4	6	+0 49	106°E
1982 Aug 10	1	-2 05	76°E
1984 Oct 13	23	-1 55	75°E
1986 Dec 19	7	+0 31	79°E
1989 Mar 12	8	+1 57	68°E
1991 Jun 14	5	+0 38	48°E
1993 Sep 7	0	-0 54	32°E
1995 Nov 16	8	-1 11	26°E
1998 Jan 21	1	-0 12	26°E
2000 Apr 6	23	+1 06	23°E
2002 Jul 3	6	+0 49	12°E
2004 Sep 27	5	-0 12	4°W

(n) MARS – URANUS

1976 Oct 18	22^h	-0°21′	11°E
1978 Oct 12	2	-0 38	27°E
1980 Oct 2	24	-0 59	43°E
1982 Sep 22	13	-1 27	62°E
1984 Sep 4	11	-2 23	88°E
1986 Mar 13	9	+0 21	90°W
1988 Feb 22	21	+0 01	63°W
1990 Feb 9	14	-0 13	43°W
1992 Jan 29	21	-0 23	24°W
1994 Jan 18	13	-0 29	6°W
1996 Jan 7	24	-0 34	13°E
1997 Dec 26	20	-0 36	32°E
1999 Dec 14	5	-0 39	52°E
2001 Nov 26	10	-0 48	77°E
2003 Jun 20	23	-3 11	116°W
2005 May 14	20	-1 11	74°W

(m) MARS – SATURN

1976 May 12	2^h	+1°17′	66°E
1978 Jun 5	0	-0 06	71°E
1980 Jun 25	13	-1 40	77°E
1982 Jul 9	24	-3 08	88°E
1984 Feb 15	13	-0 48	100°W
1986 Feb 17	24	-1 16	80°W
1988 Feb 23	13	-1 18	63°W
1990 Feb 28	17	-1 00	48°W
1992 Mar 6	13	-0 26	33°W
1994 Mar 14	10	+0 22	18°W
1996 Mar 22	20	+1 16	4°W
1998 Apr 3	12	+2 01	9°E
2000 Apr 16	23	+2 22	20°E
2002 May 4	17	+2 13	30°E
2004 May 24	23	+1 35	37°E

(o) MARS – NEPTUNE

1976 Dec 10	12^h	-1°53′	4°W
1978 Nov 26	7	-2 05	14°E
1980 Nov 10	13	-2 19	33°E
1982 Oct 25	6	-2 38	53°E
1984 Oct 3	14	-3 18	78°E
1986 Apr 8	22	-1 27	103°W
1988 Mar 7	22	-1 22	68°W
1990 Feb 17	6	-1 27	45°W
1992 Feb 1	9	-1 31	24°W
1994 Jan 16	7	-1 33	5°W
1996 Jan 1	7	-1 35	14°E
1997 Dec 15	19	-1 38	34°E
1999 Nov 28	14	-1 45	56°E
2001 Nov 4	18	-2 11	83°E
2003 May 13	14	-2 02	99°W
2005 Apr 13	0	-1 15	66°W

XX (continued)

(p) JUPITER – SATURN

1981 Jan 14	8^h	-1°09′	104°W
1981 Feb 19	7	-1 09	141°W
1981 Jul 30	22	-1 12	58°E
2000 May 31	10	+1 11	17°W

(s) SATURN – URANUS

1988 Feb 13	0^h	+1°17′	54°W
1988 Jun 27	3	+1 21	173°E
1988 Oct 18	2	+1 04	63°E

(q) JUPITER – URANUS

1983 Feb 17	14^h	+0°46′	80°W
1983 May 16	13	+0 50	167°W
1983 Sep 24	22	+0 27	64°E
1997 Feb 16	8	+0 10	22°W

(t) SATURN – NEPTUNE

1989 Mar 3	3^h	-0°14′	61°W
1989 Jun 24	15	-0 18	172°W
1989 Nov 12	22	-0 30	50°E

(r) JUPITER – NEPTUNE

1984 Jan 19	18^h	-0°52′	29°W
1997 Jan 8	18	-0 47	8°E

(u) URANUS – NEPTUNE

1993 Jan 26	1^h	-1°06′	17°W
1993 Sep 17	18	-1 09	113°E
1993 Sep 28	4	-1 08	103°E

XX A
CLOSE PLANETARY CONJUNCTIONS

Date	Planets	Conj. in right asc. UT	Δδ	Least Distance UT	Dist.
1978 Sep 13	Mercury-Saturn	14^h47^m	+236″	15^h06^m	224″
1978 Nov 29	Mercury-Mars	19 37	+306	19 14	294
1982 Nov 22	Venus-Uranus	14 41	+ 93	14 48	91
1984 Jan 25	Venus-Neptune	23 23	+ 94	23 24	94
1985 Sep 4	Mercury-Mars	21 03	- 47	20 59	46
1988 Feb 22	Mars-Uranus	20 47	+ 40	20 47	40
1989 Aug 5	Mercury-Mars	21 44	+ 52	21 53	47
1990 Aug 12	Venus-Jupiter	23 24	+153	23 34	150
1990 Oct 16	Mercury-Venus	5 18	+103	6 02	88
1991 Jul 15	Mercury-Jupiter	7 53	-282	7 12	256
1991 Sep 10	Mercury-Jupiter	10 25	-221	10 11	218
1994 Feb 14	Venus-Saturn	2 39	- 92	2 51	86
1996 Feb 11	Mercury-Neptune	14 31	+228	14 29	228
2000 Mar 4	Venus-Uranus	0 12	-244	0 35	234
2000 May 17	Venus-Jupiter	10 35	+ 44	10 30	42
2000 Aug 10	Mercury-Mars	12 59	-292	12 46	289
2003 Mar 28	Venus-Uranus	13 03	+165	12 45	157
2005 Jun 27	Mercury-Venus	20 18	-294	16 00	233

XXI

QUASI-CONJUNCTIONS

Approaches of two bright planets within 5° of each other,
without a conjunction in right ascension

1976 Feb 22	13^h	Mercury-Venus	3°59'	
1976 Aug 25	12	Mercury-Mars	2°57'	
1979 Jul 9	5	Mercury-Jupiter	1°59'	
1981 Oct 28	7	Mercury-Saturn	2°54'	
1982 Feb 23	21	Mars-Saturn	2°41'	
1984 Jan 13	21	Mercury-Jupiter	3°10'	
1984 Jun 11	1	Mars-Saturn	4°20'	
1987 Jun 11	0	Mercury-Mars	0°37'	
1996 Dec 22	15	Mercury-Jupiter	3°56'	
1997 Jul 27	14	Mercury-Venus	4°27'	
1999 Mar 6	8	Mercury-Jupiter	3°59'	
2001 Oct 29	22	Mercury-Venus	0°35'	} The separation between Mercury and
2001 Nov 4	0	Mercury-Venus	0°39'	∫ Venus will be less than 1° during eleven days
2002 Oct 10	13	Mercury-Mars	2°50'	
2002 Dec 6	12	Venus-Mars	1°33'	

XXII

HELIOCENTRIC CONJUNCTIONS BETWEEN THE OUTER PLANETS

JUPITER-SATURN

1961 Apr 16	293°41'
1981 Apr 16	187 08
2000 Jun 22	52 01

JUPITER-NEPTUNE

1958 Jul 20	213°56'
1971 May 24	241 43
1984 Mar 16	269 28
1996 Dec 24	297 19

SATURN-NEPTUNE

1953 Feb 18	202°12'
1989 Jul 18	281 05
(Next one in Dec. 2025)	

JUPITER-URANUS

1955 Jan 26	115°47'
1969 Apr 3	182 08
1983 Jun 12	247 17
1997 Mar 28	305 20

SATURN-URANUS

1988 Jun 9	268°56'
(Next one in July 2032)	

URANUS-NEPTUNE

1993 Apr 21	289°16'
(Previous one in Sept. 1821)	

URANUS-PLUTO

1966 Jan 7	166°49'
(Previous one in Dec. 1850)	

XXIII

PASSAGES OF EARTH AND SUN
THROUGH THE RING-PLANE OF SATURN

Passages of Earth			Passages of Sun		
1848 Apr 21	N → S				
1848 Sep 14	S → N		1848 Aug 31	N → S	
1849 Jan 18	N → S				
1861 Nov 22	S → N				
1862 Feb 2	N → S		1862 May 16	S → N	
1862 Aug 12	S → N				
1878 Mar 1	N → S		1878 Feb 5	N → S	
1891 Sep 22	S → N		1891 Oct 29	S → N	
1907 Apr 12	N → S				
1907 Oct 4	S → N		1907 Jul 26	N → S	
1908 Jan 7	N → S				
1920 Nov 7	S → N				
1921 Feb 22	N → S		1921 Apr 11	S → N	
1921 Aug 3	S → N				
1936 Jun 28	(a)		1936 Dec 28	N → S	
1937 Feb 20	N → S				
1950 Sep 14	S → N		1950 Sep 21	S → N	
1966 Apr 2	N → S				
1966 Oct 28	S → N		1966 Jun 15	N → S	
1966 Dec 17	N → S				
1979 Oct 27	S → N				
1980 Mar 12	N → S		1980 Mar 3	S → N	
1980 Jul 23	S → N				
1995 May 21	N → S				
1995 Aug 11	S → N		1995 Nov 19	N → S	
1996 Feb 11	N → S				
2009 Sep 4	S → N		2009 Aug 10	S → N	
2025 Mar 23	N → S		2025 May 6	N → S	
2038 Oct 15	S → N				
2039 Apr 1	N → S		2039 Jan 22	S → N	
2039 Jul 9	S → N				
2054 May 5	N → S				
2054 Aug 31	S → N		2054 Oct 9	N → S	
2055 Feb 1	N → S				

(a) The Earth did not cross the plane of the rings ; the minimum
saturnicentric latitude of the Earth was 0°0001 North.

2

OPPOSITIONS
OF
MARS

0 - 3010

OPPOSITIONS OF MARS

The list which follows has been calculated by Edwin Goffin of the Public Observatory of Antwerp, Belgium, and was first published in July 1976 as *Memoir* No. 1 of the Belgian astronomical society "Vereniging Voor Sterrenkunde".

The list gives all the oppositions of Mars taking place during the period 0 to 3010.

The first three columns give the Julian Date (JD), the calendar date (year, month, day of the month), and the Ephemeris Time (ET) of the instant of opposition, this being the instant when the true heliocentric longitudes of the Earth and Mars are equal. Hence, the times are those of the *geometric* oppositions of Mars. The *apparent* opposition occurs approximately 7 minutes later than the geometric opposition, by reason of the aberration of light and the effect of light-time.

The next column gives the geocentric declination δ of Mars, referred to the mean equinox of date at the time of opposition.

The fifth and sixth columns give the calendar date and the Ephemeris Time (ET) of the instant of Mars' shortest distance to the Earth. The next two columns contain the value of this least distance, in astronomical units and in millions of kilometers, respectively.

The next column gives the planet's apparent diameter d in seconds of arc at the instant of minimum distance, the adopted diameter at unit distance being $9''.36$.

The last column gives the time interval Δt in days between the instant of Mars' opposition and that of its shortest distance to the Earth. The Julian Date corresponding to the instant of minimum distance can be obtained by adding the values of the first and last columns of the list, that is JD + Δt.

For the computation of the heliocentric positions of the Earth and Mars, use was made of Newcomb's theories for these planets in the introduction to his *Tables of the Sun* (*Astronomical Papers American Ephemeris*, Vol. VI, Part 1) and his *Tables of Mars* (*ibid.*, Vol. VI, Part 4). In the latter case the orbital elements given by Ross in *New Elements of Mars* (*ibid.*, Vol. IX, Part 2) were adopted.

During the period 0 to 3010, the extreme values of the shortest distance Mars-Earth are 0.37198 a.u. (55.65 millions of km) at the perihelic opposition of 2729, and 0.67853 a.u. (101.51 millions of km) at the aphelic opposition of 2832.

The extreme values of Δt are +8.54 days (oppositions of 359, 722, 1527 and 2774) and −8.27 days (oppositions of 2526 and 2889). On the other hand, $|\Delta t|$ is less than 0.005 day at the oppositions of 2208 and 2232.

In the years 69, 274, 558, 763, 968, 1047, 1252, 1536, 1630, 1709, 1835, 2040, 2119, 2245, 2450, 2529, 2655, 2734, 2860 and 2939, the opposition occurs in early January, while Mars' least distance takes place at the end of December of the p r e c e d i n g year.

In 656 and 1728 the opposition of Mars took place on the bissextile day, February 29.

OPPOSITIONS OF MARS

	Mars at opposition				Mars nearest to Earth				
JD	date	ET	δ	date	ET	least distance	10^6 km	d	Δt
		h m	° '		h	a.u.		"	d
1721319.17	0 Sep 18	16 05	- 6 14	0 Sep 10	20.1	0.43428	64.97	21.55	-7.83
1722101.12	2 Nov 9	14 48	+17 34	2 Nov 3	0.8	0.56177	84.04	16.66	-6.58
1722869.78	4 Dec 17	6 49	+27 09	4 Dec 14	8.1	0.65001	97.24	14.40	-2.95
1723634.12	7 Jan 20	14 46	+25 07	7 Jan 21	12.9	0.67578	101.10	13.85	+0.92
1724399.30	9 Feb 23	19 06	+14 16	9 Feb 28	14.2	0.63584	95.12	14.72	+4.80
1725171.12	11 Apr 6	14 53	- 3 45	11 Apr 14	16.3	0.53400	79.89	17.53	+8.06
1725960.43	13 Jun 3	22 24	-25 57	13 Jun 10	6.3	0.40637	60.79	23.03	+6.33
1726770.97	15 Aug 23	11 23	-18 20	15 Aug 17	16.1	0.39320	58.82	23.80	-5.80
1727562.23	17 Oct 22	17 36	+10 08	17 Oct 15	1.4	0.51361	76.83	18.22	-7.68
1728334.63	19 Dec 4	3 00	+25 00	19 Nov 29	16.6	0.62242	93.11	15.04	-4.43
1729099.94	22 Jan 7	10 38	+27 04	22 Jan 6	20.3	0.67356	100.76	13.90	-0.60
1729864.23	24 Feb 10	17 33	+19 19	24 Feb 14	0.7	0.65933	98.64	14.20	+3.30
1730632.59	26 Mar 20	2 04	+ 3 59	26 Mar 27	1.4	0.58021	86.80	16.13	+6.97
1731413.21	28 May 8	17 03	-17 43	28 May 16	22.7	0.45312	67.79	20.66	+8.23
1732218.47	30 Jul 22	23 11	-27 43	30 Jul 21	19.9	0.37462	56.04	24.99	-1.14
1733020.55	32 Oct 2	1 18	+ 0 24	32 Sep 23	22.4	0.46219	69.14	20.25	-8.12
1733798.47	34 Nov 18	23 16	+20 57	34 Nov 13	2.8	0.58580	87.63	15.98	-5.86
1734565.63	36 Dec 25	3 05	+27 36	36 Dec 23	0.6	0.66112	98.90	14.16	-2.10
1735329.74	39 Jan 28	5 39	+23 16	39 Jan 30	0.3	0.67274	100.64	13.91	+1.78
1736095.79	41 Mar 4	7 02	+10 42	41 Mar 9	21.7	0.61856	92.54	15.13	+5.61
1736870.21	43 Apr 17	17 05	- 8 47	43 Apr 26	2.6	0.50547	75.62	18.52	+8.40
1737665.23	45 Jun 20	17 33	-28 55	45 Jun 24	23.6	0.38711	57.91	24.18	+4.25
1738474.88	47 Sep 8	9 14	-11 20	47 Sep 1	4.3	0.41379	61.90	22.62	-7.20
1739260.69	49 Nov 2	4 35	+14 47	49 Oct 26	1.7	0.54085	80.91	17.31	-7.12
1740030.85	51 Dec 12	8 31	+26 30	51 Dec 8	18.2	0.63886	95.57	14.65	-3.59
1740795.53	54 Jan 15	0 42	+26 02	54 Jan 15	7.0	0.67601	101.13	13.85	+0.26
1741560.20	56 Feb 18	16 50	+16 23	56 Feb 22	20.4	0.64756	96.87	14.45	+4.15
1742330.30	58 Mar 29	19 18	- 0 29	58 Apr 6	10.1	0.55560	83.12	16.85	+7.62
1743115.41	60 May 22	21 49	-22 44	60 May 30	9.3	0.42594	63.72	21.97	+7.48
1743924.59	62 Aug 10	2 06	-22 59	62 Aug 6	2.0	0.38126	57.04	24.55	-4.00
1744720.70	64 Oct 14	4 53	+ 6 23	64 Oct 6	4.5	0.49067	73.40	19.08	-8.02
1745495.29	66 Nov 27	18 59	+23 36	66 Nov 22	17.5	0.60708	90.82	15.42	-5.06
1746261.31	69 Jan 1	19 33	+27 27	68 Dec 31	13.4	0.66929	100.12	13.98	-1.25
1747025.44	71 Feb 4	22 40	+21 00	71 Feb 7	13.9	0.66666	99.73	14.04	+2.63
1747792.63	73 Mar 13	3 10	+ 6 48	73 Mar 19	12.5	0.59848	89.53	15.64	+6.39
1748570.19	75 Apr 29	16 29	-13 58	75 May 8	3.4	0.47654	71.29	19.64	+8.46
1749371.13	77 Jul 8	15 00	-29 22	77 Jul 10	0.1	0.37640	56.31	24.87	+1.38
1750177.37	79 Sep 22	20 56	- 4 20	79 Sep 14	21.9	0.43959	65.76	21.29	-7.96
1750958.51	81 Nov 12	0 11	+18 42	81 Nov 5	13.9	0.56670	84.78	16.52	-6.43
1751726.85	83 Dec 20	8 27	+27 21	83 Dec 17	14.1	0.65242	97.60	14.35	-2.76

	Mars at opposition					Mars nearest to Earth			
JD	date	ET	δ	date	ET	least distance	10^6 km	d	Δt
		h m	° ′		h	a.u.		″	d
1752491.11	86 Jan 22	14 40	+24 29	86 Jan 23	17.3	0.67548	101.05	13.86	+1.11
1753256.45	88 Feb 26	22 46	+13 04	88 Mar 2	22.0	0.63274	94.66	14.79	+4.97
1754028.75	90 Apr 9	6 02	- 5 17	90 Apr 17	8.4	0.52845	79.05	17.71	+8.10
1754819.07	92 Jun 7	13 42	-26 53	92 Jun 13	15.0	0.40191	60.13	23.29	+6.05
1755629.64	94 Aug 27	3 26	-16 31	94 Aug 21	1.6	0.39631	59.29	23.62	-6.08
1756419.83	96 Oct 25	7 56	+11 37	96 Oct 17	17.9	0.51881	77.61	18.04	-7.58
1757191.76	98 Dec 6	6 13	+25 33	98 Dec 2	0.5	0.62578	93.62	14.96	-4.24
1757956.93	101 Jan 9	10 26	+26 44	101 Jan 9	0.7	0.67425	100.87	13.88	-0.40
1758721.27	103 Feb 12	18 26	+18 20	103 Feb 16	5.9	0.65734	98.34	14.24	+3.48
1759489.90	105 Mar 22	9 37	+ 2 35	105 Mar 29	11.8	0.57586	86.15	16.25	+7.09
1760271.31	107 May 12	19 30	-19 08	107 May 20	23.1	0.44797	67.02	20.89	+8.15
1761077.51	109 Jul 27	0 21	-26 42	109 Jul 25	6.6	0.37512	56.12	24.95	-1.74
1761878.49	111 Oct 5	23 47	+ 2 11	111 Sep 27	19.6	0.46743	69.93	20.02	-8.18
1762655.70	113 Nov 21	4 55	+21 52	113 Nov 15	12.9	0.58994	88.25	15.87	-5.67
1763422.64	115 Dec 28	3 15	+27 35	115 Dec 26	5.1	0.66299	99.18	14.12	-1.92
1764186.74	118 Jan 30	5 49	+22 30	118 Feb 1	4.8	0.67192	100.52	13.93	+1.96
1764952.99	120 Mar 6	11 40	+ 9 25	120 Mar 12	6.0	0.61498	92.00	15.22	+5.77
1765727.94	122 Apr 20	10 38	-10 18	122 Apr 28	20.1	0.50015	74.82	18.71	+8.39
1766524.08	124 Jun 24	13 59	-29 16	124 Jun 28	7.6	0.38453	57.52	24.34	+3.73
1767333.31	126 Sep 11	19 24	- 9 24	126 Sep 4	9.5	0.41854	62.61	22.36	-7.41
1768118.19	128 Nov 4	16 38	+16 06	128 Oct 28	17.5	0.54619	81.71	17.14	-6.96
1768887.95	130 Dec 14	10 51	+26 50	130 Dec 11	0.7	0.64175	96.00	14.59	-3.42
1769652.51	133 Jan 17	0 19	+25 31	133 Jan 17	10.7	0.67619	101.16	13.84	+0.43
1770417.31	135 Feb 20	19 24	+15 16	135 Feb 25	2.6	0.64501	96.49	14.51	+4.30
1771187.80	137 Apr 1	7 06	- 1 58	137 Apr 8	23.3	0.55046	82.35	17.00	+7.67
1771973.81	139 May 27	7 31	-23 56	139 Jun 3	15.2	0.42078	62.95	22.24	+7.32
1772783.46	141 Aug 13	23 04	-21 25	141 Aug 9	13.3	0.38324	57.33	24.42	-4.40
1773578.41	143 Oct 17	21 52	+ 8 00	143 Oct 9	23.4	0.49590	74.19	18.87	-7.93
1774352.46	145 Nov 29	23 08	+24 18	145 Nov 25	2.0	0.61081	91.38	15.32	-4.88
1775118.32	148 Jan 4	19 39	+27 14	148 Jan 3	17.6	0.67047	100.30	13.96	-1.08
1775882.45	150 Feb 6	22 55	+20 05	150 Feb 9	18.0	0.66508	99.49	14.07	+2.80
1776649.87	152 Mar 15	8 59	+ 5 27	152 Mar 21	21.0	0.59449	88.93	15.74	+6.50
1777428.12	154 May 2	14 51	-15 27	154 May 11	1.2	0.47127	70.50	19.86	+8.43
1778230.16	156 Jul 12	15 44	-28 56	156 Jul 13	9.9	0.37540	56.16	24.93	+0.76
1779035.53	158 Sep 26	0 37	- 2 27	158 Sep 17	22.7	0.44463	66.52	21.05	-8.08
1779815.84	160 Nov 14	8 05	+19 47	160 Nov 8	1.9	0.57123	85.46	16.39	-6.26
1780583.87	162 Dec 22	8 58	+27 29	162 Dec 19	18.6	0.65471	97.94	14.30	-2.60
1781348.10	165 Jan 24	14 22	+23 49	165 Jan 25	20.9	0.67515	101.00	13.86	+1.27
1782113.59	167 Mar 1	2 14	+11 52	167 Mar 6	4.8	0.62967	94.20	14.86	+5.11
1782886.32	169 Apr 11	19 46	- 6 47	169 Apr 19	23.2	0.52330	78.29	17.89	+8.14
1783677.70	171 Jun 12	4 54	-27 41	171 Jun 17	20.9	0.39821	59.57	23.51	+5.67
1784488.27	173 Aug 30	18 27	-14 38	173 Aug 24	9.0	0.40027	59.88	23.38	-6.40
1785277.42	175 Oct 28	22 03	+13 03	175 Oct 21	11.7	0.52437	78.44	17.85	-7.43
1786048.90	177 Dec 8	9 30	+26 03	177 Dec 4	7.4	0.62912	94.11	14.88	-4.09
1786813.92	180 Jan 12	10 06	+26 20	180 Jan 12	3.9	0.67492	100.97	13.87	-0.26
1787578.33	182 Feb 14	19 51	+17 17	182 Feb 18	10.6	0.65526	98.03	14.28	+3.62
1788347.30	184 Mar 24	19 08	+ 1 09	184 Mar 31	23.0	0.57111	85.44	16.39	+7.16
1789129.53	186 May 16	0 49	-20 31	186 May 24	2.8	0.44238	66.18	21.16	+8.08
1789936.53	188 Jul 31	0 45	-25 31	188 Jul 28	19.0	0.37566	56.20	24.92	-2.24
1790736.38	190 Oct 8	21 05	+ 3 56	190 Sep 30	17.8	0.47261	70.70	19.80	-8.14

	Mars at opposition				Mars nearest to Earth					
JD	date	ET		δ	date	ET	least distance	10^6 km	d	Δt
		h	m	° ′		h	a.u.		″	d
1791512.95	192 Nov 23	10	42	+22 43	192 Nov 17	22.4	0.59410	88.88	15.75	-5.51
1792279.65	194 Dec 30	3	40	+27 30	194 Dec 28	9.0	0.66465	99.43	14.08	-1.78
1793043.74	197 Feb 1	5	43	+21 41	197 Feb 3	8.1	0.67078	100.35	13.95	+2.10
1793810.16	199 Mar 9	15	53	+ 8 08	199 Mar 15	12.8	0.61142	91.47	15.31	+5.87
1794585.68	201 Apr 23	4	20	-11 48	201 May 1	14.9	0.49490	74.04	18.91	+8.44
1795382.98	203 Jun 29	11	28	-29 26	203 Jul 2	15.4	0.38213	57.17	24.49	+3.16
1796191.67	205 Sep 15	4	06	- 7 27	205 Sep 7	12.6	0.42320	63.31	22.12	-7.64
1796975.60	207 Nov 8	2	21	+17 18	207 Nov 1	6.7	0.55098	82.43	16.99	-6.82
1797745.00	209 Dec 16	11	57	+27 06	209 Dec 13	5.1	0.64443	96.40	14.52	-3.28
1798509.49	212 Jan 19	23	42	+24 56	212 Jan 20	13.4	0.67639	101.19	13.84	+0.57
1799274.40	214 Feb 22	21	38	+14 07	214 Feb 27	7.9	0.64239	96.10	14.57	+4.43
1800045.26	216 Apr 3	18	13	- 3 26	216 Apr 11	12.5	0.54551	81.61	17.16	+7.76
1800832.24	218 May 30	17	44	-25 03	218 Jun 6	19.4	0.41626	62.27	22.49	+7.07
1801642.28	220 Aug 17	18	46	-19 43	220 Aug 12	22.9	0.38599	57.74	24.25	-4.83
1802436.14	222 Oct 20	15	27	+ 9 35	222 Oct 12	19.9	0.50160	75.04	18.66	-7.81
1803209.66	224 Dec 2	3	54	+24 56	224 Nov 27	9.9	0.61468	91.96	15.23	-4.75
1803975.32	227 Jan 6	19	35	+26 57	227 Jan 5	20.6	0.67162	100.47	13.94	-0.96
1804739.49	229 Feb 8	23	40	+19 08	229 Feb 11	21.8	0.66348	99.25	14.11	+2.92
1805507.18	231 Mar 18	16	22	+ 4 04	231 Mar 25	6.6	0.59024	88.30	15.86	+6.59
1806286.13	233 May 5	15	11	-16 54	233 May 14	2.1	0.46545	69.63	20.11	+8.46
1807089.20	235 Jul 17	16	47	-28 18	235 Jul 17	22.1	0.37439	56.01	25.00	+0.22
1807893.59	237 Sep 29	2	09	- 0 35	237 Sep 20	23.5	0.44962	67.26	20.82	-8.11
1808673.13	239 Nov 17	15	07	+20 47	239 Nov 11	11.9	0.57567	86.12	16.26	-6.13
1809440.91	241 Dec 24	9	49	+27 31	241 Dec 21	22.5	0.65683	98.26	14.25	-2.47
1810205.08	244 Jan 27	13	56	+23 05	244 Jan 28	23.5	0.67454	100.91	13.88	+1.40
1810970.71	246 Mar 3	4	59	+10 38	246 Mar 8	10.2	0.62647	93.72	14.94	+5.22
1811743.92	248 Apr 14	10	02	- 8 16	248 Apr 22	15.9	0.51817	77.52	18.06	+8.24
1812536.42	250 Jun 15	22	01	-28 19	250 Jun 21	2.8	0.39469	59.04	23.71	+5.20
1813346.87	252 Sep 3	8	49	-12 42	252 Aug 27	14.8	0.40415	60.46	23.16	-6.75
1814134.96	254 Oct 31	10	59	+14 26	254 Oct 24	3.2	0.52940	79.20	17.68	-7.32
1814905.99	256 Dec 10	11	41	+26 27	256 Dec 6	12.5	0.63226	94.58	14.80	-3.97
1815670.90	259 Jan 14	9	29	+25 53	259 Jan 14	6.4	0.67557	101.06	13.85	-0.13
1816435.39	261 Feb 16	21	24	+16 13	261 Feb 20	15.2	0.65312	97.71	14.33	+3.74
1817204.66	263 Mar 28	3	55	- 0 17	263 Apr 4	10.8	0.56651	84.75	16.52	+7.29
1817987.72	265 May 19	5	21	-21 48	265 May 27	4.4	0.43727	65.41	21.41	+7.96
1818795.49	267 Aug 4	23	42	-24 10	267 Aug 2	6.0	0.37695	56.39	24.83	-2.74
1819594.24	269 Oct 11	17	45	+ 5 39	269 Oct 3	16.5	0.47832	71.56	19.57	-8.05
1820370.19	271 Nov 26	16	34	+23 31	271 Nov 21	6.9	0.59840	89.52	15.64	-5.40
1821136.67	274 Jan 1	4	03	+27 21	273 Dec 30	12.3	0.66629	99.68	14.05	-1.66
1821900.74	276 Feb 4	5	40	+20 48	276 Feb 6	11.1	0.66974	100.19	13.98	+2.23
1822667.39	278 Mar 11	21	19	+ 6 48	278 Mar 17	21.1	0.60763	90.90	15.40	+5.99
1823443.53	280 Apr 26	0	45	-13 18	280 May 4	13.1	0.48908	73.16	19.14	+8.51
1824241.94	282 Jul 3	10	36	-29 23	282 Jul 6	2.5	0.37970	56.80	24.65	+2.66
1825049.96	284 Sep 18	11	07	- 5 31	284 Sep 10	16.7	0.42772	63.99	21.88	-7.77
1825833.00	286 Nov 10	11	54	+18 28	286 Nov 3	18.6	0.55568	83.13	16.84	-6.72
1826602.07	288 Dec 18	13	35	+27 17	288 Dec 15	9.7	0.64705	96.80	14.47	-3.16
1827366.47	291 Jan 21	23	24	+24 19	291 Jan 22	16.3	0.67626	101.17	13.84	+0.70
1828131.49	293 Feb 24	23	43	+12 58	293 Mar 1	13.1	0.63958	95.68	14.63	+4.56
1828902.73	295 Apr 7	5	25	- 4 54	295 Apr 15	2.9	0.54065	80.88	17.31	+7.89
1829690.71	297 Jun 3	5	09	-26 02	297 Jun 9	22.6	0.41185	61.61	22.73	+6.73

	Mars at opposition				Mars nearest to Earth				
JD	date	ET	δ	date	ET	least distance		d	Δt
		h m	° ′		h	a.u.	10^6 km	″	d
1830501.09	299 Aug 22	14 13	-17 55	299 Aug 17	6.7	0.38878	58.16	24.08	-5.31
1831293.80	301 Oct 23	7 07	+11 06	301 Oct 15	13.4	0.50680	75.82	18.47	-7.74
1832066.79	303 Dec 5	6 55	+25 29	303 Nov 30	15.6	0.61820	92.48	15.14	-4.64
1832832.29	306 Jan 8	19 01	+26 36	306 Jan 7	23.2	0.67273	100.64	13.91	-0.83
1833596.51	308 Feb 12	0 10	+18 08	308 Feb 15	1.6	0.66190	99.02	14.14	+3.06
1834364.47	310 Mar 20	23 11	+ 2 41	310 Mar 27	17.0	0.58601	87.67	15.97	+6.74
1835144.13	312 May 8	15 08	-18 18	312 May 17	0.9	0.46003	68.82	20.35	+8.41
1835948.19	314 Jul 21	16 39	-27 28	314 Jul 21	9.5	0.37419	55.98	25.01	-0.30
1836751.63	316 Oct 2	3 04	+ 1 15	316 Sep 24	1.3	0.45514	68.09	20.57	-8.07
1837530.46	318 Nov 19	23 08	+21 44	318 Nov 13	22.3	0.58040	86.83	16.13	-6.04
1838297.95	320 Dec 26	10 54	+27 30	320 Dec 24	2.8	0.65900	98.58	14.20	-2.34
1839062.07	323 Jan 29	13 45	+22 18	323 Jan 31	2.7	0.67399	100.83	13.89	+1.54
1839827.89	325 Mar 5	9 19	+ 9 22	325 Mar 10	18.1	0.62317	93.22	15.02	+5.36
1840601.61	327 Apr 18	2 45	- 9 46	327 Apr 26	11.1	0.51259	76.68	18.26	+8.35
1841395.21	329 Jun 19	17 01	-28 49	329 Jun 24	11.9	0.39100	58.49	23.94	+4.78
1842205.40	331 Sep 7	21 41	-10 45	331 Aug 31	21.5	0.40797	61.03	22.94	-7.01
1842992.44	333 Nov 2	22 32	+15 44	333 Oct 26	16.5	0.53432	79.93	17.52	-7.25
1843763.08	335 Dec 13	13 49	+26 47	335 Dec 9	17.8	0.63532	95.04	14.73	-3.83
1844527.88	338 Jan 16	9 14	+25 21	338 Jan 16	9.6	0.67596	101.12	13.85	+0.02
1845292.43	340 Feb 19	22 22	+15 07	340 Feb 23	19.9	0.65080	97.36	14.38	+3.90
1846062.02	342 Mar 30	12 33	- 1 42	342 Apr 6	23.2	0.56194	84.07	16.66	+7.44
1846845.98	344 May 22	11 32	-23 02	344 May 30	4.9	0.43231	64.67	21.65	+7.72
1847654.44	346 Aug 8	22 36	-22 41	346 Aug 5	14.8	0.37848	56.62	24.73	-3.33
1848452.05	348 Oct 14	13 18	+ 7 19	348 Oct 6	13.0	0.48362	72.35	19.35	-8.01
1849227.40	350 Nov 28	21 29	+24 13	350 Nov 23	14.4	0.60234	90.11	15.54	-5.29
1849993.66	353 Jan 3	3 46	+27 07	353 Jan 1	15.6	0.66787	99.91	14.01	-1.51
1850757.74	355 Feb 6	5 51	+19 53	355 Feb 8	15.1	0.66867	100.03	14.00	+2.38
1851524.63	357 Mar 14	3 03	+ 5 28	357 Mar 20	6.9	0.60379	90.33	15.50	+6.16
1852301.35	359 Apr 29	20 20	-14 44	359 May 8	9.2	0.48362	72.35	19.35	+8.54
1853100.87	361 Jul 7	8 47	-29 08	361 Jul 9	13.4	0.37800	56.55	24.76	+2.19
1853908.20	363 Sep 22	16 50	- 3 36	363 Sep 14	21.3	0.43285	64.75	21.62	-7.82
1854690.39	365 Nov 12	21 26	+19 34	365 Nov 6	6.3	0.56079	83.89	16.69	-6.63
1855459.13	367 Dec 21	15 08	+27 25	367 Dec 18	15.0	0.64973	97.20	14.41	-3.01
1856223.46	370 Jan 23	23 01	+23 38	370 Jan 24	19.8	0.67621	101.16	13.84	+0.86
1856988.60	372 Feb 28	2 21	+11 45	372 Mar 3	19.8	0.63681	95.27	14.70	+4.73
1857760.29	374 Apr 9	18 55	- 6 23	374 Apr 17	19.6	0.53537	80.09	17.48	+8.03
1858549.31	376 Jun 6	19 32	-26 55	376 Jun 13	5.2	0.40722	60.92	22.98	+6.40
1859359.85	378 Aug 26	8 22	-16 04	378 Aug 20	15.0	0.39164	58.59	23.90	-5.72
1860151.41	380 Oct 25	21 50	+12 34	380 Oct 18	5.1	0.51190	76.58	18.28	-7.70
1860923.94	382 Dec 7	10 27	+25 59	382 Dec 2	22.7	0.62170	93.00	15.06	-4.49
1861689.29	385 Jan 10	18 58	+26 12	385 Jan 10	3.3	0.67362	100.77	13.90	-0.65
1862453.53	387 Feb 14	0 50	+17 07	387 Feb 17	6.7	0.66003	98.74	14.18	+3.24
1863221.76	389 Mar 23	6 18	+ 1 18	389 Mar 30	4.4	0.58173	87.03	16.09	+6.92
1864002.16	391 May 12	15 54	-19 39	391 May 20	22.6	0.45482	68.04	20.58	+8.28
1864807.22	393 Jul 25	17 10	-26 26	393 Jul 24	19.1	0.37423	55.98	25.01	-0.92
1865609.63	395 Oct 6	3 10	+ 3 03	395 Sep 28	0.6	0.46040	68.87	20.33	-8.11
1866387.73	397 Nov 22	5 26	+22 35	397 Nov 16	7.4	0.58471	87.47	16.01	-5.92
1867154.95	399 Dec 29	10 45	+27 24	399 Dec 27	6.7	0.66099	98.88	14.16	-2.17
1867919.06	402 Jan 31	13 27	+21 29	402 Feb 2	6.7	0.67340	100.74	13.90	+1.72
1868685.05	404 Mar 7	13 13	+ 8 05	404 Mar 13	2.4	0.61979	92.72	15.10	+5.55

OPPOSITIONS OF MARS

	Mars at opposition				Mars nearest to Earth				
JD	date	ET	δ		date	ET	least distance	d	Δt
		h m	° ′			h	a.u.	10^6 km ″	d

JD	date	ET	δ	date	ET	least distance	a.u.	10^6 km	d ″	Δt d
1869459.28	406 Apr 20	18 38	-11 14	406 Apr 29	4.5	0.50720	75.88	18.45	+8.41
1870253.99	408 Jun 23	11 39	-29 07	408 Jun 27	21.3	0.38801	58.05	24.12	+4.40
1871063.84	410 Sep 11	8 08	- 8 48	410 Sep 4	4.4	0.41245	61.70	22.69	-7.16
1871849.94	412 Nov 5	10 30	+17 00	412 Oct 29	6.4	0.53974	80.74	17.34	-7.17
1872620.19	414 Dec 15	16 38	+27 02	414 Dec 12	0.7	0.63851	95.52	14.66	-3.66
1873384.87	417 Jan 18	8 56	+24 46	417 Jan 18	13.7	0.67636	101.18	13.84	+0.20
1874149.51	419 Feb 22	0 12	+13 59	419 Feb 26	2.2	0.64849	97.01	14.43	+4.08
1874919.50	421 Apr 1	23 53	- 3 10	421 Apr 9	13.9	0.55702	83.33	16.80	+7.58
1875704.35	423 May 26	20 21	-24 11	423 Jun 3	8.6	0.42708	63.89	21.92	+7.51
1876513.39	425 Aug 12	21 26	-21 04	425 Aug 9	0.6	0.38005	56.85	24.63	-3.87
1877309.83	427 Oct 18	7 51	+ 8 56	427 Oct 10	7.2	0.48882	73.13	19.15	-8.03
1878084.58	429 Dec 1	1 53	+24 51	429 Nov 25	22.9	0.60620	90.69	15.44	-5.13
1878850.65	432 Jan 6	3 43	+26 50	432 Jan 4	20.2	0.66921	100.11	13.99	-1.31
1879614.75	434 Feb 8	6 04	+18 56	434 Feb 10	20.0	0.66725	99.82	14.03	+2.58
1880381.83	436 Mar 16	7 59	+ 4 08	436 Mar 22	16.3	0.59985	89.74	15.60	+6.35
1881159.18	438 May 2	16 20	-16 09	438 May 11	3.8	0.47831	71.55	19.57	+8.48
1881959.85	440 Jul 11	8 29	-28 41	440 Jul 12	22.8	0.37658	56.34	24.86	+1.60
1882766.39	442 Sep 25	21 26	- 1 41	442 Sep 17	23.4	0.43782	65.50	21.38	-7.92
1883547.75	444 Nov 15	5 58	+20 36	444 Nov 8	18.0	0.56544	84.59	16.55	-6.50
1884316.16	446 Dec 23	15 53	+27 27	446 Dec 20	20.1	0.65214	97.56	14.35	-2.82
1885080.44	449 Jan 25	22 27	+22 53	449 Jan 26	23.8	0.67605	101.14	13.85	+1.06
1885845.72	451 Mar 2	5 19	+10 32	451 Mar 7	3.4	0.63388	94.83	14.77	+4.92
1886617.85	453 Apr 12	8 23	- 7 50	453 Apr 20	10.8	0.53016	79.31	17.66	+8.10
1887407.86	455 Jun 11	8 42	-27 39	455 Jun 17	12.5	0.40318	60.32	23.22	+6.16
1888218.51	457 Aug 30	0 10	-14 10	457 Aug 24	0.4	0.39514	59.11	23.69	-5.99
1889009.03	459 Oct 29	12 39	+13 59	459 Oct 21	21.5	0.51756	77.43	18.08	-7.63
1889781.08	461 Dec 9	14 01	+26 24	461 Dec 5	6.7	0.62533	93.55	14.97	-4.31
1890546.28	464 Jan 13	18 38	+25 43	464 Jan 13	7.5	0.67446	100.90	13.88	-0.46
1891310.57	466 Feb 16	1 45	+16 02	466 Feb 19	12.0	0.65819	98.46	14.22	+3.43
1892079.11	468 Mar 25	14 44	- 0 07	468 Apr 1	16.1	0.57722	86.35	16.22	+7.06
1892860.31	470 May 15	19 21	-20 58	470 May 23	23.0	0.44925	67.21	20.83	+8.15
1893666.27	472 Jul 29	18 35	-25 14	472 Jul 28	5.0	0.37436	56.00	25.00	-1.57
1894467.57	474 Oct 9	1 37	+ 4 49	474 Sep 30	21.5	0.46548	69.63	20.11	-8.17
1895244.98	476 Nov 24	11 26	+23 23	476 Nov 18	17.7	0.58887	88.09	15.89	-5.74
1896011.97	478 Dec 31	11 17	+27 15	478 Dec 29	12.0	0.66280	99.15	14.12	-1.97
1896776.06	481 Feb 2	13 27	+20 36	481 Feb 4	11.4	0.67240	100.59	13.92	+1.92
1897542.21	483 Mar 10	17 02	+ 6 47	483 Mar 16	10.6	0.61619	92.18	15.19	+5.73
1898316.97	485 Apr 23	11 17	-12 40	485 May 1	20.8	0.50194	75.09	18.65	+8.40
1899112.82	487 Jun 28	7 37	-29 13	487 Jul 2	5.9	0.38524	57.63	24.30	+3.93
1899922.26	489 Sep 14	18 16	- 6 49	489 Sep 7	10.0	0.41686	62.36	22.45	-7.35
1900707.40	491 Nov 8	21 37	+18 11	491 Nov 1	20.8	0.54475	81.49	17.18	-7.03
1901477.25	493 Dec 17	17 58	+27 14	493 Dec 14	6.5	0.64129	95.94	14.60	-3.48
1902241.84	496 Jan 21	8 03	+24 08	496 Jan 21	17.4	0.67661	101.22	13.83	+0.39
1903006.59	498 Feb 24	2 03	+12 49	498 Feb 28	8.5	0.64605	96.65	14.49	+4.27
1903776.93	500 Apr 4	10 15	- 4 36	500 Apr 12	2.3	0.55207	82.59	16.95	+7.67
1904562.66	502 May 30	3 44	-25 12	502 Jun 6	12.7	0.42228	63.17	22.17	+7.37
1905372.23	504 Aug 16	17 30	-19 22	504 Aug 12	10.8	0.38233	57.20	24.48	-4.28
1906167.56	506 Oct 21	1 23	+10 30	506 Oct 13	1.9	0.49449	73.98	18.93	-7.98
1906941.78	508 Dec 3	6 44	+25 25	508 Nov 28	8.1	0.61022	91.29	15.34	-4.94
1907707.66	511 Jan 8	3 50	+26 28	511 Jan 7	0.7	0.67055	100.31	13.96	-1.13

	Mars at opposition				*Mars nearest to Earth*				
JD	date	ET	δ	date	ET	least distance		d	Δt
		h m	° ′		h	a.u.	10^6 km	″	d
1908471.77	513 Feb 10	6 25	+17 57	513 Feb 13	0.6	0.66583	99.61	14.06	+2.76
1909239.11	515 Mar 19	14 40	+ 2 46	515 Mar 26	2.0	0.59576	89.12	15.71	+6.47
1910017.16	517 May 5	15 44	-17 33	517 May 14	1.8	0.47267	70.71	19.80	+8.42
1910818.91	519 Jul 16	9 51	-28 02	519 Jul 17	8.7	0.37515	56.12	24.95	+0.95
1911624.55	521 Sep 29	1 11	+ 0 13	521 Sep 20	23.7	0.44263	66.22	21.15	-8.06
1912405.08	523 Nov 18	13 51	+21 33	523 Nov 12	6.0	0.56996	85.26	16.42	-6.33
1913173.20	525 Dec 25	16 44	+27 25	525 Dec 23	1.4	0.65438	97.89	14.30	-2.64
1913937.43	528 Jan 28	22 13	+22 07	528 Jan 30	3.9	0.67551	101.05	13.86	+1.24
1914702.83	530 Mar 4	7 54	+ 9 18	530 Mar 9	9.9	0.63072	94.35	14.84	+5.08
1915475.38	532 Apr 14	21 12	- 9 16	532 Apr 23	0.4	0.52506	78.55	17.83	+8.13
1916266.47	534 Jun 14	23 11	-28 15	534 Jun 20	18.9	0.39934	59.74	23.44	+5.82
1917077.13	536 Sep 2	15 13	-12 13	536 Aug 27	8.3	0.39875	59.65	23.47	-6.29
1917866.59	538 Nov 1	2 14	+15 20	538 Oct 24	14.1	0.52276	78.20	17.90	-7.51
1918638.19	540 Dec 11	16 28	+26 43	540 Dec 7	13.3	0.62847	94.02	14.89	-4.13
1919403.24	543 Jan 15	17 46	+25 11	543 Jan 15	10.8	0.67517	101.00	13.86	-0.29
1920167.62	545 Feb 18	2 49	+14 57	545 Feb 21	17.0	0.65621	98.17	14.26	+3.59
1920936.47	547 Mar 28	23 16	- 1 31	547 Apr 5	2.9	0.57260	85.66	16.35	+7.15
1921718.43	549 May 18	22 14	-22 11	549 May 27	0.8	0.44405	66.43	21.08	+8.11
1922525.24	551 Aug 3	17 39	-23 54	551 Aug 1	15.9	0.37518	56.13	24.95	-2.07
1923325.47	553 Oct 11	23 15	+ 6 32	553 Oct 3	19.2	0.47109	70.47	19.87	-8.17
1924102.24	555 Nov 27	17 41	+24 07	555 Nov 22	4.2	0.59336	88.77	15.77	-5.56
1924868.99	558 Jan 2	11 46	+27 01	557 Dec 31	16.3	0.66463	99.43	14.08	-1.81
1925633.06	560 Feb 5	13 23	+19 41	560 Feb 7	15.1	0.67145	100.45	13.94	+2.07
1926399.41	562 Mar 12	21 47	+ 5 28	562 Mar 18	18.1	0.61263	91.65	15.28	+5.85
1927174.75	564 Apr 26	6 06	-14 06	564 May 4	15.9	0.49641	74.26	18.86	+8.41
1927971.75	566 Jul 2	6 06	-29 09	566 Jul 5	14.3	0.38238	57.20	24.48	+3.34
1928780.62	568 Sep 18	2 57	- 4 51	568 Sep 10	12.8	0.42120	63.01	22.22	-7.59
1929564.81	570 Nov 11	7 21	+19 18	570 Nov 4	10.3	0.54949	82.20	17.03	-6.88
1930334.31	572 Dec 19	19 32	+27 20	572 Dec 16	11.9	0.64393	96.33	14.54	-3.32
1931098.82	575 Jan 23	7 40	+23 27	575 Jan 23	20.8	0.67660	101.22	13.83	+0.55
1931863.66	577 Feb 26	3 54	+11 38	577 Mar 2	13.6	0.64328	96.23	14.55	+4.41
1932634.37	579 Apr 7	20 46	- 6 01	579 Apr 15	14.4	0.54716	81.85	17.11	+7.74
1933421.04	581 Jun 2	13 03	-26 08	581 Jun 9	17.4	0.41773	62.49	22.41	+7.18
1934231.07	583 Aug 21	13 45	-17 34	583 Aug 16	21.5	0.38483	57.57	24.32	-4.68
1935025.28	585 Oct 23	18 45	+12 01	585 Oct 15	21.6	0.49989	74.78	18.72	-7.88
1935798.93	587 Dec 6	10 26	+25 54	587 Dec 1	15.6	0.61383	91.83	15.25	-4.78
1936564.63	590 Jan 10	3 03	+26 03	590 Jan 9	3.5	0.67173	100.49	13.93	-0.98
1937328.79	592 Feb 13	6 52	+16 55	592 Feb 16	4.4	0.66435	99.39	14.09	+2.90
1938096.38	594 Mar 21	21 03	+ 1 23	594 Mar 28	10.8	0.59165	88.51	15.82	+6.57
1938875.07	596 May 8	13 43	-18 53	596 May 17	0.4	0.46725	69.90	20.03	+8.45
1939677.87	598 Jul 20	8 48	-27 12	598 Jul 20	18.4	0.37441	56.01	25.00	+0.40
1940482.60	600 Oct 2	2 27	+ 2 02	600 Sep 23	23.5	0.44798	67.02	20.89	-8.12
1941262.39	602 Nov 20	21 21	+22 26	602 Nov 14	17.3	0.57477	85.98	16.28	-6.17
1942030.24	604 Dec 27	17 50	+27 19	604 Dec 25	5.8	0.65668	98.24	14.25	-2.50
1942794.41	607 Jan 30	21 44	+21 17	607 Feb 1	6.7	0.67506	100.99	13.87	+1.37
1943559.97	609 Mar 6	11 19	+ 8 01	609 Mar 11	15.9	0.62760	93.89	14.91	+5.19
1944333.04	611 Apr 18	12 58	-10 44	611 Apr 26	17.4	0.51970	77.75	18.01	+8.18
1945125.24	613 Jun 18	17 42	-28 41	613 Jun 24	2.2	0.39538	59.15	23.67	+5.36
1945935.76	615 Sep 7	6 12	-10 13	615 Aug 31	14.7	0.40236	60.19	23.26	-6.65
1946724.13	617 Nov 3	15 12	+16 37	617 Oct 27	6.3	0.52776	78.95	17.74	-7.37

	Mars at opposition				Mars nearest to Earth			
JD	date	ET	δ	date	ET	least distance		d	Δt
		h m	° ′		h	a.u.	10⁶ km	″	d
1947495.28	619 Dec 14	18 50	+26 59	619 Dec 10	18.9	0.63159	94.48	14.82	-4.00
1948260.22	622 Jan 17	17 22	+24 36	622 Jan 17	13.7	0.67569	101.08	13.85	-0.15
1949024.67	624 Feb 21	4 08	+13 49	624 Feb 24	21.3	0.65390	97.82	14.31	+3.71
1949793.80	626 Mar 31	7 07	- 2 55	626 Apr 7	13.0	0.56806	84.98	16.48	+7.24
1950576.57	628 May 22	1 41	-23 19	628 May 30	2.2	0.43902	65.68	21.32	+8.02
1951384.21	630 Aug 7	16 59	-22 24	630 Aug 5	3.4	0.37629	56.29	24.87	-2.57
1952183.32	632 Oct 14	19 47	+ 8 12	632 Oct 6	17.3	0.47651	71.28	19.64	-8.10
1952959.44	634 Nov 29	22 35	+24 45	634 Nov 24	12.2	0.59735	89.36	15.67	-5.43
1953725.97	637 Jan 4	11 17	+26 42	637 Jan 2	18.9	0.66622	99.67	14.05	-1.68
1954490.04	639 Feb 7	12 58	+18 44	639 Feb 9	17.7	0.67046	100.30	13.96	+2.20
1955256.61	641 Mar 15	2 38	+ 4 09	641 Mar 21	1.5	0.60893	91.10	15.37	+5.95
1956032.53	643 Apr 30	0 48	-15 30	643 May 8	12.3	0.49089	73.44	19.07	+8.48
1956830.62	645 Jul 6	2 52	-28 53	645 Jul 8	22.9	0.38021	56.88	24.62	+2.84
1957638.90	647 Sep 22	9 35	- 2 55	647 Sep 14	15.8	0.42610	63.74	21.97	-7.74
1958422.22	649 Nov 13	17 19	+20 21	649 Nov 6	23.5	0.55465	82.97	16.88	-6.74
1959191.39	651 Dec 22	21 19	+27 22	651 Dec 19	16.7	0.64679	96.76	14.47	-3.19
1959955.80	654 Jan 25	7 07	+22 42	654 Jan 25	23.3	0.67668	101.23	13.83	+0.67
1960720.76	656 Feb 29	6 16	+10 25	656 Mar 4	18.7	0.64067	95.84	14.61	+4.52
1961491.87	658 Apr 10	8 56	- 7 28	658 Apr 18	4.8	0.54219	81.11	17.26	+7.83
1962279.58	660 Jun 6	1 50	-26 57	660 Jun 12	22.0	0.41294	61.77	22.67	+6.84
1963089.92	662 Aug 25	10 04	-15 40	662 Aug 20	6.0	0.38733	57.94	24.17	-5.17
1963882.93	664 Oct 26	10 21	+13 28	664 Oct 18	15.8	0.50499	75.55	18.53	-7.77
1964656.06	666 Dec 8	13 31	+26 19	666 Dec 3	21.4	0.61732	92.35	15.16	-4.67
1965421.61	669 Jan 12	2 43	+25 34	669 Jan 11	6.1	0.67271	100.64	13.91	-0.86
1966185.80	671 Feb 15	7 15	+15 51	671 Feb 18	7.8	0.66253	99.11	14.13	+3.02
1966953.63	673 Mar 24	3 05	+ 0 01	673 Mar 30	19.7	0.58739	87.87	15.93	+6.69
1967733.04	675 May 12	12 53	-20 09	675 May 20	23.4	0.46190	69.10	20.26	+8.44
1968536.87	677 Jul 24	8 55	-26 10	677 Jul 24	6.2	0.37404	55.96	25.02	-0.11
1969340.66	679 Oct 6	3 49	+ 3 51	679 Sep 28	1.2	0.45329	67.81	20.65	-8.11
1970119.69	681 Nov 23	4 39	+23 15	681 Nov 17	3.0	0.57919	86.65	16.16	-6.07
1970887.25	683 Dec 30	17 53	+27 09	683 Dec 28	8.8	0.65874	98.55	14.21	-2.38
1971651.37	686 Feb 1	20 55	+20 25	686 Feb 3	8.9	0.67458	100.92	13.88	+1.50
1972417.12	688 Mar 8	14 59	+ 6 44	688 Mar 13	22.5	0.62441	93.41	14.99	+5.31
1973190.66	690 Apr 21	3 48	-12 09	690 Apr 29	11.2	0.51436	76.95	18.20	+8.31
1973983.91	692 Jun 22	9 51	-28 56	692 Jun 27	8.1	0.39195	58.64	23.88	+4.93
1974794.25	694 Sep 10	18 00	- 8 15	694 Sep 3	19.4	0.40646	60.81	23.03	-6.94
1975581.62	696 Nov 6	2 54	+17 50	696 Oct 29	20.4	0.53310	79.75	17.56	-7.27
1976352.38	698 Dec 16	21 11	+27 09	698 Dec 13	0.2	0.63490	94.98	14.74	-3.87
1977117.20	701 Jan 19	16 50	+23 58	701 Jan 19	16.3	0.67625	101.17	13.84	-0.02
1977881.71	703 Feb 23	5 07	+12 40	703 Feb 27	1.4	0.65176	97.50	14.36	+3.85
1978651.19	705 Apr 2	16 37	- 4 21	705 Apr 10	1.6	0.56343	84.29	16.61	+7.37
1979434.90	707 May 26	9 33	-24 24	707 Jun 3	4.9	0.43367	64.88	21.58	+7.81
1980243.22	709 Aug 11	17 22	-20 45	709 Aug 8	14.1	0.37744	56.46	24.80	-3.14
1981041.15	711 Oct 18	15 39	+ 9 49	711 Oct 10	14.6	0.48167	72.06	19.43	-8.04
1981816.65	713 Dec 2	3 34	+25 19	713 Nov 26	19.6	0.60127	89.95	15.57	-5.33
1982582.97	716 Jan 7	11 16	+26 21	716 Jan 5	22.1	0.66771	99.89	14.02	-1.55
1983347.04	718 Feb 9	13 02	+17 45	718 Feb 11	21.2	0.66916	100.11	13.99	+2.34
1984113.81	720 Mar 17	7 28	+ 2 48	720 Mar 23	9.8	0.60506	90.52	15.47	+6.10
1984890.30	722 May 2	19 13	-16 51	722 May 11	8.3	0.48551	72.63	19.28	+8.54
1985689.52	724 Jul 10	0 24	-28 26	724 Jul 12	9.2	0.37839	56.61	24.74	+2.37

	Mars at opposition			Mars nearest to Earth				
JD	date	ET	δ	date	ET	least distance	d	Δt
		h m	° ′		h	a.u.	10^6 km	″	d
1986497.16	726 Sep 25	15 48	- 0 59	726 Sep 17	20.0	0.43105	64.48	21.71	-7.83
1987279.60	728 Nov 16	2 17	+21 20	728 Nov 9	10.2	0.55940	83.69	16.73	-6.67
1988048.41	730 Dec 24	21 52	+27 20	730 Dec 21	20.4	0.64926	97.13	14.42	-3.06
1988812.75	733 Jan 27	5 59	+21 55	733 Jan 28	1.5	0.67666	101.23	13.83	+0.81
1989577.85	735 Mar 3	8 21	+ 9 11	735 Mar 8	0.3	0.63796	95.44	14.67	+4.66
1990349.37	737 Apr 12	20 56	- 8 53	737 Apr 20	20.4	0.53707	80.34	17.43	+7.98
1991138.06	739 Jun 10	13 30	-27 37	739 Jun 17	2.0	0.40855	61.12	22.91	+6.52
1991948.61	741 Aug 29	2 42	-13 45	741 Aug 23	12.6	0.39035	58.40	23.98	-5.59
1992740.55	743 Oct 30	1 10	+14 51	743 Oct 22	8.1	0.51046	76.36	18.34	-7.71
1993513.23	745 Dec 10	17 25	+26 39	745 Dec 6	4.4	0.62114	92.92	15.07	-4.54
1994278.61	748 Jan 15	2 33	+25 02	748 Jan 14	9.6	0.67379	100.80	13.89	-0.71
1995042.82	750 Feb 17	7 41	+14 46	750 Feb 20	12.0	0.66088	98.87	14.16	+3.18
1995810.95	752 Mar 26	10 44	- 1 22	752 Apr 2	7.1	0.58320	87.25	16.05	+6.85
1996591.13	754 May 15	15 09	-21 24	754 May 23	23.0	0.45637	68.27	20.51	+8.32
1997395.96	756 Jul 28	11 04	-24 57	756 Jul 27	17.8	0.37370	55.90	25.05	-0.72
1998198.67	758 Oct 9	4 12	+ 5 38	758 Oct 1	1.4	0.45839	68.57	20.42	-8.12
1998976.96	760 Nov 25	10 57	+23 59	760 Nov 19	11.7	0.58344	87.28	16.04	-5.97
1999744.25	763 Jan 1	18 06	+26 54	762 Dec 30	12.7	0.66067	98.83	14.17	-2.23
2000508.36	765 Feb 3	20 38	+19 29	765 Feb 5	12.6	0.67380	100.80	13.89	+1.66
2001274.26	767 Mar 11	18 15	+ 5 27	767 Mar 17	6.0	0.62092	92.89	15.07	+5.49
2002048.28	769 Apr 23	18 38	-13 31	769 May 2	4.3	0.50904	76.15	18.39	+8.40
2002842.65	771 Jun 27	3 31	-29 01	771 Jul 1	16.8	0.38892	58.18	24.07	+4.55
2003652.72	773 Sep 14	5 16	- 6 15	773 Sep 7	2.3	0.41077	61.45	22.79	-7.13
2004439.12	775 Nov 9	14 47	+19 00	775 Nov 2	9.7	0.53817	80.51	17.39	-7.21
2005209.47	777 Dec 18	23 11	+27 16	777 Dec 15	5.7	0.63786	95.42	14.67	-3.73
2005974.16	780 Jan 22	15 51	+23 16	780 Jan 22	19.1	0.67667	101.23	13.83	+0.14
2006738.77	782 Feb 25	6 29	+11 29	782 Mar 1	6.9	0.64953	97.17	14.41	+4.02
2007508.61	784 Apr 5	2 42	- 5 45	784 Apr 12	15.5	0.55866	83.57	16.75	+7.53
2008293.16	786 May 29	15 52	-25 22	786 Jun 6	5.9	0.42865	64.13	21.84	+7.59
2009102.09	788 Aug 15	14 03	-19 02	788 Aug 11	21.6	0.37912	56.72	24.69	-3.69
2009898.90	790 Oct 21	9 42	+11 22	790 Oct 13	9.2	0.48721	72.89	19.21	-8.02
2010673.85	792 Dec 4	8 21	+25 49	792 Nov 29	3.8	0.60551	90.58	15.46	-5.19
2011439.97	795 Jan 9	11 22	+25 55	795 Jan 8	2.3	0.66927	100.12	13.99	-1.38
2012204.04	797 Feb 11	12 56	+16 43	797 Feb 14	1.3	0.66801	99.93	14.01	+2.51
2012971.04	799 Mar 20	12 54	+ 1 27	799 Mar 26	19.6	0.60128	89.95	15.57	+6.28
2013748.20	801 May 5	16 47	-18 11	801 May 14	4.6	0.47996	71.80	19.50	+8.49
2014548.58	803 Jul 15	1 59	-27 45	803 Jul 16	21.2	0.37662	56.34	24.85	+1.80
2015355.38	805 Sep 28	21 12	+ 0 56	805 Sep 20	23.6	0.43584	65.20	21.48	-7.90
2016136.96	807 Nov 19	11 03	+22 14	807 Nov 12	21.5	0.56398	84.37	16.60	-6.56
2016905.46	809 Dec 26	23 03	+27 13	809 Dec 24	1.8	0.65168	97.49	14.36	-2.89
2017669.74	812 Jan 30	5 48	+21 05	812 Jan 31	5.8	0.67636	101.18	13.84	+1.00
2018434.96	814 Mar 5	11 03	+ 7 56	814 Mar 10	7.7	0.63485	94.97	14.74	+4.86
2019206.89	816 Apr 15	9 23	-10 17	816 Apr 23	11.6	0.53192	79.57	17.60	+8.09
2019996.58	818 Jun 14	1 51	-28 09	818 Jun 20	7.9	0.40452	60.51	23.14	+6.25
2020807.31	820 Sep 1	19 23	-11 46	820 Aug 26	21.3	0.39373	58.90	23.77	-5.92
2021598.16	822 Nov 1	15 56	+16 11	822 Oct 24	23.9	0.51586	77.17	18.14	-7.67
2022370.35	824 Dec 12	20 20	+26 54	824 Dec 8	11.2	0.62453	93.43	14.99	-4.38
2023135.57	827 Jan 17	1 38	+24 26	827 Jan 16	13.0	0.67463	100.92	13.87	-0.53
2023899.84	829 Feb 19	8 08	+13 39	829 Feb 22	16.9	0.65916	98.61	14.20	+3.36
2024668.27	831 Mar 29	18 23	- 2 45	831 Apr 5	18.7	0.57880	86.59	16.17	+7.01

OPPOSITIONS OF MARS

| | Mars at opposition | | | | Mars nearest to Earth | | | | |
| JD | date | ET | δ | date | ET | least distance | d | Δt | |
		h m	° '		h	a.u.	10^6 km	"	d
2025449.19	833 May 18	16 35	-22 33	833 May 26	21.2	0.45097	67.46	20.76	+8.19
2026254.91	835 Aug 2	9 51	-23 37	835 Aug 1	1.9	0.37393	55.94	25.03	-1.33
2027056.58	837 Oct 12	1 59	+ 7 21	837 Oct 3	22.5	0.46376	69.38	20.18	-8.15
2027834.23	839 Nov 28	17 29	+24 38	839 Nov 22	22.0	0.58804	87.97	15.92	-5.81
2028601.29	842 Jan 3	18 57	+26 35	842 Jan 1	18.1	0.66275	99.15	14.12	-2.03
2029365.36	844 Feb 6	20 33	+18 32	844 Feb 8	17.0	0.67306	100.69	13.91	+1.85
2030131.44	846 Mar 13	22 34	+ 4 08	846 Mar 19	14.7	0.61751	92.38	15.16	+5.67
2030906.03	848 Apr 26	12 40	-14 55	848 May 4	22.4	0.50364	75.34	18.58	+8.40
2031701.58	850 Jul 1	1 50	-28 55	850 Jul 5	4.1	0.38582	57.72	24.26	+4.09
2032511.20	852 Sep 17	16 48	- 4 14	852 Sep 10	9.4	0.41504	62.09	22.55	-7.31
2033296.58	854 Nov 12	1 57	+20 05	854 Nov 4	23.3	0.54313	81.25	17.23	-7.11
2034066.54	856 Dec 21	0 53	+27 18	856 Dec 17	11.9	0.64070	95.85	14.61	-3.54
2034831.15	859 Jan 24	15 35	+22 31	859 Jan 24	23.6	0.67683	101.25	13.83	+0.33
2035595.84	861 Feb 27	8 11	+10 17	861 Mar 3	13.4	0.64691	96.78	14.47	+4.22
2036366.01	863 Apr 8	12 17	- 7 08	863 Apr 16	4.1	0.55372	82.84	16.90	+7.66
2037151.44	865 Jun 1	22 32	-26 13	865 Jun 9	8.8	0.42390	63.41	22.08	+7.43
2037960.94	867 Aug 20	10 35	-17 14	867 Aug 16	7.1	0.38129	57.04	24.55	-4.15
2038756.65	869 Oct 24	3 40	+12 52	869 Oct 16	3.3	0.49270	73.71	19.00	-8.01
2039531.04	871 Dec 7	12 54	+26 14	871 Dec 2	12.6	0.60929	91.15	15.36	-5.01
2040296.95	874 Jan 11	10 51	+25 25	874 Jan 10	6.4	0.67055	100.31	13.96	-1.19
2041061.04	876 Feb 14	13 00	+15 40	876 Feb 17	5.9	0.66668	99.73	14.04	+2.71
2041828.30	878 Mar 22	19 13	+ 0 05	878 Mar 29	5.7	0.59722	89.34	15.67	+6.44
2042606.10	880 May 8	14 21	-19 27	880 May 17	0.9	0.47443	70.97	19.73	+8.44
2043407.54	882 Jul 19	0 54	-26 55	882 Jul 20	5.6	0.37526	56.14	24.94	+1.19
2044213.50	884 Oct 2	0 02	+ 2 47	884 Sep 23	23.7	0.44089	65.96	21.23	-8.01
2044994.30	886 Nov 21	19 05	+23 04	886 Nov 15	9.5	0.56894	85.11	16.45	-6.40
2045762.51	888 Dec 29	0 14	+27 03	888 Dec 26	7.5	0.65424	97.87	14.31	-2.70
2046526.73	891 Feb 1	5 33	+20 12	891 Feb 2	10.0	0.67607	101.14	13.84	+1.18
2047292.07	893 Mar 7	13 44	+ 6 40	893 Mar 12	14.6	0.63194	94.54	14.81	+5.03
2048064.48	895 Apr 18	23 27	-11 41	895 Apr 27	2.4	0.52677	78.80	17.77	+8.12
2048855.28	897 Jun 17	18 43	-28 32	897 Jun 23	17.1	0.40036	59.89	23.38	+5.93
2049666.01	899 Sep 6	12 15	- 9 44	899 Aug 31	7.1	0.39716	59.41	23.57	-6.21
2050455.75	901 Nov 4	5 54	+17 27	901 Oct 27	15.9	0.52101	77.94	17.97	-7.58
2051227.46	903 Dec 15	23 09	+27 05	903 Dec 11	18.6	0.62773	93.91	14.91	-4.19
2051992.56	906 Jan 19	1 21	+23 47	906 Jan 18	17.3	0.67525	101.02	13.86	-0.33
2052756.89	908 Feb 22	9 25	+12 30	908 Feb 25	22.7	0.65694	98.28	14.25	+3.55
2053525.60	910 Apr 1	2 27	- 4 09	910 Apr 8	5.8	0.57407	85.88	16.30	+7.14
2054307.27	912 May 21	18 25	-23 37	912 May 29	21.5	0.44579	66.69	21.00	+8.13
2055113.88	914 Aug 6	9 11	-22 07	914 Aug 4	11.7	0.37461	56.04	24.99	-1.90
2055914.51	916 Oct 15	0 21	+ 9 01	916 Oct 6	19.7	0.46925	70.20	19.95	-8.20
2056691.48	918 Nov 30	23 24	+25 13	918 Nov 25	8.3	0.59226	88.60	15.80	-5.63
2057458.28	921 Jan 5	18 39	+26 13	921 Jan 3	22.2	0.66445	99.40	14.09	-1.85
2058222.34	923 Feb 8	20 06	+17 33	923 Feb 10	20.9	0.67218	100.56	13.92	+2.03
2058988.62	925 Mar 16	2 48	+ 2 48	925 Mar 21	22.5	0.61398	91.85	15.24	+5.82
2059763.75	927 Apr 30	6 00	-16 15	927 May 8	15.5	0.49817	74.53	18.79	+8.40
2060560.41	929 Jul 4	21 48	-28 37	929 Jul 8	11.5	0.38301	57.30	24.44	+3.57
2061369.50	931 Sep 22	0 04	- 2 17	931 Sep 14	12.0	0.41949	62.75	22.31	-7.50
2062153.99	933 Nov 14	11 45	+21 05	933 Nov 7	13.2	0.54828	82.02	17.07	-6.94
2062923.62	935 Dec 24	2 49	+27 15	935 Dec 20	18.1	0.64367	96.29	14.54	-3.36
2063688.13	938 Jan 26	15 11	+21 43	938 Jan 27	3.4	0.67701	101.28	13.83	+0.51

	Mars at opposition				*Mars nearest to Earth*					
JD	date	ET		δ	date	ET	least distance		d	Δt
		h m		° ′		h	a.u.	10^6 km	″	d
2064452.92	940 Mar 1	10 02		+ 9 04	940 Mar 5	19.0	0.64439	96.40	14.53	+4.37
2065223.50	942 Apr 10	23 56		- 8 33	942 Apr 18	16.9	0.54882	82.10	17.05	+7.71
2066009.91	944 Jun 5	9 45		-26 58	944 Jun 12	15.6	0.41905	62.69	22.34	+7.24
2066819.89	946 Aug 24	9 16		-15 17	946 Aug 19	19.5	0.38357	57.38	24.40	-4.57
2067614.40	948 Oct 26	21 40		+14 20	948 Oct 18	22.8	0.49803	74.50	18.79	-7.95
2068388.20	950 Dec 9	16 43		+26 34	950 Dec 4	20.6	0.61289	91.69	15.27	-4.84
2069153.94	953 Jan 13	10 30		+24 52	953 Jan 12	10.1	0.67165	100.48	13.94	-1.02
2069918.07	955 Feb 16	13 44		+14 34	955 Feb 19	10.6	0.66498	99.48	14.08	+2.87
2070685.54	957 Mar 25	1 03		- 1 16	957 Mar 31	14.4	0.59294	88.70	15.79	+6.56
2071463.97	959 May 12	11 15		-20 39	959 May 20	21.6	0.46905	70.17	19.96	+8.43
2072266.49	961 Jul 22	23 41		-25 54	961 Jul 23	14.1	0.37439	56.01	25.00	+0.60
2073071.59	963 Oct 6	2 08		+ 4 37	963 Sep 27	23.2	0.44611	66.74	20.98	-8.12
2073851.61	965 Nov 24	2 37		+23 50	965 Nov 17	21.2	0.57349	85.79	16.32	-6.22
2074619.53	968 Jan 1	0 38		+26 48	967 Dec 29	11.8	0.65634	98.19	14.26	-2.53
2075383.69	970 Feb 3	4 35		+19 17	970 Feb 4	12.9	0.67562	101.07	13.85	+1.34
2076149.20	972 Mar 9	16 47		+ 5 23	972 Mar 14	20.8	0.62886	94.08	14.88	+5.17
2076922.09	974 Apr 21	14 11		-13 03	974 Apr 29	17.9	0.52144	78.01	17.95	+8.15
2077713.93	976 Jun 21	10 24		-28 45	976 Jun 26	23.5	0.39643	59.30	23.61	+5.54
2078524.56	978 Sep 10	1 26		- 7 44	978 Sep 3	13.0	0.40077	59.95	23.36	-6.52
2079313.28	980 Nov 6	18 38		+18 38	980 Oct 30	8.4	0.52635	78.74	17.78	-7.42
2080084.57	982 Dec 18	1 45		+27 11	982 Dec 14	1.0	0.63115	94.42	14.83	-4.03
2080849.54	985 Jan 21	0 54		+23 04	985 Jan 20	20.5	0.67595	101.12	13.85	-0.18
2081613.94	987 Feb 24	10 31		+11 19	987 Feb 28	3.0	0.65491	97.97	14.29	+3.69
2082382.96	989 Apr 3	10 59		- 5 32	989 Apr 10	15.9	0.56964	85.22	16.43	+7.21
2083165.48	991 May 25	23 36		-24 37	991 Jun 3	0.6	0.44058	65.91	21.24	+8.04
2083972.97	993 Aug 10	11 19		-20 27	993 Aug 8	1.0	0.37552	56.18	24.93	-2.43
2084772.41	995 Oct 18	21 52		+10 39	995 Oct 10	17.8	0.47459	71.00	19.72	-8.17
2085548.69	997 Dec 3	4 32		+25 43	997 Nov 27	17.0	0.59621	89.19	15.70	-5.48
2086315.28	1000 Jan 8	18 38		+25 47	1000 Jan 7	1.6	0.66601	99.63	14.05	-1.71
2087079.34	1002 Feb 10	20 11		+16 31	1002 Feb 13	0.3	0.67097	100.38	13.95	+2.17
2087845.81	1004 Mar 18	7 28		+ 1 28	1004 Mar 24	5.8	0.61008	91.27	15.34	+5.93
2088621.49	1006 May 2	23 49		-17 32	1006 May 11	10.6	0.49271	73.71	19.00	+8.45
2089419.25	1008 Jul 8	17 56		-28 09	1008 Jul 11	18.7	0.38072	56.96	24.58	+3.03
2090227.82	1010 Sep 25	7 34		- 0 20	1010 Sep 17	14.5	0.42426	63.47	22.06	-7.71
2091011.42	1012 Nov 16	21 59		+22 01	1012 Nov 10	3.0	0.55324	82.76	16.92	-6.79
2091780.67	1014 Dec 26	4 02		+27 08	1014 Dec 22	22.8	0.64625	96.68	14.48	-3.22
2092545.08	1017 Jan 28	14 01		+20 53	1017 Jan 29	5.6	0.67704	101.28	13.82	+0.65
2093310.00	1019 Mar 4	12 04		+ 7 49	1019 Mar 8	23.8	0.64183	96.02	14.58	+4.49
2094080.97	1021 Apr 13	11 21		- 9 56	1021 Apr 21	6.1	0.54386	81.36	17.21	+7.78
2094868.34	1023 Jun 9	20 04		-27 35	1023 Jun 16	19.7	0.41431	61.98	22.59	+6.98
2095678.65	1025 Aug 28	3 34		-13 21	1025 Aug 23	3.8	0.38609	57.76	24.24	-4.99
2096472.03	1027 Oct 30	12 50		+15 41	1027 Oct 22	17.2	0.50344	75.31	18.59	-7.82
2097245.34	1029 Dec 11	20 09		+26 50	1029 Dec 7	3.4	0.61672	92.26	15.18	-4.70
2098010.93	1032 Jan 16	10 20		+24 16	1032 Jan 15	13.2	0.67288	100.66	13.91	-0.88
2098775.09	1034 Feb 18	14 03		+13 27	1034 Feb 21	13.9	0.66341	99.24	14.11	+2.99
2099542.82	1036 Mar 27	7 40		- 2 38	1036 Apr 2	23.2	0.58889	88.10	15.89	+6.65
2100322.01	1038 May 15	12 08		-21 49	1038 May 23	22.6	0.46367	69.36	20.19	+8.43
2101125.61	1040 Jul 27	2 43		-24 40	1040 Jul 27	3.5	0.37383	55.92	25.04	+0.03
2101929.71	1042 Oct 9	4 59		+ 6 26	1042 Oct 1	1.0	0.45136	67.52	20.74	-8.17
2102708.92	1044 Nov 26	10 11		+24 30	1044 Nov 20	7.7	0.57790	86.45	16.20	-6.11

OPPOSITIONS OF MARS

	Mars at opposition				Mars nearest to Earth				
JD	date	ET	δ		date	ET	least distance		d	Δt
		h m	° ′			h	a.u.	10^6 km	″	d
2103476.55	1047 Jan 3	1 09	+26 28		1046 Dec 31	15.3	0.65838	98.49	14.22	-2.41
2104240.68	1049 Feb 5	4 22	+18 19		1049 Feb 6	15.7	0.67494	100.97	13.87	+1.47
2105006.35	1051 Mar 12	20 30	+ 4 05		1051 Mar 18	3.1	0.62543	93.56	14.97	+5.28
2105779.68	1053 Apr 24	4 22	-14 23		1053 May 2	10.6	0.51609	77.21	18.14	+8.26
2106572.58	1055 Jun 26	2 02	-28 47		1055 Jul 1	4.4	0.39293	58.78	23.82	+5.10
2107383.08	1057 Sep 13	14 02	- 5 43		1057 Sep 6	17.7	0.40484	60.56	23.12	-6.85
2108170.79	1059 Nov 10	6 58	+19 45		1059 Nov 2	23.6	0.53158	79.52	17.61	-7.31
2108941.66	1061 Dec 20	3 55	+27 13		1061 Dec 16	6.2	0.63420	94.87	14.76	-3.91
2109706.49	1064 Jan 23	23 49	+22 19		1064 Jan 23	22.5	0.67645	101.20	13.84	-0.05
2110470.97	1066 Feb 26	11 22	+10 08		1066 Mar 2	6.7	0.65281	97.66	14.34	+3.81
2111240.34	1068 Apr 5	20 08	- 6 54		1068 Apr 13	3.7	0.56503	84.53	16.57	+7.31
2112023.73	1070 May 29	5 33	-25 31		1070 Jun 6	3.2	0.43532	65.12	21.50	+7.90
2112831.90	1072 Aug 14	9 40	-18 43		1072 Aug 11	11.5	0.37666	56.35	24.85	-2.93
2113630.21	1074 Oct 21	17 08	+12 12		1074 Oct 13	15.2	0.47998	71.80	19.50	-8.08
2114405.91	1076 Dec 5	9 43	+26 09		1076 Nov 30	1.2	0.60051	89.83	15.59	-5.36
2115172.29	1079 Jan 10	18 56	+25 16		1079 Jan 9	5.0	0.66776	99.90	14.02	-1.58
2115936.34	1081 Feb 12	20 12	+15 27		1081 Feb 15	3.5	0.66988	100.21	13.97	+2.30
2116703.02	1083 Mar 21	12 32	+ 0 08		1083 Mar 27	13.6	0.60647	90.73	15.43	+6.05
2117479.33	1085 May 5	19 50	-18 48		1085 May 14	8.1	0.48734	72.90	19.21	+8.51
2118278.28	1087 Jul 13	18 36	-27 27		1087 Jul 16	6.5	0.37870	56.65	24.72	+2.50
2119086.15	1089 Sep 28	15 43	+ 1 38		1089 Sep 20	19.4	0.42923	64.21	21.81	-7.85
2119868.81	1091 Nov 20	7 28	+22 53		1091 Nov 13	14.7	0.55798	83.47	16.77	-6.70
2120637.71	1093 Dec 28	5 09	+26 57		1093 Dec 25	2.7	0.64874	97.05	14.43	-3.10
2121402.07	1096 Jan 31	13 34	+20 00		1096 Feb 1	8.2	0.67691	101.26	13.83	+0.77
2122167.11	1098 Mar 6	14 35	+ 6 33		1098 Mar 11	5.4	0.63885	95.57	14.65	+4.62
2122938.46	1100 Apr 15	22 57	-11 17		1100 Apr 23	21.0	0.53865	80.58	17.38	+7.92
2123726.80	1102 Jun 13	7 09	-28 03		1102 Jun 19	23.0	0.40989	61.32	22.84	+6.66
2124537.38	1104 Aug 31	21 02	-11 21		1104 Aug 26	10.6	0.38904	58.20	24.06	-5.44
2125329.69	1106 Nov 2	4 35	+16 59		1106 Oct 25	11.0	0.50884	76.12	18.39	-7.73
2126102.50	1108 Dec 13	23 59	+27 01		1108 Dec 9	10.0	0.62030	92.80	15.09	-4.58
2126867.90	1111 Jan 18	9 38	+23 37		1111 Jan 17	15.6	0.67382	100.80	13.89	-0.75
2127632.09	1113 Feb 20	14 16	+12 19		1113 Feb 23	17.3	0.66176	99.00	14.14	+3.13
2128400.12	1115 Mar 30	14 53	- 4 00		1115 Apr 6	9.6	0.58472	87.47	16.01	+6.78
2129180.04	1117 May 18	12 59	-22 55		1117 May 26	22.0	0.45811	68.53	20.43	+8.38
2129984.62	1119 Aug 1	2 46	-23 18		1119 Jul 31	14.8	0.37341	55.86	25.07	-0.50
2130787.68	1121 Oct 12	4 16	+ 8 08		1121 Oct 4	1.3	0.45661	68.31	20.50	-8.13
2131566.19	1123 Nov 29	16 32	+25 05		1123 Nov 23	16.6	0.58250	87.14	16.07	-6.00
2132333.58	1126 Jan 5	1 49	+26 05		1126 Jan 2	19.3	0.66060	98.82	14.17	-2.27
2133097.67	1128 Feb 8	4 01	+17 20		1128 Feb 9	18.8	0.67439	100.89	13.88	+1.62
2133863.49	1130 Mar 14	23 51	+ 2 46		1130 Mar 20	10.0	0.62223	93.08	15.04	+5.42
2134637.36	1132 Apr 26	20 32	-15 43		1132 May 5	5.2	0.51083	76.42	18.32	+8.36
2135431.44	1134 Jun 29	22 30	-28 39		1134 Jul 4	14.3	0.38972	58.30	24.02	+4.66
2136241.67	1136 Sep 17	3 58	- 3 39		1136 Sep 10	1.3	0.40914	61.21	22.88	-7.11
2137028.32	1138 Nov 12	19 37	+20 48		1138 Nov 5	13.8	0.53663	80.28	17.44	-7.24
2137798.76	1140 Dec 22	6 15	+27 10		1140 Dec 18	11.5	0.63719	95.32	14.69	-3.78
2138563.48	1143 Jan 25	23 30	+21 31		1143 Jan 25	1.6	0.67680	101.25	13.83	+0.09
2139328.05	1145 Feb 28	13 15	+ 8 55		1145 Mar 4	12.3	0.65028	97.28	14.39	+3.96
2140097.73	1147 Apr 9	5 37	- 8 16		1147 Apr 16	17.1	0.56009	83.79	16.71	+7.48
2140881.96	1149 Jun 1	11 09	-26 18		1149 Jun 9	3.6	0.43020	64.36	21.76	+7.69
2141690.78	1151 Aug 19	6 49	-16 54		1151 Aug 15	19.2	0.37819	56.58	24.75	-3.48

Mars at opposition *Mars nearest to Earth*

JD	date	ET	δ	date	ET	least distance	10⁶ km	d	Δt
		h m	° '		h	a.u.	10⁶ km	"	d
2142488.00	1153 Oct 24	11 55	+13 41	1153 Oct 16	11.3	0.48547	72.63	19.28	-8.03
2143263.11	1155 Dec 8	14 36	+26 29	1155 Dec 3	8.7	0.60451	90.43	15.48	-5.25
2144029.27	1158 Jan 12	18 36	+24 43	1158 Jan 11	8.2	0.66914	100.10	13.99	-1.44
2144793.32	1160 Feb 15	19 39	+14 22	1160 Feb 18	6.5	0.66873	100.04	14.00	+2.45
2145560.24	1162 Mar 23	17 40	- 1 12	1162 Mar 29	22.5	0.60271	90.16	15.53	+6.20
2146337.18	1164 May 8	16 25	-20 01	1164 May 17	4.9	0.48170	72.06	19.43	+8.52
2147137.23	1166 Jul 17	17 37	-26 36	1166 Jul 19	18.0	0.37684	56.37	24.84	+2.01
2147944.33	1168 Oct 1	19 57	+ 3 30	1168 Sep 23	22.6	0.43406	64.94	21.56	-7.89
2148726.17	1170 Nov 22	16 04	+23 39	1170 Nov 16	1.5	0.56286	84.20	16.63	-6.61
2149494.77	1172 Dec 30	6 35	+26 41	1172 Dec 27	8.0	0.65147	97.46	14.37	-2.94
2150259.06	1175 Feb 2	13 20	+19 04	1175 Feb 3	11.9	0.67683	101.25	13.83	+0.94
2151024.21	1177 Mar 8	17 07	+ 5 17	1177 Mar 13	12.1	0.63604	95.15	14.72	+4.79
2151796.00	1179 Apr 19	11 55	-12 38	1179 Apr 27	13.2	0.53366	79.84	17.54	+8.05
2152585.41	1181 Jun 16	21 51	-28 22	1181 Jun 23	5.5	0.40570	60.69	23.07	+6.32
2153396.19	1183 Sep 5	16 38	- 9 17	1183 Aug 30	19.9	0.39235	58.69	23.86	-5.86
2154187.33	1185 Nov 4	20 01	+18 14	1185 Oct 28	3.2	0.51417	76.92	18.20	-7.70
2154959.63	1187 Dec 17	3 06	+27 07	1187 Dec 12	16.4	0.62370	93.30	15.01	-4.44
2155724.89	1190 Jan 20	9 21	+22 54	1190 Jan 19	19.3	0.67464	100.92	13.87	-0.59
2156489.13	1192 Feb 23	15 11	+11 08	1192 Feb 26	22.5	0.65979	98.70	14.19	+3.31
2157257.43	1194 Apr 1	22 17	- 5 21	1194 Apr 8	21.4	0.58010	86.78	16.14	+6.96
2158038.08	1196 May 21	13 50	-23 54	1196 May 29	20.2	0.45260	67.71	20.68	+8.26
2158843.57	1198 Aug 5	1 34	-21 49	1198 Aug 3	23.5	0.37346	55.87	25.06	-1.09
2159645.63	1200 Oct 15	3 02	+ 9 48	1200 Oct 6	23.8	0.46197	69.11	20.26	-8.14
2160423.48	1202 Dec 1	23 27	+25 37	1202 Nov 26	2.4	0.58693	87.80	15.95	-5.88
2161190.59	1205 Jan 7	2 13	+25 38	1205 Jan 4	23.8	0.66247	99.10	14.13	-2.10
2161954.64	1207 Feb 10	3 23	+16 18	1207 Feb 11	22.4	0.67366	100.78	13.89	+1.79
2162720.66	1209 Mar 17	3 54	+ 1 27	1209 Mar 22	18.3	0.61886	92.58	15.12	+5.60
2163495.06	1211 Apr 30	13 27	-17 00	1211 May 8	23.4	0.50538	75.60	18.52	+8.42
2164290.25	1213 Jul 3	18 03	-28 20	1213 Jul 8	0.4	0.38651	57.82	24.22	+4.26
2165100.08	1215 Sep 21	13 55	- 1 40	1215 Sep 14	7.5	0.41334	61.83	22.64	-7.27
2165885.75	1217 Nov 15	6 06	+21 45	1217 Nov 8	2.2	0.54176	81.05	17.28	-7.16
2166655.84	1219 Dec 25	8 06	+27 02	1219 Dec 21	17.6	0.64035	95.79	14.62	-3.60
2167420.47	1222 Jan 27	23 18	+20 40	1222 Jan 28	5.8	0.67721	101.31	13.82	+0.27
2168185.11	1224 Mar 2	14 37	+ 7 41	1224 Mar 6	18.2	0.64796	96.93	14.45	+4.15
2168955.15	1226 Apr 11	15 41	- 9 38	1226 Apr 19	6.6	0.55541	83.09	16.85	+7.62
2169740.34	1228 Jun 4	20 13	-26 59	1228 Jun 12	7.3	0.42537	63.63	22.00	+7.46
2170549.78	1230 Aug 23	6 38	-14 56	1230 Aug 19	5.8	0.38030	56.89	24.61	-4.03
2171345.80	1232 Oct 27	7 10	+15 08	1232 Oct 19	6.1	0.49096	73.45	19.06	-8.04
2172120.31	1234 Dec 10	19 33	+26 45	1234 Dec 5	17.5	0.60834	91.01	15.39	-5.09
2172886.28	1237 Jan 14	18 39	+24 06	1237 Jan 13	12.7	0.67044	100.30	13.96	-1.25
2173650.35	1239 Feb 17	20 21	+13 15	1239 Feb 20	11.9	0.66725	99.82	14.03	+2.65
2174417.51	1241 Mar 26	0 07	- 2 34	1241 Apr 1	9.6	0.59843	89.52	15.64	+6.39
2175195.05	1243 May 12	13 09	-21 09	1243 May 21	0.6	0.47607	71.22	19.66	+8.48
2175996.17	1245 Jul 21	16 05	-25 35	1245 Jul 23	2.8	0.37533	56.15	24.94	+1.45
2176802.48	1247 Oct 5	23 35	+ 5 21	1247 Sep 28	0.2	0.43910	65.69	21.32	-7.97
2177583.52	1249 Nov 25	0 34	+24 20	1249 Nov 18	13.2	0.56768	84.92	16.49	-6.47
2178351.81	1252 Jan 2	7 29	+26 21	1251 Dec 30	13.3	0.65380	97.81	14.32	-2.76
2179116.03	1254 Feb 4	12 37	+18 07	1254 Feb 5	15.7	0.67653	101.21	13.84	+1.13
2179881.32	1256 Mar 10	19 34	+ 4 00	1256 Mar 15	19.0	0.63317	94.72	14.78	+4.98
2180653.56	1258 Apr 22	1 33	-13 58	1258 Apr 30	4.4	0.52847	79.06	17.71	+8.12

OPPOSITIONS OF MARS

| | *Mars at opposition* | | | | *Mars nearest to Earth* | | | | |
JD	date	ET	δ	date	ET	least distance		d	Δt
		h m	° ′		h	a.u.	10^6 km	″	d
2181444.04	1260 Jun 20	12 52	−28 33	1260 Jun 26	14.1	0.40147	60.06	23.31	+6.05
2182254.83	1262 Sep 9	8 00	− 7 14	1262 Sep 3	4.8	0.39568	59.19	23.66	−6.13
2183044.89	1264 Nov 7	9 23	+19 23	1264 Oct 30	18.2	0.51948	77.71	18.02	−7.63
2183816.76	1266 Dec 19	6 12	+27 08	1266 Dec 15	0.1	0.62727	93.84	14.92	−4.25
2184581.89	1269 Jan 22	9 21	+22 07	1269 Jan 22	0.0	0.67554	101.06	13.86	−0.39
2185346.18	1271 Feb 25	16 18	+ 9 57	1271 Mar 1	4.2	0.65790	98.42	14.23	+3.50
2186114.78	1273 Apr 4	6 45	− 6 43	1273 Apr 11	9.2	0.57572	86.13	16.26	+7.10
2186896.23	1275 May 25	17 29	−24 50	1275 Jun 2	20.5	0.44745	66.94	20.92	+8.13
2187702.67	1277 Aug 9	4 09	−20 08	1277 Aug 7	10.3	0.37410	55.96	25.02	−1.74
2188503.63	1279 Oct 19	3 02	+11 27	1279 Oct 10	22.1	0.46752	69.94	20.02	−8.21
2189280.74	1281 Dec 4	5 51	+26 03	1281 Nov 28	12.9	0.59118	88.44	15.83	−5.71
2190047.60	1284 Jan 10	2 27	+25 08	1284 Jan 8	4.7	0.66422	99.37	14.09	−1.91
2190811.66	1286 Feb 12	3 45	+15 14	1286 Feb 14	3.4	0.67266	100.63	13.91	+1.98
2191577.86	1288 Mar 19	8 39	+ 0 08	1288 Mar 25	3.4	0.61504	92.01	15.22	+5.78
2192352.77	1290 May 3	6 32	−18 14	1290 May 11	16.4	0.49974	74.76	18.73	+8.41
2193149.08	1292 Jul 7	14 00	−27 50	1292 Jul 11	9.2	0.38362	57.39	24.40	+3.80
2193958.43	1294 Sep 24	22 17	+ 0 18	1294 Sep 17	12.0	0.41782	62.50	22.40	−7.43
2194743.21	1296 Nov 17	16 56	+22 38	1296 Nov 10	16.6	0.54692	81.82	17.11	−7.01
2195512.93	1298 Dec 27	10 13	+26 51	1298 Dec 24	0.3	0.64315	96.21	14.55	−3.41
2196277.44	1301 Jan 29	22 38	+19 47	1301 Jan 30	9.8	0.67732	101.33	13.82	+0.46
2197042.18	1303 Mar 5	16 22	+ 6 26	1303 Mar 10	0.3	0.64551	96.57	14.50	+4.33
2197812.63	1305 Apr 14	3 14	−10 59	1305 Apr 21	19.9	0.55050	82.35	17.00	+7.69
2198598.74	1307 Jun 9	5 52	−27 32	1307 Jun 16	13.0	0.42045	62.90	22.26	+7.30
2199408.65	1309 Aug 27	3 31	−12 57	1309 Aug 22	16.6	0.38243	57.21	24.48	−4.46
2200203.51	1311 Oct 31	0 18	+16 28	1311 Oct 23	0.4	0.49640	74.26	18.86	−8.00
2200977.48	1313 Dec 12	23 27	+26 56	1313 Dec 8	1.9	0.61228	91.60	15.29	−4.90
2201743.28	1316 Jan 17	18 37	+23 26	1316 Jan 16	17.2	0.67182	100.50	13.93	−1.06
2202507.38	1318 Feb 19	21 01	+12 06	1318 Feb 22	16.9	0.66581	99.60	14.06	+2.83
2203274.76	1320 Mar 28	6 12	− 3 54	1320 Apr 3	18.8	0.59444	88.93	15.75	+6.52
2204052.99	1322 May 15	11 44	−22 15	1322 May 23	21.7	0.47078	70.43	19.88	+8.42
2204855.28	1324 Jul 25	18 39	−24 21	1324 Jul 26	13.4	0.37444	56.02	25.00	+0.78
2205660.67	1326 Oct 9	4 07	+ 7 10	1326 Oct 1	1.3	0.44446	66.49	21.06	−8.12
2206440.88	1328 Nov 27	9 05	+24 57	1328 Nov 21	1.9	0.57231	85.62	16.35	−6.30
2207208.86	1331 Jan 4	8 32	+25 58	1331 Jan 1	18.7	0.65600	98.14	14.27	−2.58
2207973.03	1333 Feb 6	12 38	+17 06	1333 Feb 7	20.0	0.67600	101.13	13.85	+1.31
2208738.48	1335 Mar 13	23 24	+ 2 42	1335 Mar 19	2.8	0.62979	94.22	14.86	+5.14
2209511.18	1337 Apr 24	16 16	−15 16	1337 May 2	19.9	0.52294	78.23	17.90	+8.15
2210302.67	1339 Jun 25	4 06	−28 33	1339 Jun 30	21.4	0.39743	59.45	23.55	+5.72
2211113.42	1341 Sep 12	22 01	− 5 12	1341 Sep 6	12.2	0.39927	59.73	23.44	−6.41
2211902.47	1343 Nov 10	23 10	+20 27	1343 Nov 3	11.3	0.52491	78.53	17.83	−7.50
2212673.88	1345 Dec 21	9 07	+27 05	1345 Dec 17	7.4	0.63051	94.32	14.85	−4.07
2213438.86	1348 Jan 25	8 36	+21 19	1348 Jan 25	3.5	0.67607	101.14	13.84	−0.21
2214203.21	1350 Feb 27	17 07	+ 8 45	1350 Mar 3	9.0	0.65588	98.12	14.27	+3.66
2214972.14	1352 Apr 6	15 16	− 8 04	1352 Apr 13	19.7	0.57121	85.45	16.39	+7.19
2215754.40	1354 May 28	21 36	−25 40	1354 Jun 5	23.1	0.44211	66.14	21.17	+8.06
2216561.69	1356 Aug 13	4 29	−18 23	1356 Aug 10	22.0	0.37482	56.07	24.97	−2.27
2217361.49	1358 Oct 21	23 42	+12 59	1358 Oct 13	18.7	0.47287	70.74	19.79	−8.21
2218137.96	1360 Dec 6	11 06	+26 23	1360 Nov 30	22.4	0.59542	89.07	15.72	−5.53
2218904.62	1363 Jan 12	2 51	+24 33	1363 Jan 10	9.1	0.66607	99.64	14.05	−1.74
2219668.66	1365 Feb 14	3 57	+14 09	1365 Feb 16	7.5	0.67167	100.48	13.94	+2.15

	Mars at opposition			*Mars nearest to Earth*				
JD	date	ET	δ	date	ET	least distance	d	Δt
		h m	° ′		h	a.u.	10^6 km	″	d
2220435.05	1367 Mar 22	13 19	- 1 12	1367 Mar 28	11.1	0.61142	91.47	15.31	+5.91
2221210.56	1369 May 6	1 27	-19 26	1369 May 14	11.4	0.49448	73.97	18.93	+8.42
2222008.06	1371 Jul 12	13 25	-27 08	1371 Jul 15	18.2	0.38127	57.04	24.55	+3.20
2222816.85	1373 Sep 28	8 29	+ 2 18	1373 Sep 20	16.2	0.42271	63.24	22.14	-7.68
2223600.67	1375 Nov 21	4 07	+23 27	1375 Nov 14	7.5	0.55196	82.57	16.96	-6.86
2224369.99	1377 Dec 29	11 51	+26 35	1377 Dec 26	5.7	0.64578	96.61	14.49	-3.26
2225134.43	1380 Feb 1	22 15	+18 51	1380 Feb 2	13.1	0.67727	101.32	13.82	+0.62
2225899.29	1382 Mar 7	19 04	+ 5 10	1382 Mar 12	6.3	0.64265	96.14	14.56	+4.47
2226670.11	1384 Apr 16	14 42	-12 18	1384 Apr 24	9.0	0.54523	81.57	17.17	+7.76
2227457.14	1386 Jun 12	15 24	-27 56	1386 Jun 19	18.2	0.41555	62.17	22.52	+7.12
2228267.43	1388 Aug 30	22 14	-10 57	1388 Aug 26	2.3	0.38486	57.57	24.32	-4.83
2229061.19	1390 Nov 2	16 35	+17 45	1390 Oct 25	19.4	0.50188	75.08	18.65	-7.88
2229834.65	1392 Dec 15	3 32	+27 02	1392 Dec 10	10.0	0.61593	92.14	15.20	-4.73
2230600.26	1395 Jan 19	18 14	+22 42	1395 Jan 18	20.6	0.67283	100.65	13.91	-0.90
2231364.38	1397 Feb 21	21 03	+10 57	1397 Feb 24	20.4	0.66421	99.36	14.09	+2.97
2232132.03	1399 Mar 31	12 45	- 5 15	1399 Apr 7	3.6	0.59034	88.31	15.86	+6.62
2232910.99	1401 May 18	11 41	-23 16	1401 May 26	21.8	0.46531	69.61	20.12	+8.42
2233714.32	1403 Jul 30	19 35	-22 58	1403 Jul 31	0.4	0.37362	55.89	25.05	+0.20
2234518.74	1405 Oct 12	5 45	+ 8 54	1405 Oct 4	1.2	0.44959	67.26	20.82	-8.19
2235298.17	1407 Nov 30	16 08	+25 29	1407 Nov 24	12.8	0.57691	86.30	16.22	-6.14
2236065.88	1410 Jan 6	9 11	+25 30	1410 Jan 3	22.9	0.65830	98.48	14.22	-2.43
2236830.02	1412 Feb 9	12 27	+16 04	1412 Feb 10	23.3	0.67549	101.05	13.86	+1.45
2237595.61	1414 Mar 16	2 38	+ 1 24	1414 Mar 21	8.8	0.62665	93.75	14.94	+5.26
2238368.79	1416 Apr 27	6 55	-16 31	1416 May 5	11.9	0.51785	77.47	18.07	+8.21
2239161.43	1418 Jun 28	22 13	-28 22	1418 Jul 4	4.3	0.39389	58.93	23.76	+5.25
2239972.05	1420 Sep 16	13 16	- 3 07	1420 Sep 9	19.0	0.40343	60.35	23.20	-6.76
2240760.03	1422 Nov 13	12 39	+21 28	1422 Nov 6	4.0	0.53021	79.32	17.65	-7.36
2241530.99	1424 Dec 23	11 45	+26 57	1424 Dec 19	13.2	0.63356	94.78	14.77	-3.94
2242295.84	1427 Jan 27	8 10	+20 28	1427 Jan 27	6.3	0.67654	101.21	13.84	-0.08
2243060.28	1429 Mar 1	18 50	+ 7 31	1429 Mar 5	13.6	0.65351	97.76	14.32	+3.78
2243829.53	1431 Apr 10	0 45	- 9 24	1431 Apr 17	7.4	0.56624	84.71	16.53	+7.28
2244612.61	1433 Jun 1	2 39	-26 23	1433 Jun 9	2.2	0.43672	65.33	21.43	+7.98
2245420.63	1435 Aug 18	3 01	-16 32	1435 Aug 15	9.2	0.37583	56.22	24.90	-2.74
2246219.33	1437 Oct 24	19 56	+14 28	1437 Oct 16	16.9	0.47834	71.56	19.57	-8.13
2246995.20	1439 Dec 9	16 42	+26 39	1439 Dec 4	7.5	0.59958	89.70	15.61	-5.39
2247761.62	1442 Jan 14	2 53	+23 56	1442 Jan 12	12.4	0.66757	99.87	14.02	-1.60
2248525.65	1444 Feb 17	3 32	+13 02	1444 Feb 19	10.2	0.67051	100.31	13.96	+2.28
2249292.25	1446 Mar 24	17 58	- 2 31	1446 Mar 30	18.2	0.60783	90.93	15.40	+6.01
2250068.36	1448 May 8	20 41	-20 34	1448 May 17	8.1	0.48904	73.16	19.14	+8.47
2250867.00	1450 Jul 16	11 57	-26 16	1450 Jul 19	3.9	0.37898	56.69	24.70	+2.66
2251675.12	1452 Oct 1	14 52	+ 4 12	1452 Sep 23	18.8	0.42744	63.94	21.90	-7.84
2252458.03	1454 Nov 23	12 47	+24 09	1454 Nov 16	19.4	0.55679	83.29	16.81	-6.72
2253227.05	1456 Dec 31	13 06	+26 14	1456 Dec 28	10.1	0.64851	97.02	14.43	-3.13
2253991.41	1459 Feb 3	21 54	+17 53	1459 Feb 4	16.0	0.67732	101.33	13.82	+0.75
2254756.39	1461 Mar 9	21 17	+ 3 53	1461 Mar 14	11.4	0.63993	95.73	14.63	+4.59
2255527.61	1463 Apr 20	2 40	-13 36	1463 Apr 27	23.2	0.54033	80.83	17.32	+7.86
2256315.69	1465 Jun 16	4 28	-28 13	1465 Jun 22	23.1	0.41120	61.51	22.76	+6.78
2257126.29	1467 Sep 4	19 00	- 8 51	1467 Aug 30	11.5	0.38796	58.04	24.13	-5.31
2257918.91	1469 Nov 5	9 50	+18 58	1469 Oct 28	15.2	0.50744	75.91	18.45	-7.78
2258691.82	1471 Dec 18	7 36	+27 03	1471 Dec 13	16.7	0.61953	92.68	15.11	-4.62

OPPOSITIONS OF MARS

		Mars at opposition				*Mars nearest to Earth*				
JD	date	ET	δ	date	ET	least distance		d	Δt
		h m	° '		h	a.u.	10^6 km	"	d
2259457.25	1474 Jan 21	18 00	+21 56	1474 Jan 20	23.2	0.67381	100.80	13.89	-0.78
2260221.42	1476 Feb 24	22 11	+ 9 45	1476 Feb 28	0.4	0.66237	99.09	14.13	+3.09
2260989.34	1478 Apr 2	20 16	- 6 35	1478 Apr 9	13.8	0.58585	87.64	15.98	+6.73
2261768.99	1480 May 21	11 48	-24 12	1480 May 29	22.0	0.45960	68.76	20.37	+8.42
2262573.31	1482 Aug 3	19 24	-21 29	1482 Aug 3	12.4	0.37306	55.81	25.09	-0.29
2263376.73	1484 Oct 15	5 38	+10 34	1484 Oct 7	1.9	0.45488	68.05	20.58	-8.15
2264155.46	1486 Dec 2	23 04	+25 56	1486 Nov 26	22.5	0.58144	86.98	16.10	-6.03
2264922.91	1489 Jan 8	9 57	+24 59	1489 Jan 6	2.7	0.66029	98.78	14.18	-2.30
2265686.99	1491 Feb 11	11 42	+15 01	1491 Feb 13	1.7	0.67483	100.95	13.87	+1.58
2266452.75	1493 Mar 18	5 54	+ 0 06	1493 Mar 23	14.8	0.62346	93.27	15.01	+5.37
2267226.46	1495 Apr 30	23 00	-17 45	1495 May 9	6.5	0.51252	76.67	18.26	+8.31
2268020.22	1497 Jul 2	17 15	-28 01	1497 Jul 7	12.3	0.39042	58.41	23.97	+4.79
2268830.58	1499 Sep 21	1 52	- 1 04	1499 Sep 14	0.4	0.40750	60.96	22.97	-7.06
2269617.52	1501 Nov 16	0 26	+22 23	1501 Nov 8	18.1	0.53531	80.08	17.49	-7.26
2270388.08	1503 Dec 26	13 57	+26 45	1503 Dec 22	18.5	0.63680	95.26	14.70	-3.81
2271152.83	1506 Jan 29	7 58	+19 34	1506 Jan 29	9.3	0.67713	101.30	13.82	+0.06
2271917.36	1508 Mar 3	20 34	+ 6 16	1508 Mar 7	18.6	0.65126	97.43	14.37	+3.92
2272686.92	1510 Apr 12	10 02	-10 43	1510 Apr 19	19.9	0.56170	84.03	16.66	+7.41
2273470.90	1512 Jun 4	9 43	-26 59	1512 Jun 12	4.3	0.43177	64.59	21.68	+7.78
2274279.65	1514 Aug 22	3 40	-14 34	1514 Aug 18	20.3	0.37753	56.48	24.79	-3.31
2275077.18	1516 Oct 27	16 26	+15 54	1516 Oct 19	14.9	0.48402	72.41	19.34	-8.06
2275852.42	1518 Dec 11	22 04	+26 51	1518 Dec 6	15.2	0.60362	90.30	15.51	-5.29
2276618.63	1521 Jan 16	3 04	+23 15	1521 Jan 14	15.6	0.66902	100.08	13.99	-1.48
2277382.66	1523 Feb 19	3 56	+11 54	1523 Feb 21	13.8	0.66921	100.11	13.99	+2.41
2278149.50	1525 Mar 26	23 58	- 3 51	1525 Apr 2	3.5	0.60373	90.32	15.50	+6.15
2278926.23	1527 May 12	17 24	-21 40	1527 May 21	6.3	0.48316	72.28	19.37	+8.54
2279725.95	1529 Jul 20	10 45	-25 13	1529 Jul 22	15.4	0.37695	56.39	24.83	+2.20
2280533.34	1531 Oct 5	20 08	+ 6 03	1531 Sep 27	22.6	0.43234	64.68	21.65	-7.90
2281315.43	1533 Nov 25	22 17	+24 47	1533 Nov 19	7.0	0.56171	84.03	16.66	-6.64
2282084.12	1536 Jan 3	14 46	+25 50	1535 Dec 31	15.0	0.65105	97.40	14.38	-2.99
2282848.39	1538 Feb 5	21 18	+16 53	1538 Feb 6	18.8	0.67714	101.30	13.82	+0.89
2283613.49	1540 Mar 11	23 43	+ 2 37	1540 Mar 16	17.2	0.63717	95.32	14.69	+4.73
2284385.14	1542 Apr 22	15 21	-14 52	1542 Apr 30	15.4	0.53532	80.08	17.48	+8.00
2285174.26	1544 Jun 19	18 10	-28 19	1544 Jun 26	4.4	0.40677	60.85	23.01	+6.43
2285985.05	1546 Sep 8	13 13	- 6 45	1546 Sep 2	19.1	0.39097	58.49	23.94	-5.76
2286776.51	1548 Nov 8	0 10	+20 05	1548 Oct 31	6.9	0.51269	76.70	18.26	-7.72
2287548.94	1550 Dec 20	10 30	+26 59	1550 Dec 15	22.8	0.62314	93.22	15.02	-4.49
2288314.25	1553 Jan 23	17 55	+21 07	1553 Jan 23	2.8	0.67488	100.96	13.87	-0.63
2289078.46	1555 Feb 26	23 01	+ 8 33	1555 Mar 2	5.0	0.66061	98.83	14.17	+3.25
2289846.65	1557 Apr 5	3 30	- 7 54	1557 Apr 12	1.0	0.58159	87.00	16.09	+6.90
2290627.09	1559 May 25	14 08	-25 03	1559 Jun 2	21.7	0.45429	67.96	20.60	+8.31
2291432.41	1561 Aug 7	21 55	-19 48	1561 Aug 7	0.3	0.37325	55.84	25.08	-0.90
2292234.79	1563 Oct 19	6 54	+12 13	1563 Oct 11	3.2	0.46053	68.89	20.32	-8.15
2293012.79	1565 Dec 5	6 52	+26 17	1565 Nov 29	8.6	0.58597	87.66	15.97	-5.93
2293779.95	1568 Jan 11	10 45	+24 24	1568 Jan 9	7.0	0.66224	99.07	14.13	-2.16
2294543.99	1570 Feb 13	11 53	+13 55	1570 Feb 15	5.6	0.67405	100.84	13.89	+1.74
2295309.95	1572 Mar 20	10 53	- 1 13	1572 Mar 25	23.8	0.61983	92.73	15.10	+5.54
2296084.16	1574 May 3	15 56	-18 56	1574 May 12	1.8	0.50676	75.81	18.47	+8.41
2296879.01	1576 Jul 6	12 08	-27 30	1576 Jul 10	22.1	0.38704	57.90	24.18	+4.42
2297689.02	1578 Sep 24	12 33	+ 0 55	1578 Sep 17	6.9	0.41171	61.59	22.73	-7.24

	Mars at opposition				Mars nearest to Earth					
JD	date	ET		δ	date	ET	least distance	10^6 km	d	Δt
		h m		° '		h	a.u.		"	d
2298474.99	1580 Nov 18	11 44		+23 12	1580 Nov 11	7.2	0.54048	80.85	17.32	-7.19
2299245.17	1583 Jan 7	16 11		+26 28	1583 Jan 4	0.3	0.63980	95.71	14.63	-3.66
2300009.81	1585 Feb 10	7 27		+18 38	1585 Feb 10	12.7	0.67739	101.34	13.82	+0.22
2300774.40	1587 Mar 16	21 39		+ 5 02	1587 Mar 20	23.6	0.64893	97.08	14.42	+4.08
2301544.33	1589 Apr 24	19 58		-12 01	1589 May 2	9.6	0.55699	83.32	16.80	+7.57
2302329.27	1591 Jun 18	18 26		-27 28	1591 Jun 26	7.1	0.42670	63.83	21.94	+7.53
2303138.59	1593 Sep 5	2 15		-12 34	1593 Sep 1	5.0	0.37925	56.74	24.68	-3.89
2303934.95	1595 Nov 10	10 41		+17 13	1595 Nov 2	9.5	0.48934	73.20	19.13	-8.05
2304709.61	1597 Dec 24	2 41		+26 57	1597 Dec 18	23.2	0.60764	90.90	15.40	-5.14
2305475.63	1600 Jan 29	3 14		+22 31	1600 Jan 27	20.0	0.67057	100.32	13.96	-1.30
2306239.69	1602 Mar 3	4 29		+10 44	1602 Mar 5	18.6	0.66796	99.93	14.01	+2.59
2307006.75	1604 Apr 8	5 55		- 5 11	1604 Apr 14	13.8	0.59984	89.73	15.60	+6.33
2307784.11	1606 May 25	14 37		-22 40	1606 Jun 3	2.6	0.47780	71.48	19.59	+8.50
2308585.00	1608 Aug 3	11 54		-24 00	1608 Aug 5	3.2	0.37559	56.19	24.92	+1.64
2309391.59	1610 Oct 19	2 15		+ 7 53	1610 Oct 11	3.1	0.43771	65.48	21.38	-7.96
2310172.82	1612 Dec 8	7 43		+25 21	1612 Dec 1	18.9	0.56663	84.77	16.52	-6.53
2310941.16	1615 Jan 15	15 56		+25 22	1615 Jan 12	20.2	0.65344	97.75	14.32	-2.82
2311705.38	1617 Feb 17	21 07		+15 51	1617 Feb 18	22.9	0.67684	101.25	13.83	+1.07
2312470.62	1619 Mar 25	3 00		+ 1 19	1619 Mar 30	1.0	0.63403	94.85	14.76	+4.92
2313242.72	1621 May 5	5 13		-16 07	1621 May 13	7.9	0.52977	79.25	17.67	+8.11
2314032.85	1623 Jul 4	8 30		-28 16	1623 Jul 10	12.1	0.40235	60.19	23.26	+6.15
2314843.71	1625 Sep 22	5 02		- 4 41	1625 Sep 16	3.4	0.39420	58.97	23.74	-6.07
2315634.10	1627 Nov 21	14 24		+21 07	1627 Nov 13	22.5	0.51803	77.50	18.07	-7.66
2316406.09	1630 Jan 1	14 09		+26 52	1629 Dec 28	6.5	0.62662	93.74	14.94	-4.32
2317171.24	1632 Feb 5	17 42		+20 15	1632 Feb 5	7.0	0.67559	101.07	13.85	-0.45
2317935.48	1634 Mar 10	23 35		+ 7 20	1634 Mar 14	10.1	0.65870	98.54	14.21	+3.44
2318703.98	1636 Apr 17	11 35		- 9 13	1636 Apr 24	12.9	0.57726	86.36	16.21	+7.06
2319485.22	1638 Jun 7	17 11		-25 48	1638 Jun 15	21.0	0.44895	67.16	20.85	+8.16
2320291.45	1640 Aug 21	22 50		-18 02	1640 Aug 20	9.7	0.37347	55.87	25.06	-1.55
2321092.73	1642 Nov 1	5 24		+13 45	1642 Oct 24	0.7	0.46582	69.69	20.09	-8.19
2321870.02	1644 Dec 17	12 32		+26 34	1644 Dec 11	17.9	0.59030	88.31	15.86	-5.78
2322636.95	1647 Jan 23	10 54		+23 46	1647 Jan 21	11.8	0.66421	99.36	14.09	-1.96
2323401.00	1649 Feb 25	11 57		+12 48	1649 Feb 27	10.3	0.67329	100.72	13.90	+1.93
2324167.12	1651 Apr 2	14 56		- 2 32	1651 Apr 8	8.4	0.61633	92.20	15.19	+5.73
2324941.87	1653 May 16	8 55		-20 03	1653 May 24	19.1	0.50147	75.02	18.67	+8.43
2325737.91	1655 Jul 21	9 48		-26 47	1655 Jul 25	9.0	0.38437	57.50	24.35	+3.97
2326547.48	1657 Oct 7	23 33		+ 2 55	1657 Sep 30	14.0	0.41651	62.31	22.47	-7.40
2327332.49	1659 Dec 1	23 50		+23 58	1659 Nov 24	21.7	0.54576	81.64	17.15	-7.09
2328102.28	1662 Jan 9	18 38		+26 07	1662 Jan 6	7.2	0.64268	96.14	14.56	-3.48
2328866.80	1664 Feb 13	7 15		+17 39	1664 Feb 13	17.1	0.67755	101.36	13.81	+0.41
2329631.51	1666 Mar 19	0 09		+ 3 45	1666 Mar 23	6.9	0.64628	96.68	14.48	+4.28
2330401.83	1668 Apr 27	7 51		-13 18	1668 May 5	0.3	0.55176	82.54	16.96	+7.69
2331187.64	1670 Jun 22	3 20		-27 49	1670 Jun 29	11.9	0.42156	63.06	22.20	+7.36
2331997.45	1672 Sep 8	22 53		-10 32	1672 Sep 4	14.8	0.38121	57.03	24.55	-4.34
2332792.67	1674 Nov 13	4 10		+18 28	1674 Nov 5	3.6	0.49462	74.02	18.92	-8.03
2333566.80	1676 Dec 26	7 08		+26 58	1676 Dec 21	8.1	0.61153	91.48	15.31	-4.96
2334332.63	1679 Jan 31	3 08		+21 44	1679 Jan 30	0.5	0.67174	100.49	13.93	-1.11
2335096.68	1681 Mar 5	4 26		+ 9 33	1681 Mar 7	23.2	0.66649	99.71	14.04	+2.78
2335863.98	1683 Apr 11	11 34		- 6 29	1683 Apr 17	23.3	0.59590	89.15	15.71	+6.49
2336642.03	1685 May 28	12 42		-23 37	1685 Jun 5	23.0	0.47239	70.67	19.81	+8.43

| | *Mars at opposition* | | | | *Mars nearest to Earth* | | | | |
JD	date	ET	δ		date	ET	least distance	d	Δt
		h m	° ′			h	a.u.	10^6 km	
								″	d
2337444.04	1687 Aug 8	12 55	-22 37		1687 Aug 9	12.8	0.37434	56.00	25.00	+1.00
2338249.72	1689 Oct 22	5 17	+ 9 37		1689 Oct 14	3.3	0.44272	66.23	21.14	-8.08
2339030.14	1691 Dec 11	15 27	+25 48		1691 Dec 5	6.5	0.57125	85.46	16.39	-6.37
2339798.20	1694 Jan 17	16 52	+24 50		1694 Jan 15	1.8	0.65586	98.12	14.27	-2.63
2340562.38	1696 Feb 20	21 06	+14 46		1696 Feb 22	3.5	0.67653	101.21	13.84	+1.26
2341327.76	1698 Mar 27	6 18	+ 0 01		1698 Apr 1	8.7	0.63093	94.39	14.84	+5.10
2342100.31	1700 May 8	19 29	-17 19		1700 May 16	23.2	0.52465	78.49	17.84	+8.16
2342891.54	1702 Jul 9	0 53	-28 04		1702 Jul 14	21.1	0.39857	59.63	23.48	+5.84
2343702.41	1704 Sep 26	21 45	- 2 35		1704 Sep 20	13.5	0.39814	59.56	23.51	-6.34
2344491.73	1706 Nov 25	5 25	+22 05		1706 Nov 17	15.7	0.52370	78.34	17.87	-7.57
2345263.23	1709 Jan 4	17 27	+26 39		1708 Dec 31	14.4	0.62997	94.24	14.86	-4.13
2346028.23	1711 Feb 8	17 25	+19 20		1711 Feb 8	11.2	0.67620	101.16	13.84	-0.26
2346792.55	1713 Mar 14	1 06	+ 6 06		1713 Mar 17	16.0	0.65659	98.22	14.26	+3.62
2347561.37	1715 Apr 21	20 48	-10 31		1715 Apr 29	1.0	0.57241	85.63	16.35	+7.18
2348343.37	1717 Jun 11	20 57	-26 27		1717 Jun 19	23.0	0.44336	66.33	21.11	+8.09
2349150.44	1719 Aug 27	22 30	-16 11		1719 Aug 25	19.9	0.37401	55.95	25.03	-2.11
2349950.60	1721 Nov 5	2 27	+15 12		1721 Oct 27	21.0	0.47118	70.49	19.86	-8.23
2350727.28	1723 Dec 21	18 37	+26 45		1723 Dec 16	4.6	0.59456	88.94	15.74	-5.58
2351493.98	1726 Jan 26	11 32	+23 04		1726 Jan 24	16.9	0.66588	99.61	14.06	-1.77
2352257.99	1728 Feb 29	11 49	+11 40		1728 Mar 2	14.6	0.67219	100.56	13.92	+2.12
2353024.31	1730 Apr 5	19 25	- 3 50		1730 Apr 11	16.5	0.61273	91.66	15.28	+5.88
2353799.65	1732 May 20	3 33	-21 08		1732 May 28	13.2	0.49614	74.22	18.87	+8.40
2354596.86	1734 Jul 26	8 31	-25 53		1734 Jul 29	18.2	0.38168	57.10	24.52	+3.40
2355405.85	1736 Oct 12	8 30	+ 4 51		1736 Oct 4	17.8	0.42105	62.99	22.23	-7.61
2356189.91	1738 Dec 5	9 54	+24 37		1738 Nov 28	11.7	0.55076	82.39	16.99	-6.92
2356959.33	1741 Jan 12	20 00	+25 43		1741 Jan 9	12.9	0.64552	96.57	14.50	-3.29
2357723.79	1743 Feb 16	7 02	+16 39		1743 Feb 16	21.2	0.67771	101.38	13.81	+0.59
2358488.60	1745 Mar 22	2 23	+ 2 29		1745 Mar 26	13.0	0.64367	96.29	14.54	+4.44
2359259.29	1747 May 1	18 51	-14 33		1747 May 9	12.8	0.54688	81.81	17.12	+7.75
2360046.07	1749 Jun 26	13 39	-28 02		1749 Jul 3	18.2	0.41699	62.38	22.45	+7.19
2360856.34	1751 Sep 14	20 13	- 8 26		1751 Sep 10	2.7	0.38402	57.45	24.37	-4.73
2361650.43	1753 Nov 16	22 13	+19 39		1753 Nov 8	23.4	0.50063	74.89	18.70	-7.95
2362424.00	1755 Dec 30	11 58	+26 54		1755 Dec 25	17.2	0.61531	92.05	15.21	-4.78
2363189.63	1758 Feb 3	3 12	+20 54		1758 Feb 2	4.7	0.67284	100.66	13.91	-0.94
2363953.73	1760 Mar 8	5 28	+ 8 21		1760 Mar 11	4.1	0.66481	99.45	14.08	+2.94
2364721.31	1762 Apr 14	19 20	- 7 48		1762 Apr 21	9.7	0.59141	88.47	15.83	+6.60
2365500.03	1764 Jun 1	12 43	-24 29		1764 Jun 9	22.5	0.46658	69.80	20.06	+8.41
2366303.07	1766 Aug 13	13 38	-21 07		1766 Aug 13	23.0	0.37326	55.84	25.08	+0.39
2367107.81	1768 Oct 26	7 25	+11 18		1768 Oct 18	3.0	0.44785	67.00	20.90	-8.19
2367887.47	1770 Dec 14	23 11	+26 10		1770 Dec 8	18.8	0.57592	86.16	16.25	-6.18
2368655.25	1773 Jan 20	17 55	+24 15		1773 Jan 18	7.0	0.65802	98.44	14.22	-2.45
2369419.37	1775 Feb 23	20 49	+13 41		1775 Feb 25	7.2	0.67588	101.11	13.85	+1.43
2370184.88	1777 Mar 30	9 10	- 1 16		1777 Apr 4	14.9	0.62780	93.92	14.91	+5.24
2370957.92	1779 May 12	10 02	-18 29		1779 May 20	14.3	0.51951	77.72	18.02	+8.18
2371750.28	1781 Jul 12	18 43	-27 41		1781 Jul 18	4.8	0.39473	59.05	23.71	+5.42
2372561.00	1783 Oct 1	11 54	- 0 32		1783 Sep 24	20.0	0.40191	60.13	23.29	-6.66
2373349.25	1785 Nov 27	17 55	+22 57		1785 Nov 20	7.9	0.52886	79.12	17.70	-7.42
2374120.32	1788 Jan 7	19 48	+26 22		1788 Jan 3	20.5	0.63315	94.72	14.78	-3.97
2374885.21	1790 Feb 10	17 07	+18 24		1790 Feb 10	14.7	0.67685	101.26	13.83	-0.10
2375649.61	1792 Mar 16	2 44	+ 4 51		1792 Mar 19	21.1	0.65439	97.90	14.30	+3.77

	Mars at opposition			Mars nearest to Earth					
JD	date	ET	δ	date	ET	least distance	10^6 km	d	Δt
		h m	° ′		h	a.u.		″	d
2376418.75	1794 Apr 24	5 56	-11 48	1794 May 1	12.0	0.56775	84.93	16.49	+7.25
2377201.59	1796 Jun 15	2 09	-26 59	1796 Jun 23	2.5	0.43829	65.57	21.36	+8.02
2378009.49	1798 Aug 31	23 41	-14 13	1798 Aug 29	9.1	0.37534	56.15	24.94	-2.61
2378808.53	1800 Nov 9	0 48	+16 37	1800 Oct 31	20.4	0.47707	71.37	19.62	-8.18
2379584.54	1802 Dec 25	0 56	+26 52	1802 Dec 19	14.7	0.59884	89.59	15.63	-5.43
2380350.99	1805 Jan 29	11 51	+22 20	1805 Jan 27	20.7	0.66747	99.85	14.02	-1.63
2381115.01	1807 Mar 4	12 17	+10 30	1807 Mar 6	18.3	0.67102	100.38	13.95	+2.25
2381881.54	1809 Apr 9	1 03	- 5 09	1809 Apr 15	0.7	0.60881	91.08	15.37	+5.98
2382657.46	1811 May 24	23 08	-22 09	1811 Jun 2	9.8	0.49033	73.35	19.09	+8.45
2383455.79	1813 Jul 31	6 51	-24 51	1813 Aug 3	3.1	0.37907	56.71	24.69	+2.84
2384264.13	1815 Oct 17	15 05	+ 6 42	1815 Oct 9	19.7	0.42566	63.68	21.99	-7.81
2385047.30	1817 Dec 8	19 15	+25 11	1817 Dec 2	1.0	0.55564	83.12	16.85	-6.76
2385816.41	1820 Jan 16	21 47	+25 14	1820 Jan 13	18.3	0.64812	96.96	14.44	-3.15
2386580.77	1822 Feb 19	6 34	+15 37	1822 Feb 20	0.2	0.67753	101.36	13.81	+0.73
2387345.68	1824 Mar 25	4 16	+ 1 13	1824 Mar 29	17.9	0.64091	95.88	14.60	+4.57
2388116.78	1826 May 5	6 46	-15 46	1826 May 13	2.3	0.54194	81.07	17.27	+7.81
2388904.59	1828 Jul 1	2 09	-28 05	1828 Jul 7	23.8	0.41234	61.68	22.70	+6.90
2389715.18	1830 Sep 19	16 14	- 6 19	1830 Sep 14	12.1	0.38667	57.84	24.21	-5.17
2390508.10	1832 Nov 20	14 25	+20 44	1832 Nov 12	18.6	0.50597	75.69	18.50	-7.83
2391281.14	1835 Jan 2	15 15	+26 46	1834 Dec 28	23.8	0.61891	92.59	15.12	-4.64
2392046.62	1837 Feb 6	2 48	+20 02	1837 Feb 5	7.6	0.67398	100.83	13.89	-0.80
2392810.77	1839 Mar 12	6 26	+ 7 08	1839 Mar 15	8.2	0.66314	99.21	14.11	+3.07
2393578.58	1841 Apr 18	1 58	- 9 05	1841 Apr 24	18.7	0.58722	87.85	15.94	+6.70
2394358.01	1843 Jun 6	12 19	-25 15	1843 Jun 14	22.7	0.46129	69.01	20.29	+8.43
2395162.13	1845 Aug 18	15 10	-19 27	1845 Aug 18	11.6	0.37302	55.80	25.09	-0.15
2395965.89	1847 Oct 31	9 18	+12 55	1847 Oct 23	4.3	0.45355	67.85	20.64	-8.21
2396744.79	1849 Dec 18	6 56	+26 27	1849 Dec 12	5.5	0.58059	86.86	16.12	-6.06
2397512.28	1852 Jan 24	18 48	+23 36	1852 Jan 22	10.8	0.66008	98.75	14.18	-2.33
2398276.36	1854 Feb 26	20 35	+12 34	1854 Feb 28	9.9	0.67521	101.01	13.86	+1.55
2399042.06	1856 Apr 2	13 22	- 2 34	1856 Apr 7	21.5	0.62436	93.40	14.99	+5.34
2399815.61	1858 May 16	2 43	-19 36	1858 May 24	9.2	0.51381	76.86	18.22	+8.27
2400609.05	1860 Jul 17	13 11	-27 07	1860 Jul 22	12.0	0.39088	58.47	23.95	+4.95
2401419.52	1862 Oct 6	0 24	+ 1 30	1862 Sep 29	0.7	0.40576	60.70	23.07	-6.99
2402206.75	1864 Dec 1	6 07	+23 43	1864 Nov 23	23.1	0.53400	79.88	17.53	-7.29
2402977.43	1867 Jan 10	22 17	+26 01	1867 Jan 7	2.2	0.63625	95.18	14.71	-3.84
2403742.19	1869 Feb 13	16 38	+17 25	1869 Feb 13	17.4	0.67720	101.31	13.82	+0.03
2404506.66	1871 Mar 20	3 50	+ 3 36	1871 Mar 24	1.1	0.65210	97.55	14.35	+3.89
2405276.11	1873 Apr 27	14 38	-13 03	1873 May 4	23.2	0.56324	84.26	16.62	+7.36
2406059.85	1875 Jun 20	8 26	-27 24	1875 Jun 28	5.2	0.43319	64.80	21.61	+7.86
2406868.48	1877 Sep 5	23 36	-12 12	1877 Sep 2	20.4	0.37664	56.35	24.85	-3.13
2407666.34	1879 Nov 12	20 04	+17 55	1879 Nov 4	17.5	0.48242	72.17	19.40	-8.11
2408441.72	1881 Dec 27	5 12	+26 53	1881 Dec 21	21.8	0.60280	90.18	15.53	-5.31
2409207.98	1884 Feb 1	11 36	+21 32	1884 Jan 30	23.5	0.66907	100.09	13.99	-1.50
2409972.02	1886 Mar 6	12 23	+ 9 20	1886 Mar 8	21.6	0.66987	100.21	13.97	+2.38
2410738.76	1888 Apr 11	6 09	- 6 26	1888 Apr 17	8.6	0.60500	90.51	15.47	+6.10
2411515.29	1890 May 27	18 58	-23 05	1890 Jun 5	7.5	0.48494	72.55	19.30	+8.52
2412314.75	1892 Aug 4	6 07	-23 38	1892 Aug 6	14.2	0.37735	56.45	24.80	+2.34
2413122.42	1894 Oct 20	22 09	+ 8 33	1894 Oct 13	0.2	0.43099	64.47	21.72	-7.92
2413904.73	1896 Dec 11	5 35	+25 40	1896 Dec 4	13.7	0.56077	83.89	16.69	-6.66
2414673.48	1899 Jan 18	23 25	+24 42	1899 Jan 15	22.7	0.65071	97.34	14.38	-3.03

	Mars at opposition				*Mars nearest to Earth*				
JD	date	ET	δ	date	ET	least distance		d	Δt
		h m	° '		h	a.u.	10^6 km	"	d
2415437.75	1901 Feb 22	6 04	+14 32	1901 Feb 23	2.6	0.67741	101.34	13.82	+0.86
2416202.81	1903 Mar 29	7 23	- 0 05	1903 Apr 2	23.8	0.63803	95.45	14.67	+4.68
2416974.33	1905 May 8	19 59	-16 57	1905 May 16	18.5	0.53659	80.27	17.44	+7.94
2417763.14	1907 Jul 6	15 21	-27 59	1907 Jul 13	4.8	0.40757	60.97	22.97	+6.56
2418573.92	1909 Sep 24	10 02	- 4 13	1909 Sep 18	19.0	0.38947	58.26	24.03	-5.62
2419365.70	1911 Nov 25	4 52	+21 43	1911 Nov 17	11.2	0.51121	76.48	18.31	-7.73
2420138.27	1914 Jan 5	18 28	+26 34	1914 Jan 1	6.0	0.62245	93.12	15.04	-4.52
2420903.61	1916 Feb 10	2 32	+19 08	1916 Feb 9	10.5	0.67485	100.96	13.87	-0.67
2421667.78	1918 Mar 15	6 37	+ 5 55	1918 Mar 18	11.5	0.66130	98.93	14.15	+3.21
2422435.86	1920 Apr 21	8 36	-10 21	1920 Apr 28	4.6	0.58306	87.22	16.05	+6.83
2423216.08	1922 Jun 10	14 02	-25 56	1922 Jun 18	22.7	0.45594	68.21	20.53	+8.36
2424021.20	1924 Aug 23	16 55	-17 41	1924 Aug 22	23.8	0.37284	55.78	25.10	-0.71
2424823.89	1926 Nov 4	9 24	+14 27	1926 Oct 27	5.2	0.45883	68.64	20.40	-8.18
2425602.06	1928 Dec 21	13 29	+26 39	1928 Dec 15	14.6	0.58498	87.51	16.00	-5.95
2426369.29	1931 Jan 27	18 59	+22 54	1931 Jan 25	14.3	0.66214	99.06	14.14	-2.20
2427133.35	1933 Mar 1	20 21	+11 26	1933 Mar 3	13.1	0.67460	100.92	13.87	+1.70
2427899.23	1935 Apr 6	17 26	- 3 52	1935 Apr 12	5.1	0.62102	92.90	15.07	+5.48
2428673.27	1937 May 19	18 29	-20 40	1937 May 28	3.6	0.50853	76.08	18.41	+8.38
2429467.83	1939 Jul 23	7 56	-26 24	1939 Jul 27	20.9	0.38788	58.03	24.13	+4.54
2430278.03	1941 Oct 10	12 40	+ 3 30	1941 Oct 3	7.5	0.41046	61.40	22.80	-7.21
2431064.27	1943 Dec 5	18 24	+24 24	1943 Nov 28	13.3	0.53946	80.70	17.35	-7.21
2431834.53	1946 Jan 14	0 45	+25 36	1946 Jan 10	7.7	0.63935	95.65	14.64	-3.71
2432599.17	1948 Feb 17	16 08	+16 25	1948 Feb 17	20.2	0.67759	101.37	13.81	+0.17
2433363.73	1950 Mar 23	5 37	+ 2 20	1950 Mar 27	6.3	0.64972	97.20	14.41	+4.03
2434133.56	1952 May 1	1 24	-14 17	1952 May 8	13.5	0.55824	83.51	16.77	+7.50
2434918.22	1954 Jun 24	17 14	-27 41	1954 Jul 2	8.0	0.42779	64.00	21.88	+7.62
2435727.41	1956 Sep 10	21 52	-10 08	1956 Sep 7	4.9	0.37809	56.56	24.76	-3.71
2436524.10	1958 Nov 16	14 26	+19 08	1958 Nov 8	13.2	0.48770	72.96	19.19	-8.05
2437298.93	1960 Dec 30	10 14	+26 49	1960 Dec 25	5.8	0.60682	90.78	15.42	-5.19
2438064.99	1963 Feb 4	11 50	+20 42	1963 Feb 3	3.4	0.67045	100.30	13.96	-1.35
2438829.01	1965 Mar 9	12 22	+ 8 08	1965 Mar 12	1.2	0.66848	100.00	14.00	+2.54
2439595.97	1967 Apr 15	11 23	- 7 43	1967 Apr 21	17.6	0.60121	89.94	15.57	+6.26
2440373.16	1969 May 31	15 43	-23 57	1969 Jun 9	4.3	0.47954	71.74	19.52	+8.52
2441173.78	1971 Aug 10	6 46	-22 15	1971 Aug 12	2.5	0.37569	56.20	24.91	+1.82
2441980.64	1973 Oct 25	3 21	+10 18	1973 Oct 17	4.2	0.43604	65.23	21.47	-7.97
2442762.08	1975 Dec 15	13 52	+26 03	1975 Dec 9	0.2	0.56549	84.60	16.55	-6.57
2443530.50	1978 Jan 22	0 05	+24 06	1978 Jan 19	3.1	0.65320	97.72	14.33	-2.87
2444294.73	1980 Feb 25	5 36	+13 27	1980 Feb 26	6.1	0.67732	101.33	13.82	+1.02
2445059.92	1982 Mar 31	10 06	- 1 21	1982 Apr 5	6.6	0.63512	95.01	14.74	+4.85
2445831.86	1984 May 11	8 45	-18 06	1984 May 19	10.7	0.53146	79.51	17.61	+8.08
2446621.72	1986 Jul 10	5 21	-27 44	1986 Jul 16	11.0	0.40356	60.37	23.19	+6.23
2447432.64	1988 Sep 28	3 25	- 2 07	1988 Sep 22	3.3	0.39314	58.81	23.81	-6.00
2448223.35	1990 Nov 27	20 28	+22 38	1990 Nov 20	4.0	0.51692	77.33	18.11	-7.69
2448995.44	1993 Jan 7	22 37	+26 16	1993 Jan 3	13.6	0.62610	93.66	14.95	-4.38
2449760.60	1995 Feb 12	2 25	+18 10	1995 Feb 11	14.3	0.67570	101.08	13.85	-0.50
2450524.83	1997 Mar 17	7 48	+ 4 40	1997 Mar 20	16.9	0.65939	98.64	14.20	+3.38
2451293.23	1999 Apr 24	17 31	-11 37	1999 May 1	17.5	0.57846	86.54	16.18	+7.00
2452074.24	2001 Jun 13	17 39	-26 30	2001 Jun 21	23.0	0.45016	67.34	20.79	+8.22
2452880.25	2003 Aug 28	17 53	-15 49	2003 Aug 27	9.9	0.37271	55.76	25.11	-1.33
2453681.83	2005 Nov 7	7 52	+15 54	2005 Oct 30	3.4	0.46406	69.42	20.17	-8.18

	Mars at opposition				*Mars nearest to Earth*				
JD	date	ET	δ		date	ET	least distance	d	Δt
		h m	° ′			h	a.u.	10⁶ km	″	d

JD	date	ET	δ	date	ET	least distance		d	Δt
		h m	° ′		h	a.u.	10⁶ km	″	d
2454459.32	2007 Dec 24	19 41	+26 46	2007 Dec 18	23.8	0.58935	88.17	15.88	-5.83
2455226.32	2010 Jan 29	19 37	+22 09	2010 Jan 27	19.0	0.66398	99.33	14.10	-2.02
2455990.34	2012 Mar 3	20 04	+10 17	2012 Mar 5	17.0	0.67368	100.78	13.89	+1.87
2456756.37	2014 Apr 8	20 57	- 5 08	2014 Apr 14	12.9	0.61756	92.39	15.16	+5.66
2457530.97	2016 May 22	11 10	-21 39	2016 May 30	21.6	0.50321	75.28	18.60	+8.43
2458326.71	2018 Jul 27	5 07	-25 30	2018 Jul 31	7.9	0.38496	57.59	24.31	+4.11
2459136.47	2020 Oct 13	23 21	+ 5 27	2020 Oct 6	14.3	0.41491	62.07	22.56	-7.38
2459921.73	2022 Dec 8	5 37	+25 00	2022 Dec 1	2.3	0.54448	81.45	17.19	-7.14
2460691.61	2025 Jan 16	2 33	+25 07	2025 Jan 12	13.6	0.64229	96.09	14.57	-3.54
2461456.16	2027 Feb 19	15 45	+15 22	2027 Feb 20	0.2	0.67793	101.42	13.81	+0.35
2462220.82	2029 Mar 25	7 43	+ 1 04	2029 Mar 29	13.0	0.64723	96.82	14.46	+4.22
2462991.00	2031 May 4	11 57	-15 29	2031 May 12	3.8	0.55336	82.78	16.91	+7.66
2463776.56	2033 Jun 28	1 23	-27 49	2033 Jul 5	11.3	0.42301	63.28	22.13	+7.41
2464586.32	2035 Sep 15	19 34	- 8 02	2035 Sep 11	14.4	0.38040	56.91	24.61	-4.22
2465381.88	2037 Nov 19	9 05	+20 16	2037 Nov 11	8.0	0.49358	73.84	18.96	-8.04
2466156.14	2040 Jan 2	15 22	+26 41	2039 Dec 28	14.8	0.61092	91.39	15.32	-5.02
2466922.00	2042 Feb 6	11 59	+19 50	2042 Feb 5	8.0	0.67175	100.49	13.93	-1.17
2467686.03	2044 Mar 11	12 44	+ 6 56	2044 Mar 14	6.1	0.66709	99.80	14.03	+2.72
2468453.25	2046 Apr 17	18 00	- 9 00	2046 Apr 24	4.6	0.59705	89.32	15.68	+6.44
2469231.11	2048 Jun 3	14 44	-24 45	2048 Jun 12	1.7	0.47366	70.86	19.76	+8.46
2470032.82	2050 Aug 14	7 46	-20 44	2050 Aug 15	12.9	0.37404	55.96	25.02	+1.21
2470838.77	2052 Oct 28	6 30	+11 58	2052 Oct 20	5.2	0.44090	65.96	21.23	-8.05
2471619.42	2054 Dec 17	22 10	+26 20	2054 Dec 11	11.7	0.57015	85.29	16.42	-6.43
2472387.56	2057 Jan 24	1 26	+23 27	2057 Jan 21	9.1	0.65553	98.07	14.28	-2.68
2473151.73	2059 Feb 27	5 25	+12 20	2059 Feb 28	10.5	0.67682	101.25	13.83	+1.21
2473917.03	2061 Apr 2	12 47	- 2 38	2061 Apr 7	13.9	0.63199	94.54	14.81	+5.05
2474689.43	2063 May 14	22 15	-19 11	2063 May 23	1.9	0.52637	78.74	17.78	+8.15
2475480.37	2065 Jul 13	20 57	-27 19	2065 Jul 19	19.9	0.39958	59.78	23.42	+5.95
2476291.33	2067 Oct 2	19 51	- 0 01	2067 Sep 26	13.0	0.39669	59.34	23.60	-6.29
2477080.93	2069 Nov 30	10 16	+23 26	2069 Nov 22	19.3	0.52221	78.12	17.92	-7.62
2477852.54	2072 Jan 11	1 00	+25 55	2072 Jan 6	20.4	0.62940	94.16	14.87	-4.19
2478617.58	2074 Feb 14	1 53	+17 12	2074 Feb 13	18.4	0.67646	101.20	13.84	-0.31
2479381.87	2076 Mar 19	8 50	+ 3 26	2076 Mar 22	22.5	0.65743	98.35	14.24	+3.57
2480150.56	2078 Apr 27	1 32	-12 51	2078 May 4	5.2	0.57391	85.86	16.31	+7.15
2480932.35	2080 Jun 16	20 22	-26 58	2080 Jun 24	22.9	0.44497	66.57	21.04	+8.11
2481739.23	2082 Sep 1	17 35	-13 53	2082 Aug 30	19.0	0.37356	55.88	25.06	-1.94
2482539.75	2084 Nov 10	6 03	+17 16	2084 Nov 2	0.6	0.46986	70.29	19.92	-8.23
2483316.61	2086 Dec 27	2 33	+26 48	2086 Dec 21	10.9	0.59387	88.84	15.76	-5.65
2484083.35	2089 Jan 31	20 18	+21 21	2089 Jan 30	0.4	0.66578	99.60	14.06	-1.83
2484847.34	2091 Mar 6	20 10	+ 9 06	2091 Mar 8	21.7	0.67273	100.64	13.91	+2.07
2485613.59	2093 Apr 11	2 15	- 6 25	2093 Apr 16	22.4	0.61382	91.83	15.25	+5.84
2486388.76	2095 May 26	6 21	-22 36	2095 Jun 3	16.2	0.49747	74.42	18.82	+8.41
2487185.65	2097 Jul 31	3 34	-24 26	2097 Aug 3	18.3	0.38186	57.13	24.51	+3.61
2487994.84	2099 Oct 18	8 04	+ 7 20	2099 Oct 10	18.8	0.41922	62.71	22.33	-7.55
2488779.15	2101 Dec 11	15 39	+25 30	2101 Dec 4	15.9	0.54943	82.19	17.04	-6.99
2489548.67	2104 Jan 20	4 12	+24 34	2104 Jan 16	20.0	0.64505	96.50	14.51	-3.34
2490313.15	2106 Feb 22	15 29	+14 18	2106 Feb 23	4.6	0.67789	101.41	13.81	+0.55
2491077.88	2108 Mar 28	9 07	- 0 11	2108 Apr 1	18.9	0.64457	96.43	14.52	+4.41
2491848.42	2110 May 7	22 12	-16 37	2110 May 15	16.0	0.54854	82.06	17.06	+7.74
2492634.96	2112 Jul 2	11 02	-27 49	2112 Jul 9	16.8	0.41834	62.58	22.37	+7.24

OPPOSITIONS OF MARS

	Mars at opposition				*Mars nearest to Earth*				
JD	date	ET	δ		date	ET	least distance		d	Δt
		h m	° ′			h	a.u.	10⁶ km	″	d
2493445.19	2114 Sep 20	16 31	- 5 55		2114 Sep 16	1.2	0.38283	57.27	24.45	-4.64
2494239.59	2116 Nov 23	2 06	+21 19		2116 Nov 15	2.1	0.49902	74.65	18.76	-8.00
2495013.30	2119 Jan 5	19 08	+26 28		2118 Dec 31	23.0	0.61459	91.94	15.23	-4.84
2495778.98	2121 Feb 9	11 29	+18 55		2121 Feb 8	12.0	0.67297	100.67	13.91	-0.98
2496543.05	2123 Mar 15	13 19	+ 5 42		2123 Mar 18	11.0	0.66559	99.57	14.06	+2.91
2497310.53	2125 Apr 21	0 43	-10 15		2125 Apr 27	14.6	0.59283	88.69	15.79	+6.58
2498089.04	2127 Jun 8	12 52	-25 26		2127 Jun 16	22.8	0.46830	70.06	19.99	+8.41
2498891.82	2129 Aug 19	7 35	-19 06		2129 Aug 19	21.9	0.37327	55.84	25.08	+0.60
2499696.89	2131 Nov 2	9 15	+13 34		2131 Oct 25	5.3	0.44645	66.79	20.97	-8.16
2500476.77	2133 Dec 21	6 24	+26 33		2133 Dec 15	0.4	0.57510	86.03	16.28	-6.25
2501244.60	2136 Jan 28	2 25	+22 44		2136 Jan 25	14.5	0.65781	98.41	14.23	-2.50
2502008.71	2138 Mar 2	5 09	+11 12		2138 Mar 3	14.5	0.67633	101.18	13.84	+1.39
2502774.18	2140 Apr 5	16 14	- 3 54		2140 Apr 10	21.1	0.62881	94.07	14.89	+5.20
2503547.07	2142 May 18	13 40	-20 15		2142 May 26	17.7	0.52088	77.92	17.97	+8.17
2504339.13	2144 Jul 18	15 02	-26 44		2144 Jul 24	4.9	0.39535	59.14	23.67	+5.58
2505149.91	2146 Oct 7	9 52	+ 2 01		2146 Sep 30	20.1	0.40020	59.87	23.39	-6.58
2505938.45	2148 Dec 3	22 46	+24 09		2148 Nov 26	11.1	0.52737	78.89	17.75	-7.49
2506709.65	2151 Jan 14	3 39	+25 30		2151 Jan 10	3.5	0.63258	94.63	14.80	-4.01
2507474.57	2153 Feb 17	1 37	+16 11		2153 Feb 16	22.5	0.67692	101.27	13.83	-0.13
2508238.91	2155 Mar 23	9 48	+ 2 11		2155 Mar 27	3.6	0.65516	98.01	14.29	+3.74
2509007.92	2157 Apr 30	9 58	-14 03		2157 May 7	15.7	0.56936	85.17	16.44	+7.24
2509790.52	2159 Jun 22	0 26	-27 18		2159 Jun 30	1.3	0.43990	65.81	21.28	+8.04
2510598.26	2161 Sep 6	18 12	-11 51		2161 Sep 4	6.9	0.37458	56.04	24.99	-2.47
2511397.64	2163 Nov 15	3 26	+18 33		2163 Nov 6	21.9	0.47536	71.11	19.69	-8.23
2512173.81	2165 Dec 30	7 32	+26 44		2165 Dec 24	20.1	0.59794	89.45	15.65	-5.48
2512940.32	2168 Feb 4	19 48	+20 31		2168 Feb 3	3.9	0.66743	99.85	14.02	-1.66
2513704.34	2170 Mar 9	20 12	+ 7 55		2170 Mar 12	1.5	0.67171	100.49	13.93	+2.22
2514470.79	2172 Apr 14	6 59	- 7 41		2172 Apr 20	6.1	0.61009	91.27	15.34	+5.96
2515246.51	2174 May 30	0 15	-23 28		2174 Jun 7	10.6	0.49209	73.62	19.02	+8.43
2516044.54	2176 Aug 5	1 03	-23 14		2176 Aug 8	2.0	0.37960	56.79	24.66	+3.04
2516853.13	2178 Oct 22	15 11	+ 9 09		2178 Oct 14	20.9	0.42427	63.47	22.06	-7.76
2517636.58	2180 Dec 14	1 50	+25 54		2180 Dec 7	6.2	0.55472	82.98	16.87	-6.82
2518405.76	2183 Jan 22	6 10	+23 57		2183 Jan 19	1.9	0.64783	96.91	14.45	-3.18
2519170.13	2185 Feb 24	15 00	+13 13		2185 Feb 25	7.8	0.67788	101.41	13.81	+0.70
2519934.98	2187 Mar 31	11 38	- 1 27		2187 Apr 5	0.5	0.64185	96.02	14.58	+4.54
2520705.97	2189 May 10	11 15	-17 46		2189 May 18	6.1	0.54334	81.28	17.23	+7.79
2521493.49	2191 Jul 6	23 50	-27 41		2191 Jul 14	0.2	0.41331	61.83	22.65	+7.02
2522304.02	2193 Sep 24	12 34	- 3 47		2193 Sep 19	11.6	0.38521	57.63	24.30	-5.04
2523097.26	2195 Nov 26	18 15	+22 16		2195 Nov 18	20.9	0.50433	75.45	18.56	-7.89
2523870.44	2198 Jan 7	22 37	+26 11		2198 Jan 3	6.5	0.61818	92.48	15.14	-4.67
2524635.97	2200 Feb 12	11 11	+17 57		2200 Feb 11	15.5	0.67393	100.82	13.89	-0.82
2525400.08	2202 Mar 18	13 48	+ 4 29		2202 Mar 21	15.1	0.66377	99.30	14.10	+3.05
2526167.78	2204 Apr 24	6 39	-11 28		2204 Apr 30	22.8	0.58866	88.06	15.90	+6.67
2526946.99	2206 Jun 12	11 43	-26 01		2206 Jun 20	21.8	0.46305	69.27	20.21	+8.42
2527750.87	2208 Aug 24	8 59	-17 21		2208 Aug 24	9.1	0.37278	55.77	25.11	+0.00
2528554.95	2210 Nov 6	10 48	+15 06		2210 Oct 29	5.0	0.45181	67.59	20.72	-8.24
2529334.05	2212 Dec 24	13 12	+26 40		2212 Dec 18	10.9	0.57951	86.69	16.15	-6.09
2530101.61	2215 Jan 31	2 34	+21 59		2215 Jan 28	18.1	0.65989	98.72	14.18	-2.35
2530865.69	2217 Mar 5	4 40	+10 03		2217 Mar 6	17.4	0.67579	101.10	13.85	+1.53
2531631.33	2219 Apr 9	19 52	- 5 10		2219 Apr 15	3.5	0.62552	93.58	14.96	+5.32

| | *Mars at opposition* | | | | *Mars nearest to Earth* | | | | |
JD	date	ET	δ		date	ET	least distance		d	Δt
		h m	° ′			h	a.u.	10^6 km	″	d
2532404.70	2221 May 22	4 54	-21 14		2221 May 30	10.4	0.51557	77.13	18.15	+8.23
2533197.84	2223 Jul 24	8 11	-25 59		2223 Jul 29	11.3	0.39184	58.62	23.89	+5.13
2534008.44	2225 Oct 11	22 34	+ 4 01		2225 Oct 5	1.2	0.40444	60.50	23.14	-6.89
2534795.99	2227 Dec 8	11 50	+24 47		2227 Dec 1	3.8	0.53297	79.73	17.56	-7.33
2535566.77	2230 Jan 17	6 27	+25 00		2230 Jan 13	9.7	0.63586	95.12	14.72	-3.87
2536331.54	2232 Feb 21	1 01	+15 08		2232 Feb 21	1.1	0.67740	101.34	13.82	+0.00
2537095.97	2234 Mar 26	11 19	+ 0 56		2234 Mar 30	7.8	0.65295	97.68	14.34	+3.85
2537865.33	2236 May 3	19 48	-15 14		2236 May 11	3.2	0.56459	84.46	16.58	+7.31
2538648.82	2238 Jun 26	7 34	-27 31		2238 Jul 4	5.8	0.43438	64.98	21.55	+7.93
2539457.28	2240 Sep 11	18 45	- 9 46		2240 Sep 8	19.2	0.37558	56.19	24.92	-2.98
2540255.46	2242 Nov 18	22 58	+19 44		2242 Nov 10	19.1	0.48065	71.90	19.47	-8.16
2541031.01	2245 Jan 2	12 16	+26 36		2244 Dec 28	4.3	0.60188	90.04	15.55	-5.33
2541797.33	2247 Feb 7	19 55	+19 38		2247 Feb 6	7.4	0.66890	100.07	13.99	-1.52
2542561.34	2249 Mar 12	20 10	+ 6 43		2249 Mar 15	4.7	0.67035	100.28	13.96	+2.36
2543327.98	2251 Apr 18	11 28	- 8 55		2251 Apr 24	13.2	0.60627	90.70	15.44	+6.07
2544104.31	2253 Jun 2	19 25	-24 15		2253 Jun 11	7.1	0.48678	72.82	19.23	+8.49
2544903.50	2255 Aug 11	0 02	-21 52		2255 Aug 13	11.9	0.37764	56.49	24.79	+2.49
2545711.43	2257 Oct 26	22 16	+10 54		2257 Oct 19	0.1	0.42924	64.21	21.81	-7.92
2546493.97	2259 Dec 18	11 15	+26 13		2259 Dec 11	18.8	0.55953	83.70	16.73	-6.69
2547262.79	2262 Jan 25	6 53	+23 18		2262 Jan 22	5.6	0.65034	97.29	14.39	-3.05
2548027.09	2264 Feb 28	14 05	+12 06		2264 Feb 29	10.0	0.67782	101.40	13.81	+0.83
2548792.09	2266 Apr 3	14 07	- 2 43		2266 Apr 8	5.8	0.63908	95.60	14.65	+4.65
2549563.46	2268 May 13	23 01	-18 50		2268 May 21	20.2	0.53827	80.52	17.39	+7.88
2550351.97	2270 Jul 11	11 17	-27 23		2270 Jul 18	4.1	0.40886	61.16	22.89	+6.70
2551162.75	2272 Sep 29	6 04	- 1 41		2272 Sep 23	18.6	0.38834	58.09	24.10	-5.48
2551954.90	2274 Nov 30	9 34	+23 06		2274 Nov 22	15.2	0.51003	76.30	18.35	-7.76
2552727.60	2277 Jan 11	2 28	+25 49		2277 Jan 6	13.3	0.62192	93.04	15.05	-4.55
2553492.95	2279 Feb 15	10 52	+16 58		2279 Feb 14	18.1	0.67493	100.97	13.87	-0.70
2554257.10	2281 Mar 20	14 20	+ 3 15		2281 Mar 23	18.3	0.66203	99.04	14.14	+3.17
2555025.10	2283 Apr 27	14 20	-12 41		2283 May 4	8.9	0.58433	87.41	16.02	+6.77
2555805.10	2285 Jun 15	14 25	-26 32		2285 Jun 23	23.8	0.45732	68.41	20.47	+8.39
2556609.98	2287 Aug 29	11 26	-15 28		2287 Aug 28	22.5	0.37224	55.69	25.15	-0.54
2557412.97	2289 Nov 9	11 10	+16 32		2289 Nov 1	5.9	0.45695	68.36	20.48	-8.22
2558191.33	2291 Dec 27	19 52	+26 42		2291 Dec 21	20.4	0.58388	87.35	16.03	-5.98
2558958.63	2294 Feb 2	3 03	+21 10		2294 Jan 30	21.7	0.66184	99.01	14.14	-2.22
2559722.68	2296 Mar 7	4 17	+ 8 53		2296 Mar 8	20.2	0.67492	100.97	13.87	+1.66
2560488.46	2298 Apr 11	23 01	- 6 25		2298 Apr 17	9.5	0.62213	93.07	15.05	+5.44
2561262.32	2300 May 25	19 42	-22 09		2300 Jun 3	3.8	0.51038	76.35	18.34	+8.34
2562056.59	2302 Jul 29	2 11	-25 06		2302 Aug 2	18.3	0.38864	58.14	24.08	+4.67
2562866.95	2304 Oct 16	10 41	+ 5 58		2304 Oct 9	6.3	0.40878	61.15	22.90	-7.18
2563653.47	2306 Dec 11	23 17	+25 18		2306 Dec 4	17.6	0.53803	80.49	17.40	-7.24
2564423.83	2309 Jan 20	8 00	+24 27		2309 Jan 16	14.1	0.63877	95.56	14.65	-3.75
2565188.50	2311 Feb 23	23 58	+14 04		2311 Feb 24	3.2	0.67785	101.40	13.81	+0.14
2565953.03	2313 Mar 29	12 38	- 0 19		2313 Apr 2	12.2	0.65066	97.34	14.39	+3.98
2566722.72	2315 May 8	5 17	-16 22		2315 May 15	16.0	0.55981	83.75	16.72	+7.45
2567507.09	2317 Jun 30	14 11	-27 36		2317 Jul 8	7.7	0.42935	64.23	21.80	+7.73
2568316.17	2319 Sep 17	16 09	- 7 41		2319 Sep 14	4.1	0.37730	56.44	24.81	-3.50
2569113.24	2321 Nov 22	17 52	+20 50		2321 Nov 14	16.2	0.48639	72.76	19.24	-8.07
2569888.24	2324 Jan 6	17 40	+26 23		2324 Jan 1	12.3	0.60617	90.68	15.44	-5.22
2570654.33	2326 Feb 10	20 00	+18 42		2326 Feb 9	10.5	0.67042	100.29	13.96	-1.40

OPPOSITIONS OF MARS

JD	date	ET	δ	date	ET	least distance	10^6 km	d	Δt
		h m	° '		h	a.u.		"	d
2571418.34	2328 Mar 15	20 06	+ 5 30	2328 Mar 18	7.8	0.66911	100.10	13.99	+2.49
2572185.22	2330 Apr 21	17 10	-10 10	2330 Apr 27	21.8	0.60245	90.13	15.54	+6.19
2572962.21	2332 Jun 6	17 01	-24 58	2332 Jun 15	5.7	0.48107	71.97	19.46	+8.53
2573762.54	2334 Aug 16	0 59	-20 21	2334 Aug 18	0.6	0.37558	56.19	24.92	+1.98
2574569.64	2336 Oct 31	3 18	+12 35	2336 Oct 23	3.5	0.43408	64.94	21.56	-7.99
2575351.31	2338 Dec 21	19 23	+26 26	2338 Dec 15	5.0	0.56417	84.40	16.59	-6.60
2576119.83	2341 Jan 28	7 53	+22 35	2341 Jan 25	9.8	0.65274	97.65	14.34	-2.92
2576884.06	2343 Mar 3	13 30	+10 58	2343 Mar 4	12.9	0.67751	101.35	13.82	+0.97
2577649.17	2345 Apr 6	16 06	- 3 58	2345 Apr 11	11.2	0.63608	95.16	14.72	+4.80
2578420.95	2347 May 18	10 54	-19 51	2347 May 26	11.7	0.53324	79.77	17.55	+8.03
2579210.51	2349 Jul 16	0 21	-26 56	2349 Jul 22	8.8	0.40476	60.55	23.12	+6.35
2580021.48	2351 Oct 4	23 31	+ 0 24	2351 Sep 29	1.7	0.39170	58.60	23.90	-5.91
2580812.52	2353 Dec 4	0 27	+23 52	2353 Nov 26	7.4	0.51537	77.10	18.16	-7.71
2581584.72	2356 Jan 15	5 22	+25 23	2356 Jan 10	19.2	0.62535	93.55	14.97	-4.42
2582349.91	2358 Feb 18	9 57	+15 58	2358 Feb 17	20.7	0.67584	101.10	13.85	-0.55
2583114.12	2360 Mar 23	14 57	+ 2 01	2360 Mar 26	22.6	0.66025	98.77	14.18	+3.32
2583882.41	2362 Apr 30	21 44	-13 52	2362 May 7	20.2	0.57997	86.76	16.14	+6.94
2584663.15	2364 Jun 19	15 35	-26 55	2364 Jun 27	22.8	0.45193	67.61	20.71	+8.30
2585468.95	2366 Sep 3	10 41	-13 33	2366 Sep 2	8.2	0.37237	55.71	25.14	-1.10
2586270.89	2368 Nov 13	9 22	+17 52	2368 Nov 5	4.9	0.46258	69.20	20.23	-8.19
2587048.60	2370 Dec 31	2 21	+26 39	2370 Dec 25	5.5	0.58856	88.05	15.90	-5.87
2587815.65	2373 Feb 5	3 40	+20 19	2373 Feb 3	1.8	0.66385	99.31	14.10	-2.08
2588579.66	2375 Mar 11	3 44	+ 7 42	2375 Mar 12	23.3	0.67420	100.86	13.88	+1.81
2589345.63	2377 Apr 15	3 01	- 7 40	2377 Apr 20	17.3	0.61874	92.56	15.13	+5.59
2590120.07	2379 May 29	13 37	-23 02	2379 Jun 6	23.6	0.50478	75.51	18.54	+8.42
2590915.50	2381 Aug 1	23 59	-24 01	2381 Aug 6	5.9	0.38534	57.65	24.29	+4.25
2591725.41	2383 Oct 20	21 46	+ 7 53	2383 Oct 13	13.0	0.41304	61.79	22.66	-7.37
2592510.93	2385 Dec 14	10 21	+25 44	2385 Dec 7	6.2	0.54299	81.23	17.24	-7.17
2593280.91	2388 Jan 23	9 56	+23 50	2388 Jan 19	19.7	0.64170	96.00	14.59	-3.59
2594045.48	2390 Feb 25	23 34	+12 59	2390 Feb 26	6.7	0.67805	101.43	13.80	+0.30
2594810.09	2392 Mar 31	14 08	- 1 34	2392 Apr 4	17.7	0.64807	96.95	14.44	+4.15
2595580.11	2394 May 10	14 33	-17 27	2394 May 18	5.2	0.55507	83.04	16.86	+7.61
2596365.39	2396 Jul 3	21 19	-27 33	2396 Jul 11	9.2	0.42458	63.52	22.05	+7.49
2597175.07	2398 Sep 21	13 46	- 5 34	2398 Sep 17	12.2	0.37934	56.75	24.67	-4.07
2597970.99	2400 Nov 25	11 45	+21 50	2400 Nov 17	10.4	0.49187	73.58	19.03	-8.06
2598745.40	2403 Jan 8	21 37	+26 06	2403 Jan 3	19.6	0.61000	91.25	15.34	-5.08
2599511.31	2405 Feb 12	19 20	+17 45	2405 Feb 11	13.9	0.67175	100.49	13.93	-1.23
2600275.33	2407 Mar 18	19 55	+ 4 17	2407 Mar 21	11.7	0.66783	99.91	14.02	+2.66
2601042.45	2409 Apr 23	22 50	-11 22	2409 Apr 30	7.8	0.59846	89.53	15.64	+6.38
2601820.09	2411 Jun 10	14 13	-25 35	2411 Jun 19	2.2	0.47548	71.13	19.69	+8.50
2602621.50	2413 Aug 20	0 06	-18 44	2413 Aug 21	10.9	0.37420	55.98	25.01	+1.45
2603427.76	2415 Nov 4	6 19	+14 10	2415 Oct 27	5.7	0.43937	65.73	21.30	-8.03
2604208.67	2417 Dec 24	4 11	+26 34	2417 Dec 17	16.6	0.56927	85.16	16.44	-6.48
2604976.89	2420 Jan 31	9 21	+21 49	2420 Jan 28	15.5	0.65531	98.03	14.28	-2.75
2605741.04	2422 Mar 5	13 04	+ 9 48	2422 Mar 6	16.7	0.67726	101.32	13.82	+1.15
2606506.30	2424 Apr 8	19 11	- 5 13	2424 Apr 13	18.6	0.63312	94.71	14.78	+4.97
2607278.55	2426 May 21	1 18	-20 51	2426 May 29	4.4	0.52796	78.98	17.73	+8.13
2608069.20	2428 Jul 19	16 54	-26 19	2428 Jul 25	17.8	0.40042	59.90	23.38	+6.04
2608880.19	2430 Oct 8	16 41	+ 2 29	2430 Oct 2	10.8	0.39503	59.10	23.69	-6.24
2609670.08	2432 Dec 6	14 00	+24 31	2432 Nov 28	22.1	0.52054	77.87	17.98	-7.66

	Mars at opposition				*Mars nearest to Earth*				
JD	date	ET	δ		date	ET	least distance		d	Δt
		h m	° ′			h	a.u.	10⁶ km	″	d

JD	date	ET	δ	date	ET	least distance a.u.	10⁶ km	d ″	Δt d
2610441.83	2435 Jan 17	7 55	+24 54	2435 Jan 13	1.8	0.62866	94.05	14.89	-4.26
2611206.90	2437 Feb 20	9 41	+14 55	2437 Feb 20	0.8	0.67649	101.20	13.84	-0.37
2611971.15	2439 Mar 26	15 37	+ 0 47	2439 Mar 30	3.8	0.65811	98.45	14.22	+3.51
2612739.70	2441 May 3	4 54	-15 00	2441 May 10	7.4	0.57546	86.09	16.27	+7.10
2613521.24	2443 Jun 23	17 52	-27 11	2443 Jul 1	21.4	0.44671	66.83	20.95	+8.15
2614327.94	2445 Sep 7	10 40	-11 32	2445 Sep 5	16.9	0.37295	55.79	25.10	-1.74
2615128.82	2447 Nov 17	7 39	+19 07	2447 Nov 9	2.3	0.46808	70.02	20.00	-8.22
2615905.85	2450 Jan 2	8 30	+26 31	2449 Dec 27	15.2	0.59283	88.69	15.79	-5.72
2616672.65	2452 Feb 8	3 33	+19 26	2452 Feb 6	6.1	0.66563	99.58	14.06	-1.89
2617436.64	2454 Mar 13	3 22	+ 6 30	2454 Mar 15	3.4	0.67338	100.74	13.90	+2.00
2618202.82	2456 Apr 17	7 42	- 8 54	2456 Apr 23	2.4	0.61516	92.03	15.22	+5.78
2618977.79	2458 Jun 1	6 54	-23 49	2458 Jun 9	17.4	0.49927	74.69	18.75	+8.44
2619774.33	2460 Aug 5	20 01	-22 49	2460 Aug 9	15.7	0.38252	57.22	24.47	+3.82
2620583.75	2462 Oct 24	6 02	+ 9 41	2462 Oct 16	18.1	0.41772	62.49	22.41	-7.50
2621368.37	2464 Dec 16	20 48	+26 04	2464 Dec 9	19.6	0.54837	82.04	17.07	-7.05
2622137.99	2467 Jan 25	11 52	+23 10	2467 Jan 22	2.2	0.64472	96.45	14.52	-3.40
2622902.46	2469 Feb 27	23 09	+11 52	2469 Feb 28	10.8	0.67824	101.46	13.80	+0.49
2623667.16	2471 Apr 3	15 47	- 2 49	2471 Apr 7	24.0	0.64558	96.58	14.50	+4.34
2624437.58	2473 May 13	1 56	-18 32	2473 May 20	19.2	0.55009	82.29	17.02	+7.72
2625223.86	2475 Jul 8	8 40	-27 21	2475 Jul 15	15.4	0.41950	62.76	22.31	+7.28
2626034.00	2477 Sep 25	12 05	- 3 24	2477 Sep 20	22.8	0.38146	57.07	24.54	-4.55
2626828.71	2479 Nov 29	5 04	+22 44	2479 Nov 21	4.0	0.49718	74.38	18.83	-8.04
2627602.57	2482 Jan 11	1 46	+25 44	2482 Jan 6	4.1	0.61371	91.81	15.25	-4.90
2628368.30	2484 Feb 15	19 19	+16 45	2484 Feb 14	18.5	0.67288	100.66	13.91	-1.03
2629132.35	2486 Mar 20	20 25	+ 3 04	2486 Mar 23	16.9	0.66615	99.65	14.05	+2.85
2629899.70	2488 Apr 26	4 50	-12 33	2488 May 2	17.8	0.59425	88.90	15.75	+6.54
2630677.98	2490 Jun 13	11 31	-26 06	2490 Jun 21	22.0	0.47014	70.33	19.91	+8.44
2631480.49	2492 Aug 23	23 50	-17 00	2492 Aug 24	19.5	0.37320	55.83	25.08	+0.82
2632285.89	2494 Nov 7	9 23	+15 41	2494 Oct 30	5.9	0.44466	66.52	21.05	-8.14
2633065.99	2496 Dec 26	11 51	+26 37	2496 Dec 20	4.0	0.57389	85.85	16.31	-6.33
2633833.89	2499 Feb 2	9 28	+21 00	2499 Jan 30	20.2	0.65747	98.36	14.24	-2.55
2634598.01	2501 Mar 8	12 18	+ 8 39	2501 Mar 9	20.3	0.67687	101.26	13.83	+1.34
2635363.42	2503 Apr 12	22 07	- 6 27	2503 Apr 18	1.7	0.63002	94.25	14.86	+5.15
2636136.13	2505 May 24	15 13	-21 46	2505 Jun 1	19.5	0.52262	78.18	17.91	+8.18
2636927.85	2507 Jul 25	8 30	-25 34	2507 Jul 31	2.2	0.39646	59.31	23.61	+5.74
2637738.75	2509 Oct 13	5 58	+ 4 29	2509 Oct 6	18.3	0.39880	59.66	23.47	-6.49
2638527.63	2511 Dec 11	3 04	+25 05	2511 Dec 3	13.8	0.52612	78.71	17.79	-7.55
2639298.96	2514 Jan 20	11 07	+24 20	2514 Jan 16	9.7	0.63215	94.57	14.81	-4.06
2640063.89	2516 Feb 24	9 25	+13 50	2516 Feb 24	5.1	0.67715	101.30	13.82	-0.18
2640828.19	2518 Mar 29	16 40	- 0 28	2518 Apr 2	9.2	0.65607	98.15	14.27	+3.69
2641597.10	2520 May 6	14 18	-16 08	2520 May 13	19.4	0.57089	85.40	16.40	+7.21
2642379.48	2522 Jun 27	23 26	-27 20	2522 Jul 6	0.5	0.44130	66.02	21.21	+8.04
2643187.02	2524 Sep 12	12 30	- 9 27	2524 Sep 10	4.4	0.37362	55.89	25.05	-2.34
2643986.72	2526 Nov 21	5 17	+20 17	2526 Nov 12	22.8	0.47343	70.82	19.77	-8.27
2644763.07	2529 Jan 5	13 39	+26 19	2528 Dec 31	0.8	0.59688	89.29	15.68	-5.54
2645529.64	2531 Feb 11	3 27	+18 31	2531 Feb 9	10.6	0.66719	99.81	14.03	-1.70
2646293.64	2533 Mar 16	3 24	+ 5 18	2533 Mar 18	7.9	0.67217	100.56	13.93	+2.19
2647059.99	2535 Apr 21	11 44	-10 06	2535 Apr 27	10.2	0.61133	91.45	15.31	+5.93
2647835.50	2537 Jun 4	23 53	-24 31	2537 Jun 13	10.2	0.49395	73.89	18.95	+8.43
2648633.21	2539 Aug 11	17 08	-21 28	2539 Aug 14	23.5	0.38011	56.86	24.62	+3.27

OPPOSITIONS OF MARS

	Mars at opposition				*Mars nearest to Earth*				
JD	*date*	*ET*	δ		*date*	*ET*	*least distance*	*d*	Δ*t*
		h m	° ′			h	a.u.	10^6 km	
								"	d
2649442.07	2541 Oct 28	13 39	+11 26		2541 Oct 20	20.8	0.42251	63.21	22.15	-7.70
2650225.78	2543 Dec 21	6 47	+26 19		2543 Dec 14	9.3	0.55333	82.78	16.92	-6.89
2650995.04	2546 Jan 28	13 02	+22 27		2546 Jan 25	7.7	0.64732	96.84	14.46	-3.22
2651759.42	2548 Mar 2	22 10	+10 44		2548 Mar 3	14.0	0.67827	101.47	13.80	+0.66
2652524.24	2550 Apr 6	17 50	- 4 03		2550 Apr 11	5.8	0.64294	96.18	14.56	+4.50
2653295.07	2552 May 16	13 43	-19 33		2552 May 24	8.3	0.54505	81.54	17.17	+7.78
2654082.28	2554 Jul 12	18 38	-27 00		2554 Jul 19	21.6	0.41476	62.05	22.57	+7.12
2654892.77	2556 Sep 30	6 34	- 1 18		2556 Sep 25	8.7	0.38405	57.45	24.37	-4.91
2655686.38	2558 Dec 2	21 11	+23 31		2558 Nov 24	22.4	0.50292	75.24	18.61	-7.95
2656459.73	2561 Jan 14	5 37	+25 18		2561 Jan 9	12.5	0.61764	92.40	15.15	-4.71
2657225.29	2563 Feb 18	18 58	+15 44		2563 Feb 17	22.4	0.67402	100.83	13.89	-0.86
2657989.36	2565 Mar 23	20 39	+ 1 50		2565 Mar 26	21.0	0.66459	99.42	14.08	+3.01
2658756.98	2567 Apr 30	11 27	-13 43		2567 May 7	2.9	0.59013	88.28	15.86	+6.64
2659535.99	2569 Jun 17	11 43	-26 31		2569 Jun 25	21.2	0.46463	69.51	20.15	+8.40
2660339.60	2571 Aug 30	2 26	-15 08		2571 Aug 30	6.5	0.37236	55.70	25.14	+0.17
2661143.97	2573 Nov 11	11 20	+17 06		2573 Nov 3	5.2	0.44982	67.29	20.81	-8.26
2661923.28	2575 Dec 30	18 49	+26 34		2575 Dec 24	15.2	0.57825	86.51	16.19	-6.15
2662690.92	2578 Feb 5	9 59	+20 09		2578 Feb 3	0.8	0.65951	98.66	14.19	-2.38
2663455.00	2580 Mar 10	12 03	+ 7 28		2580 Mar 12	0.1	0.67613	101.15	13.84	+1.50
2664220.55	2582 Apr 15	1 16	- 7 41		2582 Apr 20	8.4	0.62658	93.73	14.94	+5.30
2664993.73	2584 May 27	5 29	-22 37		2584 Jun 4	10.4	0.51741	77.40	18.09	+8.21
2665786.54	2586 Jul 29	0 56	-24 39		2586 Aug 3	8.8	0.39288	58.77	23.82	+5.33
2666597.30	2588 Oct 16	19 06	+ 6 27		2588 Oct 10	0.1	0.40284	60.26	23.24	-6.79
2667385.16	2590 Dec 13	15 50	+25 32		2590 Dec 6	6.3	0.53146	79.50	17.61	-7.40
2668156.04	2593 Jan 22	13 00	+23 43		2593 Jan 18	15.4	0.63519	95.02	14.74	-3.90
2668920.84	2595 Feb 26	8 08	+12 45		2595 Feb 26	7.5	0.67762	101.37	13.81	-0.03
2669685.23	2597 Mar 31	17 35	- 1 42		2597 Apr 4	13.4	0.65396	97.83	14.31	+3.83
2670454.46	2599 May 9	23 06	-17 13		2599 May 17	5.9	0.56623	84.71	16.53	+7.28
2671237.66	2601 Jul 2	3 46	-27 21		2601 Jul 10	3.3	0.43607	65.24	21.46	+7.98
2672045.94	2603 Sep 18	10 38	- 7 22		2603 Sep 15	15.2	0.37483	56.07	24.97	-2.81
2672844.52	2605 Nov 25	0 25	+21 20		2605 Nov 16	19.5	0.47910	71.67	19.54	-8.21
2673620.28	2608 Jan 9	18 50	+26 01		2608 Jan 4	10.0	0.60119	89.94	15.57	-5.37
2674386.65	2610 Feb 14	3 38	+17 32		2610 Feb 12	14.4	0.66887	100.06	13.99	-1.55
2675150.63	2612 Mar 19	3 12	+ 4 05		2612 Mar 21	11.0	0.67104	100.39	13.95	+2.33
2675917.20	2614 Apr 24	16 50	-11 18		2614 Apr 30	17.7	0.60762	90.90	15.40	+6.03
2676693.35	2616 Jun 8	20 22	-25 09		2616 Jun 17	7.0	0.48848	73.08	19.16	+8.44
2677492.25	2618 Aug 16	17 57	-19 57		2618 Aug 19	9.7	0.37778	56.52	24.78	+2.66
2678300.40	2620 Nov 1	21 29	+13 08		2620 Oct 24	23.5	0.42731	63.92	21.90	-7.92
2679083.17	2622 Dec 24	16 05	+26 28		2622 Dec 17	22.5	0.55812	83.49	16.77	-6.73
2679852.08	2625 Jan 31	14 00	+21 40		2625 Jan 28	12.2	0.64983	97.21	14.40	-3.07
2680616.40	2627 Mar 6	21 37	+ 9 35		2627 Mar 7	17.0	0.67804	101.43	13.80	+0.81
2681381.33	2629 Apr 9	19 53	- 5 17		2629 Apr 14	11.0	0.64000	95.74	14.63	+4.63
2682152.52	2631 May 21	0 34	-20 29		2631 May 28	20.9	0.54003	80.79	17.33	+7.85
2682940.72	2633 Jul 17	5 13	-26 32		2633 Jul 24	1.7	0.41029	61.38	22.81	+6.85
2683751.52	2635 Oct 6	0 22	+ 0 47		2635 Sep 30	16.5	0.38703	57.90	24.18	-5.33
2684544.03	2637 Dec 6	12 37	+24 14		2637 Nov 28	16.9	0.50842	76.06	18.41	-7.82
2685316.87	2640 Jan 18	8 52	+24 48		2640 Jan 13	19.0	0.62108	92.91	15.07	-4.58
2686082.25	2642 Feb 21	17 56	+14 41		2642 Feb 21	0.6	0.67496	100.97	13.87	-0.72
2686846.37	2644 Mar 26	20 51	+ 0 37		2644 Mar 30	0.2	0.66291	99.17	14.12	+3.14
2687614.27	2646 May 3	18 34	-14 51		2646 May 10	12.1	0.58582	87.64	15.98	+6.73

	Mars at opposition				*Mars nearest to Earth*			
JD	*date*	*ET*	δ	*date*	*ET*	*least distance*	*d*	Δt
		h m	° '		h	a.u.	10^6 km	"	d
2688394.00	2648 Jun 21	12 04	-26 50	2648 Jun 29	22.1	0.45912	68.68	20.39	+8.42
2689198.60	2650 Sep 4	2 30	-13 13	2650 Sep 3	18.1	0.37199	55.65	25.16	-0.35
2690001.97	2652 Nov 15	11 12	+18 25	2652 Nov 7	5.0	0.45531	68.11	20.56	-8.26
2690780.57	2655 Jan 3	1 40	+26 27	2654 Dec 28	1.5	0.58303	87.22	16.05	-6.01
2691547.94	2657 Feb 8	10 39	+19 15	2657 Feb 6	4.6	0.66171	98.99	14.15	-2.25
2692311.98	2659 Mar 14	11 34	+ 6 16	2659 Mar 16	2.6	0.67547	101.05	13.86	+1.63
2693077.69	2661 Apr 18	4 38	- 8 53	2661 Apr 23	14.1	0.62338	93.26	15.01	+5.39
2693851.40	2663 May 31	21 34	-23 24	2663 Jun 9	4.3	0.51213	76.61	18.28	+8.28
2694645.39	2665 Aug 2	21 16	-23 34	2665 Aug 7	16.8	0.38928	58.24	24.04	+4.81
2695455.85	2667 Oct 22	8 19	+ 8 22	2667 Oct 15	5.1	0.40698	60.88	23.00	-7.13
2696242.64	2669 Dec 17	3 27	+25 54	2669 Dec 9	20.8	0.53647	80.26	17.45	-7.28
2697013.12	2672 Jan 26	14 57	+23 02	2672 Jan 22	20.4	0.63810	95.46	14.67	-3.77
2697777.81	2674 Mar 1	7 33	+11 38	2674 Mar 1	10.1	0.67791	101.41	13.81	+0.11
2698542.29	2676 Apr 3	18 58	- 2 56	2676 Apr 7	17.7	0.65142	97.45	14.37	+3.95
2699311.83	2678 May 13	8 02	-18 15	2678 May 20	17.5	0.56138	83.98	16.67	+7.39
2700095.90	2680 Jul 5	9 29	-27 14	2680 Jul 13	5.8	0.43102	64.48	21.72	+7.84
2700904.87	2682 Sep 22	8 54	- 5 15	2682 Sep 19	1.5	0.37639	56.31	24.87	-3.31
2701702.33	2684 Nov 27	20 01	+22 17	2684 Nov 19	17.1	0.48470	72.51	19.31	-8.12
2702477.49	2687 Jan 11	23 41	+25 39	2687 Jan 6	17.7	0.60516	90.53	15.47	-5.25
2703243.62	2689 Feb 16	2 58	+16 33	2689 Feb 14	16.8	0.67028	100.27	13.96	-1.43
2704007.62	2691 Mar 22	2 53	+ 2 52	2691 Mar 24	13.6	0.66984	100.21	13.97	+2.45
2704774.41	2693 Apr 26	21 56	-12 28	2693 May 3	1.4	0.60383	90.33	15.50	+6.14
2705551.17	2695 Jun 12	16 07	-25 42	2695 Jun 21	4.7	0.48291	72.24	19.38	+8.52
2706351.18	2697 Aug 20	16 17	-18 21	2697 Aug 22	20.1	0.37584	56.23	24.90	+2.16
2707158.57	2699 Nov 6	1 36	+14 42	2699 Oct 29	1.5	0.43237	64.68	21.65	-8.00
2707940.52	2701 Dec 28	0 26	+26 31	2701 Dec 21	9.6	0.56314	84.25	16.62	-6.62
2708709.14	2704 Feb 4	15 25	+20 51	2704 Feb 1	16.5	0.65250	97.61	14.34	-2.95
2709473.37	2706 Mar 9	20 58	+ 8 25	2706 Mar 10	19.3	0.67790	101.41	13.81	+0.93
2710238.42	2708 Apr 12	22 07	- 6 30	2708 Apr 17	15.9	0.63720	95.32	14.69	+4.74
2711010.08	2710 May 24	13 50	-21 25	2710 Jun 1	12.9	0.53494	80.03	17.50	+7.96
2711799.36	2712 Jul 21	20 37	-25 53	2712 Jul 28	7.9	0.40578	60.70	23.07	+6.47
2712610.33	2714 Oct 10	19 50	+ 2 54	2714 Oct 5	0.5	0.39014	58.36	23.99	-5.80
2713401.67	2716 Dec 10	4 01	+24 50	2716 Dec 2	10.2	0.51367	76.84	18.22	-7.74
2714174.00	2719 Jan 21	11 58	+24 14	2719 Jan 17	0.9	0.62448	93.42	14.99	-4.46
2714939.23	2721 Feb 24	17 30	+13 37	2721 Feb 24	3.3	0.67577	101.09	13.85	-0.59
2715703.41	2723 Mar 30	21 46	- 0 37	2723 Apr 3	4.2	0.66088	98.87	14.16	+3.27
2716471.55	2725 May 7	1 16	-15 56	2725 May 13	22.3	0.58138	86.97	16.10	+6.87
2717252.01	2727 Jun 26	12 19	-27 02	2727 Jul 4	21.2	0.45375	67.88	20.63	+8.37
2718057.60	2729 Sep 9	2 20	-11 14	2729 Sep 8	5.0	0.37198	55.65	25.16	-0.89
2718859.93	2731 Nov 20	10 14	+19 38	2731 Nov 12	5.0	0.46081	68.94	20.31	-8.22
2719637.83	2734 Jan 6	7 56	+26 14	2733 Dec 31	10.3	0.58739	87.87	15.93	-5.90
2720404.94	2736 Feb 12	10 36	+18 19	2736 Feb 10	7.7	0.66355	99.27	14.11	-2.12
2721168.94	2738 Mar 17	10 37	+ 5 05	2738 Mar 19	5.0	0.67476	100.94	13.87	+1.76
2721934.85	2740 Apr 21	8 18	-10 05	2740 Apr 26	21.0	0.62003	92.75	15.10	+5.53
2722709.09	2742 Jun 4	14 15	-24 07	2742 Jun 12	23.6	0.50660	75.79	18.48	+8.39
2723504.17	2744 Aug 7	16 05	-22 22	2744 Aug 12	1.8	0.38604	57.75	24.25	+4.40
2724314.26	2746 Oct 26	18 16	+10 11	2746 Oct 19	10.2	0.41136	61.54	22.75	-7.34
2725100.11	2748 Dec 20	14 33	+26 10	2748 Dec 13	10.1	0.54178	81.05	17.28	-7.19
2725870.22	2751 Jan 29	17 11	+22 18	2751 Jan 26	1.8	0.64131	95.94	14.60	-3.64
2726634.79	2753 Mar 4	7 05	+10 30	2753 Mar 4	13.0	0.67831	101.47	13.80	+0.25

OPPOSITIONS OF MARS

	Mars at opposition				Mars nearest to Earth				
JD	date	ET	δ	date	ET	least distance	10^6 km	d	Δt
		h m	° '		h	a.u.		"	d
2727399.35	2755 Apr 7	20 28	- 4 10	2755 Apr 11	22.5	0.64910	97.10	14.42	+4.09
2728169.25	2757 May 16	18 07	-19 15	2757 May 24	7.0	0.55671	83.28	16.81	+7.54
2728954.28	2759 Jul 10	18 46	-26 59	2759 Jul 18	8.6	0.42594	63.72	21.97	+7.58
2729763.87	2761 Sep 27	8 55	- 3 04	2761 Sep 23	10.9	0.37818	56.58	24.75	-3.92
2730560.10	2763 Dec 2	14 29	+23 09	2763 Nov 24	12.6	0.49007	73.31	19.10	-8.08
2731334.66	2766 Jan 15	3 47	+25 13	2766 Jan 10	0.6	0.60896	91.10	15.37	-5.13
2732100.62	2768 Feb 20	2 47	+15 32	2768 Feb 18	20.0	0.67157	100.47	13.94	-1.28
2732864.63	2770 Mar 25	3 01	+ 1 39	2770 Mar 27	17.4	0.66833	99.98	14.01	+2.60
2733631.63	2772 Apr 30	3 09	-13 36	2772 May 6	10.5	0.59972	89.72	15.61	+6.31
2734409.02	2774 Jun 16	12 31	-26 08	2774 Jun 25	1.4	0.47736	71.41	19.61	+8.54
2735210.13	2776 Aug 25	15 10	-16 38	2776 Aug 27	7.2	0.37434	56.00	25.00	+1.67
2736016.73	2778 Nov 10	5 38	+16 12	2778 Nov 2	4.9	0.43758	65.46	21.39	-8.03
2736797.89	2780 Dec 30	9 15	+26 29	2780 Dec 23	20.6	0.56796	84.97	16.48	-6.53
2737566.17	2783 Feb 6	16 09	+19 59	2783 Feb 3	20.9	0.65483	97.96	14.29	-2.80
2738330.33	2785 Mar 11	19 54	+ 7 15	2785 Mar 12	22.1	0.67767	101.38	13.81	+1.09
2739095.53	2787 Apr 16	0 50	- 7 43	2787 Apr 20	22.4	0.63435	94.90	14.76	+4.90
2739867.63	2789 May 27	3 03	-22 16	2789 Jun 4	5.4	0.52973	79.25	17.67	+8.10
2740657.93	2791 Jul 26	10 17	-25 06	2791 Aug 1	13.9	0.40154	60.07	23.31	+6.15
2741468.97	2793 Oct 14	11 17	+ 4 54	2793 Oct 8	7.2	0.39356	58.88	23.78	-6.17
2742259.21	2795 Dec 13	17 09	+25 19	2795 Dec 6	0.9	0.51914	77.66	18.03	-7.68
2743031.12	2798 Jan 23	14 46	+23 36	2798 Jan 19	7.3	0.62813	93.97	14.90	-4.31
2743796.22	2800 Feb 27	17 11	+12 31	2800 Feb 27	6.9	0.67668	101.23	13.83	-0.43
2744560.43	2802 Apr 1	22 13	- 1 51	2802 Apr 5	8.7	0.65903	98.59	14.20	+3.44
2745328.88	2804 May 9	9 02	-17 01	2804 May 16	9.9	0.57707	86.33	16.22	+7.04
2746110.19	2806 Jun 29	16 38	-27 07	2806 Jul 7	21.5	0.44835	67.07	20.88	+8.20
2746916.71	2808 Sep 13	4 58	- 9 08	2808 Sep 11	15.8	0.37228	55.69	25.14	-1.55
2747717.89	2810 Nov 23	9 24	+20 46	2810 Nov 15	3.8	0.46620	69.74	20.08	-8.23
2748495.09	2813 Jan 8	14 14	+25 57	2813 Jan 2	19.4	0.59164	88.51	15.82	-5.78
2749261.95	2815 Feb 14	10 51	+17 21	2815 Feb 12	11.9	0.66532	99.53	14.07	-1.96
2750025.94	2817 Mar 19	10 34	+ 3 52	2817 Mar 21	9.0	0.67377	100.79	13.89	+1.94
2750792.03	2819 Apr 24	12 40	-11 15	2819 Apr 30	5.7	0.61630	92.20	15.19	+5.71
2751566.77	2821 Jun 7	6 36	-24 45	2821 Jun 15	17.6	0.50111	74.97	18.68	+8.46
2752362.97	2823 Aug 12	11 20	-21 02	2823 Aug 16	11.5	0.38320	57.33	24.43	+4.01
2753172.65	2825 Oct 30	3 29	+11 56	2825 Oct 22	16.0	0.41605	62.24	22.50	-7.48
2753957.55	2827 Dec 24	1 12	+26 20	2827 Dec 16	22.7	0.54696	81.82	17.11	-7.11
2754727.27	2830 Jan 31	18 33	+21 32	2830 Jan 28	7.3	0.64410	96.36	14.53	-3.47
2755491.75	2832 Mar 6	5 56	+ 9 21	2832 Mar 6	16.1	0.67853	101.51	13.79	+0.42
2756256.41	2834 Apr 9	21 45	- 5 22	2834 Apr 14	4.1	0.64670	96.75	14.47	+4.26
2757026.70	2836 May 19	4 41	-20 12	2836 May 26	21.1	0.55181	82.55	16.96	+7.68
2757812.66	2838 Jul 14	3 45	-26 37	2838 Jul 21	12.4	0.42098	62.98	22.23	+7.36
2758622.71	2840 Oct 1	4 59	- 0 57	2840 Sep 26	19.0	0.38033	56.90	24.61	-4.41
2759417.80	2842 Dec 5	7 09	+23 53	2842 Nov 27	5.7	0.49561	74.14	18.89	-8.06
2760191.84	2845 Jan 17	8 11	+24 43	2845 Jan 12	9.1	0.61307	91.71	15.27	-4.96
2760957.62	2847 Feb 22	2 51	+14 28	2847 Feb 21	0.6	0.67297	100.67	13.91	-1.09
2761721.63	2849 Mar 27	3 12	+ 0 25	2849 Mar 29	22.1	0.66693	99.77	14.03	+2.79
2762488.89	2851 May 3	9 26	-14 43	2851 May 9	21.0	0.59579	89.13	15.71	+6.48
2763266.98	2853 Jun 19	11 35	-26 29	2853 Jun 27	22.6	0.47190	70.60	19.83	+8.46
2764069.24	2855 Aug 30	17 52	-14 47	2855 Aug 31	18.4	0.37308	55.81	25.09	+1.02
2764874.92	2857 Nov 13	9 58	+17 37	2857 Nov 5	6.7	0.44283	66.25	21.14	-8.13
2765655.22	2860 Jan 2	17 11	+26 22	2859 Dec 27	7.6	0.57259	85.66	16.35	-6.40

	Mars at opposition				Mars nearest to Earth					
JD	date	ET		δ	date	ET	least distance	10^6 km	d	Δt
		h m		° '		h	a.u.		"	d
2766423.20	2862 Feb 8	16 42		+19 05	2862 Feb 6	1.9	0.65703	98.29	14.25	-2.62
2767187.32	2864 Mar 13	19 34		+ 6 03	2864 Mar 15	2.2	0.67718	101.30	13.82	+1.28
2767952.66	2866 Apr 18	3 48		- 8 55	2866 Apr 23	6.0	0.63103	94.40	14.83	+5.09
2768725.18	2868 May 29	16 16		-23 02	2868 Jun 6	20.7	0.52436	78.44	17.85	+8.18
2769516.54	2870 Jul 30	1 00		-24 11	2870 Aug 4	22.2	0.39761	59.48	23.54	+5.88
2770327.57	2872 Oct 18	1 36		+ 6 52	2872 Oct 11	15.5	0.39733	59.44	23.56	-6.42
2771116.79	2874 Dec 16	6 54		+25 43	2874 Dec 8	16.3	0.52459	78.48	17.84	-7.61
2771888.24	2877 Jan 25	17 44		+22 55	2877 Jan 21	14.7	0.63141	94.46	14.82	-4.13
2772653.18	2879 Mar 1	16 18		+11 24	2879 Mar 1	10.6	0.67729	101.32	13.82	-0.24
2773417.45	2881 Apr 3	22 54		- 3 04	2881 Apr 7	13.9	0.65705	98.29	14.25	+3.62
2774186.24	2883 May 12	17 49		-18 03	2883 May 19	22.1	0.57253	85.65	16.35	+7.18
2774968.35	2885 Jul 2	20 24		-27 05	2885 Jul 10	22.2	0.44295	66.26	21.13	+8.08
2775775.68	2887 Sep 18	4 19		- 7 02	2887 Sep 16	0.5	0.37290	55.78	25.10	-2.16
2776575.75	2889 Nov 26	6 02		+21 47	2889 Nov 17	23.6	0.47175	70.57	19.84	-8.27
2777352.32	2892 Jan 11	19 38		+25 35	2892 Jan 6	5.2	0.59611	89.18	15.70	-5.60
2778118.96	2894 Feb 16	11 08		+16 21	2894 Feb 14	17.0	0.66718	99.81	14.03	-1.76
2778882.93	2896 Mar 21	10 23		+ 2 40	2896 Mar 23	13.5	0.67284	100.66	13.91	+2.13
2779649.21	2898 Apr 26	16 56		-12 24	2898 May 2	14.2	0.61276	91.67	15.28	+5.88
2780424.55	2900 Jun 11	1 12		-25 19	2900 Jun 19	11.5	0.49573	74.16	18.88	+8.43
2781221.98	2902 Aug 17	11 34		-19 32	2902 Aug 20	22.6	0.38055	56.93	24.60	+3.46
2782031.05	2904 Nov 3	13 10		+13 38	2904 Oct 26	21.3	0.42080	62.95	22.24	-7.66
2782815.00	2906 Dec 27	11 56		+26 25	2906 Dec 20	12.6	0.55191	82.57	16.96	-6.97
2783584.34	2909 Feb 3	20 14		+20 42	2909 Jan 31	13.5	0.64675	96.75	14.47	-3.28
2784348.74	2911 Mar 10	5 39		+ 8 11	2911 Mar 10	20.4	0.67846	101.50	13.80	+0.61
2785113.51	2913 Apr 13	0 10		- 6 35	2913 Apr 17	11.1	0.64379	96.31	14.54	+4.46
2785884.16	2915 May 23	15 52		-21 06	2915 May 31	10.4	0.54664	81.78	17.12	+7.77
2786671.04	2917 Jul 18	12 53		-26 06	2917 Jul 25	17.9	0.41620	62.26	22.49	+7.21
2787481.52	2919 Oct 7	0 26		+ 1 09	2919 Oct 2	5.0	0.38285	57.27	24.45	-4.81
2788275.50	2921 Dec 9	0 04		+24 32	2921 Nov 30	23.9	0.50129	74.99	18.67	-8.01
2789049.01	2924 Jan 21	12 08		+24 08	2924 Jan 16	17.5	0.61677	92.27	15.18	-4.78
2789814.59	2926 Feb 25	2 09		+13 23	2926 Feb 24	4.4	0.67398	100.83	13.89	-0.91
2790578.63	2928 Mar 30	3 06		- 0 47	2928 Apr 2	2.3	0.66542	99.55	14.07	+2.97
2791346.16	2930 May 6	15 49		-15 48	2930 May 13	6.5	0.59167	88.51	15.82	+6.61
2792124.94	2932 Jun 23	10 34		-26 44	2932 Jul 1	20.2	0.46633	69.76	20.07	+8.40
2792928.25	2934 Sep 4	17 53		-12 52	2934 Sep 5	3.2	0.37215	55.67	25.15	+0.39
2793732.95	2936 Nov 17	10 54		+18 54	2936 Nov 9	5.3	0.44808	67.03	20.89	-8.23
2794512.52	2939 Jan 6	0 28		+26 10	2938 Dec 30	19.4	0.57733	86.37	16.21	-6.21
2795280.24	2941 Feb 11	17 44		+18 08	2941 Feb 9	7.6	0.65938	98.64	14.20	-2.42
2796044.31	2943 Mar 17	19 25		+ 4 51	2943 Mar 19	6.5	0.67667	101.23	13.83	+1.46
2796809.80	2945 Apr 21	7 05		-10 06	2945 Apr 26	13.2	0.62785	93.93	14.91	+5.26
2797582.82	2947 Jun 3	7 38		-23 45	2947 Jun 11	12.3	0.51918	77.67	18.03	+8.20
2798375.35	2949 Aug 3	20 19		-23 06	2949 Aug 9	7.9	0.39379	58.91	23.77	+5.48
2799186.23	2951 Oct 23	17 26		+ 8 49	2951 Oct 16	24.0	0.40130	60.03	23.32	-6.73
2799974.35	2953 Dec 19	20 25		+26 01	2953 Dec 12	8.9	0.52991	79.27	17.66	-7.48
2800745.34	2956 Jan 29	20 04		+22 11	2956 Jan 25	21.4	0.63448	94.92	14.75	-3.94
2801510.16	2958 Mar 4	15 55		+10 16	2958 Mar 4	14.4	0.67768	101.38	13.81	-0.06
2802274.51	2960 Apr 7	0 21		- 4 17	2960 Apr 10	19.4	0.65466	97.94	14.30	+3.79
2803043.60	2962 May 16	2 26		-19 01	2962 May 23	9.1	0.56765	84.92	16.49	+7.28
2803826.50	2964 Jul 7	0 04		-26 55	2964 Jul 15	0.7	0.43767	65.47	21.39	+8.02
2804634.62	2966 Sep 23	2 51		- 4 56	2966 Sep 20	11.1	0.37402	55.95	25.03	-2.65

	Mars at opposition				*Mars nearest to Earth*				
JD	date	ET	δ	date	ET	least distance		d	Δt
		h m	° ′		h	a.u.	10⁶ km	″	d
2805433.60	2968 Nov 30	2 21	+22 41	2968 Nov 21	20.2	0.47741	71.42	19.61	-8.25
2806209.55	2971 Jan 15	1 16	+25 09	2971 Jan 9	15.2	0.60022	89.79	15.59	-5.42
2806975.96	2973 Feb 19	11 00	+15 19	2973 Feb 17	21.0	0.66866	100.03	14.00	-1.58
2807739.92	2975 Mar 25	9 58	+ 1 27	2975 Mar 27	16.9	0.67171	100.49	13.93	+2.29
2808506.42	2977 Apr 29	21 59	-13 32	2977 May 5	22.1	0.60903	91.11	15.37	+6.00
2809282.36	2979 Jun 14	20 31	-25 47	2979 Jun 23	6.9	0.49022	73.34	19.09	+8.43
2810080.91	2981 Aug 21	9 50	-17 56	2981 Aug 24	6.9	0.37808	56.56	24.76	+2.88
2810889.32	2983 Nov 7	19 36	+15 12	2983 Oct 30	22.7	0.42556	63.66	21.99	-7.87
2811672.37	2985 Dec 29	20 58	+26 24	2985 Dec 23	2.0	0.55698	83.32	16.81	-6.79
2812441.40	2988 Feb 6	21 35	+19 50	2988 Feb 3	19.1	0.64956	97.17	14.41	-3.10
2813205.72	2990 Mar 12	5 22	+ 7 00	2990 Mar 12	24.0	0.67844	101.49	13.80	+0.78
2813970.59	2992 Apr 15	2 06	- 7 47	2992 Apr 19	16.4	0.64112	95.91	14.60	+4.60
2814741.65	2994 May 26	3 40	-21 57	2994 Jun 2	23.4	0.54176	81.05	17.28	+7.82
2815529.59	2996 Jul 22	2 10	-25 25	2996 Jul 29	1.2	0.41157	61.57	22.74	+6.96
2816340.39	2998 Oct 10	21 20	+ 3 16	2998 Oct 5	16.0	0.38577	57.71	24.26	-5.22
2817133.20	3000 Dec 12	16 41	+25 04	3000 Dec 4	19.1	0.50681	75.82	18.47	-7.90
2817906.16	3003 Jan 24	15 51	+23 30	3003 Jan 20	1.1	0.62025	92.79	15.09	-4.61
2818671.58	3005 Feb 28	1 51	+12 17	3005 Feb 27	7.9	0.67488	100.96	13.87	-0.75
2819435.67	3007 Apr 3	4 04	- 2 01	3007 Apr 6	6.8	0.66351	99.26	14.11	+3.12
2820203.46	3009 May 9	23 05	-16 51	3009 May 16	16.2	0.58710	87.83	15.94	+6.71

3

EQUINOXES AND SOLSTICES

1 – 3000

EQUINOXES AND SOLSTICES

The list on the next pages gives the times of the equinoxes and solstices on the planet Earth for the years 1 to 3000. These are the instants when the apparent longitude of the Sun's center (that is, calculated by including the effects of aberration and nutation) is a multiple of 90 degrees.

The instants are given to the nearest second of time, and they are expressed in *Ephemeris Time*. For each equinox and solstice, the day of the month is given first, followed by the hours, minutes, and seconds. The tabulated times may be converted into Universal Time by subtracting the quantity $\Delta T = ET - UT$, as explained in the Introduction.

The times have been calculated by Jean Meeus on a HP-85 microcomputer. For the computation of the Sun's longitude, use was made of Newcomb's theory as given in *Astronomical Papers American Ephemeris*, Vol. VI, Part 1 (Washington, 1895). All the terms in longitude given by Newcomb have been used.

The table below, due to J. Meeus and first published in *Hemel en Dampkring* (Netherlands), Vol. 60, page 171 (1962), gives the duration of the four astronomical seasons for some epochs. The winter is the shortest astronomical season since the year +1245. Its duration is still decreasing, and the minimum value (88.71 days) will be reached by about the year +3500. It appears that the first winter (since several thousands of years) having a duration less than 89 days has been that of 1971-1972.

Duration of the astronomical seasons, in days

Year	Spring	Summer	Autumn	Winter
-3000	94.05	89.93	88.61	92.65
-2000	94.29	90.77	88.39	91.80
-1000	94.25	91.64	88.42	90.93
0	93.97	92.45	88.69	90.14
+1000	93.44	93.15	89.18	89.47
2000	92.76	93.65	89.84	88.99
3000	91.97	93.92	90.61	88.74
4000	91.17	93.93	91.40	88.74
5000	90.43	93.69	92.16	88.96
6000	89.82	93.25	92.80	89.39

After four years, the equinoxes and the solstices occur approximately 45 minutes earlier. However, there is a jump of +1 day when the 4-year period contains a non-bissextile century year. For example :

1969 March 20	19^h09^m ET		1895 March 20	20^h49^m ET	
1973 March 20	18 13		1899 March 20	19 46	
1977 March 20	17 43		1903 March 21	19 15	

After 400 years, in the Gregorian calendar, the equinoxes and the solstices occur approximately 3 hours earlier. This is a *mean* value, however. After a lapse of time of four centuries the eccentricity and the longitude of perihelion of the Earth's orbit have sensibly changed, influencing the duration of the seasons. If we compare the times for the years 1600 and 2000, we see that in the year 2000 the June solstice will occur 8 hours earlier than in 1600, but the December solstice 3 hours *later*.

In 2044 the *March equinox* will take place on March 19, for the first time since 1796. It did not occur on March 21 from 1616 to 1701, from 1744 to 1800, and from 1876 to 1899. It will not occur on March 21 from 2008 to 2101.

A *June solstice* on June 20 occurred for the last time in 1896, and will take place for the next time in 2012 (if we use ET, but in 2008 if the dates are based on UT). It occurred for the last time on June 22 in 1975, and will again take place on that date in 2203, 2207, 2211 and 2215, and then not before 2302. The June solstice will take place on June 19 in 2488, for the first time since the Gregorian calendar reform in A.D. 1582.

In 1968, for the first time since 1897, the *September equinox* took place on September 22. Since the Gregorian calendar reform, the September equinox occurred on September 24 only in 1803, 1807, 1903, 1907, 1911, 1915, 1919, 1923, 1927 and 1931. The next time this will occur in 2303, but then not again before many centuries. (In 2307, if $\Delta T > 10$ minutes, the equinox will take place on September 23 *Universal Time*). The September equinox will occur on September 21 in 2092, for the first time since the Gregorian calendar reform. It will take place on that date again in 2096, but then not again before 2464.

The *December solstice* will occur on December 20 in A.D. 2080, for the first time since 1697. It will again occur on that date in 2084, 2088, 2092 and 2096, but then not again before 2492 (for dates based on ET) or 2488 (for dates based on UT). Since the Gregorian calendar reform, the December solstice occurred only once on December 23, namely in 1903. It will again take place on that date in 2303, 2307, 2311 and 2315.

Of course, these statements are not valid if the instants are expressed in standard times of zones other than that of Greenwich.

Year	March Equinox	June Solstice	September Equinox	December Solstice
1	23^d $0^h30^m57^s$	24^d $23^h41^m56^s$	25^d $10^h26^m11^s$	23^d $3^h04^m27^s$
2	23 6 27 11	25 5 31 58	25 16 17 26	23 8 55 06
3	23 12 05 18	25 11 10 13	25 22 10 50	23 14 50 02
4	22 17 55 40	24 17 08 45	25 4 04 32	22 20 45 14
5	22 23 53 20	24 23 01 44	25 9 51 47	23 2 38 06
6	23 5 44 43	25 4 42 02	25 15 39 49	23 8 32 33
7	23 11 41 45	25 10 37 28	25 21 35 54	23 14 19 45
8	22 17 29 40	24 16 27 48	25 3 24 59	22 20 07 17
9	22 23 19 09	24 22 20 46	25 9 12 02	23 2 00 13
10	23 5 14 45	25 4 14 25	25 15 01 15	23 7 49 58
11	23 10 52 19	25 9 52 18	25 20 50 44	23 13 35 59
12	22 16 37 42	24 15 46 35	25 2 43 22	22 19 22 04
13	22 22 30 24	24 21 35 38	25 8 31 33	23 1 09 40
14	23 4 12 15	25 3 07 16	25 14 14 39	23 6 59 24
15	23 9 56 50	25 8 55 30	25 20 06 38	23 12 49 15
16	22 15 40 34	24 14 41 10	25 1 50 17	22 18 37 18
17	22 21 27 33	24 20 26 29	25 7 32 11	23 0 28 36
18	23 3 25 56	25 2 18 02	25 13 22 52	23 6 18 41
19	23 9 09 33	25 7 55 45	25 19 11 50	23 12 04 31
20	22 14 58 14	24 13 52 47	25 1 04 40	22 17 54 51
21	22 20 55 12	24 19 49 40	25 6 53 24	22 23 47 24
22	23 2 42 13	25 1 28 53	25 12 37 32	23 5 38 37
23	23 8 31 24	25 7 24 59	25 18 35 35	23 11 27 56
24	22 14 20 35	24 13 17 58	25 0 27 22	22 17 15 11
25	22 20 10 10	24 19 05 38	25 6 16 13	22 23 06 54
26	23 2 05 20	25 0 59 39	25 12 13 34	23 5 02 48
27	23 7 46 48	25 6 38 43	25 18 03 23	23 10 54 07
28	22 13 32 46	24 12 32 26	24 23 54 48	22 16 47 26
29	22 19 29 28	24 18 25 49	25 5 42 22	22 22 40 05
30	23 1 18 57	24 23 58 33	25 11 23 22	23 4 27 26
31	23 7 07 47	25 5 48 31	25 17 17 18	23 10 12 56
32	22 12 55 00	24 11 39 10	24 23 02 12	22 15 58 57
33	22 18 41 55	24 17 23 39	25 4 41 08	22 21 45 53
34	23 0 31 30	24 23 15 55	25 10 32 22	23 3 36 32
35	23 6 12 26	25 4 55 55	25 16 19 25	23 9 19 51
36	22 11 56 07	24 10 46 32	24 22 10 03	22 15 04 34
37	22 17 49 29	24 16 40 54	25 4 03 26	22 20 57 41
38	22 23 36 27	24 22 14 49	25 9 46 43	23 2 47 13
39	23 5 20 00	25 4 05 24	25 15 43 28	23 8 40 20
40	22 11 09 22	24 10 00 18	24 21 33 23	22 14 34 59
41	22 17 02 33	24 15 45 17	25 3 15 32	22 20 25 03
42	22 22 59 01	24 21 39 26	25 9 13 48	23 2 20 30
43	23 4 49 07	25 3 24 56	25 15 06 03	23 8 09 17
44	22 10 37 29	24 9 18 19	24 20 55 26	22 13 57 44
45	22 16 32 33	24 15 17 34	25 2 49 17	22 19 55 51
46	22 22 22 37	24 20 54 46	25 8 29 23	23 1 41 37
47	23 4 05 23	25 2 43 58	25 14 23 48	23 7 27 51
48	22 9 54 15	24 8 38 35	24 20 15 44	22 13 16 25
49	22 15 43 05	24 14 18 36	25 1 55 40	22 19 00 39
50	22 21 29 42	24 20 07 21	25 7 51 14	23 0 55 32

EQUINOXES AND SOLSTICES

Year	March Equinox		June Solstice		September Equinox		December Solstice	
	23^d	$3^h11^m45^s$	25^d	$1^h48^m51^s$	25^d	$13^h38^m34^s$	23^d	$6^h45^m24^s$
51	23	3 11 45	25	1 48 51	25	13 38 34	23	6 45 24
52	22	8 55 54	24	7 35 38	24	19 21 29	22	12 31 39
53	22	14 49 06	24	13 29 23	25	1 13 00	22	18 26 44
54	22	20 42 18	24	19 04 36	25	6 53 13	23	0 10 27
55	23	2 27 26	25	0 52 10	25	12 46 12	23	5 55 27
56	22	8 15 26	24	6 49 52	24	18 37 57	22	11 47 55
57	22	14 06 52	24	12 34 31	25	0 15 55	22	17 33 43
58	22	19 54 16	24	18 26 55	25	6 12 44	22	23 27 50
59	23	1 39 45	25	0 14 58	25	12 06 11	23	5 17 30
60	22	7 28 30	24	6 04 43	24	17 55 54	22	11 02 40
61	22	13 20 40	24	12 00 19	24	23 54 37	22	17 01 53
62	22	19 12 55	24	17 38 54	25	5 39 18	22	22 54 44
63	23	0 58 12	24	23 26 18	25	11 31 57	23	4 45 04
64	22	6 46 35	24	5 23 57	24	17 24 55	22	10 42 41
65	22	12 44 58	24	11 08 08	24	23 02 56	22	16 27 48
66	22	18 35 57	24	16 56 21	25	4 58 17	22	22 18 48
67	23	0 23 08	24	22 44 57	25	10 50 27	23	4 08 44
68	22	6 11 33	24	4 33 21	24	16 31 46	22	9 49 50
69	22	11 58 58	24	10 27 55	24	22 24 20	22	15 45 28
70	22	17 48 46	24	16 07 19	25	4 05 00	22	21 30 32
71	22	23 29 58	24	21 51 00	25	9 54 11	23	3 09 54
72	22	5 12 50	24	3 45 38	24	15 49 53	22	9 01 58
73	22	11 05 31	24	9 27 12	24	21 29 01	22	14 45 44
74	22	16 48 18	24	15 11 11	25	3 23 08	22	20 39 39
75	22	22 30 06	24	20 59 16	25	9 17 19	23	2 36 59
76	22	4 20 30	24	2 46 28	24	14 58 05	22	8 21 26
77	22	10 12 35	24	8 39 07	24	20 54 09	22	14 21 24
78	22	16 11 04	24	14 21 18	25	2 40 08	22	20 11 05
79	22	22 00 04	24	20 08 13	25	8 29 41	23	1 54 53
80	22	3 48 14	24	2 09 32	24	14 26 42	22	7 53 14
81	22	9 46 43	24	7 59 05	24	20 05 44	22	13 39 57
82	22	15 34 15	24	13 47 49	25	1 59 43	22	19 30 08
83	22	21 17 48	24	19 39 44	25	7 58 02	23	1 23 06
84	22	3 10 11	24	1 28 59	24	13 42 35	22	7 03 16
85	22	8 57 17	24	7 18 53	24	19 38 54	22	13 00 33
86	22	14 46 42	24	12 59 31	25	1 24 58	22	18 53 47
87	22	20 30 45	24	18 41 56	25	7 09 41	23	0 36 40
88	22	2 12 52	24	0 35 39	24	13 02 25	22	6 32 28
89	22	8 11 20	24	6 19 51	24	18 39 28	22	12 16 15
90	22	13 59 17	24	12 01 38	25	0 29 21	22	18 01 07
91	22	19 39 23	24	17 50 46	25	6 24 21	22	23 54 19
92	22	1 31 14	23	23 40 58	24	12 04 17	22	5 35 46
93	22	7 16 36	24	5 29 54	24	17 54 52	22	11 30 00
94	22	13 05 01	24	11 13 13	24	23 42 08	22	17 22 23
95	22	18 54 17	24	16 58 59	25	5 30 04	22	23 00 52
96	22	0 36 59	23	22 53 45	24	11 28 53	22	4 57 03
97	22	6 36 13	24	4 43 17	24	17 13 32	22	10 48 21
98	22	12 24 27	24	10 27 36	24	23 05 35	22	16 39 55
99	22	18 05 12	24	16 20 16	25	5 04 57	22	22 41 47
100	22	0 03 54	23	22 14 11	24	10 47 47	22	4 27 12

Year	March Equinox	June Solstice	September Equinox	December Solstice
	22^d $5^h56^m52^s$	24^d $4^h03^m19^s$	24^d $16^h41^m18^s$	22^d $10^h22^m03^s$
101	22 5 56 52	24 4 03 19	24 16 41 18	22 10 22 03
102	22 11 50 28	24 9 48 37	24 22 32 57	22 16 16 00
103	22 17 42 21	24 15 36 23	25 4 18 32	22 21 54 50
104	21 23 23 20	23 21 30 34	24 10 11 36	22 3 49 40
105	22 5 18 48	24 3 20 03	24 15 51 43	22 9 37 17
106	22 11 04 20	24 9 01 26	24 21 37 49	22 15 18 26
107	22 16 39 18	24 14 48 51	25 3 36 04	22 21 10 53
108	21 22 31 35	23 20 38 00	24 9 18 07	22 2 50 30
109	22 4 15 50	24 2 19 51	24 15 07 43	22 8 44 07
110	22 9 59 49	24 8 00 00	24 20 57 05	22 14 41 06
111	22 15 48 26	24 13 44 52	25 2 39 19	22 20 22 56
112	21 21 31 46	23 19 35 39	24 8 31 02	22 2 18 17
113	22 3 32 17	24 1 24 57	24 14 15 00	22 8 08 50
114	22 9 27 04	24 7 10 01	24 20 03 53	22 13 52 39
115	22 15 07 00	24 13 02 16	25 2 03 03	22 19 50 03
116	21 21 05 04	23 19 01 39	24 7 48 19	22 1 37 01
117	22 2 56 11	24 0 51 35	24 13 38 52	22 7 29 53
118	22 8 42 33	24 6 38 27	24 19 34 19	22 13 26 13
119	22 14 36 34	24 12 29 49	25 1 24 11	22 19 06 15
120	21 20 19 19	23 18 20 17	24 7 20 01	22 0 59 38
121	22 2 13 34	24 0 09 07	24 13 07 23	22 6 54 48
122	22 8 03 37	24 5 52 01	24 18 54 23	22 12 41 03
123	22 13 37 37	24 11 37 25	25 0 48 08	22 18 37 47
124	21 19 33 44	23 17 30 48	24 6 30 58	22 0 24 11
125	22 1 26 35	23 23 12 26	24 12 15 43	22 6 09 46
126	22 7 09 40	24 4 52 02	24 18 06 01	22 12 02 54
127	22 13 02 43	24 10 42 39	24 23 49 56	22 17 41 51
128	21 18 42 50	23 16 30 41	24 5 36 44	21 23 31 12
129	22 0 33 55	23 22 20 10	24 11 21 13	22 5 23 15
130	22 6 25 06	24 4 05 11	24 17 07 42	22 11 03 20
131	22 11 59 44	24 9 51 06	24 23 05 05	22 16 55 36
132	21 17 54 38	23 15 47 29	24 4 54 58	21 22 44 16
133	21 23 46 51	23 21 33 17	24 10 44 30	22 4 36 45
134	22 5 28 35	24 3 15 38	24 16 39 26	22 10 38 31
135	22 11 23 55	24 9 10 36	24 22 28 38	22 16 26 12
136	21 17 12 25	23 15 00 24	24 4 19 03	21 22 18 49
137	21 23 10 12	23 20 50 40	24 10 09 29	22 4 14 13
138	22 5 08 18	24 2 39 27	24 15 58 19	22 9 58 02
139	22 10 46 01	24 8 26 45	24 21 52 50	22 15 52 10
140	21 16 39 19	23 14 24 26	24 3 38 48	21 21 40 42
141	21 22 31 39	23 20 10 42	24 9 22 58	22 3 27 57
142	22 4 11 31	24 1 49 29	24 15 14 06	22 9 19 11
143	22 10 02 11	24 7 41 01	24 21 03 09	22 14 59 29
144	21 15 45 44	23 13 26 08	24 2 50 55	21 20 46 06
145	21 21 32 19	23 19 09 32	24 8 37 30	22 2 41 08
146	22 3 23 08	24 0 55 28	24 14 23 23	22 8 27 20
147	22 8 58 04	24 6 37 26	24 20 12 18	22 14 19 19
148	21 14 51 33	23 12 32 16	24 1 59 39	21 20 08 55
149	21 20 50 34	23 18 19 20	24 7 46 08	22 1 55 42
150	22 2 32 55	23 23 58 47	24 13 37 56	22 7 49 13

EQUINOXES AND SOLSTICES

Year	March Equinox	June Solstice	September Equinox	December Solstice
151	22^d $8^h26^m17^s$	24^d $5^h56^m50^s$	24^d $19^h29^m27^s$	22^d $13^h36^m26^s$
152	21 14 14 35	23 11 48 48	24 1 17 10	21 19 25 31
153	21 20 03 59	23 17 37 18	24 7 06 16	22 1 21 09
154	22 1 59 42	23 23 30 06	24 12 59 57	22 7 07 12
155	22 7 38 24	24 5 14 34	24 18 54 21	22 12 56 38
156	21 13 29 42	23 11 10 11	24 0 48 07	21 18 51 19
157	21 19 27 07	23 16 58 34	24 6 36 13	22 0 43 35
158	22 1 05 40	23 22 34 16	24 12 24 52	22 6 40 11
159	22 6 58 14	24 4 29 59	24 18 15 24	22 12 29 14
160	21 12 50 18	23 10 17 54	23 23 59 58	21 18 14 39
161	21 18 41 20	23 16 00 27	24 5 45 41	22 0 05 44
162	22 0 35 41	23 21 51 04	24 11 34 59	22 5 49 55
163	22 6 12 52	24 3 34 42	24 17 21 25	22 11 36 00
164	21 11 59 16	23 9 28 38	23 23 08 12	21 17 25 15
165	21 17 53 32	23 15 18 25	24 4 53 48	21 23 12 07
166	21 23 31 07	23 20 52 38	24 10 42 02	22 4 59 51
167	22 5 18 44	24 2 46 37	24 16 37 00	22 10 45 50
168	21 11 08 09	23 8 35 32	23 22 25 15	21 16 34 04
169	21 16 54 21	23 14 16 22	24 4 12 16	21 22 30 43
170	21 22 45 54	23 20 08 20	24 10 05 00	22 4 23 44
171	22 4 28 50	24 1 53 53	24 15 54 17	22 10 15 31
172	21 10 22 09	23 7 46 58	23 21 44 31	21 16 07 27
173	21 16 24 44	23 13 40 13	24 3 35 31	21 22 00 18
174	21 22 11 03	23 19 18 50	24 9 24 22	22 3 51 14
175	22 4 00 59	24 1 17 18	24 15 18 31	22 9 41 30
176	21 9 55 00	23 7 13 13	23 21 05 31	21 15 30 16
177	21 15 42 54	23 12 54 51	24 2 49 19	21 21 20 06
178	21 21 33 24	23 18 47 45	24 8 44 39	22 3 06 46
179	22 3 14 51	24 0 32 21	24 14 34 02	22 8 50 52
180	21 8 59 50	23 6 20 43	23 20 22 43	21 14 40 45
181	21 14 53 00	23 12 10 48	24 2 12 04	21 20 35 08
182	21 20 31 34	23 17 43 17	24 7 55 22	22 2 24 36
183	22 2 16 52	23 23 34 42	24 13 46 10	22 8 13 17
184	21 8 11 56	23 5 26 08	23 19 32 42	21 13 59 58
185	21 14 01 46	23 11 02 48	24 1 13 52	21 19 47 26
186	21 19 51 35	23 16 55 27	24 7 08 48	22 1 37 18
187	22 1 34 38	23 22 43 44	24 12 55 05	22 7 23 33
188	21 7 20 47	23 4 34 09	23 18 41 25	21 13 14 20
189	21 13 17 27	23 10 29 37	24 0 34 38	21 19 08 11
190	21 19 01 14	23 16 06 37	24 6 23 13	22 0 55 21
191	22 0 49 02	23 22 01 14	24 12 21 26	22 6 45 45
192	21 6 44 07	23 3 57 25	23 18 14 26	21 12 40 09
193	21 12 34 19	23 9 36 26	23 23 57 28	21 18 34 13
194	21 18 23 20	23 15 29 32	24 5 53 29	22 0 29 19
195	22 0 11 42	23 21 20 03	24 11 41 42	22 6 18 19
196	21 6 03 47	23 3 07 33	23 17 27 09	21 12 05 07
197	21 12 00 56	23 9 01 38	23 23 20 53	21 17 58 31
198	21 17 45 39	23 14 39 11	24 5 05 08	21 23 44 03
199	21 23 28 35	23 20 30 41	24 10 55 15	22 5 30 14
200	21 5 18 41	23 2 26 06	23 16 43 06	21 11 20 09

Year	March Equinox	June Solstice	September Equinox	December Solstice
201	21^d $11^h06^m13^s$	23^d $8^h01^m41^s$	23^d $22^h21^m28^s$	21^d $17^h03^m51^s$
202	21 16 48 57	23 13 49 02	24 4 16 29	21 22 49 58
203	21 22 32 19	23 19 37 01	24 10 06 53	22 4 37 23
204	21 4 18 16	23 1 19 23	23 15 49 33	21 10 23 54
205	21 10 07 22	23 7 10 22	23 21 42 50	21 16 23 42
206	21 15 53 25	23 12 49 17	24 3 27 44	21 22 14 28
207	21 21 40 49	23 18 39 23	24 9 18 34	22 4 03 01
208	21 3 39 08	23 0 38 13	23 15 13 12	21 9 58 01
209	21 9 36 41	23 6 18 39	23 20 54 44	21 15 45 43
210	21 15 25 08	23 12 12 11	24 2 52 01	21 21 38 02
211	21 21 13 53	23 18 09 48	24 8 44 08	22 3 29 58
212	21 3 04 53	22 23 58 57	23 14 27 03	21 9 15 37
213	21 8 56 34	23 5 54 17	23 20 24 27	21 15 11 29
214	21 14 43 21	23 11 35 48	24 2 14 18	21 20 57 30
215	21 20 27 29	23 17 23 09	24 8 06 12	22 2 42 23
216	21 2 16 34	22 23 18 13	23 14 01 13	21 8 39 04
217	21 8 06 17	23 4 53 27	23 19 38 01	21 14 27 24
218	21 13 48 00	23 10 38 05	24 1 29 28	21 20 18 09
219	21 19 33 43	23 16 28 14	24 7 17 31	22 2 06 57
220	21 1 25 48	22 22 09 22	23 12 55 14	21 7 48 04
221	21 7 16 56	23 3 59 26	23 18 48 17	21 13 41 36
222	21 13 02 28	23 9 41 36	24 0 33 42	21 19 29 03
223	21 18 47 17	23 15 30 32	24 6 19 42	22 1 12 23
224	21 0 35 54	22 21 28 33	23 12 13 57	21 7 07 47
225	21 6 30 09	23 3 09 31	23 17 54 42	21 12 53 33
226	21 12 14 47	23 8 56 32	23 23 51 16	21 18 40 59
227	21 18 00 54	23 14 52 46	24 5 49 23	22 0 36 03
228	20 23 54 54	22 20 39 22	23 11 32 28	21 6 24 07
229	21 5 43 57	23 2 31 17	23 17 29 14	21 12 25 59
230	21 11 32 55	23 8 17 38	23 23 19 56	21 18 21 08
231	21 17 24 49	23 14 06 22	24 5 07 42	22 0 04 47
232	20 23 18 16	22 20 03 18	23 11 04 56	21 6 01 14
233	21 5 16 50	23 1 45 38	23 16 45 59	21 11 48 34
234	21 11 00 32	23 7 30 36	23 22 35 26	21 17 33 38
235	21 16 42 12	23 13 26 04	24 4 28 18	21 23 27 35
236	20 22 35 00	22 19 10 57	23 10 04 15	21 5 06 57
237	21 4 18 57	23 0 56 36	23 15 55 41	21 10 57 42
238	21 10 02 59	23 6 39 54	23 21 47 05	21 16 46 02
239	21 15 48 46	23 12 23 43	24 3 31 46	21 22 25 15
240	20 21 32 56	22 18 16 05	23 9 26 46	21 4 23 37
241	21 3 25 55	22 23 56 40	23 15 06 07	21 10 14 54
242	21 9 10 09	23 5 39 32	23 20 54 07	21 16 01 11
243	21 14 55 34	23 11 34 47	24 2 50 31	21 21 56 22
244	20 20 56 07	22 17 23 09	23 8 31 26	21 3 39 23
245	21 2 46 46	22 23 12 29	23 14 25 38	21 9 34 23
246	21 8 33 29	23 5 03 09	23 20 20 43	21 15 30 06
247	21 14 25 20	23 10 55 15	24 2 06 20	21 21 12 37
248	20 20 13 18	22 16 52 18	23 8 04 01	21 3 10 20
249	21 2 09 30	22 22 37 38	23 13 49 33	21 9 00 09
250	21 7 56 01	23 4 21 17	23 19 42 04	21 14 43 41

EQUINOXES AND SOLSTICES

Year	March Equinox	June Solstice	September Equinox	December Solstice
251	21^d $13^h36^m51^s$	23^d $10^h14^m34^s$	24^d $1^h40^m37^s$	21^d $20^h39^m37^s$
252	20 19 31 37	22 16 00 56	23 7 19 49	21 2 27 10
253	21 1 18 07	22 21 44 34	23 13 08 16	21 8 21 50
254	21 7 00 24	23 3 29 17	23 18 59 44	21 14 16 45
255	21 12 54 54	23 9 16 36	24 0 41 24	21 19 54 29
256	20 18 43 06	22 15 06 19	23 6 33 34	21 1 46 42
257	21 0 37 37	22 20 50 16	23 12 15 47	21 7 36 15
258	21 6 23 05	23 2 33 43	23 18 00 15	21 13 15 43
259	21 12 00 22	23 8 27 57	23 23 54 48	21 19 09 15
260	20 17 55 53	22 14 18 15	23 5 34 45	21 0 52 46
261	20 23 42 28	22 20 02 05	23 11 25 18	21 6 40 47
262	21 5 22 50	23 1 48 35	23 17 24 48	21 12 36 21
263	21 11 16 27	23 7 39 20	23 23 12 25	21 18 19 29
264	20 17 02 04	22 13 29 29	23 5 07 02	21 0 19 29
265	20 22 56 12	22 19 15 58	23 10 54 57	21 6 19 28
266	21 4 48 02	23 1 01 07	23 16 41 25	21 12 02 50
267	21 10 32 22	23 6 54 41	23 22 39 15	21 17 59 54
268	20 16 36 32	22 12 47 49	23 4 22 53	20 23 47 04
269	20 22 28 04	22 18 33 19	23 10 10 38	21 5 36 04
270	21 4 08 45	23 0 22 50	23 16 07 20	21 11 32 47
271	21 10 02 57	23 6 16 48	23 21 50 20	21 17 12 56
272	20 15 48 21	22 12 05 41	23 3 40 19	20 23 03 39
273	20 21 38 07	22 17 49 13	23 9 28 07	21 4 54 36
274	21 3 26 08	22 23 32 10	23 15 14 51	21 10 31 31
275	21 9 01 46	23 5 20 09	23 21 09 45	21 16 24 50
276	20 14 54 41	22 11 09 12	23 2 51 35	20 22 15 08
277	20 20 41 45	22 16 50 12	23 8 35 24	21 4 04 04
278	21 2 19 13	22 22 33 50	23 14 30 45	21 10 00 07
279	21 8 16 43	23 4 26 17	23 20 15 47	21 15 41 31
280	20 14 07 27	22 10 13 46	23 2 06 07	20 21 33 00
281	20 19 58 05	22 15 59 18	23 7 54 54	21 3 29 27
282	21 1 50 21	22 21 48 56	23 13 41 52	21 9 12 30
283	21 7 30 30	23 3 41 06	23 19 35 30	21 15 05 44
284	20 13 26 58	22 9 35 38	23 1 21 51	20 20 56 51
285	20 19 21 29	22 15 22 22	23 7 11 33	21 2 43 02
286	21 0 59 39	22 21 07 49	23 13 12 30	21 8 39 38
287	21 6 55 52	23 3 04 22	23 19 02 33	21 14 27 27
288	20 12 44 48	22 8 51 22	23 0 51 13	20 20 22 10
289	20 18 32 27	22 14 34 52	23 6 39 53	21 2 21 45
290	21 0 27 17	22 20 23 00	23 12 25 24	21 8 02 34
291	21 6 09 52	23 2 10 06	23 18 16 20	21 13 50 58
292	20 12 06 04	22 8 01 32	23 0 01 16	20 19 40 11
293	20 17 58 17	22 13 46 37	23 5 44 40	21 1 23 22
294	20 23 31 17	22 19 29 24	23 11 37 48	21 7 16 03
295	21 5 22 43	23 1 25 31	23 17 23 59	21 12 59 32
296	20 11 09 57	22 7 10 34	22 23 09 18	20 18 45 31
297	20 16 53 32	22 12 50 27	23 5 00 41	21 0 40 03
298	20 22 44 41	22 18 37 59	23 10 49 37	21 6 20 49
299	21 4 23 09	23 0 23 15	23 16 40 30	21 12 13 46
300	20 10 16 18	22 6 14 46	22 22 27 42	20 18 11 24

Year	March Equinox	June Solstice	September Equinox	December Solstice
301	20^d $16^h11^m37^s$	22^d $12^h02^m17^s$	23^d $4^h13^m12^s$	21^d $0^h01^m17^s$
302	20 21 52 21	22 17 46 28	23 10 09 25	21 5 57 07
303	21 3 52 26	22 23 46 04	23 16 01 58	21 11 45 22
304	20 9 50 17	22 5 37 15	22 21 50 45	20 17 35 29
305	20 15 38 00	22 11 21 34	23 3 41 58	20 23 33 12
306	20 21 31 55	22 17 17 10	23 9 32 03	21 5 17 52
307	21 3 14 20	22 23 06 47	23 15 21 04	21 11 05 38
308	20 9 05 08	22 4 57 58	22 21 09 13	20 16 56 53
309	20 14 59 07	22 10 45 10	23 2 57 16	20 22 40 29
310	20 20 33 06	22 16 23 35	23 8 51 42	21 4 30 34
311	21 2 21 24	22 22 17 26	23 14 41 45	21 10 19 59
312	20 8 11 37	22 4 02 49	22 20 25 28	20 16 10 36
313	20 13 52 22	22 9 37 29	23 2 09 54	20 22 05 39
314	20 19 44 21	22 15 26 58	23 7 58 13	21 3 50 08
315	21 1 31 15	22 21 11 41	23 13 43 55	21 9 33 56
316	20 7 22 05	22 2 59 28	22 19 29 16	20 15 26 14
317	20 13 18 33	22 8 51 15	23 1 16 05	20 21 13 40
318	20 18 55 21	22 14 33 46	23 7 07 05	21 3 03 58
319	21 0 46 34	22 20 33 56	23 12 59 50	21 8 54 11
320	20 6 43 29	22 2 26 36	22 18 48 52	20 14 42 52
321	20 12 28 26	22 8 05 39	23 0 41 23	20 20 37 14
322	20 18 21 01	22 14 01 07	23 6 39 03	21 2 26 29
323	21 0 07 38	22 19 50 49	23 12 29 18	21 8 17 38
324	20 5 57 09	22 1 39 07	22 18 17 00	20 14 16 40
325	20 11 54 02	22 7 31 14	23 0 06 09	20 20 08 16
326	20 17 35 40	22 13 11 16	23 5 56 31	21 1 56 40
327	20 23 28 49	22 19 07 13	23 11 49 06	21 7 45 10
328	20 5 26 20	22 0 58 21	22 17 34 13	20 13 32 52
329	20 11 08 05	22 6 33 31	22 23 18 29	20 19 24 23
330	20 16 54 34	22 12 26 04	23 5 08 32	21 1 08 39
331	20 22 38 08	22 18 13 26	23 10 51 43	21 6 51 34
332	20 4 24 43	21 23 56 33	22 16 35 47	20 12 40 02
333	20 10 16 06	22 5 45 28	22 22 26 39	20 18 26 36
334	20 15 53 43	22 11 23 58	23 4 17 02	21 0 13 44
335	20 21 39 25	22 17 16 53	23 10 08 41	21 6 06 32
336	20 3 35 06	21 23 09 43	22 15 55 55	20 12 01 44
337	20 9 20 33	22 4 44 53	22 21 40 40	20 17 55 08
338	20 15 12 50	22 10 39 56	23 3 37 17	20 23 44 25
339	20 21 07 52	22 16 33 37	23 9 27 13	21 5 32 01
340	20 3 00 26	21 22 20 39	22 15 13 37	20 11 25 28
341	20 8 55 19	22 4 17 22	22 21 07 53	20 17 19 15
342	20 14 38 03	22 10 03 08	23 2 58 24	20 23 07 25
343	20 20 25 36	22 15 58 29	23 8 50 12	21 4 57 14
344	20 2 22 57	21 21 53 29	22 14 41 59	20 10 49 05
345	20 8 07 03	22 3 25 47	22 20 27 12	20 16 35 54
346	20 13 51 07	22 9 16 00	23 2 23 03	20 22 25 30
347	20 19 39 48	22 15 06 14	23 8 09 06	21 4 14 21
348	20 1 25 21	21 20 44 39	22 13 47 29	20 10 05 53
349	20 7 16 06	22 2 35 00	22 19 38 06	20 15 57 26
350	20 13 00 31	22 8 15 35	23 1 24 37	20 21 40 30

Year	March Equinox	June Solstice	September Equinox	December Solstice
351	20^d $18^h49^m05^s$	22^d $14^h05^m36^s$	23^d $7^h12^m37^s$	21^d $3^h26^m44^s$
352	20 0 45 32	21 20 00 34	22 13 01 33	20 9 19 19
353	20 6 30 13	22 1 36 00	22 18 43 02	20 15 06 29
354	20 12 13 59	22 7 29 34	23 0 37 18	20 20 54 24
355	20 18 04 19	22 13 25 55	23 6 27 01	21 2 42 15
356	19 23 54 08	21 19 07 40	22 12 10 43	20 8 30 27
357	20 5 44 10	22 1 00 47	22 18 10 28	20 14 23 57
358	20 11 28 46	22 6 46 46	23 0 03 47	20 20 14 31
359	20 17 16 09	22 12 37 48	23 5 54 44	21 2 08 41
360	19 23 11 56	21 18 34 03	22 11 47 19	20 8 09 33
361	20 5 02 29	22 0 09 58	22 17 30 55	20 14 00 33
362	20 10 52 23	22 6 00 27	22 23 25 34	20 19 47 27
363	20 16 47 04	22 11 56 35	23 5 15 34	21 1 36 57
364	19 22 40 18	21 17 38 17	22 10 53 53	20 7 23 34
365	20 4 26 45	21 23 30 13	22 16 47 07	20 13 14 42
366	20 10 09 49	22 5 18 04	22 22 35 02	20 19 00 06
367	20 15 54 39	22 11 05 18	23 4 19 17	21 0 42 54
368	19 21 45 16	21 16 58 17	22 10 12 36	20 6 35 53
369	20 3 31 03	21 22 31 37	22 15 56 11	20 12 20 14
370	20 9 11 25	22 4 17 29	22 21 49 00	20 18 07 07
371	20 14 58 44	22 10 11 59	23 3 39 22	21 0 01 52
372	19 20 49 21	21 15 50 36	22 9 15 37	20 5 50 40
373	20 2 36 42	21 21 39 37	22 15 10 06	20 11 43 54
374	20 8 26 22	22 3 29 06	22 21 03 09	20 17 32 35
375	20 14 18 29	22 9 17 56	23 2 48 51	20 23 18 13
376	19 20 12 50	21 15 15 40	22 8 45 21	20 5 18 56
377	20 2 03 41	21 20 56 32	22 14 29 25	20 11 07 45
378	20 7 48 12	22 2 46 50	22 20 21 35	20 16 55 17
379	20 13 40 36	22 8 47 05	23 2 16 48	20 22 48 48
380	19 19 35 49	21 14 28 39	22 7 57 10	20 4 33 40
381	20 1 22 10	21 20 17 42	22 13 55 11	20 10 25 32
382	20 7 06 50	22 2 07 49	22 19 49 29	20 16 17 46
383	20 12 54 43	22 7 54 05	23 1 31 26	20 22 04 38
384	19 18 43 52	21 13 46 17	22 7 23 51	20 4 04 09
385	20 0 34 22	21 19 24 00	22 13 06 25	20 9 50 57
386	20 6 20 56	22 1 08 16	22 18 54 46	20 15 31 51
387	20 12 10 06	22 7 04 22	23 0 47 27	20 21 23 50
388	19 18 04 03	21 12 45 30	22 6 22 31	20 3 07 53
389	19 23 46 07	21 18 32 19	22 12 13 51	20 8 56 12
390	20 5 27 47	22 0 24 05	22 18 06 28	20 14 46 03
391	20 11 17 18	22 6 11 07	22 23 48 47	20 20 27 05
392	19 17 04 51	21 12 01 45	22 5 45 09	20 2 22 51
393	19 22 53 37	21 17 41 52	22 11 34 51	20 8 14 45
394	20 4 39 41	21 23 27 15	22 17 25 09	20 14 01 03
395	20 10 26 22	22 5 24 21	22 23 21 02	20 20 02 46
396	19 16 26 05	21 11 09 29	22 4 59 40	20 1 54 09
397	19 22 15 46	21 16 56 12	22 10 52 59	20 7 44 24
398	20 4 05 40	21 22 51 23	22 16 51 21	20 13 38 09
399	20 10 03 22	22 4 42 21	22 22 33 27	20 19 20 43
400	19 15 53 00	21 10 35 49	22 4 27 32	20 1 18 21

Year	March Equinox	June Solstice	September Equinox	December Solstice
	$19^d\ 21^h41^m44^s$	$21^d\ 16^h20^m41^s$	$22^d\ 10^h15^m10^s$	$20^d\ 7^h09^m06^s$
401	19 21 41 44	21 16 20 41	22 10 15 10	20 7 09 06
402	20 3 27 00	21 22 07 12	22 16 00 58	20 12 48 05
403	20 9 10 42	22 4 02 00	22 21 56 15	20 18 41 09
404	19 15 05 37	21 9 43 42	22 3 35 34	20 0 24 58
405	19 20 47 15	21 15 23 46	22 9 26 00	20 6 10 01
406	20 2 25 29	21 21 13 09	22 15 22 17	20 12 05 20
407	20 8 15 59	22 2 58 49	22 20 59 23	20 17 49 08
408	19 14 01 50	21 8 44 46	22 2 49 22	19 23 46 52
409	19 19 51 25	21 14 25 38	22 8 37 22	20 5 37 27
410	20 1 42 19	21 20 10 24	22 14 22 27	20 11 16 19
411	20 7 30 35	22 2 06 36	22 20 18 09	20 17 12 49
412	19 13 29 00	21 7 54 44	22 1 58 18	19 23 03 05
413	19 19 16 15	21 13 41 55	22 7 48 22	20 4 49 57
414	20 0 58 24	21 19 38 15	22 13 48 09	20 10 46 01
415	20 6 56 15	22 1 32 08	22 19 32 15	20 16 29 10
416	19 12 45 59	21 7 21 08	22 1 28 00	19 22 24 00
417	19 18 33 12	21 13 06 29	22 7 24 14	20 4 20 02
418	20 0 23 14	21 18 54 22	22 13 11 28	20 10 03 30
419	20 6 06 39	22 0 47 17	22 19 05 48	20 16 03 01
420	19 12 03 42	21 6 32 51	22 0 45 56	19 21 55 08
421	19 17 53 10	21 12 14 03	22 6 32 59	20 3 36 37
422	19 23 34 08	21 18 03 51	22 12 29 44	20 9 28 59
423	20 5 30 52	21 23 54 55	22 18 09 10	20 15 11 14
424	19 11 16 20	21 5 38 49	21 23 54 31	19 21 01 00
425	19 16 56 47	21 11 21 30	22 5 44 33	20 2 54 22
426	19 22 46 39	21 17 09 12	22 11 27 02	20 8 30 15
427	20 4 27 52	21 22 58 17	22 17 19 38	20 14 21 37
428	19 10 22 28	21 4 44 08	21 23 04 53	19 20 12 29
429	19 16 09 44	21 10 25 56	22 4 53 39	20 1 55 42
430	19 21 46 45	21 16 16 37	22 10 52 53	20 7 55 37
431	20 3 44 10	21 22 11 07	22 16 35 27	20 13 46 19
432	19 9 36 26	21 3 57 38	21 22 24 11	19 19 40 51
433	19 15 25 06	21 9 43 41	22 4 21 34	20 1 38 07
434	19 21 24 29	21 15 37 30	22 10 09 58	20 7 18 57
435	20 3 12 38	21 21 31 01	22 16 03 45	20 13 14 17
436	19 9 08 28	21 3 22 02	21 21 50 03	19 19 09 20
437	19 14 58 26	21 9 08 40	22 3 36 36	20 0 50 41
438	19 20 35 21	21 14 59 07	22 9 34 21	20 6 44 12
439	20 2 30 22	21 20 51 58	22 15 17 56	20 12 28 08
440	19 8 18 37	21 2 33 21	21 21 05 32	19 18 16 17
441	19 13 57 01	21 8 12 20	22 2 59 50	20 0 11 05
442	19 19 46 36	21 14 00 44	22 8 42 49	20 5 53 21
443	20 1 29 29	21 19 46 45	22 14 29 07	20 11 46 37
444	19 7 21 12	21 1 30 45	21 20 12 01	19 17 40 17
445	19 13 14 54	21 7 14 41	22 1 57 15	19 23 18 46
446	19 18 54 29	21 13 02 10	22 7 52 22	20 5 10 23
447	20 0 50 50	21 18 58 42	22 13 36 43	20 10 59 10
448	19 6 42 43	21 0 45 48	21 19 22 16	19 16 48 03
449	19 12 21 52	21 6 30 26	22 1 18 22	19 22 44 24
450	19 18 16 57	21 12 27 20	22 7 07 55	20 4 27 11

Year	March Equinox	June Solstice	September Equinox	December Solstice
451	20^d $0^h04^m48^s$	21^d $18^h17^m14^s$	22^d $13^h00^m38^s$	20^d $10^h17^m45^s$
452	19 5 56 47	21 0 04 35	21 18 52 59	19 16 15 07
453	19 11 50 26	21 5 53 03	22 0 44 32	19 22 00 26
454	19 17 27 32	21 11 40 11	22 6 40 05	20 3 57 54
455	19 23 22 21	21 17 35 51	22 12 26 39	20 9 53 33
456	19 5 18 13	20 23 20 06	21 18 10 20	19 15 41 07
457	19 10 59 27	21 4 58 41	22 0 04 12	19 21 35 09
458	19 16 57 13	21 10 53 52	22 5 51 31	20 3 17 05
459	19 22 44 46	21 16 41 01	22 11 35 47	20 9 04 22
460	19 4 32 04	20 22 26 14	21 17 20 43	19 14 58 26
461	19 10 22 41	21 4 14 41	21 23 05 47	19 20 37 18
462	19 15 58 03	21 9 59 20	22 4 56 32	20 2 24 16
463	19 21 48 22	21 15 52 06	22 10 44 00	20 8 12 09
464	19 3 40 41	20 21 36 24	21 16 30 30	19 13 57 40
465	19 9 15 51	21 3 13 16	21 22 25 17	19 19 53 32
466	19 15 06 55	21 9 08 42	22 4 15 12	20 1 43 15
467	19 20 56 38	21 14 56 40	22 10 00 52	20 7 35 38
468	19 2 47 45	20 20 40 35	21 15 49 35	19 13 33 16
469	19 8 46 37	21 2 32 31	21 21 41 14	19 19 17 13
470	19 14 31 15	21 8 21 05	22 3 35 02	20 1 08 23
471	19 20 25 00	21 14 18 24	22 9 23 52	20 7 02 26
472	19 2 21 54	20 20 09 57	21 15 10 40	19 12 52 56
473	19 8 01 13	21 1 50 18	21 21 03 45	19 18 45 45
474	19 13 53 02	21 7 48 03	22 2 55 56	20 0 31 53
475	19 19 45 34	21 13 37 23	22 8 43 50	20 6 18 10
476	19 1 31 24	20 19 17 00	21 14 32 09	19 12 12 15
477	19 7 22 25	21 1 06 57	21 20 22 59	19 17 58 11
478	19 13 00 41	21 6 50 17	22 2 09 58	19 23 47 41
479	19 18 48 27	21 12 40 55	22 7 54 34	20 5 40 45
480	19 0 45 32	20 18 28 12	21 13 39 03	19 11 26 53
481	19 6 25 08	21 0 02 52	21 19 29 11	19 17 14 39
482	19 12 15 58	21 5 58 51	22 1 20 27	19 23 01 17
483	19 18 08 06	21 11 49 31	22 7 04 45	20 4 48 23
484	18 23 52 11	20 17 29 36	21 12 49 16	19 10 42 14
485	19 5 43 12	20 23 23 27	21 18 42 10	19 16 28 08
486	19 11 25 26	21 5 10 08	22 0 31 42	19 22 13 51
487	19 17 14 29	21 11 01 39	22 6 23 03	20 4 07 34
488	18 23 12 08	20 16 52 36	21 12 14 37	19 9 58 37
489	19 4 50 38	20 22 29 10	21 18 06 51	19 15 53 21
490	19 10 41 15	21 4 27 18	22 0 01 13	19 21 48 35
491	19 16 38 13	21 10 20 14	22 5 47 22	20 3 41 04
492	18 22 29 25	20 15 59 14	21 11 32 58	19 9 34 59
493	19 4 25 33	20 21 53 10	21 17 28 04	19 15 21 27
494	19 10 13 15	21 3 42 27	21 23 15 42	19 21 07 03
495	19 16 01 23	21 9 33 20	22 5 00 31	20 2 58 01
496	18 21 55 30	20 15 26 21	21 10 48 16	19 8 46 46
497	19 3 32 53	20 21 02 40	21 16 35 28	19 14 31 50
498	19 9 17 29	21 2 56 16	21 22 28 05	19 20 17 11
499	19 15 09 43	21 8 46 31	22 4 15 40	20 2 03 44
500	18 20 52 58	20 14 19 25	21 9 59 33	19 7 54 07

Year	March Equinox	June Solstice	September Equinox	December Solstice
501	19^d 2^{h}37^{m}56^s	20^d 20^{h}08^{m}41^s	21^d 15^{h}53^{m}36^s	19^d 13^{h}44^{m}19^s
502	19 8 21 32	21 1 55 39	21 21 39 44	19 19 34 31
503	19 14 08 53	21 7 40 52	22 3 21 44	20 1 26 23
504	18 20 06 31	20 13 32 27	21 9 12 44	19 7 19 12
505	19 1 53 54	20 19 11 16	21 15 02 17	19 13 06 02
506	19 7 43 36	21 1 07 59	21 20 55 58	19 18 56 43
507	19 13 41 59	21 7 07 33	22 2 46 11	20 0 50 33
508	18 19 30 58	20 12 47 16	21 8 29 43	19 6 41 57
509	19 1 19 59	20 18 43 16	21 14 28 04	19 12 31 50
510	19 7 09 24	21 0 36 28	21 20 19 16	19 18 18 26
511	19 12 58 15	21 6 22 50	22 2 06 00	20 0 07 44
512	18 18 51 33	20 12 15 22	21 8 01 37	19 6 02 17
513	19 0 32 45	20 17 53 23	21 13 50 04	19 11 51 33
514	19 6 16 26	20 23 44 29	21 19 39 38	19 17 42 55
515	19 12 11 15	21 5 38.14	22 1 27 17	19 23 35 37
516	18 18 00 41	20 11 10 29	21 7 06 17	19 5 21 11
517	18 23 48 00	20 16 59 27	21 13 01 25	19 11 07 31
518	19 5 35 17	20 22 50 19	21 18 47 05	19 16 53 08
519	19 11 21 12	21 4 33 31	22 0 25 32	19 22 40 58
520	18 17 11 02	20 10 25 12	21 6 16 38	19 4 32 41
521	18 22 52 45	20 16 05 03	21 12 03 15	19 10 17 14
522	19 4 38 26	20 21 56 20	21 17 54 35	19 16 03 10
523	19 10 32 49	21 3 52 23	21 23 49 07	19 21 57 18
524	18 16 22 54	20 9 28 48	21 5 33 01	19 3 47 51
525	18 22 08 07	20 15 19 50	21 11 31 08	19 9 42 03
526	19 3 58 10	20 21 16 21	21 17 22 31	19 15 37 33
527	19 9 52 07	21 3 01 52	21 23 03 46	19 21 27 41
528	18 15 48 11	20 8 55 33	21 5 01 19	19 3 22 56
529	18 21 38 55	20 14 40 49	21 10 52 33	19 9 10 49
530	19 3 27 31	20 20 33 22	21 16 41 28	19 14 57 44
531	19 9 20 37	21 2 32 05	21 22 34 00	19 20 53 46
532	18 15 10 30	20 8 10 00	21 4 13 38	19 2 39 25
533	18 20 52 09	20 13 57 20	21 10 07 04	19 8 24 06
534	19 2 37 54	20 19 50 31	21 15 58 54	19 14 12 37
535	19 8 26 07	21 1 29 06	21 21 36 34	19 19 54 19
536	18 14 10 00	20 7 14 40	21 3 30 13	19 1 48 41
537	18 19 52 53	20 12 56 14	21 9 17 45	19 7 38 41
538	19 1 36 53	20 18 41 55	21 15 00 20	19 13 24 57
539	19 7 29 40	21 0 36 41	21 20 52 36	19 19 21 52
540	18 13 25 22	20 6 13 52	21 2 33 16	19 1 07 07
541	18 19 12 37	20 12 02 07	21 8 27 37	19 6 53 27
542	19 1 02 09	20 18 01 41	21 14 21 33	19 12 47 34
543	19 6 55 51	20 23 48 56	21 20 00 23	19 18 33 57
544	18 12 44 07	20 5 41 35	21 1 57 21	19 0 29 28
545	18 18 31 06	20 11 31 16	21 7 52 06	19 6 20 04
546	19 0 20 43	20 17 21 17	21 13 41 07	19 12 04 11
547	19 6 12 29	20 23 17 30	21 19 40 44	19 18 03 18
548	18 12 05 19	20 4 56 48	21 1 24 45	18 23 54 10
549	18 17 48 36	20 10 42 01	21 7 17 11	19 5 44 05
550	18 23 34 06	20 16 37 10	21 13 08 55	19 11 40 06

Year	March Equinox	June Solstice	September Equinox	December Solstice
551	19^d $5^h29^m30^s$	20^d $22^h19^m01^s$	21^d $18^h43^m50^s$	19^d $17^h24^m26^s$
552	18 11 19 32	20 4 03 47	21 0 36 08	18 23 13 51
553	18 17 05 03	20 9 49 49	21 6 26 41	19 5 02 29
554	18 22 53 59	20 15 37 59	21 12 07 11	19 10 42 53
555	19 4 40 25	20 21 31 23	21 17 58 59	19 16 38 16
556	18 10 31 08	20 3 12 12	20 23 39 37	18 22 24 25
557	18 16 13 59	20 8 56 02	21 5 28 53	19 4 04 44
558	18 21 57 16	20 14 51 39	21 11 25 39	19 9 57 34
559	19 3 52 06	20 20 36 04	21 17 05 31	19 15 42 01
560	18 9 36 50	20 2 21 08	20 23 00 36	18 21 36 41
561	18 15 19 46	20 8 10 48	21 4 56 42	19 3 35 28
562	18 21 12 32	20 14 01 05	21 10 39 57	19 9 21 38
563	19 3 04 59	20 19 54 42	21 16 36 22	19 15 21 40
564	18 9 04 27	20 1 39 12	20 22 24 21	18 21 13 12
565	18 14 55 04	20 7 26 05	21 4 14 27	19 2 56 13
566	18 20 40 50	20 13 25 18	21 10 11 26	19 8 54 47
567	19 2 38 50	20 19 14 10	21 15 48 30	19 14 39 33
568	18 8 23 48	20 0 58 51	20 21 38 43	18 20 28 09
569	18 14 05 59	20 6 48 46	21 3 35 23	19 2 19 54
570	18 19 57 12	20 12 36 38	21 9 17 46	19 7 57 44
571	19 1 42 34	20 18 24 36	21 15 12 09	19 13 53 18
572	18 7 31 15	20 0 04 32	20 20 57 19	18 19 45 45
573	18 13 15 07	20 5 46 15	21 2 41 38	19 1 28 11
574	18 18 55 52	20 11 38 38	21 8 33 59	19 7 24 31
575	19 0 54 28	20 17 24 24	21 14 10 54	19 13 08 40
576	18 6 43 44	19 23 06 03	20 20 00 11	18 18 54 17
577	18 12 25 34	20 4 56 19	21 1 57 28	19 0 48 49
578	18 18 19 34	20 10 49 31	21 7 38 58	19 6 30 59
579	19 0 06 41	20 16 40 47	21 13 32 11	19 12 27 43
580	18 5 57 05	19 22 26 37	20 19 21 38	18 18 21 38
581	18 11 47 25	20 4 13 48	21 1 11 34	19 0 02 32
582	18 17 30 27	20 10 07 34	21 7 10 12	19 5 58 19
583	18 23 28 40	20 15 56 44	21 12 53 54	19 11 49 29
584	18 5 18 18	19 21 40 04	20 18 44 17	18 17 39 57
585	18 10 57 20	20 3 29 48	21 0 42 10	18 23 40 30
586	18 16 55 11	20 9 23 55	21 6 23 54	19 5 25 12
587	18 22 46 48	20 15 10 43	21 12 14 32	19 11 18 17
588	18 4 39 08	19 20 54 36	20 18 04 39	18 17 10 58
589	18 10 31 10	20 2 41 38	20 23 49 10	18 22 48 36
590	18 16 10 44	20 8 33 53	21 5 40 46	19 4 41 57
591	18 22 05 00	20 14 23 29	21 11 19 53	19 10 29 06
592	18 3 51 09	19 20 05 01	20 17 05 08	18 16 10 05
593	18 9 25 48	20 1 51 33	20 23 02 49	18 22 02 28
594	18 15 19 24	20 7 43 45	21 4 47 14	19 3 43 04
595	18 21 05 03	20 13 27 02	21 10 37 32	19 9 35 49
596	18 2 49 14	19 19 08 49	20 16 29 54	18 15 35 11
597	18 8 39 37	20 0 55 43	20 22 14 36	18 21 18 34
598	18 14 22 11	20 6 45 35	21 4 07 01	19 3 16 33
599	18 20 23 46	20 12 35 55	21 9 51 24	19 9 08 01
600	18 2 19 44	19 18 20 42	20 15 39 36	18 14 52 17

Year	March Equinox	June Solstice	September Equinox	December Solstice
601	18^d $8^h01^m13^s$	20^d $0^h12^m25^s$	20^d $21^h39^m02^s$	18^d $20^h50^m07^s$
602	18 13 59 32	20 6 12 29	21 3 24 29	19 2 36 46
603	18 19 50 47	20 12 02 12	21 9 13 41	19 8 28 53
604	18 1 36 31	19 17 47 28	20 15 07 30	18 14 24 28
605	18 7 29 52	19 23 38 27	20 20 56 15	18 20 03 13
606	18 13 11 15	20 5 26 29	21 2 49 34	19 1 54 28
607	18 19 03 29	20 11 14 12	21 8 35 24	19 7 47 14
608	18 0 52 34	19 16 56 00	20 14 20 25	18 13 32 06
609	18 6 25 39	19 22 39 24	20 20 13 41	18 19 28 18
610	18 12 20 12	20 4 33 04	21 1 56 00	19 1 13 54
611	18 18 13 08	20 10 15 47	21 7 41 19	19 7 00 29
612	17 23 57 12	19 15 55 08	20 13 32 11	18 12 53 36
613	18 5 50 18	19 21 46 41	20 19 18 22	18 18 34 54
614	18 11 31 39	20 3 34 28	21 1 05 11	19 0 24 32
615	18 17 22 00	20 9 23 44	21 6 49 37	19 6 17 58
616	17 23 16 09	19 15 11 04	20 12 36 53	18 12 00 05
617	18 4 52 27	19 20 56 29	20 18 34 21	18 17 53 09
618	18 10 48 37	20 2 55 14	21 0 25 57	18 23 42 48
619	18 16 43 00	20 8 42 43	21 6 15 48	19 5 35 14
620	17 22 25 35	19 14 24 50	20 12 10 36	18 11 36 59
621	18 4 21 01	19 20 20 25	20 18 00 25	18 17 25 05
622	18 10 08 57	20 2 09 38	20 23 49 15	18 23 16 35
623	18 16 05 09	20 7 58 29	21 5 37 36	19 5 10 24
624	17 22 03 12	19 13 47 14	20 11 25 42	18 10 52 52
625	18 3 40 02	19 19 32 06	20 17 18 02	18 16 44 43
626	18 9 31 37	20 1 29 57	20 23 04 48	18 22 32 53
627	18 15 23 20	20 7 16 25	21 4 48 06	19 4 18 19
628	17 21 01 17	19 12 53 09	20 10 38 33	18 10 10 00
629	18 2 50 56	19 18 43 55	20 16 27 36	18 15 49 48
630	18 8 33 05	20 0 27 27	20 22 14 06	18 21 36 28
631	18 14 20 06	20 6 10 08	21 4 00 00	19 3 30 49
632	17 20 10 47	19 11 56 10	20 9 46 07	18 9 17 53
633	18 1 47 17	19 17 38 28	20 15 35 46	18 15 11 33
634	18 7 41 03	19 23 34 15	20 21 24 17	18 21 02 44
635	18 13 42 07	20 5 23 53	21 3 11 56	19 2 51 10
636	17 19 27 24	19 11 03 37	20 9 04 15	18 8 45 51
637	18 1 22 14	19 17 03 18	20 14 57 47	18 14 34 17
638	18 7 12 04	19 22 56 42	20 20 45 40	18 20 23 58
639	18 13 02 02	20 4 45 21	21 2 34 08	19 2 19 04
640	17 18 57 34	19 10 38 33	20 8 26 58	18 8 05 05
641	18 0 37 02	19 16 22 43	20 14 20 54	18 13 53 38
642	18 6 26 47	19 22 17 26	20 20 13 32	18 19 45 14
643	18 12 22 33	20 4 06 12	21 2 01 49	19 1 36 48
644	17 18 00 19	19 9 39 21	20 7 48 40	18 7 31 28
645	17 23 49 07	19 15 32 11	20 13 38 41	18 13 21 04
646	18 5 40 08	19 21 18 35	20 19 20 53	18 19 05 11
647	18 11 29 08	20 2 57 49	21 1 03 24	19 0 54 40
648	17 17 23 59	19 8 48 35	20 6 52 37	18 6 39 09
649	17 23 02 11	19 14 31 46	20 12 38 34	18 12 25 04
650	18 4 48 40	19 20 26 23	20 18 26 03	18 18 15 01

EQUINOXES AND SOLSTICES

Year	March Equinox	June Solstice	September Equinox	December Solstice
651	18^d $10^h44^m09^s$	20^d $2^h18^m17^s$	21^d $0^h12^m45^s$	19^d $0^h03^m22^s$
652	17 16 24 02	19 7 53 33	20 6 01 18	18 5 52 23
653	17 22 13 00	19 13 48 50	20 11 58 10	18 11 40 04
654	18 4 04 28	19 19 40 41	20 17 47 59	18 17 28 56
655	18 9 52 18	20 1 22 23	20 23 35 02	18 23 25 39
656	17 15 44 59	19 7 16 14	20 5 30 04	18 5 20 05
657	17 21 28 56	19 13 02 49	20 11 19 26	18 11 11 49
658	18 3 21 53	19 18 56 39	20 17 10 43	18 17 04 29
659	18 9 24 17	20 0 50 33	20 23 02 04	18 22 56 11
660	17 15 09 45	19 6 27 19	20 4 50 08	18 4 46 50
661	17 20 57 48	19 12 22 46	20 10 42 56	18 10 35 21
662	18 2 48 35	19 18 16 11	20 16 27 28	18 16 23 08
663	18 8 36 01	19 23 55 18	20 22 07 51	18 22 10 56
664	17 14 24 08	19 5 45 27	20 4 00 56	18 3 56 31
665	17 20 05 31	19 11 29 34	20 9 49 03	18 9 39 47
666	18 1 49 50	19 17 16 26	20 15 36 25	18 15 27 59
667	18 7 42 16	19 23 06 58	20 21 26 05	18 21 22 14
668	17 13 22 11	19 4 39 35	20 3 09 09	18 3 12 23
669	17 19 07 22	19 10 30 52	20 9 00 48	18 9 02 21
670	18 1 03 01	19 16 24 14	20 14 48 07	18 14 50 52
671	18 6 55 41	19 22 02 34	20 20 29 53	18 20 38 56
672	17 12 47 11	19 3 56 35	20 2 26 17	18 2 30 22
673	17 18 33 27	19 9 48 30	20 8 15 45	18 8 19 05
674	18 0 21 24	19 15 40 38	20 14 03 11	18 14 09 31
675	18 6 17 45	19 21 37 58	20 19 58 47	18 20 05 28
676	17 12 03 18	19 3 15 34	20 1 47 02	18 1 52 28
677	17 17 49 33	19 9 07 44	20 7 44 35	18 7 43 00
678	17 23 44 04	19 15 03 16	20 13 36 26	18 13 35 44
679	18 5 32 54	19 20 39 42	20 19 16 10	18 19 27 08
680	17 11 20 15	19 2 30 40	20 1 10 41	18 1 21 26
681	17 17 07 02	19 8 19 39	20 6 57 02	18 7 09 21
682	17 22 57 39	19 14 05 28	20 12 40 01	18 12 54 17
683	18 4 53 30	19 19 58 08	20 18 32 42	18 18 46 27
684	17 10 38 48	19 1 35 35	20 0 15 59	18 0 31 06
685	17 16 21 22	19 7 25 46	20 6 05 41	18 6 17 05
686	17 22 10 42	19 13 22 09	20 11 54 18	18 12 07 18
687	18 3 59 15	19 18 58 45	20 17 31 48	18 17 51 16
688	17 9 43 06	19 0 47 01	19 23 28 23	17 23 39 08
689	17 15 27 52	19 6 37 14	20 5 20 19	18 5 27 02
690	17 21 15 32	19 12 22 16	20 11 05 35	18 11 14 59
691	18 3 05 44	19 18 14 46	20 17 01 18	18 17 15 44
692	17 8 52 30	18 23 54 47	19 22 48 05	17 23 09 25
693	17 14 40 45	19 5 43 55	20 4 38 50	18 4 58 32
694	17 20 37 19	19 11 41 45	20 10 32 54	18 10 54 07
695	18 2 37 11	19 17 22 49	20 16 13 22	18 16 41 07
696	17 8 25 25	18 23 14 03	19 22 09 15	17 22 32 40
697	17 14 13 31	19 5 11 24	20 4 00 52	18 4 24 21
698	17 20 04 07	19 10 59 24	20 9 41 36	18 10 08 03
699	18 1 53 49	19 16 52 44	20 15 36 55	18 16 02 33
700	17 7 40 15	18 22 33 23	19 21 24 45	17 21 47 31

Year	March Equinox	June Solstice	September Equinox	December Solstice
701	17^d $13^h23^m36^s$	19^d $4^h18^m50^s$	20^d $3^h14^m47^s$	18^d $3^h30^m07^s$
702	17 19 10 35	19 10 12 50	20 9 08 43	18 9 25 21
703	18 1 00 47	19 15 48 53	20 14 45 01	18 15 12 38
704	17 6 41 51	18 21 32 21	19 20 35 39	17 21 03 17
705	17 12 27 20	19 3 24 09	20 2 25 57	18 2 54 25
706	17 18 20 58	19 9 07 15	20 8 04 25	18 8 35 06
707	18 0 11 59	19 14 57 33	20 13 59 28	18 14 30 42
708	17 5 59 38	18 20 41 12	19 19 46 21	17 20 19 30
709	17 11 45 07	19 2 29 54	20 1 32 58	18 2 04 46
710	17 17 35 05	19 8 28 57	20 7 28 17	18 8 01 25
711	17 23 30 53	19 14 11 24	20 13 08 45	18 13 47 49
712	17 5 17 24	18 19 58 56	19 19 05 31	17 19 36 15
713	17 11 03 51	19 1 55 26	20 1 04 19	18 1 31 14
714	17 16 58 33	19 7 43 14	20 6 47 03	18 7 17 53
715	17 22 46 57	19 13 33 27	20 12 42 41	18 13 18 42
716	17 4 34 45	18 19 19 00	19 18 32 26	17 19 13 13
717	17 10 25 30	19 1 06 02	20 0 17 59	18 0 55 40
718	17 16 16 44	19 7 01 05	20 6 13 36	18 6 50 12
719	17 22 14 38	19 12 42 58	20 11 52 37	18 12 35 37
720	17 3 58 21	18 18 27 00	19 17 41 40	17 18 20 13
721	17 9 38 22	19 0 21 39	19 23 34 25	18 0 12 52
722	17 15 30 16	19 6 07 54	20 5 11 15	18 5 52 43
723	17 21 14 00	19 11 52 20	20 11 02 09	18 11 43 19
724	17 2 56 54	18 17 34 45	19 16 54 04	17 17 33 42
725	17 8 44 31	18 23 18 37	19 22 38 43	17 23 12 43
726	17 14 27 45	19 5 09 43	20 4 33 21	18 5 11 20
727	17 20 22 59	19 10 52 45	20 10 13 55	18 11 04 11
728	17 2 09 00	18 16 36 04	19 16 02 23	17 16 52 10
729	17 7 55 05	18 22 32 25	19 22 00 07	17 22 49 29
730	17 13 57 54	19 4 23 14	20 3 41 55	18 4 33 01
731	17 19 50 19	19 10 12 41	20 9 36 07	18 10 28 07
732	17 1 37 55	18 16 03 41	19 15 31 36	17 16 24 30
733	17 7 30 43	18 21 56 32	19 21 17 02	17 22 06 14
734	17 13 17 24	19 3 52 06	20 3 12 48	18 4 02 46
735	17 19 13 21	19 9 37 55	20 8 57 28	18 9 51 36
736	17 0 59 28	18 15 20 31	19 14 47 40	17 15 32 48
737	17 6 38 28	18 21 12 50	19 20 46 29	17 21 27 58
738	17 12 32 17	19 2 59 45	20 2 25 24	18 3 12 52
739	17 18 16 17	19 8 40 53	20 8 12 58	18 9 07 22
740	16 23 56 40	18 14 23 11	19 14 03 18	17 15 01 55
741	17 5 49 21	18 20 08 59	19 19 43 29	17 20 39 54
742	17 11 38 00	19 1 57 29	20 1 34 41	18 2 31 43
743	17 17 33 01	19 7 41 35	20 7 16 54	18 8 21 30
744	16 23 20 48	18 13 26 49	19 13 02 20	17 14 02 39
745	17 4 59 04	18 19 21 23	19 18 58 02	17 19 57 36
746	17 10 55 59	19 1 14 50	20 0 39 44	18 1 42 29
747	17 16 44 57	19 6 59 57	20 6 30 47	18 7 31 54
748	16 22 26 36	18 12 47 20	19 12 31 28	17 13 28 32
749	17 4 21 54	18 18 40 09	19 18 20 11	17 19 12 09
750	17 10 08 22	19 0 30 36	20 0 15 10	18 1 11 34

EQUINOXES AND SOLSTICES

Year	March Equinox	June Solstice	September Equinox	December Solstice
751	17^d 16^{h}02^{m}10^s	19^d 6^{h}17^{m}20^s	20^d 6^{h}02^{m}46^s	18^d 7^{h}11^{m}13^s
752	16 21 54 43	18 12 03 37	19 11 49 54	17 12 55 34
753	17 3 37 48	18 17 55 29	19 17 46 26	17 18 50 48
754	17 9 39 44	18 23 48 23	19 23 30 14	18 0 37 09
755	17 15 30 46	19 5 31 46	20 5 15 52	18 6 23 58
756	16 21 08 18	18 11 17 10	19 11 10 30	17 12 20 25
757	17 3 01 48	18 17 09 53	19 16 52 02	17 17 59 13
758	17 8 45 01	18 22 55 36	19 22 38 54	17 23 48 20
759	17 14 34 20	19 4 38 35	20 4 25 28	18 5 39 15
760	16 20 23 12	18 10 21 44	19 10 11 35	17 11 16 07
761	17 1 59 04	18 16 09 09	19 16 06 16	17 17 09 18
762	17 7 52 37	18 21 59 47	19 21 49 15	17 22 59 46
763	17 13 41 22	19 3 42 21	20 3 33 45	18 4 49 48
764	16 19 19 49	18 9 26 17	19 9 29 48	17 10 48 01
765	17 1 18 59	18 15 21 34	19 15 16 58	17 16 31 00
766	17 7 11 44	18 21 10 29	19 21 07 49	17 22 23 04
767	17 13 04 43	19 2 58 13	20 2 58 48	18 4 21 01
768	16 18 59 07	18 8 50 31	19 8 47 08	17 10 04 34
769	17 0 39 41	18 14 43 25	19 14 42 16	17 15 59 08
770	17 6 35 54	18 20 38 48	19 20 29 11	17 21 49 19
771	17 12 29 21	19 2 24 36	20 2 17 52	18 3 35 41
772	16 18 06 39	18 8 06 37	19 8 16 14	17 9 30 00
773	16 23 59 46	18 14 00 37	19 14 04 20	17 15 15 39
774	17 5 47 50	18 19 46 00	19 19 50 44	17 21 08 00
775	17 11 33 05	19 1 26 23	20 1 37 00	18 3 06 01
776	16 17 26 33	18 7 14 12	19 7 21 35	17 8 47 10
777	16 23 08 22	18 12 59 13	19 13 10 39	17 14 34 43
778	17 5 03 40	18 18 50 18	19 18 55 19	17 20 22 57
779	17 10 57 05	19 0 36 10	20 0 38 30	18 2 06 34
780	16 16 30 52	18 6 18 20	19 6 31 30	17 7 59 39
781	16 22 22 32	18 12 15 52	19 12 19 04	17 13 44 22
782	17 4 11 44	18 18 03 16	19 18 05 37	17 19 31 29
783	17 9 56 48	18 23 44 10	19 23 57 19	18 1 26 51
784	16 15 50 18	18 5 35 28	19 5 50 02	17 7 10 38
785	16 21 31 05	18 11 22 23	19 11 42 35	17 13 03 35
786	17 3 23 35	18 17 14 46	19 17 32 05	17 19 02 36
787	17 9 19 51	18 23 02 56	19 23 17 46	18 0 53 08
788	16 14 59 25	18 4 44 11	19 5 11 55	17 6 49 46
789	16 20 59 27	18 10 42 58	19 11 03 45	17 12 37 12
790	17 2 56 47	18 16 32 45	19 16 50 18	17 18 25 08
791	17 8 44 20	18 22 15 21	19 22 39 22	18 0 21 35
792	16 14 37 10	18 4 10 01	19 4 28 31	17 6 05 29
793	16 20 18 32	18 9 58 18	19 10 15 33	17 11 51 46
794	17 2 07 41	18 15 47 50	19 16 01 48	17 17 41 21
795	17 8 01 01	18 21 34 56	19 21 48 31	17 23 24 06
796	16 13 34 58	18 3 11 40	19 3 41 04	17 5 13 10
797	16 19 22 39	18 9 05 40	19 9 31 50	17 11 01 41
798	17 1 13 00	18 14 52 21	19 15 15 31	17 16 51 57
799	17 6 54 56	18 20 27 57	19 21 01 04	17 22 48 32
800	16 12 47 28	18 2 19 32	19 2 51 24	17 4 34 25

Year	March Equinox	June Solstice	September Equinox	December Solstice
801	16^d 18^{h}35^{m}30^s	18^d 8^{h}06^{m}34^s	19^d 8^{h}39^{m}42^s	17^d 10^{h}20^{m}34^s
802	17 0 27 57	18 13 55 20	19 14 26 50	17 16 13 13
803	17 6 24 55	18 19 48 15	19 20 15 30	17 22 03 43
804	16 12 04 16	18 1 30 39	19 2 06 28	17 3 54 45
805	16 17 54 42	18 7 29 53	19 7 59 13	17 9 45 20
806	16 23 52 57	18 13 24 19	19 13 48 14	17 15 34 24
807	17 5 38 39	18 19 02 04	19 19 38 44	17 21 27 49
808	16 11 30 34	18 0 57 20	19 1 36 22	17 3 16 37
809	16 17 16 57	18 6 46 24	19 7 25 21	17 9 05 54
810	16 23 04 40	18 12 32 37	19 13 11 08	17 15 02 27
811	17 4 59 28	18 18 23 22	19 18 58 43	17 20 53 40
812	16 10 40 04	18 0 01 23	19 0 46 35	17 2 40 41
813	16 16 31 13	18 5 55 01	19 6 37 28	17 8 27 34
814	16 22 28 38	18 11 46 45	19 12 22 42	17 14 14 27
815	17 4 10 53	18 17 21 13	19 18 05 13	17 20 04 40
816	16 9 56 57	17 23 14 50	18 23 57 38	17 1 51 03
817	16 15 41 32	18 5 03 56	19 5 41 58	17 7 34 18
818	16 21 27 34	18 10 46 45	19 11 26 37	17 13 24 09
819	17 3 19 45	18 16 36 13	19 17 18 08	17 19 11 50
820	16 8 58 19	17 22 14 23	18 23 08 13	17 1 00 03
821	16 14 45 58	18 4 07 57	19 5 01 04	17 6 53 27
822	16 20 42 24	18 10 02 12	19 10 49 27	17 12 49 34
823	17 2 30 08	18 15 38 44	19 16 34 13	17 18 44 35
824	16 8 23 10	17 21 34 09	18 22 31 31	17 0 35 16
825	16 14 19 16	18 3 29 21	19 4 21 50	17 6 22 48
826	16 20 12 51	18 9 15 56	19 10 07 06	17 12 14 44
827	17 2 06 58	18 15 12 20	19 16 00 51	17 18 07 54
828	16 7 49 27	17 20 57 21	18 21 49 36	16 23 54 59
829	16 13 35 39	18 2 50 59	19 3 40 08	17 5 42 52
830	16 19 30 52	18 8 45 18	19 9 30 12	17 11 32 55
831	17 1 15 15	18 14 17 32	19 15 14 26	17 17 18 57
832	16 6 58 08	17 20 06 32	18 21 10 02	16 23 06 37
833	16 12 44 43	18 1 56 58	19 2 57 12	17 4 55 34
834	16 18 29 40	18 7 34 07	19 8 34 32	17 10 45 34
835	17 0 17 40	18 13 22 26	19 14 24 42	17 16 39 10
836	16 6 03 22	17 19 03 04	18 20 10 41	16 22 23 02
837	16 11 51 56	18 0 51 33	19 1 58 15	17 4 09 04
838	16 17 49 40	18 6 48 25	19 7 48 55	17 10 03 14
839	16 23 37 18	18 12 25 31	19 13 30 50	17 15 51 51
840	16 5 22 35	17 18 20 04	18 19 26 44	16 21 41 54
841	16 11 14 32	18 0 18 37	19 1 18 03	17 3 31 23
842	16 17 06 19	18 6 01 47	19 7 01 34	17 9 19 22
843	16 22 56 54	18 11 55 03	19 13 01 18	17 15 13 33
844	16 4 42 41	17 17 42 01	18 18 54 47	16 21 03 26
845	16 10 29 40	17 23 31 54	19 0 44 29	17 2 55 43
846	16 16 24 04	18 5 28 05	19 6 37 31	17 8 56 07
847	16 22 14 10	18 11 03 41	19 12 18 56	17 14 45 28
848	16 4 02 06	17 16 52 27	18 18 13 14	16 20 32 34
849	16 9 55 28	17 22 47 36	19 0 02 49	17 2 20 20
850	16 15 47 11	18 4 27 22	19 5 39 24	17 8 05 52

EQUINOXES AND SOLSTICES

Year	March Equinox	June Solstice	September Equinox	December Solstice
851	16^d $21^h32^m26^s$	18^d $10^h16^m34^s$	19^d $11^h30^m37^s$	17^d $13^h56^m00^s$
852	16 3 13 44	17 16 02 35	18 17 16 50	16 19 41 06
853	16 8 59 11	17 21 49 21	18 23 00 03	17 1 23 17
854	16 14 48 57	18 3 41 52	19 4 53 04	17 7 16 20
855	16 20 36 41	18 9 17 15	19 10 36 51	17 13 01 44
856	16 2 18 45	17 15 03 23	18 16 30 37	16 18 49 29
857	16 8 06 45	17 20 59 52	18 22 23 13	17 0 45 26
858	16 13 59 43	18 2 40 54	19 4 00 24	17 6 35 17
859	16 19 48 06	18 8 30 47	19 9 55 39	17 12 30 26
860	16 1 39 07	17 14 21 58	18 15 49 46	16 18 20 38
861	16 7 33 37	17 20 12 22	18 21 36 47	17 0 05 56
862	16 13 28 04	18 2 10 37	19 3 33 39	17 6 06 01
863	16 19 20 16	18 7 53 51	19 9 19 08	17 11 56 10
864	16 1 04 56	17 13 43 23	18 15 11 09	16 17 42 36
865	16 6 54 03	17 19 42 12	18 21 06 32	16 23 36 17
866	16 12 48 29	18 1 22 26	19 2 44 15	17 5 18 31
867	16 18 31 39	18 7 06 55	19 8 38 46	17 11 09 09
868	16 0 15 21	17 12 55 04	18 14 31 23	16 16 59 33
869	16 6 01 36	17 18 38 55	18 20 10 59	16 22 43 43
870	16 11 48 47	18 0 29 31	19 2 02 18	17 4 42 49
871	16 17 39 01	18 6 07 07	19 7 43 46	17 10 30 09
872	15 23 25 57	17 11 50 45	18 13 31 41	16 16 11 23
873	16 5 15 15	17 17 46 51	18 19 25 16	16 22 03 53
874	16 11 11 11	17 23 30 18	19 1 00 56	17 3 48 17
875	16 16 54 43	18 5 17 25	19 6 52 25	17 9 38 29
876	15 22 37 55	17 11 11 06	18 12 47 12	16 15 30 12
877	16 4 29 41	17 17 00 38	18 18 30 17	16 21 11 53
878	16 10 19 04	17 22 53 19	19 0 28 41	17 3 09 24
879	16 16 10 07	18 4 36 14	19 6 19 51	17 9 01 32
880	15 21 57 13	17 10 22 56	18 12 12 19	16 14 49 00
881	16 3 43 23	17 16 19 39	18 18 09 08	16 20 50 08
882	16 9 41 40	17 22 04 18	18 23 46 58	17 2 41 58
883	16 15 30 56	18 3 48 18	19 5 37 19	17 8 31 25
884	15 21 18 06	17 9 40 02	18 11 33 13	16 14 23 36
885	16 3 15 59	17 15 30 34	18 17 13 42	16 20 03 54
886	16 9 04 04	17 21 21 17	18 23 05 12	17 1 59 24
887	16 14 51 45	18 3 05 43	19 4 51 40	17 7 49 40
888	15 20 36 56	17 8 51 19	18 10 36 16	16 13 27 47
889	16 2 18 57	17 14 44 52	18 16 30 46	16 19 19 57
890	16 8 13 56	17 20 27 33	18 22 09 27	17 1 03 21
891	16 13 56 18	18 2 07 27	19 3 59 18	17 6 48 14
892	15 19 34 18	17 7 56 43	18 9 56 20	16 12 43 56
893	16 1 26 29	17 13 45 16	18 15 35 49	16 18 28 30
894	16 7 12 57	17 19 32 00	18 21 26 16	17 0 26 54
895	16 13 03 39	18 1 15 46	19 3 17 20	17 6 21 14
896	15 18 57 27	17 7 02 53	18 9 04 45	16 12 01 15
897	16 0 45 36	17 12 58 53	18 15 02 24	16 17 59 35
898	16 6 45 29	17 18 48 15	18 20 43 03	16 23 49 59
899	16 12 32 34	18 0 33 55	19 2 31 35	17 5 37 36
900	15 18 14 57	17 6 29 02	18 8 30 52	16 11 33 37

Year	March Equinox	June Solstice	September Equinox	December Solstice
901	16^d $0^h12^m22^s$	17^d $12^h22^m34^s$	18^d $14^h13^m35^s$	16^d $17^h15^m16^s$
902	16 6 01 31	17 18 10 01	18 20 06 56	16 23 08 42
903	16 11 47 50	17 23 53 42	19 2 01 26	17 5 03 19
904	15 17 37 17	17 5 40 53	18 7 47 26	16 10 44 53
905	15 23 18 45	17 11 31 14	18 13 39 53	16 16 42 33
906	16 5 14 00	17 17 16 20	18 19 18 38	16 22 33 32
907	16 11 02 44	17 22 56 24	19 1 03 20	17 4 14 52
908	15 16 43 10	17 4 44 42	18 6 59 54	16 10 06 38
909	15 22 40 11	17 10 36 59	18 12 39 33	16 15 47 58
910	16 4 27 02	17 16 22 18	18 18 26 06	16 21 39 03
911	16 10 08 35	17 22 06 16	19 0 17 36	17 3 33 19
912	15 15 59 19	17 3 56 24	18 6 03 00	16 9 12 01
913	15 21 41 38	17 9 45 48	18 11 56 13	16 15 03 59
914	16 3 36 15	17 15 32 27	18 17 42 15	16 20 56 43
915	16 9 26 41	17 21 15 46	18 23 31 22	17 2 40 51
916	15 15 03 54	17 3 05 09	18 5 30 41	16 8 40 43
917	15 21 02 15	17 9 01 49	18 11 14 46	16 14 31 59
918	16 2 55 06	17 14 48 14	18 17 02 31	16 20 26 36
919	16 8 43 23	17 20 33 34	18 22 59 04	17 2 24 09
920	15 14 43 22	17 2 27 28	18 4 47 13	16 8 04 08
921	15 20 30 51	17 8 19 18	18 10 39 15	16 13 57 11
922	16 2 25 08	17 14 09 16	18 16 23 54	16 19 50 48
923	16 8 14 27	17 19 55 26	18 22 08 51	17 1 31 01
924	15 13 49 24	17 1 43 03	18 4 04 17	16 7 22 52
925	15 19 43 11	17 7 36 25	18 9 48 14	16 13 06 00
926	16 1 31 06	17 13 17 35	18 15 34 17	16 18 52 07
927	16 7 08 30	17 18 55 58	18 21 29 28	17 0 47 47
928	15 12 58 25	17 0 45 12	18 3 14 05	16 6 30 02
929	15 18 40 01	17 6 29 57	18 9 00 39	16 12 24 38
930	16 0 31 55	17 12 13 38	18 14 43 16	16 18 19 12
931	16 6 25 55	17 17 57 43	18 20 27 58	16 23 59 32
932	15 12 07 57	16 23 45 17	18 2 23 46	16 5 52 08
933	15 18 05 39	17 5 43 23	18 8 09 49	16 11 41 37
934	15 23 59 58	17 11 33 00	18 13 56 14	16 17 32 07
935	16 5 40 57	17 17 17 45	18 19 52 40	16 23 29 52
936	15 11 37 04	16 23 16 19	18 1 43 32	16 5 13 33
937	15 17 25 52	17 5 06 27	18 7 35 16	16 11 03 34
938	15 23 17 44	17 10 53 25	18 13 26 26	16 16 59 28
939	16 5 11 13	17 16 41 59	18 19 16 39	16 22 43 46
940	15 10 47 49	16 22 27 22	18 1 11 05	16 4 39 07
941	15 16 40 06	17 4 22 05	18 6 56 34	16 10 32 19
942	15 22 34 46	17 10 06 54	18 12 39 48	16 16 20 09
943	16 4 15 10	17 15 43 24	18 18 32 07	16 22 12 45
944	15 10 10 40	16 21 37 30	18 0 20 27	16 3 55 02
945	15 15 58 02	17 3 23 28	18 6 03 22	16 9 40 42
946	15 21 43 24	17 9 06 15	18 11 46 29	16 15 34 56
947	16 3 35 11	17 14 55 33	18 17 31 38	16 21 14 56
948	15 9 10 58	16 20 39 04	17 23 21 32	16 3 01 51
949	15 15 01 53	17 2 33 25	18 5 10 15	16 8 50 37
950	15 20 56 20	17 8 20 06	18 10 57 39	16 14 37 15

EQUINOXES AND SOLSTICES

Year	March Equinox	June Solstice	September Equinox	December Solstice
951	16^d 2^{h}33^{m}36^s	17^d 13^{h}57^{m}45^s	18^d 16^{h}53^{m}20^s	16^d 20^{h}34^{m}10^s
952	15 8 25 58	16 19 55 00	17 22 46 01	16 2 25 25
953	15 14 17 03	17 1 45 01	18 4 32 47	16 8 18 36
954	15 20 08 43	17 7 29 05	18 10 20 57	16 14 17 21
955	16 2 08 30	17 13 22 21	18 16 13 27	16 20 02 17
956	15 7 54 01	16 19 10 18	17 22 06 15	16 1 52 04
957	15 13 47 36	17 1 08 01	18 3 55 56	16 7 45 29
958	15 19 44 08	17 7 00 13	18 9 41 59	16 13 34 20
959	16 1 21 53	17 12 38 39	18 15 34 09	16 19 27 00
960	15 7 11 36	16 18 34 25	17 21 25 42	16 1 11 42
961	15 13 01 25	17 0 21 31	18 3 11 24	16 6 57 08
962	15 18 46 36	17 5 58 02	18 8 56 28	16 12 48 59
963	16 0 35 20	17 11 45 39	18 14 45 33	16 18 34 00
964	15 6 13 56	16 17 28 17	17 20 31 36	16 0 22 45
965	15 12 00 38	16 23 17 47	18 2 15 31	16 6 15 23
966	15 17 57 51	17 5 06 37	18 8 00 08	16 12 03 07
967	15 23 39 34	17 10 41 22	18 13 49 59	16 17 51 46
968	15 5 31 39	16 16 38 27	17 19 43 24	15 23 39 20
969	15 11 25 58	16 22 31 58	18 1 29 12	16 5 27 52
970	15 17 12 23	17 4 13 32	18 7 14 21	16 11 22 50
971	15 23 04 32	17 10 09 19	18 13 08 50	16 17 10 57
972	15 4 49 27	16 15 58 31	17 19 00 07	15 22 57 50
973	15 10 39 19	16 21 50 59	18 0 51 34	16 4 50 28
974	15 16 37 25	17 3 44 22	18 6 45 06	16 10 42 50
975	15 22 17 14	17 9 20 50	18 12 37 03	16 16 36 07
976	15 4 05 09	16 15 16 46	17 18 31 54	15 22 31 12
977	15 10 00 26	16 21 08 28	18 0 16 19	16 4 21 57
978	15 15 48 49	17 2 43 27	18 5 57 34	16 10 14 18
979	15 21 43 36	17 8 35 19	18 11 51 06	16 15 59 52
980	15 3 30 28	16 14 22 36	17 17 36 49	15 21 43 15
981	15 9 17 29	16 20 12 02	17 23 20 05	16 3 32 44
982	15 15 10 40	17 2 04 42	18 5 07 09	16 9 21 34
983	15 20 48 37	17 7 40 51	18 10 53 22	16 15 06 34
984	15 2 33 03	16 13 33 49	17 16 46 10	15 20 52 13
985	15 8 25 38	16 19 25 53	17 22 34 30	16 2 39 01
986	15 14 10 33	17 0 59 33	18 4 17 52	16 8 29 44
987	15 19 56 38	17 6 50 14	18 10 14 20	16 14 21 30
988	15 1 41 47	16 12 39 23	17 16 02 06	15 20 12 34
989	15 7 30 41	16 18 26 47	17 21 46 28	16 2 06 42
990	15 13 29 21	17 0 20 30	18 3 39 18	16 8 01 04
991	15 19 18 11	17 6 00 27	18 9 30 26	16 13 49 50
992	15 1 08 52	16 11 56 13	17 15 24 34	15 19 39 59
993	15 7 05 49	16 17 55 07	17 21 14 22	16 1 34 20
994	15 12 56 07	16 23 33 58	18 2 55 27	16 7 24 35
995	15 18 43 04	17 5 27 00	18 8 51 40	16 13 13 25
996	15 0 31 50	16 11 19 48	17 14 41 26	15 18 58 52
997	15 6 19 44	16 17 04 04	17 20 25 16	16 0 45 26
998	15 12 11 08	16 22 55 16	18 2 19 23	16 6 38 40
999	15 17 52 27	17 4 32 25	18 8 06 19	16 12 26 26
1000	14 23 34 47	16 10 21 34	17 13 54 57	15 18 16 17

Year	March Equinox	June Solstice	September Equinox	December Solstice
1001	15^d 5^{h}27^{m}50^s	16^d 16^{h}15^{m}16^s	17^d 19^{h}42^{m}02^s	16^d 0^{h}09^{m}08^s
1002	15 11 18 01	16 21 48 07	18 1 19 48	16 5 54 35
1003	15 17 05 24	17 3 36 34	18 7 14 57	16 11 41 19
1004	14 22 54 26	16 9 30 04	17 13 03 25	15 17 28 18
1005	15 4 42 12	16 15 15 19	17 18 42 47	15 23 15 47
1006	15 10 32 17	16 21 09 10	18 0 37 06	16 5 10 53
1007	15 16 16 40	17 2 51 28	18 6 25 45	16 10 57 09
1008	14 22 02 40	16 8 42 20	17 12 18 18	15 16 45 06
1009	15 3 58 01	16 14 39 08	17 18 13 32	15 22 39 30
1010	15 9 48 59	16 20 15 33	17 23 56 04	16 4 29 42
1011	15 15 35 05	17 2 06 05	18 5 54 11	16 10 23 57
1012	14 21 24 21	16 8 02 08	17 11 45 34	15 16 18 51
1013	15 3 17 40	16 13 47 26	17 17 25 07	15 22 08 00
1014	15 9 12 06	16 19 39 01	17 23 20 43	16 4 02 34
1015	15 15 02 25	17 1 23 16	18 5 10 34	16 9 48 55
1016	14 20 50 21	16 7 13 27	17 10 57 35	15 15 33 46
1017	15 2 41 15	16 13 10 53	17 16 48 57	15 21 28 17
1018	15 8 30 42	16 18 48 26	17 22 26 18	16 3 12 59
1019	15 14 11 53	17 0 34 28	18 4 19 06	16 8 57 25
1020	14 19 56 38	16 6 27 43	17 10 10 46	15 14 45 02
1021	15 1 45 39	16 12 08 19	17 15 49 27	15 20 27 12
1022	15 7 29 59	16 17 54 06	17 21 44 00	16 2 21 39
1023	15 13 12 46	16 23 36 30	18 3 33 48	16 8 13 55
1024	14 18 58 00	16 5 22 02	17 9 17 01	15 14 00 40
1025	15 0 49 22	16 11 16 05	17 15 09 25	15 19 59 47
1026	15 6 48 17	16 16 55 31	17 20 50 15	16 1 46 44
1027	15 12 37 24	16 22 43 15	18 2 44 43	16 7 33 42
1028	14 18 28 27	16 4 44 33	17 8 40 31	15 13 28 58
1029	15 0 24 14	16 10 33 50	17 14 19 23	15 19 15 14
1030	15 6 12 46	16 16 26 17	17 20 15 49	16 1 11 04
1031	15 12 00 07	16 22 16 10	18 2 10 20	16 7 01 38
1032	14 17 49 42	16 4 05 36	17 7 57 59	15 12 43 50
1033	14 23 39 28	16 10 00 07	17 13 55 44	15 18 41 07
1034	15 5 32 00	16 15 39 27	17 19 38 22	16 0 30 07
1035	15 11 13 59	16 21 22 29	18 1 28 40	16 6 17 52
1036	14 16 57 30	16 3 17 03	17 7 21 09	15 12 13 41
1037	14 22 52 04	16 8 59 37	17 12 55 00	15 17 56 12
1038	15 4 40 10	16 14 42 54	17 18 47 03	15 23 46 23
1039	15 10 25 37	16 20 28 16	18 0 38 17	16 5 34 43
1040	14 16 14 03	16 2 15 32	17 6 18 47	15 11 55 50
1041	14 22 00 43	16 8 08 08	17 12 10 22	15 17 11 37
1042	15 3 51 41	16 13 49 24	17 17 50 41	15 22 59 03
1043	15 9 36 51	16 19 34 24	17 23 40 35	16 4 41 08
1044	14 15 21 05	16 1 30 34	17 5 38 34	15 10 35 09
1045	14 21 18 09	16 7 18 18	17 11 19 41	15 16 20 26
1046	15 3 04 55	16 13 03 54	17 17 15 04	15 22 15 49
1047	15 8 48 41	16 18 54 24	17 23 12 48	16 4 15 12
1048	14 14 42 25	16 0 45 46	17 4 56 21	15 10 01 35
1049	14 20 34 03	16 6 38 34	17 10 51 48	15 16 01 16
1050	15 2 33 08	16 12 22 55	17 16 38 22	15 21 52 20

EQUINOXES AND SOLSTICES

Year	March Equinox	June Solstice	September Equinox	December Solstice
1051	15^d 8^{h}24^{m}38^s	16^d 18^{h}09^{m}19^s	17^d 22^{h}28^{m}04^s	16^d 3^{h}34^{m}21^s
1052	14 14 08 53	16 0 06 50	17 4 23 50	15 9 30 34
1053	14 20 05 37	16 5 56 46	17 10 01 02	15 15 15 04
1054	15 1 49 51	16 11 39 40	17 15 49 20	15 21 02 13
1055	15 7 29 07	16 17 26 48	17 21 45 20	16 2 54 13
1056	14 13 20 03	15 23 13 39	17 3 26 02	15 8 30 08
1057	14 19 03 02	16 4 58 23	17 9 17 45	15 14 24 18
1058	15 0 52 17	16 10 38 42	17 15 02 39	15 20 16 50
1059	15 6 36 54	16 16 20 15	17 20 46 52	16 1 59 22
1060	14 12 17 32	15 22 12 46	17 2 40 04	15 7 57 10
1061	14 18 17 19	16 4 01 00	17 8 17 51	15 13 42 53
1062	15 0 08 50	16 9 43 59	17 14 07 31	15 19 29 49
1063	15 5 52 51	16 15 35 14	17 20 06 42	16 1 25 50
1064	14 11 49 22	15 21 31 36	17 1 50 17	15 7 08 49
1065	14 17 37 36	16 3 23 44	17 7 43 40	15 13 06 14
1066	14 23 29 17	16 9 11 25	17 13 34 33	15 19 01 20
1067	15 5 20 51	16 14 59 53	17 19 24 28	16 0 41 56
1068	14 11 03 48	15 20 53 44	17 1 24 01	15 6 37 55
1069	14 17 01 38	16 2 43 51	17 7 07 43	15 12 27 02
1070	14 22 49 35	16 8 25 43	17 12 57 13	15 18 16 55
1071	15 4 26 31	16 14 11 57	17 18 53 33	16 0 15 28
1072	14 10 20 54	15 20 03 35	17 0 33 15	15 5 59 25
1073	14 16 11 10	16 1 47 34	17 6 20 28	15 11 50 44
1074	14 22 01 23	16 7 28 26	17 12 08 10	15 17 42 03
1075	15 3 53 57	16 13 15 33	17 17 52 16	15 23 19 17
1076	14 9 33 10	15 19 06 06	16 23 42 57	15 5 12 03
1077	14 15 27 03	16 0 56 54	17 5 22 29	15 10 59 45
1078	14 21 14 47	16 6 39 19	17 11 07 27	15 16 41 51
1079	15 2 50 18	16 12 25 37	17 17 05 42	15 22 35 10
1080	14 8 45 20	15 18 20 10	16 22 51 34	15 4 16 56
1081	14 14 33 04	16 0 05 13	17 4 42 38	15 10 10 02
1082	14 20 18 37	16 5 48 17	17 10 36 00	15 16 10 29
1083	15 2 11 37	16 11 38 38	17 16 24 00	15 21 56 08
1084	14 7 55 09	15 17 29 17	16 22 17 10	15 3 54 14
1085	14 13 56 10	15 23 21 33	17 4 03 31	15 9 47 37
1086	14 19 53 45	16 5 06 57	17 9 51 37	15 15 31 34
1087	15 1 34 08	16 10 55 32	17 15 50 13	15 21 29 27
1088	14 7 32 00	15 16 54 40	16 21 34 34	15 3 14 29
1089	14 13 21 07	15 22 41 37	17 3 20 09	15 9 04 26
1090	14 19 05 52	16 4 24 56	17 9 11 35	15 14 58 55
1091	15 0 58 12	16 10 14 54	17 14 58 46	15 20 36 16
1092	14 6 38 29	15 16 01 21	16 20 50 06	15 2 25 44
1093	14 12 29 26	15 21 48 20	17 2 34 56	15 8 16 57
1094	14 18 18 26	16 3 30 28	17 8 19 17	15 14 01 07
1095	14 23 51 12	16 9 12 08	17 14 12 01	15 19 57 14
1096	14 5 45 02	15 15 06 44	16 19 55 19	15 1 43 23
1097	14 11 38 34	15 20 50 32	17 1 40 06	15 7 30 37
1098	14 17 24 32	16 2 30 45	17 7 32 20	15 13 25 04
1099	14 23 19 42	16 8 25 02	17 13 20 57	15 19 07 23
1100	14 5 03 04	15 14 15 39	16 19 10 46	15 0 59 32

Year	March Equinox	June Solstice	September Equinox	December Solstice
	d h m s	d h m s	d h m s	d h m s
1101	14 10 54 46	15 20 07 10	17 0 57 28	15 6 54 03
1102	14 16 49 29	16 1 56 13	17 6 46 34	15 12 39 05
1103	14 22 27 08	16 7 40 16	17 12 43 32	15 18 32 03
1104	14 4 21 46	15 13 37 47	16 18 34 41	15 0 21 36
1105	14 10 17 20	15 19 25 34	17 0 23 13	15 6 12 46
1106	14 15 59 06	16 1 04 45	17 6 15 37	15 12 12 49
1107	14 21 53 24	16 6 59 52	17 12 05 10	15 18 00 34
1108	14 3 40 06	15 12 47 21	16 17 51 32	14 23 50 35
1109	14 9 34 10	15 18 34 22	16 23 37 34	15 5 42 44
1110	14 15 31 47	16 0 22 20	17 5 24 12	15 11 24 10
1111	14 21 08 22	16 6 05 01	17 11 14 49	15 17 14 16
1112	14 2 58 19	15 12 02 00	16 17 00 59	14 23 01 47
1113	14 8 50 09	15 17 49 28	16 22 43 55	15 4 46 56
1114	14 14 28 21	15 23 25 20	17 4 32 54	15 10 38 30
1115	14 20 18 59	16 5 18 15	17 10 24 40	15 16 20 19
1116	14 2 02 59	15 11 03 59	16 16 12 17	14 22 06 36
1117	14 7 50 05	15 16 48 00	16 22 00 15	15 4 02 36
1118	14 13 42 24	15 22 36 04	17 3 48 29	15 9 51 10
1119	14 19 19 22	16 4 17 41	17 9 38 51	15 15 47 02
1120	14 1 14 02	15 10 14 00	16 15 28 09	14 21 39 23
1121	14 7 15 33	15 16 04 20	16 21 15 18	15 3 28 19
1122	14 13 03 16	15 21 44 08	17 3 07 12	15 9 23 22
1123	14 18 58 47	16 3 43 57	17 9 01 34	15 15 11 55
1124	14 0 49 07	15 9 38 22	16 14 49 00	14 21 01 02
1125	14 6 38 36	15 15 25 50	16 20 35 36	15 2 54 45
1126	14 12 32 59	15 21 18 36	17 2 27 22	15 8 39 51
1127	14 18 11 48	16 3 00 56	17 8 18 58	15 14 26 31
1128	13 23 59 29	15 8 53 52	16 14 10 15	14 20 15 34
1129	14 5 53 28	15 14 42 09	16 19 56 53	15 2 05 14
1130	14 11 31 06	15 20 14 09	17 1 42 51	15 7 59 05
1131	14 17 18 12	16 2 06 10	17 7 33 11	15 13 48 01
1132	13 23 08 17	15 7 54 01	16 13 16 34	14 19 33 46
1133	14 4 57 40	15 13 32 52	16 18 58 42	15 1 22 53
1134	14 10 51 56	15 19 23 22	17 0 49 27	15 7 09 38
1135	14 16 32 22	16 1 06 35	17 6 35 33	15 12 55 52
1136	13 22 18 13	15 7 00 11	16 12 22 56	14 18 46 39
1137	14 4 15 32	15 12 54 35	16 18 10 56	15 0 36 55
1138	14 9 57 50	15 18 30 17	16 23 59 12	15 6 27 04
1139	14 15 48 07	16 0 26 48	17 5 57 50	15 12 16 27
1140	13 21 41 30	15 6 20 47	16 11 48 30	14 18 05 47
1141	14 3 30 33	15 12 02 50	16 17 34 51	15 0 01 42
1142	14 9 23 16	15 17 56 41	16 23 30 06	15 5 56 22
1143	14 15 07 24	15 23 43 07	17 5 18 37	15 11 47 12
1144	13 20 58 37	15 5 35 09	16 11 07 38	14 17 38 26
1145	14 2 59 53	15 11 28 50	16 16 58 16	14 23 28 52
1146	14 8 45 25	15 17 04 14	16 22 43 52	15 5 17 04
1147	14 14 32 19	15 22 58 55	17 4 37 30	15 11 05 34
1148	13 20 22 09	15 4 52 53	16 10 21 59	14 16 52 02
1149	14 2 07 49	15 10 30 42	16 16 01 16	14 22 39 47
1150	14 7 54 51	15 16 19 16	16 21 53 33	15 4 25 10

Year	March Equinox	June Solstice	September Equinox	December Solstice
1151	14^d $13^h35^m22^s$	15^d $22^h02^m04^s$	17^d $3^h40^m32^s$	15^d $10^h08^m42^s$
1152	13 19 20 35	15 3 48 21	16 9 27 05	14 15 55 57
1153	14 1 12 28	15 9 38 59	16 15 17 03	14 21 50 15
1154	14 6 54 40	15 15 13 24	16 21 00 55	15 3 41 47
1155	14 12 40 47	15 21 05 01	17 2 54 07	15 9 33 34
1156	13 18 37 17	15 3 01 07	16 8 43 18	14 15 24 16
1157	14 0 32 42	15 8 41 03	16 14 25 22	14 21 13 06
1158	14 6 25 46	15 14 35 56	16 20 23 17	15 3 05 40
1159	14 12 13 49	15 20 29 28	17 2 13 40	15 8 55 15
1160	13 18 02 29	15 2 22 00	16 8 00 56	14 14 45 01
1161	13 23 57 48	15 8 19 21	16 13 55 57	14 20 40 29
1162	14 5 44 27	15 13 57 52	16 19 43 56	15 2 27 23
1163	14 11 29 51	15 19 48 25	17 1 40 29	15 8 15 31
1164	13 17 21 42	15 1 43 44	16 7 32 55	14 14 07 50
1165	13 23 09 50	15 7 18 47	16 13 10 35	14 19 56 36
1166	14 4 53 38	15 13 05 43	16 19 03 35	15 1 50 57
1167	14 10 39 17	15 18 52 40	17 0 48 08	15 7 37 59
1168	13 16 27 35	15 0 35 21	16 6 27 42	14 13 21 17
1169	13 22 22 56	15 6 27 15	16 12 19 40	14 19 13 18
1170	14 4 09 49	15 12 04 57	16 18 02 22	15 0 57 55
1171	14 9 53 20	15 17 55 16	16 23 52 39	15 6 44 43
1172	13 15 43 15	14 23 53 21	16 5 42 49	14 12 36 28
1173	13 21 34 03	15 5 32 23	16 11 20 42	14 18 21 24
1174	14 3 19 32	15 11 21 20	16 17 18 10	15 0 11 14
1175	14 9 06 19	15 17 13 55	16 23 12 07	15 6 00 22
1176	13 14 55 55	14 23 00 43	16 4 57 41	14 11 48 11
1177	13 20 47 20	15 4 54 49	16 10 55 20	14 17 49 48
1178	14 2 35 42	15 10 36 48	16 16 42 47	14 23 43 30
1179	14 8 24 10	15 16 26 39	16 22 34 58	15 5 34 02
1180	13 14 19 47	14 22 24 46	16 4 29 33	14 11 29 04
1181	13 20 18 44	15 4 05 12	16 10 09 02	14 17 15 30
1182	14 2 06 13	15 9 53 09	16 16 02 44	14 23 05 07
1183	14 7 51 14	15 15 47 23	16 21 52 17	15 4 55 43
1184	13 13 41 22	14 21 34 13	16 3 30 28	14 10 37 15
1185	13 19 28 36	15 3 24 29	16 9 23 14	14 16 30 08
1186	14 1 14 47	15 9 05 05	16 15 09 57	14 22 14 48
1187	14 6 58 13	15 14 49 15	16 20 58 48	15 3 56 14
1188	13 12 43 56	14 20 43 10	16 2 53 01	14 9 50 59
1189	13 18 35 14	15 2 20 43	16 8 29 15	14 15 38 12
1190	14 0 16 43	15 8 03 52	16 14 20 18	14 21 29 44
1191	14 6 02 04	15 13 56 33	16 20 11 53	15 3 22 57
1192	13 11 57 37	14 19 42 08	16 1 51 29	14 9 04 33
1193	13 17 49 58	15 1 33 01	16 7 47 17	14 15 00 57
1194	13 23 40 41	15 7 20 03	16 13 37 37	14 20 52 35
1195	14 5 28 48	15 13 10 46	16 19 26 02	15 2 38 12
1196	13 11 17 56	14 19 10 45	16 1 23 38	14 8 37 00
1197	13 17 15 06	15 0 54 31	16 7 03 57	14 14 23 22
1198	13 23 00 43	15 6 39 27	16 12 58 59	14 20 12 14
1199	14 4 46 47	15 12 34 16	16 18 56 40	15 2 05 56
1200	13 10 40 07	14 18 20 33	16 0 36 39	14 7 49 39

Year	March Equinox	June Solstice	September Equinox	December Solstice
	d $\quad h \quad m \quad s$	d $\quad h \quad m \quad s$	d $\quad h \quad m \quad s$	d $\quad h \quad m \quad s$
1201	13 16 27 13	15 0 08 27	16 6 30 02	14 13 48 45
1202	13 22 13 35	15 5 52 12	16 12 18 10	14 19 42 18
1203	14 4 03 14	15 11 38 03	16 18 01 49	15 1 23 42
1204	13 9 52 40	14 17 31 08	15 23 55 42	14 7 17 06
1205	13 15 50 41	14 23 13 23	16 5 33 45	14 13 01 13
1206	13 21 34 56	15 4 56 32	16 11 21 42	14 18 45 31
1207	14 3 14 28	15 10 51 07	16 17 15 12	15 0 38 25
1208	13 9 06 52	14 16 39 24	15 22 52 04	14 6 18 19
1209	13 14 51 57	14 22 24 54	16 4 44 05	14 12 10 24
1210	13 20 36 18	15 4 09 18	16 10 37 43	14 18 01 43
1211	14 2 25 57	15 9 56 24	16 16 25 38	14 23 43 01
1212	13 8 10 31	14 15 48 45	15 22 22 17	14 5 42 19
1213	13 14 06 10	14 21 33 16	16 4 04 47	14 11 37 24
1214	13 19 54 02	15 3 16 33	16 9 53 15	14 17 26 03
1215	14 1 38 31	15 9 10 58	16 15 50 14	14 23 24 04
1216	13 7 42 28	14 15 03 19	15 21 31 44	14 5 07 11
1217	13 13 35 08	14 20 50 52	16 3 24 09	14 11 00 52
1218	13 19 22 16	15 2 41 09	16 9 19 14	14 16 56 57
1219	14 1 14 57	15 8 33 36	16 15 03 32	14 22 37 26
1220	13 6 59 35	14 14 26 59	15 20 57 10	14 4 32 15
1221	13 12 54 17	14 20 11 56	16 2 39 54	14 10 19 55
1222	13 18 39 57	15 1 53 17	16 8 28 02	14 15 59 27
1223	14 0 17 03	15 7 43 15	16 14 25 13	14 21 53 06
1224	13 6 10 25	14 13 31 13	15 20 04 21	14 3 36 39
1225	13 11 54 03	14 19 11 33	16 1 50 45	14 9 29 57
1226	13 17 34 02	15 0 54 30	16 7 43 12	14 15 26 51
1227	13 23 27 46	15 6 42 17	16 13 24 54	14 21 05 42
1228	13 5 15 51	14 12 30 26	15 19 17 12	14 2 59 29
1229	13 11 12 36	14 18 15 24	16 1 00 30	14 8 50 09
1230	13 17 01 48	15 0 01 00	16 6 46 15	14 14 33 24
1231	13 22 42 07	15 5 55 49	16 12 42 35	14 20 29 43
1232	13 4 39 53	14 11 50 55	15 18 24 53	14 2 15 07
1233	13 10 30 58	14 17 37 29	16 0 15 51	14 8 05 24
1234	13 16 13 44	14 23 24 40	16 6 16 37	14 14 02 21
1235	13 22 09 54	15 5 19 00	16 12 05 51	14 19 45 26
1236	13 3 56 17	14 11 08 31	15 17 59 30	14 1 43 19
1237	13 9 48 56	14 16 54 32	15 23 46 12	14 7 41 36
1238	13 15 40 40	14 22 40 08	16 5 31 35	14 13 25 16
1239	13 21 21 52	15 4 29 33	16 11 26 01	14 19 19 00
1240	13 3 21 55	14 10 21 50	15 17 08 13	14 1 03 04
1241	13 9 13 20	14 16 05 13	15 22 53 23	14 6 48 58
1242	13 14 50 13	14 21 49 21	16 4 47 34	14 12 43 58
1243	13 20 42 20	15 3 43 12	16 10 31 01	14 18 23 56
1244	13 2 25 20	14 9 28 23	15 16 17 25	14 0 12 37
1245	13 8 13 09	14 15 10 05	15 22 03 53	14 6 05 10
1246	13 14 03 56	14 20 53 59	16 3 50 05	14 11 42 43
1247	13 19 39 36	15 2 39 43	16 9 43 56	14 17 35 55
1248	13 1 34 22	14 8 32 37	15 15 28 38	13 23 27 14
1249	13 7 25 11	14 14 16 42	15 21 13 50	14 5 18 23
1250	13 13 04 47	14 20 01 10	16 3 10 51	14 11 18 43

Year	March Equinox	June Solstice	September Equinox	December Solstice
1251	13^d $19^h05^m17^s$	15^d $1^h58^m30^s$	16^d $8^h59^m41^s$	14^d $17^h03^m19^s$
1252	13 0 59 28	14 7 48 16	15 14 50 38	13 22 55 09
1253	13 6 53 23	14 13 35 53	15 20 41 31	14 4 53 08
1254	13 12 48 43	14 19 29 13	16 2 30 04	14 10 36 40
1255	13 18 28 34	15 1 20 24	16 8 23 14	14 16 30 05
1256	13 0 23 34	14 7 15 43	15 14 09 50	13 22 19 09
1257	13 6 16 30	14 13 01 11	15 19 56 22	14 4 03 20
1258	13 11 52 35	14 18 41 33	16 1 54 01	14 9 57 12
1259	13 17 44 30	15 0 35 07	16 7 42 17	14 15 40 57
1260	12 23 30 11	14 6 18 45	15 13 27 34	13 21 32 30
1261	13 5 14 09	14 11 56 21	15 19 11 56	14 3 29 14
1262	13 11 05 29	14 17 42 31	16 0 55 10	14 9 11 09
1263	13 16 47 55	14 23 26 37	16 6 42 50	14 14 58 49
1264	12 22 43 01	14 5 17 15	15 12 27 23	13 20 46 40
1265	13 4 38 43	14 11 05 30	15 18 11 33	14 2 31 38
1266	13 10 14 45	14 16 48 04	16 0 05 02	14 8 26 04
1267	13 16 07 37	14 22 48 04	16 5 55 06	14 14 12 32
1268	12 21 59 07	14 4 38 10	15 11 42 39	13 20 00 58
1269	13 3 46 02	14 10 20 10	15 17 34 54	14 1 57 12
1270	13 9 41 00	14 16 13 34	15 23 29 06	14 7 42 18
1271	13 15 23 27	14 22 01 36	16 5 22 19	14 13 34 52
1272	12 21 15 27	14 3 54 12	15 11 11 48	13 19 32 24
1273	13 3 12 04	14 9 44 06	15 16 58 46	14 1 23 48
1274	13 8 51 15	14 15 23 56	15 22 51 36	14 7 19 06
1275	13 14 48 08	14 21 21 17	16 4 43 46	14 13 06 32
1276	12 20 44 33	14 3 09 42	15 10 28 29	13 18 52 02
1277	13 2 29 42	14 8 48 08	15 16 14 27	14 0 47 16
1278	13 8 21 39	14 14 41 18	15 22 02 24	14 6 30 25
1279	13 14 01 12	14 20 27 05	16 3 46 43	14 12 15 06
1280	12 19 49 09	14 2 15 36	15 9 31 45	13 18 03 55
1281	13 1 42 33	14 8 02 51	15 15 17 50	13 23 46 51
1282	13 7 17 29	14 13 39 13	15 21 09 41	14 5 35 56
1283	13 13 05 35	14 19 33 37	16 3 01 42	14 11 24 50
1284	12 18 57 06	14 1 22 31	15 8 46 35	13 17 15 33
1285	13 0 40 29	14 6 58 35	15 14 32 14	13 23 13 34
1286	13 6 34 02	14 12 52 10	15 20 24 50	14 5 01 55
1287	13 12 23 49	14 18 41 03	16 2 13 39	14 10 49 12
1288	12 18 18 26	14 0 31 33	15 8 02 46	13 16 42 41
1289	13 0 17 14	14 6 26 50	15 13 53 02	13 22 33 39
1290	13 5 57 51	14 12 10 27	15 19 45 25	14 4 26 21
1291	13 11 48 14	14 18 09 29	16 1 38 55	14 10 16 35
1292	12 17 44 51	14 0 03 19	15 7 27 14	13 16 05 55
1293	12 23 30 43	14 5 38 24	15 13 14 16	13 21 57 06
1294	13 5 19 50	14 11 30 28	15 19 09 36	14 3 44 13
1295	13 11 05 44	14 17 18 52	16 0 56 37	14 9 31 21
1296	12 16 51 33	13 23 02 21	15 6 39 50	13 15 25 03
1297	12 22 44 29	14 4 52 30	15 12 26 40	13 21 16 13
1298	13 4 24 50	14 10 29 24	15 18 12 41	14 3 02 51
1299	13 10 14 51	14 16 21 47	16 0 03 04	14 8 48 59
1300	12 16 12 37	13 22 14 23	15 5 48 35	13 14 35 44

Year	March Equinox	June Solstice	September Equinox	December Solstice
1301	12^d $21^h56^m33^s$	14^d $3^h49^m18^s$	15^d $11^h30^m35^s$	13^d $20^h25^m52^s$
1302	13 3 42 55	14 9 43 42	15 17 24 16	14 2 13 50
1303	13 9 29 19	14 15 35 45	15 23 10 45	14 7 58 42
1304	12 15 16 44	13 21 20 04	15 4 55 53	13 13 48 33
1305	12 21 10 22	14 3 12 45	15 10 51 01	13 19 39 30
1306	13 2 51 54	14 8 53 20	15 16 43 00	14 1 28 23
1307	13 8 39 12	14 14 46 50	15 22 37 58	14 7 23 11
1308	12 14 35 52	13 20 41 37	15 4 27 09	13 13 19 19
1309	12 20 23 02	14 2 16 20	15 10 09 48	13 19 14 41
1310	13 2 15 52	14 8 10 18	15 16 06 03	14 1 05 32
1311	13 8 10 55	14 14 04 08	15 21 54 31	14 6 51 31
1312	12 14 04 18	13 19 49 21	15 3 37 33	13 12 41 07
1313	12 19 57 09	14 1 44 02	15 9 29 54	13 18 33 13
1314	13 1 38 57	14 7 28 15	15 15 16 57	14 0 19 06
1315	13 7 23 45	14 13 19 35	15 21 05 37	14 6 05 27
1316	12 13 17 09	13 19 13 16	15 2 54 43	13 11 54 27
1317	12 19 01 50	14 0 44 55	15 8 36 40	13 17 39 37
1318	13 0 44 29	14 6 33 04	15 14 32 18	13 23 26 58
1319	13 6 30 55	14 12 24 33	15 20 19 45	14 5 15 30
1320	12 12 17 14	13 18 03 28	15 1 58 21	13 11 06 01
1321	12 18 05 38	13 23 53 02	15 7 50 10	13 17 00 51
1322	12 23 52 09	14 5 36 03	15 13 38 39	13 22 47 59
1323	13 5 42 34	14 11 25 04	15 19 27 34	14 4 34 38
1324	12 11 40 16	13 17 22 23	15 1 20 03	13 10 31 01
1325	12 17 31 26	13 23 01 01	15 7 01 55	13 16 20 29
1326	12 23 16 53	14 4 54 24	15 12 57 30	13 22 11 21
1327	13 5 09 44	14 10 54 39	15 18 49 37	14 4 01 42
1328	12 11 02 36	13 16 38 03	15 0 31 35	13 9 48 18
1329	12 16 52 44	13 22 30 43	15 6 30 31	13 15 42 05
1330	12 22 38 43	14 4 17 33	15 12 23 05	13 21 30 35
1331	13 4 24 37	14 10 05 45	15 18 11 08	14 3 20 02
1332	12 10 16 20	13 16 00 14	15 0 03 02	13 9 18 54
1333	12 16 05 29	13 21 35 11	15 5 42 24	13 15 07 01
1334	12 21 51 24	14 3 21 19	15 11 34 27	13 20 53 15
1335	13 3 43 42	14 9 16 17	15 17 24 33	14 2 40 35
1336	12 9 35 55	13 14 56 28	14 23 00 02	13 8 23 50
1337	12 15 20 35	13 20 45 53	15 4 52 37	13 14 15 51
1338	12 21 02 43	14 2 33 13	15 10 40 24	13 20 01 49
1339	13 2 47 42	14 8 19 39	15 16 24 22	14 1 45 25
1340	12 8 37 40	13 14 11 53	14 22 17 49	13 7 39 05
1341	12 14 26 19	13 19 47 42	15 4 01 04	13 13 25 46
1342	12 20 10 56	14 1 34 25	15 9 55 36	13 19 14 41
1343	13 1 59 26	14 7 31 38	15 15 49 41	14 1 11 25
1344	12 7 54 30	13 13 15 14	14 21 27 40	13 7 02 00
1345	12 13 43 45	13 19 04 54	15 3 22 45	13 12 58 38
1346	12 19 35 18	14 0 56 53	15 9 17 12	13 18 49 34
1347	13 1 30 57	14 6 47 14	15 15 03 22	14 0 33 31
1348	12 7 24 44	13 12 44 26	14 20 59 29	13 6 32 05
1349	12 13 16 48	13 18 27 41	15 2 43 16	13 12 21 23
1350	12 19 00 58	14 0 15 48	15 8 33 59	13 18 06 37

EQUINOXES AND SOLSTICES

Year	March Equinox	June Solstice	September Equinox	December Solstice
1351	13^d $0^h47^m37^s$	14^d $6^h13^m11^s$	15^d $14^h27^m58^s$	13^d $23^h58^m32^s$
1352	12 6 41 51	13 11 54 32	14 20 05 15	13 5 39 55
1353	12 12 24 22	13 17 37 29	15 1 58 23	13 11 28 52
1354	12 18 05 58	13 23 24 46	15 7 51 50	13 17 19 41
1355	12 23 52 12	14 5 08 03	15 13 30 45	13 23 02 06
1356	12 5 36 47	13 10 56 07	14 19 21 07	13 5 01 46
1357	12 11 28 08	13 16 34 32	15 1 02 14	13 10 50 20
1358	12 17 15 31	13 22 17 19	15 6 49 27	13 16 32 09
1359	12 23 05 45	14 4 14 31	15 12 44 46	13 22 25 58
1360	12 5 04 29	13 10 00 48	14 18 21 40	13 4 11 08
1361	12 10 50 04	13 15 49 13	15 0 14 06	13 10 03 18
1362	12 16 34 52	13 21 44 26	15 6 10 53	13 15 56 58
1363	12 22 28 35	14 3 36 20	15 11 54 57	13 21 38 49
1364	12 4 18 23	13 9 29 00	14 17 53 00	13 3 36 30
1365	12 10 10 26	13 15 13 10	14 23 44 28	13 9 28 31
1366	12 15 57 48	13 20 59 35	15 5 35 55	13 15 14 33
1367	12 21 42 42	14 2 55 39	15 11 33 20	13 21 14 52
1368	12 3 39 56	13 8 40 41	14 17 10 13	13 3 04 25
1369	12 9 27 00	13 14 23 00	14 22 59 27	13 8 54 26
1370	12 15 12 14	13 20 12 16	15 4 54 26	13 14 45 29
1371	12 21 08 13	14 2 00 49	15 10 33 37	13 20 24 41
1372	12 2 55 29	13 7 48 21	14 16 22 27	13 2 18 10
1373	12 8 41 17	13 13 30 39	14 22 06 54	13 8 08 06
1374	12 14 27 22	13 19 16 27	15 3 50 48	13 13 46 24
1375	12 20 08 41	14 1 08 46	15 9 44 44	13 19 38 32
1376	12 2 04 39	13 6 53 45	14 15 24 02	13 1 22 42
1377	12 7 49 10	13 12 34 28	14 21 13 54	13 7 08 39
1378	12 13 28 12	13 18 24 31	15 3 12 40	13 13 05 24
1379	12 19 22 29	14 0 15 52	15 8 53 57	13 18 50 51
1380	12 1 10 15	13 6 03 32	14 14 45 05	13 0 50 16
1381	12 7 01 55	13 11 48 48	14 20 36 54	13 6 46 40
1382	12 12 58 19	13 17 38 19	15 2 25 56	13 12 27 45
1383	12 18 47 12	13 23 34 26	15 8 23 47	13 18 25 03
1384	12 0 47 54	13 5 26 28	14 14 06 28	13 0 16 20
1385	12 6 35 50	13 11 12 30	14 19 54 36	13 6 03 19
1386	12 12 15 33	13 17 05 06	15 1 53 39	13 11 59 36
1387	12 18 11 51	13 22 57 32	15 7 34 42	13 17 39 01
1388	11 23 58 09	13 4 40 59	14 13 24 05	12 23 30 28
1389	12 5 43 43	13 10 22 41	14 19 16 26	13 5 23 31
1390	12 11 32 09	13 16 08 22	15 1 00 42	13 11 02 59
1391	12 17 12 08	13 21 57 07	15 6 51 58	13 16 59 22
1392	11 23 06 22	13 3 42 14	14 12 30 21	12 22 50 04
1393	12 4 55 09	13 9 22 42	14 18 14 12	13 4 32 14
1394	12 10 35 45	13 15 09 56	15 0 10 56	13 10 24 36
1395	12 16 33 52	13 21 04 08	15 5 52 15	13 16 06 05
1396	11 22 22 16	13 2 50 29	14 11 38 56	12 21 58 09
1397	12 4 05 26	13 8 35 56	14 17 32 14	13 3 54 31
1398	12 9 58 11	13 14 28 51	14 23 19 21	13 9 34 46
1399	12 15 42 24	13 20 20 16	15 5 14 42	13 15 28 37
1400	11 21 38 37	13 2 09 22	14 11 02 32	12 21 21 26

Year	March Equinox	June Solstice	September Equinox	December Solstice
1401	12^d $3^h30^m03^s$	13^d $7^h54^m36^s$	14^d $16^h53^m20^s$	13^d $3^h07^m25^s$
1402	12 9 07 41	13 13 42 35	14 22 52 46	13 9 06 35
1403	12 15 03 59	13 19 38 08	15 4 36 50	13 14 57 49
1404	11 20 56 39	13 1 22 50	14 10 21 56	12 20 51 08
1405	12 2 42 21	13 7 04 16	14 16 15 02	13 2 47 31
1406	12 8 41 57	13 12 57 32	14 22 01 55	13 8 26 23
1407	12 14 28 36	13 18 46 47	15 3 51 18	13 14 16 52
1408	11 20 21 34	13 0 35 58	14 9 34 37	12 20 09 11
1409	12 2 10 54	13 6 22 06	14 15 18 09	13 1 49 04
1410	12 7 45 01	13 12 07 49	14 21 12 15	13 7 39 55
1411	12 13 37 59	13 18 01 36	15 2 56 14	13 13 22 39
1412	11 19 26 38	12 23 43 41	14 8 41 43	12 19 08 20
1413	12 1 04 28	13 5 21 43	14 14 36 34	13 1 04 20
1414	12 6 56 03	13 11 13 43	14 20 24 16	13 6 48 07
1415	12 12 38 59	13 17 00 10	15 2 11 56	13 12 42 42
1416	11 18 31 27	12 22 46 28	14 7 57 38	12 18 40 29
1417	12 0 27 50	13 4 33 34	14 13 44 25	13 0 23 09
1418	12 6 10 29	13 10 20 22	14 19 41 23	13 6 17 34
1419	12 12 09 30	13 16 19 05	15 1 28 41	13 12 06 44
1420	11 18 03 54	12 22 08 30	14 7 13 55	12 17 57 17
1421	11 23 45 57	13 3 52 27	14 13 09 19	12 23 55 12
1422	12 5 41 24	13 9 50 26	14 18 59 53	13 5 38 24
1423	12 11 29 44	13 15 40 09	15 0 49 55	13 11 27 02
1424	11 17 20 34	12 21 25 24	14 6 38 58	12 17 21 05
1425	11 23 13 07	13 3 13 33	14 12 27 48	12 23 04 03
1426	12 4 48 37	13 8 56 21	14 18 20 01	13 4 57 10
1427	12 10 38 29	13 14 49 41	15 0 04 51	13 10 48 25
1428	11 16 31 39	12 20 34 06	14 5 46 10	12 16 35 43
1429	11 22 11 40	13 2 09 02	14 11 37 13	12 22 28 13
1430	12 4 06 43	13 8 03 08	14 17 26 14	13 4 10 07
1431	12 9 55 22	13 13 51 07	14 23 10 55	13 9 56 49
1432	11 15 41 53	12 19 34 47	14 4 55 01	12 15 51 27
1433	11 21 33 47	13 1 25 55	14 10 42 45	12 21 34 59
1434	12 3 11 26	13 7 09 35	14 16 33 01	13 3 22 42
1435	12 9 01 55	13 13 03 38	14 22 22 21	13 9 12 58
1436	11 14 59 15	12 18 52 33	14 4 10 12	12 15 00 36
1437	11 20 38 01	13 0 29 25	14 10 05 03	12 20 57 31
1438	12 2 31 22	13 6 28 06	14 15 59 42	13 2 49 39
1439	12 8 23 25	13 12 19 13	14 21 46 28	13 8 42 36
1440	11 14 14 51	12 18 02 46	14 3 33 22	12 14 41 16
1441	11 20 14 24	12 23 55 57	14 9 25 11	12 20 26 30
1442	12 2 00 04	13 5 42 34	14 15 16 19	13 2 14 05
1443	12 7 52 15	13 11 38 40	14 21 04 27	13 8 05 19
1444	11 13 47 52	12 17 30 59	14 2 49 16	12 13 52 52
1445	11 19 24 39	12 23 07 28	14 8 38 38	12 19 43 50
1446	12 1 13 05	13 5 02 55	14 14 31 07	13 1 28 26
1447	12 7 02 23	13 10 50 49	14 20 16 15	13 7 11 59
1448	11 12 46 39	12 16 26 58	14 2 01 26	12 13 04 14
1449	11 18 35 21	12 22 14 56	14 7 51 44	12 18 49 45
1450	12 0 13 26	13 3 56 41	14 13 38 07	13 0 39 26

EQUINOXES AND SOLSTICES

Year	March Equinox			June Solstice			September Equinox			December Solstice		
1451	12^d	$6^h00^m25^s$		13^d	$9^h45^m26^s$		14^d	$19^h22^m05^s$		13^d	$6^h32^m04^s$	
1452	11	11 56 57		12	15 34 32		14	1 06 27		12	12 21 26	
1453	11	17 41 18		12	21 10 21		14	6 56 19		12	18 11 37	
1454	11	23 34 50		13	3 07 53		14	12 51 36		13	0 00 11	
1455	12	5 31 15		13	9 04 31		14	18 39 16		13	5 50 02	
1456	11	11 19 53		12	14 47 02		14	0 24 25		12	11 45 53	
1457	11	17 12 33		12	20 43 54		14	6 20 11		12	17 35 25	
1458	11	22 58 37		13	2 33 45		14	12 11 08		12	23 22 12	
1459	12	4 48 11		13	8 25 20		14	18 01 28		13	5 13 10	
1460	11	10 45 15		12	14 18 24		13	23 53 39		12	11 04 27	
1461	11	16 25 29		12	19 54 07		14	5 44 11		12	16 56 00	
1462	11	22 11 24		13	1 48 01		14	11 38 03		12	22 48 33	
1463	12	4 04 40		13	7 40 20		14	17 23 01		13	4 39 21	
1464	11	9 52 12		12	13 13 56		13	23 02 09		12	10 30 18	
1465	11	15 43 58		12	19 03 22		14	4 55 27		12	16 16 53	
1466	11	21 31 02		13	0 49 14		14	10 39 58		12	21 58 43	
1467	12	3 16 27		13	6 35 54		14	16 21 05		13	3 47 17	
1468	11	9 10 10		12	12 29 05		13	22 08 04		12	9 37 01	
1469	11	14 49 27		12	18 05 17		14	3 53 22		12	15 22 23	
1470	11	20 34 51		12	23 58 59		14	9 47 18		12	21 09 21	
1471	12	2 29 01		13	5 53 36		14	15 37 10		13	2 57 22	
1472	11	8 16 32		12	11 29 26		13	21 20 59		12	8 48 49	
1473	11	14 04 15		12	17 21 31		14	3 19 24		12	14 42 18	
1474	11	19 51 12		12	23 13 02		14	9 09 06		12	20 34 05	
1475	12	1 40 51		13	5 01 07		14	14 53 30		13	2 28 33	
1476	11	7 39 36		12	10 55 56		13	20 47 09		12	8 24 06	
1477	11	13 29 40		12	16 36 28		14	2 37 20		12	14 12 07	
1478	11	19 20 52		12	22 31 58		14	8 32 31		12	20 01 54	
1479	12	1 17 08		13	4 31 28		14	14 22 36		13	1 54 27	
1480	11	7 06 01		12	10 09 45		13	20 02 40		12	7 44 22	
1481	11	12 51 05		12	16 00 09		14	1 57 32		12	13 32 08	
1482	11	18 36 55		12	21 50 21		14	7 45 33		12	19 16 58	
1483	12	0 24 13		13	3 32 19		14	13 26 18		13	1 00 55	
1484	11	6 12 55		12	9 20 34		13	19 18 09		12	6 52 46	
1485	11	11 54 45		12	14 57 51		14	1 04 13		12	12 39 58	
1486	11	17 36 43		12	20 45 19		14	6 52 19		12	18 29 08	
1487	11	23 28 49		13	2 39 55		14	12 40 18		13	0 23 04	
1488	11	5 20 41		12	8 14 11		13	18 17 31		12	6 09 39	
1489	11	11 09 07		12	14 02 48		14	0 13 37		12	11 57 43	
1490	11	17 00 02		12	19 58 13		14	6 04 07		12	17 46 03	
1491	11	22 50 13		13	1 45 41		14	11 44 33		12	23 34 02	
1492	11	4 41 04		12	7 40 44		13	17 39 54		12	5 30 56	
1493	11	10 28 12		12	13 26 00		13	23 30 42		12	11 19 16	
1494	11	16 15 40		12	19 17 31		14	5 23 43		12	17 06 43	
1495	11	22 10 35		13	1 15 57		14	11 21 08		12	23 02 29	
1496	11	4 03 13		12	6 53 33		13	17 03 12		12	4 51 36	
1497	11	9 47 31		12	12 41 21		13	23 00 25		12	10 45 52	
1498	11	15 35 30		12	18 35 26		14	4 50 36		12	16 38 52	
1499	11	21 26 04		13	0 17 52		14	10 26 24		12	22 25 41	
1500	11	3 18 41		12	6 07 03		13	16 19 21		12	4 19 21	

Year	March Equinox	June Solstice	September Equinox	December Solstice
1501	11^d $9^h08^m11^s$	12^d $11^h49^m31^s$	13^d $22^h07^m12^s$	12^d $10^h04^m04^s$
1502	11 14 55 52	12 17 38 21	14 3 52 55	12 15 47 22
1503	11 20 45 36	12 23 35 03	14 9 43 47	12 21 41 18
1504	11 2 35 36	12 5 13 47	13 15 20 21	12 3 25 55
1505	11 8 17 08	12 · 10 59 12	13 21 12 37	12 9 10 57
1506	11 14 02 01	12 16 53 23	14 3 05 33	12 14 59 02
1507	11 19 52 25	12 22 36 03	14 8 44 17	12 20 41 08
1508	11 1 38 06	12 4 23 00	13 14 40 21	12 2 36 46
1509	11 7 22 33	12 10 07 46	13 20 32 13	12 8 29 54
1510	11 13 09 50	12 15 56 01	14 2 18 46	12 14 18 54
1511	11 19 01 40	12 21 51 42	14 8 13 29	12 20 19 22
1512	11 1 01 05	12 3 32 58	13 13 55 43	12 2 09 00
1513	11 6 51 37	12 9 19 35	13 19 49 51	12 7 55 54
1514	11 12 41 22	12 15 18 55	14 1 45 23	12 13 51 16
1515	11 18 38 20	12 21 08 44	14 7 22 41	12 19 36 04
1516	11 0 25 11	12 2 58 08	13 13 16 12	12 1 30 29
1517	11 6 11 35	12 8 47 13	13 19 09 38	12 7 20 23
1518	11 12 00 32	12 14 35 03	14 0 55 04	12 13 00 23
1519	11 17 47 59	12 20 27 30	14 6 51 02	12 18 55 49
1520	10 23 40 06	12 2 06 25	13 12 31 58	12 0 43 30
1521	11 5 21 26	12 7 47 45	13 18 20 45	12 6 29 48
1522	11 11 03 07	12 13 40 51	14 0 12 50	12 12 25 15
1523	11 16 57 37	12 19 24 45	14 5 46 18	12 18 07 32
1524	10 22 45 29	12 1 07 17	13 11 36 54	11 23 57 59
1525	11 4 32 28	12 6 54 16	13 17 30 44	12 5 48 00
1526	11 10 23 22	12 12 43 56	13 23 13 02	12 11 28 59
1527	11 16 10 07	12 18 38 03	14 5 07 21	12 17 27 30
1528	10 22 03 18	12 0 21 59	13 10 49 38	11 23 17 01
1529	11 3 49 31	12 6 07 09	13 16 40 19	12 5 01 43
1530	11 9 35 14	12 12 03 13	13 22 38 53	12 10 56 19
1531	11 15 32 52	12 17 51 53	14 4 19 18	12 16 41 03
1532	10 21 21 07	11 23 37 40	13 10 13 45	11 22 36 13
1533	11 3 04 51	12 5 27 22	13 16 11 27	12 4 34 52
1534	11 8 58 25	12 11 19 21	13 21 54 40	12 10 20 17
1535	11 14 48 26	12 17 10 17	14 3 48 04	12 16 18 50
1536	10 20 46 11	11 22 53 53	13 9 33 03	11 22 08 49
1537	11 2 37 27	12 4 38 46	13 15 20 53	12 3 49 07
1538	11 8 20 05	12 10 34 01	13 21 15 32	12 9 43 23
1539	11 14 15 28	12 16 24 09	14 2 51 20	12 15 26 27
1540	10 19 59 25	11 22 06 26	13 8 38 31	11 21 13 15
1541	11 1 37 46	12 3 52 46	13 14 34 19	12 3 04 31
1542	11 7 28 59	12 9 41 49	13 20 17 01	12 8 41 34
1543	11 13 12 23	12 15 26 42	14 2 09 13	12 14 35 22
1544	10 19 01 00	11 21 07 26	13 7 55 49	11 20 30 05
1545	11 0 47 27	12 2 49 34	13 13 40 45	12 2 13 18
1546	11 6 27 03	12 8 40 16	13 19 34 00	12 8 12 34
1547	11 12 28 30	12 14 30 56	14 1 12 41	12 14 00 01
1548	10 18 21 46	11 20 14 05	13 7 01 40	11 19 47 48
1549	11 0 07 37	12 2 05 50	13 13 02 10	12 1 44 54
1550	11 6 06 17	12 8 04 14	13 18 46 53	12 7 28 09

EQUINOXES AND SOLSTICES

Year	March Equinox		June Solstice		September Equinox		December Solstice	
1551	11^d	$11^h54^m56^s$	12^d	$13^h56^m30^s$	14^d	$0^h39^m35^s$	12^d	$13^h25^m14^s$
1552	10	17 46 38	11	19 44 10	13	6 29 57	11	19 20 29
1553	10	23 38 27	12	1 32 46	13	12 18 53	12	1 00 07
1554	11	5 19 58	12	7 24 13	13	18 16 11	12	6 54 19
1555	11	11 16 37	12	13 14 16	13	23 58 56	12	12 41 26
1556	10	17 03 42	11	18 54 57	13	5 45 46	11	18 28 52
1557	10	22 39 25	12	0 39 42	13	11 42 17	12	0 26 57
1558	11	4 32 47	12	6 31 45	13	17 22 16	12	6 09 28
1559	11	10 20 40	12	12 14 37	13	23 08 45	12	12 01 18
1560	10	16 10 23	11	17 54 16	13	4 55 52	11	17 52 31
1561	10	22 02 49	11	23 40 45	13	10 39 55	11	23 30 42
1562	11	3 43 24	12	5 30 29	13	16 30 10	12	5 23 28
1563	11	9 36 55	12	11 21 54	13	22 09 56	12	11 11 46
1564	10	15 26 59	11	17 06 46	13	3 55 31	11	16 55 45
1565	10	21 04 17	11	22 53 08	13	9 54 34	11	22 50 23
1566	11	3 01 02	12	4 50 36	13	15 42 47	12	4 33 29
1567	11	8 50 58	12	10 37 24	13	21 34 17	12	10 26 51
1568	10	14 37 43	11	16 21 21	13	3 28 32	11	16 27 41
1569	10	20 31 50	11	22 13 20	13	9 17 40	11	22 13 56
1570	11	2 15 11	12	4 03 25	13	15 10 24	12	4 11 30
1571	11	8 14 28	12	9 55 28	13	20 55 41	12	10 04 10
1572	10	14 12 24	11	15 41 16	13	2 43 13	11	15 47 40
1573	10	19 52 12	11	21 27 22	13	8 40 19	11	21 42 55
1574	11	1 48 01	12	3 26 28	13	14 25 41	12	3 27 53
1575	11	7 36 10	12	9 12 27	13	20 09 36	12	9 16 05
1576	10	13 18 06	11	14 52 24	13	1 59 09	11	15 10 46
1577	10	19 09 59	11	20 41 04	13	7 45 03	11	20 47 09
1578	11	0 48 24	12	2 24 22	13	13 33 29	12	2 35 01
1579	11	6 39 05	12	8 11 19	13	19 17 48	12	8 25 38
1580	10	12 28 45	11	13 53 55	13	1 01 41	11	14 09 49
1581	10	18 02 18	11	19 35 11	13	6 54 49	11	20 06 48
1582	10	23 56 37	12	1 31 12	13	12 39 52 J|G	22	1 54 39
1583	21	5 51 38	22	7 17 13	23	18 25 05	22	7 43 13
1584	20	11 39 51	21	12 58 00	23	0 17 48	21	13 39 13
1585	20	17 37 35	21	18 54 48	23	6 09 00	21	19 22 54
1586	20	23 22 51	22	0 46 58	23	11 59 14	22	1 15 35
1587	21	5 15 37	22	6 40 17	23	17 47 19	22	7 10 49
1588	20	11 11 23	21	12 31 16	22	23 36 33	21	12 55 58
1589	20	16 49 52	21	18 15 30	23	5 34 11	21	18 49 33
1590	20	22 43 55	22	0 13 10	23	11 26 01	22	0 37 35
1591	21	4 37 39	22	6 00 23	23	17 13 49	22	6 27 57
1592	20	10 18 21	21	11 36 17	22	23 03 25	21	12 25 42
1593	20	16 09 22	21	17 28 19	23	4 51 22	21	18 12 39
1594	20	21 55 12	21	23 14 10	23	10 35 01	22	0 00 59
1595	21	3 46 38	22	4 58 17	23	16 18 04	22	5 51 15
1596	20	9 44 02	21	10 46 26	22	22 04 04	21	11 32 26
1597	20	15 21 27	21	16 27 43	23	3 53 42	21	17 21 33
1598	20	21 10 42	21	22 25 01	23	9 40 49	21	23 09 18
1599	21	3 03 19	22	4 14 16	23	15 24 08	22	4 55 16
1600	20	8 42 46	21	9 50 19	22	21 12 58	21	10 47 48

Year	March Equinox	June Solstice	September Equinox	December Solstice
1601	20^d $14^h34^m20^s$	21^d $15^h44^m43^s$	23^d $3^h06^m36^s$	21^d $16^h31^m31^s$
1602	20 20 20 35	21 21 33 00	23 8 55 39	21 22 18 35
1603	21 2 08 52	22 3 18 18	23 14 44 05	22 4 14 46
1604	20 8 03 16	21 9 09 41	22 20 35 39	21 10 05 44
1605	20 13 42 09	21 14 52 28	23 2 27 18	21 16 01 32
1606	20 19 35 34	21 20 49 24	23 8 18 53	21 21 55 55
1607	21 1 37 23	22 2 40 31	23 14 06 00	22 3 44 54
1608	20 7 24 31	21 8 17 19	22 19 55 49	21 9 39 55
1609	20 13 19 58	21 14 15 08	23 1 49 17	21 15 27 15
1610	20 19 08 34	21 20 07 31	23 7 33 49	21 21 14 18
1611	21 0 56 55	22 1 53 14	23 13 17 31	22 3 06 15
1612	20 6 49 48	21 7 44 27	22 19 07 25	21 8 50 20
1613	20 12 28 11	21 13 25 34	23 0 57 00	21 14 35 27
1614	20 18 14 39	21 19 16 52	23 6 47 02	21 20 22 49
1615	21 0 07 48	22 1 05 35	23 12 33 10	22 2 11 19
1616	20 5 45 58	21 6 36 51	22 18 17 47	21 8 04 42
1617	20 11 32 36	21 12 28 47	23 0 09 23	21 13 54 20
1618	20 17 22 50	21 18 18 18	23 5 53 00	21 19 41 06
1619	20 23 13 43	21 23 58 29	23 11 35 44	22 1 31 18
1620	20 5 09 40	21 5 50 59	22 17 28 31	21 7 19 03
1621	20 10 52 57	21 11 37 28	22 23 17 55	21 13 07 52
1622	20 16 40 34	21 17 32 52	23 5 07 38	21 18 59 39
1623	20 22 37 41	21 23 28 56	23 10 57 52	22 0 52 37
1624	20 4 22 16	21 5 04 43	22 16 45 23	21 6 42 59
1625	20 10 11 32	21 10 59 16	22 22 43 32	21 12 32 36
1626	20 16 05 53	21 16 54 16	23 4 33 48	21 18 21 14
1627	20 21 54 34	21 22 34 37	23 10 17 30	22 0 14 43
1628	20 3 45 59	21 4 27 34	22 16 12 11	21 6 08 42
1629	20 9 29 31	21 10 12 53	22 21 59 08	21 11 58 06
1630	20 15 18 20	21 16 02 43	23 3 45 53	21 17 47 34
1631	20 21 17 36	21 21 55 12	23 9 35 02	21 23 36 51
1632	20 3 03 14	21 3 29 21	22 15 18 37	21 5 23 14
1633	20 8 48 55	21 9 22 06	22 21 11 29	21 11 11 00
1634	20 14 38 14	21 15 16 59	23 2 56 30	21 16 57 27
1635	20 20 23 58	21 20 54 54	23 8 34 16	21 22 44 22
1636	20 2 11 20	21 2 44 37	22 14 28 39	21 4 32 08
1637	20 7 53 48	21 8 29 25	22 20 17 17	21 10 16 04
1638	20 13 39 00	21 14 16 16	23 2 05 29	21 16 04 47
1639	20 19 31 46	21 20 08 04	23 7 57 09	21 22 00 01
1640	20 1 15 07	21 1 42 46	22 13 41 12	21 3 53 24
1641	20 7 02 59	21 7 34 16	22 19 35 11	21 9 46 34
1642	20 12 59 12	21 13 30 56	23 1 24 32	21 15 38 20
1643	20 18 56 39	21 19 12 14	23 7 05 54	21 21 27 19
1644	20 0 50 46	21 1 06 26	22 13 03 48	21 3 19 55
1645	20 6 39 46	21 7 00 48	22 18 54 19	21 9 09 12
1646	20 12 28 23	21 12 52 33	23 0 39 52	21 14 57 26
1647	20 18 21 58	21 18 48 59	23 6 33 45	21 20 51 37
1648	20 0 08 24	21 0 26 57	22 12 19 22	21 2 37 08
1649	20 5 52 39	21 6 15 23	22 18 14 18	21 8 23 17
1650	20 11 42 18	21 12 09 59	23 0 05 36	21 14 13 44

EQUINOXES AND SOLSTICES

Year	March Equinox	June Solstice	September Equinox	December Solstice
1651	20^d $17^h30^m28^s$	21^d $17^h45^m36^s$	23^d $5^h42^m28^s$	21^d $20^h01^m05^s$
1652	19 23 13 05	20 23 31 16	22 11 35 10	21 1 54 16
1653	20 4 57 27	21 5 19 01	22 17 21 40	21 7 43 06
1654	20 10 45 52	21 11 01 47	22 23 00 58	21 13 26 13
1655	20 16 39 44	21 16 52 29	23 4 53 32	21 19 19 59
1656	19 22 29 25	20 22 31 02	22 10 36 27	21 1 05 04
1657	20 4 13 22	21 4 20 13	22 16 26 34	21 6 52 34
1658	20 10 04 26	21 10 20 35	22 22 18 43	21 12 46 14
1659	20 15 57 28	21 16 01 47	23 3 56 48	21 18 32 01
1660	19 21 44 24	20 21 51 34	22 9 55 16	21 0 23 43
1661	20 3 32 49	21 3 45 37	22 15 50 40	21 6 13 54
1662	20 9 23 41	21 9 33 34	22 21 36 06	21 12 00 46
1663	20 15 14 50	21 15 27 08	23 3 33 27	21 18 01 52
1664	19 21 03 35	20 21 09 27	22 9 20 28	20 23 54 49
1665	20 2 50 50	21 2 57 25	22 15 10 42	21 5 44 11
1666	20 8 44 17	21 8 54 34	22 21 04 41	21 11 38 25
1667	20 14 42 38	21 14 34 39	23 2 41 48	21 17 22 06
1668	19 20 29 21	20 20 21 08	22 8 35 16	20 23 11 34
1669	20 2 13 26	21 2 14 40	22 14 25 09	21 5 01 12
1670	20 8 01 32	21 8 00 36	22 20 02 34	21 10 42 34
1671	20 13 47 29	21 13 48 24	23 1 53 38	21 16 34 41
1672	19 19 32 40	20 19 27 32	22 7 38 52	20 22 19 54
1673	20 1 17 46	21 1 11 25	22 13 26 49	21 4 01 01
1674	20 7 02 58	21 7 04 52	22 19 21 04	21 9 55 36
1675	20 12 56 18	21 12 45 25	23 0 58 13	21 15 43 38
1676	19 18 39 38	20 18 29 07	22 6 50 05	20 21 36 36
1677	20 0 25 41	21 0 23 32	22 12 44 01	21 3 32 08
1678	20 6 23 20	21 6 11 49	22 18 24 25	21 9 14 52
1679	20 12 17 08	21 12 03 09	23 0 20 54	21 15 11 50
1680	19 18 09 51	20 17 51 47	22 6 12 16	20 21 04 28
1681	19 23 59 49	20 23 43 44	22 12 01 17	21 2 49 58
1682	20 5 48 01	21 5 43 36	22 17 58 29	21 8 48 04
1683	20 11 45 47	21 11 29 25	22 23 39 15	21 14 34 29
1684	19 17 31 16	20 17 13 19	22 5 32 57	20 20 21 24
1685	19 23 14 33	20 23 06 36	22 11 31 11	21 2 14 54
1686	20 5 06 53	21 4 52 18	22 17 09 45	21 7 55 54
1687	20 10 50 42	21 10 36 01	22 23 00 32	21 13 53 53
1688	19 16 36 03	20 16 17 37	22 4 47 07	20 19 46 28
1689	19 22 23 45	20 22 01 02	22 10 28 13	21 1 26 42
1690	20 4 11 49	21 3 52 51	22 16 20 52	21 7 20 01
1691	20 10 10 31	21 9 35 35	22 21 58 30	21 13 03 47
1692	19 15 56 14	20 15 19 11	22 3 46 30	20 18 48 39
1693	19 21 36 29	20 21 14 09	22 9 41 23	21 0 42 56
1694	20 3 30 15	21 3 05 19	22 15 19 37	21 6 23 42
1695	20 9 16 49	21 8 51 33	22 21 11 55	21 12 17 15
1696	19 15 02 57	20 14 37 39	22 3 07 26	20 18 10 19
1697	19 20 54 48	20 20 26 56	22 8 56 21	20 23 52 16
1698	20 2 40 40	21 2 20 29	22 14 54 35	21 5 52 19
1699	20 8 37 29	21 8 07 05	22 20 38 03	21 11 46 58
1700	20 14 26 02	21 13 51 38	23 2 27 51	21 17 37 10

Year	March Equinox	June Solstice	September Equinox	December Solstice
1701	20^d $20^h09^m49^s$	21^d $19^h45^m02^s$	23^d $8^h25^m01^s$	21^d $23^h35^m06^s$
1702	21 2 11 45	22 1 36 44	23 14 05 46	22 5 18 03
1703	21 8 04 09	22 7 21 11	23 19 55 07	22 11 09 10
1704	20 13 48 50	21 13 07 41	23 1 47 36	21 17 03 47
1705	20 19 41 25	21 18 59 41	23 7 30 12	21 22 42 46
1706	21 1 23 54	22 0 50 09	23 13 20 59	22 4 35 40
1707	21 7 17 36	22 6 35 06	23 19 02 39	22 10 22 50
1708	20 13 03 54	21 12 16 11	23 0 49 23	21 16 01 47
1709	20 18 40 22	21 18 05 05	23 6 46 13	21 21 54 53
1710	21 0 34 06	21 23 54 47	23 12 25 58	22 3 38 02
1711	21 6 18 37	22 5 35 44	23 18 12 26	22 9 31 11
1712	20 11 58 51	21 11 19 03	23 0 06 00	21 15 29 44
1713	20 17 54 12	21 17 09 44	23 5 49 58	21 21 10 32
1714	20 23 43 19	21 22 58 53	23 11 42 45	22 3 04 55
1715	21 5 42 22	22 4 47 01	23 17 29 36	22 8 58 07
1716	20 11 34 50	21 10 35 34	22 23 17 37	21 14 42 20
1717	20 17 14 52	21 16 30 18	23 5 15 48	21 20 40 51
1718	20 23 13 01	21 22 26 30	23 10 58 49	22 2 26 11
1719	21 5 03 18	22 4 11 24	23 16 47 44	22 8 16 26
1720	20 10 46 19	21 9 56 34	22 22 46 46	21 14 12 23
1721	20 16 41 11	21 15 49 26	23 4 34 14	21 19 53 19
1722	20 22 26 20	21 21 37 19	23 10 25 32	22 1 49 01
1723	21 4 17 15	22 3 21 20	23 16 10 30	22 7 45 32
1724	20 10 07 49	21 9 06 16	22 21 54 31	21 13 28 27
1725	20 15 47 17	21 14 52 50	23 3 46 36	21 19 21 17
1726	20 21 45 50	21 20 44 38	23 9 28 00	22 1 03 59
1727	21 3 37 42	22 2 27 37	23 15 11 41	22 6 48 58
1728	20 9 14 58	21 8 10 54	22 21 05 44	21 12 43 58
1729	20 15 07 14	21 14 06 15	23 2 50 07	21 18 24 31
1730	20 20 51 29	21 19 53 17	23 8 37 46	22 0 14 33
1731	21 2 40 28	22 1 36 32	23 14 25 36	22 6 07 50
1732	20 8 32 55	21 7 23 42	22 20 15 00	21 11 48 37
1733	20 14 10 44	21 13 10 07	23 2 10 12	21 17 42 42
1734	20 20 05 22	21 19 03 51	23 7 56 46	21 23 35 41
1735	21 1 58 39	22 0 49 05	23 13 41 59	22 5 27 05
1736	20 7 37 52	21 6 31 29	22 19 37 39	21 11 27 49
1737	20 13 38 52	21 12 30 04	23 1 27 03	21 17 13 06
1738	20 19 33 16	21 18 19 15	23 7 16 17	21 23 03 27
1739	21 1 27 02	22 0 06 00	23 13 06 12	22 5 00 20
1740	20 7 22 33	21 5 59 24	22 18 54 02	21 10 43 10
1741	20 13 01 11	21 11 48 43	23 0 45 06	21 16 34 51
1742	20 18 54 06	21 17 42 51	23 6 30 17	21 22 22 17
1743	21 0 46 21	21 23 27 58	23 12 15 03	22 4 04 51
1744	20 6 21 21	21 5 05 54	22 18 10 15	21 9 57 09
1745	20 12 12 29	21 10 59 45	22 23 59 29	21 15 40 09
1746	20 17 57 45	21 16 43 52	23 5 44 01	21 21 29 49
1747	20 23 41 19	21 22 21 58	23 11 29 57	22 3 28 01
1748	20 5 33 13	21 4 09 54	22 17 15 22	21 9 11 36
1749	20 11 14 59	21 9 53 37	22 23 03 47	21 15 01 41
1750	20 17 10 49	21 15 44 11	23 4 49 09	21 20 49 56

EQUINOXES AND SOLSTICES

Year	March Equinox	June Solstice	September Equinox	December Solstice
	20^d $23^h07^m35^s$	21^d $21^h32^m52^s$	23^d $10^h33^m34^s$	22^d $2^h36^m24^s$
1751	20 23 07 35	21 21 32 52	23 10 33 34	22 2 36 24
1752	20 4 46 40	21 3 15 53	22 16 27 06	21 8 32 05
1753	20 10 40 11	21 9 16 35	22 22 18 30	21 14 19 34
1754	20 16 33 21	21 15 09 12	23 4 06 39	21 20 08 50
1755	20 22 21 40	21 20 51 04	23 9 58 13	22 2 05 15
1756	20 4 17 10	21 2 45 31	22 15 53 09	21 7 50 44
1757	20 10 00 08	21 8 33 06	22 21 45 08	21 13 41 59
1758	20 15 50 43	21 14 24 26	23 3 33 31	21 19 37 16
1759	20 21 46 08	21 20 13 44	23 9 18 57	22 1 27 34
1760	20 3 24 19	21 1 51 23	22 15 09 22	21 7 21 32
1761	20 9 18 41	21 7 46 56	22 20 59 59	21 13 07 21
1762	20 15 14 45	21 13 36 00	23 2 44 34	21 18 52 00
1763	20 21 00 03	21 19 13 13	23 8 29 14	22 0 45 11
1764	20 2 50 22	21 1 06 34	22 14 18 57	21 6 30 00
1765	20 8 30 10	21 6 52 17	22 20 02 52	21 12 14 27
1766	20 14 16 24	21 12 39 06	23 1 47 14	21 18 04 00
1767	20 20 11 18	21 18 27 15	23 7 33 22	21 23 47 48
1768	20 1 47 17	21 0 02 48	22 13 24 01	21 5 36 59
1769	20 7 36 40	21 5 58 40	22 19 17 52	21 11 26 58
1770	20 13 30 02	21 11 50 05	23 1 04 11	21 17 18 23
1771	20 19 15 16	21 17 27 30	23 6 50 29	21 23 17 37
1772	20 1 09 55	20 23 22 41	22 12 45 04	21 5 08 14
1773	20 7 01 00	21 5 13 15	22 18 34 22	21 10 56 01
1774	20 12 56 18	21 11 03 36	23 0 23 10	21 16 48 43
1775	20 18 55 42	21 16 59 40	23 6 14 00	21 22 39 29
1776	20 0 36 31	20 22 42 40	22 12 04 39	21 4 31 05
1777	20 6 25 42	21 4 40 51	22 17 58 19	21 10 20 38
1778	20 12 20 53	21 10 34 43	22 23 45 29	21 16 07 47
1779	20 18 05 34	21 16 08 45	23 5 31 11	21 21 58 31
1780	19 23 53 21	20 21 59 18	22 11 26 19	21 3 44 31
1781	20 5 36 58	21 3 45 59	22 17 12 18	21 9 31 04
1782	20 11 21 44	21 9 26 45	22 22 53 10	21 15 22 15
1783	20 17 11 57	21 15 14 30	23 4 38 36	21 21 13 24
1784	19 22 53 03	20 20 51 16	22 10 23 17	21 3 00 31
1785	20 4 42 31	21 2 42 05	22 16 13 08	21 8 46 42
1786	20 10 41 15	21 8 36 32	22 21 59 57	21 14 34 24
1787	20 16 28 09	21 14 12 42	23 3 41 56	21 20 25 25
1788	19 22 15 54	20 20 08 34	22 9 37 40	21 2 15 34
1789	20 4 04 26	21 2 03 27	22 15 25 48	21 8 02 16
1790	20 9 53 40	21 7 49 16	22 21 11 21	21 13 52 21
1791	20 15 48 18	21 13 43 23	23 3 07 33	21 19 44 38
1792	19 21 32 16	20 19 25 55	22 9 00 20	21 1 33 46
1793	20 3 19 46	21 1 19 08	22 14 55 10	21 7 27 05
1794	20 9 16 02	21 7 15 35	22 20 46 16	21 13 23 44
1795	20 15 03 38	21 12 50 19	23 2 27 50	21 19 17 41
1796	19 20 53 25	20 18 41 58	22 8 23 46	21 1 09 24
1797	20 2 47 07	21 0 34 15	22 14 10 31	21 6 53 41
1798	20 8 38 19	21 6 15 53	22 19 50 00	21 12 40 47
1799	20 14 30 28	21 12 08 37	23 1 40 37	21 18 31 37
1800	20 20 11 15	21 17 51 15	23 7 25 21	22 0 16 02

Year	March Equinox	June Solstice	September Equinox	December Solstice
1801	21^d $1^h55^m11^s$	21^d $23^h41^m29^s$	23^d $13^h12^m57^s$	22^d $6^h01^m35^s$
1802	21 7 47 40	22 5 35 08	23 19 02 02	22 11 50 28
1803	21 13 33 30	22 11 07 50	24 0 43 20	22 17 35 44
1804	20 19 16 55	21 16 55 50	23 6 39 35	21 23 23 55
1805	21 1 04 09	21 22 49 11	23 12 28 49	22 5 13 05
1806	21 6 51 56	22 4 29 43	23 18 07 46	22 11 03 46
1807	21 12 41 10	22 10 20 35	24 0 01 32	22 17 00 42
1808	20 18 29 10	21 16 05 51	23 5 51 01	21 22 49 36
1809	21 0 21 32	21 21 56 26	23 11 41 58	22 4 37 36
1810	21 6 20 13	22 3 55 19	23 17 36 35	22 10 33 52
1811	21 12 12 46	22 9 36 05	23 23 19 56	22 16 25 03
1812	20 17 58 42	21 15 28 34	23 5 15 42	21 22 16 12
1813	20 23 49 25	21 21 27 18	23 11 07 27	22 4 07 13
1814	21 5 42 31	22 3 09 17	23 16 46 19	22 9 51 15
1815	21 11 30 01	22 8 57 58	23 22 41 50	22 15 43 10
1816	20 17 15 33	21 14 44 01	23 4 32 36	21 21 30 03
1817	20 23 00 11	21 20 29 41	23 10 18 01	22 3 16 28
1818	21 4 49 24	22 2 22 46	23 16 09 07	22 9 14 05
1819	21 10 38 19	22 7 57 34	23 21 46 49	22 15 01 46
1820	20 16 23 18	21 13 42 05	23 3 37 35	21 20 48 00
1821	20 22 15 01	21 19 36 53	23 9 27 54	22 2 35 36
1822	21 4 08 54	22 1 18 29	23 15 03 17	22 8 17 56
1823	21 9 54 16	22 7 08 04	23 20 55 56	22 14 11 00
1824	20 15 38 18	21 12 58 06	23 2 46 14	21 19 59 04
1825	20 21 25 04	21 18 46 38	23 8 31 15	22 1 42 45
1826	21 3 15 48	22 0 41 28	23 14 28 06	22 7 39 17
1827	21 9 07 33	22 6 20 39	23 20 13 25	22 13 27 20
1828	20 14 52 48	21 12 07 23	23 2 09 28	21 19 18 16
1829	20 20 41 43	21 18 04 35	23 8 04 36	22 1 14 58
1830	21 2 36 24	21 23 47 53	23 13 41 14	22 7 04 43
1831	21 8 25 51	22 5 36 14	23 19 34 39	22 13 01 20
1832	20 14 16 05	21 11 26 33	23 1 27 33	21 18 51 29
1833	20 20 11 26	21 17 16 14	23 7 12 02	22 0 33 17
1834	21 2 03 52	21 23 11 13	23 13 06 30	22 6 29 45
1835	21 7 55 14	22 4 54 19	23 18 48 54	22 12 17 56
1836	20 13 38 32	21 10 40 41	23 0 37 44	21 18 02 07
1837	20 19 22 53	21 16 36 31	23 6 31 03	21 23 53 02
1838	21 1 16 33	21 22 18 14	23 12 06 44	22 5 33 07
1839	21 6 58 59	22 4 00 06	23 17 58 50	22 11 21 45
1840	20 12 40 14	21 9 47 20	22 23 52 36	21 17 12 17
1841	20 18 27 48	21 15 32 56	23 5 33 30	21 22 55 16
1842	21 0 12 37	21 21 21 33	23 11 25 17	22 4 55 09
1843	21 6 04 02	22 3 02 02	23 17 08 59	22 10 47 17
1844	20 11 53 18	21 8 45 33	22 22 57 02	21 16 30 43
1845	20 17 43 09	21 14 41 46	23 4 53 25	21 22 26 32
1846	20 23 45 09	21 20 30 15	23 10 30 52	22 4 11 57
1847	21 5 31 43	22 2 17 47	23 16 22 06	22 10 04 42
1848	20 11 17 25	21 8 13 49	22 22 19 47	21 15 59 34
1849	20 17 12 20	21 14 06 42	23 4 03 21	21 21 40 33
1850	20 23 01 36	21 19 58 28	23 10 00 00	22 3 37 16

EQUINOXES AND SOLSTICES

Year	March Equinox	June Solstice	September Equinox	December Solstice
1851	21^d $4^h53^m46^s$	22^d $1^h42^m28^s$	23^d $15^h50^m20^s$	22^d $9^h28^m24^s$
1852	20 10 40 54	21 7 28 04	22 21 40 10	21 15 12 16
1853	20 16 23 35	21 13 21 53	23 3 36 01	21 21 10 33
1854	20 22 19 27	21 19 06 55	23 9 11 46	22 2 58 04
1855	21 4 04 47	22 0 47 20	23 14 58 09	22 8 47 08
1856	20 9 48 39	21 6 35 28	22 20 53 13	21 14 38 21
1857	20 15 44 45	21 12 24 42	23 2 32 17	21 20 15 24
1858	20 21 31 25	21 18 12 09	23 8 21 55	22 2 09 38
1859	21 3 18 05	21 23 55 30	23 14 07 40	22 8 00 38
1860	20 9 04 14	21 5 41 41	22 19 52 31	21 13 41 03
1861	20 14 46 11	21 11 33 20	23 1 46 35	21 19 33 40
1862	20 20 42 19	21 17 19 11	23 7 26 03	22 1 18 33
1863	21 2 29 39	21 23 01 37	23 13 16 11	22 7 05 49
1864	20 8 09 43	21 4 51 36	22 19 16 08	21 13 03 21
1865	20 14 05 39	21 10 45 50	23 0 59 23	21 18 49 26
1866	20 19 54 22	21 16 33 47	23 6 50 26	22 0 49 16
1867	21 1 46 09	21 22 19 33	23 12 42 34	22 6 46 49
1868	20 7 43 26	21 4 09 41	22 18 31 08	21 12 27 53
1869	20 13 31 49	21 10 04 06	23 0 27 53	21 18 23 24
1870	20 19 31 44	21 15 55 59	23 6 09 27	22 0 13 01
1871	21 1 19 28	21 21 41 32	23 11 56 00	22 5 58 58
1872	20 6 57 03	21 3 31 35	22 17 53 19	21 11 53 22
1873	20 12 52 04	21 9 24 44	22 23 34 58	21 17 32 26
1874	20 18 37 37	21 15 06 54	23 5 22 43	21 23 21 57
1875	21 0 20 56	21 20 46 41	23 11 14 49	22 5 15 40
1876	20 6 09 38	21 2 31 53	22 16 58 32	21 10 54 10
1877	20 11 47 35	21 8 17 41	22 22 48 03	21 16 50 08
1878	20 17 42 21	21 14 03 26	23 4 26 26	21 22 41 08
1879	20 23 31 38	21 19 44 05	23 10 09 10	22 4 24 18
1880	20 5 13 17	21 1 31 25	22 16 06 36	21 10 18 10
1881	20 11 13 33	21 7 27 54	22 21 50 00	21 16 00 24
1882	20 17 04 18	21 13 16 22	23 3 37 37	21 21 53 31
1883	20 22 49 29	21 19 03 11	23 9 32 11	22 3 51 56
1884	20 4 44 15	21 0 58 45	22 15 21 09	21 9 33 29
1885	20 10 29 21	21 6 50 38	22 21 16 06	21 15 27 26
1886	20 16 26 11	21 12 40 59	23 3 04 29	21 21 19 50
1887	20 22 18 17	21 18 27 09	23 8 54 13	22 3 04 51
1888	20 3 55 45	21 0 14 13	22 14 53 45	21 9 03 11
1889	20 9 50 40	21 6 09 50	22 20 38 10	21 14 51 52
1890	20 15 40 51	21 11 53 46	23 2 22 36	21 20 44 54
1891	20 21 24 42	21 17 32 24	23 8 13 47	22 2 40 30
1892	20 3 21 45	20 23 23 15	22 13 59 32	21 8 19 17
1893	20 9 08 10	21 5 09 49	22 19 46 14	21 14 07 19
1894	20 14 59 06	21 10 56 35	23 1 27 26	21 19 58 10
1895	20 20 49 01	21 16 43 48	23 7 10 21	22 1 38 38
1896	20 2 23 05	20 22 27 57	22 13 03 28	21 7 29 18
1897	20 8 15 55	21 4 23 08	22 18 48 53	21 13 12 42
1898	20 14 06 15	21 10 06 55	23 0 34 31	21 18 59 05
1899	20 19 45 32	21 15 45 21	23 6 30 02	22 0 56 07
1900	21 1 38 48	21 21 39 33	23 12 20 08	22 6 41 30

Year	March Equinox	June Solstice	September Equinox	December Solstice
1901	21^d $7^h23^m24^s$	22^d $3^h27^m38^s$	23^d $18^h08^m57^s$	22^d $12^h36^m35^s$
1902	21 13 16 24	22 9 15 00	23 23 55 21	22 18 35 33
1903	21 19 14 37	22 15 04 48	24 5 43 43	23 0 20 30
1904	21 0 58 23	21 20 51 16	23 11 40 19	22 6 14 02
1905	21 6 57 27	22 2 51 18	23 17 30 06	22 12 03 47
1906	21 12 52 47	22 8 41 47	23 23 15 10	22 17 53 26
1907	21 18 33 01	22 14 22 58	24 5 09 04	22 23 51 43
1908	21 0 27 16	21 20 19 05	23 10 58 28	22 5 33 38
1909	21 6 13 02	22 2 05 40	23 16 44 46	22 11 20 04
1910	21 12 03 01	22 7 48 50	23 22 30 58	22 17 12 01
1911	21 17 54 28	22 13 35 39	24 4 17 48	22 22 53 28
1912	20 23 29 27	21 19 17 03	23 10 08 20	22 4 44 57
1913	21 5 18 07	22 1 09 39	23 15 53 05	22 10 35 07
1914	21 11 10 52	22 6 55 13	23 21 34 08	22 16 22 43
1915	21 16 51 24	22 12 29 31	24 3 24 08	22 22 16 05
1916	20 22 47 01	21 18 24 35	23 9 15 06	22 3 58 53
1917	21 4 37 25	22 0 14 32	23 15 00 32	22 9 46 02
1918	21 10 25 52	22 5 59 49	23 20 46 01	22 15 41 53
1919	21 16 19 20	22 11 53 46	24 2 35 44	22 21 27 27
1920	20 21 59 30	21 17 40 03	23 8 28 31	22 3 17 24
1921	21 3 51 17	21 23 35 55	23 14 20 09	22 9 07 49
1922	21 9 48 52	22 5 26 59	23 20 09 59	22 14 57 21
1923	21 15 29 02	22 11 03 02	24 2 04 01	22 20 53 41
1924	20 21 20 27	21 16 59 41	23 7 58 46	22 2 45 54
1925	21 3 12 30	21 22 50 19	23 13 43 55	22 8 37 09
1926	21 9 01 32	22 4 30 22	23 19 27 08	22 14 33 53
1927	21 14 59 26	22 10 22 32	24 1 17 28	22 20 19 00
1928	20 20 44 35	21 16 06 49	23 7 06 03	22 2 04 12
1929	21 2 35 12	21 22 00 59	23 12 52 49	22 7 53 16
1930	21 8 30 05	22 3 53 10	23 18 36 28	22 13 40 04
1931	21 14 06 37	22 9 28 22	24 0 23 48	22 19 30 04
1932	20 19 53 55	21 15 22 55	23 6 16 22	22 1 14 44
1933	21 1 43 26	21 21 12 07	23 12 01 37	22 6 58 01
1934	21 7 28 15	22 2 48 13	23 17 45 40	22 12 49 55
1935	21 13 18 05	22 8 38 12	23 23 38 38	22 18 37 35
1936	20 18 58 09	21 14 21 56	23 5 26 28	22 0 27 12
1937	21 0 45 24	21 20 12 19	23 11 13 24	22 6 22 12
1938	21 6 43 23	22 2 03 55	23 17 00 01	22 12 13 56
1939	21 12 28 47	22 7 39 44	23 22 49 56	22 18 06 30
1940	20 18 24 04	21 13 36 44	23 4 46 04	21 23 55 15
1941	21 0 20 44	21 19 33 37	23 10 33 17	22 5 44 43
1942	21 6 11 00	22 1 16 34	23 16 17 02	22 11 40 08
1943	21 12 03 00	22 7 12 38	23 22 12 16	22 17 29 41
1944	20 17 48 59	21 13 02 39	23 4 02 11	21 23 15 27
1945	20 23 37 38	21 18 52 27	23 9 50 21	22 5 04 13
1946	21 5 33 06	22 0 44 46	23 15 41 15	22 10 53 58
1947	21 11 13 08	22 6 19 14	23 21 29 14	22 16 43 26
1948	20 16 57 16	21 12 11 01	23 3 22 19	21 22 33 57
1949	20 22 48 32	21 18 03 12	23 9 06 30	22 4 23 36
1950	21 4 35 40	21 23 36 31	23 14 44 16	22 10 14 03

EQUINOXES AND SOLSTICES

Year	March Equinox	June Solstice	September Equinox	December Solstice
1951	21^d $10^h26^m14^s$	22^d $5^h25^m18^s$	23^d $20^h37^m31^s$	22^d $16^h00^m46^s$
1952	20 16 14 15	21 11 13 00	23 2 24 20	21 21 43 53
1953	20 22 01 00	21 17 00 23	23 8 06 35	22 3 32 09
1954	21 3 53 55	21 22 54 30	23 13 55 57	22 9 25 03
1955	21 9 35 35	22 4 31 48	23 19 41 33	22 15 11 38
1956	20 15 20 49	21 10 24 12	23 1 35 44	21 21 00 15
1957	20 21 17 01	21 16 21 01	23 7 26 48	22 2 49 24
1958	21 3 06 23	21 21 57 26	23 13 09 36	22 8 40 31
1959	21 8 55 07	22 3 50 18	23 19 09 11	22 14 35 07
1960	20 14 43 15	21 9 42 49	23 0 59 38	21 20 26 46
1961	20 20 32 43	21 15 30 40	23 6 43 15	22 2 20 18
1962	21 2 30 10	21 21 24 42	23 12 36 01	22 8 16 05
1963	21 8 20 17	22 3 04 35	23 18 24 17	22 14 02 45
1964	20 14 10 29	21 8 57 22	23 0 17 27	21 19 50 25
1965	20 20 05 24	21 14 56 17	23 6 06 51	22 1 41 18
1966	21 1 53 37	21 20 34 00	23 11 44 00	22 7 29 05
1967	21 7 37 31	22 2 23 29	23 17 38 53	22 13 17 14
1968	20 13 22 43	21 8 13 54	22 23 27 03	21 19 00 44
1969	20 19 08 50	21 13 55 45	23 5 07 49	22 0 44 40
1970	21 0 57 05	21 19 43 21	23 10 59 53	22 6 36 38
1971	21 6 38 55	22 1 20 16	23 16 45 47	22 12 24 55
1972	20 12 22 12	21 7 06 53	22 22 33 42	21 18 13 56
1973	20 18 13 16	21 13 01 21	23 4 22 07	22 0 08 45
1974	21 0 07 28	21 18 38 24	23 9 59 27	22 5 57 02
1975	21 5 57 31	22 0 27 14	23 15 56 12	22 11 46 40
1976	20 11 50 28	21 6 25 00	22 21 49 14	21 17 36 16
1977	20 17 43 12	21 12 14 37	23 3 30 19	21 23 24 18
1978	20 23 34 30	21 18 10 26	23 9 26 29	22 5 22 07
1979	21 5 22 53	21 23 57 01	23 15 17 27	22 11 10 58
1980	20 11 10 39	21 5 47 54	22 21 09 49	21 16 57 19
1981	20 17 03 52	21 11 45 36	23 3 06 21	21 22 51 46
1982	20 22 56 53	21 17 23 56	23 8 47 21	22 4 39 24
1983	21 4 39 48	21 23 09 37	23 14 42 47	22 10 31 11
1984	20 10 25 22	21 5 03 11	22 20 34 06	21 16 24 06
1985	20 16 14 49	21 10 45 07	23 2 08 39	21 22 08 59
1986	20 22 03 47	21 16 30 58	23 8 00 03	22 4 03 26
1987	21 3 53 02	21 22 11 42	23 13 46 26	22 9 47 12
1988	20 9 39 40	21 3 57 32	22 19 30 04	21 15 29 12
1989	20 15 29 22	21 9 54 01	23 1 20 50	21 21 23 20
1990	20 21 20 23	21 15 33 46	23 6 56 44	22 3 08 20
1991	21 3 03 03	21 21 19 45	23 12 49 19	22 8 54 59
1992	20 8 49 11	21 3 15 08	22 18 44 00	21 14 44 36
1993	20 14 41 49	21 9 00 46	23 0 23 44	21 20 27 12
1994	20 20 29 12	21 14 48 37	23 6 20 29	22 2 24 08
1995	21 2 15 36	21 20 35 26	23 12 14 17	22 8 18 15
1996	20 8 04 18	21 2 24 52	22 18 01 28	21 14 07 24
1997	20 13 55 58	21 8 21 06	22 23 57 11	21 20 08 34
1998	20 19 55 54	21 14 03 45	23 5 38 37	22 1 58 01
1999	21 1 47 11	21 19 50 20	23 11 32 58	22 7 45 21
2000	20 7 36 37	21 1 48 55	22 17 29 02	21 13 39 00

Year	March Equinox	June Solstice	September Equinox	December Solstice
2001	20^d $13^h32^m04^s$	21^d $7^h38^m56^s$	22^d $23^h05^m55^s$	21^d $19^h23^m04^s$
2002	20 19 17 28	21 13 25 36	23 4 56 49	22 1 15 57
2003	21 1 01 03	21 19 11 38	23 10 48 15	22 7 05 21
2004	20 6 49 57	21 0 58 04	22 16 31 16	21 12 43 11
2005	20 12 34 47	21 6 47 20	22 22 24 38	21 18 36 31
2006	20 18 26 56	21 12 27 03	23 4 04 50	22 0 23 42
2007	21 0 08 46	21 18 07 39	23 9 52 42	22 6 09 25
2008	20 5 49 41	21 0 00 34	22 15 45 57	21 12 05 22
2009	20 11 45 02	21 5 46 45	22 21 20 03	21 17 48 25
2010	20 17 33 35	21 11 29 38	23 3 10 30	21 23 40 03
2011	20 23 22 04	21 17 17 42	23 9 06 05	22 5 31 38
2012	20 5 15 46	20 23 10 02	22 14 50 27	21 11 13 15
2013	20 11 03 18	21 5 05 12	22 20 45 38	21 17 12 40
2014	20 16 58 29	21 10 52 28	23 2 30 36	21 23 04 41
2015	20 22 46 33	21 16 39 13	23 8 22 06	22 4 49 39
2016	20 4 31 40	20 22 35 29	22 14 22 43	21 10 45 54
2017	20 10 30 09	21 4 25 29	22 20 03 25	21 16 29 43
2018	20 16 16 57	21 10 08 39	23 1 55 44	21 22 24 29
2019	20 21 59 56	21 15 55 35	23 7 51 46	22 4 21 12
2020	20 3 51 07	20 21 45 02	22 13 32 16	21 10 04 08
2021	20 9 39 00	21 3 33 31	22 19 22 43	21 16 01 07
2022	20 15 34 55	21 9 15 13	23 1 05 20	21 21 49 58
2023	20 21 25 56	21 14 59 11	23 6 51 36	22 3 29 07
2024	20 3 07 55	20 20 52 20	22 12 45 16	21 9 22 19
2025	20 9 03 00	21 2 43 35	22 18 20 55	21 15 04 51
2026	20 14 47 26	21 8 25 50	23 0 06 48	21 20 51 57
2027	20 20 26 10	21 14 12 07	23 6 03 16	22 2 43 52
2028	20 2 18 36	20 20 03 17	22 11 46 51	21 8 21 24
2029	20 8 03 27	21 1 49 35	22 17 40 05	21 14 15 51
2030	20 13 53 33	21 7 32 37	22 23 28 28	21 20 11 20
2031	20 19 42 26	21 13 18 26	23 5 16 53	22 1 57 18
2032	20 1 23 23	20 19 10 05	22 11 12 28	21 7 57 42
2033	20 7 24 14	21 1 02 28	22 16 53 16	21 13 47 46
2034	20 13 19 00	21 6 45 33	22 22 41 10	21 19 35 45
2035	20 19 04 14	21 12 34 28	23 4 40 32	22 1 32 38
2036	20 1 04 21	20 18 33 35	22 10 24 56	21 7 14 42
2037	20 6 51 48	21 0 23 50	22 16 14 47	21 13 09 35
2038	20 12 42 14	21 6 10 47	22 22 03 56	21 19 04 09
2039	20 18 33 38	21 11 58 49	23 3 51 18	22 0 42 27
2040	20 0 13 20	20 17 47 49	22 9 46 37	21 6 34 45
2041	20 6 08 27	20 23 37 20	22 15 28 17	21 12 20 13
2042	20 11 54 58	21 5 17 19	22 21 13 16	21 18 05 58
2043	20 17 29 25	21 10 59 50	23 3 08 39	22 0 03 10
2044	19 23 22 12	20 16 52 36	22 8 49 36	21 5 45 33
2045	20 5 09 18	20 22 35 25	22 14 34 44	21 11 37 04
2046	20 10 59 38	21 4 16 09	22 20 23 31	21 17 30 25
2047	20 16 54 24	21 10 05 00	23 2 09 52	21 23 09 12
2048	19 22 35 35	20 15 55 28	22 8 02 27	21 5 04 17
2049	20 4 30 22	20 21 48 53	22 13 44 25	21 10 54 10
2050	20 10 21 20	21 3 34 34	22 19 30 19	21 16 40 43

EQUINOXES AND SOLSTICES

Year	March Equinox	June Solstice	September Equinox	December Solstice
2051	20^d $16^h00^m55^s$	21^d $9^h20^m13^s$	23^d $1^h29^m10^s$	21^d $22^h36^m11^s$
2052	19 21 57 53	20 15 17 49	22 7 17 34	21 4 19 17
2053	20 3 49 13	20 21 05 48	22 13 08 15	21 10 12 04
2054	20 9 36 23	21 2 48 51	22 19 01 28	21 16 12 07
2055	20 15 30 34	21 8 41 35	23 0 50 47	21 21 57 52
2056	19 21 13 06	20 14 30 01	22 6 41 36	21 3 53 57
2057	20 3 10 03	20 20 20 57	22 12 25 23	21 9 45 12
2058	20 9 07 10	21 2 05 58	22 18 10 32	21 15 27 21
2059	20 14 46 22	21 7 49 12	23 0 05 41	21 21 20 23
2060	19 20 40 37	20 13 47 36	22 5 50 24	21 3 03 51
2061	20 2 28 26	20 19 34 13	22 11 33 38	21 8 51 19
2062	20 8 09 43	21 1 13 21	22 17 22 10	21 14 45 09
2063	20 14 01 26	21 7 03 54	22 23 10 32	21 20 23 39
2064	19 19 40 56	20 12 47 45	22 4 59 27	21 2 11 20
2065	20 1 30 34	20 18 34 36	22 10 45 00	21 8 03 24
2066	20 7 22 11	21 0 18 32	22 16 29 24	21 13 48 13
2067	20 12 55 59	21 5 58 10	22 22 22 03	21 19 45 50
2068	19 18 51 24	20 11 55 59	22 4 09 13	21 1 35 23
2069	20 0 47 33	20 17 43 32	22 9 54 22	21 7 24 49
2070	20 6 37 25	20 23 24 51	22 15 47 19	21 13 22 04
2071	20 12 37 10	21 5 23 05	22 21 40 17	21 19 06 38
2072	19 18 23 36	20 11 16 08	22 3 30 26	21 0 58 50
2073	20 0 15 58	20 17 09 22	22 9 17 55	21 6 53 33
2074	20 6 11 29	20 23 00 49	22 15 06 25	21 12 38 07
2075	20 11 49 12	21 4 42 50	22 21 01 31	21 18 29 59
2076	19 17 41 31	20 10 39 15	22 2 52 58	21 0 16 20
2077	19 23 33 49	20 16 25 58	22 8 38 38	21 6 03 52
2078	20 5 13 35	20 22 00 29	22 14 27 18	21 12 01 02
2079	20 11 03 27	21 3 51 53	22 20 16 01	21 17 47 05
2080	19 16 47 00	20 9 36 44	22 1 59 20	20 23 35 57
2081	19 22 37 12	20 15 19 04	22 7 40 45	21 5 25 40
2082	20 4 33 33	20 21 05 58	22 13 25 54	21 11 07 50
2083	20 10 13 25	21 2 46 38	22 19 14 38	21 16 56 34
2084	19 16 02 28	20 8 43 23	22 1 02 08	20 22 44 32
2085	19 21 56 34	20 14 35 42	22 6 46 36	21 4 32 06
2086	20 3 38 23	20 20 12 29	22 12 35 22	21 10 26 04
2087	20 9 31 29	21 2 08 56	22 18 31 35	21 16 11 56
2088	19 15 20 18	20 7 59 46	22 0 21 38	20 21 59 45
2089	19 21 09 56	20 13 46 06	22 6 10 19	21 3 55 39
2090	20 3 05 17	20 19 39 03	22 12 02 47	21 9 47 15
2091	20 8 45 16	21 1 22 11	22 17 54 21	21 15 42 15
2092	19 14 37 03	20 7 18 15	21 23 45 19	20 21 35 39
2093	19 20 38 03	20 13 10 23	22 5 32 28	21 3 24 32
2094	20 2 25 06	20 18 45 38	22 11 20 09	21 9 17 06
2095	20 8 18 32	21 0 42 14	22 17 14 45	21 15 04 38
2096	19 14 06 32	20 6 34 13	21 22 58 24	20 20 50 07
2097	19 19 52 04	20 12 16 50	22 4 39 31	21 2 41 20
2098	20 1 44 06	20 18 06 25	22 10 27 54	21 8 24 42
2099	20 7 21 22	20 23 45 07	22 16 14 47	21 14 08 15
2100	20 13 07 23	21 5 35 42	22 22 04 14	21 19 54 51

Year	March Equinox	June Solstice	September Equinox	December Solstice
2101	20^d $19^h00^m19^s$	21^d $11^h25^m01^s$	23^d $3^h50^m24^s$	22^d $1^h43^m00^s$
2102	21 0 39 46	21 16 57 04	23 9 34 51	22 7 36 52
2103	21 6 26 59	21 22 49 25	23 15 28 23	22 13 28 12
2104	20 12 18 06	21 4 41 19	22 21 13 32	21 19 16 38
2105	20 18 10 28	21 10 22 34	23 2 56 07	22 1 08 01
2106	21 0 08 12	21 16 16 31	23 8 50 55	22 6 57 20
2107	21 5 54 06	21 22 04 46	23 14 41 09	22 12 46 38
2108	20 11 43 02	21 4 01 18	22 20 32 14	21 18 39 00
2109	20 17 40 14	21 9 59 01	23 2 23 17	22 0 31 49
2110	20 23 25 49	21 15 36 06	23 8 11 13	22 6 23 16
2111	21 5 14 53	21 21 29 41	23 14 09 48	22 12 12 17
2112	20 11 07 08	21 3 23 43	22 19 59 43	21 18 00 42
2113	20 16 55 08	21 9 01 34	23 1 39 58	21 23 51 09
2114	20 22 43 19	21 14 50 37	23 7 31 59	22 5 43 48
2115	21 4 26 36	21 20 34 43	23 13 16 50	22 11 31 28
2116	20 10 12 59	21 2 21 24	22 19 00 28	21 17 18 48
2117	20 16 10 32	21 8 13 31	23 0 48 47	21 23 07 44
2118	20 21 57 24	21 13 47 25	23 6 30 52	22 4 53 12
2119	21 3 43 14	21 19 39 34	23 12 24 16	22 10 41 05
2120	20 9 33 00	21 1 36 13	22 18 10 24	21 16 28 31
2121	20 15 20 01	21 7 15 44	22 23 48 02	21 22 15 42
2122	20 21 08 08	21 13 06 21	23 5 43 32	22 4 05 27
2123	21 2 53 08	21 18 53 50	23 11 34 16	22 9 50 45
2124	20 8 39 50	21 0 42 08	22 17 23 06	21 15 39 21
2125	20 14 33 59	21 6 36 54	22 23 18 14	21 21 36 37
2126	20 20 20 02	21 12 14 19	23 5 03 52	22 3 30 11
2127	21 2 07 17	21 18 05 47	23 11 00 05	22 9 25 27
2128	20 8 03 05	21 0 02 55	22 16 50 15	21 15 17 49
2129	20 13 59 29	21 5 42 39	22 22 29 15	21 21 06 23
2130	20 19 53 42	21 11 34 17	23 4 25 20	22 2 57 41
2131	21 1 41 28	21 17 26 19	23 10 13 41	22 8 44 56
2132	20 7 29 00	20 23 16 47	22 15 56 43	21 14 31 06
2133	20 13 20 16	21 5 11 06	22 21 48 42	21 20 23 47
2134	20 19 06 11	21 10 48 44	23 3 32 26	22 2 08 13
2135	21 0 49 41	21 16 35 00	23 9 25 41	22 7 53 04
2136	20 6 37 51	20 22 29 11	22 15 16 53	21 13 42 30
2137	20 12 26 15	21 4 05 19	22 20 52 21	21 19 28 39
2138	20 18 08 39	21 9 50 09	23 2 45 14	22 1 22 04
2139	20 23 52 55	21 15 38 41	23 8 32 40	22 7 11 51
2140	20 5 42 25	20 21 23 13	22 14 12 56	21 12 56 15
2141	20 11 36 54	21 3 14 52	22 20 06 48	21 18 50 46
2142	20 17 29 11	21 8 56 42	23 1 52 51	22 0 38 47
2143	20 23 15 58	21 14 47 14	23 7 44 56	22 6 27 34
2144	20 5 06 37	20 20 48 22	22 13 39 21	21 12 23 57
2145	20 11 02 01	21 2 31 10	22 19 16 54	21 18 09 43
2146	20 16 48 33	21 8 18 50	23 1 13 42	22 0 01 37
2147	20 22 37 55	21 14 13 21	23 7 09 11	22 5 51 42
2148	20 4 28 50	20 20 00 42	22 12 52 35	21 11 36 06
2149	20 10 18 35	21 1 52 59	22 18 48 40	21 17 35 44
2150	20 16 07 06	21 7 34 49	23 0 34 20	21 23 27 22

EQUINOXES AND SOLSTICES

Year	March Equinox			June Solstice			September Equinox			December Solstice		
2151	20^d	$21^h52^m52^s$		21^d	$13^h20^m57^s$		23^d	$6^h22^m37^s$		22^d	$5^h15^m03^s$	
2152	20	3 43 37		20	19 16 23		22	12 15 08		21	11 08 19	
2153	20	9 41 24		21	0 56 27		22	17 50 24		21	16 49 59	
2154	20	15 27 29		21	6 40 54		22	23 42 03		21	22 38 11	
2155	20	21 11 12		21	12 34 25		23	5 32 59		22	4 27 49	
2156	20	2 59 14		20	18 21 30		22	11 09 57		21	10 08 00	
2157	20	8 44 52		21	0 10 01		22	17 02 37		21	16 02 00	
2158	20	14 31 41		21	5 51 08		22	22 49 46		21	21 48 26	
2159	20	20 17 14		21	11 35 42		23	4 39 26		22	3 31 37	
2160	20	2 03 10		20	17 29 28		22	10 35 00		21	9 27 02	
2161	20	7 57 17		20	23 11 11		22	16 12 27		21	15 16 13	
2162	20	13 43 01		21	4 55 18		22	22 04 27		21	21 10 11	
2163	20	19 28 52		21	10 49 15		23	3 59 11		22	3 06 44	
2164	20	1 27 36		20	16 39 45		22	9 39 46		21	8 49 59	
2165	20	7 21 58		20	22 29 59		22	15 35 19		21	14 46 36	
2166	20	13 15 28		21	4 18 52		22	21 26 56		21	20 39 07	
2167	20	19 06 10		21	10 10 33		23	3 14 47		22	2 23 30	
2168	20	0 52 32		20	16 08 48		22	9 10 37		21	8 20 10	
2169	20	6 49 17		20	21 54 32		22	14 49 24		21	14 05 06	
2170	20	12 34 03		21	3 36 41		22	20 40 43		21	19 50 20	
2171	20	18 15 13		21	9 27 48		23	2 37 34		22	1 42 01	
2172	20	0 07 06		20	15 14 32		22	8 16 03		21	7 21 41	
2173	20	5 49 59		20	20 56 54		22	14 05 40		21	13 17 37	
2174	20	11 33 49		21	2 38 29		22	19 53 55		21	19 11 53	
2175	20	17 22 02		21	8 22 18		23	1 35 19		22	0 52 32	
2176	19	23 08 13		20	14 12 11		22	7 27 28		21	6 47 32	
2177	20	5 08 56		20	19 55 55		22	13 05 10		21	12 31 37	
2178	20	10 56 05		21	1 39 07		22	18 52 23		21	18 16 48	
2179	20	16 38 00		21	7 35 24		23	0 48 58		22	0 12 49	
2180	19	22 33 37		20	13 29 30		22	6 28 30		21	5 54 28	
2181	20	4 21 43		20	19 17 11		22	12 21 22		21	11 49 14	
2182	20	10 09 35		21	1 04 32		22	18 18 16		21	17 43 46	
2183	20	16 03 10		21	6 55 42		23	0 08 00		21	23 25 36	
2184	19	21 49 04		20	12 48 47		22	6 05 37		21	5 24 51	
2185	20	3 45 49		20	18 36 02		22	11 49 04		21	11 18 20	
2186	20	9 33 50		21	0 19 56		22	17 37 08		21	17 07 11	
2187	20	15 15 47		21	6 11 31		22	23 33 42		21	23 04 27	
2188	19	21 15 51		20	12 03 04		22	5 13 05		21	4 44 59	
2189	20	3 06 56		20	17 46 21		22	11 01 35		21	10 35 24	
2190	20	8 50 57		20	23 31 12		22	16 53 56		21	16 28 59	
2191	20	14 41 39		21	5 22 15		22	22 36 27		21	22 08 20	
2192	19	20 22 54		20	11 10 13		22	4 25 05		21	4 00 03	
2193	20	2 14 40		20	16 53 21		22	10 05 10		21	9 47 16	
2194	20	8 02 42		20	22 35 05		22	15 50 58		21	15 26 27	
2195	20	13 39 09		21	4 22 21		22	21 47 09		21	21 19 23	
2196	19	19 33 52		20	10 14 38		22	3 28 32		21	3 03 07	
2197	20	1 20 21		20	15 56 52		22	9 15 21		21	8 56 51	
2198	20	7 01 38		20	21 41 05		22	15 10 41		21	14 57 18	
2199	20	12 58 43		21	3 34 22		22	20 56 26		21	20 40 08	
2200	20	18 48 54		21	9 24 12		23	2 49 23		22	2 35 12	

Year	March Equinox	June Solstice	September Equinox	December Solstice
2201	21^d $0^h49^m10^s$	21^d $15^h13^m23^s$	23^d $8^h36^m58^s$	22^d $8^h28^m59^s$
2202	21 6 44 07	21 21 03 52	23 14 25 47	22 14 13 30
2203	21 12 24 08	22 2 57 43	23 20 23 06	22 20 11 11
2204	20 18 21 55	21 8 55 37	23 2 07 29	22 1 56 44
2205	21 0 12 14	21 14 40 29	23 7 55 02	22 7 45 01
2206	21 5 52 58	21 20 23 20	23 13 53 26	22 13 40 57
2207	21 11 46 56	22 2 15 13	23 19 40 01	22 19 20 00
2208	20 17 29 17	21 7 59 40	23 1 28 09	22 1 13 50
2209	20 23 19 11	21 13 41 39	23 7 11 07	22 7 08 35
2210	21 5 08 14	21 19 24 59	23 12 53 05	22 12 50 24
2211	21 10 46 38	22 1 10 03	23 18 43 35	22 18 43 07
2212	20 16 44 33	21 7 01 51	23 0 24 54	22 0 25 29
2213	20 22 37 44	21 12 46 04	23 6 08 36	22 6 10 26
2214	21 4 16 38	21 18 29 23	23 12 03 05	22 12 06 21
2215	21 10 10 07	22 0 27 12	23 17 49 56	22 17 48 18
2216	20 15 55 42	21 6 16 08	22 23 38 14	21 23 39 24
2217	20 21 46 04	21 12 00 59	23 5 27 40	22 5 34 09
2218	21 3 40 27	21 17 50 46	23 11 18 32	22 11 16 25
2219	21 9 20 11	21 23 38 22	23 17 15 13	22 17 11 33
2220	20 15 15 16	21 5 33 40	22 23 03 26	21 23 03 45
2221	20 21 08 56	21 11 20 51	23 4 50 15	22 4 56 19
2222	21 2 48 19	21 17 01 51	23 10 45 24	22 10 56 52
2223	21 8 46 40	21 22 58 40	23 16 34 37	22 16 42 52
2224	20 14 40 23	21 4 45 40	22 22 20 42	21 22 30 42
2225	20 20 31 43	21 10 28 06	23 4 07 03	22 4 25 05
2226	21 2 27 00	21 16 20 58	23 9 53 28	22 10 06 52
2227	21 8 04 14	21 22 07 31	23 15 41 33	22 15 56 35
2228	20 13 55 06	21 4 00 47	22 21 25 50	21 21 43 02
2229	20 19 47 26	21 9 46 02	23 3 09 17	22 3 25 02
2230	21 1 22 29	21 15 22 34	23 9 02 58	22 9 16 45
2231	21 7 13 38	21 21 16 51	23 14 53 04	22 14 59 52
2232	20 12 59 46	21 3 02 31	22 20 37 48	21 20 48 58
2233	20 18 43 52	21 8 40 58	23 2 23 49	22 2 47 42
2234	21 0 37 11	21 14 31 34	23 8 12 04	22 8 33 47
2235	21 6 20 17	21 20 16 46	23 14 01 04	22 14 24 47
2236	20 12 17 16	21 2 09 45	22 19 49 40	21 20 15 25
2237	20 18 17 04	21 8 01 28	23 1 36 34	22 2 02 54
2238	20 23 57 09	21 13 44 31	23 7 31 03	22 8 00 35
2239	21 5 51 15	21 19 45 24	23 13 23 34	22 13 48 29
2240	20 11 43 35	21 1 37 34	22 19 10 11	21 19 37 24
2241	20 17 32 40	21 7 18 11	23 0 59 18	22 1 32 33
2242	20 23 26 56	21 13 11 05	23 6 52 53	22 7 16 41
2243	21 5 09 15	21 18 57 54	23 12 42 56	22 13 05 51
2244	20 10 58 01	21 0 47 08	22 18 29 29	21 18 58 28
2245	20 16 51 47	21 6 35 54	23 0 13 57	22 0 47 29
2246	20 22 28 59	21 12 11 28	23 6 01 51	22 6 40 21
2247	21 4 21 09	21 18 05 24	23 11 51 52	22 12 25 29
2248	20 10 16 32	20 23 54 35	22 17 35 25	21 18 09 07
2249	20 16 02 32	21 5 31 08	22 23 19 10	22 0 01 31
2250	20 21 52 50	21 11 25 12	23 5 09 57	22 5 47 07

EQUINOXES AND SOLSTICES

Year	March Equinox	June Solstice	September Equinox	December Solstice
2251	21^d $3^h33^m58^s$	21^d $17^h13^m19^s$	23^d $10^h55^m44^s$	22^d $11^h33^m17^s$
2252	20 9 20 52	20 23 01 16	22 16 41 26	21 17 23 00
2253	20 15 16 13	21 4 51 55	22 22 30 47	21 23 10 14
2254	20 20 55 03	21 10 28 34	23 4 22 15	22 5 00 36
2255	21 2 44 23	21 16 23 46	23 10 17 35	22 10 52 27
2256	20 8 39 49	20 22 16 46	22 16 04 26	21 16 43 56
2257	20 14 25 23	21 3 52 51	22 21 48 48	21 22 42 39
2258	20 20 20 15	21 9 48 36	23 3 44 02	22 4 34 31
2259	21 2 11 31	21 15 39 06	23 9 32 00	22 10 21 38
2260	20 8 06 16	20 21 28 17	22 15 19 13	21 16 12 33
2261	20 14 05 28	21 3 23 44	22 21 09 17	21 22 02 18
2262	20 19 46 06	21 9 05 42	23 2 57 43	22 3 52 06
2263	21 1 33 21	21 15 01 42	23 8 49 48	22 9 40 09
2264	20 7 26 58	20 20 55 18	22 14 35 50	21 15 25 42
2265	20 13 10 56	21 2 27 57	22 20 18 26	21 21 14 35
2266	20 18 58 01	21 8 18 00	23 2 14 01	22 3 00 33
2267	21 0 41 33	21 14 05 31	23 7 59 54	22 8 45 32
2268	20 6 25 52	20 19 47 03	22 13 41 51	21 14 37 05
2269	20 12 16 21	21 1 35 56	22 19 29 26	21 20 29 28
2270	20 17 57 22	21 7 13 03	23 1 14 52	22 2 19 06
2271	20 23 47 34	21 13 03 03	23 7 05 00	22 8 06 05
2272	20 5 46 16	20 18 57 40	22 12 52 25	21 13 54 45
2273	20 11 36 53	21 0 35 47	22 18 34 24	21 19 46 29
2274	20 17 25 53	21 6 31 59	23 0 31 14	22 1 38 03
2275	20 23 15 54	21 12 29 38	23 6 21 02	22 7 26 09
2276	20 5 06 27	20 18 16 14	22 12 06 07	21 13 15 50
2277	20 11 01 04	21 0 10 48	22 18 02 53	21 19 08 30
2278	20 16 46 01	21 5 53 45	22 23 54 42	22 0 56 46
2279	20 22 32 37	21 11 45 13	23 5 48 20	22 6 47 55
2280	20 4 26 41	20 17 40 46	22 11 38 26	21 12 42 52
2281	20 10 13 39	20 23 14 41	22 17 17 59	21 18 35 06
2282	20 16 00 55	21 5 03 44	22 23 12 01	22 0 25 37
2283	20 21 53 08	21 10 56 02	23 4 59 03	22 6 10 06
2284	20 3 44 09	20 16 36 46	22 10 37 09	21 11 54 28
2285	20 9 34 20	20 22 28 07	22 16 28 27	21 17 46 21
2286	20 15 15 48	21 4 10 45	22 22 12 43	21 23 30 45
2287	20 20 58 20	21 9 58 41	23 3 59 08	22 5 16 45
2288	20 2 51 17	20 15 52 43	22 9 48 29	21 11 06 15
2289	20 8 38 25	20 21 25 51	22 15 28 30	21 16 51 41
2290	20 14 23 27	21 3 14 38	22 21 25 44	21 22 41 26
2291	20 20 12 23	21 9 09 53	23 3 16 48	22 4 31 43
2292	20 2 02 12	20 14 52 54	22 8 56 35	21 10 22 34
2293	20 7 52 40	20 20 44 46	22 14 51 35	21 16 21 14
2294	20 13 42 07	21 2 31 58	22 20 42 01	21 22 11 19
2295	20 19 35 12	21 8 22 34	23 2 32 24	22 3 58 57
2296	20 1 33 47	20 14 21 35	22 8 27 44	21 9 54 27
2297	20 7 27 07	20 20 03 06	22 14 09 39	21 15 44 18
2298	20 13 12 45	21 1 54 48	22 20 05 09	21 21 35 20
2299	20 19 01 47	21 7 53 03	23 1 56 39	22 3 24 43
2300	21 0 53 21	21 13 35 11	23 7 34 45	22 9 07 51

Year	March Equinox	June Solstice	September Equinox	December Solstice
2301	21^d $6^h39^m40^s$	21^d $19^h21^m44^s$	23^d $13^h28^m54^s$	22^d $14^h58^m38^s$
2302	21 12 23 13	22 1 05 46	23 19 18 39	22 20 45 16
2303	21 18 07 31	22 6 49 17	24 1 01 47	23 2 28 58
2304	20 23 53 38	21 12 39 28	23 6 51 19	22 8 25 13
2305	21 5 43 06	21 18 15 22	23 12 28 19	22 14 13 10
2306	21 11 27 51	21 23 58 35	23 18 18 08	22 20 00 04
2307	21 17 19 23	22 5 54 14	24 0 09 59	23 1 49 07
2308	20 23 15 47	21 11 38 02	23 5 46 00	22 7 31 35
2309	21 5 02 48	21 17 28 26	23 11 39 41	22 13 26 15
2310	21 10 48 48	21 23 20 43	23 17 31 59	22 19 16 30
2311	21 16 37 31	22 5 11 15	23 23 18 05	23 1 00 38
2312	20 22 28 31	21 11 06 51	23 5 15 42	22 6 57 56
2313	21 4 22 28	21 16 48 36	23 11 02 12	22 12 46 55
2314	21 10 08 40	21 22 35 18	23 16 57 54	22 18 36 56
2315	21 15 56 48	22 4 33 07	23 22 55 08	23 0 34 25
2316	20 21 51 48	21 10 17 23	23 4 31 31	22 6 22 15
2317	21 3 38 28	21 16 02 43	23 10 23 26	22 12 19 18
2318	21 9 27 05	21 21 50 27	23 16 14 22	22 18 08 27
2319	21 15 20 08	22 3 36 48	23 21 55 16	22 23 47 50
2320	20 21 11 39	21 9 29 10	23 3 47 13	22 5 41 59
2321	21 3 02 16	21 15 11 03	23 9 27 25	22 11 28 41
2322	21 8 45 20	21 20 56 57	23 15 14 58	22 17 12 27
2323	21 14 28 36	22 2 51 53	23 21 08 11	22 23 03 19
2324	20 20 22 46	21 8 35 30	23 2 43 53	22 4 43 07
2325	21 2 06 18	21 14 17 22	23 8 35 24	22 10 32 28
2326	21 7 48 31	21 20 05 35	23 14 30 48	22 16 23 54
2327	21 13 37 52	22 1 53 35	23 20 12 47	22 22 07 02
2328	20 19 23 37	21 7 43 22	23 2 06 06	22 4 07 51
2329	21 1 16 21	21 13 26 23	23 7 51 23	22 10 01 46
2330	21 7 07 46	21 19 12 36	23 13 41 41	22 15 47 42
2331	21 12 58 29	22 1 09 36	23 19 40 05	22 21 43 43
2332	20 19 01 14	21 7 00 30	23 1 19 50	22 3 30 17
2333	21 0 49 07	21 12 47 41	23 7 10 30	22 9 23 06
2334	21 6 32 56	21 18 41 23	23 13 07 41	22 15 18 43
2335	21 12 28 07	22 0 33 42	23 18 49 24	22 20 57 37
2336	20 18 14 40	21 6 21 33	23 0 42 11	22 2 51 43
2337	21 0 05 54	21 12 04 43	23 6 30 47	22 8 41 39
2338	21 5 52 22	21 17 48 43	23 12 18 14	22 14 23 11
2339	21 11 32 44	21 23 40 24	23 18 12 51	22 20 19 46
2340	20 17 27 37	21 5 25 33	22 23 47 52	22 2 05 58
2341	20 23 12 06	21 11 04 55	23 5 32 31	22 7 54 46
2342	21 4 54 48	21 16 51 36	23 11 27 11	22 13 46 37
2343	21 10 51 41	21 22 41 57	23 17 06 54	22 19 23 08
2344	20 16 38 53	21 4 29 12	22 22 55 59	22 1 17 08
2345	20 22 27 17	21 10 14 42	23 4 44 00	22 7 10 25
2346	21 4 15 37	21 16 03 38	23 10 30 28	22 12 51 47
2347	21 9 58 13	21 21 57 03	23 16 27 25	22 18 47 24
2348	20 15 56 38	21 3 46 04	22 22 09 29	22 0 33 39
2349	20 21 45 03	21 9 29 06	23 4 00 15	22 6 23 05
2350	21 3 26 25	21 15 18 04	23 10 00 22	22 12 20 36

EQUINOXES AND SOLSTICES

Year	March Equinox	June Solstice	September Equinox	December Solstice
2351	21^d $9^h21^m55^s$	21^d $21^h12^m00^s$	23^d $15^h43^m03^s$	22^d $18^h05^m44^s$
2352	20 15 11 04	21 2 59 25	22 21 32 22	22 0 04 31
2353	20 21 01 41	21 8 43 29	23 3 22 47	22 6 01 31
2354	21 2 58 33	21 14 33 43	23 9 10 05	22 11 41 48
2355	21 8 45 56	21 20 25 38	23 15 04 33	22 17 35 06
2356	20 14 44 44	21 2 17 07	22 20 45 12	21 23 22 41
2357	20 20 32 08	21 8 02 09	23 2 29 33	22 5 07 28
2358	21 2 08 05	21 13 50 01	23 8 25 37	22 11 00 43
2359	21 8 01 49	21 19 43 22	23 14 06 50	22 16 38 41
2360	20 13 47 11	21 1 25 23	22 19 53 35	21 22 27 20
2361	20 19 30 14	21 7 04 49	23 1 45 30	22 4 20 48
2362	21 1 20 15	21 12 52 42	23 7 31 57	22 10 00 56
2363	21 6 59 01	21 18 39 11	23 13 22 44	22 15 56 46
2364	20 12 53 04	21 0 26 28	22 19 04 03	21 21 50 22
2365	20 18 44 13	21 6 08 46	23 0 47 29	22 3 35 31
2366	21 0 25 40	21 11 54 21	23 6 44 49	22 9 31 24
2367	21 6 28 20	21 17 52 30	23 12 29 47	22 15 14 06
2368	20 12 20 19	20 23 41 03	22 18 16 32	21 21 06 47
2369	20 18 06 37	21 5 28 27	23 0 11 26	22 3 06 26
2370	21 0 02 26	21 11 25 06	23 6 00 52	22 8 48 21
2371	21 5 47 12	21 17 16 23	23 11 54 32	22 14 41 22
2372	20 11 43 28	20 23 06 11	22 17 41 50	21 20 32 31
2373	20 17 35 23	21 4 52 11	22 23 29 51	22 2 15 55
2374	20 23 11 28	21 10 36 23	23 5 26 54	22 8 11 59
2375	21 5 04 28	21 16 31 08	23 11 10 59	22 13 58 28
2376	20 10 52 43	20 22 13 59	22 16 52 55	21 19 49 26
2377	20 16 35 11	21 3 51 06	22 22 43 13	22 1 45 28
2378	20 22 31 36	21 9 41 59	23 4 29 14	22 7 23 31
2379	21 4 17 24	21 15 28 23	23 10 16 26	22 13 12 00
2380	20 10 09 05	20 21 15 24	22 15 58 24	21 19 03 01
2381	20 15 59 11	21 3 03 18	22 21 41 51	22 0 45 54
2382	20 21 34 55	21 8 46 44	23 3 34 20	22 6 36 58
2383	21 3 27 24	21 14 41 59	23 9 20 25	22 12 20 53
2384	20 9 20 10	20 20 28 38	22 15 06 36	21 18 08 16
2385	20 15 01 24	21 2 07 16	22 21 02 11	22 0 06 10
2386	20 20 56 16	21 8 04 02	23 2 54 58	22 5 53 01
2387	21 2 42 32	21 13 53 39	23 8 44 17	22 11 48 08
2388	20 8 35 41	20 19 41 32	22 14 31 05	21 17 47 14
2389	20 14 34 18	21 1 32 29	22 20 19 14	21 23 33 03
2390	20 20 18 04	21 7 17 31	23 2 14 16	22 5 25 07
2391	21 2 15 52	21 13 16 27	23 8 03 28	22 11 12 43
2392	20 8 10 57	20 19 07 24	22 13 47 40	21 17 00 48
2393	20 13 49 56	21 0 46 32	22 19 39 09	21 22 57 04
2394	20 19 42 12	21 6 42 38	23 1 30 00	22 4 39 43
2395	21 1 27 12	21 12 28 51	23 7 15 23	22 10 24 30
2396	20 7 14 36	20 18 09 36	22 13 00 32	21 16 16 28
2397	20 13 05 56	20 23 55 36	22 18 46 32	21 21 57 31
2398	20 18 39 46	21 5 34 12	23 0 34 46	22 3 48 03
2399	21 0 28 31	21 11 26 37	23 6 19 46	22 9 37 50
2400	20 6 21 15	20 17 12 51	22 12 00 29	21 15 25 50

Year	March Equinox	June Solstice	September Equinox	December Solstice
2401	20^d 12^{h}02^{m}53^s	20^d 22^{h}47^{m}39^s	22^d 17^{h}50^{m}12^s	21^d 21^{h}20^{m}53^s
2402	20　17 59 51	21　4 43 46	22　23 43 22	22　3 05 26
2403	20　23 52 30	21　10 36 13	23　5 30 19	22　8 53 20
2404	20　5 43 08	20　16 22 44	22　11 16 23	21　14 50 18
2405	20　11 38 11	20　22 18 57	22　17 07 57	21　20 37 52
2406	20　17 19 52	21　4 06 10	22　23 00 33	22　2 28 07
2407	20　23 12 02	21　10 02 25	23　4 52 48	22　8 18 11
2408	20　5 09 42	20　15 54 32	22　10 42 02	21　14 06 41
2409	20　10 50 34	20　21 30 26	22　16 35 18	21　20 02 28
2410	20　16 40 55	21　3 26 10	22　22 30 33	22　1 52 47
2411	20　22 30 31	21　9 16 32	23　4 15 36	22　7 43 25
2412	20　4 18 03	20　14 54 06	22　9 56 16	21　13 38 28
2413	20　10 12 41	20　20 43 18	22　15 44 44	21　19 24 12
2414	20　15 58 04	21　2 25 32	22　21 30 36	22　1 07 16
2415	20　21 46 52	21　8 16 30	23　3 15 03	22　6 53 52
2416	20　3 41 43	20　14 09 44	22　8 58 28	21　12 40 43
2417	20　9 19 09	20　19 44 42	22　14 44 18	21　18 30 32
2418	20　15 06 19	21　1 39 45	22　20 38 21	22　0 16 17
2419	20　20 57 08	21　7 31 22	23　2 24 43	22　6 00 25
2420	20　2 43 38	20　13 08 40	22　8 09 00	21　11 52 42
2421	20　8 34 47	20　19 00 28	22　14 04 07	21　17 42 24
2422	20　14 17 10	21　0 46 31	22　19 53 44	21　23 32 35
2423	20　20 04 59	21　6 37 41	23　1 41 36	22　5 27 49
2424	20　2 03 46	20　12 31 46	22　7 30 36	21　11 21 45
2425	20　7 50 36	20　18 08 34	22　13 20 03	21　17 14 09
2426	20　13 45 34	21　0 05 36	22　19 18 21	21　23 04 16
2427	20　19 42 55	21　6 03 23	23　1 06 21	22　4 52 52
2428	20　1 31 52	20　11 44 21	22　6 48 03	21　10 47 50
2429	20　7 22 33	20　17 37 57	22　12 41 36	21　16 36 32
2430	20　13 06 11	20　23 25 00	22　18 28 20	21　22 20 11
2431	20　18 53 41	21　5 12 34	23　0 13 32	22　4 06 30
2432	20　0 47 12	20　11 02 50	22　6 02 24	21　9 54 37
2433	20　6 27 13	20　16 36 47	22　11 48 24	21　15 42 39
2434	20　12 10 35	20　22 26 48	22　17 40 49	21　21 32 11
2435	20　18 00 52	21　4 19 43	22　23 25 31	22　3 21 38
2436	19　23 48 27	20　9 53 29	22　5 01 57	21　9 12 15
2437	20　5 38 58	20　15 42 20	22　10 55 58	21　15 00 30
2438	20　11 28 30	20　21 31 35	22　16 43 50	21　20 44 19
2439	20　17 17 30	21　3 20 31	22　22 27 25	22　2 33 00
2440	19　23 11 30	20　9 16 48	22　4 18 46	21　8 27 25
2441	20　4 56 05	20　14 57 52	22　10 07 00	21　14 16 54
2442	20　10 43 10	20　20 51 25	22　16 03 15	21　20 06 28
2443	20　16 38 59	21　2 49 47	22　21 56 47	22　1 57 45
2444	19　22 30 18	20　8 26 53	22　3 38 29	21　7 47 59
2445	20　4 17 46	20　14 17 22	22　9 36 54	21　13 42 42
2446	20　10 06 28	20　20 09 42	22　15 26 31	21　19 32 49
2447	20　15 54 08	21　1 55 03	22　21 07 06	22　1 23 24
2448	19　21 49 07	20　7 47 42	22　2 58 26	21　7 18 41
2449	20　3 38 41	20　13 26 21	22　8 44 12	21　13 03 47
2450	20　9 27 36	20　19 16 36	22　14 35 41	21　18 49 23

EQUINOXES AND SOLSTICES

Year	March Equinox	June Solstice	September Equinox	December Solstice
2451	20^d $15^h21^m18^s$	21^d $1^h14^m45^s$	22^d $20^h24^m31^s$	22^d $0^h39^m01^s$
2452	19 21 09 29	20 6 52 45	22 1 59 55	21 6 25 23
2453	20 2 52 18	20 12 41 02	22 7 53 57	21 12 13 47
2454	20 8 37 24	20 18 32 13	22 13 43 08	21 17 57 30
2455	20 14 23 46	21 0 14 45	22 19 23 15	21 23 40 32
2456	19 20 12 28	20 6 03 35	22 1 17 26	21 5 34 22
2457	20 1 56 26	20 11 42 53	22 7 05 04	21 11 23 06
2458	20 7 40 18	20 17 30 37	22 12 55 44	21 17 14 29
2459	20 13 32 10	20 23 26 38	22 18 46 52	21 23 11 14
2460	19 19 27 10	20 5 04 45	22 0 24 21	21 5 01 53
2461	20 1 18 59	20 10 52 45	22 6 20 31	21 10 52 29
2462	20 7 12 04	20 16 49 31	22 12 13 32	21 16 41 45
2463	20 13 06 31	20 22 40 27	22 17 53 49	21 22 28 37
2464	19 18 57 13	20 4 34 57	21 23 48 42	21 4 25 55
2465	20 0 45 30	20 10 21 53	22 5 38 54	21 10 14 23
2466	20 6 32 52	20 16 11 09	22 11 29 11	21 15 59 01
2467	20 12 23 53	20 22 07 16	22 17 24 27	21 21 51 45
2468	19 18 16 35	20 3 45 23	21 23 03 02	21 3 37 18
2469	19 23 58 24	20 9 28 53	22 4 56 24	21 9 27 13
2470	20 5 42 03	20 15 21 13	22 10 47 10	21 15 18 43
2471	20 11 31 03	20 21 03 56	22 16 21 14	21 21 02 25
2472	19 17 18 32	20 2 48 45	21 22 11 38	21 2 56 32
2473	19 23 08 01	20 8 31 04	22 4 00 08	21 8 42 29
2474	20 4 56 35	20 14 17 45	22 9 45 05	21 14 24 25
2475	20 10 45 25	20 20 14 10	22 15 37 51	21 20 20 50
2476	19 16 38 48	20 1 56 01	21 21 14 18	21 2 07 01
2477	19 22 21 56	20 7 41 13	22 3 06 18	21 7 55 20
2478	20 4 10 01	20 13 38 05	22 9 02 41	21 13 46 25
2479	20 10 04 29	20 19 25 39	22 14 42 30	21 19 28 36
2480	19 15 53 00	20 1 14 18	21 20 39 33	21 1 26 12
2481	19 21 40 49	20 7 02 09	22 2 34 23	21 7 20 35
2482	20 3 29 58	20 12 52 10	22 8 21 40	21 13 08 45
2483	20 9 20 04	20 18 47 15	22 14 16 41	21 19 09 21
2484	19 15 19 10	20 0 29 57	21 19 56 42	21 0 58 00
2485	19 21 09 24	20 6 14 22	22 1 48 20	21 6 43 52
2486	20 2 57 30	20 12 10 52	22 7 43 54	21 12 35 46
2487	20 8 51 58	20 18 00 54	22 13 19 00	21 18 17 08
2488	19 14 36 05	19 23 46 34	21 19 08 58	21 0 09 58
2489	19 20 18 50	20 5 32 01	22 1 00 37	21 5 58 37
2490	20 2 06 24	20 11 18 17	22 6 44 04	21 11 36 54
2491	20 7 50 30	20 17 06 00	22 12 36 49	21 17 29 47
2492	19 13 42 03	19 22 45 13	21 18 16 36	20 23 17 57
2493	19 19 25 56	20 4 25 50	22 0 03 42	21 5 03 46
2494	20 1 06 15	20 10 17 21	22 5 57 10	21 11 00 08
2495	20 7 03 05	20 16 06 41	22 11 32 33	21 16 44 28
2496	19 12 53 03	19 21 50 05	21 17 22 30	20 22 37 33
2497	19 18 43 13	20 3 39 29	21 23 19 59	21 4 30 48
2498	20 0 39 57	20 9 34 11	22 5 05 45	21 10 12 43
2499	20 6 28 22	20 15 29 57	22 11 01 18	21 16 12 13
2500	20 12 24 21	20 21 18 53	22 16 46 42	21 22 04 59

Year	March Equinox	June Solstice	September Equinox	December Solstice
2501	20^d 18^{h}13^{m}18^s	21^d 3^{h}06^{m}15^s	22^d 22^{h}37^{m}41^s	22^d 3^{h}49^{m}33^s
2502	20 23 57 07	21 9 00 55	23 4 37 13	22 9 44 33
2503	21 5 55 19	21 14 52 01	23 10 17 52	22 15 27 04
2504	20 11 41 17	20 20 33 51	22 16 07 46	21 21 19 02
2505	20 17 22 11	21 2 19 31	22 22 04 19	22 3 15 39
2506	20 23 12 40	21 8 08 45	23 3 44 38	22 8 56 52
2507	21 4 57 01	21 13 53 59	23 9 33 04	22 14 53 35
2508	20 10 51 54	20 19 34 02	22 15 13 59	21 20 41 51
2509	20 16 41 59	21 1 15 56	22 20 58 00	22 2 20 01
2510	20 22 24 22	21 7 07 58	23 2 51 18	22 8 12 50
2511	21 4 19 47	21 13 00 21	23 8 27 20	22 13 55 19
2512	20 10 05 20	20 18 44 10	22 14 13 16	21 19 43 41
2513	20 15 45 27	21 0 30 59	22 20 10 59	22 1 37 22
2514	20 21 39 43	21 6 24 59	23 1 56 45	22 7 16 07
2515	21 3 26 05	21 12 12 31	23 7 50 21	22 13 11 14
2516	20 9 17 46	20 17 57 21	22 13 40 27	21 19 07 45
2517	20 15 08 24	20 23 45 15	22 19 29 49	22 0 54 19
2518	20 20 49 53	21 5 36 50	23 1 26 24	22 6 55 06
2519	21 2 50 06	21 11 30 20	23 7 07 39	22 12 44 44
2520	20 8 44 59	20 17 14 15	22 12 55 25	21 18 33 40
2521	20 14 30 12	20 23 01 13	22 18 54 21	22 0 29 13
2522	20 20 28 40	21 4 59 20	23 0 39 00	22 6 10 51
2523	21 2 14 59	21 10 47 14	23 6 25 44	22 12 03 20
2524	20 8 02 23	20 16 30 42	22 12 12 00	21 17 57 04
2525	20 13 53 51	20 22 18 06	22 17 57 15	21 23 34 06
2526	20 19 31 42	21 4 03 36	22 23 49 30	22 5 24 11
2527	21 1 25 47	21 9 53 25	23 5 30 51	22 11 08 47
2528	20 7 12 58	20 15 33 41	22 11 14 32	21 16 53 48
2529	20 12 47 29	20 21 15 19	22 17 09 56	21 22 51 01
2530	20 18 40 50	21 3 09 39	22 22 52 33	22 4 34 14
2531	21 0 28 30	21 8 53 48	23 4 37 40	22 10 26 20
2532	20 6 19 44	20 14 35 30	22 10 27 05	21 16 21 29
2533	20 12 17 16	20 20 27 14	22 16 15 58	21 22 01 48
2534	20 18 00 18	21 2 18 47	22 22 09 24	22 3 57 11
2535	20 23 56 30	21 8 15 27	23 3 54 41	22 9 49 10
2536	20 5 49 27	20 14 03 57	22 9 41 48	21 15 36 15
2537	20 11 29 02	20 19 49 24	22 15 41 59	21 21 33 32
2538	20 17 26 28	21 1 47 31	22 21 31 39	22 3 16 30
2539	20 23 16 31	21 7 33 58	23 3 20 22	22 9 08 21
2540	20 5 03 27	20 13 14 33	22 9 10 59	21 15 06 23
2541	20 10 55 33	20 19 05 01	22 14 58 12	21 20 50 05
2542	20 16 36 30	21 0 51 33	22 20 46 30	22 2 44 05
2543	20 22 30 39	21 6 40 20	23 2 28 14	22 8 33 43
2544	20 4 26 34	20 12 24 59	22 8 11 40	21 14 15 07
2545	20 10 05 41	20 18 05 40	22 14 04 56	21 20 06 47
2546	20 15 59 11	21 0 03 46	22 19 50 18	22 1 49 39
2547	20 21 47 04	21 5 51 15	23 1 32 42	22 7 36 48
2548	20 3 28 38	20 11 30 07	22 7 20 53	21 13 31 13
2549	20 9 21 01	20 17 22 11	22 13 10 36	21 19 11 05
2550	20 15 02 20	20 23 07 45	22 19 00 43	22 0 59 41

Year	March Equinox	June Solstice	September Equinox	December Solstice
2551	20^d $20^h52^m56^s$	21^d $4^h56^m16^s$	23^d $0^h47^m48^s$	22^d $6^h51^m53^s$
2552	20 2 46 28	20 10 43 58	22 6 35 35	21 12 39 46
2553	20 8 22 46	20 16 24 31	22 12 29 52	21 18 38 20
2554	20 14 17 19	20 22 22 36	22 18 19 39	22 0 30 25
2555	20 20 14 31	21 4 10 56	23 0 04 06	22 6 19 58
2556	20 2 03 50	20 9 49 29	22 5 54 19	21 12 17 02
2557	20 8 04 52	20 15 47 48	22 11 47 17	21 18 01 11
2558	20 13 51 18	20 21 39 37	22 17 35 06	21 23 51 01
2559	20 19 42 16	21 3 32 06	22 23 21 02	22 5 44 17
2560	20 1 37 04	20 9 23 21	22 5 08 05	21 11 27 55
2561	20 7 13 53	20 15 03 28	22 11 00 48	21 17 17 51
2562	20 13 04 24	20 20 58 19	22 16 51 22	21 23 02 40
2563	20 18 55 42	21 2 45 07	22 22 35 44	22 4 48 10
2564	20 0 34 49	20 8 18 04	22 4 22 10	21 10 44 03
2565	20 6 24 09	20 14 09 38	22 10 12 17	21 16 30 23
2566	20 12 07 26	20 19 55 16	22 15 55 23	21 22 18 40
2567	20 17 57 29	21 1 38 46	22 21 38 05	22 4 10 08
2568	19 23 55 14	20 7 27 45	22 3 25 27	21 9 53 37
2569	20 5 36 25	20 13 09 14	22 9 16 11	21 15 44 18
2570	20 11 26 54	20 19 06 47	22 15 05 54	21 21 33 32
2571	20 17 21 16	21 1 00 23	22 20 51 15	22 3 23 12
2572	19 23 05 49	20 6 37 44	22 2 39 18	21 9 18 07
2573	20 4 59 02	20 12 33 42	22 8 36 08	21 15 04 50
2574	20 10 49 19	20 18 26 23	22 14 26 24	21 20 52 37
2575	20 16 39 22	21 0 12 05	22 20 13 37	22 2 47 09
2576	19 22 34 22	20 6 05 18	22 2 06 08	21 8 38 22
2577	20 4 14 28	20 11 47 28	22 7 56 00	21 14 31 36
2578	20 10 03 55	20 17 41 25	22 13 45 43	21 20 23 16
2579	20 16 02 39	20 23 32 44	22 19 30 58	22 2 10 53
2580	19 21 49 05	20 5 05 55	22 1 15 47	21 8 01 28
2581	20 3 40 53	20 11 00 21	22 7 09 32	21 13 47 38
2582	20 9 28 42	20 16 53 03	22 12 53 19	21 19 32 37
2583	20 15 13 30	20 22 35 08	22 18 32 59	22 1 22 13
2584	19 21 04 28	20 4 25 32	22 0 23 15	21 7 08 01
2585	20 2 43 29	20 10 05 05	22 6 10 34	21 12 51 52
2586	20 8 28 40	20 15 54 26	22 12 00 27	21 18 39 40
2587	20 14 22 58	20 21 44 55	22 17 47 19	22 0 28 08
2588	19 20 03 32	20 3 16 43	21 23 30 34	21 6 22 10
2589	20 1 52 14	20 9 10 16	22 5 26 02	21 12 15 16
2590	20 7 44 31	20 15 04 04	22 11 12 19	21 18 04 55
2591	20 13 38 07	20 20 46 54	22 16 54 48	21 23 57 14
2592	19 19 37 21	20 2 41 43	21 22 50 45	21 5 47 47
2593	20 1 25 26	20 8 31 17	22 4 41 30	21 11 36 36
2594	20 7 14 36	20 14 27 31	22 10 31 54	21 17 28 03
2595	20 13 10 49	20 20 25 38	22 16 22 54	21 23 20 03
2596	19 18 56 04	20 2 02 16	21 22 08 22	21 5 09 55
2597	20 0 43 51	20 7 54 07	22 4 06 27	21 10 57 53
2598	20 6 34 25	20 13 47 38	22 9 55 23	21 16 43 55
2599	20 12 21 20	20 19 25 17	22 15 34 42	21 22 33 15
2600	20 18 08 16	21 1 13 03	22 21 26 48	22 4 25 04

Year	March Equinox	June Solstice	September Equinox	December Solstice
2601	20^d $23^h49^m40^s$	21^d $6^h56^m03^s$	23^d $3^h11^m49^s$	22^d $10^h13^m10^s$
2602	21 5 34 54	21 12 40 36	23 8 53 43	22 15 59 31
2603	21 11 29 53	21 18 30 42	23 14 40 53	22 21 49 04
2604	20 17 19 08	21 0 05 13	22 20 21 50	22 3 34 26
2605	20 23 05 31	21 5 55 59	23 2 15 11	22 9 22 27
2606	21 4 56 17	21 11 55 11	23 8 03 26	22 15 11 38
2607	21 10 45 28	21 17 36 50	23 13 41 15	22 20 59 41
2608	20 16 34 45	20 23 28 39	22 19 38 27	22 2 51 43
2609	20 22 22 19	21 5 18 22	23 1 30 46	22 8 38 28
2610	21 4 10 35	21 11 07 38	23 7 19 52	22 14 26 45
2611	21 10 04 57	21 17 03 13	23 13 15 44	22 20 24 11
2612	20 15 52 43	20 22 42 10	22 19 01 19	22 2 17 16
2613	20 21 39 15	21 4 32 22	23 0 56 41	22 8 11 30
2614	21 3 33 39	21 10 30 00	23 6 47 48	22 14 04 21
2615	21 9 29 31	21 16 09 34	23 12 24 47	22 19 50 30
2616	20 15 21 30	20 21 58 49	22 18 20 27	22 1 41 55
2617	20 21 09 09	21 3 49 37	23 0 08 10	22 7 27 45
2618	21 2 54 26	21 9 37 19	23 5 48 30	22 13 12 39
2619	21 8 44 24	21 15 29 36	23 11 38 32	22 19 04 06
2620	20 14 29 29	20 21 05 55	22 17 19 40	22 0 47 19
2621	20 20 13 07	21 2 51 30	22 23 11 53	22 6 31 45
2622	21 2 00 52	21 8 45 45	23 5 03 33	22 12 20 59
2623	21 7 50 29	21 14 24 13	23 10 39 11	22 18 06 57
2624	20 13 34 00	20 20 09 24	22 16 32 51	22 0 01 36
2625	20 19 19 20	21 1 59 46	22 22 22 47	22 5 53 01
2626	21 1 10 09	21 7 46 19	23 4 03 43	22 11 38 29
2627	21 7 05 23	21 13 38 54	23 9 59 01	22 17 34 26
2628	20 13 00 08	20 19 22 57	22 15 46 13	21 23 23 19
2629	20 18 49 14	21 1 14 52	22 21 39 51	22 5 12 56
2630	21 0 39 51	21 7 17 07	23 3 35 46	22 11 09 15
2631	21 6 35 40	21 13 02 36	23 9 14 32	22 16 56 05
2632	20 12 22 24	20 18 49 16	22 15 10 49	21 22 47 47
2633	20 18 09 34	21 0 41 40	22 21 06 15	22 4 38 11
2634	20 24 00 00	21 6 27 16	23 2 46 51	22 10 19 30
2635	21 5 46 17	21 12 15 01	23 8 39 21	22 16 16 43
2636	20 11 34 25	20 17 55 55	22 14 23 03	21 22 06 50
2637	20 17 18 34	20 23 39 15	22 20 08 33	22 3 52 27
2638	20 23 06 48	21 5 33 22	23 2 00 12	22 9 45 28
2639	21 5 04 44	21 11 13 56	23 7 34 08	22 15 26 19
2640	20 10 51 14	21 16 57 23	22 13 24 58	21 21 14 15
2641	20 16 35 33	20 22 51 13	22 19 17 07	22 3 04 41
2642	20 22 24 46	21 4 40 31	23 0 54 42	22 8 45 00
2643	21 4 10 42	21 10 29 28	23 6 47 36	22 14 40 13
2644	20 9 59 44	20 16 13 07	22 12 36 58	21 20 28 49
2645	20 15 47 31	20 21 59 38	22 18 27 36	22 2 12 28
2646	20 21 34 44	21 3 55 38	23 0 26 18	22 8 10 08
2647	21 3 31 23	21 9 40 50	23 6 05 53	22 13 59 32
2648	20 9 17 21	20 15 25 19	22 11 59 03	21 19 55 20
2649	20 15 03 17	20 21 18 40	22 17 54 44	22 1 52 23
2650	20 21 00 41	21 3 08 27	22 23 33 42	22 7 35 17

EQUINOXES AND SOLSTICES

Year	March Equinox			June Solstice			September Equinox			December Solstice		
	21^d	$2^h54^m56^s$		21^d	$8^h56^m26^s$		23^d	$5^h26^m27^s$		22^d	$13^h30^m16^s$	
2651	21	2 54 56		21	8 56 26		23	5 26 27		22	13 30 16	
2652	20	8 47 25		20	14 42 51		22	11 15 59		21	19 20 52	
2653	20	14 37 55		20	20 34 05		22	17 02 08		22	1 03 27	
2654	20	20 22 14		21	2 30 03		22	22 55 53		22	6 58 24	
2655	21	2 17 45		21	8 16 03		23	4 33 30		22	12 42 05	
2656	20	8 02 11		20	13 57 01		22	10 22 37		21	18 26 10	
2657	20	13 42 03		20	19 46 35		22	16 19 15		22	0 16 56	
2658	20	19 33 41		21	1 34 22		22	21 57 24		22	5 55 22	
2659	21	1 16 24		21	7 16 16		23	3 46 30		22	11 50 37	
2660	20	7 00 01		20	12 58 16		22	9 35 43		21	17 45 40	
2661	20	12 49 30		20	18 44 28		22	15 19 18		21	23 28 10	
2662	20	18 35 49		21	0 34 53		22	21 12 29		22	5 24 07	
2663	21	0 37 31		21	6 21 28		23	2 53 27		22	11 11 19	
2664	20	6 28 04		20	12 06 27		22	8 42 11		21	16 57 52	
2665	20	12 10 18		20	18 02 02		22	14 40 20		21	22 56 20	
2666	20	18 07 53		20	23 58 00		22	20 20 28		22	4 38 09	
2667	20	23 55 36		21	5 44 08		23	2 11 10		22	10 32 24	
2668	20	5 44 12		20	11 31 32		22	8 07 54		21	16 27 17	
2669	20	11 38 15		20	17 22 31		22	13 56 29		21	22 07 38	
2670	20	17 22 56		20	23 14 06		22	19 52 20		22	4 05 11	
2671	20	23 18 54		21	5 00 47		23	1 34 31		22	9 57 04	
2672	20	5 06 05		20	10 43 43		22	7 20 33		21	15 44 09	
2673	20	10 45 50		20	16 32 31		22	13 15 08		21	21 40 13	
2674	20	16 44 10		20	22 23 48		22	18 53 23		22	3 19 09	
2675	20	22 34 15		21	4 05 32		23	0 39 15		22	9 07 37	
2676	20	4 18 17		20	9 49 31		22	6 31 53		21	15 01 00	
2677	20	10 09 19		20	15 41 49		22	12 14 53		21	20 39 37	
2678	20	15 50 17		20	21 30 48		22	18 04 45		22	2 33 00	
2679	20	21 42 59		21	3 15 47		22	23 46 52		22	8 21 31	
2680	20	3 31 48		20	8 58 54		22	5 34 09		21	14 03 19	
2681	20	9 09 53		20	14 45 40		22	11 30 49		21	19 57 03	
2682	20	15 04 54		20	20 38 49		22	17 13 09		22	1 41 37	
2683	20	20 54 08		21	2 22 58		22	23 00 01		22	7 35 35	
2684	20	2 36 08		20	8 06 43		22	4 55 42		21	13 36 41	
2685	20	8 34 12		20	14 02 25		22	10 43 10		21	19 20 50	
2686	20	14 24 46		20	19 52 11		22	16 35 10		22	1 15 44	
2687	20	20 25 00		21	1 41 24		22	22 22 58		22	7 09 09	
2688	20	2 21 07		20	7 32 23		22	4 11 05		21	12 52 51	
2689	20	8 00 15		20	13 24 13		22	10 06 44		21	18 48 59	
2690	20	13 56 01		20	19 21 37		22	15 50 07		22	0 33 02	
2691	20	19 45 24		21	1 05 49		22	21 35 35		22	6 19 27	
2692	20	1 24 22		20	6 46 06		22	3 31 42		21	12 13 35	
2693	20	7 17 31		20	12 38 24		22	9 19 14		21	17 52 22	
2694	20	12 59 05		20	18 22 09		22	15 06 15		21	23 43 54	
2695	20	18 46 51		21	0 03 22		22	20 50 10		22	5 39 37	
2696	20	0 36 27		20	5 47 13		22	2 32 38		21	11 22 04	
2697	20	6 13 27		20	11 29 43		22	8 21 58		21	17 16 12	
2698	20	12 12 02		20	17 21 47		22	14 03 31		21	22 59 08	
2699	20	18 06 20		20	23 06 05		22	19 46 18		22	4 43 56	
2700	20	23 47 38		21	4 50 09		23	1 41 17		22	10 41 12	

Year	March Equinox	June Solstice	September Equinox	December Solstice
2701	21^d 5^{h}43^{m}01^s	21^d 10^{h}50^{m}08^s	23^d 7^{h}30^{m}17^s	22^d 16^{h}24^{m}41^s
2702	21 11 30 07	21 16 41 32	23 13 19 25	22 22 16 45
2703	21 17 22 01	21 22 27 33	23 19 09 35	23 4 12 41
2704	20 23 18 07	21 4 19 25	23 1 01 38	22 9 55 48
2705	21 4 58 46	21 10 06 44	23 6 57 25	22 15 50 17
2706	21 10 53 24	21 16 02 10	23 12 45 57	22 21 41 06
2707	21 16 46 38	21 21 49 38	23 18 31 19	23 3 31 51
2708	20 22 25 15	21 3 29 01	23 0 25 07	22 9 31 28
2709	21 4 21 29	21 9 24 55	23 6 13 56	22 15 15 53
2710	21 10 13 13	21 15 11 46	23 11 59 17	22 21 03 14
2711	21 16 03 57	21 20 52 19	23 17 44 30	23 2 56 03
2712	20 21 57 37	21 2 43 54	22 23 30 54	22 8 38 25
2713	21 3 34 46	21 8 28 34	23 5 16 48	22 14 26 47
2714	21 9 23 07	21 14 19 34	23 10 59 39	22 20 12 30
2715	21 15 16 29	21 20 06 13	23 16 42 42	23 1 54 44
2716	20 20 52 21	21 1 41 37	22 22 35 00	22 7 46 19
2717	21 2 44 09	21 7 37 24	23 4 27 20	22 13 30 36
2718	21 8 32 09	21 13 25 33	23 10 13 02	22 19 20 06
2719	21 14 17 33	21 19 05 09	23 16 00 02	23 1 19 51
2720	20 20 12 18	21 0 58 08	22 21 50 48	22 7 08 23
2721	21 1 56 44	21 6 44 42	23 3 40 28	22 13 00 26
2722	21 7 53 51	21 12 38 07	23 9 29 39	22 18 51 27
2723	21 13 55 27	21 18 31 33	23 15 17 49	23 0 39 18
2724	20 19 36 38	21 0 14 07	22 21 11 02	22 6 35 53
2725	21 1 30 08	21 6 15 32	23 3 05 23	22 12 24 44
2726	21 7 22 09	21 12 08 24	23 8 51 33	22 18 12 03
2727	21 13 08 59	21 17 46 39	23 14 38 53	23 0 06 53
2728	20 19 02 12	20 23 37 39	22 20 31 22	22 5 49 57
2729	21 0 42 21	21 5 21 19	23 2 18 10	22 11 37 10
2730	21 6 30 01	21 11 08 13	23 8 02 12	22 17 27 11
2731	21 12 21 54	21 16 55 14	23 13 44 46	22 23 14 38
2732	20 17 58 48	20 22 30 00	22 19 30 49	22 5 07 02
2733	20 23 49 58	21 4 22 47	23 1 20 38	22 10 52 26
2734	21 5 45 35	21 10 13 26	23 7 04 31	22 16 36 14
2735	21 11 33 32	21 15 50 22	23 12 47 41	22 22 28 44
2736	20 17 25 18	20 21 46 08	22 18 40 45	22 4 16 01
2737	20 23 08 18	21 3 36 51	23 0 27 42	22 10 03 28
2738	21 4 56 30	21 9 26 30	23 6 14 49	22 15 54 06
2739	21 10 53 07	21 15 19 37	23 12 06 02	22 21 42 54
2740	20 16 34 44	20 20 58 41	22 17 59 01	22 3 34 44
2741	20 22 25 00	21 2 54 36	22 23 56 09	22 9 26 17
2742	21 4 20 14	21 8 49 46	23 5 45 20	22 15 18 58
2743	21 10 06 44	21 14 25 39	23 11 28 47	22 21 16 53
2744	20 15 59 15	20 20 19 03	22 17 23 45	22 3 09 42
2745	20 21 50 04	21 2 08 12	22 23 09 18	22 8 55 07
2746	21 3 42 05	21 7 53 20	23 4 52 29	22 14 42 48
2747	21 9 40 19	21 13 47 39	23 10 41 20	22 20 31 07
2748	20 15 20 34	20 19 27 59	22 16 27 00	22 2 18 51
2749	20 21 05 48	21 1 22 15	22 22 18 08	22 8 05 58
2750	21 2 58 18	21 7 15 55	23 4 03 40	22 13 51 00

EQUINOXES AND SOLSTICES

Year	March Equinox			June Solstice			September Equinox			December Solstice		
	21^d	$8^h42^m22^s$		21^d	$12^h48^m17^s$		23^d	$9^h44^m30^s$		22^d	$19^h39^m03^s$	
2751	21	8 42 22		21	12 48 17		23	9 44 30		22	19 39 03	
2752	20	14 29 19		20	18 37 50		22	15 40 23		22	1 25 41	
2753	20	20 13 44		21	0 26 46		22	21 27 10		22	7 10 39	
2754	21	1 58 29		21	6 08 55		23	3 08 52		22	13 01 36	
2755	21	7 49 50		21	11 59 37		23	8 59 11		22	18 56 02	
2756	20	13 32 38		20	17 38 54		22	14 45 42		22	0 46 41	
2757	20	19 23 31		20	23 30 21		22	20 38 37		22	6 36 46	
2758	21	1 23 56		21	5 27 14		23	2 28 52		22	12 26 49	
2759	21	7 16 04		21	11 06 07		23	8 11 23		22	18 20 06	
2760	20	13 06 29		20	17 01 41		22	14 08 26		22	0 12 24	
2761	20	18 55 50		20	22 58 38		22	19 57 18		22	6 00 34	
2762	21	0 47 06		21	4 45 03		23	1 40 00		22	11 48 26	
2763	21	6 40 21		21	10 37 31		23	7 34 55		22	17 39 42	
2764	20	12 25 14		20	16 20 24		22	13 25 02		21	23 26 21	
2765	20	18 10 45		20	22 09 32		22	19 16 26		22	5 14 55	
2766	21	0 02 29		21	4 04 10		23	1 05 54		22	11 08 06	
2767	21	5 48 55		21	9 37 36		23	6 43 06		22	16 58 39	
2768	20	11 34 24		20	15 24 38		22	12 35 59		21	22 48 44	
2769	20	17 25 14		20	21 16 38		22	18 22 29		22	4 33 03	
2770	20	23 16 40		21	2 57 36		22	23 59 43		22	10 15 57	
2771	21	5 06 51		21	8 48 49		23	5 51 25		22	16 07 50	
2772	20	10 50 09		20	14 34 21		22	11 38 05		21	21 54 18	
2773	20	16 33 55		20	20 23 32		22	17 26 01		22	3 40 39	
2774	20	22 26 08		21	2 19 20		22	23 18 52		22	9 33 19	
2775	21	4 16 13		21	7 54 53		23	4 59 55		22	15 20 08	
2776	20	10 01 48		20	13 42 41		22	10 57 41		21	21 12 05	
2777	20	15 52 45		20	19 39 10		22	16 50 04		22	3 03 03	
2778	20	21 43 11		21	1 22 24		22	22 28 35		22	8 52 30	
2779	21	3 33 50		21	7 14 25		23	4 23 45		22	14 51 50	
2780	20	9 23 37		20	13 01 46		22	10 13 37		21	20 41 56	
2781	20	15 15 57		20	18 51 24		22	16 02 30		22	2 28 16	
2782	20	21 13 16		21	0 48 52		22	21 56 58		22	8 22 16	
2783	21	3 06 39		21	6 30 09		23	3 37 03		22	14 10 07	
2784	20	8 50 59		20	12 19 35		22	9 30 16		21	19 59 54	
2785	20	14 37 57		20	18 16 35		22	15 21 14		22	1 48 14	
2786	20	20 28 19		20	23 58 22		22	20 56 54		22	7 28 59	
2787	21	2 13 46		21	5 43 38		23	2 50 20		22	13 19 42	
2788	20	7 57 18		20	11 27 46		22	8 40 11		21	19 05 19	
2789	20	13 41 20		20	17 12 01		22	14 24 19		22	0 49 26	
2790	20	19 27 17		20	23 02 24		22	20 15 20		22	6 46 07	
2791	21	1 16 46		21	4 39 07		23	1 53 08		22	12 36 19	
2792	20	7 02 58		20	10 21 53		22	7 42 32		21	18 24 27	
2793	20	12 53 51		20	16 16 41		22	13 34 50		22	0 14 54	
2794	20	18 53 27		20	22 03 29		22	19 11 29		22	5 57 32	
2795	21	0 42 26		21	3 53 56		23	1 05 10		22	11 53 03	
2796	20	6 30 10		20	9 48 33		22	6 59 27		21	17 45 06	
2797	20	12 20 42		20	15 40 51		22	12 45 48		21	23 29 07	
2798	20	18 11 23		20	21 36 37		22	18 43 39		22	5 26 17	
2799	21	0 06 17		21	3 19 35		23	0 29 39		22	11 14 55	
2800	20	5 52 34		20	9 05 26		22	6 23 59		21	17 03 20	

Year	March Equinox	June Solstice	September Equinox	December Solstice
2801	20^d $11^h38^m39^s$	20^d $15^h01^m40^s$	22^d $12^h20^m26^s$	21^d $22^h59^m09^s$
2802	20 17 32 51	20 20 46 24	22 17 55 55	22 4 44 37
2803	20 23 17 28	21 2 29 23	22 23 45 22	22 10 39 45
2804	20 5 04 07	20 8 16 28	22 5 36 56	21 16 29 48
2805	20 10 56 43	20 14 02 28	22 11 17 07	21 22 07 17
2806	20 16 45 42	20 19 52 25	22 17 08 49	22 4 01 29
2807	20 22 37 02	21 1 34 15	22 22 48 41	22 9 48 04
2808	20 4 19 25	20 7 18 37	22 4 34 45	21 15 32 22
2809	20 10 02 55	20 13 13 16	22 10 28 15	21 21 23 58
2810	20 15 57 38	20 18 58 03	22 16 03 44	22 3 03 39
2811	20 21 42 48	21 0 40 58	22 21 55 23	22 8 54 26
2812	20 3 26 45	20 6 30 07	22 3 52 33	21 14 47 21
2813	20 9 18 12	20 12 20 55	22 9 36 16	21 20 30 55
2814	20 15 05 02	20 18 11 23	22 15 30 13	22 2 32 25
2815	20 20 58 56	20 23 56 11	22 21 16 55	22 8 27 40
2816	20 2 51 16	20 5 43 18	22 3 06 49	21 14 14 09
2817	20 8 41 40	20 11 39 28	22 9 05 21	21 20 09 53
2818	20 14 44 34	20 17 31 08	22 14 44 31	22 1 54 34
2819	20 20 32 49	20 23 18 05	22 20 34 14	22 7 47 02
2820	20 2 15 21	20 5 10 20	22 2 31 00	21 13 41 20
2821	20 8 08 38	20 11 02 46	22 8 12 49	21 19 19 50
2822	20 13 53 50	20 16 48 17	22 14 03 21	22 1 12 12
2823	20 19 42 56	20 22 28 56	22 19 50 19	22 7 02 05
2824	20 1 29 58	20 4 11 47	22 1 35 31	21 12 41 38
2825	20 7 07 56	20 10 00 06	22 7 27 46	21 18 36 20
2826	20 13 02 42	20 15 46 43	22 13 02 56	22 0 21 59
2827	20 18 47 31	20 21 25 57	22 18 46 22	22 6 11 00
2828	20 0 29 57	20 3 12 40	22 0 42 01	21 12 04 33
2829	20 6 28 36	20 9 05 13	22 6 23 14	21 17 41 46
2830	20 12 17 27	20 14 53 28	22 12 12 57	21 23 36 10
2831	20 18 07 56	20 20 41 03	22 18 02 50	22 5 31 33
2832	19 23 58 47	20 2 33 02	21 23 51 15	21 11 14 20
2833	20 5 41 58	20 8 27 08	22 5 48 48	21 17 10 41
2834	20 11 41 45	20 14 18 54	22 11 33 08	21 22 57 51
2835	20 17 31 30	20 20 03 11	22 17 23 38	22 4 46 35
2836	19 23 12 30	20 1 51 57	21 23 25 18	21 10 45 07
2837	20 5 08 06	20 7 46 40	22 5 08 57	21 16 28 54
2838	20 10 54 34	20 13 31 37	22 10 56 31	21 22 27 00
2839	20 16 43 34	20 19 12 46	22 16 44 41	22 4 22 58
2840	19 22 37 57	20 1 00 06	21 22 28 51	21 10 01 54
2841	20 4 24 14	20 6 49 09	22 4 20 11	21 15 52 54
2842	20 10 21 40	20 12 38 47	22 9 59 05	21 21 38 09
2843	20 16 09 02	20 18 24 08	22 15 41 57	22 3 22 21
2844	19 21 44 27	20 0 10 20	21 21 37 03	21 9 15 26
2845	20 3 37 47	20 6 04 48	22 3 19 19	21 14 53 28
2846	20 9 23 51	20 11 47 30	22 9 05 20	21 20 42 11
2847	20 15 07 50	20 17 27 16	22 14 57 58	22 2 36 44
2848	19 20 59 37	19 23 17 27	21 20 46 01	21 8 17 58
2849	20 2 39 55	20 5 05 13	22 2 37 55	21 14 14 30
2850	20 8 34 49	20 10 54 37	22 8 21 10	21 20 08 53

EQUINOXES AND SOLSTICES

Year	March Equinox	June Solstice	September Equinox	December Solstice
2851	20^d $14^h28^m11^s$	20^d $16^h40^m32^s$	22^d $14^h07^m03^s$	22^d $1^h57^m00^s$
2852	19 20 11 02	19 22 26 28	21 20 05 26	21 7 53 43
2853	20 2 13 32	20 4 25 50	22 1 53 17	21 13 37 59
2854	20 8 07 09	20 10 14 49	22 7 39 28	21 19 29 46
2855	20 13 52 25	20 15 59 44	22 13 32 41	22 1 29 48
2856	19 19 48 43	19 21 56 13	21 19 21 06	21 7 10 50
2857	20 1 31 27	20 3 44 32	22 1 11 04	21 13 01 10
2858	20 7 26 10	20 9 33 24	22 6 56 44	21 18 50 37
2859	20 13 17 44	20 15 18 46	22 12 42 34	22 0 32 13
2860	19 18 52 31	19 21 00 44	21 18 37 37	21 6 26 18
2861	20 0 44 07	20 2 54 56	22 0 21 44	21 12 11 14
2862	20 6 31 22	20 8 37 52	22 6 02 37	21 18 00 50
2863	20 12 12 50	20 14 13 41	22 11 51 42	21 23 57 16
2864	19 18 09 11	19 20 05 19	21 17 39 01	21 5 36 15
2865	19 23 55 28	20 1 52 05	21 23 25 55	21 11 24 17
2866	20 5 48 33	20 7 40 43	22 5 10 03	21 17 16 44
2867	20 11 40 48	20 13 31 43	22 10 55 36	21 23 01 06
2868	19 17 17 34	19 19 16 47	21 16 50 30	21 4 55 13
2869	19 23 11 38	20 1 14 05	21 22 39 46	21 10 40 48
2870	20 5 05 06	20 7 01 56	22 4 26 49	21 16 30 04
2871	20 10 48 27	20 12 39 45	22 10 21 24	21 22 28 03
2872	19 16 42 52	19 18 35 33	21 16 14 12	21 4 14 45
2873	19 22 29 52	20 0 25 44	21 22 02 24	21 10 08 36
2874	20 4 21 52	20 6 11 47	22 3 47 22	21 16 06 11
2875	20 10 19 29	20 12 02 46	22 9 34 32	21 21 52 00
2876	19 16 02 32	19 17 45 28	21 15 26 37	21 3 42 15
2877	19 21 58 36	19 23 42 31	21 21 14 53	21 9 27 33
2878	20 3 53 07	20 5 33 14	22 2 57 14	21 15 13 44
2879	20 9 31 21	20 11 10 34	22 8 46 16	21 21 08 37
2880	19 15 21 59	19 17 05 58	21 14 37 03	21 2 50 56
2881	19 21 06 47	19 22 52 52	21 20 22 04	21 8 35 02
2882	20 2 53 47	20 4 33 08	22 2 06 22	21 14 25 58
2883	20 8 45 53	20 10 21 27	22 7 54 58	21 20 09 28
2884	19 14 21 40	19 16 01 13	21 13 44 16	21 2 00 02
2885	19 20 09 41	19 21 54 02	21 19 31 50	21 7 51 50
2886	20 2 03 59	20 3 42 17	22 1 13 52	21 13 41 10
2887	20 7 46 07	20 9 16 15	22 7 02 27	21 19 38 01
2888	19 13 45 00	19 15 13 30	21 12 57 19	21 1 24 08
2889	19 19 39 09	19 21 06 56	21 18 44 19	21 7 11 31
2890	20 1 31 19	20 2 54 29	22 0 30 25	21 13 08 38
2891	20 7 27 18	20 8 51 53	22 6 22 48	21 18 57 14
2892	19 13 09 25	19 14 39 28	21 12 14 46	21 0 46 49
2893	19 19 00 48	19 20 34 42	21 18 06 09	21 6 35 32
2894	20 0 57 33	20 2 26 44	21 23 54 25	21 12 22 28
2895	20 6 37 43	20 8 00 54	22 5 44 35	21 18 16 02
2896	19 12 26 15	19 13 54 51	21 11 39 34	21 0 04 43
2897	19 18 13 51	19 19 44 44	21 17 23 12	21 5 52 59
2898	19 23 59 59	20 1 21 24	21 23 02 47	21 11 47 30
2899	20 5 53 10	20 7 10 01	22 4 51 32	21 17 33 27
2900	20 11 37 36	20 12 52 00	22 10 37 55	21 23 17 30

Year	March Equinox	June Solstice	September Equinox	December Solstice
2901	20^d $17^h26^m53^s$	20^d $18^h41^m52^s$	22^d $16^h22^m36^s$	22^d $5^h03^m37^s$
2902	20 23 21 17	21 0 34 53	22 22 06 33	22 10 52 09
2903	21 5 01 13	21 6 10 19	23 3 51 10	22 16 42 24
2904	20 10 48 06	20 12 04 24	22 9 45 31	21 22 29 08
2905	20 16 40 44	20 17 59 04	22 15 33 24	22 4 14 38
2906	20 22 29 25	20 23 37 26	22 21 17 17	22 10 07 12
2907	21 4 21 56	21 5 31 02	23 3 14 35	22 15 58 51
2908	20 10 06 40	20 11 19 03	22 9 05 07	21 21 49 21
2909	20 15 54 55	20 17 10 23	22 14 53 14	22 3 43 58
2910	20 21 53 20	20 23 05 12	22 20 42 33	22 9 38 22
2911	21 3 40 39	21 4 41 58	23 2 30 12	22 15 29 57
2912	20 9 34 21	20 10 36 57	22 8 27 14	21 21 18 36
2913	20 15 31 15	20 16 35 00	22 14 15 17	22 3 05 44
2914	20 21 19 38	20 22 15 09	22 19 54 26	22 8 57 43
2915	21 3 08 08	21 4 08 11	23 1 48 45	22 14 47 33
2916	20 8 51 23	20 9 55 10	22 7 35 14	21 20 30 11
2917	20 14 36 27	20 15 40 12	22 13 19 00	22 2 15 59
2918	20 20 29 14	20 21 29 05	22 19 06 59	22 8 03 24
2919	21 2 08 57	21 3 01 34	23 0 50 31	22 13 50 26
2920	20 7 52 53	20 8 51 00	22 6 42 44	21 19 39 43
2921	20 13 42 54	20 14 44 32	22 12 28 12	22 1 29 15
2922	20 19 31 35	20 20 20 21	22 18 04 44	22 7 20 42
2923	21 1 23 01	21 2 09 58	23 0 00 04	22 13 11 24
2924	20 7 14 28	20 8 01 28	22 5 50 08	21 18 56 36
2925	20 13 05 48	20 13 51 53	22 11 34 36	22 0 45 34
2926	20 19 00 52	20 19 49 48	22 17 27 56	22 6 41 30
2927	21 0 47 18	21 1 32 58	22 23 16 24	22 12 31 48
2928	20 6 35 04	20 7 26 28	22 5 13 12	21 18 21 42
2929	20 12 30 15	20 13 25 18	22 11 07 04	22 0 12 15
2930	20 18 22 14	20 19 03 38	22 16 48 20	22 6 01 56
2931	21 0 09 04	21 0 52 41	22 22 46 13	22 11 55 19
2932	20 5 55 27	20 6 43 51	22 4 36 11	21 17 45 10
2933	20 11 41 52	20 12 27 07	22 10 14 22	21 23 33 06
2934	20 17 32 55	20 18 15 49	22 16 03 15	22 5 28 08
2935	20 23 22 29	20 23 53 11	22 21 46 08	22 11 11 58
2936	20 5 09 51	20 5 39 38	22 3 34 36	21 16 55 21
2937	20 11 02 53	20 11 37 42	22 9 23 31	21 22 44 32
2938	20 16 52 19	20 17 16 48	22 14 57 36	22 4 30 15
2939	20 22 35 16	20 23 05 16	22 20 51 51	22 10 19 57
2940	20 4 21 33	20 4 58 06	22 2 42 45	21 16 05 01
2941	20 10 09 35	20 10 42 41	22 8 23 29	21 21 48 00
2942	20 15 59 13	20 16 32 40	22 14 18 57	22 3 43 19
2943	20 21 45 52	20 22 14 50	22 20 08 31	22 9 33 15
2944	20 3 31 11	20 4 03 39	22 2 00 03	21 15 24 55
2945	20 9 23 35	20 10 01 45	22 7 54 16	21 21 23 31
2946	20 15 20 06	20 15 42 26	22 13 32 16	22 3 14 07
2947	20 21 11 30	20 21 30 27	22 19 29 20	22 9 06 54
2948	20 3 05 15	20 3 27 03	22 1 23 35	21 14 55 53
2949	20 8 58 54	20 9 16 49	22 7 02 25	21 20 41 24
2950	20 14 48 41	20 15 08 48	22 12 54 48	22 2 37 06

EQUINOXES AND SOLSTICES

Year	March Equinox	June Solstice	September Equinox	December Solstice
2951	20^d $20^h34^m52^s$	20^d $20^h53^m17^s$	22^d $18^h42^m13^s$	22^d $8^h23^m50^s$
2952	20 2 21 20	20 2 41 03	22 0 30 02	21 14 06 27
2953	20 8 09 54	20 8 34 37	22 6 23 31	21 19 57 23
2954	20 14 02 08	20 14 13 12	22 12 00 45	22 1 41 32
2955	20 19 43 40	20 19 55 13	22 17 52 25	22 7 30 39
2956	20 1 26 20	20 1 47 06	21 23 44 02	21 13 21 54
2957	20 7 15 32	20 7 31 09	22 5 17 53	21 19 05 01
2958	20 13 02 33	20 13 15 39	22 11 08 12	22 1 00 08
2959	20 18 52 46	20 18 59 08	22 16 57 43	22 6 47 35
2960	20 0 43 41	20 0 47 26	21 22 44 22	21 12 30 17
2961	20 6 33 25	20 6 44 56	22 4 39 00	21 18 27 25
2962	20 12 29 06	20 12 31 01	22 10 18 38	22 0 16 45
2963	20 18 14 36	20 18 17 51	22 16 12 03	22 6 06 47
2964	20 0 02 17	20 0 14 42	21 22 10 48	21 12 00 28
2965	20 5 58 37	20 6 03 33	22 3 50 20	21 17 41 48
2966	20 11 46 12	20 11 49 37	22 9 44 48	21 23 38 47
2967	20 17 34 51	20 17 37 06	22 15 38 44	22 5 32 07
2968	19 23 23 05	19 23 25 23	21 21 23 34	21 11 17 26
2969	20 5 10 39	20 5 18 40	22 3 16 59	21 17 16 33
2970	20 11 08 35	20 11 00 45	22 8 54 57	21 23 04 00
2971	20 16 57 39	20 16 43 26	22 14 44 06	22 4 48 32
2972	19 22 44 37	19 22 37 46	21 20 38 49	21 10 39 05
2973	20 4 39 00	20 4 28 30	22 2 13 13	21 16 18 30
2974	20 10 22 17	20 10 13 20	22 8 01 17	21 22 10 47
2975	20 16 04 58	20 15 59 06	22 13 53 50	22 4 00 04
2976	19 21 53 02	19 21 46 42	21 19 37 31	21 9 37 53
2977	20 3 37 26	20 3 35 31	22 1 32 00	21 15 32 12
2978	20 9 30 46	20 9 17 26	22 7 14 06	21 21 21 11
2979	20 15 15 47	20 15 00 07	22 13 03 41	22 3 09 44
2980	19 20 57 30	19 20 52 08	21 18 59 37	21 9 07 46
2981	20 2 54 39	20 2 42 51	22 0 36 25	21 14 53 57
2982	20 8 46 24	20 8 26 26	22 6 25 14	21 20 47 51
2983	20 14 36 33	20 14 14 22	22 12 22 13	22 2 41 19
2984	19 20 35 04	19 20 10 39	21 18 08 23	21 8 22 31
2985	20 2 23 21	20 2 05 15	22 0 02 24	21 14 20 41
2986	20 8 18 41	20 7 54 42	22 5 47 35	21 20 13 14
2987	20 14 07 28	20 13 41 42	22 11 36 48	22 1 56 48
2988	19 19 49 21	19 19 33 44	21 17 34 49	21 7 50 10
2989	20 1 46 13	20 1 24 26	21 23 14 03	21 13 30 37
2990	20 7 31 15	20 7 04 39	22 5 01 16	21 19 20 24
2991	20 13 10 28	20 12 48 09	22 10 56 26	22 1 15 31
2992	19 19 00 33	19 18 38 01	21 16 37 20	21 6 55 59
2993	20 0 43 24	20 0 22 15	21 22 24 36	21 12 51 20
2994	20 6 37 15	20 6 03 21	22 4 07 18	21 18 42 11
2995	20 12 29 07	20 11 46 35	22 9 52 18	22 0 21 21
2996	19 18 11 19	19 17 37 04	21 15 46 35	21 6 15 56
2997	20 0 08 42	19 23 31 07	21 21 23 50	21 11 59 07
2998	20 5 55 06	20 5 15 10	22 3 08 39	21 17 48 34
2999	20 11 37 27	20 11 03 01	22 9 07 15	21 23 43 57
3000	20 17 33 33	20 16 58 49	22 14 54 14	22 5 23 12

4

PHASES OF THE MOON

1951 – 2050

PHASES OF THE MOON

The list which follows gives the instants of the phases of the Moon for the years 1951 to 2050, with an accuracy of one second of time. These instants are expressed in *Ephemeris Time*. For their conversion to Universal Time or to other standard times, see the Introduction.

The times of New Moon, First Quarter, Full Moon and Last Quarter are the times at which the excess of the apparent longitude of the Moon (i.e. affected by the nutation) over the apparent longitude of the Sun is 0, 90, 180 and 270 degrees, respectively.

Cycles

The Metonic Cycle is a cycle of 19 years, or 235 lunations (synodic months), after which time the phases of the Moon are repeated on the same days of the year, or approximately so. For example :

New Moon	1942 Aug 12
New Moon	1961 Aug 11
New Moon	1980 Aug 10
New Moon	1999 Aug 11

There are other interesting cycles. After 2 years, the *preceding* lunar phase occurs at nearly the same calendar date. For example :

New Moon	1983 Aug 8
Last Quarter	1985 Aug 8
Full Moon	1987 Aug 9
First Quarter	1989 Aug 9
New Moon	1991 Aug 10
Last Quarter	1993 Aug 10
Full Moon	1995 Aug 10

Thus, after 8 years the same lunar phases repeat, occurring one or two days later in the year :

New Moon	1983 Aug 8
New Moon	1991 Aug 10
New Moon	1999 Aug 11
New Moon	2007 Aug 12

Combining the cycle of 19 years with that of 2 years, we obtain cycles of 17 and 21 years, after which respectively the following and the preceding lunar phase occurs near the same calendar date :

New Moon	1945 Jul 9	New Moon	1945 Jul 9
First Quarter	1962 Jul 9	Last Quarter	1966 Jul 10
Full Moon	1979 Jul 9	Full Moon	1987 Jul 11
Last Quarter	1996 Jul 7	First Quarter	2008 Jul 10

In the Gregorian Calendar, 372 years is an excellent cycle of long period :

New Moon	1627 Aug 11
New Moon	1999 Aug 11
New Moon	2371 Aug 11
New Moon	2743 Aug 12

During the period 1800–2100, the following lunar phases take place on the bissextile day, February 29 :

1820	Full Moon	1972	Full Moon
1824	New Moon	1976	New Moon
1856	Last Quarter	2008	Last Quarter
1860	First Quarter	2048	Full Moon
1936	First Quarter	2088	First Quarter

Some years there are only three lunar phases in February. Here is the complete list of these cases during the period 1800–2100. For each case the missing phase is given. For instance, there was no First Quarter in February 1974. The sign B refers to a bissextile year. The list is based on the Universal Time ; different results are obtained for dates based on another standard time.

1805	NM	1849	FQ	1911	NM	1957	NM	1999	FM	2054	LQ
1807	LQ	1866	FM	1915	FM	1959	LQ	B 2012	FQ	2058	FQ
1809	FM	1870	NM	1919	NM	1961	FM	2014	NM	2067	FM
1811	FQ	1879	FQ	1921	LQ	1970	LQ	2018	FM	2071	NM
1826	LQ	1883	LQ	1925	FQ	1974	FQ	2031	FQ	2077	FQ
1830	FQ	1885	FM	1934	FM	1978	LQ	2033	NM	2081	LQ
1843	NM	1889	NM	1938	NM	1993	FQ	2035	LQ	2090	NM
1845	LQ	1900	NM	B 1940	LQ	1995	NM	2037	FM	B 2092	LQ
1847	FM	1902	LQ	1955	FQ	1997	LQ	B 2052	NM	2094	FM

On 2031 December 21, First Quarter will occur at $0^h01^m41^s$ Ephemeris Time. If in 2031 the difference $\Delta T = ET - UT$ is larger than 101 seconds, then that lunar phase will take place on the *previous* UT date.

Year	New Moon		First Quarter		Full Moon		Last Quarter	
		h m s		h m s		h m s		
1951							Jan 1	$5^h11^m34^s$
	Jan 7	20 10 22	Jan 15	0 23 08	Jan 23	4 47 25	Jan 30	15 14 01
	Feb 6	7 54 10	Feb 13	20 55 41	Feb 21	21 12 38	Feb 28	22 59 54
	Mar 7	20 50 58	Mar 15	17 40 14	Mar 23	10 50 15	Mar 30	5 35 15
	Apr 6	10 52 05	Apr 14	12 56 00	Apr 21	21 30 24	Apr 28	12 18 05
	May 6	1 36 01	May 14	5 32 14	May 21	5 45 09	May 27	20 17 17
	Jun 4	16 40 48	Jun 12	18 52 19	Jun 19	12 36 22	Jun 26	6 21 52
	Jul 4	7 48 26	Jul 12	4 56 49	Jul 18	19 17 51	Jul 25	18 59 48
	Aug 2	22 39 40	Aug 10	12 22 54	Aug 17	2 59 53	Aug 24	10 20 38
	Sep 1	12 50 02	Sep 8	18 16 36	Sep 15	12 38 49	Sep 23	4 13 35
	Oct 1	1 57 07	Oct 8	0 00 32	Oct 15	0 51 20	Oct 22	23 55 50
	Oct 30	13 54 58	Nov 6	6 59 04	Nov 13	15 52 31	Nov 21	20 01 42
	Nov 29	1 00 44	Dec 5	16 20 55	Dec 13	9 30 53	Dec 21	14 37 51
	Dec 28	11 43 47						
		h m s						
1952			Jan 4	$4^h42^m35^s$	Jan 12	$4^h55^m26^s$	Jan 20	$6^h09^m34^s$
	Jan 26	22 26 47	Feb 2	20 01 51	Feb 11	0 28 38	Feb 18	18 01 36
	Feb 25	9 16 26	Mar 3	13 43 48	Mar 11	18 14 24	Mar 19	2 40 06
	Mar 25	20 13 03	Apr 2	8 48 54	Apr 10	8 53 37	Apr 17	9 07 55
	Apr 24	7 27 47	May 2	3 58 06	May 9	20 16 15	May 16	14 39 32
	May 23	19 28 16	May 31	21 46 53	Jun 8	5 07 21	Jun 14	20 28 10
	Jun 22	8 45 31	Jun 30	13 11 52	Jul 7	12 33 57	Jul 14	3 42 39
	Jul 21	23 31 02	Jul 30	1 51 19	Aug 5	19 40 28	Aug 12	13 27 24
	Aug 20	15 20 48	Aug 28	12 03 41	Sep 4	3 19 43	Sep 11	2 36 15
	Sep 19	7 22 12	Sep 26	20 31 13	Oct 3	12 15 39	Oct 10	19 33 06
	Oct 18	22 42 43	Oct 26	4 04 31	Nov 1	23 10 11	Nov 9	15 43 22
	Nov 17	12 56 11	Nov 24	11 34 45	Dec 1	12 41 49	Dec 9	13 22 17
	Dec 17	2 02 53	Dec 23	19 51 59	Dec 31	5 06 04		
		h m s		h m s		h m s		
1953							Jan 8	$10^h09^m23^s$
	Jan 15	14 08 45	Jan 22	5 43 12	Jan 29	23 44 39	Feb 7	4 09 38
	Feb 14	1 10 44	Feb 20	17 44 56	Feb 28	18 59 11	Mar 8	18 26 51
	Mar 15	11 05 28	Mar 22	8 11 02	Mar 30	12 55 20	Apr 7	4 58 51
	Apr 13	20 09 20	Apr 21	0 41 00	Apr 29	4 20 54	May 6	12 21 13
	May 13	5 06 09	May 20	18 20 33	May 28	17 03 25	Jun 4	17 35 59
	Jun 11	14 55 12	Jun 19	12 01 21	Jun 27	3 29 45	Jul 3	22 03 53
	Jul 11	2 28 36	Jul 19	4 47 43	Jul 26	12 21 05	Aug 2	3 16 39
	Aug 9	16 10 25	Aug 17	20 08 27	Aug 24	20 21 11	Aug 31	10 46 51
	Sep 8	7 48 01	Sep 16	9 49 50	Sep 23	4 16 06	Sep 29	21 51 51
	Oct 8	0 40 58	Oct 15	21 44 47	Oct 22	12 56 11	Oct 29	13 09 51
	Nov 6	17 58 10	Nov 14	7 52 41	Nov 20	23 12 35	Nov 28	8 16 34
	Dec 6	10 48 25	Dec 13	16 30 43	Dec 20	11 43 57	Dec 28	5 43 53
1954	Jan 5	$2^h21^m48^s$	Jan 12	$0^h22^m22^s$	Jan 19	$2^h37^m15^s$	Jan 27	$3^h28^m31^s$
	Feb 3	15 55 52	Feb 10	8 29 42	Feb 17	19 17 47	Feb 25	23 29 22
	Mar 5	3 11 44	Mar 11	17 52 07	Mar 19	12 42 54	Mar 27	16 14 15
	Apr 3	12 25 28	Apr 10	5 05 53	Apr 18	5 48 57	Apr 26	4 57 50
	May 2	20 22 45	May 9	18 17 48	May 17	21 47 25	May 25	13 49 35
	Jun 1	4 03 21	Jun 8	9 14 08	Jun 16	12 06 10	Jun 23	19 46 20
	Jun 30	12 26 11	Jul 8	1 33 41	Jul 16	0 29 30	Jul 23	0 14 23
	Jul 29	22 20 12	Aug 6	18 51 06	Aug 14	11 03 31	Aug 21	4 51 25
	Aug 28	10 21 20	Sep 5	12 28 53	Sep 12	20 19 55	Sep 19	11 11 27
	Sep 27	0 50 50	Oct 5	5 31 19	Oct 12	5 10 15	Oct 18	20 30 35
	Oct 26	17 47 24	Nov 3	20 55 17	Nov 10	14 29 34	Nov 17	9 33 06
	Nov 25	12 30 50	Dec 3	9 56 25	Dec 10	0 57 03	Dec 17	2 21 52
	Dec 25	7 33 37						

Year	New Moon	First Quarter	Full Moon	Last Quarter
	h m s	Jan 1 $20^h29^m08^s$	Jan 8 $12^h44^m34^s$	Jan 15 $22^h14^m01^s$
1955	Jan 24 1 07 05	Jan 31 5 05 59	Feb 7 1 43 14	Feb 14 19 40 17
	Feb 22 15 54 46	Mar 1 12 40 56	Mar 8 15 41 43	Mar 16 16 36 28
	Mar 24 3 42 57	Mar 30 20 10 23	Apr 7 6 35 20	Apr 15 11 01 10
	Apr 22 13 06 40	Apr 29 4 23 17	May 6 22 14 32	May 15 1 42 51
	May 21 20 59 01	May 28 14 01 52	Jun 5 14 08 53	Jun 13 12 37 30
	Jun 20 4 12 15	Jun 27 1 44 26	Jul 5 5 29 03	Jul 12 20 31 10
	Jul 19 11 35 02	Jul 26 16 00 03	Aug 3 19 30 39	Aug 11 2 33 24
	Aug 17 19 58 30	Aug 25 8 52 09	Sep 2 7 59 45	Sep 9 7 59 48
	Sep 16 6 19 43	Sep 24 3 41 04	Oct 1 19 17 53	Oct 8 14 04 21
	Oct 15 19 32 48	Oct 23 23 05 07	Oct 31 6 04 17	Nov 6 21 56 23
	Nov 14 12 01 59	Nov 22 17 29 21	Nov 29 16 50 18	Dec 6 8 35 54
	Dec 14 7 07 32	Dec 22 9 39 53	Dec 29 3 44 24	
	h m s	h m s	h m s	Jan 4 $22^h41^m14^s$
1956	Jan 13 3 01 30	Jan 20 22 58 41	Jan 27 14 40 46	Feb 3 16 08 19
	Feb 11 21 38 12	Feb 19 9 21 34	Feb 26 1 42 05	Mar 4 11 53 41
	Mar 12 13 37 05	Mar 19 17 14 08	Mar 26 13 11 37	Apr 3 8 06 38
	Apr 11 2 39 18	Apr 17 23 28 24	Apr 25 1 41 03	May 3 2 55 59
	May 10 13 04 52	May 17 5 15 47	May 24 15 26 21	Jun 1 19 13 38
	Jun 8 21 29 38	Jun 15 11 56 29	Jun 23 6 14 01	Jul 1 8 41 08
	Jul 8 4 38 07	Jul 14 20 47 06	Jul 22 21 29 29	Jul 30 19 31 37
	Aug 6 11 25 25	Aug 13 8 45 27	Aug 21 12 38 16	Aug 29 4 13 22
	Sep 4 18 57 40	Sep 12 0 13 22	Sep 20 3 19 51	Sep 27 11 25 45
	Oct 4 4 25 01	Oct 11 18 44 45	Oct 19 17 25 06	Oct 26 18 02 39
	Nov 2 16 44 07	Nov 10 15 09 34	Nov 18 6 45 04	Nov 25 1 13 03
	Dec 2 8 13 00	Dec 10 11 51 27	Dec 17 19 06 48	Dec 24 10 10 09
1957	Jan 1 $2^h14^m03^s$	Jan 9 $7^h06^m36^s$	Jan 16 $6^h21^m54^s$	Jan 22 $21^h48^m24^s$
	Jan 30 21 25 07	Feb 7 23 23 44	Feb 14 16 38 18	Feb 21 12 19 07
	Mar 1 16 12 48	Mar 9 11 50 35	Mar 16 2 22 18	Mar 23 5 04 57
	Mar 31 9 19 26	Apr 7 20 33 04	Apr 14 12 09 51	Apr 21 23 01 01
	Apr 29 23 54 17	May 7 2 29 59	May 13 22 34 46	May 21 17 03 43
	May 29 11 39 31	Jun 5 7 10 15	Jun 12 10 02 24	Jun 20 10 22 48
	Jun 27 20 53 58	Jul 4 12 09 46	Jul 11 22 50 12	Jul 20 2 17 54
	Jul 27 4 28 35	Aug 2 18 55 31	Aug 10 13 08 45	Aug 18 16 16 40
	Aug 25 11 33 03	Sep 1 4 35 06	Sep 9 4 55 34	Sep 17 4 02 28
	Sep 23 19 18 58	Sep 30 17 49 44	Oct 8 21 42 43	Oct 16 13 44 30
	Oct 23 4 43 50	Oct 30 10 48 20	Nov 7 14 32 28	Nov 14 21 59 58
	Nov 21 16 19 45	Nov 29 6 58 08	Dec 7 6 16 23	Dec 14 5 45 32
	Dec 21 6 12 19	Dec 29 4 52 36		
	h m s	h m s	Jan 5 $20^h09^m19^s$	Jan 12 $14^h01^m26^s$
1958	Jan 19 22 08 17	Jan 28 2 16 32	Feb 4 8 05 59	Feb 10 23 34 15
	Feb 18 15 38 40	Feb 26 20 52 03	Mar 5 18 28 26	Mar 12 10 48 15
	Mar 20 9 50 16	Mar 28 11 19 02	Apr 4 3 45 17	Apr 10 23 50 18
	Apr 19 3 23 58	Apr 26 21 36 17	May 3 12 23 47	May 10 14 38 12
	May 18 19 00 52	May 26 4 38 34	Jun 1 20 55 37	Jun 9 6 59 22
	Jun 17 8 00 00	Jun 24 9 45 04	Jul 1 6 05 01	Jul 9 0 21 24
	Jul 16 18 33 53	Jul 23 14 20 02	Jul 30 16 47 22	Aug 7 17 50 01
	Aug 15 3 33 45	Aug 21 19 45 13	Aug 29 5 53 55	Sep 6 10 24 34
	Sep 13 12 02 36	Sep 20 3 18 01	Sep 27 21 44 13	Oct 6 1 20 32
	Oct 12 20 52 25	Oct 19 14 07 21	Oct 27 15 41 29	Nov 4 14 19 40
	Nov 11 6 34 15	Nov 18 4 59 58	Nov 26 10 17 11	Dec 4 1 24 35
	Dec 10 17 23 37	Dec 17 23 52 56	Dec 26 3 54 34	

Year	New Moon $h\ m\ s$	First Quarter $h\ m\ s$	Full Moon $h\ m\ s$	Last Quarter $h\ m\ s$
1959				Jan 2 $10^h51^m01^s$
	Jan 9 5 34 17	Jan 16 21 27 13	Jan 24 19 33 01	Jan 31 19 06 59
	Feb 7 19 22 35	Feb 15 19 20 30	Feb 23 8 54 20	Mar 2 2 54 36
	Mar 9 10 51 36	Mar 17 15 10 31	Mar 24 20 02 57	Mar 31 11 06 59
	Apr 8 3 29 31	Apr 16 7 33 06	Apr 23 5 13 51	Apr 29 20 38 53
	May 7 20 11 58	May 15 20 09 20	May 22 12 56 21	May 29 8 14 09
	Jun 6 11 53 54	Jun 14 5 23 03	Jun 20 20 00 17	Jun 27 22 12 38
	Jul 6 2 00 55	Jul 13 12 01 54	Jul 20 3 33 47	Jul 27 14 22 18
	Aug 4 14 34 23	Aug 11 17 10 16	Aug 18 12 51 04	Aug 26 8 03 28
	Sep 3 1 56 06	Sep 9 22 07 26	Sep 17 0 52 06	Sep 25 2 22 31
	Oct 2 12 31 22	Oct 9 4 22 56	Oct 16 15 59 03	Oct 24 20 22 18
	Oct 31 22 41 32	Nov 7 13 24 05	Nov 15 9 42 15	Nov 23 13 03 29
	Nov 30 8 46 30	Dec 7 2 12 12	Dec 15 4 49 19	Dec 23 3 28 34
	Dec 29 19 09 27			
1960	$h\ m\ s$	Jan 5 $18^h53^m29^s$	Jan 13 $23^h51^m07^s$	Jan 21 $15^h01^m09^s$
	Jan 28 6 15 39	Feb 4 14 26 33	Feb 12 17 24 27	Feb 19 23 47 58
	Feb 26 18 24 00	Mar 5 11 06 04	Mar 13 8 26 20	Mar 20 6 40 55
	Mar 27 7 37 50	Apr 4 7 05 03	Apr 11 20 27 49	Apr 18 12 57 06
	Apr 25 21 44 48	May 4 1 00 51	May 11 5 42 52	May 17 19 54 31
	May 25 12 26 38	Jun 2 16 01 35	Jun 9 13 02 18	Jun 16 4 36 00
	Jun 24 3 27 24	Jul 2 3 48 34	Jul 8 19 37 08	Jul 15 15 43 07
	Jul 23 18 31 16	Jul 31 12 38 55	Aug 7 2 41 26	Aug 14 5 37 07
	Aug 22 9 15 46	Aug 29 19 22 42	Sep 5 11 19 15	Sep 12 22 19 42
	Sep 20 23 13 00	Sep 28 1 13 19	Oct 4 22 16 36	Oct 12 17 25 31
	Oct 20 12 02 47	Oct 27 7 34 07	Nov 3 11 58 04	Nov 11 13 47 59
	Nov 18 23 46 43	Nov 25 15 42 08	Dec 3 4 24 53	Dec 11 9 38 46
	Dec 18 10 47 10	Dec 25 2 30 03		
1961	$h\ m\ s$	$h\ m\ s$	Jan 1 $23^h06^m25^s$	Jan 10 $3^h02^m43^s$
	Jan 16 21 30 28	Jan 23 16 13 59	Jan 31 18 47 14	Feb 8 16 49 47
	Feb 15 8 10 52	Feb 22 8 35 01	Mar 2 13 35 10	Mar 10 2 57 50
	Mar 16 18 51 08	Mar 24 2 48 30	Apr 1 5 47 40	Apr 8 10 16 17
	Apr 15 5 37 59	Apr 22 21 49 33	Apr 30 18 40 59	May 7 15 57 51
	May 14 16 54 41	May 22 16 18 54	May 30 4 37 33	Jun 5 21 19 04
	Jun 13 5 16 42	Jun 21 9 01 49	Jun 28 12 38 03	Jul 5 3 32 47
	Jul 12 19 11 33	Jul 20 23 13 48	Jul 27 19 50 37	Aug 3 11 47 41
	Aug 11 10 36 16	Aug 19 10 51 34	Aug 26 3 13 48	Sep 1 23 05 52
	Sep 10 2 50 03	Sep 17 20 23 54	Sep 24 11 33 53	Oct 1 14 10 22
	Oct 9 18 53 00	Oct 17 4 35 00	Oct 23 21 31 01	Oct 31 8 58 57
	Nov 8 9 58 56	Nov 15 12 12 59	Nov 22 9 44 13	Nov 30 6 18 56
	Dec 7 23 52 12	Dec 14 20 06 00	Dec 22 0 42 20	Dec 30 3 57 18
1962	Jan 6 $12^h35^m48^s$	Jan 13 $5^h01^m44^s$	Jan 20 $18^h16^m50^s$	Jan 28 $23^h36^m54^s$
	Feb 5 0 10 26	Feb 11 15 43 14	Feb 19 13 18 25	Feb 27 15 50 18
	Mar 6 10 31 24	Mar 13 4 39 05	Mar 21 7 55 57	Mar 29 4 11 16
	Apr 4 19 45 17	Apr 11 19 50 38	Apr 20 0 33 50	Apr 27 12 59 33
	May 4 4 25 09	May 11 12 44 36	May 19 14 32 23	May 26 19 05 53
	Jun 2 13 27 17	Jun 10 6 21 33	Jun 18 2 02 45	Jun 24 23 43 01
	Jul 1 23 52 44	Jul 9 23 39 42	Jul 17 11 41 05	Jul 24 4 18 37
	Jul 31 12 24 13	Aug 8 15 55 23	Aug 15 20 09 44	Aug 22 10 26 57
	Aug 30 3 09 26	Sep 7 6 45 00	Sep 14 4 11 50	Sep 20 19 36 11
	Sep 28 19 39 52	Oct 6 19 54 48	Oct 13 12 33 28	Oct 20 8 47 40
	Oct 28 13 05 17	Nov 5 7 15 07	Nov 11 22 03 57	Nov 19 2 09 43
	Nov 27 6 30 01	Dec 4 16 48 23	Dec 11 9 27 54	Dec 18 22 42 39
	Dec 26 22 59 15			

PHASES OF THE MOON

Year	New Moon	First Quarter	Full Moon	Last Quarter
1963	$h \quad m \quad s$	Jan 3 $\quad 1^h02^m08^s$	Jan 9 $\quad 23^h08^m39^s$	Jan 17 $\quad 20^h34^m32^s$
	Jan 25 13 42 26	Feb 1 8 50 24	Feb 8 14 52 11	Feb 16 17 39 01
	Feb 24 2 06 12	Mar 2 17 17 31	Mar 10 7 49 07	Mar 18 12 08 07
	Mar 25 12 10 12	Apr 1 3 15 10	Apr 9 0 57 15	Apr 17 2 52 59
	Apr 23 20 29 14	Apr 30 15 08 07	May 8 17 23 43	May 16 13 36 33
	May 23 4 00 21	May 30 4 55 49	Jun 7 8 31 09	Jun 14 20 53 42
	Jun 21 11 46 18	Jun 28 20 24 04	Jul 6 21 55 58	Jul 14 1 57 37
	Jul 20 20 43 16	Jul 28 13 13 13	Aug 5 9 31 24	Aug 12 6 21 34
	Aug 19 7 35 03	Aug 27 6 54 16	Sep 3 19 34 03	Sep 10 11 42 42
	Sep 17 20 51 16	Sep 26 0 38 54	Oct 3 4 44 26	Oct 9 19 27 34
	Oct 17 12 43 12	Oct 25 17 20 35	Nov 1 13 55 43	Nov 8 6 37 09
	Nov 16 6 50 49	Nov 24 7 56 15	Nov 30 23 54 34	Dec 7 21 34 26
	Dec 16 2 06 47	Dec 23 19 54 54	Dec 30 11 04 23	
1964	$h \quad m \quad s$	$h \quad m \quad s$	$h \quad m \quad s$	Jan 6 $\quad 15^h58^m22^s$
	Jan 14 20 44 01	Jan 22 5 29 08	Jan 28 23 23 18	Feb 5 12 42 44
	Feb 13 13 01 42	Feb 20 13 24 35	Feb 27 12 39 46	Mar 6 10 00 13
	Mar 14 2 14 20	Mar 20 20 39 35	Mar 28 2 48 54	Apr 5 5 45 33
	Apr 12 12 37 58	Apr 19 4 09 31	Apr 26 17 50 07	May 4 22 20 20
	May 11 21 02 12	May 18 12 42 32	May 26 9 29 20	Jun 3 11 07 50
	Jun 10 4 22 52	Jun 16 23 02 21	Jun 25 1 08 33	Jul 2 20 31 21
	Jul 9 11 31 22	Jul 16 11 47 30	Jul 24 15 58 21	Aug 1 3 29 53
	Aug 7 19 17 16	Aug 15 3 19 44	Aug 23 5 25 30	Aug 30 9 15 36
	Sep 6 4 34 31	Sep 13 21 24 09	Sep 21 17 31 18	Sep 28 15 01 56
	Oct 5 16 20 12	Oct 13 16 56 57	Oct 21 4 45 35	Oct 27 21 59 06
	Nov 4 7 16 51	Nov 12 12 20 41	Nov 19 15 43 17	Nov 26 7 11 00
	Dec 4 1 18 45	Dec 12 6 01 54	Dec 19 2 41 45	Dec 25 19 27 23
1965	Jan 2 $\quad 21^h07^m24^s$	Jan 10 $\quad 21^h00^m03^s$	Jan 17 $\quad 13^h37^m52^s$	Jan 24 $\quad 11^h07^m28^s$
	Feb 1 16 36 05	Feb 9 8 53 10	Feb 16 0 27 10	Feb 23 5 39 42
	Mar 3 9 56 19	Mar 10 17 52 55	Mar 17 11 24 23	Mar 25 1 37 05
	Apr 2 0 21 06	Apr 9 0 40 19	Apr 15 23 02 40	Apr 23 21 07 13
	May 1 11 56 16	May 8 6 20 07	May 15 11 52 32	May 23 14 40 47
	May 30 21 13 07	Jun 6 12 11 31	Jun 14 1 59 53	Jun 22 5 36 42
	Jun 29 4 52 57	Jul 5 19 36 46	Jul 13 17 01 57	Jul 21 17 53 34
	Jul 28 11 45 27	Aug 4 5 47 31	Aug 12 8 22 43	Aug 20 3 50 47
	Aug 26 18 51 01	Sep 2 19 27 58	Sep 10 23 32 27	Sep 18 11 58 42
	Sep 25 3 18 12	Oct 2 12 37 49	Oct 10 14 14 08	Oct 17 19 00 07
	Oct 24 14 11 46	Nov 1 8 26 25	Nov 9 4 15 43	Nov 16 1 54 13
	Nov 23 4 10 25	Dec 1 5 24 49	Dec 8 17 21 41	Dec 15 9 52 18
	Dec 22 21 03 21	Dec 31 1 46 42		
1966	$h \quad m \quad s$	$h \quad m \quad s$	Jan 7 $\quad 5^h16^m40^s$	Jan 13 $\quad 20^h00^m14^s$
	Jan 21 15 46 43	Jan 29 19 48 48	Feb 5 15 58 28	Feb 12 8 53 15
	Feb 20 10 49 34	Feb 28 10 15 32	Mar 7 1 45 50	Mar 14 0 19 30
	Mar 22 4 46 40	Mar 29 20 43 49	Apr 5 11 13 43	Apr 12 17 28 46
	Apr 20 20 35 33	Apr 28 3 49 36	May 4 21 01 04	May 12 11 19 25
	May 20 9 42 50	May 27 8 50 49	Jun 3 7 40 49	Jun 11 4 58 33
	Jun 18 20 09 24	Jun 25 13 23 03	Jul 2 19 36 55	Jul 10 21 43 20
	Jul 18 4 30 45	Jul 24 19 00 21	Aug 1 9 05 51	Aug 9 12 56 06
	Aug 16 11 48 19	Aug 23 3 02 19	Aug 31 0 14 12	Sep 8 2 07 36
	Sep 14 19 13 50	Sep 21 14 25 10	Sep 29 16 47 51	Oct 7 13 08 39
	Oct 14 3 52 11	Oct 21 5 35 01	Oct 29 10 00 36	Nov 5 22 18 33
	Nov 12 14 26 54	Nov 20 0 20 32	Nov 28 2 40 56	Dec 5 6 22 36
	Dec 12 3 13 50	Dec 19 21 41 23	Dec 27 17 43 41	

Year	New Moon	First Quarter	Full Moon	Last Quarter
	h m s	h m s	h m s	h m s
1967	Jan 10 18 06 29	Jan 18 19 41 50	Jan 26 6 41 00	Jan 3 $14^h19^m25^s$
	Feb 9 10 44 25	Feb 17 15 56 38	Feb 24 17 43 53	Feb 1 23 03 23
	Mar 11 4 30 18	Mar 19 8 31 33	Mar 26 3 21 23	Mar 3 9 10 50
	Apr 9 22 20 43	Apr 17 20 48 09	Apr 24 12 04 01	Apr 1 20 58 46
	May 9 14 55 55	May 17 5 18 15	May 23 20 22 52	May 1 10 33 06
	Jun 8 5 14 04	Jun 15 11 12 21	Jun 22 4 57 23	May 31 1 52 25
	Jul 7 17 00 41	Jul 14 15 53 29	Jul 21 14 39 46	Jun 29 18 40 00
	Aug 6 2 48 57	Aug 12 20 44 52	Aug 20 2 27 17	Jul 29 12 14 34
	Sep 4 11 37 33	Sep 11 3 06 11	Sep 18 17 00 02	Aug 28 5 35 31
	Oct 3 20 24 21	Oct 10 12 11 19	Oct 18 10 11 22	Sep 26 21 44 18
	Nov 2 5 48 55	Nov 9 1 00 10	Nov 17 4 53 07	Oct 26 12 04 14
	Dec 1 16 10 16	Dec 8 17 57 34	Dec 16 23 21 54	Nov 25 0 23 37
	Dec 31 3 39 00			Dec 24 10 48 29
	h m s			
1968	Jan 29 16 29 47	Jan 7 $14^h23^m10^s$	Jan 15 $16^h11^m51^s$	Jan 22 $19^h38^m27^s$
	Feb 28 6 56 07	Feb 6 12 20 55	Feb 14 6 43 29	Feb 21 3 28 22
	Mar 28 22 48 36	Mar 7 9 20 52	Mar 14 18 52 55	Mar 21 11 08 03
	Apr 27 15 21 35	Apr 6 3 28 04	Apr 13 4 52 17	Apr 19 19 35 25
	May 27 7 30 19	May 5 17 54 48	May 12 13 05 25	May 19 5 44 56
	Jun 25 22 25 04	Jun 4 4 47 16	Jun 10 20 13 41	Jun 17 18 14 13
	Jul 25 11 50 02	Jul 3 12 42 13	Jul 10 3 18 15	Jul 17 9 11 57
	Aug 23 23 57 09	Aug 1 18 34 42	Aug 8 11 32 54	Aug 16 2 13 52
	Sep 22 11 08 43	Aug 30 23 35 05	Sep 6 22 07 50	Sep 14 20 31 33
	Oct 21 21 44 50	Sep 29 5 07 08	Oct 6 11 46 42	Oct 14 15 05 53
	Nov 20 8 02 09	Oct 28 12 40 13	Nov 5 4 25 14	Nov 13 8 53 42
	Dec 19 18 19 12	Nov 26 23 30 54	Dec 4 23 07 52	Dec 13 0 49 59
		Dec 26 14 14 58		
	h m s	h m s		
1969	Jan 18 4 59 16	Jan 25 8 23 47	Jan 3 $18^h28^m11^s$	Jan 11 $14^h00^m59^s$
	Feb 16 16 25 40	Feb 24 4 30 44	Feb 2 12 56 25	Feb 10 0 08 49
	Mar 18 4 51 58	Mar 26 0 48 38	Mar 4 5 17 41	Mar 11 7 45 01
	Apr 16 18 16 22	Apr 24 19 45 09	Apr 2 18 45 31	Apr 9 13 58 45
	May 16 8 26 33	May 24 12 15 32	May 2 5 13 59	May 8 20 12 17
	Jun 14 23 09 11	Jun 23 1 44 40	May 31 13 18 49	Jun 7 3 39 59
	Jul 14 14 11 59	Jul 22 12 10 06	Jun 29 20 04 22	Jul 6 13 17 30
	Aug 13 5 17 03	Aug 20 20 03 31	Jul 29 2 45 38	Aug 5 1 39 01
	Sep 11 19 56 25	Sep 19 2 25 05	Aug 27 10 33 02	Sep 3 16 58 22
	Oct 11 9 39 48	Oct 18 8 32 22	Sep 25 20 21 28	Oct 3 11 05 35
	Nov 9 22 12 01	Nov 16 15 45 33	Oct 25 8 44 47	Nov 2 7 14 16
	Dec 9 9 42 35	Dec 16 1 09 42	Nov 23 23 54 21	Dec 2 3 50 56
			Dec 23 17 35 33	Dec 31 22 52 47
1970	Jan 7 $20^h36^m07^s$	Jan 14 $13^h18^m49^s$	Jan 22 $12^h55^m58^s$	Jan 30 $14^h39^m06^s$
	Feb 6 7 13 29	Feb 13 4 11 00	Feb 21 8 19 19	Mar 1 2 33 47
	Mar 7 17 43 06	Mar 14 21 16 16	Mar 23 1 53 05	Mar 30 11 05 11
	Apr 6 4 10 00	Apr 13 15 44 14	Apr 21 16 21 57	Apr 28 17 18 52
	May 5 14 51 34	May 13 10 26 42	May 21 3 38 14	May 27 22 32 18
	Jun 4 2 21 53	Jun 12 4 07 04	Jun 19 12 28 04	Jun 26 4 01 56
	Jul 3 15 18 27	Jul 11 19 43 39	Jul 18 19 59 08	Jul 25 11 00 12
	Aug 2 5 58 47	Aug 10 8 50 31	Aug 17 3 16 00	Aug 23 20 35 00
	Aug 31 22 01 52	Sep 8 19 38 55	Sep 15 11 10 04	Sep 22 9 42 42
	Sep 30 14 32 08	Oct 8 4 43 29	Oct 14 20 21 52	Oct 22 2 47 56
	Oct 30 6 28 32	Nov 6 12 47 33	Nov 13 7 28 32	Nov 20 23 13 43
	Nov 28 21 14 48	Dec 5 20 36 14	Dec 12 21 03 58	Dec 20 21 09 31
	Dec 28 10 43 19			

Year	New Moon	First Quarter	Full Moon	Last Quarter
1971	$h \quad m \quad s$	Jan 4 $\quad 4^h55^m34^s$	Jan 11 $\quad 13^h20^m56^s$	Jan 19 $\quad 18^h08^m54^s$
	Jan 26 $\quad$ 22 55 46	Feb 2 $\quad$ 14 31 15	Feb 10 $\quad$ 7 42 02	Feb 18 $\quad$ 12 14 16
	Feb 25 $\quad$ 9 49 13	Mar 4 $\quad$ 2 01 42	Mar 12 $\quad$ 2 34 14	Mar 20 $\quad$ 2 30 38
	Mar 26 $\quad$ 19 24 09	Apr 2 $\quad$ 15 46 28	Apr 10 $\quad$ 20 10 37	Apr 18 $\quad$ 12 58 31
	Apr 25 $\quad$ 4 02 21	May 2 $\quad$ 7 34 34	May 10 $\quad$ 11 24 13	May 17 $\quad$ 20 15 34
	May 24 $\quad$ 12 32 28	Jun 1 $\quad$ 0 42 55	Jun 9 $\quad$ 0 04 17	Jun 16 $\quad$ 1 24 48
	Jun 22 $\quad$ 21 58 00	Jun 30 $\quad$ 18 11 39	Jul 8 $\quad$ 10 37 17	Jul 15 $\quad$ 5 47 23
	Jul 22 $\quad$ 9 15 39	Jul 30 $\quad$ 11 07 43	Aug 6 $\quad$ 19 43 01	Aug 13 $\quad$ 10 55 47
	Aug 20 $\quad$ 22 54 01	Aug 29 $\quad$ 2 56 45	Sep 5 $\quad$ 4 03 05	Sep 11 $\quad$ 18 23 49
	Sep 19 $\quad$ 14 42 52	Sep 27 $\quad$ 17 17 58	Oct 4 $\quad$ 12 20 07	Oct 11 $\quad$ 5 29 56
	Oct 19 $\quad$ 7 59 35	Oct 27 $\quad$ 5 54 54	Nov 2 $\quad$ 21 20 02	Nov 9 $\quad$ 20 51 58
	Nov 18 $\quad$ 1 46 26	Nov 25 $\quad$ 16 37 25	Dec 2 $\quad$ 7 48 47	Dec 9 $\quad$ 16 03 03
	Dec 17 $\quad$ 19 03 26	Dec 25 $\quad$ 1 35 33	Dec 31 $\quad$ 20 20 12	
1972	$h \quad m \quad s$	$h \quad m \quad s$	$h \quad m \quad s$	Jan 8 $\quad 13^h31^m24^s$
	Jan 16 $\quad$ 10 53 04	Jan 23 $\quad$ 9 29 23	Jan 30 $\quad$ 10 58 52	Feb 7 $\quad$ 11 12 10
	Feb 15 $\quad$ 0 29 27	Feb 21 $\quad$ 17 20 56	Feb 29 $\quad$ 3 12 36	Mar 8 $\quad$ 7 06 07
	Mar 15 $\quad$ 11 35 23	Mar 22 $\quad$ 2 12 22	Mar 29 $\quad$ 20 06 10	Apr 6 $\quad$ 23 45 06
	Apr 13 $\quad$ 20 31 34	Apr 20 $\quad$ 12 45 52	Apr 28 $\quad$ 12 45 08	May 6 $\quad$ 12 27 02
	May 13 $\quad$ 4 08 42	May 20 $\quad$ 1 17 00	May 28 $\quad$ 4 28 17	Jun 4 $\quad$ 21 22 13
	Jun 11 $\quad$ 11 30 38	Jun 18 $\quad$ 15 41 40	Jun 26 $\quad$ 18 46 51	Jul 4 $\quad$ 3 26 08
	Jul 10 $\quad$ 19 39 27	Jul 18 $\quad$ 7 46 24	Jul 26 $\quad$ 7 24 14	Aug 2 $\quad$ 8 03 00
	Aug 9 $\quad$ 5 26 30	Aug 17 $\quad$ 1 09 49	Aug 24 $\quad$ 18 22 20	Aug 31 $\quad$ 12 48 51
	Sep 7 $\quad$ 17 29 06	Sep 15 $\quad$ 19 13 35	Sep 23 $\quad$ 4 07 26	Sep 29 $\quad$ 19 16 42
	Oct 7 $\quad$ 8 08 36	Oct 15 $\quad$ 12 55 25	Oct 22 $\quad$ 13 25 43	Oct 29 $\quad$ 4 41 28
	Nov 6 $\quad$ 1 21 57	Nov 14 $\quad$ 5 01 24	Nov 20 $\quad$ 23 07 12	Nov 27 $\quad$ 17 45 17
	Dec 5 $\quad$ 20 24 43	Dec 13 $\quad$ 18 36 17	Dec 20 $\quad$ 9 45 45	Dec 27 $\quad$ 10 28 05
1973	Jan 4 $\quad 15^h43^m10^s$	Jan 12 $\quad 5^h27^m42^s$	Jan 18 $\quad 21^h29^m06^s$	Jan 26 $\quad 6^h05^m34^s$
	Feb 3 $\quad$ 9 23 31	Feb 10 $\quad$ 14 06 04	Feb 17 $\quad$ 10 07 40	Feb 25 $\quad$ 3 10 59
	Mar 5 $\quad$ 0 08 07	Mar 11 $\quad$ 21 26 26	Mar 18 $\quad$ 23 34 07	Mar 26 $\quad$ 23 46 52
	Apr 3 $\quad$ 11 45 46	Apr 10 $\quad$ 4 28 42	Apr 17 $\quad$ 13 51 12	Apr 25 $\quad$ 17 59 16
	May 2 $\quad$ 20 55 36	May 9 $\quad$ 12 07 23	May 17 $\quad$ 4 58 45	May 25 $\quad$ 8 40 33
	Jun 1 $\quad$ 4 34 58	Jun 7 $\quad$ 21 11 30	Jun 15 $\quad$ 20 35 16	Jun 23 $\quad$ 19 45 58
	Jun 30 $\quad$ 11 39 28	Jul 7 $\quad$ 8 26 23	Jul 15 $\quad$ 11 56 35	Jul 23 $\quad$ 3 58 15
	Jul 29 $\quad$ 18 59 30	Aug 5 $\quad$ 22 27 26	Aug 14 $\quad$ 2 17 12	Aug 21 $\quad$ 10 22 52
	Aug 28 $\quad$ 3 26 00	Sep 4 $\quad$ 15 22 37	Sep 12 $\quad$ 15 16 58	Sep 19 $\quad$ 16 11 14
	Sep 26 $\quad$ 13 54 32	Oct 4 $\quad$ 10 32 39	Oct 12 $\quad$ 3 09 39	Oct 18 $\quad$ 22 33 23
	Oct 26 $\quad$ 3 17 12	Nov 3 $\quad$ 6 29 59	Nov 10 $\quad$ 14 27 24	Nov 17 $\quad$ 6 35 07
	Nov 24 $\quad$ 19 55 47	Dec 3 $\quad$ 1 29 17	Dec 10 $\quad$ 1 35 18	Dec 16 $\quad$ 17 13 20
	Dec 24 $\quad$ 15 07 42			
1974	$h \quad m \quad s$	Jan 1 $\quad 18^h07^m01^s$	Jan 8 $\quad 12^h37^m01^s$	Jan 15 $\quad 7^h04^m14^s$
	Jan 23 $\quad$ 11 02 35	Jan 31 $\quad$ 7 39 56	Feb 6 $\quad$ 23 24 55	Feb 14 $\quad$ 0 04 23
	Feb 22 $\quad$ 5 34 27	Mar 1 $\quad$ 18 03 14	Mar 8 $\quad$ 10 03 38	Mar 15 $\quad$ 19 15 55
	Mar 23 $\quad$ 21 24 40	Mar 31 $\quad$ 1 44 49	Apr 6 $\quad$ 21 01 02	Apr 14 $\quad$ 14 58 09
	Apr 22 $\quad$ 10 17 13	Apr 29 $\quad$ 7 40 01	May 6 $\quad$ 8 55 16	May 14 $\quad$ 9 29 23
	May 21 $\quad$ 20 35 00	May 28 $\quad$ 13 03 55	Jun 4 $\quad$ 22 10 21	Jun 13 $\quad$ 1 45 58
	Jun 20 $\quad$ 4 56 24	Jun 26 $\quad$ 19 20 46	Jul 4 $\quad$ 12 41 01	Jul 12 $\quad$ 15 28 41
	Jul 19 $\quad$ 12 07 14	Jul 26 $\quad$ 3 51 40	Aug 3 $\quad$ 3 57 43	Aug 11 $\quad$ 2 46 33
	Aug 17 $\quad$ 19 02 17	Aug 24 $\quad$ 15 39 09	Sep 1 $\quad$ 19 25 33	Sep 9 $\quad$ 12 01 38
	Sep 16 $\quad$ 2 45 57	Sep 23 $\quad$ 7 08 34	Oct 1 $\quad$ 10 38 40	Oct 8 $\quad$ 19 46 16
	Oct 15 $\quad$ 12 25 30	Oct 23 $\quad$ 1 53 46	Oct 31 $\quad$ 1 19 41	Nov 7 $\quad$ 2 47 53
	Nov 14 $\quad$ 0 53 55	Nov 21 $\quad$ 22 40 11	Nov 29 $\quad$ 15 10 46	Dec 6 $\quad$ 10 10 40
	Dec 13 $\quad$ 16 25 27	Dec 21 $\quad$ 19 44 02	Dec 29 $\quad$ 3 51 34	

Year	New Moon			First Quarter			Full Moon			Last Quarter		

1975 — New Moon h m s; First Quarter h m s; Full Moon h m s; Last Quarter starts Jan 4 $19^h04^m46^s$

Year	New Moon		First Quarter		Full Moon		Last Quarter	
1975							Jan 4	19ʰ04ᵐ46ˢ
	Jan 12	10 20 19	Jan 20	15 15 02	Jan 27	15 10 10	Feb 3	6 23 29
	Feb 11	5 17 34	Feb 19	7 39 12	Feb 26	1 15 14	Mar 4	20 20 30
	Mar 12	23 48 10	Mar 20	20 05 20	Mar 27	10 36 47	Apr 3	12 25 35
	Apr 11	16 39 44	Apr 19	4 41 54	Apr 25	19 55 40	May 3	5 44 24
	May 11	7 05 38	May 18	10 29 43	May 25	5 51 20	Jun 1	23 23 12
	Jun 9	18 49 51	Jun 16	14 59 03	Jun 23	16 54 57	Jul 1	16 37 53
	Jul 9	4 10 53	Jul 15	19 47 27	Jul 23	5 29 11	Jul 31	8 49 08
	Aug 7	11 58 07	Aug 14	2 24 15	Aug 21	19 48 21	Aug 29	23 20 20
	Sep 5	19 19 26	Sep 12	11 59 35	Sep 20	11 51 06	Sep 28	11 46 33
	Oct 5	3 24 01	Oct 12	1 16 01	Oct 20	5 06 29	Oct 27	22 07 35
	Nov 3	13 05 29	Nov 10	18 21 27	Nov 18	22 28 58	Nov 26	6 52 34
	Dec 3	0 50 46	Dec 10	14 39 47	Dec 18	14 40 12	Dec 25	14 52 37
1976	Jan 1	14ʰ40ᵐ46ˢ	Jan 9	12ʰ40ᵐ19ˢ	Jan 17	4ʰ47ᵐ41ˢ	Jan 23	23ʰ05ᵐ05ˢ
	Jan 31	6 20 54	Feb 8	10 05 35	Feb 15	16 44 01	Feb 22	8 16 36
	Feb 29	23 25 30	Mar 9	4 38 41	Mar 16	2 53 30	Mar 22	18 55 08
	Mar 30	17 08 41	Apr 7	19 02 28	Apr 14	11 49 31	Apr 21	7 14 48
	Apr 29	10 20 14	May 7	5 18 01	May 13	20 04 49	May 20	21 22 46
	May 29	1 47 32	Jun 5	12 20 40	Jun 12	4 15 44	Jun 19	13 16 05
	Jun 27	14 50 43	Jul 4	17 29 13	Jul 11	13 09 38	Jul 19	6 30 04
	Jul 27	1 39 29	Aug 2	22 07 15	Aug 9	23 44 14	Aug 18	0 13 15
	Aug 25	11 01 19	Sep 1	3 36 10	Sep 8	12 52 48	Sep 16	17 20 57
	Sep 23	19 55 37	Sep 30	11 12 58	Oct 8	4 56 08	Oct 16	8 59 21
	Oct 23	5 10 23	Oct 29	22 05 48	Nov 6	23 15 19	Nov 14	22 39 30
	Nov 21	15 11 23	Nov 28	12 59 29	Dec 6	18 15 16	Dec 14	10 14 58
	Dec 21	2 08 31	Dec 28	7 48 26				
1977		h m s		h m s	Jan 5	12ʰ11ᵐ07ˢ	Jan 12	19ʰ55ᵐ45ˢ
	Jan 19	14 11 50	Jan 27	5 12 11	Feb 4	3 57 01	Feb 11	4 07 57
	Feb 18	3 37 38	Feb 26	2 50 47	Mar 5	17 13 59	Mar 12	11 35 21
	Mar 19	18 33 24	Mar 27	22 27 25	Apr 4	4 09 40	Apr 10	19 15 24
	Apr 18	10 36 12	Apr 26	14 42 58	May 3	13 04 12	May 10	4 09 03
	May 18	2 52 10	May 26	3 20 53	Jun 1	20 31 39	Jun 8	15 07 48
	Jun 16	18 23 30	Jun 24	12 44 31	Jul 1	3 24 49	Jul 8	4 40 01
	Jul 16	8 37 16	Jul 23	19 39 01	Jul 30	10 53 01	Aug 6	20 40 53
	Aug 14	21 31 47	Aug 22	1 05 11	Aug 28	20 10 50	Sep 5	14 33 38
	Sep 13	9 23 33	Sep 20	6 18 46	Sep 27	8 18 09	Oct 5	9 21 28
	Oct 12	20 31 27	Oct 19	12 46 30	Oct 26	23 36 06	Nov 4	3 59 05
	Nov 11	7 10 08	Nov 17	21 53 11	Nov 25	17 32 10	Dec 3	21 16 46
	Dec 10	17 33 30	Dec 17	10 37 30	Dec 25	12 49 47		
1978		h m s		h m s		h m s	Jan 2	12ʰ08ᵐ02ˢ
	Jan 9	4 00 29	Jan 16	3 03 37	Jan 24	7 56 17	Jan 31	23 51 47
	Feb 7	14 54 46	Feb 14	22 11 24	Feb 23	1 27 07	Mar 2	8 34 48
	Mar 9	2 37 00	Mar 16	18 21 42	Mar 24	16 20 47	Mar 31	15 11 34
	Apr 7	15 15 56	Apr 15	13 56 22	Apr 23	4 11 45	Apr 29	21 02 47
	May 7	4 47 30	May 15	7 40 15	May 22	13 17 29	May 29	3 31 01
	Jun 5	19 02 15	Jun 13	22 44 47	Jun 20	20 31 19	Jun 27	11 44 48
	Jul 5	9 51 07	Jul 13	10 49 44	Jul 20	3 05 37	Jul 26	22 31 50
	Aug 4	1 01 31	Aug 11	20 07 13	Aug 18	10 14 47	Aug 25	12 18 21
	Sep 2	16 09 41	Sep 10	3 21 00	Sep 16	19 01 53	Sep 24	5 08 16
	Oct 2	6 41 26	Oct 9	9 38 43	Oct 16	6 10 18	Oct 24	0 34 38
	Oct 31	20 07 18	Nov 7	16 18 54	Nov 14	20 00 54	Nov 22	21 24 59
	Nov 30	8 20 13	Dec 7	0 34 38	Dec 14	12 31 25	Dec 22	17 42 17
	Dec 29	19 36 41						

PHASES OF THE MOON

1979

Year	New Moon		First Quarter		Full Moon		Last Quarter	
	Date	h m s	Date	h m s	Date	h m s	Date	h m s
1979			Jan 5	$11^h15^m36^s$	Jan 13	$7^h09^m24^s$	Jan 21	$11^h24^m07^s$
	Jan 28	6 20 20	Feb 4	0 37 10	Feb 12	2 40 11	Feb 20	1 17 56
	Feb 26	16 46 01	Mar 5	16 23 37	Mar 13	21 15 04	Mar 21	11 22 55
	Mar 28	3 00 17	Apr 4	9 58 03	Apr 12	13 15 50	Apr 19	18 31 05
	Apr 26	13 15 33	May 4	4 26 08	May 12	2 01 41	May 18	23 57 39
	May 26	0 01 12	Jun 2	22 38 16	Jun 10	11 56 05	Jun 17	5 01 49
	Jun 24	11 58 39	Jul 2	15 24 25	Jul 9	20 00 05	Jul 16	10 59 31
	Jul 24	1 41 27	Aug 1	5 58 08	Aug 8	3 21 55	Aug 14	19 02 40
	Aug 22	17 11 15	Aug 30	18 09 51	Sep 6	10 59 23	Sep 13	6 16 10
	Sep 21	9 47 23	Sep 29	4 20 32	Oct 5	19 36 08	Oct 12	21 24 36
	Oct 21	2 23 50	Oct 28	13 06 40	Nov 4	5 47 50	Nov 11	16 24 47
	Nov 19	18 04 25	Nov 26	21 09 23	Dec 3	18 08 26	Dec 11	13 59 40
	Dec 19	8 24 10	Dec 26	5 11 44				

1980

Year	New Moon		First Quarter		Full Moon		Last Quarter	
	Date	h m s	Date	h m s	Date	h m s	Date	h m s
1980					Jan 2	$9^h02^m55^s$	Jan 10	$11^h50^m19^s$
	Jan 17	21 20 12	Jan 24	13 58 56	Feb 1	2 22 02	Feb 9	7 35 57
	Feb 16	8 51 42	Feb 23	0 14 48	Mar 1	21 00 28	Mar 9	23 49 29
	Mar 16	18 56 42	Mar 23	12 32 07	Mar 31	15 14 41	Apr 8	12 07 15
	Apr 15	3 47 03	Apr 22	3 00 15	Apr 30	7 36 07	May 7	20 51 26
	May 14	12 01 01	May 21	19 16 47	May 29	21 28 26	Jun 6	2 54 01
	Jun 12	20 39 08	Jun 20	12 32 36	Jun 28	9 02 57	Jul 5	7 27 59
	Jul 12	6 46 29	Jul 20	5 51 24	Jul 27	18 54 28	Aug 3	12 01 08
	Aug 10	19 10 07	Aug 18	22 28 42	Aug 26	3 43 02	Sep 1	18 08 23
	Sep 9	10 01 00	Sep 17	13 55 07	Sep 24	12 08 32	Oct 1	3 18 44
	Oct 9	2 50 26	Oct 17	3 47 59	Oct 23	20 52 41	Oct 30	16 33 37
	Nov 7	20 43 20	Nov 15	15 47 35	Nov 22	6 39 40	Nov 29	9 59 22
	Dec 7	14 35 51	Dec 15	1 47 40	Dec 21	18 08 49	Dec 29	6 32 46

1981

Year	New Moon		First Quarter		Full Moon		Last Quarter	
	Date	h m s	Date	h m s	Date	h m s	Date	h m s
1981	Jan 6	$7^h24^m49^s$	Jan 13	$10^h10^m48^s$	Jan 20	$7^h39^m46^s$	Jan 28	$4^h19^m36^s$
	Feb 4	22 14 35	Feb 11	17 49 58	Feb 18	22 59 03	Feb 27	1 14 51
	Mar 6	10 31 49	Mar 13	1 51 21	Mar 20	15 23 17	Mar 28	19 34 53
	Apr 4	20 20 16	Apr 11	11 11 24	Apr 19	7 59 45	Apr 27	10 15 18
	May 4	4 20 00	May 10	22 22 40	May 19	0 04 25	May 26	21 01 00
	Jun 2	11 32 32	Jun 9	11 34 06	Jun 17	15 05 18	Jun 25	4 25 54
	Jul 1	19 03 56	Jul 9	2 40 12	Jul 17	4 39 41	Jul 24	9 40 44
	Jul 31	3 52 47	Aug 7	19 26 54	Aug 15	16 37 28	Aug 22	14 16 23
	Aug 29	14 44 15	Sep 6	13 26 23	Sep 14	3 09 32	Sep 20	19 48 14
	Sep 28	4 08 02	Oct 6	7 46 02	Oct 13	12 50 15	Oct 20	3 41 19
	Oct 27	20 14 06	Nov 5	1 09 45	Nov 11	22 27 20	Nov 18	14 54 33
	Nov 26	14 39 10	Dec 4	16 23 02	Dec 11	8 42 11	Dec 18	5 48 05
	Dec 26	10 10 50						

1982

Year	New Moon		First Quarter		Full Moon		Last Quarter	
	Date	h m s	Date	h m s	Date	h m s	Date	h m s
1982			Jan 3	$4^h46^m20^s$	Jan 9	$19^h53^m39^s$	Jan 16	$23^h58^m49^s$
	Jan 25	4 56 47	Feb 1	14 28 39	Feb 8	7 57 57	Feb 15	20 21 45
	Feb 23	21 13 57	Mar 2	22 15 53	Mar 9	20 45 58	Mar 17	17 15 29
	Mar 25	10 18 20	Apr 1	5 09 03	Apr 8	10 19 10	Apr 16	12 43 04
	Apr 23	20 29 32	Apr 30	12 08 08	May 8	0 45 31	May 16	5 12 05
	May 23	4 41 10	May 29	20 07 26	Jun 6	16 00 13	Jun 14	18 06 46
	Jun 21	11 52 37	Jun 28	5 57 17	Jul 6	7 32 28	Jul 14	3 47 37
	Jul 20	18 57 29	Jul 27	18 22 31	Aug 4	22 34 36	Aug 12	11 09 19
	Aug 19	2 45 43	Aug 26	9 50 12	Sep 3	12 29 04	Sep 10	17 19 30
	Sep 17	12 09 56	Sep 25	4 07 50	Oct 3	1 09 22	Oct 9	23 27 13
	Oct 17	0 04 57	Oct 25	0 08 38	Nov 1	12 57 30	Nov 8	6 38 34
	Nov 15	15 10 45	Nov 23	20 06 23	Dec 1	0 21 33	Dec 7	15 54 02
	Dec 15	9 18 54	Dec 23	14 17 24	Dec 30	11 33 26		

Year	New Moon		First Quarter		Full Moon		Last Quarter	
		$h \quad m \quad s$		$h \quad m \quad s$		$h \quad m \quad s$	Jan 6	$4^h00^m49^s$
1983	Jan 14	5 08 37	Jan 22	5 34 15	Jan 28	22 27 08	Feb 4	19 17 37
	Feb 13	0 32 40	Feb 20	17 32 51	Feb 27	8 58 58	Mar 6	13 16 30
	Mar 14	17 44 17	Mar 22	2 26 19	Mar 28	19 27 31	Apr 5	8 39 09
	Apr 13	7 59 11	Apr 20	8 58 37	Apr 27	6 31 32	May 5	3 43 49
	May 12	19 25 58	May 19	14 17 51	May 26	18 48 33	Jun 3	21 08 17
	Jun 11	4 38 24	Jun 17	19 46 42	Jun 25	8 32 39	Jul 3	12 12 51
	Jul 10	12 19 22	Jul 17	2 51 19	Jul 24	23 27 45	Aug 2	0 53 11
	Aug 8	19 18 48	Aug 15	12 47 50	Aug 23	15 00 10	Aug 31	11 23 09
	Sep 7	2 35 50	Sep 14	2 24 36	Sep 22	6 37 02	Sep 29	20 06 06
	Oct 6	11 16 31	Oct 13	19 43 06	Oct 21	21 54 07	Oct 29	3 37 38
	Nov 4	22 21 59	Nov 12	15 50 02	Nov 20	12 30 13	Nov 27	10 50 56
	Dec 4	12 26 43	Dec 12	13 09 53	Dec 20	2 01 08	Dec 26	18 53 18
1984	Jan 3	$5^h16^m31^s$	Jan 11	$9^h49^m09^s$	Jan 18	$14^h06^m01^s$	Jan 25	$4^h48^m37^s$
	Feb 1	23 47 17	Feb 10	4 00 29	Feb 17	0 41 56	Feb 23	17 12 48
	Mar 2	18 31 45	Mar 10	18 28 21	Mar 17	10 10 36	Mar 24	7 59 10
	Apr 1	12 10 26	Apr 9	4 52 17	Apr 15	19 11 39	Apr 23	0 27 04
	May 1	3 46 11	May 8	11 50 34	May 15	4 29 29	May 22	17 45 44
	May 30	16 48 44	Jun 6	16 42 33	Jun 13	14 42 41	Jun 21	11 10 31
	Jun 29	3 19 21	Jul 5	21 05 00	Jul 13	2 20 37	Jul 21	4 02 15
	Jul 28	11 52 07	Aug 4	2 33 41	Aug 11	15 44 05	Aug 19	19 41 24
	Aug 26	19 26 23	Sep 2	10 30 28	Sep 10	7 01 56	Sep 18	9 31 50
	Sep 25	3 11 30	Oct 1	21 53 21	Oct 9	23 59 02	Oct 17	21 14 42
	Oct 24	12 09 01	Oct 31	13 08 18	Nov 8	17 43 28	Nov 16	6 59 59
	Nov 22	22 57 32	Nov 30	8 01 23	Dec 8	10 54 13	Dec 15	15 26 06
	Dec 22	11 47 28	Dec 30	5 28 11				
1985		$h \quad m \quad s$		$h \quad m \quad s$	Jan 7	$2^h17^m00^s$	Jan 13	$23^h27^m35^s$
	Jan 21	2 29 08	Jan 29	3 29 53	Feb 5	15 19 39	Feb 12	7 57 39
	Feb 19	18 43 34	Feb 27	23 41 31	Mar 7	2 13 59	Mar 13	17 34 54
	Mar 21	11 59 35	Mar 29	16 12 01	Apr 5	11 33 19	Apr 12	4 42 22
	Apr 20	5 22 51	Apr 28	4 26 02	May 4	19 53 43	May 11	17 34 40
	May 19	21 42 01	May 27	12 56 38	Jun 3	3 51 20	Jun 10	8 19 58
	Jun 18	11 58 50	Jun 25	18 53 54	Jul 2	12 09 14	Jul 10	0 50 16
	Jul 17	23 57 15	Jul 24	23 39 53	Jul 31	21 41 50	Aug 8	18 29 41
	Aug 16	10 06 25	Aug 23	4 36 54	Aug 30	9 28 03	Sep 7	12 16 36
	Sep 14	19 20 48	Sep 21	11 03 49	Sep 29	0 09 21	Oct 7	5 04 59
	Oct 14	4 34 11	Oct 20	20 13 34	Oct 28	17 38 25	Nov 5	20 07 23
	Nov 12	14 21 13	Nov 19	9 04 24	Nov 27	12 42 37	Dec 5	9 02 08
	Dec 12	0 55 19	Dec 19	1 58 41	Dec 27	7 31 11		
1986		$h \quad m \quad s$		$h \quad m \quad s$		$h \quad m \quad s$	Jan 3	$19^h48^m06^s$
	Jan 10	12 22 34	Jan 17	22 14 10	Jan 26	0 31 56	Feb 2	4 41 40
	Feb 9	0 56 12	Feb 16	19 56 00	Feb 24	15 03 07	Mar 3	12 18 09
	Mar 10	14 52 27	Mar 18	16 39 29	Mar 26	3 02 41	Apr 1	19 30 48
	Apr 9	6 09 02	Apr 17	10 35 43	Apr 24	12 47 16	May 1	3 22 49
	May 8	22 10 31	May 17	1 00 35	May 23	20 45 39	May 30	12 55 26
	Jun 7	14 01 18	Jun 15	12 00 43	Jun 22	3 42 41	Jun 29	0 54 03
	Jul 7	4 55 44	Jul 14	20 10 49	Jul 21	10 41 07	Jul 28	15 35 06
	Aug 5	18 36 39	Aug 13	2 22 22	Aug 19	18 55 05	Aug 27	8 39 25
	Sep 4	7 11 17	Sep 11	7 41 44	Sep 18	5 34 37	Sep 26	3 18 14
	Oct 3	18 55 39	Oct 10	13 29 14	Oct 17	19 22 32	Oct 25	22 26 25
	Nov 2	6 03 01	Nov 8	21 11 26	Nov 16	12 12 26	Nov 24	16 51 14
	Dec 1	16 43 29	Dec 8	8 02 20	Dec 16	7 05 18	Dec 24	9 17 56
	Dec 31	3 10 40						

Year	New Moon	First Quarter	Full Moon	Last Quarter
	h m s			
1987		Jan 6 $22^h35^m20^s$	Jan 15 $2^h31^m17^s$	Jan 22 $22^h46^m16^s$
	Jan 29 13 45 24	Feb 5 16 21 38	Feb 13 20 58 48	Feb 21 8 56 43
	Feb 28 0 51 32	Mar 7 11 59 00	Mar 15 13 13 35	Mar 22 16 22 34
	Mar 29 12 46 27	Apr 6 7 48 34	Apr 14 2 31 53	Apr 20 22 16 06
	Apr 28 1 35 17	May 6 2 26 32	May 13 12 51 09	May 20 4 03 18
	May 27 15 14 16	Jun 4 18 53 34	Jun 11 20 49 43	Jun 18 11 03 25
	Jun 26 5 37 45	Jul 4 8 35 04	Jul 11 3 33 31	Jul 17 20 17 44
	Jul 25 20 38 26	Aug 2 19 24 53	Aug 9 10 18 13	Aug 16 8 25 54
	Aug 24 11 59 31	Sep 1 3 48 32	Sep 7 18 13 46	Sep 14 23 45 22
	Sep 23 3 09 07	Sep 30 10 39 56	Oct 7 4 13 09	Oct 14 18 06 29
	Oct 22 17 28 46	Oct 29 17 11 10	Nov 5 16 46 44	Nov 13 14 39 05
	Nov 21 6 33 47	Nov 28 0 37 54	Dec 5 8 01 53	Dec 13 11 42 07
	Dec 20 18 26 09	Dec 27 10 01 36		
	h m s	h m s		
1988			Jan 4 $1^h41^m16^s$	Jan 12 $7^h04^m42^s$
	Jan 19 5 26 30	Jan 25 21 54 26	Feb 2 20 52 24	Feb 10 23 01 42
	Feb 17 15 55 09	Feb 24 12 16 03	Mar 3 16 01 59	Mar 11 10 57 13
	Mar 18 2 03 14	Mar 25 4 42 26	Apr 2 9 22 12	Apr 9 19 21 53
	Apr 16 12 00 55	Apr 23 22 32 53	May 1 23 41 42	May 9 1 23 43
	May 15 22 11 28	May 23 16 49 47	May 31 10 54 25	Jun 7 6 22 27
	Jun 14 9 14 44	Jun 22 10 24 06	Jun 29 19 46 41	Jul 6 11 37 12
	Jul 13 21 53 57	Jul 22 2 15 13	Jul 29 3 26 08	Aug 4 18 23 17
	Aug 12 12 31 54	Aug 20 15 52 23	Aug 27 10 56 46	Sep 3 3 51 16
	Sep 11 4 50 03	Sep 19 3 19 06	Sep 25 19 08 00	Oct 2 16 59 23
	Oct 10 21 49 33	Oct 18 13 01 47	Oct 25 4 36 26	Nov 1 10 12 22
	Nov 9 14 20 38	Nov 16 21 36 11	Nov 23 15 53 59	Dec 1 6 50 20
	Dec 9 5 36 53	Dec 16 5 41 12	Dec 23 5 29 48	Dec 31 4 57 29
1989	Jan 7 $19^h23^m07^s$	Jan 14 $13^h59^m16^s$	Jan 21 $21^h34^m26^s$	Jan 30 $2^h03^m02^s$
	Feb 6 7 37 52	Feb 12 23 15 37	Feb 20 15 32 46	Feb 28 20 08 50
	Mar 7 18 19 35	Mar 14 10 11 44	Mar 22 9 58 51	Mar 30 10 22 22
	Apr 6 3 33 37	Apr 12 23 13 43	Apr 21 3 14 12	Apr 28 20 46 54
	May 5 11 47 18	May 12 14 20 27	May 20 18 17 10	May 28 4 01 40
	Jun 3 19 53 42	Jun 11 6 59 49	Jun 19 6 58 14	Jun 26 9 09 49
	Jul 3 4 59 58	Jul 11 0 19 53	Jul 18 17 42 32	Jul 25 13 32 11
	Aug 1 16 06 31	Aug 9 17 29 27	Aug 17 3 07 32	Aug 23 18 41 06
	Aug 31 5 45 26	Sep 8 9 50 07	Sep 15 11 51 29	Sep 22 2 10 45
	Sep 29 21 47 45	Oct 8 0 53 03	Oct 14 20 32 46	Oct 21 13 19 42
	Oct 29 15 27 57	Nov 6 14 11 48	Nov 13 5 52 18	Nov 20 4 44 31
	Nov 28 9 41 34	Dec 6 1 26 39	Dec 12 16 30 46	Dec 19 23 55 25
	Dec 28 3 20 30			
	h m s			
1990		Jan 4 $10^h41^m14^s$	Jan 11 $4^h57^m44^s$	Jan 18 $21^h18^m10^s$
	Jan 26 19 20 57	Feb 2 18 33 20	Feb 9 19 16 40	Feb 17 18 48 33
	Feb 25 8 55 19	Mar 4 2 05 59	Mar 11 10 59 26	Mar 19 14 31 21
	Mar 26 19 49 15	Apr 2 10 24 56	Apr 10 3 19 25	Apr 18 7 03 29
	Apr 25 4 28 24	May 1 20 18 46	May 9 19 31 43	May 17 19 45 54
	May 24 11 48 04	May 31 8 11 41	Jun 8 11 02 04	Jun 16 4 48 38
	Jun 22 18 55 32	Jun 29 22 08 25	Jul 8 1 24 26	Jul 15 11 04 42
	Jul 22 2 55 14	Jul 29 14 02 18	Aug 6 14 20 22	Aug 13 15 55 19
	Aug 20 12 39 56	Aug 28 7 35 03	Sep 5 1 46 32	Sep 11 20 54 09
	Sep 19 0 47 16	Sep 27 2 06 34	Oct 4 12 02 45	Oct 11 3 32 19
	Oct 18 15 37 36	Oct 26 20 27 19	Nov 2 21 49 12	Nov 9 13 02 42
	Nov 17 9 05 29	Nov 25 13 12 25	Dec 2 7 50 35	Dec 9 2 04 52
	Dec 17 4 22 35	Dec 25 3 16 43	Dec 31 18 36 06	

Year	New Moon		First Quarter		Full Moon		Last Quarter	
		$h\ \ m\ \ s$		$h\ \ m\ \ s$		$h\ \ m\ \ s$		
1991							Jan 7	$18^h36^m25^s$
	Jan 15	23 50 34	Jan 23	14 22 28	Jan 30	6 10 38	Feb 6	13 53 09
	Feb 14	17 32 38	Feb 21	22 59 09	Feb 28	18 25 31	Mar 8	10 32 52
	Mar 16	8 11 26	Mar 23	6 03 40	Mar 30	7 18 20	Apr 7	6 46 16
	Apr 14	19 38 34	Apr 21	12 39 37	Apr 28	20 59 27	May 7	0 47 08
	May 14	4 36 46	May 20	19 46 47	May 28	11 37 32	Jun 5	15 30 57
	Jun 12	12 06 53	Jun 19	4 20 20	Jun 27	2 59 23	Jul 5	2 51 13
	Jul 11	19 07 03	Jul 18	15 11 40	Jul 26	18 25 18	Aug 3	11 26 27
	Aug 10	2 28 39	Aug 17	5 01 42	Aug 25	9 08 06	Sep 1	18 17 20
	Sep 8	11 01 45	Sep 15	22 02 16	Sep 23	22 40 54	Oct 1	0 30 30
	Oct 7	21 39 45	Oct 15	17 33 48	Oct 23	11 09 06	Oct 30	7 11 28
	Nov 6	11 11 56	Nov 14	14 02 28	Nov 21	22 57 14	Nov 28	15 22 01
	Dec 6	3 57 14	Dec 14	9 32 51	Dec 21	10 24 07	Dec 28	1 55 57
1992	Jan 4	$23^h10^m31^s$	Jan 13	$2^h32^m54^s$	Jan 19	$21^h29^m25^s$	Jan 26	$15^h28^m04^s$
	Feb 3	19 00 35	Feb 11	16 15 40	Feb 18	8 05 04	Feb 25	7 56 42
	Mar 4	13 23 06	Mar 12	2 36 50	Mar 18	18 18 45	Mar 26	2 30 56
	Apr 3	5 02 25	Apr 10	10 06 56	Apr 17	4 43 26	Apr 24	21 40 53
	May 2	17 45 25	May 9	15 44 29	May 16	16 03 32	May 24	15 54 25
	Jun 1	3 57 34	Jun 7	20 47 46	Jun 15	4 50 43	Jun 23	8 12 03
	Jun 30	12 19 00	Jul 7	2 44 27	Jul 14	19 07 16	Jul 22	22 13 16
	Jul 29	19 36 09	Aug 5	10 59 32	Aug 13	10 28 09	Aug 21	10 02 10
	Aug 28	2 42 50	Sep 3	22 39 39	Sep 12	2 17 36	Sep 19	19 53 55
	Sep 26	10 41 06	Oct 3	14 12 50	Oct 11	18 03 58	Oct 19	4 13 01
	Oct 25	20 34 42	Nov 2	9 12 10	Nov 10	9 21 04	Nov 17	11 39 58
	Nov 24	9 12 15	Dec 2	6 17 56	Dec 9	23 41 37	Dec 16	19 13 55
	Dec 24	0 43 48						
1993		$h\ \ m\ \ s$	Jan 1	$3^h39^m17^s$	Jan 8	$12^h38^m12^s$	Jan 15	$4^h02^m13^s$
	Jan 22	18 27 49	Jan 30	23 20 47	Feb 6	23 56 22	Feb 13	14 57 44
	Feb 21	13 06 03	Mar 1	15 47 31	Mar 8	9 46 51	Mar 15	4 17 32
	Mar 23	7 15 24	Mar 31	4 10 40	Apr 6	18 44 14	Apr 13	19 39 38
	Apr 21	23 49 57	Apr 29	12 41 29	May 6	3 34 44	May 13	12 20 40
	May 21	14 07 30	May 28	18 22 22	Jun 4	13 03 12	Jun 12	5 36 44
	Jun 20	1 53 26	Jun 26	22 44 20	Jul 3	23 46 06	Jul 11	22 49 35
	Jul 19	11 25 02	Jul 26	3 26 00	Aug 2	12 10 36	Aug 10	15 20 22
	Aug 17	19 29 16	Aug 24	9 58 20	Sep 1	2 33 54	Sep 9	6 27 22
	Sep 16	3 11 16	Sep 22	19 33 03	Sep 30	18 54 48	Oct 8	19 36 19
	Oct 15	11 36 55	Oct 22	8 53 01	Oct 30	12 38 36	Nov 7	6 36 48
	Nov 13	21 35 19	Nov 21	2 04 11	Nov 29	6 31 43	Dec 6	15 49 49
	Dec 13	9 27 52	Dec 20	22 26 59	Dec 28	23 06 21		
1994		$h\ \ m\ \ s$		$h\ \ m\ \ s$		$h\ \ m\ \ s$	Jan 5	$0^h01^m30^s$
	Jan 11	23 11 19	Jan 19	20 27 35	Jan 27	13 23 50	Feb 3	8 07 06
	Feb 10	14 30 52	Feb 18	17 48 17	Feb 26	1 16 01	Mar 4	16 54 17
	Mar 12	7 05 35	Mar 20	12 15 17	Mar 27	11 10 32	Apr 3	2 55 48
	Apr 11	0 18 05	Apr 19	2 35 12	Apr 25	19 45 58	May 2	14 33 25
	May 10	17 07 34	May 18	12 50 53	May 25	3 40 22	Jun 1	4 03 24
	Jun 9	8 27 28	Jun 16	19 57 30	Jun 23	11 34 05	Jun 30	19 31 49
	Jul 8	21 38 18	Jul 16	1 12 35	Jul 22	20 16 52	Jul 30	12 41 00
	Aug 7	8 46 08	Aug 14	5 58 21	Aug 21	6 47 52	Aug 29	6 41 41
	Sep 5	18 33 45	Sep 12	11 34 58	Sep 19	20 01 30	Sep 28	0 24 21
	Oct 5	3 56 08	Oct 11	19 18 18	Oct 19	12 18 45	Oct 27	16 45 16
	Nov 3	13 36 28	Nov 10	6 14 53	Nov 18	6 57 59	Nov 26	7 04 35
	Dec 2	23 55 02	Dec 9	21 07 14	Dec 18	2 18 02	Dec 25	19 07 24

Year	New Moon	First Quarter	Full Moon	Last Quarter
1995	Jan 1 $10^h56^m37^s$	Jan 8 $15^h47^m25^s$	Jan 16 $20^h27^m20^s$	Jan 24 $4^h59^m01^s$
	Jan 30 22 48 43	Feb 7 12 55 05	Feb 15 12 16 22	Feb 22 13 04 47
	Mar 1 11 48 46	Mar 9 10 14 36	Mar 17 1 26 37	Mar 23 20 10 52
	Mar 31 2 09 37	Apr 8 5 35 52	Apr 15 12 09 13	Apr 22 3 19 13
	Apr 29 17 37 20	May 7 21 44 50	May 14 20 49 08	May 21 11 36 41
	May 29 9 28 07	Jun 6 10 26 38	Jun 13 4 04 30	Jun 19 22 01 49
	Jun 28 0 50 58	Jul 5 20 03 24	Jul 12 10 50 19	Jul 19 11 10 37
	Jul 27 15 13 58	Aug 4 3 17 01	Aug 10 18 16 33	Aug 18 3 04 53
	Aug 26 4 31 59	Sep 2 9 04 10	Sep 9 3 37 39	Sep 16 21 10 22
	Sep 24 16 55 45	Oct 1 14 36 35	Oct 8 15 52 50	Oct 16 16 26 53
	Oct 24 4 37 12	Oct 30 21 17 39	Nov 7 7 21 29	Nov 15 11 40 47
	Nov 22 15 43 48	Nov 29 6 29 18	Dec 7 1 27 52	Dec 15 5 32 06
	Dec 22 2 23 25	Dec 28 19 07 26		
1996		h m s	h m s	Jan 5 $20^h51^m53^s$	Jan 13 $20^h46^m04^s$
	Jan 20 12 51 28	Jan 27 11 14 42	Feb 4 15 58 38	Feb 12 8 38 06
	Feb 18 23 31 09	Feb 26 5 53 26	Mar 5 9 23 42	Mar 12 17 15 54
	Mar 19 10 45 44	Mar 27 1 31 55	Apr 4 0 07 57	Apr 10 23 36 45
	Apr 17 22 49 47	Apr 25 20 41 21	May 3 11 49 15	May 10 5 04 43
	May 17 11 47 12	May 25 14 14 00	Jun 1 20 47 56	Jun 8 11 06 31
	Jun 16 1 36 48	Jun 24 5 24 17	Jul 1 3 59 10	Jul 7 18 55 48
	Jul 15 16 15 53	Jul 23 17 49 57	Jul 30 10 36 12	Aug 6 5 25 46
	Aug 14 7 34 51	Aug 22 3 37 28	Aug 28 17 53 12	Sep 4 19 07 00
	Sep 12 23 08 18	Sep 20 11 23 44	Sep 27 2 51 47	Oct 4 12 05 07
	Oct 12 14 15 26	Oct 19 18 10 05	Oct 26 14 12 10	Nov 3 7 51 08
	Nov 11 4 17 08	Nov 18 1 09 51	Nov 25 4 10 53	Dec 3 5 06 33
	Dec 10 16 57 18	Dec 17 9 31 53	Dec 24 20 41 58	
1997		h m s		h m s		h m s	Jan 2 $1^h46^m04^s$
	Jan 9 4 26 43	Jan 15 20 03 00	Jan 23 15 11 42	Jan 31 19 41 16
	Feb 7 15 07 16	Feb 14 8 58 33	Feb 22 10 27 40	Mar 2 9 38 34
	Mar 9 1 15 36	Mar 16 0 07 16	Mar 24 4 46 15	Mar 31 19 39 19
	Apr 7 11 02 55	Apr 14 17 00 56	Apr 22 20 34 31	Apr 30 2 38 05
	May 6 20 47 33	May 14 10 56 01	May 22 9 14 26	May 29 7 52 17
	Jun 5 7 04 35	Jun 13 4 52 31	Jun 20 19 09 51	Jun 27 12 43 10
	Jul 4 18 40 47	Jul 12 21 44 39	Jul 20 3 21 20	Jul 26 18 29 09
	Aug 3 8 14 56	Aug 11 12 43 27	Aug 18 10 56 29	Aug 25 2 24 33
	Sep 1 23 52 36	Sep 10 1 32 08	Sep 16 18 51 31	Sep 23 13 36 11
	Oct 1 16 52 32	Oct 9 12 23 03	Oct 16 3 46 41	Oct 23 4 49 16
	Oct 31 10 02 04	Nov 7 21 44 24	Nov 14 14 12 38	Nov 21 23 58 59
	Nov 30 2 15 01	Dec 7 6 10 19	Dec 14 2 38 06	Dec 21 21 44 05
	Dec 29 16 57 33			
1998		h m s	Jan 5 $14^h19^m07^s$	Jan 12 $17^h24^m42^s$	Jan 20 $19^h41^m09^s$
	Jan 28 6 01 52	Feb 3 22 54 27	Feb 11 10 23 39	Feb 19 15 27 51
	Feb 26 17 26 57	Mar 5 8 41 54	Mar 13 4 35 10	Mar 21 7 38 57
	Mar 28 3 14 40	Apr 3 20 19 24	Apr 11 22 24 30	Apr 19 19 53 43
	Apr 26 11 42 25	May 3 10 04 44	May 11 14 30 23	May 19 4 36 21
	May 25 19 33 17	Jun 2 1 46 04	Jun 10 4 19 20	Jun 17 10 39 10
	Jun 24 3 51 20	Jul 1 18 43 44	Jul 9 16 01 56	Jul 16 15 14 30
	Jul 23 13 44 49	Jul 31 12 06 13	Aug 8 2 10 40	Aug 14 19 49 31
	Aug 22 2 04 08	Aug 30 5 07 32	Sep 6 11 22 24	Sep 13 1 58 57
	Sep 20 17 02 31	Sep 28 21 11 52	Oct 5 20 12 57	Oct 12 11 11 54
	Oct 20 10 10 24	Oct 28 11 47 10	Nov 4 5 19 13	Nov 11 0 29 03
	Nov 19 4 27 46	Nov 27 0 23 36	Dec 3 15 20 11	Dec 10 17 54 33
	Dec 18 22 43 22	Dec 26 10 47 02		

Year	New Moon	First Quarter	Full Moon	Last Quarter
	h m s	h m s	h m s	h m s
1999			Jan 2 $2^h50^m34^s$	Jan 9 $14^h22^m37^s$
	Jan 17 15 47 06	Jan 24 19 16 10	Jan 31 16 07 33	Feb 8 11 58 48
	Feb 16 6 39 44	Feb 23 2 43 52	Mar 2 6 59 30	Mar 10 8 41 16
	Mar 17 18 48 59	Mar 24 10 18 52	Mar 31 22 49 57	Apr 9 2 51 38
	Apr 16 4 22 51	Apr 22 19 02 36	Apr 30 14 55 39	May 8 17 29 35
	May 15 12 06 05	May 22 5 35 04	May 30 6 40 56	Jun 7 4 21 00
	Jun 13 19 03 55	Jun 20 18 13 54	Jun 28 21 38 31	Jul 6 11 57 50
	Jul 13 2 25 01	Jul 20 9 01 14	Jul 28 11 25 50	Aug 4 17 27 40
	Aug 11 11 09 33	Aug 19 1 47 54	Aug 26 23 48 52	Sep 2 22 18 21
	Sep 9 22 03 18	Sep 17 20 06 48	Sep 25 10 52 07	Oct 2 4 03 01
	Oct 9 11 35 27	Oct 17 15 00 47	Oct 24 21 03 26	Oct 31 12 04 38
	Nov 8 3 54 03	Nov 16 9 04 07	Nov 23 7 04 42	Nov 29 23 19 35
	Dec 7 22 32 39	Dec 16 0 51 18	Dec 22 17 32 20	Dec 29 14 05 21
2000	Jan 6 $18^h14^m40^s$	Jan 14 $13^h35^m13^s$	Jan 21 $4^h41^m29^s$	Jan 28 $7^h57^m43^s$
	Feb 5 13 04 18	Feb 12 23 22 27	Feb 19 16 27 43	Feb 27 3 54 32
	Mar 6 5 17 45	Mar 13 6 59 52	Mar 20 4 45 24	Mar 28 0 21 42
	Apr 4 18 13 03	Apr 11 13 31 25	Apr 18 17 42 34	Apr 26 19 31 07
	May 4 4 13 08	May 10 20 01 34	May 18 7 35 29	May 26 11 56 05
	Jun 2 12 15 01	Jun 9 3 30 17	Jun 16 22 28 06	Jun 25 1 01 07
	Jul 1 19 20 59	Jul 8 12 53 51	Jul 16 13 56 17	Jul 24 11 03 07
	Jul 31 2 26 13	Aug 7 1 02 42	Aug 15 5 13 42	Aug 22 18 51 44
	Aug 29 10 20 21	Sep 5 16 28 24	Sep 13 19 37 52	Sep 21 1 29 24
	Sep 27 19 54 00	Oct 5 11 00 14	Oct 13 8 54 00	Oct 20 8 00 08
	Oct 27 7 59 03	Nov 4 7 27 52	Nov 11 21 15 39	Nov 18 15 25 30
	Nov 25 23 12 20	Dec 4 3 56 23	Dec 11 9 03 50	Dec 18 0 42 20
	Dec 25 17 22 39			
2001		h m s	Jan 9 $20^h25^m27^s$	Jan 16 $12^h35^m39^s$
	Jan 24 13 07 52	Jan 2 $22^h32^m34^s$	Feb 8 7 12 42	Feb 15 3 24 34
	Feb 23 8 22 08	Feb 1 14 03 26	Mar 9 17 24 07	Mar 16 20 46 13
	Mar 25 1 22 03	Mar 3 2 04 14	Apr 8 3 22 55	Apr 15 15 32 10
	Apr 23 15 26 40	Apr 1 10 50 18	May 7 13 53 36	May 15 10 11 40
	May 23 2 47 06	Apr 30 17 08 39	Jun 6 1 40 24	Jun 14 3 29 20
	Jun 21 11 58 50	May 29 22 10 12	Jul 5 15 04 50	Jul 13 18 46 14
	Jul 20 19 45 29	Jun 28 3 20 36	Aug 4 5 56 47	Aug 12 7 54 13
	Aug 19 2 56 20	Jul 27 10 09 14	Sep 2 21 44 04	Sep 10 19 00 35
	Sep 17 10 28 25	Aug 25 19 55 52	Oct 2 13 49 51	Oct 10 4 20 40
	Oct 16 19 24 21	Sep 24 9 31 54	Nov 1 5 42 03	Nov 8 12 22 18
	Nov 15 6 41 01	Oct 24 2 59 23	Nov 30 20 50 05	Dec 7 19 52 44
	Dec 14 20 48 25	Nov 22 23 21 45	Dec 30 10 41 35	
		Dec 22 20 57 27		
2002	h m s	h m s	h m s	Jan 6 $3^h55^m45^s$
	Jan 13 13 29 38	Jan 21 17 47 33	Jan 28 22 51 33	Feb 4 13 34 05
	Feb 12 7 41 55	Feb 20 12 02 47	Feb 27 9 17 44	Mar 6 1 25 36
	Mar 14 2 03 35	Mar 22 2 29 17	Mar 28 18 25 57	Apr 4 15 30 05
	Apr 12 19 22 12	Apr 20 12 49 23	Apr 27 3 00 59	May 4 7 17 04
	May 12 10 46 09	May 19 19 43 08	May 26 11 52 19	Jun 3 0 06 12
	Jun 10 23 47 35	Jun 18 0 30 21	Jun 24 21 43 25	Jul 2 17 20 20
	Jul 10 10 27 06	Jul 17 4 48 11	Jul 24 9 08 04	Aug 1 10 23 07
	Aug 8 19 16 15	Aug 15 10 13 20	Aug 22 22 30 18	Aug 31 2 32 18
	Sep 7 3 11 22	Sep 13 18 09 06	Sep 21 14 00 18	Sep 29 17 03 55
	Oct 6 11 18 37	Oct 13 5 34 05	Oct 21 7 21 04	Oct 29 5 28 42
	Nov 4 20 35 29	Nov 11 20 53 11	Nov 20 1 34 42	Nov 27 15 47 22
	Dec 4 7 35 24	Dec 11 15 49 35	Dec 19 19 11 12	Dec 27 0 32 02

Year	New Moon	First Quarter	Full Moon	Last Quarter
2003	Jan 2 $20^h23^m52^s$	Jan 10 $13^h15^m49^s$	Jan 18 $10^h48^m41^s$	Jan 25 $8^h34^m12^s$
	Feb 1 10 49 25	Feb 9 11 12 12	Feb 16 23 52 13	Feb 23 16 46 57
	Mar 3 2 36 01	Mar 11 7 16 19	Mar 18 10 35 37	Mar 25 1 52 12
	Apr 1 19 19 46	Apr 9 23 41 15	Apr 16 19 36 46	Apr 23 12 19 27
	May 1 12 15 52	May 9 11 54 17	May 16 3 37 02	May 23 0 31 45
	May 31 4 20 57	Jun 7 20 28 44	Jun 14 11 16 59	Jun 21 14 46 04
	Jun 29 18 39 45	Jul 7 2 33 25	Jul 13 19 22 28	Jul 21 7 02 18
	Jul 29 6 53 48	Aug 5 7 28 40	Aug 12 4 49 17	Aug 20 0 49 15
	Aug 27 17 27 25	Sep 3 12 35 09	Sep 10 16 37 18	Sep 18 19 03 56
	Sep 26 3 10 15	Oct 2 19 10 21	Oct 10 7 28 31	Oct 18 12 32 19
	Oct 25 12 51 21	Nov 1 4 25 38	Nov 9 1 14 24	Nov 17 4 15 56
	Nov 23 23 00 00	Nov 30 17 17 01	Dec 8 20 37 43	Dec 16 17 43 11
	Dec 23 9 44 03	Dec 30 10 04 20		
2004			Jan 7 $15^h41^m12^s$	Jan 15 $4^h46^m40^s$
	Jan 21 21 05 57	Jan 29 6 04 16	Feb 6 8 47 57	Feb 13 13 40 37
	Feb 20 9 18 44	Feb 28 3 25 04	Mar 6 23 15 20	Mar 13 21 01 46
	Mar 20 22 42 24	Mar 28 23 48 55	Apr 5 11 03 49	Apr 12 3 47 08
	Apr 19 13 22 16	Apr 27 17 33 32	May 4 20 34 31	May 11 11 05 10
	May 19 4 53 00	May 27 7 58 06	Jun 3 4 20 39	Jun 9 20 03 26
	Jun 17 20 27 51	Jun 25 19 09 02	Jul 2 11 09 59	Jul 9 7 34 39
	Jul 17 11 24 50	Jul 25 3 38 15	Jul 31 18 06 15	Aug 7 22 02 17
	Aug 16 1 24 57	Aug 23 10 12 54	Aug 30 2 23 19	Sep 6 15 11 35
	Sep 14 14 30 06	Sep 21 15 54 39	Sep 28 13 10 20	Oct 6 10 12 52
	Oct 14 2 49 18	Oct 20 21 59 44	Oct 28 3 08 25	Nov 5 5 54 27
	Nov 12 14 28 13	Nov 19 5 51 25	Nov 26 20 08 17	Dec 5 0 53 46
	Dec 12 1 30 04	Dec 18 16 40 39	Dec 26 15 07 21	
2005				Jan 3 $17^h46^m43^s$
	Jan 10 12 03 52	Jan 17 6 58 31	Jan 25 10 33 23	Feb 2 7 27 53
	Feb 8 22 29 05	Feb 16 0 17 03	Feb 24 4 54 46	Mar 3 17 37 31
	Mar 10 9 11 25	Mar 17 19 20 06	Mar 25 20 59 35	Apr 2 0 51 29
	Apr 8 20 33 05	Apr 16 14 38 32	Apr 24 10 07 33	May 1 6 25 16
	May 8 8 46 31	May 16 8 57 39	May 23 20 19 14	May 30 11 48 24
	Jun 6 21 56 10	Jun 15 1 23 19	Jun 22 4 14 54	Jun 28 18 24 31
	Jul 6 12 03 34	Jul 14 15 20 57	Jul 21 11 01 19	Jul 28 3 20 00
	Aug 5 3 05 49	Aug 13 2 39 37	Aug 19 17 54 04	Aug 26 15 19 07
	Sep 3 18 46 28	Sep 11 11 37 42	Sep 18 2 01 50	Sep 25 6 41 48
	Oct 3 10 28 56	Oct 10 19 01 52	Oct 17 12 14 41	Oct 25 1 17 45
	Nov 2 1 25 39	Nov 9 1 58 07	Nov 16 0 58 35	Nov 23 22 12 17
	Dec 1 15 01 59	Dec 8 9 37 19	Dec 15 16 16 34	Dec 23 19 37 09
	Dec 31 3 12 48			
2006		Jan 6 $18^h57^m32^s$	Jan 14 $9^h49^m10^s$	Jan 22 $15^h14^m50^s$
	Jan 29 14 15 42	Feb 5 6 29 52	Feb 13 4 45 17	Feb 21 7 17 45
	Feb 28 0 31 51	Mar 6 20 16 51	Mar 14 23 36 30	Mar 22 19 11 35
	Mar 29 10 16 20	Apr 5 12 01 44	Apr 13 16 41 12	Apr 21 3 29 31
	Apr 27 19 44 58	May 5 5 14 05	May 13 6 52 08	May 20 9 21 38
	May 27 5 26 42	Jun 3 23 06 49	Jun 11 18 04 12	Jun 18 14 09 22
	Jun 25 16 06 21	Jul 3 16 37 48	Jul 11 3 02 58	Jul 17 19 13 38
	Jul 25 4 31 59	Aug 2 8 46 52	Aug 9 10 55 03	Aug 16 1 51 50
	Aug 23 19 10 50	Aug 31 22 57 35	Sep 7 18 43 08	Sep 14 11 16 25
	Sep 22 11 46 06	Sep 30 11 04 55	Oct 7 3 13 53	Oct 14 0 26 38
	Oct 22 5 15 06	Oct 29 21 26 26	Nov 5 12 59 19	Nov 12 17 46 24
	Nov 20 22 19 02	Nov 28 6 30 02	Dec 5 0 25 52	Dec 12 14 32 43
	Dec 20 14 01 50	Dec 27 14 48 51		

Year	New Moon	First Quarter	Full Moon	Last Quarter
2007	$h \quad m \quad s$	$h \quad m \quad s$	Jan 3 $13^h58^m25^s$	Jan 11 $12^h45^m38^s$
	Jan 19 4 01 45	Jan 25 23 02 30	Feb 2 5 46 21	Feb 10 9 52 06
	Feb 17 16 15 22	Feb 24 7 56 51	Mar 3 23 18 10	Mar 12 3 55 10
	Mar 19 2 43 39	Mar 25 18 17 12	Apr 2 17 16 06	Apr 10 18 05 16
	Apr 17 11 37 06	Apr 24 6 36 36	May 2 10 10 30	May 10 4 28 07
	May 16 19 28 19	May 23 21 03 36	Jun 1 1 04 45	Jun 8 11 43 45
	Jun 15 3 14 14	Jun 22 13 16 21	Jun 30 13 49 48	Jul 7 16 54 42
	Jul 14 12 04 52	Jul 22 6 30 10	Jul 30 0 48 49	Aug 5 21 20 43
	Aug 12 23 03 35	Aug 20 23 55 17	Aug 28 10 36 10	Sep 4 2 33 27
	Sep 11 12 45 17	Sep 19 16 48 58	Sep 26 19 46 13	Oct 3 10 06 56
	Oct 11 5 01 42	Oct 19 8 34 01	Oct 26 4 52 36	Nov 1 21 19 14
	Nov 9 23 04 07	Nov 17 22 33 37	Nov 24 14 30 53	Dec 1 12 45 06
	Dec 9 17 41 27	Dec 17 10 18 35	Dec 24 1 16 39	Dec 31 7 51 52
2008	Jan 8 $11^h38^m13^s$	Jan 15 $19^h46^m45^s$	Jan 22 $13^h35^m45^s$	Jan 30 $5^h03^m56^s$
	Feb 7 3 45 35	Feb 14 3 34 33	Feb 21 3 31 35	Feb 29 2 19 23
	Mar 7 17 15 18	Mar 14 10 46 40	Mar 21 18 41 01	Mar 29 21 48 22
	Apr 6 3 56 23	Apr 12 18 32 50	Apr 20 10 26 27	Apr 28 14 13 14
	May 5 12 19 19	May 12 3 48 01	May 20 2 12 27	May 28 2 57 40
	Jun 3 19 23 40	Jun 10 15 04 41	Jun 18 17 31 33	Jun 26 12 10 52
	Jul 3 2 19 41	Jul 10 4 35 56	Jul 18 8 00 10	Jul 25 18 42 39
	Aug 1 10 13 38	Aug 8 20 21 18	Aug 16 21 17 32	Aug 23 23 50 38
	Aug 30 19 59 07	Sep 7 14 05 09	Sep 15 9 14 27	Sep 22 5 05 25
	Sep 29 8 13 18	Oct 7 9 05 16	Oct 14 20 03 31	Oct 21 11 55 41
	Oct 28 23 14 54	Nov 6 4 04 26	Nov 13 6 18 23	Nov 19 21 31 47
	Nov 27 16 55 38	Dec 5 21 26 40	Dec 12 16 38 13	Dec 19 10 30 18
	Dec 27 12 23 26			
2009	$h \quad m \quad s$	Jan 4 $11^h57^m17^s$	Jan 11 $3^h27^m50^s$	Jan 18 $2^h46^m47^s$
	Jan 26 7 56 21	Feb 2 23 14 11	Feb 9 14 50 14	Feb 16 21 38 14
	Feb 25 1 36 08	Mar 4 7 46 54	Mar 11 2 38 49	Mar 18 17 48 26
	Mar 26 16 07 00	Apr 2 14 34 48	Apr 9 14 56 54	Apr 17 13 37 27
	Apr 25 3 23 39	May 1 20 45 19	May 9 4 02 30	May 17 7 27 07
	May 24 12 12 06	May 31 3 23 18	Jun 7 18 12 49	Jun 15 22 15 41
	Jun 22 19 36 04	Jun 29 11 29 31	Jul 7 9 22 31	Jul 15 9 54 15
	Jul 22 2 35 42	Jul 28 22 00 52	Aug 6 0 55 57	Aug 13 18 56 17
	Aug 20 10 02 40	Aug 27 11 43 03	Sep 4 16 03 41	Sep 12 2 16 49
	Sep 18 18 45 23	Sep 26 4 50 45	Oct 4 6 11 15	Oct 11 8 56 54
	Oct 18 5 34 09	Oct 26 0 43 17	Nov 2 19 14 59	Nov 9 15 56 53
	Nov 16 19 14 47	Nov 24 21 40 17	Dec 2 7 31 31	Dec 9 0 14 24
	Dec 16 12 03 09	Dec 24 17 37 01	Dec 31 19 13 50	
2010	$h \quad m \quad s$	$h \quad m \quad s$	$h \quad m \quad s$	Jan 7 $10^h40^m32^s$
	Jan 15 7 12 26	Jan 23 10 54 27	Jan 30 6 18 40	Feb 5 23 49 30
	Feb 14 2 52 23	Feb 22 0 43 27	Feb 28 16 38 57	Mar 7 15 42 52
	Mar 15 21 02 10	Mar 23 11 01 09	Mar 30 2 26 30	Apr 6 9 37 54
	Apr 14 12 30 00	Apr 21 18 20 52	Apr 28 12 19 33	May 6 4 16 00
	May 14 1 05 29	May 20 23 43 46	May 27 23 08 24	Jun 4 22 14 13
	Jun 12 11 15 41	Jun 19 4 30 35	Jun 26 11 31 27	Jul 4 14 36 19
	Jul 11 19 41 34	Jul 18 10 11 39	Jul 26 1 37 38	Aug 3 4 59 43
	Aug 10 3 09 15	Aug 16 18 15 08	Aug 24 17 05 39	Sep 1 17 22 52
	Sep 8 10 30 55	Sep 15 5 50 50	Sep 23 9 18 17	Oct 1 3 53 12
	Oct 7 18 45 33	Oct 14 21 28 28	Oct 23 1 37 36	Oct 30 12 46 57
	Nov 6 4 52 51	Nov 13 16 39 38	Nov 21 17 28 25	Nov 28 20 37 30
	Dec 5 17 36 46	Dec 13 13 59 46	Dec 21 8 14 32	Dec 28 4 19 31

Year	New Moon	First Quarter	Full Moon	Last Quarter
2011	Jan 4 $9^h03^m41^s$	Jan 12 $11^h32^m29^s$	Jan 19 $21^h22^m30^s$	Jan 26 $12^h58^m19^s$
	Feb 3 2 31 44	Feb 11 7 19 20	Feb 18 8 36 45	Feb 24 23 27 26
	Mar 4 20 46 57	Mar 12 23 46 03	Mar 19 18 11 09	Mar 26 12 08 26
	Apr 3 14 33 25	Apr 11 12 06 29	Apr 18 2 45 05	Apr 25 2 47 59
	May 3 6 51 48	May 10 20 34 01	May 17 11 09 45	May 24 18 53 16
	Jun 1 21 03 43	Jun 9 2 11 41	Jun 15 20 14 41	Jun 23 11 49 21
	Jul 1 8 55 02	Jul 8 6 30 30	Jul 15 6 40 42	Jul 23 5 03 02
	Jul 30 18 40 55	Aug 6 11 09 25	Aug 13 18 58 36	Aug 21 21 55 33
	Aug 29 3 05 11	Sep 4 17 40 24	Sep 12 9 27 44	Sep 20 13 39 40
	Sep 27 11 09 46	Oct 4 3 16 11	Oct 12 2 06 48	Oct 20 3 31 29
	Oct 26 19 56 52	Nov 2 16 39 14	Nov 10 20 17 11	Nov 18 15 10 08
	Nov 25 6 10 45	Dec 2 9 53 19	Dec 10 14 37 27	Dec 18 0 48 45
	Dec 24 18 07 27			
2012		Jan 1 $6^h15^m38^s$	Jan 9 $7^h31^m10^s$	Jan 16 $9^h09^m00^s$
	Jan 23 7 40 21	Jan 31 4 10 42	Feb 7 21 54 50	Feb 14 17 04 57
	Feb 21 22 35 41	Mar 1 1 22 35	Mar 8 9 40 35	Mar 15 1 26 14
	Mar 22 14 38 12	Mar 30 19 41 46	Apr 6 19 19 47	Apr 13 10 50 43
	Apr 21 7 19 30	Apr 29 9 58 36	May 6 3 36 12	May 12 21 47 50
	May 20 23 48 08	May 28 20 17 07	Jun 4 11 12 41	Jun 11 10 42 28
	Jun 19 15 03 13	Jun 27 3 31 29	Jul 3 18 53 00	Jul 11 1 48 58
	Jul 19 4 25 09	Jul 26 8 57 17	Aug 2 3 28 34	Aug 9 18 56 05
	Aug 17 15 55 32	Aug 24 13 54 40	Aug 31 13 59 12	Sep 8 13 16 07
	Sep 16 2 11 44	Sep 22 19 41 55	Sep 30 3 19 42	Oct 8 7 34 21
	Oct 15 12 03 36	Oct 22 3 33 04	Oct 29 19 50 31	Nov 7 0 36 44
	Nov 13 22 09 05	Nov 20 14 32 31	Nov 28 14 47 00	Dec 6 15 32 31
	Dec 13 8 42 42	Dec 20 5 20 08	Dec 28 10 22 15	
2013				Jan 5 $3^h58^m51^s$
	Jan 11 19 44 42	Jan 18 23 46 06	Jan 27 4 39 28	Feb 3 13 57 24
	Feb 10 7 21 12	Feb 17 20 31 40	Feb 25 20 27 10	Mar 4 21 53 54
	Mar 11 19 52 06	Mar 19 17 27 40	Mar 27 9 28 25	Apr 3 4 37 41
	Apr 10 9 36 24	Apr 18 12 31 58	Apr 25 19 58 14	May 2 11 15 17
	May 10 0 29 30	May 18 4 35 39	May 25 4 26 03	May 31 18 59 15
	Jun 8 15 57 27	Jun 16 17 24 49	Jun 23 11 33 23	Jun 30 4 54 39
	Jul 8 7 15 24	Jul 16 3 19 25	Jul 22 18 16 39	Jul 29 17 44 27
	Aug 6 21 51 49	Aug 14 10 57 08	Aug 21 1 45 43	Aug 28 9 36 03
	Sep 5 11 37 15	Sep 12 17 09 31	Sep 19 11 13 56	Sep 27 3 56 33
	Oct 5 0 35 37	Oct 11 23 03 25	Oct 18 23 38 45	Oct 26 23 41 34
	Nov 3 12 51 03	Nov 10 5 58 16	Nov 17 15 16 48	Nov 25 19 28 48
	Dec 3 0 23 27	Dec 9 15 12 49	Dec 17 9 29 08	Dec 25 13 48 45
2014	Jan 1 $11^h15^m15^s$	Jan 8 $3^h40^m22^s$	Jan 16 $4^h53^m15^s$	Jan 24 $5^h20^m03^s$
	Jan 30 21 39 37	Feb 6 19 23 08	Feb 14 23 54 06	Feb 22 17 16 21
	Mar 1 8 00 46	Mar 8 13 27 50	Mar 16 17 09 27	Mar 24 1 47 12
	Mar 30 18 45 47	Apr 7 8 31 44	Apr 15 7 43 25	Apr 22 7 52 49
	Apr 29 6 15 27	May 7 3 15 59	May 14 19 17 02	May 21 13 00 18
	May 28 18 41 20	Jun 5 20 39 54	Jun 13 4 12 37	Jun 19 18 39 51
	Jun 27 8 09 35	Jul 5 11 59 56	Jul 12 11 26 02	Jul 19 2 09 28
	Jul 26 22 42 53	Aug 4 0 50 47	Aug 10 18 10 29	Aug 17 12 26 53
	Aug 25 14 13 52	Sep 2 11 12 17	Sep 9 1 39 17	Sep 16 2 05 59
	Sep 24 6 14 52	Oct 1 19 33 40	Oct 8 10 51 41	Oct 15 19 13 07
	Oct 23 21 57 45	Oct 31 2 49 24	Nov 6 22 23 54	Nov 14 15 16 35
	Nov 22 12 33 21	Nov 29 10 07 24	Dec 6 12 27 49	Dec 14 12 52 02
	Dec 22 1 36 56	Dec 28 18 32 29		

Year	New Moon		First Quarter		Full Moon		Last Quarter	
		$h\quad m\quad s$		$h\quad m\quad s$	Jan 5	$4^h54^m21^s$	Jan 13	$9^h47^m32^s$
2015	Jan 20	13 14 48	Jan 27	4 49 29	Feb 3	23 10 01	Feb 12	3 50 56
	Feb 18	23 48 22	Feb 25	17 15 06	Mar 5	18 06 30	Mar 13	17 49 01
	Mar 20	9 37 18	Mar 27	7 43 43	Apr 4	12 06 41	Apr 12	3 45 35
	Apr 18	18 57 58	Apr 25	23 56 17	May 4	3 43 12	May 11	10 37 16
	May 18	4 14 21	May 25	17 19 59	Jun 2	16 20 09	Jun 9	15 42 54
	Jun 16	14 06 28	Jun 24	11 03 40	Jul 2	2 20 43	Jul 8	20 25 05
	Jul 16	1 25 28	Jul 24	4 05 06	Jul 31	10 44 02	Aug 7	2 03 49
	Aug 14	14 54 31	Aug 22	19 32 07	Aug 29	18 36 19	Sep 5	9 55 09
	Sep 13	6 42 22	Sep 21	9 00 11	Sep 28	2 51 37	Oct 4	21 07 10
	Oct 13	0 06 48	Oct 20	20 32 27	Oct 27	12 06 13	Nov 3	12 24 57
	Nov 11	17 48 14	Nov 19	6 28 28	Nov 25	22 45 20	Dec 3	7 41 26
	Dec 11	10 30 30	Dec 18	15 15 25	Dec 25	11 12 34		
		$h\quad m\quad s$		$h\quad m\quad s$		$h\quad m\quad s$	Jan 2	$5^h31^m31^s$
2016	Jan 10	1 31 38	Jan 16	23 27 27	Jan 24	1 46 50	Feb 1	3 28 55
	Feb 8	14 40 02	Feb 15	7 47 35	Feb 22	18 20 58	Mar 1	23 11 47
	Mar 9	1 55 37	Mar 15	17 04 01	Mar 23	12 01 57	Mar 31	15 17 58
	Apr 7	11 24 47	Apr 14	4 00 27	Apr 22	5 24 43	Apr 30	3 29 51
	May 6	19 30 38	May 13	17 03 16	May 21	21 15 33	May 29	12 13 07
	Jun 5	3 00 43	Jun 12	8 10 56	Jun 20	11 03 27	Jun 27	18 19 47
	Jul 4	11 02 08	Jul 12	0 52 56	Jul 19	22 57 42	Jul 26	23 00 48
	Aug 2	20 45 40	Aug 10	18 21 57	Aug 18	9 27 41	Aug 25	3 41 57
	Sep 1	9 04 13	Sep 9	11 49 59	Sep 16	19 06 12	Sep 23	9 57 10
	Oct 1	0 12 28	Oct 9	4 34 06	Oct 16	4 24 12	Oct 22	19 14 53
	Oct 30	17 39 16	Nov 7	19 52 12	Nov 14	13 53 09	Nov 21	8 34 17
	Nov 29	12 19 19	Dec 7	9 04 06	Dec 14	0 06 39	Dec 21	1 56 49
	Dec 29	6 54 17						
		$h\quad m\quad s$	Jan 5	$19^h48^m06^s$	Jan 12	$11^h35^m04^s$	Jan 19	$22^h14^m32^s$
2017	Jan 28	0 08 09	Feb 4	4 19 59	Feb 11	0 33 59	Feb 18	19 34 14
	Feb 26	14 59 30	Mar 5	11 33 29	Mar 12	14 54 54	Mar 20	15 59 20
	Mar 28	2 58 21	Apr 3	18 40 33	Apr 11	6 09 13	Apr 19	9 57 53
	Apr 26	12 17 17	May 3	2 47 59	May 10	21 43 37	May 19	0 33 57
	May 25	19 45 36	Jun 1	12 43 15	Jun 9	13 10 43	Jun 17	11 33 51
	Jun 24	2 31 51	Jul 1	0 52 14	Jul 9	4 07 43	Jul 16	19 26 47
	Jul 23	9 46 43	Jul 30	15 24 13	Aug 7	18 11 45	Aug 15	1 16 12
	Aug 21	18 31 18	Aug 29	8 14 06	Sep 6	7 03 56	Sep 13	6 26 08
	Sep 20	5 30 59	Sep 28	2 54 38	Oct 5	18 41 14	Oct 12	12 26 33
	Oct 19	19 13 09	Oct 27	22 23 10	Nov 4	5 24 01	Nov 10	20 37 30
	Nov 18	11 43 13	Nov 26	17 04 01	Dec 3	15 48 05	Dec 10	7 52 27
	Dec 18	6 31 31	Dec 26	9 21 13				
		$h\quad m\quad s$		$h\quad m\quad s$	Jan 2	$2^h25^m12^s$	Jan 8	$22^h26^m21^s$
2018	Jan 17	2 18 21	Jan 24	22 21 29	Jan 31	13 27 51	Feb 7	15 55 02
	Feb 15	21 06 19	Feb 23	8 10 17	Mar 2	0 52 28	Mar 9	11 20 54
	Mar 17	13 12 42	Mar 24	15 36 18	Mar 31	12 37 58	Apr 8	7 18 40
	Apr 16	1 58 16	Apr 22	21 46 44	Apr 30	0 59 19	May 8	2 09 45
	May 15	11 48 55	May 22	3 50 18	May 29	14 20 42	Jun 6	18 32 50
	Jun 13	19 44 23	Jun 20	10 52 01	Jun 28	4 54 07	Jul 6	7 51 51
	Jul 13	2 49 01	Jul 19	19 53 22	Jul 27	20 21 29	Aug 4	18 19 08
	Aug 11	9 58 52	Aug 18	7 49 39	Aug 26	11 57 18	Sep 3	2 38 32
	Sep 9	18 02 35	Sep 16	23 16 05	Sep 25	2 53 32	Oct 2	9 46 33
	Oct 9	3 47 58	Oct 16	18 02 46	Oct 24	16 46 18	Oct 31	16 41 26
	Nov 7	16 03 09	Nov 15	14 55 17	Nov 23	5 40 19	Nov 30	0 19 56
	Dec 7	7 21 27	Dec 15	11 50 21	Dec 22	17 49 42	Dec 29	9 35 25

Year	New Moon	First Quarter	Full Moon	Last Quarter
2019	Jan 6 $1^h29^m18^s$	Jan 14 $6^h46^m38^s$	Jan 21 $5^h17^m12^s$	Jan 27 $21^h11^m28^s$
	Feb 4 21 04 42	Feb 12 22 27 21	Feb 19 15 54 42	Feb 26 11 28 50
	Mar 6 16 05 06	Mar 14 10 28 13	Mar 21 1 44 00	Mar 28 4 10 49
	Apr 5 8 51 37	Apr 12 19 07 02	Apr 19 11 13 19	Apr 26 22 19 26
	May 4 22 46 39	May 12 1 13 23	May 18 21 12 30	May 26 16 34 44
	Jun 3 10 03 06	Jun 10 6 00 27	Jun 17 8 31 48	Jun 25 9 47 33
	Jul 2 19 17 22	Jul 9 10 56 00	Jul 16 21 39 22	Jul 25 1 19 10
	Aug 1 3 13 04	Aug 7 17 32 06	Aug 15 12 30 23	Aug 23 14 57 14
	Aug 30 10 38 16	Sep 6 3 11 33	Sep 14 4 33 53	Sep 22 2 42 02
	Sep 28 18 27 29	Oct 5 16 48 13	Oct 13 21 08 59	Oct 21 12 40 25
	Oct 28 3 39 35	Nov 4 10 24 11	Nov 12 13 35 31	Nov 19 21 12 03
	Nov 26 15 06 42	Dec 4 6 59 19	Dec 12 5 13 22	Dec 19 4 58 12
	Dec 26 5 14 14			
2020	_h m s_	Jan 3 $4^h46^m31^s$	Jan 10 $19^h22^m25^s$	Jan 17 $12^h59^m32^s$
	Jan 24 21 43 07	Feb 2 1 42 47	Feb 9 7 34 24	Feb 15 22 18 20
	Feb 23 15 33 08	Mar 2 19 58 31	Mar 9 17 48 52	Mar 16 9 35 20
	Mar 24 9 29 21	Apr 1 10 22 24	Apr 8 2 36 13	Apr 14 22 57 17
	Apr 23 2 27 00	Apr 30 20 39 30	May 7 10 46 22	May 14 14 03 51
	May 22 17 40 00	May 30 3 31 04	Jun 5 19 13 32	Jun 13 6 24 49
	Jun 21 6 42 36	Jun 28 8 16 50	Jul 5 4 45 33	Jul 12 23 30 09
	Jul 20 17 34 06	Jul 27 12 33 42	Aug 3 15 59 54	Aug 11 16 45 55
	Aug 19 2 42 48	Aug 25 17 58 47	Sep 2 5 23 12	Sep 10 9 26 51
	Sep 17 11 01 20	Sep 24 1 56 00	Oct 1 21 06 23	Oct 10 0 40 39
	Oct 16 19 32 10	Oct 23 13 24 03	Oct 31 14 50 16	Nov 8 13 47 12
	Nov 15 5 08 18	Nov 22 4 46 08	Nov 30 9 30 47	Dec 8 0 37 43
	Dec 14 16 17 41	Dec 21 23 42 19	Dec 30 3 29 19	
2021	_h m s_	_h m s_	_h m s_	Jan 6 $9^h38^m20^s$
	Jan 13 5 01 17	Jan 20 21 02 43	Jan 28 19 17 21	Feb 4 17 38 14
	Feb 11 19 06 48	Feb 19 18 48 27	Feb 27 8 18 28	Mar 6 1 31 21
	Mar 13 10 22 17	Mar 21 14 41 33	Mar 28 18 49 19	Apr 4 10 03 36
	Apr 12 2 31 59	Apr 20 7 00 05	Apr 27 3 32 42	May 3 19 51 15
	May 11 19 00 56	May 19 19 13 47	May 26 11 15 02	Jun 2 7 25 33
	Jun 10 10 53 47	Jun 18 3 55 26	Jun 24 18 40 51	Jul 1 21 11 47
	Jul 10 1 17 46	Jul 17 10 11 48	Jul 24 2 38 03	Jul 31 13 17 09
	Aug 8 13 51 16	Aug 15 15 20 45	Aug 22 12 03 06	Aug 30 7 14 22
	Sep 7 0 52 54	Sep 13 20 40 30	Sep 20 23 55 50	Sep 29 1 58 16
	Oct 6 11 06 31	Oct 13 3 26 16	Oct 20 14 57 48	Oct 28 20 06 17
	Nov 4 21 15 43	Nov 11 12 47 10	Nov 19 8 58 34	Nov 27 12 28 47
	Dec 4 7 44 09	Dec 11 1 36 40	Dec 19 4 36 37	Dec 27 2 24 52
2022	Jan 2 $18^h34^m37^s$	Jan 9 $18^h12^m23^s$	Jan 17 $23^h49^m33^s$	Jan 25 $13^h42^m02^s$
	Feb 1 5 47 09	Feb 8 13 51 14	Feb 16 16 57 38	Feb 23 22 33 35
	Mar 2 17 35 55	Mar 10 10 46 31	Mar 18 7 18 43	Mar 25 5 38 25
	Apr 1 6 25 33	Apr 9 6 48 46	Apr 16 18 56 12	Apr 23 11 57 31
	Apr 30 20 29 15	May 9 0 22 33	May 16 4 15 19	May 22 18 44 16
	May 30 11 31 26	Jun 7 14 49 39	Jun 14 11 52 55	Jun 21 3 11 59
	Jun 29 2 53 25	Jul 7 2 15 18	Jul 13 18 38 47	Jul 20 14 19 47
	Jul 28 17 56 10	Aug 5 11 07 42	Aug 12 1 36 53	Aug 19 4 37 15
	Aug 27 8 18 16	Sep 3 18 08 53	Sep 10 10 00 11	Sep 17 21 53 07
	Sep 25 21 55 42	Oct 3 0 15 11	Oct 9 20 56 06	Oct 17 17 16 13
	Oct 25 10 49 49	Nov 1 6 38 14	Nov 8 11 03 15	Nov 16 13 28 10
	Nov 23 22 58 20	Nov 30 14 37 40	Dec 8 4 09 17	Dec 16 8 57 12
	Dec 23 10 18 00	Dec 30 1 21 39		

Year	New Moon	First Quarter	Full Moon	Last Quarter

2023

New Moon	First Quarter	Full Moon	Last Quarter
h m s	h m s	Jan 6 $23^h09^m00^s$	Jan 15 $2^h11^m26^s$
Jan 21 20 54 22	Jan 28 15 19 53	Feb 5 18 29 41	Feb 13 16 01 51
Feb 20 7 06 58	Feb 27 8 06 45	Mar 7 12 41 29	Mar 15 2 09 25
Mar 21 17 24 17	Mar 29 2 33 30	Apr 6 4 35 40	Apr 13 9 12 33
Apr 20 4 13 41	Apr 27 21 21 04	May 5 17 35 12	May 12 14 29 27
May 19 15 54 26	May 27 15 23 26	Jun 4 3 42 53	Jun 10 19 32 32
Jun 18 4 38 18	Jun 26 7 50 52	Jul 3 11 39 51	Jul 10 1 49 03
Jul 17 18 32 58	Jul 25 22 07 56	Aug 1 18 32 49	Aug 8 10 29 35
Aug 16 9 39 19	Aug 24 9 58 23	Aug 31 1 36 46	Sep 6 22 22 14
Sep 15 1 40 56	Sep 22 19 32 55	Sep 29 9 58 40	Oct 6 13 48 50
Oct 14 17 56 16	Oct 22 3 30 34	Oct 28 20 25 10	Nov 5 8 37 55
Nov 13 9 28 31	Nov 20 10 51 01	Nov 27 9 17 25	Dec 5 5 50 22
Dec 12 23 33 09	Dec 19 18 40 21	Dec 27 0 34 19	

2024

New Moon	First Quarter	Full Moon	Last Quarter
h m s	h m s	h m s	Jan 4 $3^h31^m33^s$
Jan 11 11 58 32	Jan 18 3 53 44	Jan 25 17 55 07	Feb 2 23 19 05
Feb 9 23 00 18	Feb 16 15 02 04	Feb 24 12 31 33	Mar 3 15 24 37
Mar 10 9 01 35	Mar 17 4 11 52	Mar 25 7 01 27	Apr 2 3 15 52
Apr 8 18 22 01	Apr 15 19 14 16	Apr 23 23 50 07	May 1 11 28 25
May 8 3 23 05	May 15 11 49 09	May 23 13 54 17	May 30 17 13 50
Jun 6 12 38 54	Jun 14 5 19 35	Jun 22 1 09 02	Jun 28 21 54 34
Jul 5 22 58 33	Jul 13 22 49 57	Jul 21 10 18 18	Jul 28 2 52 43
Aug 4 11 14 12	Aug 12 15 19 55	Aug 19 18 26 57	Aug 26 9 27 00
Sep 3 1 56 43	Sep 11 6 06 47	Sep 18 2 35 36	Sep 24 18 51 01
Oct 2 18 50 23	Oct 10 18 56 15	Oct 17 11 27 32	Oct 24 8 04 12
Nov 1 12 48 15	Nov 9 5 56 35	Nov 15 21 29 38	Nov 23 1 29 02
Dec 1 6 22 32	Dec 8 15 27 44	Dec 15 9 02 48	Dec 22 22 19 17
Dec 30 22 27 54			

2025

New Moon	First Quarter	Full Moon	Last Quarter
h m s	Jan 6 $23^h57^m25^s$	Jan 13 $22^h28^m01^s$	Jan 21 $20^h31^m54^s$
Jan 29 12 37 06	Feb 5 8 03 17	Feb 12 13 54 31	Feb 20 17 33 39
Feb 28 0 45 57	Mar 6 16 32 46	Mar 14 6 55 46	Mar 22 11 30 34
Mar 29 10 58 59	Apr 5 2 15 50	Apr 13 0 23 24	Apr 21 1 36 42
Apr 27 19 32 18	May 4 13 52 54	May 12 16 57 05	May 20 11 59 55
May 27 3 03 30	Jun 3 3 42 07	Jun 11 7 44 59	Jun 18 19 20 16
Jun 25 10 32 46	Jul 2 19 31 20	Jul 10 20 37 56	Jul 18 0 38 49
Jul 24 19 12 21	Aug 1 12 42 27	Aug 9 7 56 13	Aug 16 5 13 23
Aug 23 6 07 41	Aug 31 6 26 19	Sep 7 18 10 02	Sep 14 10 34 06
Sep 21 19 55 15	Sep 29 23 54 56	Oct 7 3 48 45	Oct 13 18 13 49
Oct 21 12 26 17	Oct 29 16 21 55	Nov 5 13 20 26	Nov 12 5 29 16
Nov 20 6 48 22	Nov 28 6 59 54	Dec 4 23 15 12	Dec 11 20 52 48
Dec 20 1 44 27	Dec 27 19 10 58		

2026

New Moon	First Quarter	Full Moon	Last Quarter
h m s	h m s	Jan 3 $10^h04^m02^s$	Jan 10 $15^h49^m30^s$
Jan 18 19 53 06	Jan 26 4 48 31	Feb 1 22 10 22	Feb 9 12 44 13
Feb 17 12 02 17	Feb 24 12 28 45	Mar 3 11 39 02	Mar 11 9 39 38
Mar 19 1 24 37	Mar 25 19 18 51	Apr 2 2 13 06	Apr 10 4 52 47
Apr 17 11 52 57	Apr 24 2 32 54	May 1 17 24 19	May 9 21 11 36
May 16 20 02 12	May 23 11 12 07	May 31 8 46 21	Jun 8 10 01 40
Jun 15 2 55 19	Jun 21 21 56 34	Jun 29 23 57 50	Jul 7 19 30 08
Jul 14 9 44 46	Jul 21 11 06 45	Jul 29 14 36 52	Aug 6 2 22 38
Aug 12 17 37 54	Aug 20 2 47 29	Aug 28 4 19 40	Sep 4 7 52 22
Sep 11 3 28 08	Sep 18 20 44 54	Sep 26 16 50 10	Oct 3 13 26 12
Oct 10 15 51 12	Oct 18 16 13 48	Oct 26 4 12 56	Nov 1 20 29 35
Nov 9 7 03 14	Nov 17 11 48 56	Nov 24 14 54 41	Dec 1 6 09 47
Dec 9 0 52 57	Dec 17 5 43 46	Dec 24 1 29 21	Dec 30 19 00 37

PHASES OF THE MOON

Year	New Moon		First Quarter		Full Moon		Last Quarter	
2027	Jan 7	$20^h25^m30^s$	Jan 15	$20^h35^m38^s$	Jan 22	$12^h18^m30^s$	Jan 29	$10^h56^m35^s$
	Feb 6	15 57 14	Feb 14	7 59 37	Feb 20	23 24 47	Feb 28	5 17 38
	Mar 8	9 30 37	Mar 15	16 26 21	Mar 22	10 44 56	Mar 30	0 55 05
	Apr 6	23 52 18	Apr 13	22 57 48	Apr 20	22 28 19	Apr 28	20 19 01
	May 6	10 59 47	May 13	4 45 03	May 20	11 00 11	May 28	13 59 05
	Jun 4	19 41 31	Jun 11	10 57 18	Jun 19	0 45 30	Jun 27	4 55 30
	Jul 4	3 03 15	Jul 10	18 40 11	Jul 18	15 46 04	Jul 26	16 56 04
	Aug 2	10 06 23	Aug 9	4 55 20	Aug 17	7 29 50	Aug 25	2 28 30
	Aug 31	17 42 20	Sep 7	18 32 27	Sep 15	23 04 39	Sep 23	10 21 36
	Sep 30	2 37 13	Oct 7	11 48 32	Oct 15	13 48 08	Oct 22	17 30 17
	Oct 29	13 37 41	Nov 6	8 01 08	Nov 14	3 27 03	Nov 21	0 49 20
	Nov 28	3 25 33	Dec 6	5 23 15	Dec 13	16 09 55	Dec 20	9 12 02
	Dec 27	20 13 27						
2028		$h\ m\ s$	Jan 5	$1^h41^m35^s$	Jan 12	$4^h04^m13^s$	Jan 18	$19^h27^m07^s$
	Jan 26	15 13 38	Feb 3	19 11 38	Feb 10	15 04 54	Feb 17	8 09 08
	Feb 25	10 38 33	Mar 4	9 03 32	Mar 11	1 07 13	Mar 17	23 24 00
	Mar 26	4 32 28	Apr 2	19 16 38	Apr 9	10 27 47	Apr 16	16 38 05
	Apr 24	19 48 05	May 2	2 26 55	May 8	19 50 06	May 16	10 44 16
	May 24	8 17 28	May 31	7 37 47	Jun 7	6 09 58	Jun 15	4 28 34
	Jun 22	18 28 43	Jun 29	12 11 47	Jul 6	18 11 58	Jul 14	20 57 44
	Jul 22	3 02 52	Jul 28	17 41 19	Aug 5	8 10 58	Aug 13	11 46 28
	Aug 20	10 45 01	Aug 27	1 37 02	Sep 3	23 48 44	Sep 12	0 46 54
	Sep 18	18 24 54	Sep 25	13 11 11	Oct 3	16 26 07	Oct 11	11 58 00
	Oct 18	2 57 57	Oct 25	4 54 19	Nov 2	9 18 29	Nov 9	21 27 00
	Nov 16	13 19 09	Nov 24	0 15 46	Dec 2	1 41 21	Dec 9	5 40 04
	Dec 16	2 07 26	Dec 23	21 46 04	Dec 31	16 49 39		
2029		$h\ m\ s$		$h\ m\ s$		$h\ m\ s$	Jan 7	$13^h27^m31^s$
	Jan 14	17 25 38	Jan 22	19 24 21	Jan 30	6 04 46	Feb 5	21 53 21
	Feb 13	10 32 39	Feb 21	15 11 02	Feb 28	17 11 24	Mar 7	7 52 45
	Mar 15	4 20 23	Mar 23	7 34 17	Mar 30	2 27 34	Apr 5	19 52 45
	Apr 13	21 41 -20	Apr 21	19 51 18	Apr 28	10 37 58	May 5	9 49 16
	May 13	13 43 19	May 21	4 17 19	May 27	18 38 40	Jun 4	1 20 02
	Jun 12	3 51 43	Jun 19	9 55 18	Jun 26	3 23 30	Jul 3	17 58 41
	Jul 11	15 52 13	Jul 18	14 15 30	Jul 25	13 36 55	Aug 2	11 16 33
	Aug 10	1 56 57	Aug 16	18 56 29	Aug 24	1 52 22	Sep 1	4 34 13
	Sep 8	10 45 32	Sep 15	1 30 26	Sep 22	16 30 27	Sep 30	20 57 59
	Oct 7	19 15 42	Oct 14	11 10 01	Oct 22	9 28 42	Oct 30	11 33 25
	Nov 6	4 25 16	Nov 13	0 36 25	Nov 21	4 04 05	Nov 28	23 48 56
	Dec 5	14 53 16	Dec 12	17 50 37	Dec 20	22 47 38	Dec 28	9 50 14
2030	Jan 4	$2^h50^m41^s$	Jan 11	$14^h07^m11^s$	Jan 19	$15^h55^m30^s$	Jan 26	$18^h15^m35^s$
	Feb 2	16 08 39	Feb 10	11 50 34	Feb 18	6 20 57	Feb 25	1 58 52
	Mar 4	6 35 49	Mar 12	8 48 45	Mar 19	17 57 39	Mar 26	9 52 33
	Apr 2	22 03 42	Apr 11	2 58 03	Apr 18	3 21 12	Apr 24	18 40 10
	May 2	14 13 19	May 10	17 12 39	May 17	11 20 19	May 24	4 58 39
	Jun 1	6 22 31	Jun 9	3 36 55	Jun 15	18 42 11	Jun 22	17 20 53
	Jun 30	21 35 39	Jul 8	11 03 03	Jul 15	2 13 09	Jul 22	8 08 45
	Jul 30	11 12 10	Aug 6	16 43 53	Aug 13	10 45 33	Aug 21	1 16 36
	Aug 28	23 08 31	Sep 4	21 56 45	Sep 11	21 19 05	Sep 19	19 57 40
	Sep 27	9 55 50	Oct 4	3 57 19	Oct 11	10 47 59	Oct 19	14 51 30
	Oct 26	20 18 08	Nov 2	11 57 09	Nov 10	3 31 26	Nov 18	8 33 28
	Nov 25	6 47 37	Dec 1	22 58 01	Dec 9	22 41 35	Dec 18	0 02 20
	Dec 24	17 33 22	Dec 31	13 37 18				

2031

Year	New Moon	First Quarter	Full Moon	Last Quarter
	h m s	*h m s*		
			Jan 8 $18^h26^m57^s$	Jan 16 $12^h48^m24^s$
	Jan 23 4 32 08	Jan 30 7 44 15	Feb 7 12 47 21	Feb 14 22 50 58
	Feb 21 15 50 01	Mar 1 4 03 23	Mar 9 4 30 48	Mar 16 6 36 59
	Mar 23 3 50 15	Mar 31 0 33 25	Apr 7 17 22 30	Apr 14 12 59 03
	Apr 21 16 58 15	Apr 29 19 20 37	May 7 3 41 03	May 13 19 08 09
	May 21 7 18 23	May 29 11 20 46	Jun 5 11 59 43	Jun 12 2 21 50
	Jun 19 22 25 52	Jun 28 0 20 11	Jul 4 19 02 33	Jul 11 11 51 01
	Jul 19 13 41 25	Jul 27 10 35 57	Aug 3 1 46 47	Aug 10 0 24 50
	Aug 18 4 33 28	Aug 25 18 40 53	Sep 1 9 21 41	Sep 8 16 15 34
	Sep 16 18 48 02	Sep 24 1 20 56	Sep 30 18 59 01	Oct 8 10 51 23
	Oct 16 8 21 53	Oct 23 7 37 36	Oct 30 7 33 50	Nov 7 7 03 23
	Nov 14 21 10 46	Nov 21 14 46 03	Nov 28 23 19 36	Dec 7 3 20 49
	Dec 14 9 06 57	Dec 21 0 01 41	Dec 28 17 34 09	

2032

Year	New Moon	First Quarter	Full Moon	Last Quarter
	h m s	*h m s*	*h m s*	
				Jan 5 $22^h05^m20^s$
	Jan 12 20 07 51	Jan 19 12 15 28	Jan 27 12 53 38	Feb 4 13 50 06
	Feb 11 6 25 25	Feb 18 3 30 08	Feb 26 7 44 23	Mar 5 1 48 09
	Mar 11 16 25 51	Mar 18 20 57 50	Mar 27 0 47 28	Apr 3 10 11 25
	Apr 10 2 40 41	Apr 17 15 25 33	Apr 25 15 10 51	May 2 16 02 57
	May 9 13 36 55	May 17 9 44 34	May 25 2 38 25	May 31 20 52 24
	Jun 8 1 33 17	Jun 16 3 01 03	Jun 23 11 33 44	Jun 30 2 13 01
	Jul 7 14 42 41	Jul 15 18 33 13	Jul 22 18 52 45	Jul 29 9 26 33
	Aug 6 5 12 43	Aug 14 7 52 10	Aug 21 1 48 02	Aug 27 19 34 35
	Sep 4 20 57 47	Sep 12 18 50 22	Sep 19 9 31 26	Sep 26 9 13 37
	Oct 4 13 27 31	Oct 12 3 48 45	Oct 18 18 59 17	Oct 26 2 30 01
	Nov 3 5 46 05	Nov 10 11 34 25	Nov 17 6 43 12	Nov 24 22 48 55
	Dec 2 20 54 01	Dec 9 19 09 41	Dec 16 20 50 09	Dec 24 20 40 21

2033

Year	New Moon	First Quarter	Full Moon	Last Quarter
	Jan 1 $10^h18^m10^s$	Jan 8 $3^h35^m32^s$	Jan 15 $13^h08^m15^s$	Jan 23 $17^h46^m53^s$
	Jan 30 22 01 01	Feb 6 13 35 23	Feb 14 7 05 21	Feb 22 11 54 15
	Mar 1 8 24 42	Mar 8 1 28 15	Mar 16 1 38 35	Mar 24 1 50 58
	Mar 30 17 52 50	Apr 6 15 15 06	Apr 14 19 18 33	Apr 22 11 43 28
	Apr 29 2 47 22	May 6 6 46 28	May 14 10 43 54	May 21 18 30 11
	May 28 11 37 44	Jun 4 23 40 05	Jun 12 23 20 24	Jun 19 23 30 44
	Jun 26 21 08 14	Jul 4 17 13 17	Jul 12 9 29 46	Jul 19 4 08 26
	Jul 26 8 13 45	Aug 3 10 26 54	Aug 10 18 08 54	Aug 17 9 44 08
	Aug 24 21 40 58	Sep 2 2 24 50	Sep 9 2 21 45	Sep 15 17 34 45
	Sep 23 13 40 56	Oct 1 16 33 55	Oct 8 10 59 18	Oct 15 4 48 38
	Oct 23 7 29 34	Oct 31 4 47 30	Nov 6 20 33 15	Nov 13 20 09 57
	Nov 22 1 40 16	Nov 29 15 16 23	Dec 6 7 23 14	Dec 13 15 29 07
	Dec 21 18 47 37	Dec 29 0 21 15		

2034

Year	New Moon	First Quarter	Full Moon	Last Quarter
	h m s	*h m s*	Jan 4 $19^h48^m16^s$	Jan 12 $13^h18^m12^s$
	Jan 20 10 02 42	Jan 27 8 33 08	Feb 3 10 05 42	Feb 11 11 10 08
	Feb 18 23 11 25	Feb 25 16 35 21	Mar 5 2 11 19	Mar 13 6 45 37
	Mar 20 10 15 45	Mar 27 1 19 41	Apr 3 19 20 03	Apr 11 22 46 20
	Apr 18 19 27 05	Apr 25 11 35 53	May 3 12 16 52	May 11 10 57 24
	May 18 3 13 47	May 24 23 58 46	Jun 2 3 55 11	Jun 9 19 45 09
	Jun 16 10 27 08	Jun 23 14 36 21	Jul 1 17 45 44	Jul 9 2 00 15
	Jul 15 18 16 27	Jul 23 7 06 13	Jul 31 5 55 41	Aug 7 6 51 30
	Aug 14 3 54 15	Aug 22 0 44 31	Aug 29 16 50 28	Sep 5 11 42 35
	Sep 12 16 14 59	Sep 20 18 40 40	Sep 28 2 58 00	Oct 4 18 05 53
	Oct 12 7 33 49	Oct 20 12 04 08	Oct 27 12 43 40	Nov 3 3 28 29
	Nov 11 1 17 23	Nov 19 4 02 37	Nov 25 22 33 20	Dec 2 16 47 35
	Dec 10 20 15 33	Dec 18 17 46 06	Dec 25 8 55 39	

2035

New Moon	h m s	First Quarter	h m s	Full Moon	h m s	Last Quarter	h m s
						Jan 1	$10^h02^m10^s$
Jan 9	15 04 15	Jan 17	4 46 34	Jan 23	20 17 46	Jan 31	6 03 40
Feb 8	8 23 18	Feb 15	13 18 10	Feb 22	8 55 05	Mar 2	3 02 13
Mar 9	23 10 38	Mar 16	20 16 11	Mar 23	22 43 18	Mar 31	23 07 49
Apr 8	10 58 59	Apr 15	2 56 03	Apr 22	13 21 56	Apr 30	16 55 08
May 7	20 05 06	May 14	10 29 40	May 22	4 26 59	May 30	7 31 55
Jun 6	3 21 55	Jun 12	19 51 15	Jun 20	19 38 40	Jun 28	18 43 57
Jul 5	10 00 32	Jul 12	7 34 05	Jul 20	10 38 03	Jul 28	2 56 48
Aug 3	17 13 01	Aug 10	21 53 40	Aug 19	1 01 29	Aug 26	9 09 11
Sep 2	2 00 44	Sep 9	14 48 24	Sep 17	14 24 42	Sep 24	14 40 49
Oct 1	13 07 59	Oct 9	9 50 27	Oct 17	2 36 44	Oct 23	20 58 00
Oct 31	2 59 52	Nov 8	5 51 37	Nov 15	13 50 03	Nov 22	5 17 32
Nov 29	19 38 44	Dec 8	1 06 22	Dec 15	0 34 21	Dec 21	16 29 42
Dec 29	14 32 05						

2036

New Moon	h m s	First Quarter	h m s	Full Moon	h m s	Last Quarter	h m s
		Jan 6	$17^h49^m12^s$	Jan 13	$11^h17^m17^s$	Jan 20	$6^h47^m32^s$
Jan 28	10 18 21	Feb 5	7 02 00	Feb 11	22 09 53	Feb 18	23 48 08
Feb 27	5 00 26	Mar 5	16 50 09	Mar 12	9 10 40	Mar 19	18 39 52
Mar 27	20 57 55	Apr 4	0 04 35	Apr 10	20 23 45	Apr 18	14 07 02
Apr 26	9 34 25	May 3	5 55 34	May 10	8 10 45	May 18	8 40 41
May 25	19 18 11	Jun 1	11 35 40	Jun 8	21 03 11	Jun 17	1 04 25
Jun 24	3 10 51	Jun 30	18 14 18	Jul 8	11 20 30	Jul 16	14 40 51
Jul 23	10 18 13	Jul 30	2 57 30	Aug 7	2 50 07	Aug 15	1 37 05
Aug 21	17 36 33	Aug 28	14 44 38	Sep 5	18 46 45	Sep 13	10 30 23
Sep 20	1 52 46	Sep 27	6 13 43	Oct 5	10 16 19	Oct 12	18 10 35
Oct 19	11 51 11	Oct 27	1 14 54	Nov 4	0 45 26	Nov 11	1 29 50
Nov 18	0 15 40	Nov 25	22 29 17	Dec 3	14 09 41	Dec 10	9 19 41
Dec 17	15 35 37	Dec 25	19 45 43				

2037

New Moon	h m s	First Quarter	h m s	Full Moon	h m s	Last Quarter	h m s
				Jan 2	$2^h36^m28^s$	Jan 8	$18^h30^m19^s$
Jan 16	9 35 34	Jan 24	14 56 17	Jan 31	14 05 20	Feb 7	5 44 40
Feb 15	4 55 15	Feb 23	6 42 11	Mar 2	0 29 21	Mar 8	19 26 25
Mar 16	23 37 24	Mar 24	18 40 49	Mar 31	9 54 44	Apr 7	11 26 14
Apr 15	16 08 54	Apr 23	3 12 44	Apr 29	18 55 05	May 7	4 57 17
May 15	5 55 33	May 22	9 09 47	May 29	4 25 21	Jun 5	22 50 17
Jun 13	17 11 28	Jun 20	13 46 33	Jun 27	15 21 08	Jul 5	16 01 43
Jul 13	2 33 01	Jul 19	18 32 17	Jul 27	4 16 20	Aug 4	7 52 18
Aug 11	10 42 48	Aug 18	1 00 54	Aug 25	19 10 31	Sep 2	22 04 28
Sep 9	18 26 41	Sep 16	10 37 20	Sep 24	11 32 52	Oct 2	10 30 26
Oct 9	2 35 43	Oct 16	0 16 41	Oct 24	4 37 41	Oct 31	21 07 45
Nov 7	12 04 18	Nov 14	17 59 56	Nov 22	21 36 24	Nov 30	6 07 43
Dec 6	23 39 31	Dec 14	14 43 11	Dec 22	13 39 48	Dec 29	14 06 13

2038

New Moon	h m s	First Quarter	h m s	Full Moon	h m s	Last Quarter	h m s
Jan 5	$13^h42^m31^s$	Jan 13	$12^h35^m04^s$	Jan 21	$4^h01^m08^s$	Jan 27	$22^h01^m42^s$
Feb 4	5 53 21	Feb 12	9 30 53	Feb 19	16 10 36	Feb 26	6 57 09
Mar 5	23 16 11	Mar 14	3 42 51	Mar 21	2 10 47	Mar 27	17 37 16
Apr 4	16 44 11	Apr 12	18 03 08	Apr 19	10 37 13	Apr 26	6 16 32
May 4	9 20 51	May 12	4 19 16	May 18	18 24 45	May 25	20 44 41
Jun 3	0 25 26	Jun 10	11 12 37	Jun 17	2 31 45	Jun 24	12 40 50
Jul 2	13 33 22	Jul 9	16 01 49	Jul 16	11 49 28	Jul 24	5 41 07
Aug 1	0 41 32	Aug 7	20 22 42	Aug 14	22 58 01	Aug 22	23 13 17
Aug 30	10 13 58	Sep 6	1 51 59	Sep 13	12 25 36	Sep 21	16 28 13
Sep 28	18 58 47	Oct 5	9 53 28	Oct 13	4 23 01	Oct 21	8 24 36
Oct 28	3 54 03	Nov 3	21 24 57	Nov 11	22 28 21	Nov 19	22 11 13
Nov 26	13 47 54	Dec 3	12 47 22	Dec 11	17 31 34	Dec 19	9 30 12
Dec 26	1 03 09						

Year	New Moon		First Quarter		Full Moon		Last Quarter	
2039		h m s	Jan 2	$7^h38^m04^s$	Jan 10	$11^h46^m41^s$	Jan 17	$18^h42^m53^s$
	Jan 24	13 37 17	Feb 1	4 46 18	Feb 9	3 40 25	Feb 16	2 37 12
	Feb 23	3 18 42	Mar 3	2 16 13	Mar 10	16 36 11	Mar 17	10 08 52
	Mar 24	18 00 42	Apr 1	21 55 52	Apr 9	2 53 58	Apr 15	18 08 13
	Apr 23	9 35 57	May 1	14 08 34	May 8	11 21 13	May 15	3 18 13
	May 23	1 39 14	May 31	2 25 41	Jun 6	18 48 54	Jun 13	14 17 42
	Jun 21	17 22 39	Jun 29	11 18 31	Jul 6	2 04 42	Jul 13	3 39 29
	Jul 21	7 55 17	Jul 28	17 50 56	Aug 4	9 57 52	Aug 11	19 37 09
	Aug 19	20 51 44	Aug 26	23 17 36	Sep 2	19 24 45	Sep 10	13 46 49
	Sep 18	8 24 13	Sep 25	4 53 45	Oct 2	7 24 28	Oct 10	9 00 39
	Oct 17	19 10 12	Oct 24	11 51 44	Oct 31	22 37 26	Nov 9	3 47 13
	Nov 16	5 47 16	Nov 22	21 17 51	Nov 30	16 50 44	Dec 8	20 45 44
	Dec 15	16 33 13	Dec 22	10 02 49	Dec 30	12 38 54		
2040		h m s		h m s		h m s	Jan 7	$11^h06^m43^s$
	Jan 14	3 26 27	Jan 21	2 22 21	Jan 29	7 55 51	Feb 5	22 33 39
	Feb 12	14 25 38	Feb 19	21 34 50	Feb 28	1 00 44	Mar 6	7 19 59
	Mar 13	1 47 18	Mar 20	18 00 16	Mar 28	15 12 54	Apr 4	14 07 26
	Apr 11	14 01 24	Apr 19	13 38 40	Apr 27	2 39 05	May 3	20 00 57
	May 11	3 29 05	May 19	7 01 45	May 26	11 48 18	Jun 2	2 18 35
	Jun 9	18 04 16	Jun 17	21 33 34	Jun 24	19 20 31	Jul 1	10 18 57
	Jul 9	9 15 56	Jul 17	9 17 28	Jul 24	2 06 50	Jul 30	21 06 59
	Aug 8	0 27 36	Aug 15	18 37 11	Aug 22	9 10 53	Aug 29	11 17 44
	Sep 6	15 14 52	Sep 14	2 08 46	Sep 20	17 43 59	Sep 28	4 42 36
	Oct 6	5 27 05	Oct 13	8 42 25	Oct 20	4 51 02	Oct 28	0 28 03
	Nov 4	18 57 10	Nov 11	15 24 34	Nov 18	19 07 19	Nov 26	21 08 37
	Dec 4	7 34 20	Dec 10	23 30 57	Dec 18	12 16 54	Dec 26	17 03 35
2041	Jan 2	$19^h08^m57^s$	Jan 9	$10^h06^m57^s$	Jan 17	$7^h12^m32^s$	Jan 25	$10^h34^m21^s$
	Feb 1	5 44 05	Feb 7	23 41 26	Feb 16	2 22 26	Feb 24	0 29 55
	Mar 2	15 40 31	Mar 9	15 52 22	Mar 17	20 20 19	Mar 25	10 33 08
	Apr 1	1 30 39	Apr 8	9 39 45	Apr 16	12 01 51	Apr 23	17 25 19
	Apr 30	11 47 32	May 8	3 55 20	May 16	0 53 35	May 22	22 27 08
	May 29	22 57 11	Jun 6	21 41 51	Jun 14	10 59 55	Jun 21	3 13 24
	Jun 28	11 18 07	Jul 6	14 13 41	Jul 13	19 02 02	Jul 20	9 14 24
	Jul 28	1 03 31	Aug 5	4 53 55	Aug 12	2 05 54	Aug 18	17 44 19
	Aug 26	16 17 21	Sep 3	17 20 00	Sep 10	9 25 03	Sep 17	5 34 06
	Sep 25	8 42 27	Oct 3	3 33 52	Oct 9	18 03 52	Oct 16	21 06 23
	Oct 25	1 31 29	Nov 1	12 06 08	Nov 8	4 44 40	Nov 15	16 07 32
	Nov 23	17 37 54	Nov 30	19 49 55	Dec 7	17 43 23	Dec 15	13 33 53
	Dec 23	8 07 33	Dec 30	3 46 51				
2042		h m s		h m s	Jan 6	$8^h55^m06^s$	Jan 14	$11^h25^m36^s$
	Jan 21	20 43 16	Jan 28	12 49 46	Feb 5	1 58 58	Feb 13	7 17 32
	Feb 20	7 40 06	Feb 26	23 30 37	Mar 6	20 11 07	Mar 14	23 22 18
	Mar 21	17 24 13	Mar 28	12 00 58	Apr 5	14 17 07	Apr 13	11 10 21
	Apr 20	2 20 32	Apr 27	2 20 31	May 5	6 49 47	May 12	19 19 17
	May 19	10 56 04	May 26	18 19 19	Jun 3	20 49 30	Jun 11	1 01 25
	Jun 17	19 49 23	Jun 25	11 30 02	Jul 3	8 10 36	Jul 10	5 39 28
	Jul 17	5 53 05	Jul 25	5 02 51	Aug 1	17 34 32	Aug 8	10 35 56
	Aug 15	18 02 30	Aug 23	21 56 41	Aug 31	2 03 33	Sep 6	17 10 09
	Sep 14	8 51 23	Sep 22	13 21 53	Sep 29	10 35 30	Oct 6	2 36 03
	Oct 14	2 04 18	Oct 22	2 54 18	Oct 28	19 49 35	Nov 4	15 52 32
	Nov 12	20 29 36	Nov 20	14 32 31	Nov 27	6 07 11	Dec 4	9 20 00
	Dec 12	14 30 50	Dec 20	0 28 45	Dec 26	17 43 55		

PHASES OF THE MOON

Year	New Moon				First Quarter				Full Moon				Last Quarter			
		h	m	s		h	m	s		h	m	s		h	m	s
2043													Jan 3	6^h	09^m	19^s
	Jan 11	6	54	27	Jan 18	9	06	07	Jan 25	6	57	49	Feb 2	4	15	56
	Feb 9	21	08	55	Feb 16	17	01	23	Feb 23	21	59	09	Mar 4	1	08	36
	Mar 11	9	10	31	Mar 18	1	04	23	Mar 25	14	27	30	Apr 2	18	57	27
	Apr 9	19	07	51	Apr 16	10	10	08	Apr 24	7	24	14	May 2	9	00	07
	May 9	3	22	27	May 15	21	06	18	May 23	23	38	11	May 31	19	25	56
	Jun 7	10	36	17	Jun 14	10	19	59	Jun 22	14	21	51	Jun 30	2	54	16
	Jul 6	17	52	10	Jul 14	1	48	03	Jul 22	3	25	25	Jul 29	8	24	00
	Aug 5	2	23	58	Aug 12	18	58	24	Aug 20	15	05	44	Aug 27	13	10	24
	Sep 3	13	18	33	Sep 11	13	02	14	Sep 19	1	48	17	Sep 25	18	41	33
	Oct 3	3	13	21	Oct 11	7	06	08	Oct 18	11	56	53	Oct 25	2	28	45
	Nov 1	19	58	38	Nov 10	0	14	17	Nov 16	21	53	40	Nov 23	13	46	51
	Dec 1	14	38	08	Dec 9	15	28	41	Dec 16	8	03	12	Dec 23	5	05	36
	Dec 31	9	49	15												
		h	m	s												
2044					Jan 8	4^h	03^m	06^s	Jan 14	18^h	52^m	20^s	Jan 21	23^h	48^m	14^s
	Jan 30	4	05	42	Feb 6	13	47	23	Feb 13	6	42	59	Feb 20	20	21	21
	Feb 28	20	13	35	Mar 6	21	18	24	Mar 13	19	42	23	Mar 21	16	53	38
	Mar 29	9	27	13	Apr 5	3	46	15	Apr 12	9	40	22	Apr 20	11	49	44
	Apr 27	19	43	21	May 4	10	29	18	May 12	0	17	53	May 20	4	03	13
	May 27	3	40	51	Jun 2	18	34	43	Jun 10	15	17	21	Jun 18	17	01	07
	Jun 25	10	25	34	Jul 2	4	49	44	Jul 10	6	23	14	Jul 18	2	47	58
	Jul 24	17	11	44	Jul 31	17	41	37	Aug 8	21	15	07	Aug 16	10	04	32
	Aug 23	1	07	13	Aug 30	9	19	51	Sep 7	11	25	38	Sep 14	15	58	58
	Sep 21	11	04	41	Sep 29	3	31	48	Oct 7	0	31	16	Oct 13	21	53	45
	Oct 20	23	37	38	Oct 28	23	28	52	Nov 5	12	27	56	Nov 12	5	10	39
	Nov 19	14	58	59	Nov 27	19	37	34	Dec 4	23	35	06	Dec 11	14	53	29
	Dec 19	8	54	21	Dec 27	14	01	13								
		h	m	s		h	m	s								
2045									Jan 3	10^h	21^m	40^s	Jan 10	3^h	33^m	19^s
	Jan 18	4	26	41	Jan 26	5	10	14	Feb 1	21	06	44	Feb 8	19	04	28
	Feb 16	23	52	20	Feb 24	16	38	12	Mar 3	7	53	52	Mar 10	12	51	09
	Mar 18	17	16	05	Mar 26	0	57	35	Apr 1	18	44	13	Apr 9	7	53	28
	Apr 17	7	27	55	Apr 24	7	13	17	May 1	5	53	24	May 9	2	52	07
	May 16	18	27	51	May 23	12	39	39	May 30	17	53	39	Jun 7	20	24	24
	Jun 15	3	06	04	Jun 21	18	29	48	Jun 29	7	17	02	Jul 7	11	31	54
	Jul 14	10	29	38	Jul 21	1	53	33	Jul 28	22	11	49	Aug 5	23	58	18
	Aug 12	17	40	30	Aug 19	11	56	28	Aug 27	14	08	53	Sep 4	10	04	44
	Sep 11	1	28	57	Sep 18	1	31	26	Sep 26	6	12	45	Oct 3	18	32	53
	Oct 10	10	38	05	Oct 17	18	56	35	Oct 25	21	32	24	Nov 2	2	10	53
	Nov 8	21	50	08	Nov 16	15	27	05	Nov 24	11	44	45	Dec 1	9	47	40
	Dec 8	11	42	34	Dec 16	13	09	34	Dec 24	0	50	28	Dec 30	18	12	32
2046	Jan 7	4^h	25^m	14^s	Jan 15	9^h	43^m	46^s	Jan 22	12^h	52^m	31^s	Jan 29	4^h	12^m	39^s
	Feb 5	23	10	54	Feb 14	3	21	40	Feb 20	23	45	31	Feb 27	16	24	10
	Mar 7	18	16	33	Mar 15	17	14	03	Mar 22	9	28	10	Mar 29	6	58	29
	Apr 6	11	52	54	Apr 14	3	22	46	Apr 20	18	22	20	Apr 27	23	31	39
	May 6	2	57	19	May 13	10	26	08	May 20	3	16	31	May 27	17	07	50
	Jun 4	15	23	42	Jun 11	15	28	56	Jun 18	13	11	16	Jun 26	10	41	13
	Jul 4	1	40	08	Jul 10	19	54	40	Jul 18	0	56	16	Jul 26	3	20	39
	Aug 2	10	26	44	Aug 9	1	16	57	Aug 16	14	51	20	Aug 24	18	37	30
	Aug 31	18	26	40	Sep 7	9	08	13	Sep 15	6	40	35	Sep 23	8	16	56
	Sep 30	2	26	44	Oct 6	20	42	18	Oct 14	23	42	30	Oct 22	20	08	56
	Oct 29	11	18	15	Nov 5	12	29	50	Nov 13	17	05	39	Nov 21	6	11	38
	Nov 27	21	51	18	Dec 5	7	57	35	Dec 13	9	56	39	Dec 20	14	44	16
	Dec 27	10	40	06												

Year	New Moon	First Quarter	Full Moon	Last Quarter
		h m s		
2047		Jan 4 $5^h32^m10^s$	Jan 12 $1^h22^m33^s$	Jan 18 $22^h33^m47^s$
	Jan 26 1 44 58	Feb 3 3 10 14	Feb 10 14 41 03	Feb 17 6 43 47
	Feb 24 18 27 06	Mar 4 22 53 06	Mar 12 1 38 13	Mar 18 16 12 00
	Mar 26 11 45 24	Apr 3 15 12 08	Apr 10 10 36 31	Apr 17 3 31 26
	Apr 25 4 41 03	May 3 3 27 37	May 9 18 25 42	May 16 16 47 05
	May 24 20 28 47	Jun 1 11 55 39	Jun 8 2 06 08	Jun 15 7 46 25
	Jun 23 10 37 04	Jun 30 17 38 14	Jul 7 10 35 04	Jul 15 0 10 46
	Jul 22 22 50 34	Jul 29 22 04 06	Aug 5 20 39 43	Aug 13 17 35 19
	Aug 21 9 17 29	Aug 28 2 50 49	Sep 4 8 55 24	Sep 12 11 19 32
	Sep 19 18 32 39	Sep 26 9 30 08	Oct 3 23 43 15	Oct 12 4 23 02
	Oct 19 3 29 10	Oct 25 19 14 04	Nov 2 16 59 21	Nov 10 19 40 36
	Nov 17 13 00 09	Nov 24 8 42 00	Dec 2 11 56 17	Dec 10 8 30 12
	Dec 16 23 39 27	Dec 24 1 52 10		
	h m s	h m s		
2048			Jan 1 $6^h58^m17^s$	Jan 8 $18^h50^m44^s$
	Jan 15 11 33 39	Jan 22 21 57 21	Jan 31 0 15 30	Feb 7 3 17 55
	Feb 14 0 32 44	Feb 21 19 23 39	Feb 29 14 39 20	Mar 7 10 46 20
	Mar 14 14 28 52	Mar 22 16 04 42	Mar 30 2 05 37	Apr 5 18 11 50
	Apr 13 5 20 56	Apr 21 10 03 25	Apr 28 11 14 12	May 5 2 23 40
	May 12 20 59 25	May 21 0 17 28	May 27 18 58 29	Jun 3 12 05 58
	Jun 11 12 51 10	Jun 19 10 50 49	Jun 26 2 09 20	Jul 2 23 59 05
	Jul 11 4 05 26	Jul 18 18 32 55	Jul 25 9 35 09	Aug 1 14 31 44
	Aug 9 18 00 09	Aug 17 0 33 07	Aug 23 18 08 23	Aug 31 7 43 08
	Sep 8 6 25 48	Sep 15 6 05 09	Sep 22 4 47 45	Sep 30 2 46 42
	Oct 7 17 46 26	Oct 14 12 21 19	Oct 21 18 26 09	Oct 29 22 15 40
	Nov 6 4 39 34	Nov 12 20 30 03	Nov 20 11 20 42	Nov 28 16 34 45
	Dec 5 15 31 20	Dec 12 7 30 17	Dec 20 6 40 23	Dec 28 8 32 45
2049	Jan 4 $2^h25^m42^s$	Jan 10 $21^h57^m00^s$	Jan 19 $2^h30^m17^s$	Jan 26 $21^h34^m16^s$
	Feb 2 13 17 03	Feb 9 15 39 32	Feb 17 20 48 37	Feb 25 7 37 31
	Mar 4 0 12 44	Mar 11 11 27 22	Mar 19 12 24 31	Mar 26 15 11 30
	Apr 2 11 40 25	Apr 10 7 28 40	Apr 18 1 05 56	Apr 24 21 12 25
	May 2 0 12 15	May 10 1 58 46	May 17 11 15 00	May 24 2 55 21
	May 31 14 01 21	Jun 8 17 57 43	Jun 15 19 28 02	Jun 22 9 42 27
	Jun 30 4 51 33	Jul 8 7 11 23	Jul 15 2 30 53	Jul 21 18 49 50
	Jul 29 20 08 29	Aug 6 17 52 58	Aug 13 9 20 49	Aug 20 7 12 00
	Aug 28 11 19 53	Sep 5 2 29 27	Sep 11 17 05 36	Sep 18 23 04 59
	Sep 27 2 06 18	Oct 4 9 40 05	Oct 11 2 54 29	Oct 18 17 56 46
	Oct 26 16 16 19	Nov 2 16 20 23	Nov 9 15 39 07	Nov 17 14 33 20
	Nov 25 5 36 54	Dec 1 23 40 56	Dec 9 7 29 14	Dec 17 11 15 52
	Dec 24 17 52 58	Dec 31 8 53 59		
	h m s	h m s		
2050			Jan 8 $1^h40^m13^s$	Jan 16 $6^h18^m47^s$
	Jan 23 4 58 09	Jan 29 20 49 05	Feb 6 20 48 41	Feb 14 22 11 56
	Feb 21 15 04 42	Feb 28 11 30 54	Mar 8 15 24 27	Mar 16 10 09 12
	Mar 23 0 42 08	Mar 30 4 18 37	Apr 7 8 13 20	Apr 14 18 25 07
	Apr 21 10 26 58	Apr 28 22 09 45	May 6 22 27 19	May 14 0 05 04
	May 20 20 52 15	May 28 16 05 29	Jun 5 9 52 23	Jun 12 4 40 37
	Jun 19 8 23 05	Jun 27 9 18 04	Jul 4 18 52 11	Jul 11 9 47 14
	Jul 18 21 18 03	Jul 27 1 06 34	Aug 3 2 21 42	Aug 9 16 49 35
	Aug 17 11 48 41	Aug 25 14 57 34	Sep 1 9 32 04	Sep 8 2 52 15
	Sep 16 3 50 31	Sep 24 2 35 21	Sep 30 17 32 59	Oct 7 16 33 19
	Oct 15 20 49 49	Oct 23 12 11 51	Oct 30 3 17 07	Nov 6 9 58 15
	Nov 14 13 42 36	Nov 21 20 26 33	Nov 28 15 10 57	Dec 6 6 28 42
	Dec 14 5 19 17	Dec 21 4 16 22	Dec 28 5 16 45	

   With the tables on the next pages, it is possible to find the instant of any lunar phase between the years −1500 and +2999 with an accuracy of about 10 minutes.

   Write down, from Table 1, the values of "Time", A, B, and C for the century of the given year.
(Thus, for instance, for the year 1967 take the values for 1900). The time is expressed in days
and decimals.

   Write down, from Tables 2 and 3, the values of the same quantities for the year and the date
of the year.

   Add up the numbers obtained, in order to find the date (of the *mean* lunar phase), and the
values of the arguments A, B, C. If the value of any of the quantities A, B, C exceeds 1000,
subtract 1000 or a convenient multiple.

   Compute the quantities $A + B$ and $A - B$. If $A + B$ > 1000, subtract 1000. If $A - B$ < 0, add 1000.

   With the arguments A, B, C, $A + B$, $A - B$, find the five corrections from Tables 4 to 8, respectively. Note that Table 4 is in a certain sense a double-entry table, the value of the correction
being given for the years −1500, 0, +1500, and +3000. Hence, a "double" interpolation should be
performed here.

   From Tables 5 − 8, take the correction from the appropriate column, according to the lunar
phase (NM = New Moon, FQ = First Quarter, FM = Full Moon, LQ = Last Quarter).

   If the final instant should be expressed in UT instead of ET, take the correction from Table 9.

   Add the five (or six) corrections to the instant of the mean phase, in order to obtain the
time of the true phase, expressed in ET (or UT). The fraction of the day can then be converted
into hours and minutes, if this is necessary.

NOTES

• It will be clear to the reader that *no* interpolation is involved in Tables 1, 2, and 3. On the
other hand, in Tables 4 − 9 the corrections *should* be interpolated ; the tabular intervals in these
tables have been taken so small that linear interpolation is always sufficient for our purpose.

• In the case of a negative year, the year number should be split in such a way as to have the
last two figures positive. For example, the year −328 should be split up as −400 + 72.   Then
Table 1 should be consulted for −400, and Table 2 for +72.

• After the addition of the several times and corrections, the final instant may exceed the
number of days in the month. It should be remembered that, for instance, May 32 is the same as
June 1.   In Example II, we obtain April 78 ; this is the same as May 48, or as June 17.

• Care must be given to choose the correct line in Table 3. In Example II, for instance, the sum of the two first times (25.632 and 25.011 days) is already as large as 50 days. Consequently, in order to obtain the New Moon taking place in mid-June, the line corresponding to April 28 should be taken from Table 3. (The line June 26 would give the New Moon of August).

• In the case of a bissextile year, for the months January and February the lines indicated by *(B)* should be used in Table 3.

• If an accuracy of one hour is sufficient, it's not necessary to use Tables 6, 7, and 8. In that case, it's not necessary to calculate the arguments C, $A + B$, $A - B$.

EXAMPLE I. — Find the time of the Full Moon of July 1953.

Table	Argument	Time	A	B	C
1	1900	1.259	998	850	118
2	53	14.066	36	44	777
3	July	10.949	526	966	108
4	560	-0.062	1560	1860	1003
5	860	+0.298	= 560	= 860	= 3
6	3	+0.000			
7	420	-0.002	$A + B$ = 1420 or 420		
8	700	+0.007			
9	1953	0.000	$A - B$ = -300 or 700		
	Sum	26.515			

Hence, the Full Moon took place on 1953 July 26.515, that is on 1953 July 26 at approximately $12^h 22^m$ Universal Time. The instant given by the *American Ephemeris* is $12^h 20^m$ UT.

EXAMPLE II. — Find the time of the New Moon of June 1433.

Table	Argument	Time	A	B	C
1	1400	25.632	113	445	586
2	33	25.011	67	331	690
3	April	28.122	323	287	682
4	503	-0.003	503	1063	1958
5	63	-0.146		= 63	=958
6	958	-0.003			
7	566	+0.002	$A + B$ = 566		
8	440	-0.003	$A - B$ = 440		
	Sum	78.612			

Hence, New Moon took place on 1433 April 78.612 ET, that is 1433 June 17, at approximately $14^h 41^m$ Ephemeris Time.

TABLE 1
VALUES FOR BEGINNING OF CENTURIES

Julian Calendar						Gregorian Calendar				
Year	Time	A	B	C		Year	Time	A	B	C
-1500	18.^{d}082	169	179	797		1500	10.^{d}439	42	83	190
-1400	22.412	178	887	572		1600	14.776	51	793	965
-1300	26.742	188	595	347		1700	20.114	60	502	740
-1200	1.542	116	232	951		1800	25.452	69	212	514
-1100	5.873	125	940	727		1900	1.259	998	850	118
-1000	10.204	134	648	502		2000	5.598	7	560	893
- 900	14.535	144	356	277		2100	10.936	16	270	667
- 800	18.867	153	64	52		2200	16.275	26	980	442
- 700	23.198	162	773	827		2300	21.614	35	690	216
- 600	27.530	171	481	602		2400	25.954	44	400	991
- 500	2.332	100	118	206		2500	1.763	972	38	595
- 400	6.665	109	826	981		2600	7.103	982	749	369
- 300	10.997	118	535	756		2700	12.443	991	459	144
- 200	15.330	127	243	531		2800	16.783	0	169	918
- 100	19.664	137	952	306		2900	22.123	9	880	692
0	23.997	146	661	81						
+ 100	28.331	155	369	855						
200	3.134	83	7	460						
300	7.469	93	715	235						
400	11.803	102	424	10						
500	16.138	111	133	784						
600	20.473	120	842	559						
700	24.809	130	552	334						
800	29.144	139	261	109						
900	3.949	67	898	713						
1000	8.285	76	607	488						
1100	12.622	86	317	262						
1200	16.958	95	26	37						
1300	21.295	104	736	812						
1400	25.632	113	445	586						
1500	0.439	42	83	190						

TABLE 2
VALUES FOR ADDITIONAL YEARS

Year	Time	A	B	C
0 B?	0^d.000	0	0	0
1	18.898	51	932	215
2	8.265	21	793	260
3	27.162	72	725	475
4 B	15.529	42	586	520
5	4.896	13	446	564
6	23.794	64	379	779
7	13.161	34	239	824
8 B	1.528	4	100	869
9	20.426	55	32	84
10	9.793	25	892	129
11	28.691	76	825	344
12 B	17.058	46	685	388
13	6.425	17	546	433
14	25.322	68	478	648
15	14.689	38	339	693
16 B	3.057	8	199	738
17	21.954	59	132	953
18	11.321	29	992	997
19	0.688	999	853	42
20 B	18.586	50	785	257
21	7.953	21	646	302
22	26.851	72	578	517
23	16.218	42	438	562
24 B	4.585	12	299	606
25	23.482	63	231	821
26	12.850	33	92	866
27	2.217	3	952	911
28 B	20.114	54	885	126
29	9.481	25	745	171
30	28.379	76	677	386
31	17.746	46	538	430
32 B	6.113	16	399	475
33	25.011	67	331	690
34	14.378	37	191	735
35	3.745	7	52	780
36 B	21.643	58	984	995
37	11.010	28	845	39
38	0.377	999	705	84
39	19.274	50	638	299

Year	Time	A	B	C
40 B	7^d.641	20	498	344
41	26.539	71	431	559
42	15.906	41	291	604
43	5.273	11	152	648
44 B	23.171	62	84	863
45	12.538	32	945	908
46	1.905	3	805	953
47	20.803	54	737	168
48 B	9.170	24	598	213
49	28.067	75	530	428
50	17.434	45	391	472
51	6.801	15	251	517
52 B	24.699	66	184	732
53	14.066	36	44	777
54	3.433	7	905	822
55	22.331	58	837	37
56 B	10.698	28	698	81
57	0.065	998	558	126
58	18.963	49	490	341
59	8.330	19	351	386
60 B	26.227	70	283	601
61	15.594	40	144	646
62	4.962	11	4	690
63	23.859	62	937	906
64 B	12.226	32	797	950
65	1.593	2	658	995
66	20.491	53	590	210
67	9.858	23	451	255
68 B	27.756	74	383	470
69	17.123	44	243	515
70	6.490	15	104	559
71	25.387	66	36	774
72 B	13.755	36	897	819
73	3.122	6	757	864
74	22.019	57	690	79
75	11.386	27	550	123
76 B	29.284	78	483	339
77	18.651	48	343	383
78	8.018	19	204	428
79	26.916	70	136	643

Year	Time	A	B	C
80 B	15^d.283	40	997	688
81	4.650	10	857	732
82	23.548	61	789	948
83	12.915	31	650	992
84 B	1.282	1	510	37
85	20.179	52	443	252
86	9.546	22	303	297
87	28.444	74	236	512
88 B	16.811	44	96	557
89	6.178	14	957	601
90	25.076	65	889	816
91	14.443	35	750	861
92 B	2.810	5	610	906
93	21.708	56	542	121
94	11.075	26	403	166
95	0.442	997	264	210
96 B	18.339	48	196	425
97	7.706	18	56	470
98	26.604	69	989	685
99	15.971	39	849	730

B = Leap (Bissextile) Year

All the century years are bissextile, except 1700, 1800, 1900, 2100, 2200, 2300, 2500, etc., which are common years.

TABLE   3

VALUES FOR DATES OF THE YEAR

Lunar Phase	Month	Time	A	B	C
NM	Jan.	$0^d.000$	0	0	0
NM	Jan. (B)	1.000	0	0	0
FQ	Jan.	7.383	20	268	543
FQ	Jan. (B)	8.383	20	268	543
FM	Jan.	14.765	40	536	85
FM	Jan. (B)	15.765	40	536	85
LQ	Jan.	22.148	61	804	628
LQ	Jan. (B)	23.148	61	804	628
NM	Jan.	29.531	81	72	170
NM	Jan. (B)	30.531	81	72	170
FQ	Feb.	5.913	101	340	713
FQ	Feb. (B)	6.913	101	340	713
FM	Feb.	13.296	121	608	256
FM	Feb. (B)	14.296	121	608	256
LQ	Feb.	20.679	141	875	798
LQ	Feb. (B)	21.679	141	875	798
NM	March	0.061	162	143	341
FQ	March	7.444	182	411	883
FM	March	14.826	202	679	426
LQ	March	22.209	222	947	969
NM	March	29.592	243	215	511
FQ	April	5.974	263	483	54
FM	April	13.357	283	751	596
LQ	April	20.740	303	19	139
NM	April	28.122	323	287	682
FQ	May	5.505	344	555	224
FM	May	12.888	364	823	767
LQ	May	20.270	384	91	309

Lunar Phase	Month	Time	A	B	C
NM	May	$27^d.653$	404	359	852
FQ	June	4.036	424	626	395
FM	June	11.418	445	894	937
LQ	June	18.801	465	162	480
NM	June	26.184	485	430	22
FQ	July	3.566	505	698	565
FM	July	10.949	526	966	108
LQ	July	18.331	546	234	650
NM	July	25.714	566	502	193
FQ	Aug.	2.097	586	770	735
FM	Aug.	9.479	606	38	278
LQ	Aug.	16.862	627	306	821
NM	Aug.	24.245	647	574	363
FQ	Sept.	0.627	667	842	906
FM	Sept.	8.010	687	110	448
LQ	Sept.	15.393	707	377	991
NM	Sept.	22.775	728	645	534
FQ	Oct.	0.158	748	913	76
FM	Oct.	7.541	768	181	619
LQ	Oct.	14.923	788	449	161
NM	Oct.	22.306	808	717	704
FQ	Oct.	29.689	829	985	247
FM	Nov.	6.071	849	253	789
LQ	Nov.	13.454	869	521	332
NM	Nov.	20.836	889	789	874
FQ	Nov.	28.219	910	57	417
FM	Dec.	5.602	930	325	960
LQ	Dec.	12.984	950	593	502
NM	Dec.	20.367	970	861	45
FQ	Dec.	27.750	990	128	587
FM	Dec.	35.132	11	396	130

(B) = Leap (Bissextile) Year

NM = New Moon

FQ = First Quarter

FM = Full Moon

LQ = Last Quarter

TABLE 4

FIRST CORRECTION

A	Year -1500	Year 0	Year 1500	Year 3000	A	Year -1500	Year 0	Year 1500	Year 3000
0	0^d.000	0^d.000	0^d.000	0^d.000	500	0^d.000	0^d.000	0^d.000	0^d.000
10	+0.012	+0.012	+0.011	+0.011	510	−0.011	−0.011	−0.011	−0.010
20	+0.024	+0.023	+0.022	+0.022	520	−0.023	−0.022	−0.021	−0.021
30	+0.036	+0.035	+0.033	+0.032	530	−0.034	−0.033	−0.032	−0.031
40	+0.047	+0.046	+0.044	+0.043	540	−0.045	−0.044	−0.042	−0.041
50	+0.059	+0.057	+0.055	+0.053	550	−0.056	−0.054	−0.053	−0.051
60	+0.070	+0.068	+0.066	+0.063	560	−0.067	−0.065	−0.063	−0.061
70	+0.081	+0.078	+0.076	+0.073	570	−0.078	−0.075	−0.073	−0.070
80	+0.091	+0.089	+0.086	+0.083	580	−0.088	−0.085	−0.082	−0.079
90	+0.102	+0.098	+0.095	+0.092	590	−0.098	−0.095	−0.092	−0.088
100	+0.111	+0.108	+0.104	+0.101	600	−0.107	−0.104	−0.100	−0.097
110	+0.121	+0.117	+0.113	+0.109	610	−0.117	−0.113	−0.109	−0.105
120	+0.130	+0.125	+0.121	+0.117	620	−0.125	−0.121	−0.117	−0.113
130	+0.138	+0.134	+0.129	+0.125	630	−0.134	−0.129	−0.125	−0.121
140	+0.146	+0.141	+0.136	+0.132	640	−0.141	−0.137	−0.132	−0.128
150	+0.153	+0.148	+0.143	+0.138	650	−0.149	−0.144	−0.139	−0.134
160	+0.159	+0.154	+0.149	+0.144	660	−0.155	−0.150	−0.145	−0.140
170	+0.165	+0.160	+0.155	+0.149	670	−0.161	−0.156	−0.151	−0.146
180	+0.170	+0.165	+0.159	+0.154	680	−0.167	−0.161	−0.156	−0.151
190	+0.175	+0.169	+0.164	+0.158	690	−0.172	−0.166	−0.161	−0.155
200	+0.178	+0.173	+0.167	+0.161	700	−0.176	−0.170	−0.165	−0.159
210	+0.181	+0.176	+0.170	+0.164	710	−0.179	−0.174	−0.168	−0.162
220	+0.184	+0.178	+0.172	+0.166	720	−0.182	−0.176	−0.171	−0.165
230	+0.185	+0.179	+0.174	+0.168	730	−0.184	−0.178	−0.172	−0.167
240	+0.186	+0.180	+0.174	+0.168	740	−0.186	−0.180	−0.174	−0.168
250	+0.186	+0.180	+0.174	+0.168	750	−0.186	−0.180	−0.174	−0.168
260	+0.186	+0.180	+0.174	+0.168	760	−0.186	−0.180	−0.174	−0.168
270	+0.184	+0.178	+0.172	+0.167	770	−0.185	−0.179	−0.174	−0.168
280	+0.182	+0.176	+0.171	+0.165	780	−0.184	−0.178	−0.172	−0.166
290	+0.179	+0.174	+0.168	+0.162	790	−0.181	−0.176	−0.170	−0.164
300	+0.176	+0.170	+0.165	+0.159	800	−0.178	−0.173	−0.167	−0.161
310	+0.172	+0.166	+0.161	+0.155	810	−0.175	−0.169	−0.164	−0.158
320	+0.167	+0.161	+0.156	+0.151	820	−0.170	−0.165	−0.159	−0.154
330	+0.161	+0.156	+0.151	+0.146	830	−0.165	−0.160	−0.155	−0.149
340	+0.155	+0.150	+0.145	+0.140	840	−0.159	−0.154	−0.149	−0.144
350	+0.149	+0.144	+0.139	+0.134	850	−0.153	−0.148	−0.143	−0.138
360	+0.141	+0.137	+0.132	+0.128	860	−0.146	−0.141	−0.136	−0.132
370	+0.134	+0.129	+0.125	+0.121	870	−0.138	−0.134	−0.129	−0.125
380	+0.125	+0.121	+0.117	+0.113	880	−0.130	−0.125	−0.121	−0.117
390	+0.117	+0.113	+0.109	+0.105	890	−0.121	−0.117	−0.113	−0.109
400	+0.107	+0.104	+0.100	+0.097	900	−0.111	−0.108	−0.104	−0.101
410	+0.098	+0.095	+0.092	+0.088	910	−0.102	−0.098	−0.095	−0.092
420	+0.088	+0.085	+0.082	+0.079	920	−0.091	−0.089	−0.086	−0.083
430	+0.078	+0.075	+0.073	+0.070	930	−0.081	−0.078	−0.076	−0.073
440	+0.067	+0.065	+0.063	+0.061	940	−0.070	−0.068	−0.066	−0.063
450	+0.056	+0.054	+0.053	+0.051	950	−0.059	−0.057	−0.055	−0.053
460	+0.045	+0.044	+0.042	+0.041	960	−0.047	−0.046	−0.044	−0.043
470	+0.034	+0.033	+0.032	+0.031	970	−0.036	−0.035	−0.033	−0.032
480	+0.023	+0.022	+0.021	+0.021	980	−0.024	−0.023	−0.022	−0.022
490	+0.011	+0.011	+0.011	+0.010	990	−0.012	−0.012	−0.011	−0.011
500	0.000	0.000	0.000	0.000	1000	0.000	0.000	0.000	0.000

PHASES OF THE MOON

TABLE 5
SECOND CORRECTION

B	NM/FM	FQ/LQ	B	NM/FM	FQ/LQ
0	$0^d.000$	$0^d.000$	500	$0^d.000$	$0^d.000$
10	-0.024	-0.038	510	+0.028	+0.041
20	-0.047	-0.077	520	+0.055	+0.081
30	-0.071	-0.115	530	+0.082	+0.121
40	-0.094	-0.152	540	+0.109	+0.161
50	-0.117	-0.189	550	+0.135	+0.200
60	-0.139	-0.225	560	+0.161	+0.238
70	-0.161	-0.261	570	+0.186	+0.275
80	-0.183	-0.295	580	+0.210	+0.310
90	-0.204	-0.329	590	+0.233	+0.345
100	-0.224	-0.361	600	+0.255	+0.378
110	-0.244	-0.392	610	+0.275	+0.409
120	-0.263	-0.421	620	+0.295	+0.439
130	-0.281	-0.449	630	+0.313	+0.467
140	-0.298	-0.475	640	+0.329	+0.493
150	-0.314	-0.500	650	+0.345	+0.517
160	-0.329	-0.522	660	+0.358	+0.538
170	-0.343	-0.543	670	+0.370	+0.558
180	-0.356	-0.561	680	+0.380	+0.575
190	-0.367	-0.578	690	+0.389	+0.590
200	-0.377	-0.592	700	+0.396	+0.602
210	-0.386	-0.604	710	+0.401	+0.612
220	-0.393	-0.613	720	+0.405	+0.620
230	-0.399	-0.620	730	+0.407	+0.625
240	-0.404	-0.625	740	+0.408	+0.627
250	-0.406	-0.628	750	+0.406	+0.628
260	-0.408	-0.627	760	+0.404	+0.625
270	-0.407	-0.625	770	+0.399	+0.620
280	-0.405	-0.620	780	+0.393	+0.613
290	-0.401	-0.612	790	+0.386	+0.604
300	-0.396	-0.602	800	+0.377	+0.592
310	-0.389	-0.590	810	+0.367	+0.578
320	-0.380	-0.575	820	+0.356	+0.561
330	-0.370	-0.558	830	+0.343	+0.543
340	-0.358	-0.538	840	+0.329	+0.522
350	-0.345	-0.517	850	+0.314	+0.500
360	-0.329	-0.493	860	+0.298	+0.475
370	-0.313	-0.467	870	+0.281	+0.449
380	-0.295	-0.439	880	+0.263	+0.421
390	-0.275	-0.409	890	+0.244	+0.392
400	-0.255	-0.378	900	+0.224	+0.361
410	-0.233	-0.345	910	+0.204	+0.329
420	-0.210	-0.310	920	+0.183	+0.295
430	-0.186	-0.275	930	+0.161	+0.261
440	-0.161	-0.238	940	+0.139	+0.225
450	-0.135	-0.200	950	+0.117	+0.189
460	-0.109	-0.161	960	+0.094	+0.152
470	-0.082	-0.121	970	+0.071	+0.115
480	-0.055	-0.081	980	+0.047	+0.077
490	-0.028	-0.041	990	+0.024	+0.038
500	0.000	0.000	1000	0.000	0.000

TABLE 6		
THIRD CORRECTION		
C	NM/FM	FQ/LQ
0	0.000^d	0.000^d
20	+0.001	+0.001
40	+0.003	+0.002
60	+0.004	+0.003
80	+0.005	+0.004
100	+0.006	+0.005
120	+0.007	+0.005
140	+0.008	+0.006
160	+0.009	+0.007
180	+0.009	+0.007
200	+0.010	+0.008
220	+0.010	+0.008
240	+0.010	+0.008
260	+0.010	+0.008
280	+0.010	+0.008
300	+0.010	+0.008
320	+0.009	+0.007
340	+0.009	+0.007
360	+0.008	+0.006
380	+0.007	+0.005
400	+0.006	+0.005
420	+0.005	+0.004
440	+0.004	+0.003
460	+0.003	+0.002
480	+0.001	+0.001
500	0.000	0.000
520	-0.001	-0.001
540	-0.003	-0.002
560	-0.004	-0.003
580	-0.005	-0.004
600	-0.006	-0.005
620	-0.007	-0.005
640	-0.008	-0.006
660	-0.009	-0.007
680	-0.009	-0.007
700	-0.010	-0.008
720	-0.010	-0.008
740	-0.010	-0.008
760	-0.010	-0.008
780	-0.010	-0.008
800	-0.010	-0.008
820	-0.009	-0.007
840	-0.009	-0.007
860	-0.008	-0.006
880	-0.007	-0.005
900	-0.006	-0.005
920	-0.005	-0.004
940	-0.004	-0.003
960	-0.003	-0.002
980	-0.001	-0.001
1000	0.000	0.000

TABLE 7			
FOURTH CORRECTION			
$A + B$	NM/FM	FQ	LQ
0	0.000^d	$+0.003^d$	-0.003^d
20	-0.001	+0.001	-0.004
40	-0.001	-0.000	-0.006
60	-0.002	-0.002	-0.007
80	-0.002	-0.003	-0.009
100	-0.003	-0.004	-0.010
120	-0.003	-0.005	-0.011
140	-0.004	-0.006	-0.012
160	-0.004	-0.007	-0.013
180	-0.005	-0.008	-0.014
200	-0.005	-0.009	-0.014
220	-0.005	-0.009	-0.014
240	-0.005	-0.009	-0.015
260	-0.005	-0.009	-0.015
280	-0.005	-0.009	-0.014
300	-0.005	-0.009	-0.014
320	-0.005	-0.008	-0.014
340	-0.004	-0.007	-0.013
360	-0.004	-0.006	-0.012
380	-0.003	-0.005	-0.011
400	-0.003	-0.004	-0.010
420	-0.002	-0.003	-0.009
440	-0.002	-0.002	-0.007
460	-0.001	-0.000	-0.006
480	-0.001	+0.001	-0.004
500	0.000	+0.003	-0.003
520	+0.001	+0.004	-0.001
540	+0.001	+0.006	+0.000
560	+0.002	+0.007	+0.002
580	+0.002	+0.009	+0.003
600	+0.003	+0.010	+0.004
620	+0.003	+0.011	+0.005
640	+0.004	+0.012	+0.006
660	+0.004	+0.013	+0.007
680	+0.005	+0.014	+0.008
700	+0.005	+0.014	+0.009
720	+0.005	+0.014	+0.009
740	+0.005	+0.015	+0.009
760	+0.005	+0.015	+0.009
780	+0.005	+0.014	+0.009
800	+0.005	+0.014	+0.009
820	+0.005	+0.014	+0.009
840	+0.004	+0.013	+0.007
860	+0.004	+0.012	+0.006
880	+0.003	+0.011	+0.005
900	+0.003	+0.010	+0.004
920	+0.002	+0.009	+0.003
940	+0.002	+0.007	+0.002
960	+0.001	+0.006	+0.000
980	+0.001	+0.004	-0.001
1000	0.000	+0.003	-0.003

TABLE 8		
FIFTH CORRECTION		
$A - B$	NM/FM	FQ/LQ
0	0.000^d	0.000^d
20	-0.001	-0.001
40	-0.002	-0.001
60	-0.003	-0.002
80	-0.004	-0.002
100	-0.004	-0.003
120	-0.005	-0.003
140	-0.006	-0.004
160	-0.006	-0.004
180	-0.007	-0.004
200	-0.007	-0.004
220	-0.007	-0.005
240	-0.007	-0.005
260	-0.007	-0.005
280	-0.007	-0.005
300	-0.007	-0.004
320	-0.007	-0.004
340	-0.006	-0.004
360	-0.006	-0.004
380	-0.005	-0.003
400	-0.004	-0.003
420	-0.004	-0.002
440	-0.003	-0.002
460	-0.002	-0.001
480	-0.001	-0.001
500	0.000	0.000
520	+0.001	+0.001
540	+0.002	+0.001
560	+0.003	+0.002
580	+0.004	+0.002
600	+0.004	+0.003
620	+0.005	+0.003
640	+0.006	+0.004
660	+0.006	+0.004
680	+0.007	+0.004
700	+0.007	+0.004
720	+0.007	+0.005
740	+0.007	+0.005
760	+0.007	+0.005
780	+0.007	+0.005
800	+0.007	+0.004
820	+0.007	+0.004
840	+0.006	+0.004
860	+0.006	+0.004
880	+0.005	+0.003
900	+0.004	+0.003
920	+0.004	+0.002
940	+0.003	+0.002
960	+0.002	+0.001
980	+0.001	+0.001
1000	0.000	0.000

TABLE 9

CORRECTION FOR REDUCTION

FROM EPHEMERIS TIME TO UNIVERSAL TIME

Year	corr.	Year	corr.	Year	corr.
-1500	$-0^{d}.373$	-500	$-0^{d}.180$	1400	$-0^{d}.005$
-1450	-0.361	-400	-0.164	1500	-0.002
-1400	-0.350	-300	-0.150	1600	-0.001
-1350	-0.339	-200	-0.136	1640	0.000
-1300	-0.328	-100	-0.122	1970	0.000
-1250	-0.318	0	-0.110	1975	-0.001
-1200	-0.307	+100	-0.098	2000	-0.001
-1150	-0.297	200	-0.086	2050	-0.002
-1100	-0.287	300	-0.076	2100	-0.003
-1050	-0.277	400	-0.066	2200	-0.006
-1000	-0.268	500	-0.057	2300	-0.009
- 950	-0.258	600	-0.048	2400	-0.013
- 900	-0.249	700	-0.040	2500	-0.018
- 850	-0.239	800	-0.033	2600	-0.023
- 800	-0.230	900	-0.027	2700	-0.029
- 750	-0.222	1000	-0.021	2800	-0.036
- 700	-0.213	1100	-0.016	2900	-0.043
- 650	-0.204	1200	-0.011	3000	-0.051
- 600	-0.196	1300	-0.008		

5

OCCULTATIONS OF PLANETS AND BRIGHT STARS BY THE MOON

1980 - 2000

OCCULTATIONS 1980 – 2000

GENERAL DATA

Table I, which begins on page 5-23, gives the list of all occultations of first–magnitude stars and of planets by the Moon, which take place during the period 1980-2000. These events are listed chronologically. For reason of completeness, *all* these occultations are given, even those which occur at a small angular distance from the Sun and thus are not observable. For example, the occultation of Neptune on 1992 January 5 will not be visible.

The second column of the Table gives the nearest integer hour T_0 (Ephemeris Time) of the geocentric least angular distance between the center of the Moon and that of the body. The stellar magnitude of the occulted star or planet is given in the fourth column.

The fifth column contains the elongation of the occulted body from the Sun, at the time T_0, in degrees :

E = East from the Sun = visible in the *evening* ;
W = West from the Sun = visible in the *morning*.

In the next column we give the least distance γ from the axis of the "shadow" to the center of the Earth, expressed in units of the Earth's equatorial radius. In fact, γ is the minimum value of the expression

$$\sqrt{x^2 + y^2}$$

where x and y are the Besselian elements which will be defined further. The quantity γ is positive or negative, according to whether the axis of the shadow passes north or south of the center of the Earth. There is occultation for some places on the Earth's surface only when $|\gamma| < 1.270$.

In the last column, the region of visibility on the Earth's surface is briefly described. However, it's important to note that no distinction is made between events taking place during the night and events in full daylight. The descriptions are, of course, necessarily short and not many details can be given ; for example, an occultation stated to be visible in Africa is not necessarily visible from each point of this continent. The letters N, E, S and W signify "northern part of", etc.

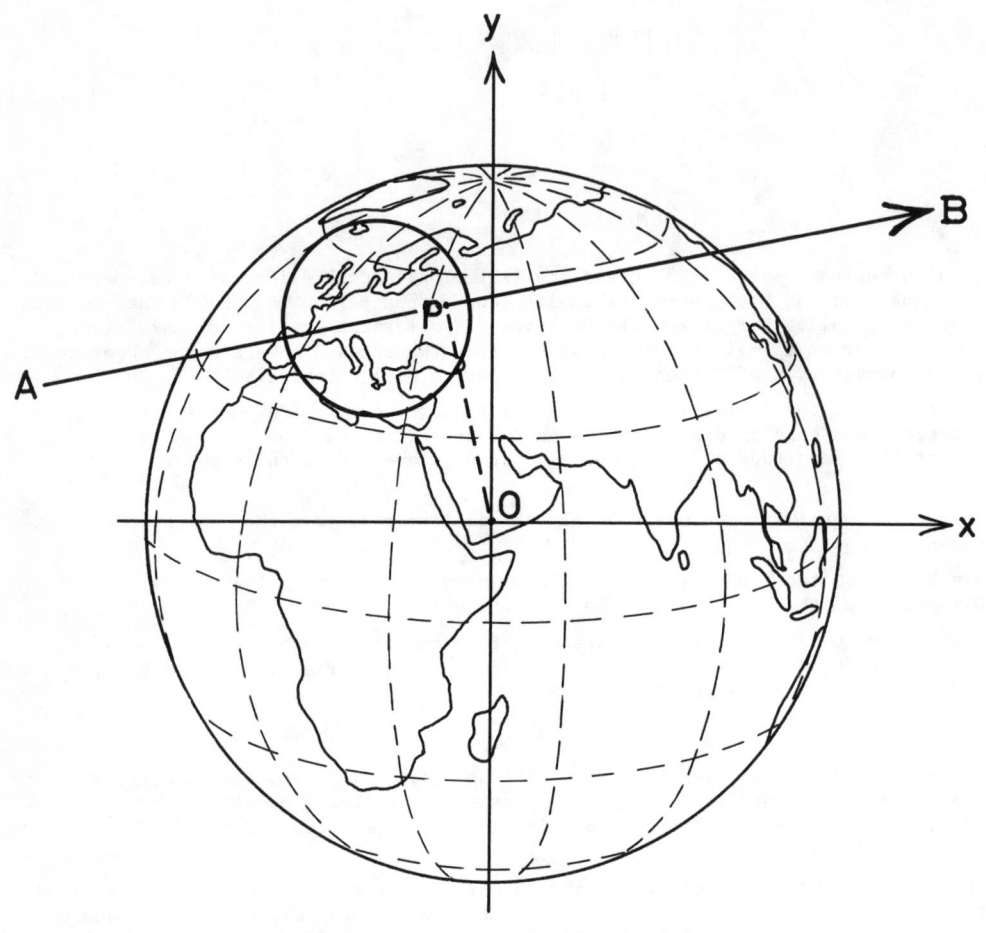

The Earth and the Moon as seen from the occulted body

The Moon's center is moving from A to B. The position of the Moon (the smaller circle) is given for the reference time T_0.

O is the center of the Earth's disk. In the fundamental plane, the coordinates of the center of the Moon (x, y) and those of the observer (ξ, η) are referred to the axes Ox, Oy.

OP is the least distance γ from the Moon's center to the center of the Earth.

BESSELIAN ELEMENTS

The Besselian elements are given in Table II for the occultations of the stars, and in Table III for the occultations of the planets. Their geometric significance is as follows.

Let us imagine a plane passing through the center of the Earth, perpendicular to the line joining the star (or planet) and the center of the Moon. The latter line is the axis of the Moon's "shadow", and the plane is called the *fundamental plane* or plane of xy. We take the intersection of this plane with that of the Earth's equator as the axis of x, and the center of the Earth as the origin of coordinates. The axis of y is perpendicular to that of x, and directed toward the north. Hence, as seen from the occulted body, the x-axis passes through the two points of intersection of the Earth's equator with the "horizon" (the Earth's limb), while the poles project on the y-axis.

Then, at any instant, x and y are the coordinates of the point in which the axis of the "shadow" intersects the fundamental plane. In other words, they are the coordinates of the center of the Moon's disk with respect to the disk of the Earth, as seen from the occulted body. These coordinates are here expressed in terms of the Earth's equatorial radius as unity.

The angle δ is the declination of the occulted body, while H_0 denotes the Greenwich hour angle of this body.

In Tables II and III, T_0 is an *integer* hour (Ephemeris Time), the same as in Table I. The values of δ and H_0 are given for that instant, and they are expressed in degrees and decimals. In the case of the occultation of a star, the declination may be considered as constant during the whole event (that is, its variation δ' is zero), while the hourly variation of H_0 is +15.04107. This is not the case for the planets, since their right ascension and declination are varying : therefore, the hourly variations δ' and H'_0 are given in Table III in degrees, and the value of δ at time T_0 is designated δ_0.

In the case of an occultation of a planet, the "shadow" is not a cylinder but a cone. In order to take this into account, Table III gives for each occultation a quantity k which is a little smaller than 1 ; in the case of an occultation of a star, we have $k = 1$.

The values of x and y at time T_0 are denoted by X and Y. They are given in the last two columns of Tables II and III, together with their hourly variations x', y' and the coefficients x'', y'' of their second-order terms. At any instant t during the occultation, measured in hours from T_0, the values of x and y can be calculated from

$$x = X + x't + x''t^2 \qquad\qquad y = Y + y't + y''t^2$$

EPHEMERIS TIME AND UNIVERSAL TIME

In the present work the uniform time known as *Ephemeris Time* (ET) is used. The Universal Time (UT), necessary for civil life and for astronomical calculations where hour angles are involved, is based on the rotation of the Earth. Because this rotation is slowing down — and, moreover, with unpredictable irregularities — UT is not a uniform time.

The exact value of the difference $\Delta T = ET - UT$ can be deduced only from observations. For the near future, and for prediction purposes, the value $\Delta T = +53$ seconds of time can be used for the year 1982, increasing with one second per year.

See also the *Note on Time Reckoning* at the beginning of this book.

LEAST GEOCENTRIC DISTANCE

The least geocentric distance between the occulted body and the center of the Moon occurs when $x^2 + y^2$ is a minimum, that is when $xx' + yy' = 0$. The instant of this least separation can be obtained, with sufficient accuracy, from

$$t = -\frac{Xx' + Yy' + 3t^2(x'x'' + y'y'')}{x'^2 + y'^2 + 2(Xx'' + Yy'')} \tag{1}$$

where t is measured in hours from the instant T_0. This equation can be solved by iteration, first putting $t = 0$ in the second member. Once t is known, the value of the least distance γ between the axis of the shadow and the center of the Earth, expressed in units of the Earth's equatorial radius, is found from

$$\gamma = \pm \sqrt{x^2 + y^2}$$

where x are y are the values for time t, to be calculated from

$$x = X + x't + x''t^2 \qquad\qquad y = Y + y't + y''t^2$$

and where γ has the same sign as y.

If no high accuracy is required, one may use the approximate formulae

$$t = -\frac{Xx' + Yy'}{x'^2 + y'^2} \qquad \text{and} \qquad \gamma = \frac{Yx' - Xy'}{\sqrt{x'^2 + y'^2}} \tag{2}$$

Example 1. — Occultation of Mercury, 1981 April 3.

From Table III we take the following values :

$$T_0 = 7\,\text{h} \qquad\qquad X = +0.42\,147 \qquad\qquad Y = -0.92\,552$$
$$x' = +0.51\,560 \qquad\qquad y' = +0.18\,016$$
$$x'' = -0.00\,002 \qquad\qquad y'' = +0.00\,011$$

Putting these values in formula (1), we find

$$t = -\frac{0.050\,568 + 0.000\,0285\,t^2}{0.298\,081}$$

Putting $t = 0$ in the second member, we find $t = -0.169\,6452$. Putting now this latter value in the second member, we obtain $t = -0.169\,6479$ which is the correct value. Consequently, the time of least geocentric separation is

$$T_c = T_0 + t = 7^\text{h} - 0^\text{h}.169\,6479$$
$$= 6^\text{h}.830\,3521$$
$$= 6^\text{h}\,49^\text{m}\,49^\text{s}\,\text{ET}$$
$$= 6^\text{h}\,48^\text{m}\,57^\text{s}\,\text{UT}$$

For $t = -0.169\,6479$, we find $x = +0.33\,400$, $y = -0.95\,608$, whence $\gamma = -1.01\,274$.

The approximate formulae (2) give $t = -0.16\,952$, whence $T_c = 6^\text{h}\,49^\text{m}\,50^\text{s}\,\text{ET}$, and $\gamma = -1.01\,274$.

GEOCENTRIC CONJUNCTION IN RIGHT ASCENSION

At the time of the geocentric conjunction in right ascension between the occulted body and the center of the Moon, we have $x = 0$. The instant of the geocentric conjunction in right ascension can be found from

$$t = - \frac{X + x''t^2}{x'} \qquad (3)$$

or, when no high accuracy is required, from $t = -X/x'$. Again, equation (3) can be solved by iteration, by first putting $t = 0$ in the second member.

Example 2. — Occultation of Mercury, 1981 April 3. As in Example 1, we have

$$T_o = 7\text{ h} \qquad X = +0.42\,147 \qquad Y = -0.92\,552$$
$$x' = +0.51\,560 \qquad y' = +0.18\,016$$
$$x'' = -0.00\,002 \qquad y'' = +0.00\,011$$

Putting these values into formula (3), we find

$$t = - \frac{0.42147 - 0.00002\,t^2}{0.51560}$$

Putting $t = 0$ in the second member, this formula gives $t = -0.817\,436$. Putting this latter value in the second member, we find $t = -0.817\,410$. A new iteration gives the value $-0.817\,410$ again. Hence, this is the correct value, and the time of geocentric conjunction in right ascension is

$$
\begin{aligned}
T_\alpha = T_o + t &= 7^\text{h} - 0^\text{h}.817\,410 \\
&= 6^\text{h}.182\,590 \\
&= 6^\text{h}\,10^\text{m}\,57^\text{s}\ \text{ET} \\
&= 6^\text{h}\,10^\text{m}\,05^\text{s}\ \text{UT}
\end{aligned}
$$

The value of y at the instant T_α is found by putting $t = -0.817\,410$ in the formula $y = Y + y't + y''t^2$. This gives $y_\alpha = -1.07\,271$.

An approximate value of y_α is given by $y_\alpha \simeq \dfrac{Yx' - Xy'}{x'}$, which gives $y_\alpha = -1.07\,279$ in this case.

LOCAL CIRCUMSTANCES

Let λ be the observer's geographical longitude, in degrees, measured westward from Greenwich, and ϕ his geographical latitude, positive (negative) in the northern (southern) hemisphere. Let z be the observer's height above sea-level, expressed in *meters*.

If not already available, calculate the geocentric rectangular coordinates of the observer as follows :

$$\kappa = 0.996\,647\,19 \qquad\qquad \tan w = \kappa \tan \phi$$

$$\rho \sin \phi' = \kappa \sin w + \frac{z}{6\,378\,140} \sin \phi$$

$$\rho \cos \phi' = \cos w + \frac{z}{6\ 378\ 140} \cos \phi$$

For an assumed value of t (take $t = 0$ as a first approximation), calculate

$\delta = \delta_0 + \delta't$ (in the case of an occultation of a *planet* only)

$H = H_0 + H'_0 t - \lambda - \Delta T/239.345$

where ΔT is the difference ET − UT expressed in seconds of time. Remember that, in the case of the occultation of a star, we have $H'_0 = +15.04107$. H, the local hour angle of the occulted body, is then expressed in degrees.

$$x = X + x't + x''t^2$$
$$y = Y + y't + y''t^2$$

$$\xi = \rho \cos \phi' \sin H$$
$$\eta = \rho \sin \phi' \cos \delta - \rho \cos \phi' \cos H \sin \delta$$

$$\xi' = 0.26252\ \rho \cos \phi' \cos H$$
$$\eta' = 0.26252\ \xi \sin \delta$$

$$u = x - \xi \qquad\qquad u' = x' - \xi'$$
$$v = y - \eta \qquad\qquad v' = y' - \eta'$$

$$n^2 = u'^2 + v'^2 \qquad (n > 0)$$

$$\tau = -\frac{uu' + vv'}{n^2}$$

Then, τ is the correction to t. That is, add τ algebraically to t in order to obtain a better value of t.

Repeat the whole calculation again and again, until t not longer varies, that is until τ is negligible, say less than 0.000 001. In practice, two or three iterations will suffice.

The time of least distance between the body and the Moon's center, as seen from the given place, is then $T_0 + t$, in hours (Ephemeris Time), where t is the final value for t. Subtract ΔT in order to obtain the instant in Universal Time.

The position angle P of the star at the time of nearest approach may be found from

$$\tan P = u/v , \tag{4}$$

$\cos P$ having the opposite sign of v, and where u and v are taken from the last iteration. This position angle P is measured in the classical way, namely from the North Point of the Moon's disk, towards the East, South and West.

The value of the least distance Δ between the body and the center of the Moon's disk, the Moon's radius being taken as unity, is then

$$\Delta = \frac{uv' - u'v}{n} \times \frac{k}{0.272\ 495} \tag{5}$$

where, again, u, v, u', etc. are taken from the last iteration. Remember that $k = 1$ in the case of the occultation of a star.

If $|\Delta| > 1$, there is no occultation for the given place.

If there is an occultation for the given place, *approximate* values of t corresponding to the immersion and to the emersion are obtained by subtracting or adding, to the final value of t found above, the quantity

$$\frac{0.272\ 495}{k} \times \frac{\sqrt{1 - \Delta^2}}{n} \tag{6}$$

To calculate the exact times of immersion or emersion, start from the approximate value of t just found, for the immersion and the emersion *separately*. Then calculate, as above, the values of

$$\delta, \quad H, \quad x, \quad y, \quad \xi, \quad \eta, \quad \xi', \quad \eta', \quad u, \quad v, \quad u', \quad v', \quad n \text{ and } \Delta.$$

The correction τ to t is then

$$\tau = -\frac{uu' + vv'}{n^2} \mp \frac{0.272\,495}{k} \frac{\sqrt{1 - \Delta^2}}{n} \tag{7}$$

where the upper sign is to be taken for immersion, the lower sign for emersion.

Repeat the calculation until t not longer varies. The Ephemeris Time of immersion or emersion is then $T_0 + t$, in hours, where t is the final value for t. Again, subtract ΔT in order to obtain the instant in Universal Time.

The position angle P of ingress or egress can be found by means of formula (4).

The altitude h of the body above the observer's horizon, at any instant t, is given by

$$\sin h = \sin \delta \sin \phi + \cos \delta \cos \phi \cos H, \tag{8}$$

where δ and H are obtained as a function of t, as indicated above. Of course, for the altitude at the time of immersion, least distance or emersion, take t from the last iteration. Evidently, the event is not visible when h is negative. Moreover, it is not visible if it occurs in daylight, except when the body is bright enough and not too near to the Sun.

Finally, it should be pointed out that, in the case of the occultation of a planet, the calculated instants and distances refer to the *center* of that body.

Example 3. — Occultation of Saturn, 1980 June 19.

Calculate the circumstances at the Uccle Observatory, Belgium, for which we have

$$\lambda = -4°21'29'' = -4°.3581$$
$$\phi = +50°47'55'' = +50°.7986$$

$$\rho \sin \phi' = +0.771\,306$$
$$\rho \cos \phi' = +0.633\,333$$

We take the elements of the occultation from Table III :

$T_0 = 19\,h$	$H_0 = 20°.6677$	$X = -0.10\,627$	$Y = +0.39\,517$
$\delta_0 = +5°.6463$	$H'_0 = +15°.0393$	$x' = +0.51\,663$	$y' = -0.17\,514$
$\delta' = -0°.0009$	$k = 0.99972$	$x'' = 0$	$y'' = -0.00\,001$

We will use the value $\Delta T = +51$ seconds.

Starting with the value $t = 0$, we find successively :

δ	+ 5°.6463	η	+0.71\,100	u'	+0.36\,572
H	+24°.8127	ξ'	+0.15\,091	v'	-0.18\,200
x	-0.10\,627	η'	+0.00\,686	n^2	0.166\,875
y	+0.39\,517	u	-0.37\,205	τ	+0.470\,921
ξ	+0.26\,578	v	-0.31\,583		

Hence, the improved value of t is

$$t = 0 + 0.470\,921 = +0.470\,921$$

Starting with this new value, we obtain :

δ	+5°.6459	η	+0.71 466	u'	+0.37 547
H	+31°.8950	ξ'	+0.14 116	v'	−0.18 378
x	+0.13 702	η'	+0.00 864	n^2	0.174 754
y	+0.31 269	u	−0.19 761	τ	+0.001 834
ξ	+0.33 463	v	−0.40 197		

Consequently, the new improved value of t is

$$t = +0.470\,921 + 0.001\,834 = +0.472\,755$$

A new calculation, performed with this new value of t, gives the correction $\tau = +0.000\,032$, so that the final value of t is

$$t = +0.472\,755 + 0.000\,032 = +0.472\,787$$

The time of least distance is thus

$$t_m = T_0 + t = 19^h.472\,787$$
$$= 19^h\,28^m\,22^s \text{ ET}$$
$$= 19^h\,27^m\,31^s \text{ UT}$$

The values taken from the last iteration are

u	−0.19 692
v	−0.40 231
u'	+0.37 551
v'	−0.18 379
n	0.41 808

whence, by formula (4), $P = 26°.08$. Finally, formula (5) gives $\Delta = +1.6433$. Because $|\Delta| > 1$, Saturn is not occulted by the Moon as seen from Uccle. Because Δ is positive, as seen from Uccle Saturn passes to the north of the center of the Moon's disk.

Formula (8) gives $h = +38°$, which shows that the conjunction is visible at Uccle, although it occurs half an hour before sunset.

Example 4. — Occultation of Aldebaran, 1997 October 19, at Palomar Mountain Observatory, for which we have

$$\lambda = +116°.8640 \qquad \rho \sin \phi' = +0.54\,686$$
$$\phi = +33°.3562 \qquad \rho \cos \phi' = +0.83\,634$$
$$z = 1706 \text{ m}$$

The elements of the occultations are

$T_0 = 9$ h	$H'_0 = +15°.04107$	$X = -0.08\,570$	$Y = +0.34\,383$
$\delta = +16°.5035$	$k = 1$	$x' = +0.58\,737$	$y' = +0.07\,656$
$H_0 = 93°.9237$	$\Delta T = +68$ s	$x'' = -0.00\,002$	$y'' = -0.00\,002$

Proceeding as in Example 3, we find that the value of t corresponding to the least topocentric distance is $t = -0.6027$, and that the least separation is $\Delta = +0.0982$. Because $|\Delta|$ is less than 1, there is an occultation of Aldebaran for the given place.

The value of n taken from the last iteration is 0.41652. The approximate value of the semiduration, given by formula (6), is then 0.6511 hour. Consequently, approximate values of t for the immersion and the emersion are :

immersion : $-0.6027 - 0.6511 = -1.2538$
emersion : $-0.6027 + 0.6511 = +0.0484$

Let us calculate the circumstances of the immersion. We thus start with the value $t = -1.2538$, and we find successively :

H	$-42°.0829$	ξ'	$+0.16\ 295$	v'	$+0.11\ 836$
x	$-0.82\ 218$	η'	$-0.04\ 180$	n^2	$+0.194\ 141$
y	$+0.24\ 781$	u	$-0.26\ 166$	n	$+0.44\ 061$
ξ	$-0.56\ 052$	v	$-0.10\ 019$	Δ	$+0.096\ 220$
η	$+0.34\ 800$	u'	$+0.42\ 442$		

Formula (7) then gives $\tau = +0.017\ 528$, whence

$$t = -1.2538 + 0.017\ 528 = -1.236\ 272$$

Starting with this improved value of t, and proceeding in the same manner as above, we obtain $\tau = +0.000\ 014$. Consequently, the correct value of t is

$$t = -1.236\ 272 + 0.000\ 014 = -1.236\ 258$$

and the time of immersion is

$$T_o + t = 9^h - 1^h.236258 = 7^h.763742$$
$$= 7^h\ 45^m\ 49^s\ \text{ET}$$
$$= 7^h\ 44^m\ 41^s\ \text{UT}$$

The values from the last iteration are

$$H = -41°.8193$$
$$u = -0.254\ 222$$
$$v = -0.098\ 122$$

Using formulae (4) and (8), we find that the star's position angle and altitude, at the instant of immersion, are

$$P = 68°.90 \qquad \text{and} \qquad h = 48°.85$$

Starting from $t = +0.0484$ we find, in a similar manner, the circumstances of the emersion :

$$9^h\ 02^m\ 44^s\ \text{UT} \qquad P = 260°.64 \qquad h = 63°.82$$

GRAZING OCCULTATIONS

If γ is between -0.725 and $+0.725$, the region of visibility of the occultation on the Earth's surface is limited by a *northern limit* and a *southern limit*. An observer situated exactly on one of these lines will see the star grazing the northern or the southern limb of the Moon.

These limits run approximately from West to East, over thousands of kilometers over the Earth's surface. On one side of such a line an occultation of the star or planet takes place, while on the other side the body is not occulted by the Moon.

If γ is between $+0.725$ and $+1.270$, only the southern limit does exist, the northern one being replaced by the curve "star on the horizon". Similarly, only the northern limit does exist if γ is between -1.270 and -0.725.

Points of the northern or southern limit of an occultation can be calculated as follows, using the Besselian elements given in the present publication.

For a given longitude λ, start from approximate values for the latitude ϕ and for the time t, the latter being, as before, measured in hours from the time T_0. Good choices are $\phi = 0°$ and $t = 0$. Then calculate

$$H = H_0 + H'_0 t - \lambda - \Delta T/239.345$$
$$\delta = \delta_0 + \delta' t$$

where, as before, $\Delta T = ET - UT$ is expressed in seconds of time. In the case of an occultation of a star, δ is already given by Table II. Then calculate further :

$$\boldsymbol{x} = X + x't + x''t^2$$
$$y = Y + y't + y''t^2$$

$$\tan w = 0.996\,647\,19 \tan \phi$$

$$\rho \sin \phi' = 0.996\,647\,19 \sin w$$
$$\rho \cos \phi' = \cos w$$

$$\xi = \rho \cos \phi' \sin H$$
$$\eta = \rho \sin \phi' \cos \delta - \rho \cos \phi' \cos H \sin \delta$$

$$\xi' = 0.26252 \, \rho \cos \phi' \cos H$$
$$\eta' = 0.26252 \, \xi \sin \delta$$

$$u = x - \xi \qquad u' = x' - \xi'$$
$$v = y - \eta \qquad v' = y' - \eta' \qquad n^2 = u'^2 + v'^2 \quad (n > 0)$$

$$\Delta = \frac{uv' - u'v}{n} \times \frac{k}{0.272\,495}$$

$$q = \frac{0.064\,311}{n}(v'\rho \sin \phi' \sin H + u'\rho \sin \phi' \cos H \sin \delta$$
$$+ 0.99328 \, u' \rho \cos \phi' \cos \delta)$$

Then the correction to the assumed value of t is

$$\tau = -\frac{uu' + vv'}{n^2} \quad \text{hours}$$

and the correction to the latitude ϕ is

$$\Delta \phi = \frac{\Delta_0 - \Delta}{q} \quad \text{degrees}$$

where $\Delta_0 = +1$ for a northern limit, and $\Delta_0 = -1$ for a southern limit.

Thus, the new values for the instant and the latitude are $t + \tau$ and $\phi + \Delta\phi$, respectively.

The whole calculation is then repeated with these new values, until the corrections τ and $\Delta\phi$ can be neglected, for instance until they are less than 0.0001 hour and 0.0001 degree, respectively.

After the last iteration, the position angle P of the point of graze is given by

$$\tan P = -v'/u' \tag{9}$$

$\cos P$ having the same sign as Δ_0, while the altitude above the horizon can be found by means of formula (8). For the calculation of these quantities, use the values of u', v', ϕ, δ and H from the last iteration.

Similar calculations can then be performed for other longitudes λ.

In the case of the occultation of a planet, the limit calculated in this manner refers to the *center* of this body. It should further be pointed out that the calculated limits refer to sea level.

Example 5. — Southern limit of the occultation of Venus, 1980 October 5.

We will make the calculation for the longitude $\lambda = -1°$. The elements of the occultation are :

T_0 = 7 h	H_0 = 326°.5667	X = +0.25 048	Y = +0.78 208
δ_0 = +11°.5787	H'_0 = +14°.9944	x' = +0.48 482	y' = -0.13 388
δ' = -0°.0143	k = 0.99737	x'' = -0.00 003	y'' = -0.00 011

We use the value ΔT = +51 seconds. Starting with t = 0 and ϕ = 0°, we find :

H	327°.3536	ξ	-0.53 945	v'	-0.10 546
δ	+11°.5787	η	-0.16 900	n^2	0.080 698
x	+0.25 048	ξ'	+0.22 105	n	0.284 073
y	+0.78 208	η'	-0.02 842	Δ	-4.30 566
w	0.00 000	u	+0.78 993	q	+0.058 107
$\rho \sin \phi'$	0.00 000	v	+0.95 108	τ	-1.33 916
$\rho \cos \phi'$	1.00 000	u'	+0.26 377	$\Delta\phi$	+56°.8891

Hence, improved values for the time and for the latitude are

$$t = 0 - 1.33916 = -1^h.33916$$
$$\phi = 0 + 56.8891 = +56°.8891$$

Starting with these new values, still better ones are obtained, and so on. The values obtained after the successive iterations are as follows :

t	ϕ
0^h	$0°$
-1.33916	+56.8891
-1.28806	+52.5508
-1.36963	+52.6603
-1.36921	+52.6565
-1.36927	+52.6565

Hence, for $\lambda = -1°$ we have $t = -1^h.36927$, which corresponds to

$$T_0 + t = 5^h.63073 = 5^h 37^m 51^s \text{ ET} = 5^h 37^m 00^s \text{ UT} ,$$

and $\phi = +52°.6565$.

The values taken from the last iteration are

u' = +0.38 918	v' = -0.10 820	δ = +11°.5983	H = 306°.8222

and formulae (9) and (8) then give $P = 195°.54$ and $h = 31°.06$.

In a similar manner, other points of the southern limit can be calculated. For instance, the following values are found :

Longitude	Latitude	UT	Pos. Angle of contact	Altitude of Venus
			°	°
+2°	+52°.6383	$5^h 33^m 51^s$	195.05	28.94
+1	+52.6508	5 34 52	195.22	29.65
0	+52.6569	5 35 55	195.38	30.35
-1	+52.6565	5 37 00	195.54	31.06
-2	+52.6494	5 38 06	195.71	31.77
-3	+52.6355	5 39 14	195.88	32.48
-4	+52.6148	5 40 24	196.05	33.19
-5	+52.5871	5 41 35	196.23	33.89

CALCULATION OF THE BESSELIAN ELEMENTS

For the years 1980 to 1984, the elements of the occultations given in the present work have been calculated on the base of the lunar and planetary ephemerides published in the Soviet almanac *Astronomicheskii Ezhegodnik*. For the years 1985 to 2000 use was made of : *(i)* positions of the Moon obtained from the Vienna Planetarium, Austria, which are based on data from Dr. J. Vondrák, Astronomical Institute, Prague, Czechoslovakia ; *(ii)* unpublished planetary ephemerides calculated in 1976 by Edwin Goffin, Public Observatory of Antwerp, Belgium.

If α_* and δ_* are the apparent right ascension and declination of the star, α and δ the same geocentric coordinates of the Moon, and π the Moon's equatorial horizontal parallax, then we have

$$x = \frac{\cos \delta \sin (\alpha - \alpha_*)}{\sin \pi}$$

$$y = \frac{\sin \delta \cos \delta_* - \cos \delta \sin \delta_* \cos (\alpha - \alpha_*)}{\sin \pi}$$

$$H_0 = \theta - \alpha_*$$

where θ is the *ephemeris sidereal time*, that is the hour angle of the vernal equinox referred to the *ephemeris* meridian ; its value at time T Ephemeris Time is numerically equal to the value of the sidereal time at time T Universal Time for the Greenwich meridian.

If the occulted body is a planet, $\sin \pi$ in the formulae above should be replaced by $\sin (\pi - \pi')$, where π' is the planet's horizontal parallax. Moreover, it will then be necessary to take into account the variations of the planet's right ascension and declination. Finally, in the case of an occultation of a planet, one should also calculate the quantity

$$k = \frac{\pi - \pi'}{\pi}$$

For the present work, the values of x and y have been calculated for three instants at intervals of two hours, namely for $T_0 - 2$ hours, T_0 and $T_0 + 2$ hours. If the values of x for these three instants are designated by A, B, C, respectively, we have

$$x' = \frac{C - A}{4} \qquad x'' = \frac{A + C - 2B}{8}$$

with similar formulae for y' and y''.

Finally, it should be pointed out that the Besselian elements given in the present publication are "improved" elements in comparison with the "classical" ones which were published, for example, in the *American Ephemeris* before 1960. Our elements are improved by reason of the inclusion of the second-order terms x'' and y'' and, in the case of a planet, of the quantities δ', H'_0 and k.

A further difference is that in the classical elements T_0 is the instant of geocentric conjunction in right ascension, while here T_0 is any *integer* hour (however, choosen as being the integer hour nearest to the middle of the occultation).

HP-67 Programs

The enclosed programs A and B have been written by Jean Meeus for the programmable pocket calculator HP-67. With these programs it is possible to calculate local circumstances, using the Besselian elements given in the present publication.

Program A should be used for the occultation of a star, and program B for the occultation of a planet.

In each case, store in advance :

- in Register 0, the observer's geographical latitude ϕ (in degrees) ;
- in Register 1, the quantity $\rho \sin \phi'$;
- in Register 2, the quantity $\rho \cos \phi'$;
- in Register 3, the observer's geographical longitude λ (in degrees), negative if East of Greenwich ;
- in Register 4, the value of ΔT ($= ET - UT$) in seconds of time.

Then, enter the Besselian elements as follows :

Program A (stars)				Program B (planets)		
T_0	press key A			T_0	press key A	
δ	R/S			δ_0	R/S	
H_0	R/S			δ'	R/S	
X	.			H_0		
x'	:			H_0'	:	
x''	.			k	.	
Y				X		
y'				x'		
y''				x''		
				Y		
				y'		
				y''		

Finally, the number 3.6698 is displayed. Then press :

> key B for the immersion,
> key C for the least distance,
> key D for the emersion.

The machine then displays successively :

- the instant (UT), in hours, minutes, seconds ;
- in the case C only, the least distance in units of the Moon's radius ;
- the position angle P, in degrees ;
- the star's or planet's altitude h, in degrees.

If one of the keys B, C or D has been pressed, after the calculation another of these keys can be pressed ; it is not necessary to enter the data again. Similarly, the calculation of the same occultation can be performed for another place by simply changing the contents of the Registers 0 to 3.

PROGRAM A

001 f LBL A	046 0	091 RCL 2	136 RCL 6	181 STO + 5
002 STO B	047 7	092 ×	137 RCL 9	182 f LBL 8
003 R/S	048 ×	093 RCL A	138 g $\to$ P	183 f DSZ
004 STO A	049 RCL 3	094 f COS	139 h $1/x$	184 GTO 7
005 7	050 −	095 RCL 1	140 f P$\leftrightarrow$S	185 RCL B
006 h ST I	051 RCL 4	096 ×	141 STO 0	186 RCL 5
007 f P$\leftrightarrow$S	052 2	097 −	142 f P$\leftrightarrow$S	187 +
008 f LBL 4	053 3	098 f P$\leftrightarrow$S	143 g x^2	188 RCL 4
009 R/S	054 9	099 RCL 3	144 RCL 7	189 6
010 STO (i)	055 .	100 +	145 RCL 9	190 0
011 f DSZ	056 3	101 RCL 2	146 ×	191 g x^2
012 GTO 4	057 4	102 RCL D	147 RCL 8	192 :
013 f P$\leftrightarrow$S	058 5	103 ×	148 RCL 6	193 −
014 3	059 :	104 +	149 ×	194 g $\to$H.MS
015 .	060 −	105 RCL D	150 +	195 f -x-
016 6	061 f P$\leftrightarrow$S	106 g x^2	151 ×	196 RCL 7
017 6	062 RCL 7	107 RCL 1	152 STO − 5	197 RCL 8
018 9	063 +	108 ×	153 h F? 1	198 g $\to$ P
019 7	064 STO C	109 +	154 GTO 8	199 RCL E
020 9	065 f SIN	110 f P$\leftrightarrow$S	155 RCL 7	200 ×
021 STO E	066 f P$\leftrightarrow$S	111 STO 8	156 RCL 6	201 h F? 1
022 h RTN	067 RCL 2	112 f P$\leftrightarrow$S	157 ×	202 f -x-
023 f LBL C	068 ×	113 RCL 5	158 RCL 9	203 h $\downarrow$
024 h SF 1	069 CHS	114 f P$\leftrightarrow$S	159 RCL 8	204 1
025 GTO 3	070 STO 6	115 RCL 2	160 ×	205 8
026 f LBL B	071 RCL 5	116 .	161 −	206 0
027 h SF 0	072 STO D	117 2	162 RCL E	207 +
028 h CF 1	073 f P$\leftrightarrow$S	118 6	163 ×	208 f -x-
029 GTO 3	074 RCL 5	119 2	164 f P$\leftrightarrow$S	209 RCL C
030 f LBL D	075 ×	120 5	165 RCL 0	210 f COS
031 h CF 0	076 +	121 2	166 f P$\leftrightarrow$S	211 RCL A
032 h CF 1	077 RCL 6	122 STO × 6	167 ×	212 f COS
033 f LBL 3	078 +	123 ×	168 g x^2	213 ×
034 0	079 RCL D	124 RCL C	169 1	214 RCL 0
035 STO 5	080 g x^2	125 f COS	170 h $x\leftrightarrow y$	215 f COS
036 3	081 RCL 4	126 ×	171 −	216 ×
037 h ST I	082 ×	127 −	172 f $\sqrt{}$	217 RCL A
038 f LBL 7	083 +	128 STO 9	173 h F? 0	218 f SIN
039 RCL 5	084 f P$\leftrightarrow$S	129 RCL A	174 CHS	219 RCL 0
040 1	085 STO 7	130 f SIN	175 f P$\leftrightarrow$S	220 f SIN
041 5	086 RCL A	131 STO × 6	176 RCL 0	221 ×
042 .	087 f SIN	132 f P$\leftrightarrow$S	177 ×	222 +
043 0	088 RCL C	133 RCL 2	178 RCL E	223 g SIN^{-1}
044 4	089 f COS	134 f P$\leftrightarrow$S	179 :	224 h RTN
045 1	090 ×	135 STO + 6	180 f P$\leftrightarrow$S	

PROGRAM B

001	f LBL 5	046	f LBL 3	091	f P↔S	136	h 1/x	181	f P↔S
002	RCL A	047	0	092	RCL 9	137	f P↔S	182	RCL 9
003	×	048	STO 5	093	RCL 8	138	STO 6	183	+
004	+	049	4	094	RCL 7	139	f P↔S	184	f P↔S
005	RCL A	050	h ST I	095	f GSB 5	140	g x^2	185	RCL 4
006	×	051	f LBL 7	096	RCL D	141	RCL 0	186	6
007	+	052	RCL 5	097	f COS	142	RCL E	187	0
008	h RTN	053	f P↔S	098	RCL 1	143	×	188	g x^2
009	f LBL A	054	RCL 5	099	×	144	RCL A	189	:
010	9	055	×	100	−	145	RCL B	190	−
011	h ST I	056	RCL 6	101	RCL C	146	×	191	g →H.MS
012	h ↓	057	+	102	f COS	147	+	192	f -x-
013	f P↔S	058	f P↔S	103	RCL 2	148	×	193	RCL E
014	f LBL 0	059	RCL 3	104	×	149	f P↔S	194	RCL A
015	STO (i)	060	−	105	STO 6	150	STO − 5	195	g → P
016	R/S	061	RCL 4	106	RCL D	151	h F? 1	196	f P↔S
017	f DSZ	062	2	107	f SIN	152	GTO 8	197	RCL 4
018	GTO 0	063	3	108	×	153	f P↔S	198	×
019	f P↔S	064	9	109	+	154	RCL E	199	h F? 1
020	STO 9	065	.	110	STO A	155	RCL B	200	f -x-
021	R/S	066	3	111	RCL 8	156	×	201	h ↓
022	STO 8	067	4	112	RCL B	157	RCL 0	202	1
023	R/S	068	5	113	3	158	RCL A	203	8
024	STO 7	069	:	114	.	159	×	204	0
025	f P↔S	070	−	115	8	160	−	205	+
026	3	071	STO C	116	1	161	RCL 4	206	f P↔S
027	.	072	f SIN	117	:	162	×	207	f -x-
028	6	073	RCL 2	118	RCL D	163	f P↔S	208	RCL C
029	6	074	×	119	f SIN	164	RCL 6	209	f COS
030	9	075	STO B	120	×	165	×	210	RCL D
031	7	076	RCL 5	121	−	166	g SIN⁻¹	211	f COS
032	9	077	STO A	122	STO B	167	f COS	212	×
033	STO × 4	078	f P↔S	123	RCL 6	168	RCL 6	213	RCL 0
034	f P↔S	079	RCL 7	124	3	169	×	214	f COS
035	h RTN	080	×	125	.	170	f P↔S	215	×
036	f LBL C	081	RCL 8	126	8	171	RCL 4	216	RCL 0
037	h SF 1	082	+	127	1	172	:	217	f SIN
038	GTO 3	083	STO D	128	:	173	f P↔S	218	RCL D
039	f LBL B	084	RCL 3	129	f P↔S	174	h F? 0	219	f SIN
040	h SF 0	085	RCL 2	130	RCL 2	175	CHS	220	×
041	h CF 1	086	RCL 1	131	h x↔y	176	STO + 5	221	+
042	GTO 3	087	f GSB 5	132	−	177	f LBL 8	222	g SIN⁻¹
043	f LBL D	088	RCL B	133	STO 0	178	f DSZ	223	h RTN
044	h CF 0	089	−	134	RCL B	179	GTO 7		
045	h CF 1	090	STO E	135	g → P	180	RCL 5		

HP-41C PROGRAM

The following program, written by Lionel Ancelet (Paris, France) for the HP-41C calculator, enables you to calculate, for a star or a planet : the distance of nearest approach
as seen from a given place, the points of the northern or southern limit for a grazing
occultation, and the local circumstances of an ordinary occultation. In the three cases,
the UT instant of the phenomenon, the position angle P and the altitude h above the horizon are given.

The procedure is as follows :

1. Load the program into the calculator (at least 2 memory modules are required to run
   this program).

2. Verify that SIZE $\geqslant$ 036, that the calculator is in DEG mode, and that no functions or
   programs are assigned to the following keys : $\Sigma+$, $1/x$, $\sqrt{x}$, LOG, LN, COS, TAN.

3. Select the occulted body : XEQ "OCS" for a star, XEQ "OCP" for a planet. The calculator
   will then ask for the following data : T_o, ΔT, H_o, H', δ_o, δ', X, x', x'', Y, y', y'', k.
   (In the case of a star, the quantities H', δ' and k are not asked for by the program,
   which will take $H' = 15.04107$, $\delta' = 0$ and $k = 1$ by itself). Notice that, as the calculator
   is unable to display characters such as Δ or δ, the exact display contents will be :
   T0=?, DT=?, H0=?, H1=?, d0=?, d1=?, X=?, X1=?, X2=?, Y=?, Y1=?, Y2=?, K=?.

4. Once all the required data have been entered, it is possible to make five different
   computations : least distance, immersion, emersion, northern limit, southern limit.

   a) *Least distance for a given place :* Enter the latitude ϕ (°.'") and press A, enter
   the longitude λ (°.'") and press B. Then press C. After a while the machine will display Δ, preceded by "L.D.=", and expressed in lunar radii, positive if the occulted
   body passes to the north of the center of the Moon, negative if it passes to the south.
   If $|\Delta|$ is larger than 1, there is no occultation for the given place.
     Pressing R/S will then give the Universal Time of the nearest approach. Pressing R/S
   again will give the corresponding position angle, and pressing R/S once more will give
   the altitude of the body above the local horizon.

   b) *Immersion or Emersion for a given place :* After having entered ϕ and λ as above,
   press key D to calculate the circumstances of the immersion, or key E for the emersion.
   After completing three iterations, the calculator then gives the Universal Time of the
   immersion or emersion and, as above, pressing R/S twice will provide the position angle
   and the altitude of the occulted body.

   c) *Northern and southern limits :* Enter the chosen longitude (negative if East from
   Greenwich), press key B. Then enter the initial value of the latitude, and press A.
   (This input is optional).
     Press I to compute the corresponding latitude of the northern limit, or J for the
   southern limit. After a while the machine will display LAT = ..., thus giving the answer,
   unless the limit doesn't exist, in which case the machine will display "ERROR".
     After the latitude has been displayed, press R/S three times to get UT, P, and h.

Remarks

• After computing the local circumstances of nearest approach, immersion or emersion, it's
  not necessary to re-enter the latitude and the longitude of the given place to make
  another calculation for that place. However, the user must be aware that pressing I or J
  will modify the latitude, since the aim of the calculation is precisely to find a
  latitude !
    Anyway, it will never be necessary to re-enter the elements T_o, ΔT, H_o, etc., as
  long as you keep working on one specific occultation, whose elements were entered once
  and for all.

- It should be noted that, as written here, the program will *exactly* fit on six tracks : adding even a single byte will make the machine ask for a seventh track. So, think twice before altering the program !

- The program runs with or without the printer attached. If the printer is attached, it should be set to NORM mode.

- As the program uses flags 0 to 3, the display annunciators enable you to control the running calculation :

  flag 0 is set for a planet,
  flag 1 is set for the calculation of a least distance,
  flag 2 is set for a grazing occultation,
  flag 3 is set for the calculation of an immersion, or for the calculation of a southern limit.

01	LBL "OCS"	38	"Y=?"	75	RTN	112	RCL 05	149	STO 22	186	RCL 22
02	CF 00	39	PROMPT	76	LBL C	113	RCL 17	150	RCL 15	187	RCL 24
03	GTO 00	40	STO 09	77	SF 01	114	×	151	RCL 20	188	—
04	LBL "OCP"	41	"Y1=?"	78	CF 02	115	RCL 04	152	SIN	189	STO 28
05	SF 00	42	PROMPT	79	CF 03	116	+	153	×	190	RCL 07
06	LBL 00	43	STO 10	80	GTO 00	117	STO 19	154	STO 23	191	RCL 25
07	"T0=?"	44	"Y2=?"	81	LBL D	118	RCL 03	155	RCL 16	192	—
08	PROMPT	45	PROMPT	82	CF 01	119	RCL 17	156	RCL 19	193	STO 29
09	STO 00	46	STO 11	83	CF 02	120	×	157	COS	194	RCL 10
10	"DT=?"	47	1	84	SF 03	121	RCL 14	158	×	195	RCL 26
11	PROMPT	48	"K=?"	85	GTO 00	122	—	159	RCL 15	196	—
12	STO 01	49	FS? 00	86	LBL E	123	RCL 01	160	RCL 20	197	STO 30
13	"H0=?"	50	PROMPT	87	CF 01	124	239.345	161	COS	198	R-P
14	PROMPT	51	STO 12	88	CF 02	125	/	162	×	199	STO 31
15	STO 02	52	RTN	89	CF 03	126	—	163	RCL 19	200	RCL 27
16	15.04107	53	LBL A	90	GTO 00	127	RCL 02	164	SIN	201	RCL 30
17	"H1=?"	54	HR	91	LBL I	128	+	165	×	202	×
18	FS? 00	55	STO 13	92	CF 01	129	STO 20	166	—	203	RCL 28
19	PROMPT	56	RTN	93	SF 02	130	RCL 08	167	STO 24	204	RCL 29
20	STO 03	57	LBL B	94	CF 03	131	RCL 17	168	RCL 20	205	×
21	"d0=?"	58	HR	95	GTO 00	132	×	169	COS	206	—
22	PROMPT	59	STO 14	96	LBL J	133	RCL 07	170	RCL 15	207	RCL 31
23	STO 04	60	RTN	97	CF 01	134	+	171	×	208	/
24	0	61	LBL 14	98	SF 02	135	RCL 17	172	.26252	209	RCL 12
25	"d1=?"	62	RCL 13	99	SF 03	136	×	173	×	210	.272495
26	FS? 00	63	TAN	100	LBL 00	137	RCL 06	174	STO 25	211	/
27	PROMPT	64	.99664719	101	0	138	+	175	LASTX	212	STO 34
28	STO 05	65	×	102	STO 17	139	STO 21	176	RCL 23	213	×
29	"X=?"	66	LASTX	103	3	140	RCL 11	177	×	214	STO 32
30	PROMPT	67	X<>Y	104	FS? 02	141	RCL 17	178	RCL 19	215	RCL 27
31	STO 06	68	ATAN	105	X↑2	142	×	179	SIN	216	RCL 29
32	"X1=?"	69	1	106	STO 18	143	RCL 10	180	×	217	×
33	PROMPT	70	P-R	107	FC? 02	144	+	181	STO 26	218	RCL 28
34	STO 07	71	STO 15	108	XEQ 14	145	RCL 17	182	RCL 21	219	RCL 30
35	"X2=?"	72	RDN	109	LBL 99	146	×	183	RCL 23	220	×
36	PROMPT	73	×	110	FS? 02	147	RCL 09	184	—	221	+
37	STO 08	74	STO 16	111	XEQ 14	148	+	185	STO 27	222	RCL 31

223	X↑2	250	RCL 19	277	1/X	304	RCL 28	331	TONE 9	358	"P="
224	/	251	1	278	1	305	SIGN	332	PROMPT	359	FIX 0
225	CHS	252	P-R	279	FS? 03	306	1	333	LBL 00	360	ARCL 35
226	STO 33	253	RDN	280	CHS	307	+	334	FC? 02	361	TONE 9
227	FC? 01	254	×	281	RCL 32	308	90	335	GTO 00	362	PROMPT
228	FS? 02	255	RCL 16	282	−	309	×	336	"LAT="	363	RCL 19
229	GTO 00	256	×	283	×	310	+	337	RCL 13	364	1
230	1	257	RCL 29	284	ST+ 13	311	STO 35	338	FIX 4	365	P-R
231	RCL 32	258	×	285	ABS	312	LBL 00	339	HMS	366	X<>Y
232	X↑2	259	X<>Y	286	1 E-5	313	FC? 02	340	ARCL X	367	RCL 13
233	−	260	RCL 16	287	X>Y?	314	GTO 00	341	TONE 9	368	1
234	SQRT	261	×	288	GTO 01	315	0	342	PROMPT	369	P-R
235	RCL 31	262	RCL 30	289	LBL 00	316	FS? 03	343	LBL 00	370	R↑
236	RCL 34	263	×	290	DSE 18	317	180	344	RCL 00	371	×
237	×	264	+	291	GTO 99	318	RCL 30	345	RCL 17	372	RCL 20
238	/	265	R↑	292	FC? 02	319	RCL 29	346	+	373	COS
239	FS? 03	266	RCL 15	293	GTO 01	320	/	347	RCL 01	374	×
240	CHS	267	×	294	"ERROR"	321	CHS	348	60	375	X<>Z
241	ST+ 33	268	RCL 29	295	TONE 0	322	ATAN	349	X↑2	376	×
242	LBL 00	269	×	296	PROMPT	323	+	350	/	377	+
243	RCL 33	270	.99328	297	LBL 01	324	STO 35	351	−	378	ASIN
244	ST+ 17	271	×	298	FS? 02	325	LBL 00	352	HMS	379	"H="
245	FC? 02	272	+	299	GTO 00	326	FC? 01	353	"UT="	380	FIX 1
246	GTO 00	273	RCL 31	300	RCL 27	327	GTO 00	354	FIX 4	381	ARCL X
247	RCL 20	274	/	301	RCL 28	328	"L.D.="	355	ARCL X	382	TONE 9
248	1	275	.064311	302	/	329	FIX 3	356	TONE 9	383	PROMPT
249	P-R	276	×	303	ATAN	330	ARCL 32	357	PROMPT	384	END

TI-59 PROGRAM

This program, written by Lionel Ancelet (Paris, France) for a TI-59 pocket calculator, enables you to calculate, for a star or a planet : the distance of nearest approach as seen from a given place, the points of the northern or southern limit for a grazing occultation, and the local circumstances of an ordinary occultation. In the three cases, the UT instant of the phenomenon, the position angle P and the altitude h above the horizon are given.

The procedure is as follows :

1. Change the partioning : enter 4, press 2nd Op 17. The display should be 639.39.

2. Load the tracks 1 and 3 of card I and the track 2 of card II.

3. Select the occulted body : press A for a star, B for a planet.

4. Press key D, and then enter the following data, each followed by R/S :

T_0 ΔT H_0 (H'_0) δ_0 (δ') X x' x'' Y y' y'' (k)

The data between parentheses are not to be entered in the case of the occultation of a star : the program will automatically skip the corresponding data entries, and take $H'_0 = 15.04107$, $\delta' = 0$ and $k = 1$ by itself.

5. Load the track 1 of card II.

6a. For the *least distance* :
   Enter ϕ (°.' ") and press key A.
   Enter λ (°.' ") (negative if East from Greenwich) and press key B.
   Then press *2nd E'*.

   The machine will then display Δ expressed in lunar radii, positive if the star (or planet) passes north of the Moon's center, negative if the body passes south of the center of the Moon. If $|\Delta| > 1$, there is no occultation for the given place. Pressing R/S will then give the Universal Time of the nearest approach. Pressing R/S again will give the corresponding position angle, and pressing R/S once more will provide the altitude of the body above the local horizon.

6b. For an *ordinary occultation* :
   Enter ϕ and λ as above.
   Press key C to indicate to the calculator that it has to calculate the circum-
      stances of an ordinary occultation.
   Then press D for an immersion, or E for an emersion.
   The machine will then display the Universal Time of contact.
   Then press R/S in order to obtain the position angle P of contact.
   Then press R/S in order to obtain the altitude h of the body.

6c. For a *grazing occultation*, press *2nd C'*.
   Enter the choosen longitude, and press key B.
   Enter the initial value of ϕ for the iterations (0° for instance, or +50° for
      Western Europe), and press key A.
   Then press D for the southern limit, or E for the northern limit.
   The machine will display the latitude ϕ (°.' "). Press R/S to obtain successively
      the instant (UT), the position angle and the altitude.

Remarks

After any calculation (6a or 6b or 6c), it is possible to make any other computation, with other values of ϕ and λ. Just re-enter the modified value(s) of ϕ or λ.
If the case of another occultation is wanted to be studied, re-load track 1 of card I, enter the new data, re-load track 1 of card II, and go on with 6a, 6b, 6c.

   It may be interesting to calculate first the least distance for a given place, and then the immersion and/or emersion if $|\Delta| < 1$.

   In the case of a grazing occultation, the display is flashing if after nine iterations the correction $\Delta\phi$ is still larger than 0.00001 in absolute value.

Part I

The following steps 000 - 087 are saved on track 1 of card I

000	LBL	018	R/S	036	IF FLG	054	09	072	5
001	A	019	STO	037	1	055	R/S	073	.
002	ST FLG	020	01	038	CP	056	STO	074	0
003	1	021	R/S	039	STO	057	10	075	4
004	INV SBR	022	STO	040	05	058	R/S	076	1
005	LBL	023	02	041	R/S	059	STO	077	0
006	B	024	R/S	042	LBL	060	11	078	7
007	INV	025	IF FLG	043	CP	061	IF FLG	079	STO
008	ST FLG	026	1	044	STO	062	1	080	03
009	1	027	LOG	045	06	063	PGM	081	1
010	INV SBR	028	STO	046	R/S	064	R/S	082	STO
011	LBL	029	03	047	STO	065	STO	083	32
012	D	030	R/S	048	07	066	32	084	CLR
013	INV	031	LBL	049	R/S	067	CLR	085	STO
014	FIX	032	LOG	050	STO	068	INV SBR	086	05
015	R/S	033	STO	051	08	069	LBL	087	INV SBR
016	STO	034	04	052	R/S	070	PGM		
017	12	035	R/S	053	STO	071	1		

Part II

Steps 000 to 239 are saved on track 1, card II,
240 to 479   —   on track 2, card II,
480 to 599   —   on track 3, card I.

000	LBL	042	33	084	4	126	=	168	x^2	210	17
001	A	043	×	085	LBL	127	STO	169	=	211	SIN
002	INV	044	RCL	086	LN x	128	17	170	STO	212	×
003	FIX	045	34	087	INV	129	RCL	171	19	213	RCL
004	D.MS	046	SIN	088	FIX	130	02	172	RCL	214	18
005	STO	047	=	089	CLR	131	+	173	09	215	COS
006	13	048	STO	090	STO	132	RCL	174	+	216	=
007	INV SBR	049	15	091	35	133	03	175	RCL	217	STO
008	LBL	050	INV SBR	092	3	134	×	176	10	218	22
009	B	051	LBL	093	STO	135	RCL	177	×	219	.
010	INV	052	C	094	00	136	35	178	RCL	220	2
011	FIX	053	ST FLG	095	INV	137	−	179	35	221	6
012	D.MS	054	2	096	IF FLG	138	RCL	180	+	222	2
013	STO	055	INV SBR	097	2	139	14	181	RCL	223	5
014	14	056	LBL	098	CLR	140	−	182	11	224	2
015	INV SBR	057	C'	099	SBR	141	RCL	183	×	225	STO
016	LBL	058	INV	100	INV	142	01	184	RCL	226	33
017	INV	059	ST FLG	101	LBL	143	:	185	35	227	×
018	RCL	060	2	102	CLR	144	2	186	x^2	228	RCL
019	13	061	INV SBR	103	IF FLG	145	3	187	=	229	16
020	TAN	062	LBL	104	2	146	9	188	STO	230	×
021	×	063	E'	105	$x \leftrightarrow t$	147	.	189	20	231	RCL
022	.	064	C	106	9	148	3	190	RCL	232	18
023	9	065	INV	107	STO	149	4	191	16	233	COS
024	9	066	ST FLG	108	00	150	5	192	×	234	=
025	6	067	4	109	LBL	151	=	193	RCL	235	STO
026	6	068	GTO	110	$x \leftrightarrow t$	152	STO	194	18	236	23
027	4	069	LN x	111	IF FLG	153	18	195	SIN	237	RCL
028	7	070	LBL	112	2	154	RCL	196	=	238	33
029	1	071	D	113	x^2	155	06	197	STO	239	×
030	9	072	INV	114	SBR	156	+	198	21	240	RCL
031	STO	073	ST FLG	115	INV	157	RCL	199	RCL	241	21
032	33	074	3	116	LBL	158	07	200	15	242	×
033	=	075	GTO	117	x^2	159	×	201	×	243	RCL
034	INV	076	CE	118	RCL	160	RCL	202	RCL	244	17
035	TAN	077	LBL	119	04	161	35	203	17	245	SIN
036	STO	078	E	120	+	162	+	204	COS	246	=
037	34	079	ST FLG	121	RCL	163	RCL	205	−	247	STO
038	COS	080	3	122	05	164	08	206	RCL	248	24
039	STO	081	LBL	123	×	165	×	207	16	249	RCL
040	16	082	CE	124	RCL	166	RCL	208	×	250	19
041	RCL	083	ST FLG	125	35	167	35	209	RCL	251	−

#	Code	#	Code	#	Code	#	Code	#	Code	#	Code
252	RCL	310	×	368	RCL	426	=	484	1	542	INV
253	21	311	RCL	369	17	427	IF FLG	485	IF FLG	543	D.MS
254	=	312	32	370	COS	428	3	486	3	544	FIX
255	STO	313	=	371	)	429	EE	487	+/−	545	4
256	25	314	STO	372	=	430	+/−	488	+/−	546	R/S
257	RCL	315	36	373	STO	431	LBL	489	LBL	547	LBL
258	20	316	IF FLG	374	37	432	EE	490	+/−	548	.
259	−	317	2	375	LBL	433	SUM	491	−	549	RCL
260	RCL	318	$\sqrt{x}$	376	$\sqrt{x}$	434	38	492	RCL	550	12
261	22	319	.	377	RCL	435	LBL	493	36	551	+
262	=	320	0	378	25	436	y^x	494	=	552	RCL
263	STO	321	6	379	×	437	RCL	495	:	553	35
264	26	322	4	380	RCL	438	38	496	RCL	554	−
265	RCL	323	3	381	27	439	SUM	497	37	555	RCL
266	07	324	1	382	+	440	35	498	=	556	01
267	−	325	1	383	RCL	441	INV	499	SUM	557	:
268	RCL	326	:	384	26	442	IF FLG	500	13	558	3
269	23	327	RCL	385	×	443	2	501	$\lvert x \rvert$	559	6
270	=	328	29	386	RCL	444	1/x	502	x ↔ t	560	0
271	STO	329	×	387	28	445	DSZ	503	5	561	0
272	27	330	(	388	=	446	0	504	INV	562	=
273	x ↔ t	331	RCL	389	+/−	447	x ↔ t	505	LOG	563	INV
274	RCL	332	28	390	:	448	RCL	506	1/x	564	FIX
275	10	333	×	391	RCL	449	25	507	x ⩾ t	565	INV
276	−	334	RCL	392	29	450	:	508	=	566	D.MS
277	RCL	335	15	393	x^2	451	RCL	509	DSZ	567	FIX
278	24	336	×	394	=	452	26	510	0	568	4
279	=	337	RCL	395	STO	453	=	511	x ↔ t	569	R/S
280	STO	338	18	396	38	454	INV	512	GTO	570	RCL
281	28	339	SIN	397	INV	455	TAN	513	P → R	571	39
282	INV	340	+	398	IF FLG	456	+	514	LBL	572	FIX
283	P → R	341	RCL	399	2	457	9	515	=	573	0
284	x ↔ t	342	27	400	y^x	458	0	516	RCL	574	R/S
285	STO	343	×	401	INV	459	+	517	28	575	RCL
286	29	344	RCL	402	IF FLG	460	9	518	:	576	17
287	RCL	345	15	403	4	461	0	519	RCL	577	SIN
288	25	346	×	404	y^x	462	×	520	27	578	×
289	×	347	RCL	405	1	463	RCL	521	=	579	RCL
290	RCL	348	18	406	−	464	26	522	+/−	580	13
291	28	349	COS	407	RCL	465	Op	523	INV	581	SIN
292	−	350	×	408	36	466	10	524	TAN	582	+
293	RCL	351	RCL	409	x^2	467	=	525	+	583	RCL
294	27	352	17	410	=	468	STO	526	9	584	17
295	×	353	SIN	411	$\sqrt{x}$	469	39	527	0	585	COS
296	RCL	354	+	412	:	470	IF FLG	528	−	586	×
297	26	355	.	413	RCL	471	4	529	9	587	RCL
298	=	356	9	414	32	472	.	530	0	588	13
299	:	357	9	415	:	473	RCL	531	IF FLG	589	COS
300	RCL	358	3	416	RCL	474	36	532	3	590	×
301	29	359	2	417	29	475	FIX	533	SBR	591	RCL
302	:	360	8	418	×	476	3	534	+/−	592	18
303	.	361	×	419	.	477	R/S	535	LBL	593	COS
304	2	362	RCL	420	2	478	INV	536	SBR	594	=
305	7	363	27	421	7	479	FIX	537	=	595	INV
306	2	364	×	422	2	480	GTO	538	STO	596	SIN
307	4	365	RCL	423	4	481	.	539	39	597	FIX
308	9	366	16	424	9	482	LBL	540	RCL	598	1
309	5	367	×	425	5	483	1/x	541	13	599	INV SBR

TABLE I

Date	$T_\circ$	Body	Magn.	Elong.	γ	Area of Visibility
	h					
1980 Jan 6	7	Regulus	+1.3	136°W	−0.5526	S Pacific Ocean, South America, S Atlantic Ocean
1980 Jan 7	4	Jupiter	−1.9	126°W	−0.2769	N South America, Atlantic Ocean, W and S Africa, Madagascar
1980 Jan 8	15	Saturn	+1.1	110°W	+0.1791	Pacific Ocean
1980 Jan 20	12	Venus	−3.5	36°E	+0.9476	NE Atlantic Ocean, Iceland, Europe, N Africa, W Asia
1980 Jan 27	5	Aldebaran	+1.1	123°E	+0.3177	Pacific Ocean, North America, Central America
1980 Feb 2	15	Regulus	+1.3	163°W	−0.4776	Indonesia, Australia, New Zealand, S Pacific Ocean
1980 Feb 3	7	Jupiter	−2.0	156°W	−0.5071	Pacific Ocean, S South America
1980 Feb 4	21	Saturn	+1.0	139°W	+0.1017	NE Africa, S Asia, Indonesia, New Guinea, Australia
1980 Feb 23	11	Aldebaran	+1.1	95°E	+0.2644	Asia, N Pacific Ocean
1980 Feb 29	21	Regulus	+1.3	169°E	−0.4885	S Atlantic Ocean, Africa, Indian Ocean
1980 Mar 1	7	Jupiter	−2.1	175°E	−0.9198	S Pacific Ocean, S South America, Antarctica
1980 Mar 2	24	Saturn	+0.9	168°W	−0.1570	N South America, Atlantic Ocean, Africa, Madagascar, S Indian Oc.
1980 Mar 21	18	Aldebaran	+1.1	68°E	+0.3492	SE North America, Centr. America, N South America, N Atlantic Oc., Europe, N Africa, W Asia
1980 Mar 28	3	Regulus	+1.3	142°E	−0.4434	Pacific Ocean, South America
1980 Mar 28	7	Jupiter	−2.0	144°E	−1.1734	S New Zealand, S Pacific Ocean, Antarctica
1980 Mar 30	1	Saturn	+0.9	163°E	−0.3746	Central America, South America, S Atlantic Ocean, S Africa
1980 Apr 13	9	Mercury	+0.2	26°W	−0.0172	S Atlantic Ocean, Africa, SW Asia, Indian Ocean
1980 Apr 18	4	Aldebaran	+1.1	42°E	+0.5103	Asia, N Pacific Ocean, W North America
1980 Apr 24	9	Regulus	+1.3	115°E	−0.2605	SE Asia, NE Australia, Pacific Oc.

Table I   (cont.)

Date	$T_\circ$	Body	Magn.	Elong.	γ	Area of Visibility
	h					
1980 Apr 24	11	Jupiter	-1.8	116°E	-1.0518	S Indian Ocean, SW Australia, Antarctica
1980 Apr 26	4	Saturn	+1.0	135°E	-0.3554	Pacific Ocean, South America
1980 May 15	14	Aldebaran	+1.1	16°E	+0.6321	E North America, Greenland, N Atlantic Ocean, Europe, W Asia
1980 May 21	17	Regulus	+1.3	89°E	+0.0164	N Atlantic Ocean, S Spain, Africa, Indian Ocean
1980 May 21	20	Jupiter	-1.6	90°E	-0.5983	South America, S Atlantic Ocean, S Africa
1980 May 22	6	Mars	+0.7	95°E	-0.4408	New Guinea, Australia, Pacific Oc.
1980 May 23	10	Saturn	+1.1	108°E	-0.0760	S Asia, Indonesia, NE Australia, SW Pacific, N New Zealand
1980 Jun 11	23	Aldebaran	+1.1	13°W	+0.6450	E Asia, N Pacific Ocean, W North America
1980 Jun 18	1	Regulus	+1.3	63°E	+0.2699	N Pacific Ocean, S North America, Central America, NW South America
1980 Jun 18	10	Jupiter	-1.5	67°E	+0.0144	E Africa, S Asia, Indonesia, New Guinea, N Australia
1980 Jun 19	19	Saturn	+1.3	82°E	+0.3401	North America, N Atlantic Ocean, S Spain, Africa, S Arabia
1980 Jul 9	7	Aldebaran	+1.1	38°W	+0.5698	N Atlantic Ocean, Europe, N Africa, Asia
1980 Jul 9	19	Venus	-4.1	32°W	+0.1677	Pacific Ocean, North America, Central America, NW South America
1980 Jul 15	9	Regulus	+1.3	37°E	+0.4084	Europe, NE Africa, Asia, N Indonesia, New Guinea
1980 Jul 16	3	Jupiter	-1.3	45°E	+0.6396	NE Asia, Arctic regions, N Pacific Ocean, W North America
1980 Jul 17	7	Saturn	+1.4	58°E	+0.7482	N Asia, Arctic regions, N Pacific Ocean, Alaska
1980 Aug 5	13	Aldebaran	+1.1	64°W	+0.4968	North America, N Atlantic Ocean, Greenland, Iceland, W Europe, NW Africa
1980 Aug 7	2	Venus	-4.1	45°W	+0.3061	NE Africa, Asia, W Pacific Ocean

Table I   (cont.)

Date	T_o	Body	Magn.	Elong.	γ	Area of Visibility
	h					
1980 Aug 11	17	Regulus	+1.3	10°E	+0.4312	N America, N Atlantic Ocean, NW Africa
1980 Aug 12	21	Jupiter	−1.2	24°E	+1.2092	Arctic regions, Greenland, Iceland
1980 Aug 13	20	Saturn	+1.4	34°E	+1.0716	Arctic regions, NE Canada, Greenland, Iceland, N Atlantic Ocean, W Europe
1980 Sep 1	18	Aldebaran	+1.1	90°W	+0.5185	E Asia, N Pacific Ocean, North America
1980 Sep 5	10	Venus	−3.9	45°W	+0.0417	Central America, N South America, Atlantic Ocean, S Spain, Africa
1980 Sep 7	24	Regulus	+1.3	16°W	+0.4269	Asia, Pacific Ocean
1980 Sep 29	1	Aldebaran	+1.1	116°W	+0.6543	NE Atlantic Ocean, Iceland, Europe, NW Africa, Asia
1980 Oct 5	5	Regulus	+1.3	43°W	+0.5114	Iceland, Europe, NW Africa, Asia
1980 Oct 5	7	Venus	−3.6	42°W	+0.8205	Greenland, Iceland, Arctic regions, N and E Europe, Asia
1980 Oct 26	11	Aldebaran	+1.1	143°W	+0.8283	NE Asia, N Pacific Ocean, Alaska, Canada, Arctic regions, Greenland, Iceland
1980 Nov 1	11	Regulus	+1.3	70°W	+0.7339	N North America, Arctic regions, Greenland, Iceland, N Atlantic Ocean, Europe, N Africa
1980 Nov 22	22	Aldebaran	+1.1	170°W	+0.9319	NE Atlantic Ocean, W and N Europe, N Asia, Greenland, Arctic regions
1980 Nov 28	19	Regulus	+1.3	97°W	+1.0215	N Siberia, Arctic regions, E North America
1980 Dec 20	8	Aldebaran	+1.1	160°E	+0.9196	Arctic regions, Greenland, NE Asia, N North America
1980 Dec 26	4	Regulus	+1.3	125°W	+1.2369	N Siberia
1981 Jan 16	16	Aldebaran	+1.1	133°E	+0.8548	Europe, NW Africa, Greenland, Arctic regions, Siberia
1981 Feb 5	21	Mars	+1.4	12°E	−0.6255	S Pacific Ocean, New Zealand, South America
1981 Feb 12	22	Aldebaran	+1.1	105°E	+0.8547	North America, Greenland, Arctic regions, N Europe

Table I   (cont.)

Date	T_o	Body	Magn.	Elong.	γ	Area of Visibility
	h					
1981 Mar 5	20	Venus	-3.4	8°W	-0.9535	Antarctica, S South America
1981 Mar 12	4	Aldebaran	+1.1	78°E	+0.9755	Arctic regions, NE Asia, Alaska, Canada
1981 Apr 3	7	Mercury	-0.0	22°W	-1.0127	Antarctica, SW Australia
1981 Apr 8	11	Aldebaran	+1.1	51°E	+1.1594	Greenland, Iceland, Spitsbergen, Arctic regions
1981 Sep 24	5	Mars	+1.8	47°W	+0.0368	Africa, Turkey, S Asia, Indonesia, Indian Ocean
1982 Jan 21	24	Neptune	+7.8	36°W	+1.2316	Extreme NE Siberia, Bering Sea
1982 Feb 18	9	Neptune	+7.8	63°W	+1.0606	N Atlantic Ocean, Europe, NW Africa
1982 Mar 17	18	Neptune	+7.8	90°W	+0.7747	Japan, NE Siberia, N Pacific Ocean, Alaska
1982 Apr 14	2	Neptune	+7.7	117°W	+0.4754	W and S Europe, Africa, Indian Ocean, SW Asia
1982 May 11	8	Neptune	+7.7	144°W	+0.2955	Mexico, Central America, South America, Atlantic Ocean
1982 Jun 7	13	Neptune	+7.7	170°W	+0.2862	Philippines, New Guinea, Pacific Ocean
1982 Jun 20	1	Mercury	+1.2	21°W	+1.0325	USSR, Arctic regions, N Alaska
1982 Jul 4	18	Neptune	+7.7	163°E	+0.3762	NE Africa, S Asia, Indian Ocean, Indonesia
1982 Jul 18	19	Venus	-3.3	28°W	-0.5559	Pacific Ocean
1982 Jul 20	9	Mercury	-1.6	6°W	-0.3959	SE Atlantic Ocean, Africa, Indian Ocean
1982 Jul 31	23	Neptune	+7.7	136°E	+0.4259	Antilles, N South America, Atlantic Ocean, Africa
1982 Aug 28	5	Neptune	+7.7	110°E	+0.3276	Pacific Ocean, Mexico, Central America
1982 Sep 24	13	Neptune	+7.8	83°E	+0.0801	Africa, Indian Ocean, Indonesia, NW Australia
1982 Oct 21	22	Neptune	+7.8	57°E	-0.2161	Pacific Ocean, South America
1982 Nov 18	8	Neptune	+7.8	30°E	-0.4423	S Africa, Madagascar, S Indian Oc., S Australia, New Zealand

Table I   (cont.)

Date	T_o	Body	Magn.	Elong.	γ	Area of Visibility
	h					
1982 Nov 19	21	Mars	+1.2	47°E	+0.5602	NE Pacific Ocean, S North America, Central America, N South America, NW Atlantic Ocean
1982 Dec 15	16	Neptune	+7.8	3°E	−0.5630	SE Pacific Ocean, S South America, S Atlantic Ocean
1982 Dec 16	8	Venus	−3.4	10°E	+0.8450	E Europe, Asia
1983 Jan 12	1	Neptune	+7.8	24°W	−0.6585	S Indian Ocean, S New Zealand, S Pacific Ocean, Antarctica
1983 Feb 8	8	Neptune	+7.8	50°W	−0.8441	S South America, S Atlantic Ocean, S Indian Ocean, Antarctica
1983 Mar 6	3	Jupiter	−1.7	95°W	+1.0153	N and E Europe, W Asia
1983 Mar 7	16	Neptune	+7.8	77°W	−1.1462	Antarctica
1983 Apr 2	13	Jupiter	−1.9	121°W	+0.6299	NE Pacific Ocean, S North America, Central America
1983 Apr 29	19	Jupiter	−2.0	150°W	+0.5761	Asia, S Japan, Indonesia, W Pacific Ocean
1983 May 26	21	Jupiter	−2.1	179°W	+0.8516	Europe, W Asia
1983 Jun 9	10	Mercury	+0.7	24°W	+0.7828	E North America, N Atlantic Ocean, W and N Europe, NW Asia, Arctic regions
1983 Jun 11	1	Mars	+1.7	2°W	−1.1938	New Zealand, SW Pacific Ocean
1983 Jun 22	22	Jupiter	−2.1	152°E	+1.2397	N Scandinavia, NW Soviet Union
1983 Jul 9	19	Mars	+1.8	10°W	+0.3430	N Pacific Ocean, North America, NW Atlantic Ocean
1983 Jul 10	14	Mercury	−1.7	2°E	+0.6096	North America, Greenland, Iceland, N Atlantic Ocean, Europe, NE Africa, SW Asia
1983 Sep 12	19	Jupiter	−1.6	75°E	+0.9385	E Canada, Greenland, Iceland, N Atlantic Ocean, W Europe, N Africa

Table I   (cont.)

Date	T_0	Body	Magn.	Elong.	γ	Area of Visibility
	h					
1983 Oct 10	8	Uranus	+6.0	50°E	+0.9792	Asia, NW Pacific Ocean
1983 Oct 10	12	Jupiter	-1.5	52°E	+0.3861	SE Europe, S Asia, Africa, Indian Ocean, Indonesia
1983 Nov 4	15	Saturn	+0.8	4°W	+1.0606	NE Canada, Greenland, Iceland, N Atlantic Ocean, W Europe, NW Africa
1983 Nov 6	20	Uranus	+6.0	24°E	+0.7224	NE Pacific Ocean, North America, Central America, N South America
1983 Nov 7	7	Jupiter	-1.4	29°E	-0.1983	E Africa, Madagascar, Indian Ocean, Australia, New Zealand
1983 Dec 2	5	Saturn	+0.8	29°W	+0.8432	Soviet Union, Asia
1983 Dec 4	8	Uranus	+6.0	2°W	+0.5594	SE Europe, NE Africa, S Asia, Indian Ocean
1983 Dec 5	3	Jupiter	-1.3	7°E	-0.7672	S Indian Ocean, S Pacific Ocean, Antarctica
1983 Dec 6	3	Mercury	-0.4	19°E	+0.9434	E Asia, N Pacific Ocean
1983 Dec 29	16	Saturn	+0.8	54°W	+0.5546	NE Pacific Ocean, SW North America, Central America, N South America
1983 Dec 30	19	Venus	-3.6	40°W	-0.6547	E Australia, New Zealand, S Pacific Ocean, S South America, Antarctica
1983 Dec 31	18	Uranus	+6.0	28°W	+0.4050	Pacific Ocean, Central America, NW South America
1984 Jan 26	1	Saturn	+0.8	80°W	+0.1552	SW Asia, NE Africa, Indian Ocean, Indonesia, Australia
1984 Jan 28	3	Uranus	+6.0	55°W	+0.1621	E Africa, Arabia, Indian Ocean, S India, Ceylon, Indonesia, N Australia
1984 Feb 22	9	Saturn	+0.7	107°W	-0.2714	E Pacific Ocean, South America, S Atlantic Ocean
1984 Feb 22	14	Mars	+0.5	104°W	+0.3183	Pacific Ocean, NW South America
1984 Feb 24	10	Uranus	+6.0	82°W	-0.1676	E Pacific Ocean, South America, S Atlantic Ocean, South Africa
1984 Mar 20	17	Saturn	+0.5	134°W	-0.5591	Indian Ocean, Australia, New Zealand, S Pacific Ocean, part of Antarctica

Table I   (cont.)

Date	T_o	Body	Magn.	Elong.	γ	Area of Visibility
	h					
1984 Mar 21	12	Mars	−0.2	124°W	−0.3921	Pacific Ocean, S South America
1984 Mar 22	18	Uranus	+5.9	109°W	−0.4665	Indonesia, Australia, New Zealand, S Pacific Ocean
1984 Apr 17	1	Saturn	+0.4	163°W	−0.5984	South America, S Atlantic Ocean, part of Antarctica
1984 Apr 17	23	Mars	−1.1	151°W	+0.0427	Africa, Indian Ocean
1984 Apr 19	3	Uranus	+5.9	136°W	−0.6092	South America, S Atlantic Ocean, S Indian Ocean, Antarctica
1984 May 14	8	Saturn	+0.3	168°E	−0.4300	Pacific Ocean, S South America
1984 May 16	11	Uranus	+5.8	164°W	−0.5750	Australia, New Zealand, S Pacific Ocean, S South America, part of Antarctica
1984 May 28	17	Mercury	+0.2	24°W	+0.9660	Extreme E of Siberia, N Pacific Ocean, Alaska, NW Canada, Arctic regions, Greenland, Iceland
1984 May 30	8	Venus	−3.4	5°W	+0.2722	SE Europe, Africa, Asia
1984 Jun 10	13	Saturn	+0.5	141°E	−0.2164	S Asia, Indonesia, Indian Ocean, Australia, New Zealand
1984 Jun 12	18	Uranus	+5.8	169°E	−0.4622	S Africa, S Indian Ocean, S Australia, New Zealand
1984 Jul 7	17	Saturn	+0.7	114°E	−0.1282	Africa, S Indian Ocean
1984 Jul 9	23	Uranus	+5.8	142°E	−0.4149	South America, S Atlantic Ocean, South Africa
1984 Aug 3	23	Saturn	+0.8	88°E	−0.2366	Pacific Ocean, South America, S Atlantic Ocean
1984 Aug 6	4	Uranus	+5.9	116°E	−0.5261	New Zealand, S Pacific Ocean, S South America, part of Antarctica
1984 Aug 31	9	Saturn	+0.9	63°E	−0.4931	E Africa, Indian Ocean, SW Australia, Tasmania, New Zealand
1984 Sep 2	10	Uranus	+5.9	90°E	−0.7774	Indian Ocean, Antarctica
1984 Sep 27	22	Saturn	+0.8	39°E	−0.7980	New Zealand, S Pacific Ocean, Antarctica

Table I   (cont.)

Date	T_o	Body	Magn.	Elong.	γ	Area of Visibility
	h					
1984 Sep 29	18	Uranus	+6.0	64°E	-1.0644	S Pacific Ocean, Antarctica
1984 Oct 1	0	Mars	+0.4	79°E	-0.2791	Pacific Ocean, South America
1984 Oct 25	13	Saturn	+0.8	15°E	-1.0868	S South America, Antarctica
1984 Oct 27	1	Venus	-3.4	35°E	+0.3247	Pacific Ocean
1984 Nov 24	14	Mercury	-0.2	22°E	+0.1022	South America, Atlantic Ocean, Africa
1985 Apr 22	13	Mars	+1.7	25°E	+0.3947	Atlantic Ocean, Europe, NW Africa, Asia
1985 Oct 15	6	Mercury	-0.3	15°E	+1.1181	NE Asia, Japan
1985 Nov 11	10	Venus	-3.4	17°W	-0.7430	South America, S Atlantic Ocean, Antarctica
1985 Nov 14	3	Mercury	+0.2	22°E	-0.4401	Indian Ocean, Australia, New Zealand, S Pacific Ocean
1985 Dec 8	11	Mars	+1.8	50°W	+0.0129	SE U.S.A., Central America, N and NE South America, Atlantic Ocean, S Africa
1986 Mar 11	16	Venus	-3.4	12°E	-1.2243	S Atlantic - Indian Ocean
1986 Mar 30	13	Antares	+1.2	120°W	+1.2161	Part of Canada and U.S.A.
1986 Apr 26	22	Antares	+1.2	147°W	+1.1738	Central Asia
1986 May 24	8	Antares	+1.2	172°W	+1.1952	Part of Canada and U.S.A.
1986 Jun 20	19	Antares	+1.2	160°E	+1.1802	W and Central Asia
1986 Jun 23	14	Mars	-2.1	160°W	-0.5255	SE Indian Ocean, SW Australia, Tasmania, New Zealand, S Pacific Ocean, part of Antarctica
1986 Jul 18	4	Antares	+1.2	134°E	+1.0667	North America
1986 Jul 20	13	Mars	-2.4	166°E	+0.9053	E Asia, Japan, NW Pacific Ocean
1986 Aug 14	12	Antares	+1.2	108°E	+0.8733	Central and E Asia, Japan, W Pacific Ocean
1986 Aug 16	16	Mars	-1.8	138°E	+0.5014	Africa, S and SE Asia, Indian Ocean
1986 Aug 21	12	Jupiter	-2.4	158°W	-1.2575	Off S South America

Table I (cont.)

Date	T_o	Body	Magn.	Elong.	γ	Area of Visibility
	h					
1986 Sep 10	17	Antares	+1.2	82°E	+0.6875	W and S Europe, SW Asia, N and Central Africa
1986 Sep 13	10	Mars	-1.2	118°E	-0.9293	S Indian Ocean, S Pacific Ocean, Antarctica
1986 Oct 5	8	Mercury	-0.1	20°E	+0.3489	Soviet Union, Egypt, W and S Asia, Indian Ocean, Indonesia, New Guinea, Australia
1986 Oct 7	23	Antares	+1.2	55°E	+0.5984	E Pacific Ocean, W and S of North America, Central America, N South America
1986 Nov 3	14	Mercury	+0.8	18°E	-0.7572	South America, S Atlantic Ocean, S Indian Ocean, Antarctica
1986 Nov 4	7	Antares	+1.2	28°E	+0.6144	Central and S Asia, Indonesia, New Guinea, W Pacific Ocean
1986 Dec 1	17	Antares	+1.2	5°E	+0.6482	North America, Central America, N South America, Atlantic Ocean
1986 Dec 29	5	Antares	+1.2	28°W	+0.5914	Egypt, S Asia, Indian Ocean, Indonesia
1987 Jan 4	20	Jupiter	-1.8	64°E	-1.2189	S Atlantic Ocean
1987 Jan 25	15	Antares	+1.2	56°W	+0.4166	E Pacific Ocean, California, Mexico, Central America, N South America
1987 Feb 1	15	Jupiter	-1.7	40°E	-0.6162	S South America, S Atlantic Ocean, Africa, Madagascar, Indian Ocean, Antarctica
1987 Feb 3	11	Mars	+1.1	64°E	+0.2822	Africa, Madagascar, Indian Ocean, Asia, Japan
1987 Feb 18	13	Spica	+1.2	126°W	+1.0679	Canada, U.S.A.
1987 Feb 21	22	Antares	+1.2	83°W	+0.2099	S Asia, Indonesia, New Guinea, Australia, W Pacific Ocean
1987 Mar 1	12	Jupiter	-1.6	19°E	+0.0050	E South America, Atlantic Ocean, Africa, SE and S Asia
1987 Mar 17	19	Spica	+1.2	153°W	+0.9763	N and E Asia, Japan, NW Pacific Ocean
1987 Mar 21	3	Antares	+1.2	110°W	+0.0928	Atlantic Ocean, W and S Africa, Madagascar, Indian Ocean

Table I (cont.)

Date	T_0	Body	Magn.	Elong.	γ	Area of Visibility
	h					
1987 Mar 29	9	Jupiter	-1.6	2°W	+0.6310	Atlantic Ocean, Africa, Europe, W and N Asia, Arctic regions
1987 Apr 14	3	Spica	+1.2	178°W	+0.9725	Greenland, Iceland, N Atlantic Ocean, Europe, N Africa
1987 Apr 17	9	Antares	+1.2	137°W	+0.1031	Pacific Ocean, South America
1987 Apr 25	12	Venus	-3.4	31°W	+0.8751	North and Central America, Atlantic Ocean, Greenland, Iceland, N Scandinavia, Arctic regions
1987 May 11	12	Spica	+1.2	153°E	+0.9441	NE Asia, N Pacific Ocean, Alaska
1987 May 14	17	Antares	+1.2	163°W	+0.1670	India, Indian Ocean, Indonesia, New Guinea, Australia, W Pacific Ocean
1987 Jun 7	22	Spica	+1.2	127°E	+0.8091	Labrador, Greenland, Iceland, N Atlantic Ocean, Europe, N Africa, SW Asia
1987 Jun 11	4	Antares	+1.2	169°E	+0.1763	E Pacific Ocean, Mexico, Central and South America, Atlantic Ocean
1987 Jul 5	7	Spica	+1.2	101°E	+0.5746	NE Asia, Japan, Pacific Ocean
1987 Jul 8	14	Antares	+1.2	143°E	+0.0749	India, Indian Ocean, Indonesia, New Guinea, Australia
1987 Aug 1	13	Spica	+1.2	75°E	+0.3278	W, S and Central Europe, N and E Africa, Arabia, India, Indian Ocean, Indonesia
1987 Aug 4	23	Antares	+1.2	117°E	-0.1052	E Pacific Ocean, South America, S Atlantic Ocean, S Africa
1987 Aug 24	22	Mercury	-1.2	5°E	+0.7688	E Siberia, North and Central America, Greenland, Arctic regions
1987 Aug 28	19	Spica	+1.2	49°E	+0.1713	North and Central America, N South America, Atlantic Ocean
1987 Sep 1	6	Antares	+1.2	91°E	-0.2659	Indonesia, New Guinea, Australia, New Zealand, S Pacific Ocean
1987 Sep 22	8	Mars	+2.0	9°W	+0.1360	Europe except N, N and E Africa, SW Asia, India, Indian Ocean, Indonesia, W Australia
1987 Sep 25	1	Spica	+1.2	22°E	+0.1373	Japan, Pacific Ocean

Table I   (cont.)

Date	T_0	Body	Magn.	Elong.	γ	Area of Visibility
	h					
1987 Sep 25	4	Mercury	+0.1	24°E	-0.2800	S Asia, Indian Ocean, Indonesia, Australia, New Zealand, S Pacific
1987 Sep 28	12	Antares	+1.2	65°E	-0.3144	S Africa, Madagascar, S Indian Ocean, Australia
1987 Oct 22	8	Spica	+1.2	5°W	+0.1531	W and S Europe, Africa, Madagascar, Arabia, Indian Ocean, W Australia
1987 Oct 23	13	Mercury	+1.8	11°E	-0.9873	SE Pacific Ocean, S South America, Antarctica
1987 Oct 25	18	Antares	+1.2	38°E	-0.2466	SE Pacific Ocean, South America, S Atlantic Ocean, S Africa
1987 Nov 18	17	Spica	+1.2	32°W	+0.0946	Pacific Ocean, Hawaii, South America
1987 Nov 22	2	Antares	+1.2	11°E	-0.1626	Indonesia, New Guinea, Australia, New Zealand, Pacific Ocean
1987 Dec 16	3	Spica	+1.2	60°W	-0.1074	Africa, Madagascar, Arabia, Indian Ocean, Australia
1987 Dec 19	12	Antares	+1.2	18°W	-0.1818	South America, S Atlantic Ocean, S Africa, Madagascar
1988 Jan 12	12	Spica	+1.2	88°W	-0.3950	Pacific Ocean, S South America, S Atlantic Ocean
1988 Jan 15	22	Antares	+1.2	46°W	-0.3322	Indonesia, Australia, New Zealand, S Pacific Ocean
1988 Jan 21	19	Venus	-3.5	37°E	-0.0582	Pacific Ocean, South America, Atlantic Ocean, W Africa
1988 Feb 8	18	Spica	+1.2	115°W	-0.6240	Indian Ocean, Sumatra, Java, Australia, Antarctica
1988 Feb 12	8	Antares	+1.2	73°W	-0.5188	South America, S Atlantic Ocean, S Indian Ocean, Antarctica
1988 Mar 7	0	Spica	+1.2	143°W	-0.7012	E South America, S Atlantic Ocean, extreme S Africa, Antarctica
1988 Mar 10	14	Antares	+1.2	101°W	-0.6079	E Australia, New Zealand, S Pacific Ocean, S South America, Antarctica
1988 Mar 16	5	Mercury	+0.2	26°W	-0.4691	S Indian Ocean, Indonesia, New Guinea, Australia, part Antarctica

Table I   (cont.)

Date	T_o	Body	Magn.	Elong.	γ	Area of Visibility
	h					
1988 Apr 3	6	Spica	+1.2	170°W	−0.6743	Pacific Ocean, S South America, Antarctica
1988 Apr 6	20	Antares	+1.2	127°W	−0.5540	SE Africa, Madagascar, S Indian Ocean, Tasmania, New Zealand, Antarctica
1988 Apr 19	24	Venus	−4.1	45°E	+1.0360	N and NE Asia, Japan, Arctic regions, Greenland, Iceland, Scandinavia
1988 Apr 30	14	Spica	+1.2	163°E	−0.6751	Indian Ocean, W Australia, Antarctica
1988 May 4	2	Antares	+1.2	154°W	−0.4378	South America, S Atlantic Ocean, part of Antarctica, S Indian Ocean
1988 May 9	6	Mars	+0.3	87°W	−0.7272	Extreme S South America, S Atlantic Ocean, S Africa, Madagascar, Indian Ocean, Antarctica
1988 May 27	22	Spica	+1.2	137°E	−0.7978	SE Pacific Ocean, South America, S Atlantic Ocean, Antarctica
1988 May 31	10	Antares	+1.2	175°E	−0.3815	Australia, New Guinea, New Zealand, S Pacific Ocean, S South America
1988 Jun 19	19	Regulus	+1.3	61°E	+1.1788	NE Scandinavia, Soviet Union
1988 Jun 24	7	Spica	+1.2	111°E	−1.0325	SW Australia, Antarctica
1988 Jun 27	19	Antares	+1.2	153°E	−0.4466	S Atlantic Ocean, S Africa, S Indian Ocean, W Australia, part Antarctica
1988 Jul 17	2	Regulus	+1.3	35°E	+0.9934	North America, Greenland, Arctic
1988 Jul 25	5	Antares	+1.2	127°E	−0.5966	E Australia, Tasmania, New Zealand, S Pacific Ocean, S South America, Antarctica
1988 Aug 13	8	Regulus	+1.3	9°E	+0.9368	Greenland, Arctic regions, extreme N Norway, NE half of Asia, Japan
1988 Aug 13	12	Mercury	−0.8	11°E	−0.4026	E South America, Atlantic Ocean, Africa, S Indian Ocean
1988 Aug 21	14	Antares	+1.2	101°E	−0.7280	S Atlantic Ocean, S Africa, Tasmania, Antarctica
1988 Sep 9	14	Regulus	+1.3	17°W	+0.9511	Arctic regions, Greenland, Iceland, Europe, N Africa, Turkey
1988 Sep 13	15	Mercury	+0.3	27°E	−0.6268	South America, S Atlantic Ocean, Antarctica

Table I   (cont.)

Date	T_o	Body	Magn.	Elong.	γ	Area of Visibility
	h					
1988 Sep 17	20	Antares	+1.2	75°E	-0.7429	S Pacific Ocean, extreme S South America, S Atlantic Ocean, Antarctica
1988 Oct 6	21	Regulus	+1.3	44°W	+0.9142	North America, Greenland, Arctic regions, N Asia
1988 Oct 7	3	Venus	-3.6	41°W	+0.6298	N and NE Europe, Asia, Japan, W Pacific Ocean, New Guinea, Arctic regions
1988 Oct 10	23	Mercury	+3.0	2°E	-1.1978	SE Indian Ocean, part Antarctica
1988 Oct 15	2	Antares	+1.2	48°E	-0.6311	Australia, New Zealand, S Pacific Ocean, S South America, Antarctica
1988 Nov 3	3	Regulus	+1.3	71°W	+0.7353	Arctic regions, Greenland, Iceland, N Europe, Asia except SW, Japan
1988 Nov 11	8	Antares	+1.2	21°E	-0.4932	S Africa, S Indian Ocean, SE Australia, New Zealand, Antarctica
1988 Nov 30	11	Regulus	+1.3	99°W	+0.4411	NE Pacific Ocean, North America, N South America, Atlantic Ocean
1988 Dec 8	16	Antares	+1.2	9°W	-0.4581	S Pacific Ocean, S South America, S Atlantic Ocean, part Antarctica
1988 Dec 27	20	Regulus	+1.3	127°W	+0.1649	Asia, SW Pacific Ocean, New Guinea, NE Australia
1989 Jan 5	1	Antares	+1.2	35°W	-0.5614	Indian Ocean, New Zealand, S Pacific Ocean, Antarctica
1989 Jan 24	4	Regulus	+1.3	154°W	+0.0262	SE North America, Central America, Atlantic Ocean, W and S Africa
1989 Feb 1	11	Antares	+1.2	63°W	-0.7083	SE Pacific Ocean, S South America, S Atlantic Ocean, Antarctica
1989 Feb 20	11	Regulus	+1.3	178°E	+0.0164	Japan, Pacific Ocean, Hawaii
1989 Feb 28	19	Antares	+1.2	91°W	-0.7599	Indian Ocean, S Pacific, Antarctica
1989 Mar 6	4	Mercury	+0.0	23°W	+0.6743	E Africa, Indian Ocean, Asia, Japan
1989 Mar 19	17	Regulus	+1.3	151°E	+0.0108	SE Europe, NE Africa, SW and S Asia, Indian Ocean, Indonesia, Australia

Table I   (cont.)

Date	T_o	Body	Magn.	Elong.	γ	Area of Visibility
	h					
1989 Mar 28	2	Antares	+1.2	118°W	-0.6632	S South America, S Atlantic Oc., S Indian Ocean, extreme SW Australia, Antarctica
1989 Apr 15	23	Regulus	+1.3	124°E	-0.1101	S North America, Central America, N South America, Atlantic Ocean, S Africa
1989 Apr 24	7	Antares	+1.2	144°W	-0.4971	S Pacific Ocean, S South America, S Atlantic Ocean, Antarctica
1989 May 13	6	Regulus	+1.3	97°E	-0.3553	New Guinea, Pacific Ocean
1989 May 21	13	Antares	+1.2	170°W	-0.3908	SE Indian Ocean, Australia, New Zealand, S Pacific Ocean
1989 Jun 9	14	Regulus	+1.3	71°E	-0.6293	S Atlantic Ocean, S Africa, Madagascar, S Indian Ocean, part of Antarctica
1989 Jun 17	21	Antares	+1.2	162°E	-0.4109	E South America, S Atlantic Ocean, extreme S Africa, S Indian Ocean
1989 Jul 5	4	Venus	-3.3	24°E	+0.1097	SE Asia, Japan, Pacific Ocean
1989 Jul 5	12	Mars	+2.0	28°E	+0.0885	Atlantic Ocean, W and S Europe, Africa, Turkey, Arabia, Indian Ocean
1989 Jul 6	22	Regulus	+1.3	45°E	-0.8212	S Pacific Ocean
1989 Jul 15	5	Antares	+1.2	137°E	-0.5183	SE Australia, Tasmania, New Zealand, S Pacific Ocean, S South America, Antarctica
1989 Aug 3	7	Regulus	+1.3	19°E	-0.8886	SE Africa, Madagascar, Indian Ocean, Antarctica
1989 Aug 11	14	Antares	+1.2	111°E	-0.6056	SE Atlantic Ocean, S Africa, SE Australia, New Zealand, Antarctica
1989 Aug 30	14	Regulus	+1.3	8°W	-0.8852	SE Pacific Ocean, South America, S Atlantic Ocean, Antarctica
1989 Sep 2	15	Mercury	+0.6	27°E	-0.5840	South America, S Atlantic Ocean, part of Antarctica
1989 Sep 7	22	Antares	+1.2	84°E	-0.5761	S Pacific Ocean, S South America, S Atlantic Ocean, Antarctica

Table I   (cont.)

Date	T_o	Body	Magn.	Elong.	γ	Area of Visibility
	h					
1989 Sep 26	20	Regulus	+1.3	34°W	−0.9235	E Australia, New Zealand, SW Pacific Ocean, Antarctica
1989 Oct 5	5	Antares	+1.2	58°E	−0.4189	E Indian Ocean, Australia, New Zealand, S Pacific Ocean
1989 Oct 24	1	Regulus	+1.3	61°W	−1.0900	S Indian Ocean, part of Antarctica
1989 Nov 1	11	Antares	+1.2	31°E	−0.2311	S Atlantic Ocean, S Africa, S Indian Ocean, W Australia
1989 Nov 2	22	Venus	−4.0	47°E	−0.7165	New Zealand, S Pacific Ocean, S South America, S Atlantic Ocean, Antarctica
1989 Nov 28	17	Antares	+1.2	6°E	−0.1436	Pacific Ocean, South America, S Atlantic Ocean
1989 Dec 2	7	Venus	−4.3	45°E	+0.8411	Asia, Japan
1989 Dec 25	24	Antares	+1.2	25°W	−0.1978	E Indian Ocean, Indonesia, Australia, New Zealand, S Pacific
1990 Jan 22	8	Antares	+1.2	53°W	−0.2994	South America, S Atlantic Ocean, S Africa, Madagascar, Indian Ocean
1990 Feb 18	16	Antares	+1.2	80°W	−0.3056	Australia, New Zealand, Pacific Ocean, SW South America
1990 Mar 18	0	Antares	+1.2	107°W	−0.1611	Africa, Madagascar, Indian Ocean, Australia
1990 Mar 22	18	Mars	+1.3	54°W	+0.4099	Pacific Ocean, North America
1990 Apr 14	7	Antares	+1.2	134°W	+0.0548	E Pacific Ocean, South America, Atlantic Ocean
1990 May 11	13	Antares	+1.2	160°W	+0.2091	New Guinea, Pacific Ocean
1990 Jun 7	19	Antares	+1.2	172°E	+0.2313	Africa, Madagascar, Indian Ocean, Indonesia
1990 Jul 5	2	Antares	+1.2	147°E	+0.1620	E Pacific Ocean, South America, Atlantic Ocean, W Africa
1990 Jul 21	18	Jupiter	−1.4	5°W	+0.9809	Greenland, Iceland, W and N Europe, N Asia, Arctic regions
1990 Aug 1	9	Antares	+1.2	121°E	+0.1136	Indonesia, New Guinea, Australia, Pacific Ocean

Table I   (cont.)

Date	T_o	Body	Magn.	Elong.	γ	Area of Visibility
	h					
1990 Aug 18	13	Jupiter	-1.4	25°W	+0.3831	North America, Greenland, Atlantic Ocean, SW Europe, Africa
1990 Aug 18	24	Venus	-3.3	20°W	-0.4724	Indonesia, New Guinea, Australia, New Zealand, Pacific Ocean
1990 Aug 22	11	Mercury	+1.0	24°E	-0.1587	Atlantic Ocean, Africa, Madagascar, Indian Ocean
1990 Aug 28	17	Antares	+1.2	94°E	+0.1838	NE South America, Atlantic Ocean, Africa, Madagascar, Indian Ocean
1990 Sep 15	6	Jupiter	-1.5	47°W	-0.2653	Africa, Madagascar, Arabia, Indian Ocean
1990 Sep 25	1	Antares	+1.2	68°E	+0.3819	Pacific Ocean, Hawaii, Central America, NW South America
1990 Oct 12	19	Jupiter	-1.6	69°W	-0.9478	Australia, New Zealand, SW Pacific
1990 Oct 22	9	Antares	+1.2	41°E	+0.6133	Asia, Indonesia, Japan
1990 Oct 25	18	Saturn	+0.8	78°E	-1.1693	S Indian Ocean, Antarctica
1990 Nov 18	15	Antares	+1.2	14°E	+0.7489	E North America, Atlantic Ocean, W Europe, NW Africa
1990 Nov 22	4	Saturn	+0.8	52°E	-0.6908	S Indian Ocean, S Pacific Ocean, Antarctica
1990 Dec 15	21	Antares	+1.2	15°W	+0.7403	Pacific Ocean, Hawaii, W North Amer.
1990 Dec 18	6	Venus	-3.4	11°E	-1.0720	S Pacific Ocean, Antarctica
1990 Dec 19	16	Saturn	+0.8	27°E	-0.2507	SE Pacific Ocean, South America, Atlantic Ocean, Africa, Madagascar
1991 Jan 12	4	Antares	+1.2	42°W	+0.6754	Asia
1991 Jan 16	4	Saturn	+0.7	2°E	+0.1179	Indian Ocean, Indonesia, Australia, New Guinea, Pacific Ocean
1991 Feb 8	11	Antares	+1.2	70°W	+0.7003	E North America, N South America, Atlantic Ocean, Iberia, NW Africa
1991 Feb 11	5	Uranus	+6.1	40°W	-1.2266	S Pacific Ocean, part of Antarctica
1991 Feb 12	18	Saturn	+0.8	23°W	+0.4808	Pacific Ocean, Hawaii, North and Central America

Table I   (cont.)

Date	T_o	Body	Magn.	Elong.	γ	Area of Visibility
	h					
1991 Mar 7	19	Antares	+1.2	97°W	+0.8768	E Asia, Japan, NW Pacific Ocean
1991 Mar 10	15	Uranus	+6.1	66°W	−0.9553	S Atlantic Ocean, Antarctica
1991 Mar 12	7	Saturn	+0.9	48°W	+0.9128	Europe, N Africa, Turkey, W Siberia
1991 Mar 22	17	Mars	+1.1	83°E	+0.6601	NE North America, Greenland, Iceland, N Atlantic Ocean, Europe except S, Asia, Arctic
1991 Apr 4	3	Antares	+1.2	124°W	+1.1291	Europe, SW Asia
1991 Apr 7	0	Uranus	+6.0	93°W	−0.6172	S Atlantic Ocean, S Indian Ocean, Australia, New Zealand, Antarctica
1991 Apr 19	24	Mars	+1.5	70°E	−0.6018	Pacific Ocean, South America
1991 May 4	8	Uranus	+6.0	120°W	−0.3342	Pacific Ocean, South America, Atlantic Ocean, SW Africa
1991 May 31	13	Uranus	+5.9	146°W	−0.2178	Australia, New Zealand, Pacific
1991 May 31	20	Neptune	+7.7	144°W	−1.2069	S Pacific Ocean, part of Antarctica
1991 Jun 27	17	Uranus	+5.9	173°W	−0.2739	SE Africa, Madagascar, Indian Oc., Australia, Pacific Ocean
1991 Jun 28	1	Neptune	+7.7	170°W	−1.2316	SE Indian Ocean, part of Antarctica
1991 Jul 24	21	Uranus	+5.9	160°E	−0.3896	South America, S Atlantic Ocean, S Africa, Madagascar, Indian Ocean
1991 Aug 11	9	Mercury	+1.6	18°E	+0.5629	Europe, Svalbard, Asia, Borneo, W Pacific Ocean
1991 Aug 21	2	Uranus	+6.0	133°E	−0.4100	S Pacific Ocean, South America, Atlantic Ocean
1991 Sep 17	9	Uranus	+6.0	106°E	−0.2467	Indian Ocean, Australia, New Zealand, Pacific Ocean
1991 Sep 17	18	Neptune	+7.7	110°E	−1.1160	SE Indian Ocean, Antarctica
1991 Oct 4	15	Venus	−4.3	42°W	−0.2127	Pacific Ocean, Hawaii, South America
1991 Oct 14	18	Uranus	+6.1	79°E	+0.0646	South America, Atlantic Ocean, Africa, Arabia
1991 Oct 15	2	Neptune	+7.8	83°E	−0.8004	S Pacific Ocean, South America, Antarctica

Table I   (cont.)

Date	T_o	Body	Magn.	Elong.	γ	Area of Visibility
	h					
1991 Nov 8	4	Mercury	-0.2	20°E	-0.7092	Indian Ocean, New Zealand, S Pacific Ocean, Antarctica
1991 Nov 11	3	Uranus	+6.1	53°E	+0.4000	New Guinea, Pacific Ocean, Hawaii
1991 Nov 11	11	Neptune	+7.8	56°E	-0.4763	S Atlantic Ocean, S Africa, S Indian Ocean, Australia, New Zealand, part of Antarctica
1991 Dec 8	13	Uranus	+6.1	26°E	+0.6522	Atlantic Ocean, Europe, Africa, SW Asia
1991 Dec 8	20	Neptune	+7.8	29°E	-0.2565	Pacific Ocean, South America, Atlantic Ocean
1992 Jan 3	10	Mars	+1.6	17°W	-0.9069	S South America, S Atlantic Ocean, S Indian Ocean, Antarctica
1992 Jan 4	23	Uranus	+6.2	0°E	+0.8204	Japan, Pacific Ocean, Hawaii, Kamchatka, NE North America
1992 Jan 5	4	Neptune	+7.8	2°E	-0.1428	Indian Ocean, Australia, New Zealand, Pacific Ocean
1992 Jan 31	17	Venus	-3.4	33°W	-1.1266	S Atlantic Ocean, Antarctica
1992 Feb 1	8	Uranus	+6.1	26°W	+1.0010	Europe, N Africa, W Asia
1992 Feb 1	12	Neptune	+7.8	24°W	-0.0287	South America, Atlantic Oc., Africa
1992 Feb 28	21	Neptune	+7.8	51°W	+0.1877	Australia, New Guinea, Pacific Ocean, Hawaii
1992 Mar 27	5	Neptune	+7.8	78°W	+0.4966	Atlantic Ocean, Africa, Asia
1992 Apr 23	13	Neptune	+7.7	105°W	+0.7832	Pacific Ocean, Hawaii, North America
1992 May 20	21	Neptune	+7.7	131°W	+0.9323	Asia
1992 Jun 1	5	Mercury	-1.9	1°E	+1.0237	Greenland, Iceland, Europe except S, N Asia, Alaska, Arctic regions
1992 Jun 17	3	Neptune	+7.7	158°W	+0.9228	E North America, Greenland, Iceland, Atlantic Ocean, W Europe
1992 Jul 14	9	Neptune	+7.7	175°E	+0.8438	Pacific Ocean, Hawaii, North America
1992 Aug 10	13	Neptune	+7.7	149°E	+0.8323	Asia, Japan, NW Pacific Ocean
1992 Sep 6	18	Neptune	+7.7	122°E	+0.9763	Europe, N Africa, W Asia

Table I   (cont.)

Date	T_o	Body	Magn.	Elong.	γ	Area of Visibility
	h					
1992 Sep 20	9	Mars	+0.6	83°W	-0.8785	South America, S Atlantic Ocean
1992 Oct 4	1	Neptune	+7.7	95°E	+1.2499	Canada
1992 Oct 27	15	Mercury	-0.0	23°E	+0.5473	E North America, Cuba, Haiti, Atlantic Ocean, Africa, S Italy, Greece
1992 Oct 28	15	Venus	-3.4	35°E	+0.4049	N South America, Atlantic Ocean, Africa, Arabia
1993 Feb 25	5	Venus	-4.3	40°E	-0.5246	SE Indian Ocean, Australia, Pacific
1993 Apr 19	16	Venus	-4.0	26°W	+0.4968	E Pacific Ocean, Hawaii, North America, Atlantic Ocean, Greenland, Iceland
1993 May 22	5	Mercury	-1.5	7°E	-0.9185	S Indian Ocean, Australia
1993 Oct 15	14	Spica	+1.2	3°E	-1.2321	Part of Antarctica
1993 Nov 14	18	Mars	+1.6	12°E	+0.5104	E Pacific Ocean, the Americas, Atlantic Ocean
1993 Dec 12	11	Mercury	-0.5	12°W	+0.1507	South America, Atlantic Ocean, Africa, Madagascar, Arabia
1993 Dec 12	18	Venus	-3.4	8°W	+0.4492	Pacific Ocean, Hawaii, the Americas, Atlantic Ocean
1994 Jan 5	16	Spica	+1.2	81°W	-1.1804	Antarctica
1994 Feb 1	21	Spica	+1.2	109°W	-0.9420	S Indian Ocean, Antarctica
1994 Mar 1	5	Spica	+1.2	137°W	-0.7025	SE Pacific Ocean, South America, S Atlantic Ocean, Antarctica
1994 Mar 28	14	Spica	+1.2	164°W	-0.5790	E Indian Ocean, Java, Australia, New Zealand, S Pacific Ocean, part of Antarctica
1994 Apr 12	23	Venus	-3.3	21°E	+1.1257	NE Asia, Greenland, Arctic regions
1994 Apr 25	2	Spica	+1.2	169°E	-0.5739	SE Pacific Ocean, South America, S Atlantic Ocean, part Antarctica
1994 May 22	12	Spica	+1.2	143°E	-0.5930	Indian Ocean, Australia, New Zealand, SW Pacific Ocean, part Antarctica
1994 Jun 18	20	Spica	+1.2	116°E	-0.5308	South America, S Atlantic Ocean, S Indian Ocean, part of Antarctica

Table I (cont.)

Date	T_o	Body	Magn.	Elong.	γ	Area of Visibility
	h					
1994 Jul 5	5	Mars	+1.4	42°W	-0.3731	Africa, Madagascar, Indian Ocean, extreme S Asia, Indonesia, W Australia
1994 Jul 16	2	Spica	+1.2	90°E	-0.3506	Pacific Ocean, South America
1994 Aug 12	7	Spica	+1.2	64°E	-0.1117	Indian Ocean, S Asia, Indonesia, Australia, New Zealand, SW Pacific
1994 Sep 8	14	Spica	+1.2	38°E	+0.0781	N South America, Atlantic Ocean, Africa, Madagascar, Indian Ocean
1994 Oct 5	23	Spica	+1.2	11°E	+0.1522	Japan, Pacific Ocean
1994 Oct 7	12	Jupiter	-1.3	32°E	-0.7148	S Atlantic Ocean, S Indian Ocean, Antarctica
1994 Nov 2	10	Spica	+1.2	16°W	+0.1443	Atlantic Ocean, Africa, Madagascar, Indian Ocean
1994 Nov 4	8	Jupiter	-1.3	11°E	-0.1056	Africa, Madagascar, Indian Ocean, Australia, New Guinea
1994 Nov 29	21	Spica	+1.2	44°W	+0.1671	Japan, E China, Pacific Ocean
1994 Dec 2	5	Jupiter	-1.3	11°W	+0.4644	Asia, Indonesia, New Guinea, W Pacific Ocean
1994 Dec 27	6	Spica	+1.2	71°W	+0.3192	Atlantic Ocean, Spain, Africa, Arabia, Madagascar, Indian Ocean, India, Ceylon
1994 Dec 30	1	Jupiter	-1.3	34°W	+1.0503	NE Asia, Japan, NW Pacific Ocean
1995 Jan 23	12	Spica	+1.2	99°W	+0.5840	NE Pacific Ocean, North America, Central America, N South America, Atlantic Ocean
1995 Jan 27	12	Venus	-3.9	46°W	+0.1758	E Pacific Ocean, South America, Atlantic Ocean, W Africa
1995 Feb 19	17	Spica	+1.2	127°W	+0.8332	NE Asia, Japan, Pacific Ocean, Hawaii, Alaska
1995 Mar 19	0	Spica	+1.2	154°W	+0.9563	Iceland, Europe except SW, western half of Asia
1995 Apr 15	10	Spica	+1.2	178°E	+0.9654	NE Siberia, NE Pacific Ocean, North America

Table I   (cont.)

Date	T_0	Body	Magn.	Elong.	γ	Area of Visibility
	h					
1995 May 12	21	Spica	+1.2	152°E	+0.9673	Iceland, Europe except SW, western half of Asia
1995 May 27	7	Venus	-3.3	23°W	+0.9171	Greenland, Iceland, Europe except SW, extreme NW Africa, N Asia, Arctic regions
1995 Jun 9	6	Spica	+1.2	126°E	+1.0675	Extreme NE Siberia, North America except E
1995 Jun 26	2	Mercury	+1.0	22°W	+0.6667	NE Europe, Asia, Japan, N Pacific, NW North America, Arctic regions
1995 Aug 30	3	Mars	+1.6	48°E	-0.1544	Indonesia, New Guinea, Australia, New Zealand, Pacific Ocean
1996 Feb 22	5	Venus	-3.7	42°E	-0.0576	Indonesia, Australia, New Guinea, Pacific Ocean, Hawaii
1996 Apr 17	5	Mars	+1.4	9°W	+0.5958	SE Europe, NE Africa, Asia, Japan
1996 Jun 14	0	Mercury	+0.5	23°W	-0.4656	Australia, New Guinea, Pacific Ocean
1996 Jul 12	9	Venus	-4.2	37°W	+0.4106	Atlantic Ocean, Europe, N Africa, Asia
1996 Aug 8	7	Aldebaran	+1.1	66°W	+1.1260	NE North America, Greenland, Arctic regions, N Asia
1996 Aug 16	18	Mercury	+0.4	27°E	-0.3421	Pacific, South America, S Atlantic
1996 Sep 4	14	Aldebaran	+1.1	93°W	+0.9313	NE Asia, Arctic regions, Alaska, Canada, Greenland, Iceland
1996 Oct 1	22	Aldebaran	+1.1	119°W	+0.8589	Europe, N Africa, northern half of Asia, Greenland, Iceland, Arctic
1996 Oct 29	8	Aldebaran	+1.1	146°W	+0.9138	NE Siberia, N Pacific Ocean, N North America, Greenland, Iceland, W Europe except Iberia, N Europe, Arctic regions
1996 Nov 25	16	Aldebaran	+1.1	172°W	+0.9928	N and E Europe, N Asia, Alaska, Arctic regions
1996 Dec 22	24	Aldebaran	+1.1	158°E	+0.9666	North America, Greenland, Iceland, Scandinavia, NW Asia, Arctic
1997 Jan 19	5	Aldebaran	+1.1	130°E	+0.7991	NE Asia, N Pacific Ocean, N North America, Greenland, Iceland, Arctic regions

Table I   (cont.)

Date	T_o	Body	Magn.	Elong.	γ	Area of Visibility
	h					
1997 Feb 15	11	Aldebaran	+1.1	103°E	+0.5909	E Europe, Asia, N Pacific, Alaska
1997 Mar 14	19	Aldebaran	+1.1	76°E	+0.4827	SE North America, Central America, Atlantic Ocean, S Greenland, Iceland, Europe, N Africa, W Asia
1997 Apr 7	1	Saturn	+1.0	6°W	+1.0429	Asia, N Alaska, Arctic regions
1997 Apr 7	13	Venus	-3.5	2°E	-0.6483	S South America, Atlantic Ocean, Africa, Madagascar, Antarctica
1997 Apr 11	4	Aldebaran	+1.1	49°E	+0.5178	SE and E Asia, N Pacific Ocean, North America
1997 May 4	15	Saturn	+1.0	30°W	+0.7580	NE Pacific Ocean, North America, N Atlantic Ocean, Greenland, Iceland, British Isles, Arctic
1997 May 5	17	Mercury	+1.9	15°W	-1.1536	S South America, S Atlantic Ocean, Antarctica
1997 May 8	14	Aldebaran	+1.1	22°E	+0.6170	E North America, Atlantic Ocean, Greenland, Iceland, Europe, Svalbard, western half of Asia
1997 Jun 1	3	Saturn	+1.0	53°W	+0.4511	Africa, Asia, Japan
1997 Jun 4	22	Aldebaran	+1.1	7°W	+0.6623	E Asia, N Pacific Ocean, North America, Greenland
1997 Jun 13	16	Mars	+0.6	95°E	-0.2906	Atlantic Ocean, Africa, Madagascar, Indian Ocean
1997 Jun 28	12	Saturn	+0.9	77°W	+0.1605	Pacific Ocean, Central America, Florida, NW South America, Atlantic Ocean, Iberia, NW Africa
1997 Jul 2	5	Aldebaran	+1.1	31°W	+0.5894	Europe, N Africa, Asia except S, Japan
1997 Jul 25	19	Saturn	+0.8	103°W	-0.0185	Indian Ocean, Indonesia, Australia, New Guinea, Pacific Ocean, Hawaii
1997 Jul 29	11	Aldebaran	+1.1	57°W	+0.4301	North and Central America, Atlantic, Europe except N, N Africa, Turkey
1997 Aug 5	20	Mercury	+0.7	27°E	+1.0596	N Canada, N Atlantic Ocean, Greenland, Iceland, British Isles, Arctic regions

Table I (cont.)

Date	T_o	Body	Magn.	Elong.	γ	Area of Visibility
	h					
1997 Aug 22	2	Saturn	+0.6	129°W	-0.0071	E South America, Atlantic Ocean, Africa, S Asia
1997 Aug 25	17	Aldebaran	+1.1	83°W	+0.2887	Pacific Ocean, Hawaii, North and Central America
1997 Sep 18	10	Saturn	+0.5	157°W	+0.1741	Pacific Ocean, Hawaii, North and Central America
1997 Sep 21	24	Aldebaran	+1.1	109°W	+0.2609	Africa, Asia, Japan
1997 Oct 15	18	Saturn	+0.4	173°E	+0.3811	Africa, Indian Ocean, Asia, Japan
1997 Oct 19	9	Aldebaran	+1.1	136°W	+0.3520	Pacific Ocean, Hawaii, North America, Atlantic Ocean
1997 Nov 12	1	Saturn	+0.6	145°E	+0.4238	The Americas, Atlantic Ocean, Europe, N Africa
1997 Nov 15	20	Aldebaran	+1.1	163°W	+0.4672	Africa, S and SE Europe, Asia, Japan, W Pacific Ocean
1997 Dec 9	6	Saturn	+0.8	116°E	+0.2046	Australia, New Guinea, Pacific Ocean, Hawaii, North America
1997 Dec 13	5	Aldebaran	+1.1	167°E	+0.4885	Pacific Ocean, Hawaii, North America, S Greenland, Atlantic Ocean, extreme W Europe
1998 Jan 5	12	Saturn	+0.9	89°E	-0.1997	S Africa, Madagascar, Indian Ocean, SE Asia, Indonesia, New Guinea
1998 Jan 9	13	Aldebaran	+1.1	140°E	+0.3809	Indian Ocean, Asia, Japan, Pacific
1998 Feb 1	21	Saturn	+0.1	63°E	-0.6150	SE Pacific Ocean, South America, Atlantic Ocean
1998 Feb 5	18	Aldebaran	+1.1	113°E	+0.2324	Atlantic Ocean, N Africa, S Europe, SW Asia
1998 Feb 27	23	Mars	+1.4	17°E	-0.6955	New Zealand, Pacific Ocean, South America, Antarctica
1998 Mar 1	10	Saturn	+0.9	38°E	-0.9300	S Indian Ocean, Australia, Antarctica
1998 Mar 5	0	Aldebaran	+1.1	86°E	+0.1744	Pacific Ocean, Hawaii, S North America, Central America, Atlantic
1998 Mar 24	19	Venus	-4.1	46°W	+0.0915	NE Australia, Pacific Ocean, S North America, Central America

Table I   (cont.)

Date	T_o	Body	Magn.	Elong.	γ	Area of Visibility
	h					
1998 Mar 26	11	Jupiter	-1.6	24°W	+0.7756	NE North America, Atlantic Ocean, Greenland, Iceland, Europe, NW Africa, W Soviet Union
1998 Mar 29	1	Saturn	+0.8	13°E	-1.1606	S Pacific Ocean, Antarctica
1998 Apr 1	8	Aldebaran	+1.1	59°E	+0.2509	Indian Ocean, S and SE Asia, Japan, Pacific Ocean
1998 Apr 23	7	Jupiter	-1.7	45°W	+0.2051	Atlantic Ocean, Africa, Asia
1998 Apr 23	8	Venus	-3.8	45°W	-0.0754	Atlantic Ocean, Africa, Madagascar, Indian Ocean, Arabia, India
1998 Apr 24	19	Mercury	+1.1	24°W	-0.8117	S Pacific, South America, Antarctica
1998 Apr 28	18	Aldebaran	+1.1	32°E	+0.3863	North and Central America, Atlantic Ocean, Europe except N, N Africa
1998 May 20	23	Jupiter	-1.8	67°W	-0.3415	Indian Ocean, Australia, New Guinea, Pacific Ocean
1998 May 26	4	Aldebaran	+1.1	7°E	+0.4683	Asia, Japan, N Pacific Ocean, extreme SW Alaska
1998 Jun 1	3	Regulus	+1.3	79°E	-1.0417	S Pacific Ocean, New Zealand
1998 Jun 17	11	Jupiter	-2.0	89°W	-0.7695	S Pacific Ocean, South America, Atlantic Ocean
1998 Jun 22	14	Aldebaran	+1.1	22°W	+0.4401	North and Central America, Atlantic Ocean, S Greenland, S Iceland, Europe, N Africa, Turkey
1998 Jun 28	12	Regulus	+1.3	53°E	-0.7956	S Africa, Madagascar, S Indian Ocean, part of Antarctica
1998 Jul 14	19	Jupiter	-2.2	114°W	-0.9526	New Zealand, S Pacific, Antarctica
1998 Jul 19	21	Aldebaran	+1.1	47°W	+0.3344	SE Asia, Japan, Pacific Ocean, SW North America
1998 Jul 25	20	Regulus	+1.3	27°E	-0.6740	Pacific Ocean, S South America
1998 Aug 11	0	Jupiter	-2.4	141°W	-0.8261	S Atlantic, Indian Ocean, Antarctica
1998 Aug 16	3	Aldebaran	+1.1	73°W	+0.2469	Africa, Asia, Japan
1998 Aug 22	4	Regulus	+1.3	1°E	-0.6636	Indian Ocean, Australia, New Zealand

Table I   (cont.)

Date	T_0	Body	Magn.	Elong.	γ	Area of Visibility
	h					
1998 Sep 7	5	Jupiter	-2.5	170°W	-0.4879	SE Pacific Ocean, South America, Atlantic Ocean, W Africa
1998 Sep 12	8	Aldebaran	+1.1	100°W	+0.2645	S and SE North America, Central America, N South America, Atlantic, W and S Europe, Africa, Turkey
1998 Sep 18	11	Regulus	+1.3	26°W	-0.6641	South America, Atlantic Ocean, S Africa
1998 Sep 19	18	Venus	-3.4	11°W	-0.0798	Pacific Ocean, Hawaii, South America
1998 Sep 20	7	Mercury	-1.3	5°W	+0.1381	Italy, Greece, NE Africa, S Asia, Indian Ocean, Indonesia, Australia
1998 Oct 4	10	Jupiter	-2.4	160°E	-0.1993	Australia, New Zealand, Pacific Ocean, Mexico
1998 Oct 9	16	Aldebaran	+1.1	126°W	+0.3914	SE Asia, Borneo, Japan, Pacific Ocean, W North America
1998 Oct 15	16	Regulus	+1.3	52°W	-0.5580	Pacific Ocean, South America
1998 Oct 16	3	Mars	+1.8	47°W	-1.0386	S Indian Ocean, Antarctica
1998 Oct 31	16	Jupiter	-2.3	130°E	-0.2140	SE Atlantic Ocean, S Africa, Madagascar, Indian Ocean, SE Asia, Indonesia
1998 Nov 6	2	Aldebaran	+1.1	153°W	+0.5389	E North America, N South America, Atlantic Ocean, S Greenland, Iceland Europe, extreme NW Africa, Asia
1998 Nov 11	22	Regulus	+1.3	79°W	-0.3095	Indian Ocean, S Asia, Indonesia, New Guinea, Australia, New Zealand
1998 Nov 13	18	Mars	+1.7	59°W	+0.5511	NE Asia, Japan, Pacific Ocean, North and Central America
1998 Nov 28	1	Jupiter	-2.1	103°E	-0.5873	S Pacific Ocean, South America, Atlantic Ocean
1998 Dec 3	13	Aldebaran	+1.1	174°E	+0.5998	Asia, Japan, N Pacific Ocean, North America
1998 Dec 9	6	Regulus	+1.3	107°W	-0.0104	Florida, N South America, Cuba, Atlantic Ocean, Africa
1998 Dec 25	11	Jupiter	-1.9	78°E	-1.1690	S New Zealand, Antarctica

Table I   (cont.)

Date	T_0	Body	Magn.	Elong.	γ	Area of Visibility
	h					
1998 Dec 30	23	Aldebaran	+1.1	150°E	+0.5447	North and Central America, Atlantic, Greenland, Iceland, Europe, W Asia
1999 Jan 5	15	Regulus	+1.3	135°W	+0.1974	E Asia, Japan, Pacific Ocean, Hawaii
1999 Jan 27	7	Aldebaran	+1.1	123°E	+0.4543	E Asia, Japan, Pacific Ocean, North America
1999 Feb 2	1	Regulus	+1.3	163°W	+0.2582	E North America, Atlantic Ocean, SW Europe, Africa, Arabia, Indian Ocean
1999 Feb 17	2	Mercury	−1.1	10°E	−0.2398	Australia, New Zealand, Pacific Ocean
1999 Feb 23	12	Aldebaran	+1.1	95°E	+0.4473	SE Europe, Africa, Asia, Japan, NW Pacific Ocean
1999 Mar 1	10	Regulus	+1.3	169°E	+0.2511	Japan, Pacific Ocean, Hawaii, Mexico, Central America
1999 Mar 22	18	Aldebaran	+1.1	68°E	+0.5636	North and Central America, Atlantic, Greenland, Iceland, Europe, W Asia
1999 Mar 28	16	Regulus	+1.3	142°E	+0.3093	Europe, N Africa, Asia, Borneo, New Guinea
1999 Apr 10	9	Neptune	+7.8	76°W	+1.1946	Labrador, N Atlantic Ocean, Greenland, Iceland
1999 Apr 11	7	Uranus	+6.2	65°W	+1.0663	NE Atlantic Ocean, Iceland, Europe
1999 Apr 14	5	Mercury	+0.7	27°W	−1.0281	SE Indian Ocean, SE Australia, Tasmania, Antarctica
1999 Apr 19	2	Aldebaran	+1.1	41°E	+0.7296	Asia, Japan, N Pacific Ocean, North America, Arctic regions
1999 Apr 24	22	Regulus	+1.3	116°E	+0.4973	North America, Greenland, Iceland, Atlantic Ocean, Europe except N and NE, Africa, Turkey, Arabia
1999 May 7	16	Neptune	+7.7	102°W	+0.9151	Japan, extreme E Siberia, Pacific Ocean, NW North America
1999 May 8	16	Uranus	+6.1	91°W	+0.7738	Pacific Ocean, Hawaii, extreme E Siberia, North America
1999 May 16	12	Aldebaran	+1.1	16°E	+0.8391	NE North America, Atlantic Ocean, Greenland, Iceland, British Isles, N Europe, N Asia, Arctic regions

Table I   (cont.)

Date	T_0	Body	Magn.	Elong.	γ	Area of Visibility
	h					
1999 May 22	4	Regulus	+1.3	89°E	+0.7597	NE Asia, North America, Greenland, Arctic regions
1999 Jun 3	22	Neptune	+7.7	129°W	+0.6976	NE Africa, Asia, Japan
1999 Jun 4	22	Uranus	+6.1	117°W	+0.5413	E Africa, Indian Ocean, Asia, Japan
1999 Jun 12	23	Aldebaran	+1.1	13°W	+0.8448	Asia, Japan, North America, Arctic
1999 Jun 18	12	Regulus	+1.3	63°E	+0.9815	Greenland, Iceland, Scandinavia, N and NE Asia, Japan, Arctic
1999 Jul 1	2	Neptune	+7.7	155°W	+0.6276	Newfoundland, Atlantic Ocean, Europe, Africa, Turkey, Arabia
1999 Jul 2	3	Uranus	+6.0	144°W	+0.4623	N South America, Atlantic Ocean, Africa, SE Europe, Turkey, Arabia
1999 Jul 10	9	Aldebaran	+1.1	38°W	+0.7854	NE North America, Atlantic Ocean, Greenland, Iceland, Europe except S, Asia, Arctic regions
1999 Jul 15	21	Regulus	+1.3	37°E	+1.0883	Arctic regions, Greenland, Iceland, N Atlantic Ocean, British Isles, W Scandinavia
1999 Jul 28	8	Neptune	+7.7	178°E	+0.6862	Pacific Ocean, Hawaii, North America
1999 Jul 29	7	Uranus	+6.0	171°W	+0.5328	Pacific Ocean, North and Central America, Atlantic Ocean
1999 Aug 6	16	Aldebaran	+1.1	64°W	+0.7516	NE Asia, N Pacific Ocean, North America, Greenland, Iceland, Arctic regions
1999 Aug 10	3	Mercury	+0.9	18°W	+1.1580	NW North America, Arctic regions
1999 Aug 12	7	Regulus	+1.3	11°E	+1.0980	Arctic regions, NE Siberia
1999 Aug 24	14	Neptune	+7.7	151°E	+0.7590	Asia, Japan, W Pacific Ocean
1999 Aug 25	13	Uranus	+6.0	162°E	+0.6405	E Asia, Japan, N Pacific Ocean
1999 Sep 2	22	Aldebaran	+1.1	90°W	+0.8184	N and E Europe, Asia, Alaska, Arctic
1999 Sep 8	15	Regulus	+1.3	16°W	+1.1029	Arctic regions, Greenland, Iceland, Europe except SW
1999 Sep 20	22	Neptune	+7.7	124°E	+0.7166	E North America, Atlantic Ocean, Iceland, Europe except NE, NW Africa

Table I   (cont.)

Date	T_o	Body	Magn.	Elong.	γ	Area of Visibility
	h					
1999 Sep 21	20	Uranus	+6.1	135°E	+0.6392	Atlantic Ocean, Europe except NW and N, Africa, W Asia
1999 Sep 30	4	Aldebaran	+1.1	117°W	+0.9816	NE North America, Greenland, Iceland, Scandinavia, N Asia, Arctic regions
1999 Oct 5	22	Regulus	+1.3	42°W	+1.2041	Canada, Arctic regions
1999 Oct 18	7	Neptune	+7.7	97°E	+0.5064	S Japan, Pacific Ocean, Hawaii
1999 Oct 19	4	Uranus	+6.1	108°E	+0.4501	Pacific Ocean, Hawaii, North America
1999 Oct 27	11	Aldebaran	+1.1	143°W	+1.1538	E Siberia, Arctic regions
1999 Nov 14	15	Neptune	+7.8	70°E	+0.1961	Atlantic Ocean, Africa, Madagascar, Indian Ocean, S Asia
1999 Nov 15	13	Uranus	+6.2	80°E	+0.1208	Africa, Madagascar, Indian Ocean, SE Asia, Indonesia
1999 Nov 23	22	Aldebaran	+1.1	170°W	+1.2377	NE Canada
1999 Dec 11	22	Neptune	+7.8	43°E	-0.0780	Pacific Ocean, South America
1999 Dec 12	18	Mars	+1.1	52°E	+0.6066	The Americas, Atlantic Ocean, W Europe, NW Africa
1999 Dec 12	21	Uranus	+6.2	54°E	-0.2112	Pacific Ocean, South America, Atlantic Ocean
1999 Dec 21	9	Aldebaran	+1.1	160°E	+1.2170	Siberia, Arctic regions
2000 Jan 8	6	Neptune	+7.8	16°E	-0.2343	Madagascar, Indian Ocean, Australia, New Guinea, Pacific
2000 Jan 9	5	Uranus	+6.2	27°E	-0.4393	Indian Ocean, Australia, New Zealand, Pacific Ocean
2000 Jan 17	19	Aldebaran	+1.1	133°E	+1.1773	NE Canada, Greenland, Arctic regions
2000 Feb 4	14	Neptune	+7.8	11°W	-0.3219	Pacific Ocean, South America, Atlantic Ocean, Africa
2000 Feb 5	14	Uranus	+6.2	1°E	-0.5822	SE Pacific Ocean, S South America, S Atlantic Ocean, Africa, part of Antarctica
2000 Feb 14	3	Aldebaran	+1.1	105°E	+1.2296	Siberia
2000 Mar 2	24	Neptune	+7.8	37°W	-0.4610	Indian Ocean, S Australia, New Zealand, Pacific Ocean

Table I   (cont.)

Date	T_o	Body	Magn.	Elong.	γ	Area of Visibility
	h					
2000 Mar 4	1	Venus	−3.3	25°W	−0.6785	S Indian Ocean, Tasmania, New Zealand, S Pacific, Antarctica
2000 Mar 4	1	Uranus	+6.2	25°W	−0.7463	S Indian Ocean, New Zealand, S Pacific Ocean, Antarctica
2000 Mar 30	10	Neptune	+7.8	64°W	−0.7167	SE Pacific Ocean, S South America, S Atlantic Ocean, S Africa, Antarctica
2000 Mar 31	12	Uranus	+6.2	51°W	−1.0070	S Atlantic Ocean, Antarctica
2000 Apr 26	19	Neptune	+7.7	90°W	−1.0363	S Pacific Ocean, Antarctica
2000 Jul 29	17	Mercury	+0.1	20°W	+0.8078	N Pacific Ocean, Alaska, Canada, Atlantic Ocean, Greenland, Iceland, W and N Europe, E Siberia, Arctic
2000 Jul 30	12	Mars	+1.9	9°W	−0.6138	South America, Atlantic Ocean, S Africa, Madagascar
2000 Aug 1	2	Venus	−3.3	14°E	+1.0385	N Asia, North America, Arctic
2000 Aug 13	17	Neptune	+7.7	164°E	−1.2515	S Pacific Ocean
2000 Aug 28	3	Mars	+2.0	18°W	+0.8929	N and NE Europe, N Asia, Pacific, Alaska, Arctic regions

REMARKS

1980 Aug 12   Last of a series of occultations of Jupiter, which began on 1979 November 13.

1980 Aug 13   Last of a series of occultations of Saturn, which began on 1979 September 20.

1980 Dec 26   Last of a series of occultations of Regulus, which began on 1979 November 12.

1981 Apr 8   Last of a series of occultations of Aldebaran, which began on 1978 January 19. There will be no other occultation of a first-magnitude star by the Moon until 1986 March 30.

1982 Jan 21   First of a series of occultations of Neptune.

1983 Mar 6   First of a series of occultations of Jupiter.

1983 Mar 7   Last of a series of occultations of Neptune.

1983 Oct 10   First of a series of occultations of Uranus.

1983 Nov 4   First of a series of occultations of Saturn.

1983 Dec 5   Last of a series of occultations of Jupiter.

1984 Sep 29   Last of a series of occultations of Uranus.

1984 Oct 25   Last of a series of occultations of Saturn.

1986 Mar 30   First of a series of occultations of Antares, which will end on 1991 April 4.
               First occultation of Antares since 1972 September 14.

1986 Aug 21   First of a series of occultations of Jupiter. In fact, this small occultation
               is an "isolated" case, the other occultations of this series taking place
               from January to March 1987.

1987 Feb 18   First of a series of occultations of Spica.

1987 Mar 29   Last of a short series of occultations of Jupiter.

1987 Apr 14   This occultation of Spica occurs during a *penumbral eclipse* of the Moon, as
               was announced by G.P. Können and J. Meeus, *Journal of the British Astrono-
               mical Association*, Vol. 85, page 23 (1974).

1988 Jun 19   First of a series of occultations of Regulus. — In a time interval of eight
               days, three first-magnitude stars will be occulted by the Moon, though not
               all in the same countries : Regulus on June 19, Spica on June 24, and
               Antares on June 27. This happens for the first time since December 1969 –
               January 1970. Next time in January 2007.

1988 Jun 24   Last of a series of occultations of Spica.

1989 Feb 20   In a part of the Pacific Ocean, the occultation of Regulus occurs during the
               initial *penumbral* phase preceding the total lunar eclipse of that date.

1989 Oct 24   Last of a series of occultations of Regulus.

1990 Jul 21   First of a series of occultations of Jupiter.

1990 Oct 12   Last of a series of occultations of Jupiter.

1990 Oct 25   First of a series of occultations of Saturn.

1991 Feb 11   First of a series of occultations of Uranus.

1991 Mar 12   Last of a series of occultations of Saturn.

1991 Apr 4   Last of a series of occultations of Antares, which began on 1986 March 30.
               There will be no other occultation of Antares by the Moon until the be-
               ginning of a new series on 2005 January 7.

1991 May 31   First of a series of occultations of Neptune.

1992 Feb 1   Last of a series of occultations of Uranus.

1992 Oct 4   Last of a series of occultations of Neptune.

1993 Oct 15   First of a series of occultations of Spica, though this series will be
               interrupted in November-December 1993.

1994 Oct 7   First of a series of occultations of Jupiter.

1994 Dec 30   Last of a series of occultations of Jupiter.

1995 Apr 15   In a part of North America, the occultation of Spica occurs during the
               initial *penumbral* phase preceding the partial lunar eclipse of that date.

1995 Jun 9   Last of a series of occultations of Spica.

1996 Aug 8   First of a series of occultations of Aldebaran, which will end on 2000 Feb-
               ruary 14.

1997 Apr 7   First of a series of occultations of Saturn.

1998 Mar 26   First of a series of occultations of Jupiter.

1998 Mar 29   Last of a series of occultations of Saturn.

1998 Apr 23   In a part of the Atlantic Ocean, of Africa, of Arabia, of the Indian Ocean and of India, Jupiter and Venus will simultaneously be hidden by the Moon.

1998 Jun 1   First of a series of occultations of Regulus.

1998 Dec 25   Last of a series of occultations of Jupiter.

1999 Apr 10   First of a series of occultations of Neptune.

1999 Apr 11   First of a series of occultations of Uranus.

1999 Jul 28   In a part of the U.S.A., the emersion of Neptune occurs during the *penumbral* phase preceding the partial lunar eclipse of that day.

1999 Oct 5   Last of a series of occultations of Regulus.

2000 Feb 14   Last of a series of occultations of Aldebaran, which began on 1996 August 8. There will be no other occultation of a first-magnitude star by the Moon until 2005 January 7, when a new series of occultations of Antares will begin. The next occultation of Aldebaran will take place on 2015 Jan. 29.

2000 Mar 4   In a part of the southern Indian Ocean, in New Zealand, in a part of the southern Pacific Ocean and in Antarctica, Venus and Uranus will simultaneously be hidden by the Moon.

2000 Mar 31   Last of a series of occultations of Uranus.

2000 Aug 13   Last of a series of occultations of Neptune, though that series was interrupted from May to July 2000.

Let us also mention the following near-misses :

1987 Jan 5	13^h	Mars	$\gamma = -1.2708$
1993 Nov 12	1	Spica	-1.2700
2000 Sep 9	23	Neptune	-1.2714

There will be no occultation on these dates, but the 'shadow' of the Moon will pass very close to the Earth's surface.

TABLE II — OCCULTATIONS OF STARS

Date Star	T_0	δ	H_0	X x' x''	Y y' y''
	h	$\circ$	$\circ$		
1980 Jan 6 Regulus	7	+12.0635	58.1989	−0.15 341 +0.53 132 −0.00 002	−0.53 097 −0.14 606 +0.00 003
1980 Jan 27 Aldebaran	5	+16.4678	131.9487	−0.24 277 +0.56 903 +0.00 003	+0.28 222 +0.09 291 −0.00 001
1980 Feb 2 Regulus	15	+12.0628	205.1376	+0.11 189 +0.53 319 −0.00 002	−0.52 555 −0.14 546 +0.00 003
1980 Feb 23 Aldebaran	11	+16.4677	248.8092	−0.00 154 +0.57 029 +0.00 000	+0.26 767 +0.09 343 −0.00 001
1980 Feb 29 Regulus	21	+12.0625	321.9954	−0.17 492 +0.53 205 +0.00 001	−0.45 865 −0.14 505 +0.00 001
1980 Mar 21 Aldebaran	18	+16.4675	20.7109	−0.17 168 +0.57 922 −0.00 001	+0.32 564 +0.09 526 −0.00 002
1980 Mar 28 Regulus	3	+12.0627	78.8543	−0.14 523 +0.52 944 +0.00 003	−0.41 994 −0.14 525 +0.00 001
1980 Apr 18 Aldebaran	4	+16.4674	197.7353	+0.15 513 +0.58 925 −0.00 003	+0.54 319 +0.09 817 −0.00 003
1980 Apr 24 Regulus	9	+12.0632	195.7142	−0.25 767 +0.52 855 +0.00 003	−0.19 906 −0.14 615 +0.00 000
1980 May 15 Aldebaran	14	+16.4675	14.7587	−0.03 525 +0.59 364 −0.00 001	+0.63 516 +0.10 112 −0.00 003
1980 May 21 Regulus	17	+12.0637	342.6567	+0.18 516 +0.53 099 −0.00 001	−0.03 433 −0.14 723 +0.00 000
1980 Jun 11 Aldebaran	23	+16.4678	176.7397	−0.31 459 +0.59 037 +0.00 002	+0.59 988 +0.10 293 −0.00 003
1980 Jun 18 Regulus	1	+12.0641	129.5991	+0.14 091 +0.53 542 −0.00 002	+0.24 104 −0.14 789 −0.00 000

Table II   (cont.)

Date Star	$T_\circ$	δ	$H_\circ$	X x' x''	Y y' y''
	h	$\circ$	$\circ$		
1980 Jul 9 Aldebaran	7	+16.4684	323.6786	−0.05 032 +0.58 285 +0.00 002	+0.56 977 +0.10 323 −0.00 002
1980 Jul 15 Regulus	9	+12.0644	276.5410	−0.03 046 +0.53 909 −0.00 002	+0.43 186 −0.14 792 −0.00 001
1980 Aug 5 Aldebaran	13	+16.4692	80.5344	+0.05 564 +0.57 722 +0.00 001	+0.51 452 +0.10 288 −0.00 002
1980 Aug 11 Regulus	17	+12.0645	63.4820	+0.12 998 +0.53 980 −0.00 001	+0.41 146 −0.14 766 −0.00 001
1980 Sep 1 Aldebaran	18	+16.4699	182.3486	−0.23 911 +0.57 865 +0.00 001	+0.48 401 +0.10 330 −0.00 002
1980 Sep 7 Regulus	24	+12.0642	195.3808	+0.34 203 +0.53 741 −0.00 000	+0.34 874 −0.14 755 −0.00 002
1980 Sep 29 Aldebaran	1	+16.4703	314.2449	−0.26 742 +0.58 751 −0.00 000	+0.61 673 +0.10 546 −0.00 003
1980 Oct 5 Regulus	5	+12.0636	297.1964	−0.05 640 +0.53 410 +0.00 002	+0.54 627 −0.14 787 −0.00 003
1980 Oct 26 Aldebaran	11	+16.4705	131.2649	+0.04 684 +0.59 821 −0.00 002	+0.85 047 +0.10 899 −0.00 005
1980 Nov 1 Regulus	11	+12.0625	54.0522	+0.01 977 +0.53 362 +0.00 001	+0.75 637 −0.14 883 −0.00 004
1980 Nov 22 Aldebaran	22	+16.4705	323.3270	+0.02 543 +0.60 300 −0.00 001	+0.95 263 +0.11 232 −0.00 005

5-56

Table II   (cont.)

Date Star	T_o	δ	H_o	X x' x''	Y y' y''
	h	$\circ$	$\circ$		
1980 Nov 28 Regulus	19	+12.0611	200.9897	+0.47 577 +0.53 816 −0.00 003	+0.92 770 −0.15 026 −0.00 004
1980 Dec 20 Aldebaran	8	+16.4703	140.3492	−0.27 097 +0.59 817 +0.00 003	+0.88 456 +0.11 369 −0.00 004
1980 Dec 26 Regulus	4	+12.0598	2.9684	+0.58 547 +0.54 552 −0.00 005	+1.12 109 −0.15 157 −0.00 004
1981 Jan 16 Aldebaran	16	+16.4702	287.2907	−0.33 695 +0.58 798 +0.00 004	+0.80 573 +0.11 292 −0.00 004
1981 Feb 12 Aldebaran	22	+16.4701	44.1510	−0.22 515 +0.58 171 +0.00 002	+0.82 704 +0.11 195 −0.00 004
1981 Mar 12 Aldebaran	4	+16.4699	161.0116	−0.00 129 +0.58 528 −0.00 001	+0.99 322 +0.11 289 −0.00 005
1981 Apr 8 Aldebaran	11	+16.4698	292.9131	−0.34 132 +0.59 582 −0.00 000	+1.11 465 +0.11 618 −0.00 006
1986 Mar 30 Antares	13	−26.4041	135.4806	+0.07 808 +0.58 871 −0.00 002	+1.22 989 −0.13 531 −0.00 006
1986 Apr 26 Antares	22	−26.4046	297.4594	+0.42 006 +0.59 953 −0.00 004	+1.10 776 −0.13 806 −0.00 006
1986 May 24 Antares	8	−26.4050	114.4805	+0.17 070 +0.60 622 −0.00 001	+1.18 705 −0.13 880 −0.00 006
1986 Jun 20 Antares	19	−26.4053	306.5438	+0.28 964 +0.60 451 −0.00 000	+1.14 452 −0.13 668 −0.00 006

Table II   (cont.)

Date Star	$T_\circ$	δ	$H_\circ$	X x' x''	Y y' y''
	h	°	°		
1986 Jul 18 Antares	4	−26.4056	108.5263	+0.01 164 +0.59 595 +0.00 002	+1.09 021 −0.13 266 −0.00 006
1986 Aug 14 Antares	12	−26.4057	255.4686	+0.46 801 +0.58 655 −0.00 001	+0.79 111 −0.12 925 −0.00 004
1986 Sep 10 Antares	17	−26.4056	357.2882	+0.13 659 +0.58 322 −0.00 001	+0.67 393 −0.12 808 −0.00 003
1986 Oct 7 Antares	23	−26.4053	114.1486	+0.24 011 +0.58 887 −0.00 003	+0.55 985 −0.12 949 −0.00 003
1986 Nov 4 Antares	7	−26.4049	261.0904	+0.23 185 +0.59 942 −0.00 003	+0.57 815 −0.13 146 −0.00 003
1986 Dec 1 Antares	17	−26.4047	78.1131	−0.09 503 +0.60 627 +0.00 001	+0.68 384 −0.13 137 −0.00 004
1986 Dec 29 Antares	5	−26.4048	285.2164	+0.25 738 +0.60 324 +0.00 000	+0.54 993 −0.12 832 −0.00 003
1987 Jan 25 Antares	15	−26.4052	102.2363	+0.32 648 +0.59 232 +0.00 000	+0.35 734 −0.12 383 −0.00 002
1987 Feb 18 Spica	13	−11.0961	141.8276	+0.61 411 +0.51 501 −0.00 005	+0.88 592 −0.26 132 −0.00 003
1987 Feb 21 Antares	22	−26.4059	234.1323	+0.25 249 +0.58 236 −0.00 001	+0.16 202 −0.12 072 −0.00 001
1987 Mar 17 Spica	19	−11.0972	258.6839	+0.54 592 +0.52 025 −0.00 004	+0.81 765 −0.26 483 −0.00 003

Table II (cont.)

Date Star	$T_\circ$	δ	$H_\circ$	X x' x''	Y y' y''
	h	$\circ$	$\circ$		
1987 Mar 21 Antares	3	-26.4065	335.9461	-0.01 371 +0.58 113 -0.00 001	+0.09 761 -0.12 038 -0.00 000
1987 Apr 14 Spica	3	-11.0978	45.6234	+0.56 934 +0.52 545 -0.00 003	+0.80 141 -0.26 760 -0.00 004
1987 Apr 17 Antares	9	-26.4071	92.8015	-0.08 737 +0.58 834 -0.00 002	+0.12 336 -0.12 169 -0.00 000
1987 May 11 Spica	12	-11.0980	207.6051	+0.27 678 +0.52 604 +0.00 000	+0.91 826 -0.26 680 -0.00 005
1987 May 14 Antares	17	-26.4075	239.7399	-0.22 887 +0.59 728 +0.00 000	+0.21 740 -0.12 234 -0.00 001
1987 Jun 7 Spica	22	-11.0979	24.6288	+0.34 046 +0.52 120 +0.00 000	+0.73 451 -0.26 273 -0.00 004
1987 Jun 11 Antares	4	-26.4079	71.8028	+0.32 992 +0.60 101 -0.00 001	+0.11 344 -0.12 092 -0.00 000
1987 Jul 5 Spica	7	-11.0976	186.6122	+0.44 575 +0.51 371 -0.00 001	+0.41 916 -0.25 778 -0.00 002
1987 Jul 8 Antares	14	-26.4081	248.8259	+0.04 237 +0.59 673 +0.00 002	+0.06 795 -0.11 765 -0.00 000
1987 Aug 1 Spica	13	-11.0972	303.4725	-0.06 543 +0.50 810 -0.00 001	+0.39 956 -0.25 476 -0.00 001
1987 Aug 4 Antares	23	-26.4083	50.8090	-0.10 989 +0.58 690 +0.00 002	-0.08 581 -0.11 405 +0.00 000

Table II   (cont.)

Date Star	$T_\circ$	δ	$H_\circ$	X x' x''	Y y' y''
	h	$\circ$	$\circ$		
1987 Aug 28 Spica	19	−11.0968	60.3326	+0.08 827 +0.50 755 −0.00 002	+0.14 732 −0.25 512 +0.00 000
1987 Sep 1 Antares	6	−26.4082	182.7106	−0.17 953 +0.57 786 +0.00 001	−0.23 609 −0.11 175 +0.00 001
1987 Sep 25 Spica	1	−11.0965	177.1919	+0.16 998 +0.51 156 −0.00 002	+0.06 817 −0.25 756 +0.00 000
1987 Sep 28 Antares	12	−26.4080	299.5712	+0.09 275 +0.57 580 −0.00 002	−0.33 820 −0.11 133 +0.00 002
1987 Oct 22 Spica	8	−11.0966	309.0912	−0.02 147 +0.51 606 +0.00 000	+0.18 216 −0.25 919 −0.00 001
1987 Oct 25 Antares	18	−26.4076	56.4313	+0.20 376 +0.58 181 −0.00 003	−0.29 035 −0.11 197 +0.00 002
1987 Nov 18 Spica	17	−11.0972	111.0715	−0.11 432 +0.51 643 +0.00 003	+0.16 271 −0.25 762 −0.00 001
1987 Nov 22 Antares	2	−26.4073	203.3723	+0.26 303 +0.59 033 −0.00 002	−0.21 535 −0.11 183 +0.00 002
1987 Dec 16 Spica	3	−11.0983	288.0917	−0.04 730 +0.51 134 +0.00 002	−0.09 645 −0.25 316 −0.00 000
1987 Dec 19 Antares	12	−26.4072	20.3940	+0.15 801 +0.59 328 +0.00 001	−0.21 414 −0.10 970 +0.00 001
1988 Jan 12 Spica	12	−11.0999	90.0703	+0.02 279 +0.50 418 +0.00 000	−0.45 169 −0.24 865 +0.00 002

5-60

Table II (cont.)

Date Star	$T_\circ$	δ	$H_\circ$	X x' x''	Y y' y''
	h	°	°		
1988 Jan 15 Antares	22	−26.4075	197.4143	−0.31 616 +0.58 709 +0.00 004	−0.28 039 −0.10 614 +0.00 001
1988 Feb 8 Spica	18	−11.1014	206.9257	−0.47 079 +0.50 023 +0.00 000	−0.46 350 −0.24 700 +0.00 003
1988 Feb 12 Antares	8	−26.4081	14.4337	+0.11 465 +0.57 615 +0.00 001	−0.54 747 −0.10 290 +0.00 003
1988 Mar 7 Spica	0	−11.1026	323.7817	−0.37 542 +0.50 148 −0.00 001	−0.59 656 −0.24 812 +0.00 004
1988 Mar 10 Antares	14	−26.4087	131.2885	−0.35 780 +0.56 863 +0.00 001	−0.55 368 −0.10 141 +0.00 003
1988 Apr 3 Spica	6	−11.1034	80.6386	−0.42 654 +0.50 532 +0.00 001	−0.54 133 −0.24 964 +0.00 003
1988 Apr 6 Antares	20	−26.4093	248.1436	−0.06 325 +0.56 917 −0.00 002	−0.55 144 −0.10 123 +0.00 003
1988 Apr 30 Spica	14	−11.1038	227.5788	−0.09 403 +0.50 788 +0.00 002	−0.70 585 −0.24 927 +0.00 003
1988 May 4 Antares	2	−26.4098	4.9995	−0.10 085 +0.57 535 −0.00 001	−0.42 675 −0.10 110 +0.00 003
1988 May 27 Spica	22	−11.1038	14.5201	−0.36 141 +0.50 694 +0.00 004	−0.71 142 −0.24 675 +0.00 002
1988 May 31 Antares	10	−26.4101	151.9388	+0.01 374 +0.58 113 +0.00 000	−0.38 949 −0.09 985 +0.00 002

Table II (cont.)

Date Star	T_o	δ	H_o	X x' x''	Y y' y''
	h	$\circ$	$\circ$		
1988 Jun 19 Regulus	19	+12.0256	41.2779	+0.71 413 +0.49 401 −0.00 005	+0.96 596 −0.23 421 −0.00 003
1988 Jun 24 Spica	7	−11.1035	176.5032	−0.30 095 +0.50 301 +0.00 003	−1.00 139 −0.24 328 +0.00 004
1988 Jun 27 Antares	19	−26.4104	313.9203	−0.19 264 +0.58 166 +0.00 003	−0.42 056 −0.09 749 +0.00 002
1988 Jul 17 Regulus	2	+12.0258	173.1787	+0.53 470 +0.49 482 −0.00 004	+0.84 554 −0.23 600 −0.00 003
1988 Jul 25 Antares	5	−26.4106	130.9441	−0.04 906 +0.57 593 +0.00 002	−0.59 656 −0.09 476 +0.00 003
1988 Aug 13 Regulus	8	+12.0257	290.0375	+0.28 137 +0.49 498 −0.00 002	+0.90 400 −0.23 743 −0.00 003
1988 Aug 21 Antares	14	−26.4106	292.9277	+0.10 189 +0.56 701 +0.00 000	−0.75 424 −0.09 260 +0.00 003
1988 Sep 9 Regulus	14	+12.0254	46.8952	+0.26 853 +0.49 415 +0.00 000	+0.92 615 −0.23 752 −0.00 004
1988 Sep 17 Antares	20	−26.4104	49.7884	−0.39 961 +0.56 049 +0.00 001	−0.68 747 −0.09 153 +0.00 003
1988 Oct 6 Regulus	21	+12.0246	178.7929	+0.63 582 +0.49 295 −0.00 000	+0.70 893 −0.23 643 −0.00 004
1988 Oct 15 Antares	2	−26.4100	166.6487	−0.14 567 +0.56 050 −0.00 002	−0.61 573 −0.09 100 +0.00 003

Table II (cont.)

Date Star	$T_\circ$	δ	$H_\circ$	X x' x''	Y y' y''
	h	$\circ$	$\circ$		
1988 Nov 3 Regulus	3	+12.0234	295.6486	+0.07 229 +0.49 278 +0.00 001	+0.78 033 -0.23 537 -0.00 004
1988 Nov 11 Antares	8	-26.4096	283.5080	-0.05 586 +0.56 589 -0.00 001	-0.49 057 -0.09 021 +0.00 003
1988 Nov 30 Regulus	11	+12.0219	82.5861	-0.02 099 +0.49 464 -0.00 001	+0.49 884 -0.23 611 -0.00 002
1988 Dec 8 Antares	16	-26.4094	70.4481	+0.17 948 +0.57 083 -0.00 000	-0.49 140 -0.08 842 +0.00 003
1988 Dec 27 Regulus	20	+12.0205	244.5648	+0.14 047 +0.49 761 -0.00 003	+0.11 558 -0.23 858 +0.00 000
1989 Jan 5 Antares	1	-26.4096	232.4278	-0.02 294 +0.56 994 +0.00 002	-0.56 430 -0.08 592 +0.00 002
1989 Jan 24 Regulus	4	+12.0194	31.5031	+0.04 560 +0.49 944 -0.00 002	+0.00 709 -0.24 095 +0.00 001
1989 Feb 1 Antares	11	-26.4101	49.4475	+0.13 529 +0.56 312 +0.00 001	-0.73 615 -0.08 360 +0.00 003
1989 Feb 20 Regulus	11	+12.0189	163.4015	+0.02 544 +0.49 863 +0.00 000	+0.00 587 -0.24 135 -0.00 000
1989 Feb 28 Antares	19	-26.4107	196.3845	-0.11 489 +0.55 533 +0.00 000	-0.75 114 -0.08 221 +0.00 003
1989 Mar 19 Regulus	17	+12.0189	280.2601	-0.01 718 +0.49 632 +0.00 002	+0.02 025 -0.23 980 -0.00 001

Table II   (cont.)

Date Star	T_o	δ	H_o	X x' x''	Y y' y''
	h	$\circ$	$\circ$		
1989 Mar 28 Antares	2	−26.4113	328.2805	+0.09 049 +0.55 189 −0.00 002	−0.68 373 −0.08 142 +0.00 003
1989 Apr 15 Regulus	23	+12.0192	37.1198	−0.05 630 +0.49 531 +0.00 002	−0.09 506 −0.23 822 −0.00 000
1989 Apr 24 Antares	7	−26.4118	70.0950	−0.26 812 +0.55 382 −0.00 001	−0.46 329 −0.08 059 +0.00 002
1989 May 13 Regulus	6	+12.0196	169.0211	−0.06 925 +0.49 733 +0.00 001	−0.36 079 −0.23 850 +0.00 001
1989 May 21 Antares	13	−26.4121	186.9516	−0.27 731 +0.55 792 +0.00 000	−0.35 538 −0.07 909 +0.00 002
1989 Jun 9 Regulus	14	+12.0199	315.9636	−0.16 286 +0.50 146 −0.00 001	−0.61 989 −0.24 084 +0.00 003
1989 Jun 17 Antares	21	−26.4124	333.8916	+0.15 156 +0.56 026 +0.00 001	−0.43 563 −0.07 711 +0.00 002
1989 Jul 6 Regulus	22	+12.0202	102.9057	−0.52 780 +0.50 521 +0.00 000	−0.65 710 −0.24 379 +0.00 004
1989 Jul 15 Antares	5	−26.4126	120.8329	−0.08 055 +0.55 865 +0.00 003	−0.51 208 −0.07 529 +0.00 002
1989 Aug 3 Regulus	7	+12.0202	264.8881	−0.16 886 +0.50 652 −0.00 001	−0.90 558 −0.24 531 +0.00 004
1989 Aug 11 Antares	14	−26.4127	282.8162	+0.03 009 +0.55 364 +0.00 001	−0.61 505 −0.07 398 +0.00 002

Table II (cont.)

Date Star	$T_\circ$	δ	$H_\circ$	X x' x''	Y y' y''
	h	$\circ$	$\circ$		
1989 Aug 30 Regulus	14	+12.0200	36.7873	−0.25 463 +0.50 498 +0.00 002	−0.86 022 −0.24 459 +0.00 003
1989 Sep 7 Antares	22	−26.4125	69.7590	−0.03 278 +0.54 814 −0.00 000	−0.57 687 −0.07 318 +0.00 002
1989 Sep 26 Regulus	20	+12.0193	153.6443	−0.29 212 +0.50 214 +0.00 003	−0.88 448 −0.24 225 +0.00 003
1989 Oct 5 Antares	5	−26.4122	201.6606	+0.08 442 +0.54 568 −0.00 003	−0.43 379 −0.07 237 +0.00 002
1989 Oct 24 Regulus	1	+12.0183	255.4593	−0.70 914 +0.50 100 +0.00 005	−0.86 862 −0.24 058 +0.00 003
1989 Nov 1 Antares	11	−26.4118	318.5203	+0.19 014 +0.54 743 −0.00 003	−0.25 776 −0.07 109 +0.00 002
1989 Nov 28 Antares	17	−26.4115	75.3788	+0.21 359 +0.55 079 −0.00 001	−0.17 156 −0.06 921 +0.00 001
1989 Dec 25 Antares	24	−26.4115	207.2769	+0.16 945 +0.55 183 +0.00 001	−0.21 994 −0.06 721 +0.00 001
1990 Jan 22 Antares	8	−26.4118	354.2148	+0.03 508 +0.54 923 +0.00 002	−0.30 575 −0.06 578 +0.00 001
1990 Feb 18 Antares	16	−26.4124	141.1520	−0.27 988 +0.54 521 +0.00 002	−0.27 432 −0.06 512 +0.00 001
1990 Mar 18 Antares	0	−26.4130	288.0889	−0.23 818 +0.54 292 −0.00 000	−0.13 390 −0.06 463 +0.00 001

Table II (cont.)

Date Star	$T_\circ$	δ	$H_\circ$	X x' x''	Y y' y''
	h	$\circ$	$\circ$		
1990 Apr 14 Antares	7	−26.4135	59.9853	−0.15 373 +0.54 340 −0.00 002	+0.07 318 −0.06 360 +0.00 000
1990 May 11 Antares	13	−26.4139	176.8415	−0.19 320 +0.54 531 −0.00 001	+0.23 238 −0.06 188 −0.00 000
1990 Jun 7 Antares	19	−26.4142	293.6989	−0.20 261 +0.54 658 +0.00 001	+0.25 487 −0.05 995 −0.00 001
1990 Jul 5 Antares	2	−26.4144	65.5986	+0.00 857 +0.54 622 +0.00 002	+0.16 203 −0.05 853 −0.00 001
1990 Aug 1 Antares	9	−26.4145	197.4995	−0.23 340 +0.54 450 +0.00 003	+0.13 901 −0.05 790 −0.00 001
1990 Aug 28 Antares	17	−26.4144	344.4421	−0.22 047 +0.54 280 +0.00 001	+0.20 826 −0.05 775 −0.00 001
1990 Sep 25 Antares	1	−26.4141	131.3849	−0.15 068 +0.54 246 −0.00 001	+0.39 992 −0.05 733 −0.00 001
1990 Oct 22 Antares	9	−26.4137	278.3271	+0.27 875 +0.54 380 −0.00 003	+0.58 780 −0.05 612 −0.00 002
1990 Nov 18 Antares	15	−26.4133	35.1861	+0.03 068 +0.54 552 −0.00 001	+0.74 949 −0.05 419 −0.00 002
1990 Dec 15 Antares	21	−26.4132	152.0437	−0.03 512 +0.54 554 +0.00 001	+0.74 702 −0.05 230 −0.00 003
1991 Jan 12 Antares	4	−26.4134	283.9410	+0.27 438 +0.54 391 +0.00 001	+0.65 258 −0.05 127 −0.00 003

Table II (cont.)

Date Star	T_0	δ	H_0	X x' x"	Y y' y"
	h	$\circ$	$\circ$		
1991 Feb 8 Antares	11	−26.4139	55.8373	+0.15 269 +0.54 290 +0.00 001	+0.68 901 −0.05 112 −0.00 003
1991 Mar 7 Antares	19	−26.4145	202.7743	+0.15 797 +0.54 435 −0.00 001	+0.86 579 −0.05 119 −0.00 003
1991 Apr 4 Antares	3	−26.4150	349.7114	+0.06 075 +0.54 743 −0.00 002	+1.12 828 −0.05 054 −0.00 004
1993 Oct 15 Spica	14	−11.1281	32.8974	−0.55 867 +0.59 156 +0.00 003	−1.11 451 −0.19 038 +0.00 007
1994 Jan 5 Spica	16	−11.1308	143.7949	−0.38 157 +0.56 864 +0.00 003	−1.11 726 −0.18 178 +0.00 005
1994 Feb 1 Spica	21	−11.1323	245.6092	−0.52 926 +0.57 125 +0.00 001	−0.81 981 −0.18 253 +0.00 005
1994 Mar 1 Spica	5	−11.1335	32.5471	+0.00 158 +0.58 204 −0.00 002	−0.73 742 −0.18 434 +0.00 005
1994 Mar 28 Spica	14	−11.1344	194.5270	−0.44 511 +0.59 195 +0.00 001	−0.46 741 −0.18 521 +0.00 003
1994 Apr 25 Spica	2	−11.1347	41.6312	+0.07 982 +0.59 363 +0.00 000	−0.62 557 −0.18 391 +0.00 003
1994 May 22 Spica	12	−11.1347	218.6544	−0.12 852 +0.58 650 +0.00 003	−0.58 098 −0.18 129 +0.00 003
1994 Jun 18 Spica	20	−11.1344	5.5963	−0.19 362 +0.57 609 +0.00 003	−0.49 572 −0.17 864 +0.00 002

Table II   (cont.)

Date Star	$T_\circ$	δ	$H_\circ$	X x' x''	Y y' y''
	h	$\circ$	$\circ$		
1994 Jul 16 Spica	2	−11.1340	122.4566	−0.13 374 +0.56 980 +0.00 001	−0.32 558 −0.17 717 +0.00 002
1994 Aug 12 Spica	7	−11.1335	224.2758	−0.26 645 +0.57 199 −0.00 000	−0.03 429 −0.17 735 +0.00 001
1994 Sep 8 Spica	14	−11.1330	356.1767	−0.06 260 +0.58 093 −0.00 002	+0.10 090 −0.17 853 +0.00 000
1994 Oct 5 Spica	23	−11.1327	158.1588	−0.16 512 +0.58 974 +0.00 000	+0.20 928 −0.17 940 −0.00 001
1994 Nov 2 Spica	10	−11.1329	350.2219	−0.18 626 +0.59 107 +0.00 002	+0.20 711 −0.17 886 −0.00 001
1994 Nov 29 Spica	21	−11.1336	182.2838	−0.06 425 +0.58 284 +0.00 003	+0.19 414 −0.17 674 −0.00 002
1994 Dec 27 Spica	6	−11.1349	344.2627	+0.25 746 +0.57 086 +0.00 000	+0.25 523 −0.17 404 −0.00 002
1995 Jan 23 Spica	12	−11.1364	101.1180	+0.35 170 +0.56 472 −0.00 002	+0.50 326 −0.17 237 −0.00 002
1995 Feb 19 Spica	17	−11.1377	202.9325	+0.17 656 +0.56 882 −0.00 003	+0.81 708 −0.17 251 −0.00 003
1995 Mar 19 Spica	0	−11.1387	334.8299	+0.07 563 +0.57 820 −0.00 002	+0.97 579 −0.17 368 −0.00 004
1995 Apr 15 Spica	10	−11.1392	151.8516	+0.32 808 +0.58 430 −0.00 002	+0.90 950 −0.17 457 −0.00 005

Table II   (cont.)

Date Star	T_0	δ	H_0	X x' x''	Y y' y''
	h	$\circ$	$\circ$		
1995 May 12 Spica	21	-11.1393	343.9155	+0.53 134 +0.58 228 -0.00 001	+0.85 067 -0.17 429 -0.00 005
1995 Jun 9 Spica	6	-11.1391	145.8982	+0.04 325 +0.57 341 +0.00 002	+1.10 178 -0.17 266 -0.00 006
1996 Aug 8 Aldebaran	7	+16.5006	353.1372	-0.04 430 +0.55 187 +0.00 002	+1.12 723 +0.05 566 -0.00 004
1996 Sep 4 Aldebaran	14	+16.5013	125.0335	-0.00 513 +0.55 569 -0.00 000	+0.93 560 +0.05 647 -0.00 004
1996 Oct 1 Aldebaran	22	+16.5017	271.9709	-0.26 162 +0.56 241 -0.00 001	+0.83 677 +0.05 719 -0.00 004
1996 Oct 29 Aldebaran	8	+16.5018	88.9909	+0.12 092 +0.56 790 -0.00 003	+0.93 103 +0.05 822 -0.00 004
1996 Nov 25 Aldebaran	16	+16.5016	235.9299	-0.35 390 +0.56 851 +0.00 001	+0.96 096 +0.05 996 -0.00 004
1996 Dec 22 Aldebaran	24	+16.5014	22.8702	+0.08 030 +0.56 409 +0.00 001	+0.98 112 +0.06 174 -0.00 004
1997 Jan 19 Aldebaran	5	+16.5012	124.6885	-0.33 721 +0.55 921 +0.00 004	+0.76 603 +0.06 325 -0.00 003
1997 Feb 15 Aldebaran	11	+16.5011	241.5488	-0.19 920 +0.55 985 +0.00 002	+0.57 189 +0.06 427 -0.00 002
1997 Mar 14 Aldebaran	19	+16.5008	28.4916	+0.15 134 +0.56 702 -0.00 002	+0.50 330 +0.06 523 -0.00 002

Table II (cont.)

Date Star	T_o	δ	H_o	X x' x''	Y y' y''
	h	$\circ$	$\circ$		
1997 Apr 11 Aldebaran	4	+16.5007	190.4751	+0.04 410 +0.57 582 -0.00 003	+0.52 638 +0.06 660 -0.00 003
1997 May 8 Aldebaran	14	+16.5007	7.4988	+0.19 442 +0.58 030 -0.00 003	+0.64 423 +0.06 840 -0.00 003
1997 Jun 4 Aldebaran	22	+16.5009	154.4392	-0.24 162 +0.57 842 +0.00 001	+0.63 782 +0.07 034 -0.00 003
1997 Jul 2 Aldebaran	5	+16.5014	286.3373	-0.22 706 +0.57 266 +0.00 003	+0.56 550 +0.07 182 -0.00 002
1997 Jul 29 Aldebaran	11	+16.5021	43.1933	-0.00 137 +0.56 824 +0.00 002	+0.43 343 +0.07 271 -0.00 002
1997 Aug 25 Aldebaran	17	+16.5028	160.0487	+0.20 982 +0.56 965 -0.00 000	+0.31 812 +0.07 341 -0.00 001
1997 Sep 21 Aldebaran	24	+16.5033	291.9450	+0.10 011 +0.57 751 -0.00 002	+0.27 594 +0.07 456 -0.00 002
1997 Oct 19 Aldebaran	9	+16.5035	93.9237	-0.08 570 +0.58 737 -0.00 002	+0.34 383 +0.07 656 -0.00 002
1997 Nov 15 Aldebaran	20	+16.5035	285.9854	+0.21 762 +0.59 252 -0.00 002	+0.50 037 +0.07 909 -0.00 002
1997 Dec 13 Aldebaran	5	+16.5033	87.9663	-0.22 298 +0.58 938 +0.00 002	+0.46 238 +0.08 127 -0.00 002
1998 Jan 9 Aldebaran	13	+16.5031	234.9074	+0.05 757 +0.58 108 +0.00 002	+0.39 283 +0.08 230 -0.00 001

Table II   (cont.)

Date Star	T_0	δ	H_0	X x' x''	Y y' y''
	h	$\circ$	$\circ$		
1998 Feb 5 Aldebaran	18	+16.5030	336.7264	−0.29 214 +0.57 573 +0.00 003	+0.19 287 +0.08 263 −0.00 001
1998 Mar 5 Aldebaran	0	+16.5028	93.5870	−0.15 253 +0.57 939 +0.00 000	+0.15 421 +0.08 339 −0.00 001
1998 Apr 1 Aldebaran	8	+16.5026	240.5296	−0.04 063 +0.58 985 −0.00 002	+0.24 764 +0.08 534 −0.00 001
1998 Apr 28 Aldebaran	18	+16.5025	57.5537	+0.02 354 +0.59 933 −0.00 002	+0.39 391 +0.08 819 −0.00 002
1998 May 26 Aldebaran	4	+16.5027	234.5766	−0.26 100 +0.60 168 +0.00 001	+0.43 418 +0.09 090 −0.00 002
1998 Jun 1 Regulus	3	+11.9750	142.3304	−0.35 644 +0.53 633 +0.00 001	−0.98 199 −0.14 974 +0.00 004
1998 Jun 22 Aldebaran	14	+16.5031	51.5984	+0.14 528 +0.59 636 +0.00 001	+0.46 788 +0.09 243 −0.00 002
1998 Jun 28 Regulus	12	+11.9754	304.3138	−0.03 276 +0.54 196 −0.00 002	−0.81 653 −0.15 033 +0.00 004
1998 Jul 19 Aldebaran	21	+16.5038	183.4958	−0.00 729 +0.58 795 +0.00 002	+0.33 743 +0.09 270 −0.00 001
1998 Jul 25 Regulus	20	+11.9756	91.2554	−0.34 622 +0.54 550 −0.00 000	−0.60 376 −0.15 007 +0.00 003
1998 Aug 16 Aldebaran	3	+16.5045	300.3514	+0.21 307 +0.58 318 +0.00 000	+0.28 384 +0.09 250 −0.00 001

Table II   (cont.)

Date Star	T_o	δ	H_o	X x' x''	Y y' y''
	h	°	°		
1998 Aug 22 Regulus	4	+11.9756	238.1960	−0.23 095 +0.54 504 +0.00 001	−0.62 478 −0.14 925 +0.00 003
1998 Sep 12 Aldebaran	8	+16.5051	42.1655	−0.16 688 +0.58 658 +0.00 000	+0.24 132 +0.09 321 −0.00 001
1998 Sep 18 Regulus	11	+11.9752	10.0945	+0.03 636 +0.54 143 +0.00 002	−0.69 867 −0.14 865 +0.00 002
1998 Oct 9 Aldebaran	16	+16.5055	189.1030	+0.10 945 +0.59 686 −0.00 003	+0.41 388 +0.09 557 −0.00 002
1998 Oct 15 Regulus	16	+11.9743	111.9097	−0.28 872 +0.53 811 +0.00 003	−0.49 909 −0.14 889 +0.00 001
1998 Nov 6 Aldebaran	2	+16.5056	6.1233	+0.09 438 +0.60 700 −0.00 002	+0.56 143 +0.09 912 −0.00 003
1998 Nov 11 Regulus	22	+11.9731	228.7653	−0.25 126 +0.53 938 +0.00 002	−0.25 133 −0.15 009 +0.00 001
1998 Dec 3 Aldebaran	13	+16.5055	198.1858	−0.08 292 +0.60 924 +0.00 001	+0.59 429 +0.10 216 −0.00 003
1998 Dec 9 Regulus	6	+11.9717	15.7028	−0.01 506 +0.54 631 −0.00 001	−0.00 659 −0.15 175 +0.00 000
1998 Dec 30 Aldebaran	23	+16.5053	15.2085	−0.24 294 +0.60 170 +0.00 003	+0.51 102 +0.10 318 −0.00 002
1999 Jan 5 Regulus	15	+11.9705	177.6816	−0.20 567 +0.55 467 −0.00 001	+0.26 149 −0.15 295 −0.00 000

Table II (cont.)

Date Star	$T_\circ$	δ	$H_\circ$	X x' x''	Y y' y''
	h	$\circ$	$\circ$		
1999 Jan 27 Aldebaran	7	+16.5051	162.1504	−0.02 084 +0.59 087 +0.00 002	+0.45 744 +0.10 240 −0.00 002
1999 Feb 2 Regulus	1	+11.9697	354.7024	−0.08 058 +0.55 842 −0.00 001	+0.28 981 −0.15 311 −0.00 001
1999 Feb 23 Aldebaran	12	+16.5050	263.9698	−0.35 096 +0.58 630 +0.00 002	+0.39 302 +0.10 187 −0.00 002
1999 Mar 1 Regulus	10	+11.9694	156.6833	+0.22 148 +0.55 549 −0.00 000	+0.19 952 −0.15 258 −0.00 001
1999 Mar 22 Aldebaran	18	+16.5048	20.8304	−0.27 005 +0.59 203 −0.00 000	+0.52 500 +0.10 328 −0.00 003
1999 Mar 28 Regulus	16	+11.9696	273.5422	−0.02 974 +0.54 960 +0.00 002	+0.32 914 −0.15 214 −0.00 002
1999 Apr 19 Aldebaran	2	+16.5047	167.7727	−0.30 245 +0.60 298 −0.00 001	+0.68 749 +0.10 661 −0.00 004
1999 Apr 24 Regulus	22	+11.9701	30.4021	+0.31 664 +0.54 694 +0.00 000	+0.42 793 −0.15 262 −0.00 003
1999 May 16 Aldebaran	12	+16.5048	344.7961	−0.40 878 +0.61 060 +0.00 001	+0.77 885 +0.11 027 −0.00 004
1999 May 22 Regulus	4	+11.9706	147.2624	+0.30 601 +0.55 081 −0.00 001	+0.70 326 −0.15 399 −0.00 003
1999 Jun 12 Aldebaran	23	+16.5052	176.8593	−0.26 716 +0.60 984 +0.00 002	+0.80 974 +0.11 240 −0.00 004

Table II   (cont.)

Date Star	T_o	δ	H_o	X x' x''	Y y' y''
	h	°	°		
1999 Jun 18 Regulus	12	+11.9710	294.2049	+0.41 134 +0.55 922 −0.00 004	+0.90 426 −0.15 571 −0.00 004
1999 Jul 10 Aldebaran	9	+16.5058	353.8803	−0.02 609 +0.60 166 +0.00 002	+0.79 405 +0.11 231 −0.00 004
1999 Jul 15 Regulus	21	+11.9713	96.1878	+0.19 372 +0.56 693 −0.00 003	+1.07 560 −0.15 698 −0.00 005
1999 Aug 6 Aldebaran	16	+16.5065	125.7771	−0.28 216 +0.59 178 +0.00 003	+0.71 175 +0.11 095 −0.00 003
1999 Aug 12 Regulus	7	+11.9714	273.2110	+0.35 816 +0.56 929 −0.00 002	+1.04 013 −0.15 759 −0.00 005
1999 Sep 2 Aldebaran	22	+16.5072	242.6324	−0.08 484 +0.58 737 +0.00 000	+0.81 671 +0.11 017 −0.00 004
1999 Sep 8 Regulus	15	+11.9711	60.1509	+0.03 110 +0.56 535 +0.00 001	+1.13 623 −0.15 744 −0.00 006
1999 Sep 30 Aldebaran	4	+16.5077	359.4877	+0.09 657 +0.59 198 −0.00 002	+1.01 715 +0.11 163 −0.00 006
1999 Oct 5 Regulus	22	+11.9705	192.0486	+0.22 922 +0.55 851 +0.00 001	+1.18 631 −0.15 714 −0.00 006
1999 Oct 27 Aldebaran	11	+16.5079	131.3844	−0.33 720 +0.60 243 −0.00 000	+1.11 015 +0.11 540 −0.00 006
1999 Nov 23 Aldebaran	22	+16.5078	323.4465	+0.05 944 +0.61 027 −0.00 001	+1.27 268 +0.11 921 −0.00 007

Table II   (cont.)

Date Star	T_o	δ	H_o	X x' x''	Y y' y''
	h	$\circ$	$\circ$		
1999 Dec 21 Aldebaran	9	+16.5077	155.5098	−0.27 125 +0.60 853 +0.00 003	+1.18 690 +0.12 091 −0.00 006
2000 Jan 17 Aldebaran	19	+16.5076	332.5334	−0.44 917 +0.59 787 +0.00 005	+1.11 075 +0.11 964 −0.00 006
2000 Feb 14 Aldebaran	3	+16.5074	119.4759	−0.16 970 +0.58 673 +0.00 002	+1.21 998 +0.11 735 −0.00 006

TABLE III - OCCULTATIONS OF PLANETS

Date / Planet	T_o	δ_o / δ'	H_o / H_o'	k	X / x' / x''	Y / y' / y''
	h	$\circ$	$\circ$			
1980 Jan 7 Jupiter	4	+ 8.8779 + 0.0008	3.8866 15.0424	0.99 943	−0.12 983 +0.52 534 −0.00 001	−0.24 973 −0.16 238 +0.00 002
1980 Jan 8 Saturn	15	+ 3.2023 + 0.0002	154.1907 15.0411	0.99 970	+0.29 219 +0.51 641 −0.00 003	+0.08 950 −0.17 627 +0.00 000
1980 Jan 20 Venus	12	−10.9510 + 0.0198	321.3484 14.9935	0.99 804	−0.48 793 +0.54 035 −0.00 002	+0.84 669 +0.15 297 +0.00 003
1980 Feb 3 Jupiter	7	+ 9.7569 + 0.0018	77.5286 15.0453	0.99 940	−0.38 343 +0.53 211 +0.00 000	−0.41 380 −0.16 088 +0.00 003
1980 Feb 4 Saturn	21	+ 3.6036 + 0.0010	271.6784 15.0429	0.99 969	+0.29 140 +0.52 019 −0.00 003	+0.00 838 −0.17 672 +0.00 001
1980 Mar 1 Jupiter	7	+11.0631 + 0.0020	107.3240 15.0462	0.99 938	−0.39 171 +0.53 474 +0.00 002	−0.84 382 −0.15 618 +0.00 004
1980 Mar 2 Saturn	24	+ 4.3769 + 0.0013	345.0034 15.0440	0.99 968	+0.17 734 +0.52 326 −0.00 002	−0.22 558 −0.17 674 +0.00 001
1980 Mar 28 Jupiter	7	+12.1508 + 0.0012	136.8304 15.0445	0.99 941	−0.40 407 +0.53 212 +0.00 004	−1.10 515 −0.15 077 +0.00 004
1980 Mar 30 Saturn	1	+ 5.2267 + 0.0012	28.5840 15.0439	0.99 968	−0.36 029 +0.52 283 +0.00 003	−0.27 398 −0.17 610 +0.00 001
1980 Apr 13 Mercury	9	− 3.0761 + 0.0218	337.3514 14.9874	0.99 769	+0.13 359 +0.53 272 −0.00 001	+0.02 721 +0.18 084 +0.00 002
1980 Apr 24 Jupiter	11	+12.5508 − 0.0000	224.8520 15.0413	0.99 945	−0.22 057 +0.52 845 +0.00 002	−1.03 054 −0.14 814 +0.00 004
1980 Apr 26 Saturn	4	+ 5.8356 + 0.0006	101.8453 15.0428	0.99 969	−0.34 865 +0.51 991 +0.00 003	−0.25 758 −0.17 520 +0.00 000

Table III (cont.)

Date Planet	T_0	δ δ'	H_0 H_0'	k	X x' x''	Y y' y''
	h	$\circ$	$\circ$			
1980 May 21 Jupiter	20	+12.1408 − 0.0012	25.9407 15.0382	0.99 949	−0.19 679 +0.52 648 +0.00 000	−0.56 611 −0.14 951 +0.00 002
1980 May 22 Mars	6	+10.6324 − 0.0070	171.6304 15.0262	0.99 775	−0.01 244 +0.51 051 −0.00 001	−0.45 596 −0.15 106 −0.00 000
1980 May 23 Saturn	10	+ 5.9932 − 0.0001	219.2739 15.0411	0.99 970	−0.17 189 +0.51 746 +0.00 002	−0.02 220 −0.17 479 −0.00 000
1980 Jun 18 Jupiter	10	+11.0115 − 0.0022	260.3315 15.0357	0.99 953	−0.01 441 +0.52 525 −0.00 002	+0.01 926 −0.15 380 −0.00 000
1980 Jun 19 Saturn	19	+ 5.6463 − 0.0009	20.6677 15.0393	0.99 972	−0.10 627 +0.51 663 −0.00 000	+0.39 517 −0.17 514 −0.00 001
1980 Jul 9 Venus	19	+17.9045 − 0.0016	137.3109 15.0361	0.99 321	+0.23 211 +0.58 381 −0.00 002	+0.20 282 +0.08 405 −0.00 003
1980 Jul 16 Jupiter	3	+ 9.3209 − 0.0029	178.4592 15.0340	0.99 956	+0.16 140 +0.52 303 −0.00 003	+0.61 945 −0.15 960 −0.00 003
1980 Jul 17 Saturn	7	+ 4.8549 − 0.0015	226.1184 15.0379	0.99 973	+0.18 380 +0.51 651 −0.00 003	+0.72 785 −0.17 619 −0.00 002
1980 Aug 7 Venus	2	+19.1057 + 0.0025	255.3538 15.0067	0.99 551	+0.06 345 +0.54 761 +0.00 000	+0.31 015 +0.03 047 −0.00 012
1980 Aug 12 Jupiter	21	+ 7.2393 − 0.0033	110.7645 15.0331	0.99 958	+0.15 761 +0.51 932 −0.00 003	+1.21 908 −0.16 591 −0.00 005
1980 Aug 13 Saturn	20	+ 3.7380 − 0.0019	85.7996 15.0369	0.99 974	+0.27 484 +0.51 592 −0.00 003	+1.03 874 −0.17 790 −0.00 004
1980 Sep 5 Venus	10	+18.6260 − 0.0049	15.4435 14.9957	0.99 672	+0.13 497 +0.51 901 −0.00 002	+0.02 696 −0.05 765 −0.00 013

Table III   (cont.)

Date Planet	T_0	δ δ'	H_0 H_0'	k	X x' x''	Y y' y''
	h	$\circ$	$\circ$			
1980 Oct 5 Venus	7	+11.5787 − 0.0143	326.5667 14.9944	0.99 737	+0.25 048 +0.48 482 −0.00 003	+0.78 208 −0.13 388 −0.00 011
1981 Feb 5 Mars	21	−12.7457 + 0.0117	119.2698 15.0095	0.99 893	+0.27 022 +0.54 484 −0.00 003	−0.57 476 +0.14 788 +0.00 010
1981 Mar 5 Venus	20	−10.2188 + 0.0188	124.2695 14.9918	0.99 855	+0.39 939 +0.52 928 −0.00 002	−0.87 532 +0.16 033 +0.00 014
1981 Apr 3 Mercury	7	− 5.5253 + 0.0253	303.1842 14.9807	0.99 791	+0.42 147 +0.51 560 −0.00 002	−0.92 552 +0.18 016 +0.00 011
1981 Sep 24 Mars	5	+17.8041 − 0.0070	301.2656 15.0152	0.99 879	−0.00 037 +0.54 261 −0.00 000	+0.03 763 −0.11 233 −0.00 008
1982 Jan 21 Neptune	24	−22.1188 − 0.0000	215.4630 15.0396	0.99 991	+0.28 565 +0.54 416 +0.00 001	+1.20 694 −0.06 233 −0.00 004
1982 Feb 18 Neptune	9	−22.1287 − 0.0000	16.6463 15.0401	0.99 991	+0.15 046 +0.54 294 +0.00 001	+1.05 055 −0.05 882 −0.00 004
1982 Mar 17 Neptune	18	−22.1225 + 0.0000	178.2117 15.0408	0.99 991	+0.17 237 +0.54 243 −0.00 001	+0.76 084 −0.05 684 −0.00 003
1982 Apr 14 Neptune	2	−22.1060 + 0.0000	325.1788 15.0414	0.99 991	+0.24 970 +0.54 345 −0.00 003	+0.45 160 −0.05 748 −0.00 002
1982 May 11 Neptune	8	−22.0841 + 0.0000	82.4681 15.0420	0.99 991	+0.06 869 +0.54 522 −0.00 003	+0.28 967 −0.06 047 −0.00 001
1982 Jun 7 Neptune	13	−22.0612 + 0.0000	184.9890 15.0423	0.99 991	+0.04 421 +0.54 609 −0.00 001	+0.28 292 −0.06 439 −0.00 002
1982 Jun 20 Mercury	1	+17.5921 + 0.0062	215.9755 15.0217	0.99 662	−0.41 492 +0.58 510 −0.00 001	+0.96 357 +0.13 612 −0.00 013

Table III   (cont.)

Date Planet	T_o	δ δ'	H_o H_o'	k	X x' x''	Y y' y''
	h	$\circ$	$\circ$			
1982 Jul 4 Neptune	18	-22.0421 + 0.0000	287.5878 15.0422	0.99 991	+0.28 075 +0.54 523 +0.00 000	+0.34 426 -0.06 766 -0.00 002
1982 Jul 18 Venus	19	+22.5615 + 0.0027	133.7143 14.9870	0.99 833	-0.07 926 +0.56 484 +0.00 005	-0.56 815 +0.06 276 -0.00 017
1982 Jul 20 Mercury	9	+22.8632 - 0.0088	319.3810 14.9454	0.99 816	-0.09 099 +0.53 181 -0.00 003	-0.39 099 -0.03 331 -0.00 031
1982 Jul 31 Neptune	23	-22.0332 + 0.0000	30.0521 15.0419	0.99 991	+0.21 486 +0.54 330 +0.00 001	+0.40 190 -0.06 939 -0.00 002
1982 Aug 28 Neptune	5	-22.0407 - 0.0000	147.2466 15.0413	0.99 991	-0.04 365 +0.54 188 +0.00 002	+0.33 588 -0.06 973 -0.00 001
1982 Sep 24 Neptune	13	-22.0667 - 0.0001	294.1083 15.0406	0.99 991	-0.06 471 +0.54 218 -0.00 000	+0.08 904 -0.06 938 -0.00 000
1982 Oct 21 Neptune	22	-22.1069 - 0.0001	95.5824 15.0400	0.99 991	-0.13 669 +0.54 400 -0.00 002	-0.20 059 -0.06 867 +0.00 001
1982 Nov 18 Neptune	8	-22.1514 - 0.0001	271.7430 15.0396	0.99 991	+0.20 489 +0.54 584 -0.00 003	-0.47 089 -0.06 717 +0.00 003
1982 Nov 19 Mars	21	-24.0481 + 0.0035	87.8985 15.0064	0.99 854	-0.22 744 +0.51 232 -0.00 002	+0.55 761 +0.00 594 +0.00 010
1982 Dec 15 Neptune	16	-22.1886 - 0.0000	57.6148 15.0394	0.99 991	-0.27 825 +0.54 613 +0.00 001	-0.53 415 -0.06 430 +0.00 003
1982 Dec 16 Venus	8	-24.1492 + 0.0003	289.8380 14.9837	0.99 838	-0.09 309 +0.49 160 +0.00 000	+0.85 303 -0.03 273 +0.00 017

Table III   (cont.)

Date / Planet	T_o	δ_o / δ'	H_o / H_o'	k	X / x' / x''	Y / y' / y''
	h	°	°			
1983 Jan 12 Neptune	1	−22.2107 − 0.0000	218.5006 15.0395	0.99 991	+0.16 337 +0.54 461 +0.00 001	−0.68 046 −0.05 990 +0.00 003
1983 Feb 8 Neptune	8	−22.2158 + 0.0000	349.4722 15.0399	0.99 991	−0.30 577 +0.54 330 +0.00 004	−0.81 728 −0.05 543 +0.00 004
1983 Mar 6 Jupiter	3	−21.1077 − 0.0003	319.5869 15.0381	0.99 949	+0.13 981 +0.53 852 −0.00 001	+1.00 976 −0.12 457 −0.00 003
1983 Mar 7 Neptune	16	−22.2080 + 0.0000	135.8097 15.0405	0.99 991	−0.34 407 +0.54 490 +0.00 002	−1.11 829 −0.05 256 +0.00 005
1983 Apr 2 Jupiter	13	−21.1719 + 0.0001	135.8725 15.0418	0.99 945	+0.07 888 +0.54 865 −0.00 002	+0.62 811 −0.12 490 −0.00 003
1983 Apr 29 Jupiter	19	−20.9288 + 0.0006	254.3855 15.0452	0.99 941	+0.08 471 +0.55 554 −0.00 002	+0.57 212 −0.13 357 −0.00 004
1983 May 26 Jupiter	21	−20.4489 + 0.0008	314.3744 15.0467	0.99 940	+0.05 677 +0.55 499 −0.00 001	+0.86 567 −0.14 604 −0.00 006
1983 Jun 9 Mercury	10	+15.4859 + 0.0140	354.3608 14.9996	0.99 704	−0.15 965 +0.52 579 +0.00 003	+0.77 182 +0.17 446 −0.00 015
1983 Jun 11 Mars	1	+23.3598 + 0.0031	197.3810 15.0100	0.99 904	+0.06 607 +0.56 866 +0.00 003	−1.20 293 +0.10 912 −0.00 006
1983 Jun 22 Jupiter	22	−19.9334 + 0.0007	359.3955 15.0454	0.99 941	+0.43 360 +0.54 798 −0.00 001	+1.16 558 −0.15 590 −0.00 007
1983 Jul 9 Mars	19	+24.0001 − 0.0012	114.4201 15.0106	0.99 906	−0.15 303 +0.58 796 +0.00 003	+0.33 725 +0.02 327 −0.00 015
1983 Jul 10 Mercury	14	+23.6025 −0.0089	27.2732 14.9437	0.99 820	−0.03 892 +0.53 099 +0.00 001	+0.61 230 −0.02 669 −0.00 039

Table III (cont.)

Date Planet	T_o	δ_o δ'	H_o H_o'	k	X x' x''	Y y' y''
	h	$\circ$	$\circ$			
1983 Sep 12 Jupiter	19	−20.4210 − 0.0011	34.0424 15.0356	0.99 953	+0.35 308 +0.54 357 −0.00 003	+0.87 598 −0.15 084 −0.00 002
1983 Oct 10 Uranus	8	−21.3143 − 0.0003	253.7975 15.0391	0.99 987	+0.15 563 +0.55 481 −0.00 003	+0.97 168 −0.14 638 −0.00 003
1983 Oct 10 Jupiter	12	−21.2237 − 0.0013	311.9120 15.0333	0.99 956	+0.36 318 +0.54 959 −0.00 004	+0.30 670 −0.13 845 +0.00 002
1983 Nov 4 Saturn	15	−12.0255 − 0.0016	52.1592 15.0362	0.99 977	+0.24 179 +0.54 418 −0.00 001	+1.04 887 −0.22 880 −0.00 004
1983 Nov 6 Uranus	20	−21.5556 − 0.0004	99.3968 15.0385	0.99 987	−0.04 524 +0.55 901 −0.00 001	+0.75 754 −0.14 373 −0.00 002
1983 Nov 7 Jupiter	7	−22.0410 − 0.0011	258.5314 15.0317	0.99 958	−0.04 358 +0.55 416 −0.00 002	−0.19 345 −0.11 988 +0.00 005
1983 Dec 2 Saturn	5	−13.0229 − 0.0014	286.2063 15.0365	0.99 976	+0.27 064 +0.53 925 +0.00 001	+0.80 032 −0.22 108 −0.00 004
1983 Dec 4 Uranus	8	−21.8181 − 0.0004	304.7412 15.0383	0.99 987	−0.05 173 +0.55 826 +0.00 001	+0.58 899 −0.13 798 −0.00 002
1983 Dec 5 Jupiter	3	−22.6916 − 0.0008	219.3918 15.0309	0.99 959	−0.25 635 +0.55 524 +0.00 001	−0.73 438 −0.09 553 +0.00 007
1983 Dec 6 Mercury	3	−25.7957 + 0.0016	206.7976 14.9778	0.99 776	+0.08 257 +0.50 498 −0.00 000	+0.93 982 −0.04 149 +0.00 019
1983 Dec 29 Saturn	16	−13.8005 − 0.0010	115.5599 15.0375	0.99 975	+0.13 569 +0.53 535 +0.00 002	+0.54 291 −0.21 287 −0.00 002
1983 Dec 30 Venus	19	−17.6716 − 0.0119	146.9342 14.9901	0.99 764	−0.07 833 +0.49 836 +0.00 005	−0.66 304 −0.16 229 +0.00 016
1983 Dec 31 Uranus	18	−22.0579 − 0.0003	120.0241 15.0386	0.99 987	−0.17 453 +0.55 404 +0.00 003	+0.45 697 −0.13 009 −0.00 001

Table III   (cont.)

Date Planet	T_{0}	δ δ'	H_{0} H_{0}'	k	X x' x''	Y y' y''
	h	$\circ$	$\circ$			
1984 Jan 26 Saturn	1	−14.2713 − 0.0005	275.6473 15.0390	0.99 974	+0.09 774 +0.53 765 +0.00 001	+0.12 854 −0.20 845 −0.00 000
1984 Jan 28 Uranus	3	−22.2429 − 0.0002	280.5595 15.0392	0.99 987	+0.22 786 +0.55 128 +0.00 002	+0.11 533 −0.12 287 −0.00 000
1984 Feb 22 Saturn	9	−14.3852 + 0.0001	61.8252 15.0409	0.99 974	−0.07 528 +0.54 734 −0.00 001	−0.26 183 −0.21 044 +0.00 002
1984 Feb 22 Mars	14	−15.9599 − 0.0043	134.3796 15.0257	0.99 767	+0.02 441 +0.53 394 −0.00 000	+0.33 054 −0.19 854 +0.00 003
1984 Feb 24 Uranus	10	−22.3543 − 0.0001	51.5443 15.0402	0.99 986	−0.04 136 +0.55 460 +0.00 001	−0.16 251 −0.11 933 +0.00 001
1984 Mar 20 Saturn	17	−14.1461 + 0.0006	209.2565 15.0427	0.99 973	−0.29 689 +0.55 911 −0.00 001	−0.48 440 −0.21 777 +0.00 003
1984 Mar 21 Mars	12	−18.1037 − 0.0021	124.0815 15.0341	0.99 695	−0.38 090 +0.55 506 +0.00 001	−0.28 776 −0.18 190 +0.00 004
1984 Mar 22 Uranus	18	−22.3851 + 0.0000	198.2454 15.0412	0.99 986	−0.05 481 +0.56 386 −0.00 001	−0.46 536 −0.12 067 +0.00 002
1984 Apr 17 Saturn	1	−13.6386 + 0.0009	357.7273 15.0439	0.99 972	−0.27 941 +0.56 494 +0.00 000	−0.53 276 −0.22 606 +0.00 002
1984 Apr 17 Mars	23	−18.7675 + 0.0002	315.9481 15.0481	0.99 601	−0.13 448 +0.57 656 −0.00 000	+0.08 857 −0.18 734 −0.00 003
1984 Apr 19 Uranus	3	−22.3373 + 0.0001	0.6746 15.0422	0.99 986	+0.11 284 +0.57 349 −0.00 003	−0.64 833 −0.12 554 +0.00 003
1984 May 14 Saturn	8	−13.0416 + 0.0009	131.6135 15.0441	0.99 972	−0.01 661 +0.56 125 +0.00 001	−0.45 812 −0.23 097 +0.00 001

Table III   (cont.)

Date Planet	T_0	δ δ'	H_0 H_0'	k	X x' x''	Y y' y''
	h	$\circ$	$\circ$			
1984 May 16 Uranus	11	−22.2254 + 0.0002	148.5872 15.0428	0.99 986	−0.10 243 +0.57 757 −0.00 001	−0.56 643 −0.13 115 +0.00 002
1984 May 28 Mercury	17	+13.2781 + 0.0209	98.8858 14.9857	0.99 731	−0.51 369 +0.46 815 +0.00 006	+0.83 429 +0.18 919 −0.00 014
1984 May 30 Venus	8	+20.6181 + 0.0109	305.3218 14.9876	0.99 850	+0.02 385 +0.50 203 +0.00 007	+0.29 027 +0.14 525 −0.00 017
1984 Jun 10 Saturn	13	−12.5800 + 0.0005	235.1296 15.0431	0.99 973	+0.03 263 +0.55 182 +0.00 002	−0.24 831 −0.23 140 +0.00 000
1984 Jun 12 Uranus	18	−22.0800 + 0.0002	281.6759 15.0429	0.99 986	−0.05 692 +0.57 432 +0.00 001	−0.46 138 −0.13 501 +0.00 001
1984 Jul 7 Saturn	17	−12.4289 − 0.0001	322.7340 15.0415	0.99 973	−0.23 039 +0.54 419 +0.00 003	−0.04 190 −0.22 971 −0.00 000
1984 Jul 9 Uranus	23	−21.9462 + 0.0002	24.5359 15.0424	0.99 986	−0.19 390 +0.56 724 +0.00 003	−0.38 009 −0.13 648 +0.00 001
1984 Aug 3 Saturn	23	−12.6550 − 0.0006	79.2855 15.0397	0.99 975	−0.29 778 +0.54 375 +0.00 001	−0.13 140 −0.22 869 +0.00 001
1984 Aug 6 Uranus	4	−21.8697 + 0.0001	126.9450 15.0415	0.99 986	+0.02 613 +0.56 245 +0.00 002	−0.54 774 −0.13 661 +0.00 002
1984 Aug 31 Saturn	9	−13.2248 − 0.0011	254.8756 15.0381	0.99 976	−0.01 845 +0.55 036 −0.00 001	−0.52 637 −0.22 876 +0.00 004
1984 Sep 2 Uranus	10	−21.8804 − 0.0001	243.7638 15.0405	0.99 987	+0.09 788 +0.56 428 −0.00 000	−0.82 358 −0.13 668 +0.00 004
1984 Sep 27 Saturn	22	−14.0384 − 0.0013	114.6562 15.0369	0.99 977	−0.14 619 +0.55 970 −0.00 001	−0.80 207 −0.22 792 +0.00 006

Table III   (cont.)

Date Planet	$T_\circ$	δ δ'	$H_\circ$ $H_\circ'$	k	X x' x''	Y y' y''
	h	$\circ$	$\circ$			
1984 Sep 29 Uranus	18	-21.9827 $-\ 0.0002$	29.9975 15.0395	$0.99\ 987$	$-0.31\ 633$ $+0.57\ 231$ $-0.00\ 000$	$-1.01\ 882$ $-0.13\ 663$ $+0.00\ 006$
1984 Oct 1 Mars	0	-25.5846 $-\ 0.0003$	103.1321 15.0097	$0.99\ 782$	$-0.10\ 728$ $+0.54\ 585$ $-0.00\ 001$	$-0.26\ 918$ $-0.05\ 838$ $+0.00\ 014$
1984 Oct 25 Saturn	13	-14.9668 $-\ 0.0014$	3.8825 15.0362	$0.99\ 978$	$-0.62\ 324$ $+0.56\ 637$ $+0.00\ 003$	$-0.92\ 255$ $-0.22\ 332$ $+0.00\ 007$
1984 Oct 27 Venus	1	-22.9736 $-\ 0.0100$	164.2788 14.9871	$0.99\ 807$	$+0.31\ 407$ $+0.53\ 319$ $-0.00\ 001$	$+0.25\ 360$ $-0.13\ 901$ $+0.00\ 019$
1984 Nov 24 Mercury	14	-25.7917 $-\ 0.0004$	10.2206 14.9909	$0.99\ 761$	$-0.10\ 028$ $+0.54\ 641$ $+0.00\ 004$	$+0.11\ 708$ $-0.07\ 561$ $+0.00\ 019$
1985 Apr 22 Mars	13	$+20.0347$ $+\ 0.0070$	350.6071 15.0110	$0.99\ 887$	$+0.10\ 344$ $+0.49\ 188$ $+0.00\ 002$	$+0.45\ 167$ $+0.16\ 636$ $-0.00\ 011$
1985 Oct 15 Mercury	6	-14.6838 $-\ 0.0252$	259.5054 14.9812	$0.99\ 823$	$+0.30\ 019$ $+0.50\ 807$ $+0.00\ 004$	$+1.09\ 216$ $-0.23\ 242$ $+0.00\ 007$
1985 Nov 11 Venus	10	-10.9319 $-\ 0.0186$	349.9482 14.9914	$0.99\ 851$	$-0.29\ 990$ $+0.51\ 384$ $+0.00\ 008$	$-0.68\ 018$ $-0.24\ 697$ $+0.00\ 011$
1985 Nov 14 Mercury	3	-24.9589 $+\ 0.0004$	206.3055 15.0141	$0.99\ 733$	$-0.31\ 254$ $+0.58\ 269$ $+0.00\ 010$	$-0.38\ 352$ $-0.12\ 388$ $+0.00\ 012$
1985 Dec 8 Mars	11	$-\ 9.0245$ $-\ 0.0095$	37.5467 15.0168	$0.99\ 884$	$+0.26\ 507$ $+0.52\ 137$ $+0.00\ 004$	$-0.11\ 774$ $-0.25\ 987$ $+0.00\ 003$
1986 Mar 11 Venus	16	$+\ 0.1164$ $+\ 0.0214$	45.6506 14.9939	$0.99\ 842$	$+0.48\ 929$ $+0.45\ 142$ $-0.00\ 005$	$-1.12\ 712$ $+0.24\ 341$ $+0.00\ 004$
1986 Jun 23 Mars	14	-26.1225 $-\ 0.0041$	187.0944 15.0486	$0.99\ 454$	$+0.33\ 115$ $+0.61\ 654$ $-0.00\ 001$	$-0.47\ 869$ $+0.09\ 987$ $-0.00\ 001$

Table III (cont.)

Date Planet	T_0	δ δ'	H_0 H_0'	k	X x' x''	Y y' y''
	h	$\circ$	$\circ$			
1986 Jul 20 Mars	13	−28.4223 − 0.0020	206.2700 15.0533	0.99 403	+0.02 821 +0.62 652 +0.00 001	+0.91 280 +0.06 360 −0.00 011
1986 Aug 16 Mars	16	−28.3977 + 0.0017	281.6562 15.0384	0.99 475	−0.29 127 +0.60 566 +0.00 002	+0.48 199 +0.04 306 −0.00 001
1986 Aug 21 Jupiter	12	− 5.1424 − 0.0020	157.7243 15.0454	0.99 937	+0.31 874 +0.52 512 −0.00 002	−1.25 149 +0.27 269 +0.00 005
1986 Sep 13 Mars	10	−26.4823 + 0.0040	212.0169 15.0240	0.99 581	+0.26 801 +0.57 967 −0.00 001	−0.90 234 +0.07 499 +0.00 013
1986 Oct 5 Mercury	8	−13.4049 − 0.0251	284.4704 14.9858	0.99 803	+0.23 982 +0.48 993 +0.00 005	+0.27 204 −0.23 426 +0.00 009
1986 Nov 3 Mercury	14	−22.3708 + 0.0080	16.2981 15.0479	0.99 685	−0.04 084 +0.59 546 +0.00 010	−0.78 080 −0.18 694 −0.00 003
1987 Jan 4 Jupiter	20	− 5.7924 + 0.0028	54.3376 15.0347	0.99 953	+0.63 998 +0.51 514 −0.00 006	−1.04 114 +0.26 733 +0.00 005
1987 Feb 1 Jupiter	15	− 3.6485 + 0.0036	1.8753 15.0330	0.99 956	+0.38 553 +0.51 589 −0.00 005	−0.49 264 +0.27 579 +0.00 002
1987 Feb 3 Mars	11	+ 7.2486 + 0.0117	281.4508 15.0143	0.99 841	+0.04 382 +0.48 221 −0.00 002	+0.34 181 +0.25 333 −0.00 006
1987 Mar 1 Jupiter	12	− 1.1223 + 0.0039	338.6016 15.0321	0.99 958	−0.05 720 +0.51 343 −0.00 002	−0.02 549 +0.28 017 −0.00 000
1987 Mar 29 Jupiter	9	+ 1.5449 + 0.0040	314.9338 15.0318	0.99 958	−0.46 887 +0.50 844 +0.00 002	+0.46 274 +0.27 812 −0.00 003
1987 Apr 25 Venus	12	− 0.0954 + 0.0193	29.2121 14.9947	0.99 814	−0.23 692 +0.46 551 +0.00 001	+0.87 010 +0.26 025 −0.00 004

Table III   (cont.)

Date Planet	T_o	δ δ'	H_o H_o'	k	X x' x''	Y y' y''
	h	$\circ$	$\circ$			
1987 Aug 24 Mercury	22	+10.8894 - 0.0313	144.3816 14.9665	0.99 804	+0.24 341 +0.41 708 -0.00 003	+0.73 873 -0.21 217 -0.00 014
1987 Sep 22 Mars	8	+ 5.1544 - 0.0106	309.9381 15.0165	0.99 900	+0.27 465 +0.46 746 -0.00 001	+0.00 549 -0.25 425 -0.00 003
1987 Sep 25 Mercury	4	-11.6088 - 0.0243	220.3723 14.9925	0.99 771	-0.31 635 +0.46 453 +0.00 008	-0.15 619 -0.22 906 +0.00 008
1987 Oct 23 Mercury	13	-16.9448 + 0.0223	9.5978 15.0771	0.99 628	-0.66 642 +0.57 712 +0.00 007	-0.78 752 -0.24 941 -0.00 012
1988 Jan 21 Venus	19	-10.1026 + 0.0202	65.6778 14.9939	0.99 802	-0.21 379 +0.50 321 -0.00 006	-0.17 478 +0.25 746 +0.00 007
1988 Mar 16 Mercury	5	-13.0929 + 0.0148	276.5953 14.9880	0.99 771	+0.01 666 +0.50 703 -0.00 006	-0.51 547 +0.25 173 +0.00 013
1988 Apr 19 Venus	24	+26.6993 + 0.0055	135.3834 15.0035	0.99 556	+0.05 201 +0.53 710 -0.00 001	+1.05 129 +0.06 932 -0.00 021
1988 May 9 Mars	6	-16.0598 + 0.0077	352.7473 15.0138	0.99 770	+0.06 988 +0.52 333 -0.00 003	-0.76 345 +0.22 951 +0.00 011
1988 Aug 13 Mercury	12	+12.3847 - 0.0302	348.1280 14.9683	0.99 797	+0.00 769 +0.41 468 -0.00 004	-0.45 429 -0.20 821 -0.00 009
1988 Sep 13 Mercury	15	- 9.0315 - 0.0221	22.6120 15.0021	0.99 731	-0.31 723 +0.44 983 +0.00 006	-0.54 209 -0.22 641 +0.00 006
1988 Oct 7 Venus	3	+10.8137 - 0.0150	265.9050 14.9943	0.99 741	+0.18 293 +0.44 042 -0.00 003	+0.61 405 -0.22 603 -0.00 010
1988 Oct 10 Mercury	23	- 9.0052 + 0.0312	168.3788 15.0826	0.99 594	-0.60 512 +0.54 203 -0.00 003	-1.03 519 -0.28 426 -0.00 002

Table III   (cont.)

Date Planet	T_0	δ δ'	H_0 H_0'	k	X x' x''	Y y' y''
	h	$\circ$	$\circ$			
1989 Mar 6 Mercury	4	−15.4900 + 0.0169	257.8285 14.9805	0.99 796	−0.14 576 +0.50 134 −0.00 003	+0.67 697 +0.23 727 +0.00 011
1989 Jul 5 Venus	4	+20.0462 − 0.0126	213.1772 14.9889	0.99 829	+0.17 073 +0.48 946 −0.00 010	+0.05 401 −0.18 022 −0.00 014
1989 Jul 5 Mars	12	+18.4318 − 0.0074	328.8664 15.0148	0.99 896	−0.18 613 +0.50 665 −0.00 005	+0.16 815 −0.19 898 −0.00 007
1989 Sep 2 Mercury	15	− 5.3943 − 0.0178	21.6458 15.0151	0.99 689	−0.47 356 +0.46 135 +0.00 006	−0.41 473 −0.23 371 +0.00 002
1989 Nov 2 Venus	22	−26.8532 − 0.0018	105.0437 14.9928	0.99 625	+0.13 439 +0.51 103 −0.00 002	−0.71 196 +0.01 937 +0.00 022
1989 Dec 2 Venus	7	−24.0321 + 0.0087	239.0807 15.0076	0.99 467	−0.40 303 +0.51 815 −0.00 001	+0.76 302 +0.13 821 +0.00 009
1990 Mar 22 Mars	18	−19.2303 + 0.0075	139.0435 15.0092	0.99 857	+0.09 172 +0.51 183 +0.00 000	+0.47 213 +0.19 197 +0.00 008
1990 Jul 21 Jupiter	18	+21.5550 − 0.0016	93.3622 15.0312	0.99 961	−0.02 869 +0.57 472 −0.00 001	+1.02 416 −0.15 593 −0.00 009
1990 Aug 18 Jupiter	13	+20.4229 − 0.0018	39.3371 15.0318	0.99 959	−0.07 959 +0.56 037 +0.00 001	+0.42 700 −0.17 736 −0.00 006
1990 Aug 18 Venus	24	+19.2693 − 0.0111	198.3503 14.9883	0.99 843	−0.05 548 +0.50 994 −0.00 006	−0.48 333 −0.18 929 −0.00 013
1990 Aug 22 Mercury	11	− 0.5885 − 0.0105	323.2386 15.0330	0.99 653	−0.31 637 +0.50 459 +0.00 006	−0.02 069 −0.24 934 −0.00 002
1990 Sep 15 Jupiter	6	+19.2097 − 0.0018	315.9041 15.0332	0.99 957	+0.09 350 +0.54 697 +0.00 001	−0.31 426 −0.19 292 −0.00 002

Table III   (cont.)

Date Planet	T_o	δ δ'	H_o H_o'	k	X x' x''	Y y' y''
	h	$\circ$	$\circ$			
1990 Oct 12 Jupiter	19	+18.1264 − 0.0014	173.4746 15.0353	0.99 954	−0.34 438 +0.53 929 +0.00 002	−0.88 304 −0.20 437 +0.00 002
1990 Oct 25 Saturn	18	−22.0807 + 0.0003	12.6749 15.0387	0.99 974	+0.40 695 +0.53 359 −0.00 003	−1.10 548 +0.12 567 +0.00 005
1990 Nov 22 Saturn	4	−21.8197 + 0.0005	187.6244 15.0372	0.99 975	+0.04 903 +0.52 922 −0.00 003	−0.70 000 +0.13 339 +0.00 004
1990 Dec 18 Venus	6	−24.1107 + 0.0015	258.4129 14.9838	0.99 837	+0.22 043 +0.48 776 −0.00 004	−1.05 056 +0.07 630 +0.00 024
1990 Dec 19 Saturn	16	−21.3917 + 0.0008	31.8037 15.0362	0.99 975	+0.29 381 +0.52 781 −0.00 004	−0.17 915 +0.14 522 +0.00 002
1991 Jan 16 Saturn	4	−20.8300 + 0.0009	235.5458 15.0359	0.99 976	−0.04 525 +0.52 767 −0.00 000	+0.10 951 +0.15 793 +0.00 001
1991 Feb 11 Uranus	5	−23.2222 + 0.0002	292.6347 15.0388	0.99 987	+0.32 864 +0.53 989 +0.00 001	−1.18 697 +0.09 803 +0.00 006
1991 Feb 12 Saturn	18	−20.1973 + 0.0010	109.3505 15.0362	0.99 976	+0.09 359 +0.52 662 +0.00 001	+0.53 489 +0.16 873 −0.00 000
1991 Mar 10 Uranus	15	−23.1252 + 0.0001	108.4331 15.0396	0.99 986	+0.12 806 +0.53 758 +0.00 001	−0.94 799 +0.10 149 +0.00 005
1991 Mar 12 Saturn	7	−19.5919 + 0.0008	328.5228 15.0371	0.99 975	−0.47 883 +0.52 384 +0.00 004	+0.80 195 +0.17 622 −0.00 001
1991 Mar 22 Mars	17	+25.3152 + 0.0006	351.1995 15.0177	0.99 827	+0.27 758 +0.57 680 −0.00 003	+0.65 140 −0.01 892 −0.00 014
1991 Apr 7 Uranus	0	−23.0785 + 0.0000	269.7877 15.0406	0.99 986	−0.09 577 +0.53 625 +0.00 000	−0.64 695 +0.10 316 +0.00 003

Table III (cont.)

Date Planet	T_o	δ δ'	H_o H_o'	k	X x' x''	Y y' y''
	h	$\circ$	$\circ$			
1991 Apr 19 Mars	24	+24.9035 − 0.0019	107.4524 15.0158	0.99 856	+0.13 992 +0.57 103 −0.00 004	−0.63 184 −0.09 142 −0.00 007
1991 May 4 Uranus	8	−23.0954 − 0.0001	56.7813 15.0417	0.99 986	+0.13 329 +0.53 697 −0.00 003	−0.31 466 +0.10 359 +0.00 001
1991 May 31 Uranus	13	−23.1677 − 0.0001	159.2695 15.0425	0.99 985	−0.14 772 +0.53 927 −0.00 001	−0.24 980 +0.10 255 +0.00 000
1991 May 31 Neptune	20	−21.6384 − 0.0001	261.2409 15.0420	0.99 991	+0.22 033 +0.53 613 −0.00 003	−1.18 711 +0.11 475 +0.00 004
1991 Jun 27 Uranus	17	−23.2686 − 0.0002	247.1238 15.0429	0.99 985	−0.20 984 +0.54 137 +0.00 000	−0.31 717 +0.09 965 +0.00 001
1991 Jun 28 Neptune	1	−21.7147 − 0.0001	3.7553 15.0422	0.99 991	+0.31 856 +0.53 805 −0.00 002	−1.19 156 +0.11 341 +0.00 004
1991 Jul 24 Uranus	21	−23.3646 − 0.0001	335.0664 15.0427	0.99 985	−0.16 219 +0.54 178 +0.00 002	−0.42 409 +0.09 521 +0.00 001
1991 Aug 11 Mercury	9	+ 5.4653 − 0.0006	298.7249 15.0549	0.99 621	+0.50 616 +0.56 693 +0.00 004	+0.38 922 −0.25 571 −0.00 004
1991 Aug 21 Uranus	2	−23.4300 − 0.0001	77.7930 15.0421	0.99 986	−0.13 741 +0.54 049 +0.00 003	−0.43 884 +0.09 071 +0.00 002
1991 Sep 17 Uranus	9	−23.4519 + 0.0000	210.0693 15.0411	0.99 986	−0.01 646 +0.53 880 +0.00 001	−0.25 273 +0.08 822 +0.00 001
1991 Sep 17 Neptune	18	−21.9257 − 0.0000	341.0435 15.0413	0.99 991	+0.35 292 +0.53 616 +0.00 000	−1.06 818 +0.10 466 +0.00 004
1991 Oct 4 Venus	15	+ 8.9385 − 0.0029	88.2834 15.0136	0.99 482	−0.20 001 +0.52 326 −0.00 001	−0.14 240 −0.23 927 −0.00 003

Table III   (cont.)

Date Planet	T_o	δ δ'	H_o H_o'	k	X x' x''	Y y' y''
	h	$\circ$	$\circ$			
1991 Oct 14 Uranus	18	−23.4262 + 0.0001	11.7562 15.0401	0.99 986	+0.18 491 +0.53 789 −0.00 002	+0.09 614 +0.08 924 −0.00 000
1991 Oct 15 Neptune	2	−21.9338 + 0.0000	127.8987 15.0406	0.99 991	+0.22 315 +0.53 517 −0.00 002	−0.77 192 +0.10 455 +0.00 003
1991 Nov 8 Mercury	4	−23.4687 − 0.0129	224.0893 14.9810	0.99 782	−0.29 463 +0.49 955 +0.00 001	−0.68 374 −0.04 893 +0.00 024
1991 Nov 11 Uranus	3	−23.3525 + 0.0001	172.7906 15.0392	0.99 987	−0.26 260 +0.53 794 −0.00 002	+0.36 028 +0.09 385 −0.00 001
1991 Nov 11 Neptune	11	−21.8978 + 0.0001	289.3638 15.0400	0.99 991	+0.20 984 +0.53 533 −0.00 003	−0.44 370 +0.10 728 +0.00 002
1991 Dec 8 Uranus	13	−23.2333 + 0.0002	348.3598 15.0386	0.99 987	−0.34 114 +0.53 781 −0.00 001	+0.59 978 +0.10 052 −0.00 002
1991 Dec 8 Neptune	20	−21.8200 + 0.0001	90.4782 15.0395	0.99 991	+0.18 335 +0.53 584 −0.00 003	−0.22 384 +0.11 169 +0.00 001
1992 Jan 3 Mars	10	−23.8530 − 0.0015	347.1379 15.0075	0.99 888	−0.17 227 +0.51 808 +0.00 000	−0.91 946 +0.03 231 +0.00 016
1992 Jan 4 Uranus	23	−23.0782 + 0.0003	163.6577 15.0384	0.99 987	−0.24 875 +0.53 633 +0.00 001	+0.78 688 +0.10 708 −0.00 002
1992 Jan 5 Neptune	4	−21.7104 + 0.0002	236.3519 15.0394	0.99 991	−0.14 185 +0.53 525 +0.00 000	−0.17 679 +0.11 585 +0.00 001
1992 Jan 31 Venus	17	−22.3553 + 0.0009	106.6763 14.9857	0.99 793	+0.02 291 +0.48 659 −0.00 002	−1.13 826 +0.08 093 +0.00 022
1992 Feb 1 Uranus	8	−22.9087 + 0.0003	323.9195 15.0386	0.99 987	−0.46 121 +0.53 369 +0.00 004	+0.92 596 +0.11 212 −0.00 002

Table III   (cont.)

Date Planet	T_0	δ δ'	H_0 H_0'	k	X x' x''	Y y' y''
	h	$\circ$	$\circ$			
1992 Feb 1 Neptune	12	−21.5867 + 0.0002	22.2101 15.0395	0.99 991	−0.25 146 +0.53 321 +0.00 003	−0.08 535 +0.11 854 +0.00 001
1992 Feb 28 Neptune	21	−21.4724 + 0.0002	183.2794 15.0399	0.99 991	+0.17 935 +0.53 152 +0.00 002	+0.23 290 +0.12 001 −0.00 000
1992 Mar 27 Neptune	5	−21.3921 + 0.0001	329.6346 15.0405	0.99 991	−0.05 635 +0.53 209 +0.00 001	+0.49 648 +0.12 121 −0.00 002
1992 Apr 23 Neptune	13	−21.3632 + 0.0000	116.4070 15.0411	0.99 991	−0.29 913 +0.53 518 +0.00 000	+0.73 502 +0.12 254 −0.00 003
1992 May 20 Neptune	21	−21.3905 − 0.0001	263.6036 15.0417	0.99 991	−0.18 751 +0.53 909 −0.00 001	+0.91 350 +0.12 320 −0.00 004
1992 Jun 1 Mercury	5	+22.8759 + 0.0192	254.9596 14.9434	0.99 813	−0.22 342 +0.51 046 +0.00 005	+1.01 942 +0.01 037 −0.00 038
1992 Jun 17 Neptune	3	−21.4634 − 0.0001	21.0547 15.0422	0.99 991	−0.47 073 +0.54 165 +0.00 000	+0.83 978 +0.12 221 −0.00 004
1992 Jul 14 Neptune	9	−21.5600 − 0.0002	138.6779 15.0423	0.99 991	−0.04 124 +0.54 178 +0.00 001	+0.85 485 +0.11 908 −0.00 004
1992 Aug 10 Neptune	13	−21.6541 − 0.0001	226.1806 15.0420	0.99 991	−0.34 695 +0.54 016 +0.00 003	+0.77 707 +0.11 501 −0.00 003
1992 Sep 6 Neptune	18	−21.7230 − 0.0001	328.4897 15.0416	0.99 991	−0.40 175 +0.53 904 +0.00 004	+0.91 378 +0.11 179 −0.00 004
1992 Sep 20 Mars	9	+23.4646 + 0.0002	39.5676 15.0164	0.99 793	+0.00 616 +0.56 277 +0.00 001	−0.88 629 −0.07 105 −0.00 005
1992 Oct 4 Neptune	1	−21.7509 − 0.0000	100.5021 15.0409	0.99 991	−0.28 333 +0.54 045 +0.00 002	+1.21 779 +0.11 116 −0.00 005

Table III   (cont.)

Date Planet	T_o	δ δ'	H_o H_o'	k	X x' x''	Y y' y''
	h	$\circ$	$\circ$			
1992 Oct 27 Mercury	15	−22.2474 − 0.0134	26.4900 14.9908	0.99 769	−0.08 714 +0.53 577 +0.00 001	+0.56 473 −0.07 457 +0.00 015
1992 Oct 28 Venus	15	−23.4719 − 0.0090	13.2246 14.9869	0.99 797	+0.19 446 +0.52 848 −0.00 005	+0.39 698 −0.02 236 +0.00 019
1993 Feb 25 Venus	5	+10.9143 + 0.0141	216.9322 15.0248	0.99 363	+0.30 245 +0.49 945 +0.00 000	−0.44 919 +0.17 754 −0.00 001
1993 Apr 19 Venus	16	+ 5.5483 − 0.0108	85.9082 15.0417	0.99 178	−0.26 669 +0.51 532 −0.00 001	+0.42 686 +0.21 649 −0.00 003
1993 May 22 Mercury	5	+22.9090 + 0.0201	248.2676 14.9465	0.99 797	+0.08 022 +0.47 900 +0.00 006	−0.91 623 +0.01 732 −0.00 025
1993 Nov 14 Mars	18	−21.3594 − 0.0060	81.9560 15.0094	0.99 901	+0.18 176 +0.57 817 −0.00 002	+0.49 603 −0.05 246 +0.00 009
1993 Dec 12 Mercury	11	−21.3492 − 0.0151	359.6611 14.9745	0.99 818	−0.15 929 +0.54 244 +0.00 001	+0.15 849 −0.02 590 +0.00 022
1993 Dec 12 Venus	18	−21.8666 − 0.0085	100.6018 14.9853	0.99 854	−0.00 880 +0.55 220 −0.00 000	+0.44 964 −0.01 563 +0.00 018
1994 Apr 12 Venus	23	+15.8759 + 0.0170	144.6212 14.9909	0.99 826	−0.19 917 +0.48 305 +0.00 002	+1.10 820 +0.09 750 −0.00 016
1994 Jul 5 Mars	5	+19.9632 + 0.0066	299.1249 15.0104	0.99 867	+0.05 941 +0.51 613 +0.00 004	−0.36 877 +0.05 800 −0.00 007
1994 Oct 7 Jupiter	12	−15.9180 − 0.0025	331.7584 15.0327	0.99 961	−0.00 912 +0.59 501 −0.00 001	−0.72 710 −0.11 875 +0.00 007
1994 Nov 4 Jupiter	8	−17.5609 − 0.0024	293.2779 15.0318	0.99 962	+0.25 873 +0.60 463 −0.00 002	−0.15 016 −0.10 070 +0.00 004

Table III   (cont.)

Date Planet	T_o	δ δ'	H_o H_o'	k	X x' x''	Y y' y''
	h	$\circ$	$\circ$			
1994 Dec 2 Jupiter	5	-19.0550 - 0.0020	269.4603 15.0316	0.99 962	+0.27 018 +0.60 758 -0.00 000	+0.43 266 -0.08 064 +0.00 001
1994 Dec 30 Jupiter	1	-20.2619 - 0.0016	230.7393 15.0322	0.99 961	+0.32 444 +0.60 307 +0.00 001	+1.02 289 -0.06 079 -0.00 003
1995 Jan 27 Venus	12	-20.0588 - 0.0051	46.1139 14.9930	0.99 686	-0.11 650 +0.56 222 +0.00 003	+0.17 489 +0.00 453 +0.00 015
1995 May 27 Venus	7	+14.1863 + 0.0162	308.8370 14.9916	0.99 825	-0.18 250 +0.48 667 +0.00 003	+0.89 878 +0.10 115 -0.00 014
1995 Jun 26 Mercury	2	+18.5725 + 0.0088	231.8888 15.0125	0.99 645	-0.10 293 +0.51 858 -0.00 005	+0.66 309 +0.02 064 -0.00 012
1995 Aug 30 Mars	3	- 9.6515 - 0.0104	180.0504 15.0157	0.99 875	-0.29 767 +0.53 777 +0.00 001	-0.07 490 -0.15 497 +0.00 005
1996 Feb 22 Venus	5	+ 6.3120 + 0.0213	212.3755 14.9975	0.99 747	+0.19 238 +0.51 514 -0.00 004	-0.00 224 +0.15 499 -0.00 007
1996 Apr 17 Mars	5	+ 6.5187 + 0.0125	263.5611 15.0116	0.99 893	-0.15 094 +0.52 960 +0.00 001	+0.57 682 +0.15 935 -0.00 006
1996 Jun 14 Mercury	0	+17.0811 + 0.0153	203.9410 14.9931	0.99 702	-0.17 604 +0.50 650 -0.00 003	-0.49 228 +0.06 562 -0.00 012
1996 Jul 12 Venus	9	+17.6830 + 0.0008	352.6973 15.0259	0.99 357	+0.18 060 +0.54 018 -0.00 001	+0.42 580 +0.04 174 -0.00 006
1996 Aug 16 Mercury	18	+ 2.4191 - 0.0253	64.1720 14.9996	0.99 729	-0.20 274 +0.48 255 +0.00 003	-0.29 608 -0.14 617 -0.00 001
1997 Apr 7 Saturn	1	+ 2.3913 + 0.0020	199.2480 15.0363	0.99 977	-0.05 787 +0.57 855 +0.00 001	+1.07 690 +0.18 601 -0.00 006

Table III   (cont.)

Date Planet	T_0	δ δ'	H_0 H_0'	k	X x' x''	Y y' y''
	h	$\circ$	$\circ$			
1997 Apr 7 Venus	13	+ 6.3549 + 0.0204	12.8373 14.9933	0.99 859	+0.09 438 +0.53 639 -0.00 001	-0.64 859 +0.16 116 -0.00 002
1997 May 4 Saturn	15	+ 3.6527 + 0.0018	73.3450 15.0366	0.99 976	-0.35 721 +0.57 162 +0.00 004	+0.68 162 +0.18 286 -0.00 003
1997 May 5 Mercury	17	+ 9.9254 - 0.0127	90.2793 15.0469	0.99 585	+0.51 185 +0.58 688 -0.00 008	-1.04 941 +0.18 076 +0.00 004
1997 Jun 1 Saturn	3	+ 4.6816 + 0.0013	277.7756 15.0375	0.99 975	-0.12 537 +0.56 422 +0.00 003	+0.43 347 +0.17 985 -0.00 002
1997 Jun 13 Mars	16	+ 1.4516 - 0.0077	323.9779 15.0260	0.99 761	+0.00 037 +0.50 415 +0.00 000	-0.30 578 -0.16 444 +0.00 001
1997 Jun 28 Saturn	12	+ 5.3587 + 0.0007	77.8235 15.0388	0.99 974	-0.01 764 +0.56 188 +0.00 001	+0.16 284 +0.17 836 -0.00 001
1997 Jul 25 Saturn	19	+ 5.5981 + 0.0000	208.8013 15.0405	0.99 973	-0.13 813 +0.56 761 -0.00 000	-0.06 302 +0.17 938 -0.00 000
1997 Aug 5 Mercury	20	+ 5.9087 - 0.0209	92.9997 15.0109	0.99 683	+0.52 813 +0.49 235 +0.00 000	+0.95 252 -0.13 772 -0.00 007
1997 Aug 22 Saturn	2	+ 5.3687 - 0.0007	340.9454 15.0423	0.99 972	-0.18 277 +0.57 948 -0.00 001	-0.06 518 +0.18 293 -0.00 000
1997 Sep 18 Saturn	10	+ 4.7393 - 0.0012	129.2024 15.0437	0.99 972	-0.00 381 +0.59 031 -0.00 001	+0.18 142 +0.18 757 -0.00 001
1997 Oct 15 Saturn	18	+ 3.9237 - 0.0012	278.0651 15.0441	0.99 971	+0.05 903 +0.59 201 +0.00 000	+0.41 942 +0.19 077 -0.00 002
1997 Nov 12 Saturn	1	+ 3.2367 - 0.0008	51.7459 15.0433	0.99 972	+0.06 588 +0.58 273 +0.00 001	+0.46 750 +0.19 091 -0.00 001

Table III   (cont.)

Date Planet	T_o	δ δ'	H_o H_o'	k	X x' x''	Y y' y''
	h	$\circ$	$\circ$			
1997 Dec 9 Saturn	6	+ 2.9522 − 0.0000	154.5080 15.0416	0.99 972	−0.29 596 +0.57 001 +0.00 003	+0.11 746 +0.18 896 +0.00 000
1998 Jan 5 Saturn	12	+ 3.1880 + 0.0008	271.1187 15.0397	0.99 973	−0.13 422 +0.56 448 +0.00 000	−0.25 500 +0.18 739 +0.00 001
1998 Feb 1 Saturn	21	+ 3.9096 + 0.0014	71.6589 15.0381	0.99 975	+0.19 547 +0.56 950 −0.00 003	−0.58 308 +0.18 756 +0.00 002
1998 Feb 27 Mars	23	− 2.1334 + 0.0132	145.6436 15.0115	0.99 896	+0.35 473 +0.55 884 −0.00 004	−0.61 607 +0.18 147 +0.00 004
1998 Mar 1 Saturn	10	+ 4.9886 + 0.0018	291.4015 15.0369	0.99 976	+0.48 695 +0.57 943 −0.00 005	−0.81 953 +0.18 868 +0.00 003
1998 Mar 24 Venus	19	−13.7209 + 0.0077	147.7264 15.0001	0.99 625	−0.02 086 +0.54 869 +0.00 002	+0.08 907 +0.12 513 +0.00 010
1998 Mar 26 Jupiter	11	− 7.9480 + 0.0036	5.0100 15.0321	0.99 959	−0.33 136 +0.57 996 +0.00 003	+0.70 959 +0.17 834 −0.00 002
1998 Mar 29 Saturn	1	+ 6.2557 + 0.0020	180.6113 15.0362	0.99 977	+0.12 251 +0.58 639 −0.00 002	−1.17 992 +0.18 906 +0.00 006
1998 Apr 23 Jupiter	7	− 5.6544 + 0.0032	326.7648 15.0332	0.99 957	−0.12 041 +0.57 234 +0.00 003	+0.17 670 +0.18 466 +0.00 001
1998 Apr 23 Venus	8	− 5.2895 + 0.0153	341.5955 14.9976	0.99 727	+0.25 209 +0.53 841 +0.00 001	+0.00 194 +0.17 341 +0.00 005
1998 Apr 24 Mercury	19	+ 2.5076 − 0.0009	127.1190 15.0243	0.99 654	+0.02 532 +0.57 138 −0.00 004	−0.85 104 +0.19 980 +0.00 001
1998 May 20 Jupiter	23	− 3.7470 + 0.0025	229.3314 15.0349	0.99 954	−0.16 436 +0.56 419 +0.00 003	−0.41 486 +0.18 816 +0.00 003

Table III (cont

Date Planet	T_0	δ δ'	H_0 H_0'	k	X x' x''	Y y' y''
	h	$\circ$	$\circ$			
1998 Jun 17 Jupiter	11	− 2.4588 + 0.0014	73.1565 15.0374	0.99 949	−0.01 576 +0.56 069 +0.00 000	−0.81 788 +0.19 019 +0.00 005
1998 Jul 14 Jupiter	19	− 2.0027 − 0.0000	218.7049 15.0406	0.99 945	+0.13 525 +0.56 528 −0.00 002	−0.96 027 +0.19 245 +0.00 005
1998 Aug 11 Jupiter	0	− 2.4821 − 0.0014	321.3011 15.0439	0.99 941	+0.08 016 +0.57 669 −0.00 002	−0.84 511 +0.19 540 +0.00 004
1998 Sep 7 Jupiter	5	− 3.7142 − 0.0022	65.7547 15.0460	0.99 939	+0.44 899 +0.58 707 −0.00 003	−0.36 395 +0.19 715 +0.00 002
1998 Sep 19 Venus	18	+ 6.7960 − 0.0198	100.7421 14.9929	0.99 841	−0.16 661 +0.47 867 −0.00 002	−0.03 171 −0.14 899 −0.00 004
1998 Sep 20 Mercury	7	+ 4.6344 − 0.0323	290.2933 14.9700	0.99 801	+0.17 877 +0.44 998 −0.00 003	+0.08 889 −0.14 026 −0.00 004
1998 Oct 4 Jupiter	10	− 5.1048 − 0.0019	170.8036 15.0456	0.99 940	+0.29 780 +0.58 693 −0.00 001	−0.11 106 +0.19 510 +0.00 001
1998 Oct 16 Mars	3	+10.8841 − 0.0090	271.8502 15.0171	0.99 876	−0.28 109 +0.50 855 +0.00 001	−0.99 993 −0.14 851 −0.00 001
1998 Oct 31 Jupiter	16	− 5.9315 − 0.0006	289.7946 15.0428	0.99 942	−0.16 232 +0.57 482 +0.00 002	−0.27 918 +0.19 022 +0.00 002
1998 Nov 13 Mars	18	+ 4.5480 − 0.0093	149.1791 15.0188	0.99 862	+0.04 033 +0.49 816 +0.00 001	+0.56 764 −0.16 653 −0.00 004
1998 Nov 28 Jupiter	1	− 5.7992 + 0.0010	91.7474 15.0392	0.99 946	+0.25 604 +0.55 950 −0.00 001	−0.53 380 +0.18 629 +0.00 003
1998 Dec 25 Jupiter	11	− 4.7136 + 0.0023	266.4947 15.0361	0.99 950	+0.24 749 +0.55 059 −0.00 002	−1.15 026 +0.18 587 +0.00 006

Table III   (cont.)

Date Planet	T_o	δ δ'	H_o H_o'	k	X x' x''	Y y' y''
	h	$\circ$	$\circ$			
1999 Feb 17 Mercury	2	- 9.6506 + 0.0336	196.3656 14.9715	0.99 803	+0.20 591 +0.50 007 -0.00 001	-0.19 168 +0.13 910 +0.00 012
1999 Apr 10 Neptune	9	-18.9314 + 0.0001	26.7181 15.0404	0.99 991	+0.08 080 +0.55 930 +0.00 001	+1.22 083 +0.08 537 -0.00 005
1999 Apr 11 Uranus	7	-16.6255 + 0.0004	345.2717 15.0397	0.99 987	-0.08 789 +0.56 071 +0.00 002	+1.07 203 +0.12 169 -0.00 004
1999 Apr 14 Mercury	5	- 3.2314 + 0.0093	279.4026 15.0059	0.99 703	+0.46 531 +0.53 694 -0.00 004	-0.92 537 +0.19 221 +0.00 005
1999 May 7 Neptune	16	-18.8860 - 0.0000	158.4189 15.0411	0.99 991	-0.30 519 +0.55 391 +0.00 000	+0.87 881 +0.08 551 -0.00 004
1999 May 8 Uranus	16	-16.4621 + 0.0001	146.6548 15.0406	0.99 987	-0.06 124 +0.55 368 +0.00 000	+0.77 897 +0.12 259 -0.00 003
1999 Jun 3 Neptune	22	-18.9353 - 0.0001	275.4872 15.0417	0.99 991	-0.03 399 +0.55 346 -0.00 002	+0.70 040 +0.08 406 -0.00 004
1999 Jun 4 Uranus	22	-16.4803 - 0.0002	263.5141 15.0415	0.99 987	-0.13 826 +0.55 124 -0.00 001	+0.52 382 +0.12 183 -0.00 003
1999 Jul 1 Neptune	2	-19.0618 - 0.0002	2.8079 15.0421	0.99 991	-0.36 702 +0.55 710 -0.00 000	+0.58 052 +0.08 174 -0.00 003
1999 Jul 2 Uranus	3	-16.6645 - 0.0004	5.8915 15.0423	0.99 986	+0.17 816 +0.55 428 -0.00 003	+0.51 163 +0.12 000 -0.00 003
1999 Jul 28 Neptune	8	-19.2304 - 0.0003	120.3984 15.0422	0.99 991	+0.09 933 +0.56 101 -0.00 000	+0.70 704 +0.07 910 -0.00 003
1999 Jul 29 Uranus	7	-16.9568 - 0.0005	93.6190 15.0427	0.99 986	-0.07 007 +0.55 954 -0.00 000	+0.52 972 +0.11 771 -0.00 003
1999 Aug 10 Mercury	3	+17.9303 + 0.0075	242.0549 15.0195	0.99 687	+0.21 592 +0.57 955 -0.00 011	+1.13 932 -0.07 813 -0.00 011

Table III   (cont.)

Date Planet	$T_\circ$	δ δ'	$H_\circ$ $H_\circ'$	k	X x' x''	Y y' y''
	h	°	°			
1999 Aug 24 Neptune	14	−19.3950 − 0.0002	237.9760 15.0421	0.99 991	−0.22 715 +0.56 137 +0.00 003	+0.73 499 +0.07 697 −0.00 003
1999 Aug 25 Uranus	13	−17.2706 − 0.0004	211.5481 15.0426	0.99 986	+0.05 558 +0.56 200 +0.00 001	+0.66 527 +0.11 523 −0.00 003
1999 Sep 20 Neptune	22	−19.5130 − 0.0001	25.4239 15.0416	0.99 991	−0.16 316 +0.55 699 +0.00 003	+0.70 100 +0.07 553 −0.00 003
1999 Sep 21 Uranus	20	−17.5124 − 0.0003	344.3158 15.0421	0.99 986	−0.15 644 +0.55 872 +0.00 003	+0.62 053 +0.11 298 −0.00 002
1999 Oct 18 Neptune	7	−19.5527 + 0.0000	187.5488 15.0410	0.99 991	+0.19 538 +0.55 012 −0.00 000	+0.53 763 +0.07 480 −0.00 002
1999 Oct 19 Uranus	4	−17.6116 − 0.0000	131.6519 15.0412	0.99 987	−0.31 438 +0.55 098 +0.00 003	+0.39 570 +0.11 129 −0.00 001
1999 Nov 14 Neptune	15	−19.5008 + 0.0001	334.2058 15.0403	0.99 991	+0.22 170 +0.54 501 −0.00 002	+0.22 835 +0.07 481 −0.00 001
1999 Nov 15 Uranus	13	−17.5344 + 0.0003	293.4134 15.0402	0.99 987	+0.10 430 +0.54 349 −0.00 001	+0.14 448 +0.11 064 −0.00 001
1999 Dec 11 Neptune	22	−19.3622 + 0.0003	105.4312 15.0398	0.99 991	−0.01 480 +0.54 446 −0.00 002	−0.08 080 +0.07 582 +0.00 000
1999 Dec 12 Mars	18	−18.2575 + 0.0096	35.4784 15.0087	0.99 847	−0.33 985 +0.50 963 −0.00 002	+0.55 215 +0.09 850 +0.00 006
1999 Dec 12 Uranus	21	−17.2859 + 0.0005	79.5301 15.0394	0.99 987	+0.17 066 +0.54 056 −0.00 003	−0.18 037 +0.11 174 +0.00 001
2000 Jan 8 Neptune	6	−19.1574 + 0.0003	251.4210 15.0395	0.99 991	+0.18 406 +0.54 706 −0.00 002	−0.21 033 +0.07 820 +0.00 001
2000 Jan 9 Uranus	5	−16.9013 + 0.0007	225.1816 15.0389	0.99 987	+0.08 109 +0.54 238 −0.00 002	−0.43 185 +0.11 504 +0.00 002

Table III   (cont.)

Date Planet	$T_\circ$	δ δ'	$H_\circ$ $H_\circ'$	k	X x' x''	Y y' y''
	h	$\circ$	$\circ$			
2000 Feb 4 Neptune	14	−18.9204 + 0.0004	37.3018 15.0395	0.99 991	−0.13 836 +0.54 904 +0.00 001	−0.34 612 +0.08 182 +0.00 002
2000 Feb 5 Uranus	14	−16.4377 + 0.0007	25.6228 15.0387	0.99 987	−0.08 886 +0.54 525 +0.00 000	−0.61 577 +0.12 019 +0.00 003
2000 Mar 2 Neptune	24	−18.6920 + 0.0003	213.3362 15.0397	0.99 991	+0.10 459 +0.54 806 +0.00 002	−0.45 028 +0.08 590 +0.00 003
2000 Mar 4 Venus	1	−16.0231 + 0.0144	216.1016 14.9896	0.99 822	+0.10 257 +0.49 525 +0.00 001	−0.67 242 +0.11 149 +0.00 016
2000 Mar 4 Uranus	1	−15.9660 + 0.0007	216.1396 15.0388	0.99 987	+0.14 190 +0.54 552 +0.00 001	−0.73 318 +0.12 602 +0.00 004
2000 Mar 30 Neptune	10	−18.5153 + 0.0002	29.6122 15.0402	0.99 991	+0.24 300 +0.54 474 +0.00 001	−0.68 652 +0.08 910 +0.00 004
2000 Mar 31 Uranus	12	−15.5641 + 0.0005	46.8947 15.0394	0.99 987	+0.13 701 +0.54 239 +0.00 002	−1.00 283 +0.13 080 +0.00 005
2000 Apr 26 Neptune	19	−18.4235 + 0.0001	191.2099 15.0408	0.99 991	+0.32 117 +0.54 153 −0.00 001	−0.99 705 +0.09 042 +0.00 004
2000 Jul 29 Mercury	17	+20.8276 + 0.0043	94.0459 14.9892	0.99 742	−0.08 564 +0.57 275 −0.00 006	+0.81 264 −0.02 651 −0.00 022
2000 Jul 30 Mars	12	+21.3682 − 0.0054	7.0717 15.0126	0.99 909	−0.00 781 +0.59 013 −0.00 001	−0.61 682 −0.06 689 −0.00 007
2000 Aug 1 Venus	2	+15.2172 − 0.0167	194.0987 14.9909	0.99 855	+0.20 821 +0.54 827 −0.00 007	+1.01 855 −0.13 835 −0.00 017
2000 Aug 13 Neptune	17	−18.8778 − 0.0003	270.4066 15.0422	0.99 991	+0.13 850 +0.54 548 +0.00 001	−1.24 493 +0.08 378 +0.00 005
2000 Aug 28 Mars	3	+16.7181 − 0.0080	241.4438 15.0146	0.99 906	−0.00 685 +0.57 339 −0.00 001	+0.91 736 −0.13 063 −0.00 013

6

SUNSPOT   ACTIVITY

1749 – 1981

SUNSPOT ACTIVITY

In 1848 R. Wolf introduced the quantity $(10g + f)$ as a measure of the sunspot activity, where g is the number of groups of sunspots and f is the number of individual spots. However, an empirical constant k has been introduced in order to remove systematic differences of values determined by different observers. Using the usual notation, therefore, the *Zürich relative sunspot number* is defined by

$$R = k(10g + f)$$

These sunspot numbers have been found useful as a broad indication, particularly for statistical studies. Until December 1980, they were regularly calculated and published by the Swiss Federal Observatory at Zürich, Switzerland. They are now issued by the Sunspot Index Data Center, Brussels, Belgium.

Table I provides the yearly means of the definitive Zürich sunspot numbers, from 1749 to 1981. Maxima and minima are indicated by M and m, respectively. The diagrams on page 6-4 show the yearly means of the sunspot numbers from 1764 to 1981.

Table II contains the monthly means of the definitive Zürich sunspot numbers, for the same period.

In Table III we give, for the same period, the smoothed monthly means, based on the definitive Zürich sunspot numbers, but calculated with the formula advocated by J. Meeus [1]. In this formula, a greater weight is given to the central months. The diagrams on pages 6-15 to 6-17 show the smoothed monthly means from 1902 to 1981. Secondary variations are well shown. The sunspot maxima of 1905-1907 and 1968-1970 are much broader than the maximum of 1957-1958.

In Table IV we find the following data :
— the epochs of maxima and minima of the solar activity, to the nearest tenth of a year, as calculated by the method used at the Swiss Federal Observatory ;
— the same epochs as deduced from the data of Table III, considering the month having the highest (or lowest) smoothed mean as the epoch of the sunspot maximum (or minimum). The corresponding smoothed means are also given ;
— the month or months with the highest or lowest monthly mean.

Table V gives the monthly and yearly numbers of spotless days, *i.e.* days with $R = 0$, in the period 1850-1981. For the years not mentioned in the list, there were no spotless days.

The synodic rotations of the Sun are numbered in continuation of Carrington's
Greenwich photoheliographic series, of which No. 1 commenced on 1853 November 9.
The *mean* synodic period is 27.2752 days ; the true value varies between 27.20 and
27.34 days, due to the non-uniform speed of the Earth in its orbit.

Table VI gives the dates of beginning of the Sun's synodic rotations, as seen
from the Earth, from 1952 to 2014, to the nearest 0.01 day (Universal Time). The
times from 1952 to 1980 have been taken from the *Astronomical Ephemeris* ; from
1981 onwards they have been calculated by Mrs. G. Evrard, of the Royal Observatory
of Uccle, Belgium.

The definitive Zürich sunspot numbers (Tables I and II) are reproduced by kind
permission of Dr. M. Waldmeier. The values for the years 1749-1960 are taken
from (2), and those for the period 1961-1975 from (3). Dr. A. Koeckelenbergh, of
the Sunspot Index Data Center, Brussels, supplied the data for 1981.

REFERENCES

(1) J. MEEUS, Une formule d'adoucissement pour l'activité solaire, *Ciel et Terre*,
     Vol. 74, pages 445-449 (November-December 1958).

(2) M. WALDMEIER, *The Sunspot-Activity in the Years 1610-1960*, Zürich, Schulthess
     & Co. AG (1961).

(3) M. WALDMEIER, The Sunspot-Activity in the Years 1961-1975, *Astronomische Mit-
     teilungen der Eidgenössischen Sternwarte Zürich*, No. 346.

I

YEARLY MEANS OF SUNSPOT NUMBERS

Year	R	Year	R	Year	R	Year	R	Year	R
1749	80.9	1796	16.0	1843	10.7 m	1890	7.1	1937	114.4 M
1750	83.4 M	1797	6.4	1844	15.0	1891	35.6	1938	109.6
1751	47.7	1798	4.1 m	1845	40.1	1892	73.0	1939	88.8
1752	47.8	1799	6.8	1846	61.5	1893	85.1 M	1940	67.8
1753	30.7	1800	14.5	1847	98.5	1894	78.0	1941	47.5
1754	12.2	1801	34.0	1848	124.7 M	1895	64.0	1942	30.6
1755	9.6 m	1802	45.0	1849	96.3	1896	41.8	1943	16.3
1756	10.2	1803	43.1	1850	66.6	1897	26.2	1944	9.6 m
1757	32.4	1804	47.5 M	1851	64.5	1898	26.7	1945	33.2
1758	47.6	1805	42.2	1852	54.1	1899	12.1	1946	92.6
1759	54.0	1806	28.1	1853	39.0	1900	9.5	1947	151.6 M
1760	62.9	1807	10.1	1854	20.6	1901	2.7 m	1948	136.3
1761	85.9 M	1808	8.1	1855	6.7	1902	5.0	1949	134.7
1762	61.2	1809	2.5	1856	4.3 m	1903	24.4	1950	83.9
1763	45.1	1810	0.0 m	1857	22.7	1904	42.0	1951	69.4
1764	36.4	1811	1.4	1858	54.8	1905	63.5 M	1952	31.5
1765	20.9	1812	5.0	1859	93.8	1906	53.8	1953	13.9
1766	11.4 m	1813	12.2	1860	95.8 M	1907	62.0	1954	4.4 m
1767	37.8	1814	13.9	1861	77.2	1908	48.5	1955	38.0
1768	69.8	1815	35.4	1862	59.1	1909	43.9	1956	141.7
1769	106.1 M	1816	45.8 M	1863	44.0	1910	18.6	1957	190.2 M
1770	100.8	1817	41.1	1864	47.0	1911	5.7	1958	184.8
1771	81.6	1818	30.1	1865	30.5	1912	3.6	1959	159.0
1772	66.5	1819	23.9	1866	16.3	1913	1.4 m	1960	112.3
1773	34.8	1820	15.6	1867	7.3 m	1914	9.6	1961	53.9
1774	30.6	1821	6.6	1868	37.6	1915	47.4	1962	37.5
1775	7.0 m	1822	4.0	1869	74.0	1916	57.1	1963	27.9
1776	19.8	1823	1.8 m	1870	139.0 M	1917	103.9 M	1964	10.2 m
1777	92.5	1824	8.5	1871	111.2	1918	80.6	1965	15.1
1778	154.4 M	1825	16.6	1872	101.6	1919	63.6	1966	47.0
1779	125.9	1826	36.3	1873	66.2	1920	37.6	1967	93.8
1780	84.8	1827	49.6	1874	44.7	1921	26.1	1968	105.9 M
1781	68.1	1828	64.2	1875	17.0	1922	14.2	1969	105.5
1782	38.5	1829	67.0	1876	11.3	1923	5.8 m	1970	104.5
1783	22.8	1830	70.9 M	1877	12.4	1924	16.7	1971	66.6
1784	10.2 m	1831	47.8	1878	3.4 m	1925	44.3	1972	68.9
1785	24.1	1832	27.5	1879	6.0	1926	63.9	1973	38.0
1786	82.9	1833	8.5 m	1880	32.3	1927	69.0	1974	34.5
1787	132.0 M	1834	13.2	1881	54.3	1928	77.8 M	1975	15.5
1788	130.9	1835	56.9	1882	59.7	1929	64.9	1976	12.6 m
1789	118.1	1836	121.5	1883	63.7 M	1930	35.7	1977	27.5
1790	89.9	1837	138.3 M	1884	63.5	1931	21.2	1978	92.5
1791	66.6	1838	103.2	1885	52.2	1932	11.1	1979	155.4 M
1792	60.0	1839	85.7	1886	25.4	1933	5.7 m	1980	154.6
1793	46.9	1840	64.6	1887	13.1	1934	8.7	1981	140.5
1794	41.0	1841	36.7	1888	6.8	1935	36.1		
1795	21.3	1842	24.2	1889	6.3 m	1936	79.7		

SUNSPOT ACTIVITY

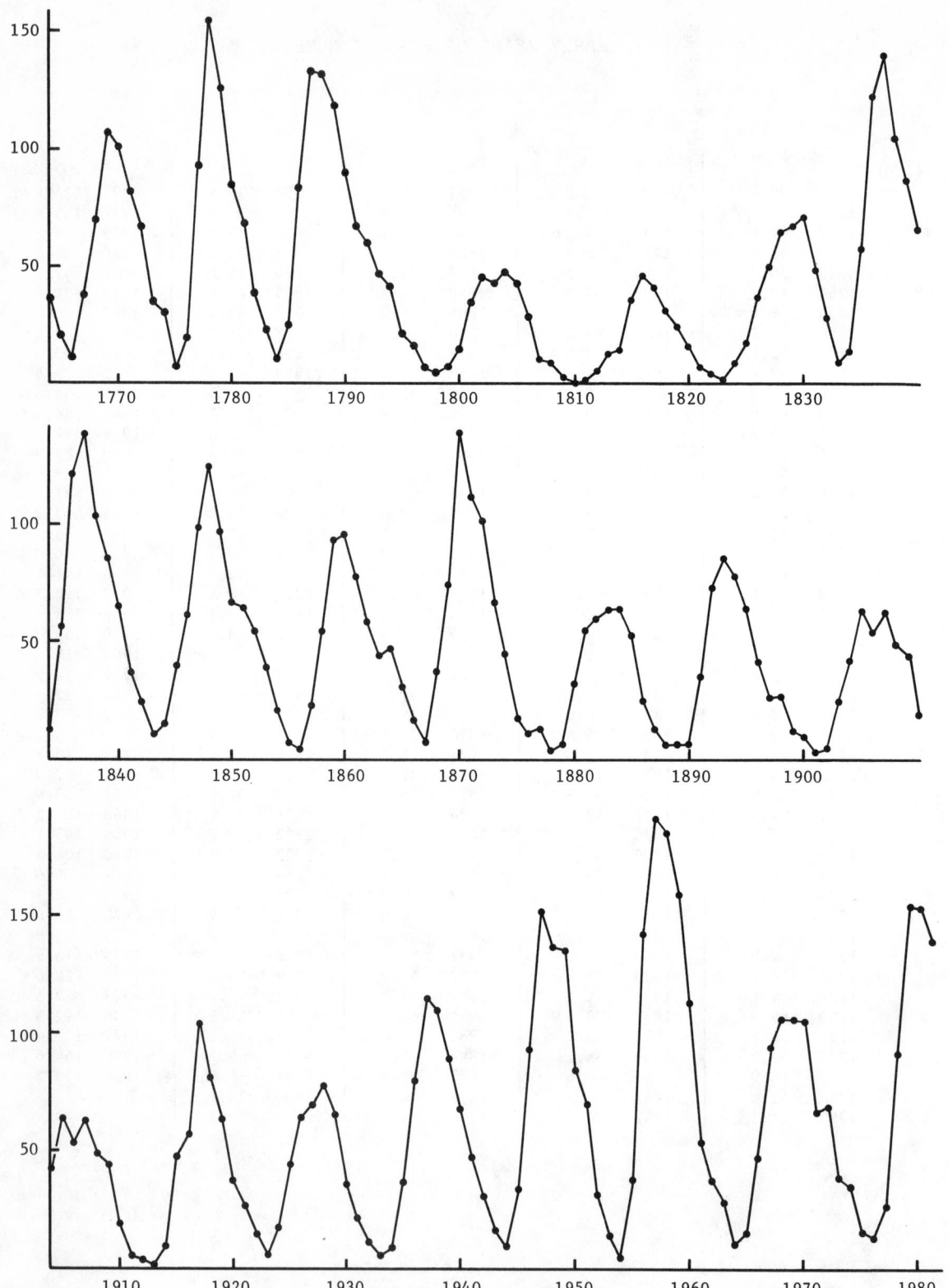

II

MONTHLY MEANS OF SUNSPOT NUMBERS

Year	Jan	Feb	Mar	Apr	May	Jun	Jul	Aug	Sep	Oct	Nov	Dec
1749	58.0	62.6	70.0	55.7	85.0	83.5	94.8	66.3	75.9	75.5	158.6	85.2
1750	73.3	75.9	89.2	88.3	90.0	100.0	85.4	103.0	91.2	65.7	63.3	75.4
1751	70.0	43.5	45.3	56.4	60.7	50.7	66.3	59.8	23.5	23.2	28.5	44.0
1752	35.0	50.0	71.0	59.3	59.7	39.6	78.4	29.3	27.1	46.6	37.6	40.0
1753	44.0	32.0	45.7	38.0	36.0	31.7	22.0	39.0	28.0	25.0	20.0	6.7
1754	0.0	3.0	1.7	13.7	20.7	26.7	18.8	12.3	8.2	24.1	13.2	4.2
1755	10.2	11.2	6.8	6.5	0.0	0.0	8.6	3.2	17.8	23.7	6.8	20.0
1756	12.5	7.1	5.4	9.4	12.5	12.9	3.6	6.4	11.8	14.3	17.0	9.4
1757	14.1	21.2	26.2	30.0	38.1	12.8	25.0	51.3	39.7	32.5	64.7	33.5
1758	37.6	52.0	49.0	72.3	46.4	45.0	44.0	38.7	62.5	37.7	43.0	43.0
1759	48.3	44.0	46.8	47.0	49.0	50.0	51.0	71.3	77.2	59.7	46.3	57.0
1760	67.3	59.5	74.7	58.3	72.0	48.3	66.0	75.6	61.3	50.6	59.7	61.0
1761	70.0	91.0	80.7	71.7	107.2	99.3	94.1	91.1	100.7	88.7	89.7	46.0
1762	43.8	72.8	45.7	60.2	39.9	77.1	33.8	67.7	68.5	69.3	77.8	77.2
1763	56.5	31.9	34.2	32.9	32.7	35.8	54.2	26.5	68.1	46.3	60.9	61.4
1764	59.7	59.7	40.2	34.4	44.3	30.0	30.0	30.0	28.2	28.0	26.0	25.7
1765	24.0	26.0	25.0	22.0	20.2	20.0	27.0	29.7	16.0	14.0	14.0	13.0
1766	12.0	11.0	36.6	6.0	26.8	3.0	3.3	4.0	4.3	5.0	5.7	19.2
1767	27.4	30.0	43.0	32.9	29.8	33.3	21.9	40.8	42.7	44.1	54.7	53.3
1768	53.5	66.1	46.3	42.7	77.7	77.4	52.6	66.8	74.8	77.8	90.6	111.8
1769	73.9	64.2	64.3	96.7	73.6	94.4	118.6	120.3	148.8	158.2	148.1	112.0
1770	104.0	142.5	80.1	51.0	70.1	83.3	109.8	126.3	104.4	103.6	132.2	102.3
1771	36.0	46.2	46.7	64.9	152.7	119.5	67.7	58.5	101.4	90.0	99.7	95.7
1772	100.9	90.8	31.1	92.2	38.0	57.0	77.3	56.2	50.5	78.6	61.3	64.0
1773	54.6	29.0	51.2	32.9	41.1	28.4	27.7	12.7	29.3	26.3	40.9	43.2
1774	46.8	65.4	55.7	43.8	51.3	28.5	17.5	6.6	7.9	14.0	17.7	12.2
1775	4.4	0.0	11.6	11.2	3.9	12.3	1.0	7.9	3.2	5.6	15.1	7.9
1776	21.7	11.6	6.3	21.8	11.2	19.0	1.0	24.2	16.0	30.0	35.0	40.0
1777	45.0	36.5	39.0	95.5	80.3	80.7	95.0	112.0	116.2	106.5	146.0	157.3
1778	177.3	109.3	134.0	145.0	238.9	171.6	153.0	140.0	171.7	156.3	150.3	105.0
1779	114.7	165.7	118.0	145.0	140.0	113.7	143.0	112.0	111.0	124.0	114.0	110.0
1780	70.0	98.0	98.0	95.0	107.2	88.0	86.0	86.0	93.7	77.0	60.0	58.7
1781	98.7	74.7	53.0	68.3	104.7	97.7	73.5	66.0	51.0	27.3	67.0	35.2
1782	54.0	37.5	37.0	41.0	54.3	38.0	37.0	44.0	34.0	23.2	31.5	30.0
1783	28.0	38.7	26.7	28.3	23.0	25.2	32.2	20.0	18.0	8.0	15.0	10.5
1784	13.0	8.0	11.0	10.0	6.0	9.0	6.0	10.0	10.0	8.0	17.0	14.0
1785	6.5	8.0	9.0	15.7	20.7	26.3	36.3	20.0	32.0	47.2	40.2	27.3

SUNSPOT ACTIVITY

Monthly means

Year	Jan	Feb	Mar	Apr	May	Jun	Jul	Aug	Sep	Oct	Nov	Dec
1786	37.2	47.6	47.7	85.4	92.3	59.0	83.0	89.7	111.5	112.3	116.0	112.7
1787	134.7	106.0	87.4	127.2	134.8	99.2	128.0	137.2	157.3	157.0	141.5	174.0
1788	138.0	129.2	143.3	108.5	113.0	154.2	141.5	136.0	141.0	142.0	94.7	129.5
1789	114.0	125.3	120.0	123.3	123.5	120.0	117.0	103.0	112.0	89.7	134.0	135.5
1790	103.0	127.5	96.3	94.0	93.0	91.0	69.3	87.0	77.3	84.3	82.0	74.0
1791	72.7	62.0	74.0	77.2	73.7	64.2	71.0	43.0	66.5	61.7	67.0	66.0
1792	58.0	64.0	63.0	75.7	62.0	61.0	45.8	60.0	59.0	59.0	57.0	56.0
1793	56.0	55.0	55.5	53.0	52.3	51.0	50.0	29.3	24.0	47.0	44.0	45.7
1794	45.0	44.0	38.0	28.4	55.7	41.5	41.0	40.0	11.1	28.5	67.4	51.4
1795	21.4	39.9	12.6	18.6	31.0	17.1	12.9	25.7	13.5	19.5	25.0	18.0
1796	22.0	23.8	15.7	31.7	21.0	6.7	26.9	1.5	18.4	11.0	8.4	5.1
1797	14.4	4.2	4.0	4.0	7.3	11.1	4.3	6.0	5.7	6.9	5.8	3.0
1798	2.0	4.0	12.4	1.1	0.0	0.0	0.0	3.0	2.4	1.5	12.5	9.9
1799	1.6	12.6	21.7	8.4	8.2	10.6	2.1	0.0	0.0	4.6	2.7	8.6
1800	6.9	9.3	13.9	0.0	5.0	23.7	21.0	19.5	11.5	12.3	10.5	40.1
1801	27.0	29.0	30.0	31.0	32.0	31.2	35.0	38.7	33.5	32.6	39.8	48.2
1802	47.8	47.0	40.8	42.0	44.0	46.0	48.0	50.0	51.8	38.5	34.5	50.0
1803	50.0	50.8	29.5	25.0	44.3	36.0	48.3	34.1	45.3	54.3	51.0	48.0
1804	45.3	48.3	48.0	50.6	33.4	34.8	29.8	43.1	53.0	62.3	61.0	60.0
1805	61.0	44.1	51.4	37.5	39.0	40.5	37.6	42.7	44.4	29.4	41.0	38.3
1806	39.0	29.6	32.7	27.7	26.4	25.6	30.0	26.3	24.0	27.0	25.0	24.0
1807	12.0	12.2	9.6	23.8	10.0	12.0	12.7	12.0	5.7	8.0	2.6	0.0
1808	0.0	4.5	0.0	12.3	13.5	13.5	6.7	8.0	11.7	4.7	10.5	12.3
1809	7.2	9.2	0.9	2.5	2.0	7.7	0.3	0.2	0.4	0.0	0.0	0.0
1810	0.0	0.0	0.0	0.0	0.0	0.0	0.0	0.0	0.0	0.0	0.0	0.0
1811	0.0	0.0	0.0	0.0	0.0	0.0	6.6	0.0	2.4	6.1	0.8	1.1
1812	11.3	1.9	0.7	0.0	1.0	1.3	0.5	15.6	5.2	3.9	7.9	10.1
1813	0.0	10.3	1.9	16.6	5.5	11.2	18.3	8.4	15.3	27.8	16.7	14.3
1814	22.2	12.0	5.7	23.8	5.8	14.9	18.5	2.3	8.1	19.3	14.5	20.1
1815	19.2	32.2	26.2	31.6	9.8	55.9	35.5	47.2	31.5	33.5	37.2	65.0
1816	26.3	68.8	73.7	58.8	44.3	43.6	38.8	23.2	47.8	56.4	38.1	29.9
1817	36.4	57.9	96.2	26.4	21.2	40.0	50.0	45.0	36.7	25.6	28.9	28.4
1818	34.9	22.4	25.4	34.5	53.1	36.4	28.0	31.5	26.1	31.6	10.9	25.8
1819	32.8	20.7	3.7	20.2	19.6	35.0	31.4	26.1	14.9	27.5	25.1	30.6
1820	19.2	26.6	4.5	19.4	29.3	10.8	20.6	25.9	5.2	8.9	7.9	9.1
1821	21.5	4.2	5.7	9.2	1.7	1.8	2.5	4.8	4.4	18.8	4.4	0.2
1822	0.0	0.9	16.1	13.5	1.5	5.6	7.9	2.1	0.0	0.4	0.0	0.0
1823	0.0	0.0	0.6	0.0	0.0	0.0	0.5	0.0	0.0	0.0	0.0	20.4
1824	21.7	10.8	0.0	19.4	2.8	0.0	0.0	1.4	20.5	25.2	0.0	0.8
1825	5.0	15.5	22.4	3.8	15.5	15.4	30.9	25.7	15.7	15.6	11.7	22.0
1826	17.7	18.2	36.7	24.0	32.4	37.1	52.5	39.6	18.9	50.6	39.5	68.1
1827	34.6	47.4	57.8	46.0	56.3	56.7	42.3	53.7	49.6	56.1	48.2	46.1
1828	52.8	64.4	65.0	61.1	89.1	98.0	54.2	76.4	50.4	54.7	57.0	46.9
1829	43.0	49.4	72.3	95.0	67.4	73.9	90.8	77.6	52.8	57.2	67.6	56.5
1830	52.2	72.1	84.6	106.3	66.3	65.1	43.9	50.7	62.1	84.4	81.2	82.1
1831	47.5	50.1	93.4	54.5	38.1	33.4	45.2	55.0	37.9	46.3	43.5	28.9
1832	30.9	55.6	55.1	26.9	41.3	26.7	14.0	8.9	8.2	21.1	14.3	27.5
1833	11.3	14.9	11.8	2.8	12.9	1.0	7.0	5.7	11.6	7.5	5.9	9.9
1834	4.9	18.1	3.9	1.4	8.8	7.8	8.7	4.0	11.5	24.8	30.5	34.5
1835	7.5	24.5	19.7	61.5	43.6	33.2	59.8	59.0	100.8	95.2	100.0	77.5

Monthly means

Year	Jan	Feb	Mar	Apr	May	Jun	Jul	Aug	Sep	Oct	Nov	Dec
1836	88.6	107.6	98.2	142.9	111.4	124.7	116.7	107.8	95.1	137.4	120.9	206.2
1837	188.0	175.6	134.6	138.2	111.7	158.0	162.8	134.0	96.3	123.7	107.0	129.8
1838	144.9	84.8	140.8	126.6	137.6	94.5	108.2	78.8	73.6	90.8	77.4	79.8
1839	105.6	102.5	77.7	61.8	53.8	54.6	84.8	131.2	132.7	90.9	68.8	63.7
1840	81.2	87.7	67.8	65.9	69.2	48.5	60.7	57.8	74.0	55.0	54.3	53.7
1841	24.1	29.9	29.7	40.2	67.5	55.7	30.8	39.3	36.5	28.5	19.8	38.8
1842	20.4	22.1	21.7	26.9	24.9	20.5	12.6	26.6	18.4	38.1	40.5	17.6
1843	13.3	3.5	8.3	9.5	21.1	10.5	9.5	11.8	4.2	5.3	19.1	12.7
1844	9.4	14.7	13.6	20.8	11.6	3.7	21.2	23.9	7.0	21.5	10.7	21.6
1845	25.7	43.6	43.3	57.0	47.8	31.1	30.6	32.3	29.6	40.7	39.4	59.7
1846	38.7	51.0	63.9	69.3	59.9	65.1	46.5	54.8	107.1	55.9	60.4	65.5
1847	62.6	44.9	85.7	44.7	75.4	85.3	52.2	140.6	160.9	180.4	138.9	109.6
1848	159.1	111.8	108.6	107.1	102.2	129.0	139.2	132.6	100.3	132.4	114.6	159.5
1849	157.0	131.7	96.2	102.5	80.6	81.1	78.0	67.7	93.7	71.5	99.0	97.0
1850	78.0	89.4	82.6	44.1	61.6	70.0	39.1	61.6	86.2	71.0	54.8	61.0
1851	75.5	105.4	64.6	56.5	62.6	63.2	36.1	57.4	67.9	62.5	51.0	71.4
1852	68.4	66.4	61.2	65.4	54.9	46.9	42.1	39.7	37.5	67.3	54.3	45.4
1853	41.1	42.9	37.7	47.6	34.7	40.0	45.9	50.4	33.5	42.3	28.8	23.4
1854	15.4	20.0	20.7	26.5	24.0	21.1	18.7	15.8	22.4	12.6	28.2	21.6
1855	12.3	11.4	17.4	4.4	9.1	5.3	0.4	3.1	0.0	9.6	4.2	3.1
1856	0.5	4.9	0.4	6.5	0.0	5.2	4.6	5.9	4.4	4.5	7.7	7.2
1857	13.7	7.4	5.2	11.1	28.6	16.0	22.2	16.9	42.4	40.6	31.4	37.2
1858	39.0	34.9	57.5	38.3	41.4	44.5	56.7	55.3	80.1	91.2	51.9	66.9
1859	83.7	87.6	90.3	85.7	91.0	87.1	95.2	106.8	105.8	114.6	97.2	81.0
1860	82.4	88.3	98.9	71.4	107.1	108.6	116.7	100.3	92.2	90.1	97.9	95.6
1861	62.3	77.7	101.0	98.5	56.8	88.1	78.0	82.5	79.9	67.2	53.7	80.5
1862	63.1	64.5	43.6	53.7	64.4	84.0	73.4	62.5	66.6	41.9	50.6	40.9
1863	48.3	56.7	66.4	40.6	53.8	40.8	32.7	48.1	22.0	39.9	37.7	41.2
1864	57.7	47.1	66.3	35.8	40.6	57.8	54.7	54.8	28.5	33.9	57.6	28.6
1865	48.7	39.3	39.5	29.4	34.5	33.6	26.8	37.8	21.6	17.1	24.6	12.8
1866	31.6	38.4	24.6	17.6	12.9	16.5	9.3	12.7	7.3	14.1	9.0	1.5
1867	0.0	0.7	9.2	5.1	2.9	1.5	5.0	4.8	9.8	13.5	9.6	25.2
1868	15.6	15.7	26.5	36.6	26.7	31.1	29.0	34.4	47.2	61.6	59.1	67.6
1869	60.9	59.9	52.7	41.0	103.9	108.4	59.2	79.6	80.6	59.3	78.1	104.3
1870	77.3	114.9	157.6	160.0	176.0	135.6	132.4	153.8	136.0	146.4	147.5	130.0
1871	88.3	125.3	143.2	162.4	145.5	91.7	103.0	110.1	80.3	89.0	105.4	90.4
1872	79.5	120.1	88.4	102.1	107.6	109.9	105.5	92.9	114.6	102.6	112.0	83.9
1873	86.7	107.0	98.3	76.2	47.9	44.8	66.9	68.2	47.1	47.1	55.4	49.2
1874	60.8	64.2	46.4	32.0	44.6	38.2	67.8	61.3	28.0	34.3	28.9	29.3
1875	14.6	21.5	33.8	29.1	11.5	23.9	12.5	14.6	2.4	12.7	17.7	9.9
1876	14.3	15.0	30.6	2.3	5.1	1.6	15.2	8.8	9.9	14.3	9.9	8.2
1877	24.4	8.7	11.9	15.8	21.6	14.2	6.0	6.3	16.9	6.7	14.2	2.2
1878	3.3	6.6	7.8	0.1	5.9	6.4	0.1	0.0	5.3	1.1	4.1	0.5
1879	1.0	0.6	0.0	6.2	2.4	4.8	7.5	10.7	6.1	12.3	13.1	7.3
1880	24.0	27.2	19.3	19.5	23.5	34.1	21.9	48.1	66.0	43.0	30.7	29.6
1881	36.4	53.2	51.5	51.6	43.5	60.5	76.9	58.4	53.2	64.4	54.8	47.3
1882	45.0	69.5	66.8	95.8	64.1	45.2	45.4	40.4	57.7	59.2	84.4	41.8
1883	60.6	46.9	42.8	82.1	31.5	76.3	80.6	46.0	52.6	83.8	84.5	75.9
1884	91.5	86.9	87.5	76.1	66.5	51.2	53.1	55.8	61.9	47.8	36.6	47.2
1885	42.8	71.8	49.8	55.0	73.0	83.7	66.5	50.0	39.6	38.7	30.9	21.7

SUNSPOT ACTIVITY

Monthly means

Year	Jan	Feb	Mar	Apr	May	Jun	Jul	Aug	Sep	Oct	Nov	Dec
1886	29.9	25.9	57.3	43.7	30.7	27.1	30.3	16.9	21.4	8.6	0.3	13.0
1887	10.3	13.2	4.2	6.9	20.0	15.7	23.3	21.4	7.4	6.6	6.9	20.7
1888	12.7	7.1	7.8	5.1	7.0	7.1	3.1	2.8	8.8	2.1	10.7	6.7
1889	0.8	8.5	6.7	4.3	2.4	6.4	9.4	20.6	6.5	2.1	0.2	6.7
1890	5.3	0.6	5.1	1.6	4.8	1.3	11.6	8.5	17.2	11.2	9.6	7.8
1891	13.5	22.2	10.4	20.5	41.1	48.3	58.8	33.0	53.8	51.5	41.9	32.5
1892	69.1	75.6	49.9	69.6	79.6	76.3	76.5	101.4	62.8	70.5	65.4	78.6
1893	75.0	73.0	65.7	88.1	84.7	89.9	88.6	129.2	77.9	80.0	75.1	93.8
1894	83.2	84.6	52.3	81.6	101.2	98.9	106.0	70.3	65.9	75.5	56.6	60.0
1895	63.3	67.2	61.0	76.9	67.5	71.5	47.8	68.9	57.7	67.9	47.2	70.7
1896	29.0	57.4	52.0	43.8	27.7	49.0	45.0	27.2	61.3	28.7	38.0	42.6
1897	40.6	29.4	29.1	31.0	20.0	11.3	27.6	21.8	48.1	14.3	8.4	33.3
1898	30.2	36.4	38.3	14.5	25.8	22.3	9.0	31.4	34.8	34.4	30.9	12.6
1899	19.5	9.2	18.1	14.2	7.7	20.5	13.5	2.9	8.4	13.0	7.8	10.5
1900	9.4	13.6	8.6	16.0	15.2	12.1	8.3	4.3	8.3	12.9	4.5	0.3
1901	0.2	2.4	4.5	0.0	10.2	5.8	0.7	1.0	0.6	3.7	3.8	0.0
1902	5.5	0.0	12.4	0.0	2.8	1.4	0.9	2.3	7.6	16.3	10.3	1.1
1903	8.3	17.0	13.5	26.1	14.6	16.3	27.9	28.8	11.1	38.9	44.5	45.6
1904	31.6	24.5	37.2	43.0	39.5	41.9	50.6	58.2	30.1	54.2	38.0	54.6
1905	54.8	85.8	56.5	39.3	48.0	49.0	73.0	58.8	55.0	78.7	107.2	55.5
1906	45.5	31.3	64.5	55.3	57.7	63.2	103.6	47.7	56.1	17.8	38.9	64.7
1907	76.4	108.2	60.7	52.6	42.9	40.4	49.7	54.3	85.0	65.4	61.5	47.3
1908	39.2	33.9	28.7	57.6	40.8	48.1	39.5	90.5	86.9	32.3	45.5	39.5
1909	56.7	46.6	66.3	32.3	36.0	22.6	35.8	23.1	38.8	58.4	55.8	54.2
1910	26.4	31.5	21.4	8.4	22.2	12.3	14.1	11.5	26.2	38.3	4.9	5.8
1911	3.4	9.0	7.8	16.5	9.0	2.2	3.5	4.0	4.0	2.6	4.2	2.2
1912	0.3	0.0	4.9	4.5	4.4	4.1	3.0	0.3	9.5	4.6	1.1	6.4
1913	2.3	2.9	0.5	0.9	0.0	0.0	1.7	0.2	1.2	3.1	0.7	3.8
1914	2.8	2.6	3.1	17.3	5.2	11.4	5.4	7.7	12.7	8.2	16.4	22.3
1915	23.0	42.3	38.8	41.3	33.0	68.8	71.6	69.6	49.5	53.5	42.5	34.5
1916	45.3	55.4	67.0	71.8	74.5	67.7	53.5	35.2	45.1	50.7	65.6	53.0
1917	74.7	71.9	94.8	74.7	114.1	114.9	119.8	154.5	129.4	72.2	96.4	129.3
1918	96.0	65.3	72.2	80.5	76.7	59.4	107.6	101.7	79.9	85.0	83.4	59.2
1919	48.1	79.5	66.5	51.8	88.1	111.2	64.7	69.0	54.7	52.8	42.0	34.9
1920	51.1	53.9	70.2	14.8	33.3	38.7	27.5	19.2	36.3	49.6	27.2	29.9
1921	31.5	28.3	26.7	32.4	22.2	33.7	41.9	22.8	17.8	18.2	17.8	20.3
1922	11.8	26.4	54.7	11.0	8.0	5.8	10.9	6.5	4.7	6.2	7.4	17.5
1923	4.5	1.5	3.3	6.1	3.2	9.1	3.5	0.5	13.2	11.6	10.0	2.8
1924	0.5	5.1	1.8	11.3	20.8	24.0	28.1	19.3	25.1	25.6	22.5	16.5
1925	5.5	23.2	18.0	31.7	42.8	47.5	38.5	37.9	60.2	69.2	58.6	98.6
1926	71.8	69.9	62.5	38.5	64.3	73.5	52.3	61.6	60.8	71.5	60.5	79.4
1927	81.6	93.0	69.6	93.5	79.1	59.1	54.9	53.8	68.4	63.1	67.2	45.2
1928	83.5	73.5	85.4	80.6	77.0	91.4	98.0	83.8	89.7	61.4	50.3	59.0
1929	68.9	62.8	50.2	52.8	58.2	71.9	70.2	65.8	34.4	54.0	81.1	108.0
1930	65.3	49.9	35.0	38.2	36.8	28.8	21.9	24.9	32.1	34.4	35.6	25.8
1931	14.6	43.1	30.0	31.2	24.6	15.3	17.4	13.0	19.0	10.0	18.7	17.8
1932	12.1	10.6	11.2	11.2	17.9	22.2	9.6	6.8	4.0	8.9	8.2	11.0
1933	12.3	22.2	10.1	2.9	3.2	5.2	2.8	0.2	5.1	3.0	0.6	0.3
1934	3.4	7.8	4.3	11.3	19.7	6.7	9.3	8.3	4.0	5.7	8.7	15.4
1935	18.6	20.5	23.1	12.2	27.3	45.7	33.9	30.1	42.1	53.2	64.2	61.5

Monthly means

Year	Jan	Feb	Mar	Apr	May	Jun	Jul	Aug	Sep	Oct	Nov	Dec
1936	62.8	74.3	77.1	74.9	54.6	70.0	52.3	87.0	76.0	89.0	115.4	123.4
1937	132.5	128.5	83.9	109.3	116.7	130.3	145.1	137.7	100.7	124.9	74.4	88.8
1938	98.4	119.2	86.5	101.0	127.4	97.5	165.3	115.7	89.6	99.1	122.2	92.7
1939	80.3	77.4	64.6	109.1	118.3	101.0	97.6	105.8	112.6	88.1	68.1	42.1
1940	50.5	59.4	83.3	60.7	54.4	83.9	67.5	105.5	66.5	55.0	58.4	68.3
1941	45.6	44.5	46.4	32.8	29.5	59.8	66.9	60.0	65.9	46.3	38.4	33.7
1942	35.6	52.8	54.2	60.7	25.0	11.4	17.7	20.2	17.2	19.2	30.7	22.5
1943	12.4	28.9	27.4	26.1	14.1	7.6	13.2	19.4	10.0	7.8	10.2	18.8
1944	3.7	0.5	11.0	0.3	2.5	5.0	5.0	16.7	14.3	16.9	10.8	28.4
1945	18.5	12.7	21.5	32.0	30.6	36.2	42.6	25.9	34.9	68.8	46.0	27.4
1946	47.6	86.2	76.6	75.7	84.9	73.5	116.2	107.2	94.4	102.3	123.8	121.7
1947	115.7	133.4	129.8	149.8	201.3	163.9	157.9	188.8	169.4	163.6	128.0	116.5
1948	108.5	86.1	94.8	189.7	174.0	167.8	142.2	157.9	143.3	136.3	95.8	138.0
1949	119.1	182.3	157.5	147.0	106.2	121.7	125.8	123.8	145.3	131.6	143.5	117.6
1950	101.6	94.8	109.7	113.4	106.2	83.6	91.0	85.2	51.3	61.4	54.8	54.1
1951	59.9	59.9	55.9	92.9	108.5	100.6	61.5	61.0	83.1	51.6	52.4	45.8
1952	40.7	22.7	22.0	29.1	23.4	36.4	39.3	54.9	28.2	23.8	22.1	34.3
1953	26.5	3.9	10.0	27.8	12.5	21.8	8.6	23.5	19.3	8.2	1.6	2.5
1954	0.2	0.5	10.9	1.8	0.8	0.2	4.8	8.4	1.5	7.0	9.2	7.6
1955	23.1	20.8	4.9	11.3	28.9	31.7	26.7	40.7	42.7	58.5	89.2	76.9
1956	73.6	124.0	118.4	110.7	136.6	116.6	129.1	169.6	173.2	155.3	201.3	192.1
1957	165.0	130.2	157.4	175.2	164.6	200.7	187.2	158.0	235.8	253.8	210.9	239.4
1958	202.5	164.9	190.7	196.0	175.3	171.5	191.4	200.2	201.2	181.5	152.3	187.6
1959	217.4	143.1	185.7	163.3	172.0	168.7	149.6	199.6	145.2	111.4	124.0	125.0
1960	146.3	106.0	102.2	122.0	119.6	110.2	121.7	134.1	127.2	82.8	89.6	85.6
1961	57.9	46.1	53.0	61.4	51.0	77.4	70.2	55.9	63.6	37.7	32.6	40.0
1962	38.7	50.3	45.6	46.4	43.7	42.0	21.8	21.8	51.3	39.5	26.9	23.2
1963	19.8	24.4	17.1	29.3	43.0	35.9	19.6	33.2	38.8	35.3	23.4	14.9
1964	15.3	17.7	16.5	8.6	9.5	9.1	3.1	9.3	4.7	6.1	7.4	15.1
1965	17.5	14.2	11.7	6.8	24.1	15.9	11.9	8.9	16.8	20.1	15.8	17.0
1966	28.2	24.4	25.3	48.7	45.3	47.7	56.7	51.2	50.2	57.2	57.2	70.4
1967	110.9	93.6	111.8	69.5	86.5	67.3	91.5	107.2	76.8	88.2	94.3	126.4
1968	121.8	111.9	92.2	81.2	127.2	110.3	96.1	109.3	117.2	107.7	86.0	109.8
1969	104.4	120.5	135.8	106.8	120.0	106.0	96.8	98.0	91.3	95.7	93.5	97.9
1970	111.5	127.8	102.9	109.5	127.5	106.8	112.5	93.0	99.5	86.6	95.2	83.5
1971	91.3	79.0	60.7	71.8	57.5	49.8	81.0	61.4	50.2	51.7	63.2	82.2
1972	61.5	88.4	80.1	63.2	80.5	88.0	76.5	76.8	64.0	61.3	41.6	45.3
1973	43.4	42.9	46.0	57.7	42.4	39.5	23.1	25.6	59.3	30.7	23.9	23.3
1974	27.6	26.0	21.3	40.3	39.5	36.0	55.8	33.6	40.2	47.1	25.0	20.5
1975	18.9	11.5	11.5	5.1	9.0	11.4	28.2	39.7	13.9	9.1	19.4	7.8
1976	8.1	4.3	21.9	18.8	12.4	12.2	1.9	16.4	13.5	20.6	5.2	15.3
1977	16.4	23.1	8.7	12.9	18.6	38.5	21.4	30.1	44.0	43.8	29.1	43.2
1978	51.9	93.6	76.5	99.7	82.7	95.1	70.4	58.1	138.2	125.1	97.9	122.7
1979	166.6	137.5	138.0	101.5	134.4	149.5	159.4	142.2	188.4	186.2	183.3	176.3
1980	159.6	155.0	126.2	164.1	179.9	157.3	136.3	135.4	155.0	164.7	147.9	174.4
1981	114.0	141.3	135.5	156.4	127.5	90.9	143.8	158.7	167.3	162.4	137.5	150.1

III

SMOOTHED MONTHLY MEANS OF SUNSPOT NUMBERS

Year	Jan	Feb	Mar	Apr	May	Jun	Jul	Aug	Sep	Oct	Nov	Dec
1749							81.8	84.5	87.2	88.6	90.1	89.9
1750	90.2	90.0	90.3	90.0	88.8	87.9	86.9	84.9	81.3	76.9	72.3	67.7
1751	63.7	60.5	58.0	55.6	53.3	50.7	48.0	45.5	43.4	42.4	42.2	43.2
1752	45.1	47.7	50.1	51.4	51.8	50.4	49.0	46.4	44.0	42.1	40.7	39.8
1753	38.8	37.9	37.5	36.2	34.7	32.9	30.3	27.5	24.0	20.5	17.2	14.6
1754	12.9	12.0	11.7	12.4	13.5	14.5	15.2	15.3	14.8	13.9	12.4	10.9
1755	9.5	8.3	7.3	6.8	6.8	7.3	8.4	9.5	10.6	11.5	11.9	12.2
1756	11.9	11.3	10.5	9.9	9.7	9.8	9.9	10.6	11.5	12.8	14.6	16.5
1757	18.5	20.7	23.3	25.6	28.3	31.0	33.6	36.1	38.3	40.6	43.2	45.2
1758	47.1	48.1	48.7	49.5	49.1	48.7	48.2	47.2	46.5	45.6	45.5	45.5
1759	45.9	46.6	48.0	49.9	52.1	54.1	56.1	57.9	59.3	60.5	61.3	61.9
1760	62.2	62.4	62.9	63.1	63.5	63.2	62.9	62.9	63.5	64.6	66.6	69.6
1761	73.3	77.2	81.3	85.3	88.8	90.4	90.2	88.2	85.1	80.6	75.4	69.8
1762	65.0	61.1	58.1	56.9	56.6	58.4	60.1	61.8	62.3	62.0	60.3	57.3
1763	53.6	49.5	45.7	43.0	41.8	41.9	43.9	46.3	49.5	51.2	52.6	52.6
1764	51.3	49.2	46.2	42.7	39.3	36.0	33.2	30.9	29.3	28.1	26.9	25.9
1765	25.2	24.6	24.1	23.6	23.0	22.2	21.2	20.1	18.9	18.1	17.3	16.8
1766	16.2	15.3	14.7	13.5	12.4	10.8	9.9	9.6	10.3	12.2	15.1	18.6
1767	22.3	25.5	28.5	30.9	33.0	35.0	36.9	39.5	42.2	44.9	47.7	50.7
1768	53.5	55.8	57.8	59.9	62.4	65.5	68.9	72.2	75.0	77.5	79.1	80.2
1769	81.3	83.2	85.8	90.5	96.5	103.8	111.2	118.2	123.7	125.3	123.7	119.3
1770	112.9	106.6	100.6	96.3	94.7	95.8	97.4	97.9	97.2	95.0	91.8	88.4
1771	84.8	82.1	81.1	81.8	83.7	85.8	88.8	91.0	91.7	90.5	88.1	84.7
1772	81.7	78.5	73.7	69.9	66.1	64.0	62.4	61.4	60.6	59.4	57.2	54.7
1773	51.1	47.0	43.1	38.7	35.3	32.6	31.4	31.4	33.0	35.2	38.3	41.3
1774	43.6	44.5	43.4	40.5	36.5	31.4	26.3	21.2	16.9	13.8	11.5	10.0
1775	9.2	8.6	8.1	7.5	7.0	7.0	7.1	7.8	8.3	9.0	10.1	11.0
1776	12.0	12.7	13.3	14.1	14.8	16.4	18.3	21.1	24.2	28.2	33.3	39.1
1777	45.6	53.0	60.9	69.5	78.0	87.4	97.3	107.1	115.7	122.9	130.6	138.2
1778	145.3	150.4	155.4	158.8	161.8	161.5	160.0	157.2	153.9	149.0	144.0	139.8
1779	137.1	135.5	133.1	131.4	129.6	127.7	125.2	120.9	116.7	112.5	108.3	104.9
1780	101.4	98.8	96.6	94.7	92.7	90.2	88.0	85.6	82.6	79.5	77.2	76.0
1781	76.0	76.1	76.3	75.9	74.9	72.8	69.1	64.8	60.4	55.5	51.3	47.5
1782	45.5	43.9	43.2	42.3	41.2	39.7	38.2	36.7	35.2	33.7	32.2	31.0
1783	30.1	29.4	28.5	27.6	26.2	24.4	22.5	20.3	18.3	16.2	14.5	12.9
1784	11.5	10.5	9.8	9.4	9.1	9.1	9.3	9.5	9.8	10.1	10.6	11.2
1785	12.3	13.8	15.5	17.9	20.8	23.8	26.8	29.6	32.4	35.3	38.9	43.0

Smoothed monthly means

Year	Jan	Feb	Mar	Apr	May	Jun	Jul	Aug	Sep	Oct	Nov	Dec
1786	47.6	53.1	59.2	65.9	73.0	80.1	87.5	94.1	99.8	104.1	108.0	111.1
1787	113.4	114.9	116.5	119.1	122.4	126.4	131.5	136.0	140.4	143.1	143.4	143.1
1788	141.6	139.3	137.1	135.3	134.2	133.2	132.3	131.6	130.2	128.7	126.6	124.8
1789	122.7	121.5	120.2	118.9	117.8	116.9	116.0	115.2	114.9	114.2	113.6	112.3
1790	110.1	107.3	103.2	99.0	94.3	90.0	86.4	83.5	80.8	79.0	77.4	75.8
1791	74.6	73.1	71.4	69.7	68.1	66.6	65.5	64.2	63.6	63.0	63.1	63.4
1792	63.5	63.5	63.1	62.5	61.5	60.6	59.5	58.7	57.8	57.1	56.5	56.1
1793	55.7	54.7	53.1	51.0	48.9	46.8	44.9	43.4	42.2	41.5	41.1	41.3
1794	41.5	41.8	41.7	40.6	39.8	39.3	38.8	38.4	38.0	37.2	36.1	34.6
1795	32.5	30.4	27.9	25.6	23.0	21.2	20.3	20.0	19.8	20.1	20.5	20.7
1796	20.9	20.8	20.3	19.6	18.5	17.0	15.7	14.0	12.7	11.1	9.9	9.0
1797	8.2	7.5	7.1	6.7	6.5	6.4	6.2	5.9	5.8	5.6	5.3	4.9
1798	4.5	4.0	3.6	3.1	2.8	2.8	3.0	3.4	4.2	5.3	6.6	7.9
1799	8.8	9.4	9.4	8.9	8.0	7.0	6.0	5.3	4.8	4.8	5.1	5.8
1800	6.9	8.2	9.6	10.9	12.0	13.4	14.8	16.5	18.1	20.1	22.0	24.0
1801	25.7	27.7	29.5	31.0	32.1	33.1	34.2	35.7	37.2	38.7	40.1	41.5
1802	42.6	43.6	44.4	45.1	45.4	45.4	45.4	45.5	45.5	45.0	44.1	43.2
1803	42.1	41.2	40.3	39.9	40.3	40.8	41.8	43.0	44.6	46.0	46.9	47.0
1804	46.5	45.6	44.4	43.6	43.2	43.8	45.1	47.0	49.1	50.9	52.2	52.5
1805	51.9	50.1	48.0	45.5	43.4	41.7	40.4	39.5	38.8	37.9	37.1	35.8
1806	34.5	33.0	31.6	30.2	29.1	28.0	27.0	25.9	24.6	23.2	21.7	20.1
1807	18.4	16.9	15.4	14.3	13.0	11.8	10.5	9.2	7.8	6.4	5.4	5.0
1808	5.0	5.4	6.1	7.1	7.9	8.7	9.2	9.4	9.4	9.0	8.4	7.7
1809	7.0	6.3	5.3	4.5	3.5	2.7	2.0	1.4	1.0	0.7	0.4	0.2
1810	0.0	0.0	0.0	0.0	0.0	0.0	0.0	0.0	0.0	0.0	0.0	0.0
1811	0.1	0.2	0.4	0.7	1.1	1.4	1.9	2.4	2.8	3.0	3.1	3.1
1812	2.9	2.8	2.9	3.0	3.3	3.8	4.3	5.0	5.5	6.1	6.5	6.9
1813	7.3	7.9	8.4	9.5	10.7	12.0	13.4	14.6	15.5	16.1	16.3	16.2
1814	15.8	15.3	14.4	13.7	13.0	12.8	12.9	13.3	14.5	15.9	17.5	19.6
1815	21.9	24.7	27.3	29.8	31.6	34.1	35.8	37.8	39.9	42.7	45.3	47.9
1816	49.3	50.7	50.8	50.4	49.1	47.1	45.0	43.2	42.6	43.2	43.9	44.7
1817	45.6	46.0	46.1	44.8	43.6	41.9	40.0	37.7	35.1	33.4	32.7	32.1
1818	31.8	31.7	32.3	32.8	33.0	32.4	31.6	30.5	28.7	26.8	24.6	23.1
1819	22.3	22.0	21.9	22.2	22.8	23.6	24.1	24.6	24.8	24.5	23.8	22.9
1820	21.7	21.1	20.3	19.5	18.4	17.2	16.2	15.2	14.0	12.9	11.6	10.4
1821	9.4	8.1	7.2	6.6	6.1	5.7	5.3	5.2	5.2	5.5	5.7	5.9
1822	6.1	6.2	6.2	6.0	5.6	5.1	4.4	3.5	2.5	1.6	1.0	0.6
1823	0.3	0.1	0.1	0.1	0.1	0.4	1.1	2.3	3.6	5.1	6.6	8.0
1824	8.7	8.8	8.5	8.5	8.1	7.7	7.4	7.5	8.2	8.8	9.3	10.0
1825	10.9	12.1	13.3	14.5	16.0	17.2	18.2	18.5	18.8	19.4	20.0	21.0
1826	22.5	24.7	27.3	29.8	32.4	34.9	37.3	39.3	41.0	43.0	44.5	46.4
1827	47.5	48.8	50.0	50.7	51.1	51.1	51.0	51.4	51.6	52.3	53.5	55.9
1828	58.9	62.0	65.0	67.2	68.6	68.8	67.2	65.0	61.8	59.8	58.4	58.1
1829	59.7	62.5	65.6	68.5	70.4	71.5	71.3	70.0	68.2	67.0	67.0	67.6
1830	68.3	69.1	69.5	69.8	69.2	68.9	68.2	67.4	66.7	66.8	66.7	66.2
1831	64.6	62.2	59.3	55.4	52.1	49.0	46.6	44.7	43.1	42.2	41.8	41.5
1832	40.6	38.9	36.7	33.7	30.5	27.0	23.8	20.8	18.3	16.8	15.5	14.7
1833	13.7	12.7	11.5	10.1	9.0	8.0	7.6	7.4	7.6	7.8	7.9	8.0
1834	7.9	7.9	7.6	7.7	8.4	9.7	11.4	13.2	15.4	18.1	21.0	24.0
1835	27.0	30.8	34.9	40.5	46.8	53.8	61.4	68.7	75.9	82.2	88.5	94.3

SUNSPOT ACTIVITY

Smoothed monthly means

Year	Jan	Feb	Mar	Apr	May	Jun	Jul	Aug	Sep	Oct	Nov	Dec
1836	99.6	104.1	107.3	110.4	112.7	116.0	120.3	125.6	131.3	137.7	142.8	147.9
1837	150.9	152.8	151.7	148.8	144.1	139.1	134.2	130.5	127.2	125.4	123.9	122.7
1838	121.6	120.1	118.9	115.6	111.7	106.2	101.0	96.5	92.8	89.5	86.8	84.5
1839	82.5	81.1	80.9	81.6	82.8	84.3	85.9	87.4	88.4	88.1	86.5	83.5
1840	79.4	74.7	70.6	67.8	65.9	63.9	61.8	58.9	56.1	52.8	49.8	47.5
1841	45.2	43.6	42.4	41.8	41.4	40.6	39.6	37.9	35.5	32.9	30.1	28.0
1842	26.2	25.0	23.9	23.3	23.2	23.5	23.7	24.0	23.5	22.8	21.5	19.8
1843	18.0	16.0	14.2	12.5	11.3	10.8	10.7	10.8	10.9	11.2	11.6	12.0
1844	12.5	13.1	13.7	14.2	14.4	14.7	15.3	16.2	17.7	20.1	23.2	27.0
1845	30.4	33.6	36.0	37.7	38.5	38.7	38.8	38.7	39.3	41.0	43.4	46.8
1846	50.0	52.9	55.8	58.6	60.7	62.4	63.7	64.4	65.0	64.7	64.6	64.5
1847	64.7	65.4	68.6	74.2	83.0	92.8	103.0	114.0	122.2	128.4	131.3	131.1
1848	129.9	126.7	123.4	120.8	119.8	120.5	122.1	124.7	126.8	128.6	128.4	126.8
1849	122.8	117.4	110.9	104.0	96.9	91.1	86.8	84.4	84.1	83.7	83.2	82.1
1850	79.9	76.8	73.3	70.0	67.3	65.0	63.9	64.3	65.6	67.3	68.8	69.7
1851	69.8	69.1	67.1	65.3	63.3	61.6	60.0	59.4	59.5	60.6	61.5	62.2
1852	62.2	61.4	59.3	57.1	54.9	52.8	50.8	49.3	48.1	47.5	46.7	46.0
1853	45.1	44.4	43.5	42.7	41.8	41.1	40.0	38.4	36.1	33.6	30.8	28.3
1854	25.9	23.9	22.5	21.6	21.0	20.7	20.5	20.1	19.6	18.7	17.8	16.6
1855	15.2	13.5	11.6	9.7	8.0	6.4	5.3	4.6	4.0	3.7	3.5	3.4
1856	3.3	3.4	3.4	3.5	3.7	4.1	4.6	5.2	5.7	6.3	7.2	8.3
1857	9.6	11.0	12.9	15.3	18.1	20.9	24.0	26.9	30.1	32.9	35.2	37.3
1858	39.3	41.1	43.3	46.0	49.5	52.9	56.8	60.8	64.9	69.0	72.6	76.2
1859	79.2	82.2	85.1	88.0	91.4	94.2	95.9	96.9	97.1	96.6	95.4	94.3
1860	93.7	93.5	94.2	94.9	96.5	97.8	98.5	97.5	96.0	94.2	91.9	89.2
1861	86.8	85.2	84.0	82.9	81.2	79.8	78.1	76.4	73.8	70.9	68.4	66.5
1862	64.7	63.9	63.4	63.6	63.5	63.1	62.0	60.4	58.9	56.9	54.9	52.8
1863	51.3	50.1	49.0	47.4	45.8	43.9	42.2	41.1	40.5	41.3	42.3	43.8
1864	45.7	47.2	48.6	48.7	48.6	48.0	46.8	45.7	44.4	43.2	42.2	40.6
1865	39.4	38.0	36.9	35.4	33.5	31.6	29.6	28.0	26.8	25.9	25.2	24.3
1866	23.7	22.6	21.3	19.8	18.0	16.0	13.8	11.6	9.8	8.5	7.3	6.2
1867	5.3	4.6	4.3	4.3	4.7	5.5	6.8	8.2	9.9	12.0	14.4	17.0
1868	19.5	22.0	24.5	27.3	30.4	34.0	38.0	42.4	46.5	50.1	53.5	57.4
1869	61.0	64.2	67.2	69.7	72.2	74.4	76.3	78.4	81.3	85.8	91.9	99.8
1870	108.9	118.7	127.8	135.3	141.1	144.1	144.8	143.3	140.2	137.4	135.2	133.7
1871	131.9	130.2	127.7	124.5	120.5	115.8	111.1	106.2	101.6	98.1	96.2	96.3
1872	97.2	98.5	99.7	101.4	102.6	103.4	103.7	103.1	102.7	101.4	99.3	95.6
1873	91.4	86.8	81.4	75.7	70.1	65.3	61.3	58.2	56.0	54.7	53.8	52.8
1874	51.7	50.7	50.0	49.0	47.9	46.2	44.3	41.7	38.8	36.2	33.1	30.2
1875	27.3	24.9	23.1	21.4	19.5	18.0	16.3	15.1	14.0	13.6	13.4	13.2
1876	13.1	12.7	12.4	11.6	11.1	10.4	10.3	10.3	10.6	11.2	12.1	12.9
1877	13.5	13.6	13.8	13.6	13.3	12.8	11.9	11.0	10.2	9.1	8.1	7.2
1878	6.5	5.8	5.1	4.4	3.9	3.5	3.2	2.8	2.5	2.2	2.1	2.0
1879	2.1	2.5	3.0	3.7	4.6	5.7	6.9	8.4	10.1	11.9	13.7	15.6
1880	17.7	19.8	22.5	25.7	28.9	31.9	34.4	36.7	38.8	40.5	41.8	42.9
1881	44.5	46.5	48.5	51.0	53.7	55.9	57.2	57.6	57.7	58.3	59.2	60.0
1882	60.5	61.1	61.0	60.8	60.1	59.4	58.4	57.3	56.4	55.7	56.0	56.0
1883	56.9	57.5	57.6	58.5	59.3	61.6	64.0	66.9	70.1	73.2	75.9	77.7
1884	78.1	77.3	75.6	72.4	68.2	63.9	59.7	56.1	53.6	51.9	51.4	52.3
1885	54.0	56.2	57.8	59.2	59.6	58.4	55.5	51.4	46.9	43.1	39.6	36.8

Smoothed monthly means

Year	Jan	Feb	Mar	Apr	May	Jun	Jul	Aug	Sep	Oct	Nov	Dec
1886	35.2	34.3	33.9	32.8	31.0	28.4	25.3	21.8	18.3	15.1	12.8	11.6
1887	10.9	11.2	11.8	12.6	13.5	13.9	14.2	14.1	13.6	13.0	12.0	11.1
1888	10.1	9.1	8.3	7.6	6.9	6.3	5.8	5.6	5.7	5.7	5.7	5.7
1889	5.7	5.9	6.2	6.6	6.9	7.1	7.2	7.2	6.7	6.2	5.5	4.9
1890	4.3	4.0	4.2	4.9	5.8	6.7	7.8	8.9	10.1	11.1	12.4	14.4
1891	17.1	20.5	24.3	28.9	33.5	37.5	41.1	44.1	47.0	49.3	51.7	54.2
1892	57.3	61.2	65.0	68.5	71.4	73.7	74.9	75.4	75.2	75.0	74.6	74.8
1893	75.4	77.1	79.8	82.8	85.1	87.3	88.5	89.5	88.7	87.3	85.4	84.2
1894	83.4	83.3	83.3	84.0	84.1	83.1	81.3	78.7	75.8	72.7	69.6	67.4
1895	65.9	65.4	65.3	65.3	64.8	64.4	63.2	61.8	59.9	58.2	55.7	53.7
1896	51.2	49.2	46.9	45.3	43.6	42.5	41.8	41.0	40.3	39.2	38.3	36.5
1897	34.4	32.2	30.1	28.4	26.8	25.4	24.8	24.6	25.3	25.7	26.3	27.0
1898	27.0	26.7	26.3	26.0	26.2	25.9	25.5	25.2	24.7	24.1	23.0	21.3
1899	19.8	17.8	15.9	14.3	13.0	12.3	11.5	11.0	10.6	10.3	10.4	10.6
1900	10.9	11.3	11.3	11.3	11.1	10.5	9.7	8.6	7.5	6.4	5.3	4.6
1901	4.1	3.8	3.5	3.3	3.3	3.2	3.1	3.0	2.8	2.9	2.9	3.0
1902	3.2	3.2	3.3	3.4	3.8	4.2	4.8	5.5	6.4	7.7	9.1	10.5
1903	12.1	13.9	15.5	17.2	19.1	21.7	24.2	26.6	28.6	30.8	32.7	34.5
1904	35.8	37.3	38.8	40.1	41.3	42.5	44.0	46.0	47.9	49.9	51.2	52.7
1905	53.7	54.9	55.6	56.3	57.8	59.6	61.4	62.5	63.4	63.9	63.7	62.2
1906	61.2	60.3	60.2	59.6	58.8	58.0	58.2	57.9	58.5	58.5	59.3	60.0
1907	60.1	60.5	60.3	60.3	59.6	58.6	57.6	56.4	55.6	54.5	53.2	51.1
1908	48.7	46.9	46.5	47.5	49.0	51.2	52.9	54.4	55.0	54.4	53.3	51.2
1909	49.0	45.9	43.0	40.8	39.8	39.0	38.9	39.0	39.4	39.5	38.8	37.2
1910	34.2	30.6	26.7	23.2	20.7	18.7	17.6	16.8	16.0	15.2	14.0	12.8
1911	11.3	10.1	8.7	7.6	6.8	6.2	5.5	4.7	3.9	3.3	2.9	2.8
1912	2.8	2.9	3.0	3.3	3.5	3.7	3.9	4.0	4.0	3.8	3.6	3.2
1913	2.8	2.4	1.8	1.4	1.2	1.1	1.1	1.3	1.5	2.0	2.6	3.4
1914	4.3	5.2	6.0	7.0	7.7	8.7	9.7	11.3	13.5	16.2	19.4	23.4
1915	28.1	33.5	38.8	43.9	48.0	51.3	52.9	53.5	53.2	53.1	53.1	53.5
1916	54.5	55.9	57.3	58.3	58.6	58.3	57.5	56.7	56.9	58.1	61.0	65.5
1917	71.7	79.1	87.7	95.0	101.5	106.8	110.6	112.1	110.9	107.8	103.5	97.9
1918	92.4	88.0	84.9	83.5	83.1	82.5	82.2	81.7	80.7	78.5	75.9	74.0
1919	72.5	71.7	71.6	71.8	71.5	70.5	68.1	64.9	61.1	57.5	53.4	49.5
1920	46.8	44.1	41.8	39.4	37.7	35.8	34.3	33.1	32.2	32.1	31.8	31.3
1921	31.0	30.8	30.2	29.4	28.6	27.8	26.6	25.0	24.1	23.4	22.7	22.1
1922	21.3	20.4	19.5	17.6	15.6	13.4	11.6	9.8	8.2	7.4	7.1	6.9
1923	6.5	6.0	5.7	5.6	5.6	5.8	6.1	6.3	6.5	6.4	6.5	6.9
1924	7.7	9.2	10.9	13.2	15.6	18.0	19.8	20.8	21.2	21.1	21.0	21.5
1925	22.6	24.6	27.3	31.1	35.4	40.5	46.1	51.6	56.6	60.8	63.4	65.6
1926	66.2	66.1	65.0	63.6	62.6	62.0	62.2	63.9	65.9	68.8	71.5	74.0
1927	75.5	76.3	75.7	74.4	72.2	69.5	66.9	65.0	64.4	64.7	66.1	68.3
1928	71.6	74.7	77.9	80.3	81.8	82.0	80.9	78.5	75.6	71.8	68.1	64.9
1929	62.4	61.0	60.2	59.8	60.0	60.9	62.1	63.5	64.3	64.7	63.8	61.9
1930	58.3	53.9	48.9	43.8	38.8	34.6	31.7	30.3	29.8	29.6	29.5	29.4
1931	28.9	28.3	26.8	25.2	23.3	21.5	19.9	18.1	16.6	15.4	14.7	14.4
1932	14.2	14.0	13.7	13.2	12.7	12.0	11.4	11.0	10.7	10.5	10.3	10.1
1933	9.9	9.5	8.7	7.7	6.5	5.3	4.2	3.3	2.9	3.0	3.5	4.2
1934	5.2	6.3	7.3	8.0	8.6	8.9	9.2	9.4	10.1	10.9	12.1	14.0
1935	16.5	19.2	22.1	25.2	28.7	32.7	36.8	41.3	46.2	51.2	55.9	59.7

SUNSPOT ACTIVITY

Smoothed monthly means

Year	Jan	Feb	Mar	Apr	May	Jun	Jul	Aug	Sep	Oct	Nov	Dec
1936	62.8	65.3	67.1	68.5	70.0	72.9	76.8	82.5	88.5	94.7	100.8	106.2
1937	110.8	114.6	117.0	119.1	119.8	119.3	117.9	116.0	113.3	110.3	106.7	104.3
1938	103.1	104.3	106.2	108.5	111.1	112.6	113.1	111.2	108.4	104.5	100.7	97.4
1939	94.8	93.6	94.3	95.9	96.8	96.9	95.7	92.7	88.5	83.0	77.0	71.9
1940	68.4	66.5	66.4	67.2	68.6	70.3	70.6	70.1	67.7	64.8	61.0	56.8
1941	53.1	50.3	48.9	48.9	49.4	50.0	50.4	50.5	50.3	49.6	48.7	46.8
1942	44.3	41.9	39.0	36.0	32.7	29.6	26.5	23.9	22.1	21.4	21.5	21.8
1943	21.7	21.4	20.5	19.4	17.8	16.3	15.0	13.6	12.2	11.0	9.9	9.0
1944	7.8	6.9	6.5	6.5	7.0	7.9	9.4	11.3	12.9	14.8	16.7	18.7
1945	20.7	22.9	25.0	27.6	30.6	33.4	35.9	38.8	42.2	45.9	49.9	54.5
1946	59.6	65.8	72.0	77.7	83.8	89.7	95.5	100.1	104.9	109.8	115.6	122.2
1947	128.8	136.1	143.9	151.2	157.2	160.4	161.2	158.9	153.1	146.5	139.9	135.3
1948	133.1	133.2	135.3	139.5	142.9	145.5	145.8	145.3	143.6	141.5	139.0	138.3
1949	137.4	137.4	136.4	135.9	134.8	133.7	131.8	129.4	126.9	124.6	122.4	119.2
1950	115.2	110.8	106.0	100.6	95.1	89.2	83.5	77.7	71.9	67.4	64.9	64.9
1951	66.5	68.9	71.9	74.8	76.2	76.0	73.9	70.1	64.9	58.4	51.7	45.4
1952	40.7	37.1	34.6	33.2	32.5	32.5	32.6	32.4	31.2	29.5	27.2	24.8
1953	22.3	20.1	18.7	17.9	16.9	16.0	14.7	13.5	11.9	10.0	8.0	6.3
1954	4.9	3.9	3.4	3.2	3.5	3.9	4.6	5.7	6.8	8.1	9.7	11.5
1955	13.5	15.7	18.3	21.6	25.8	31.2	37.2	44.5	53.2	62.5	72.3	82.0
1956	91.3	100.7	109.7	118.3	126.9	135.7	144.7	152.1	158.0	162.6	165.9	167.7
1957	169.1	169.3	170.3	173.6	178.3	184.6	192.2	199.2	204.7	207.8	208.4	206.8
1958	203.2	199.2	195.0	191.0	188.2	186.2	185.8	185.8	185.1	183.9	182.4	180.7
1959	178.6	175.8	174.1	171.2	168.1	163.7	158.3	153.4	147.1	140.6	134.7	129.1
1960	124.9	121.3	119.5	118.8	117.8	116.0	113.3	109.1	103.5	96.2	88.3	80.5
1961	73.6	68.0	64.0	61.9	60.1	58.6	56.9	54.9	52.6	49.8	47.4	45.4
1962	43.7	42.5	41.3	40.7	39.8	38.5	36.7	34.8	32.9	30.9	29.4	28.4
1963	27.9	27.7	27.7	28.3	29.3	30.1	30.3	29.8	28.6	26.8	24.3	21.7
1964	19.2	16.6	14.1	12.0	10.4	9.3	8.6	8.5	8.7	9.2	10.0	11.1
1965	12.1	12.9	13.5	13.9	14.4	14.6	15.0	15.6	16.6	18.1	20.1	22.8
1966	26.2	30.1	34.2	38.3	42.2	46.0	50.0	54.5	59.7	65.0	70.3	75.2
1967	79.7	83.1	86.0	87.0	87.9	88.3	89.5	91.5	94.0	97.0	99.8	102.6
1968	104.3	105.2	105.9	106.4	106.6	106.0	105.6	105.8	106.5	107.7	108.4	109.7
1969	110.8	111.6	111.6	110.5	109.2	106.8	104.2	102.3	101.0	100.9	102.0	104.0
1970	106.2	108.4	109.6	109.9	109.2	107.3	104.7	101.2	97.7	93.7	89.6	85.0
1971	80.8	77.0	73.1	69.6	66.4	64.1	62.9	62.3	63.2	64.6	66.3	68.8
1972	70.8	73.1	74.9	75.9	75.6	74.0	71.3	67.8	63.5	59.5	55.6	52.0
1973	48.9	46.3	44.3	42.9	41.3	39.7	37.8	35.8	33.9	31.9	30.7	30.1
1974	30.3	31.2	32.2	33.9	35.9	37.2	37.7	36.9	35.2	32.3	28.5	24.6
1975	20.8	18.1	16.6	15.7	15.8	16.4	16.9	17.0	16.7	16.2	15.4	14.4
1976	13.5	12.6	12.4	12.6	12.6	12.7	12.8	13.2	13.3	13.6	13.8	14.7
1977	15.7	16.9	18.4	20.7	23.2	26.0	28.9	32.6	37.1	42.5	48.5	55.1
1978	61.7	68.0	73.3	79.3	84.2	89.0	93.9	99.3	105.7	111.7	117.5	122.8
1979	127.9	131.7	135.0	138.3	143.3	148.5	153.9	158.8	163.3	165.9	167.1	166.6
1980	164.7	162.2	159.0	156.4	154.5	153.4	152.8	151.9	151.0	149.9	147.9	146.0
1981	143.0	141.0	138.9	138.1	138.2	139.1						

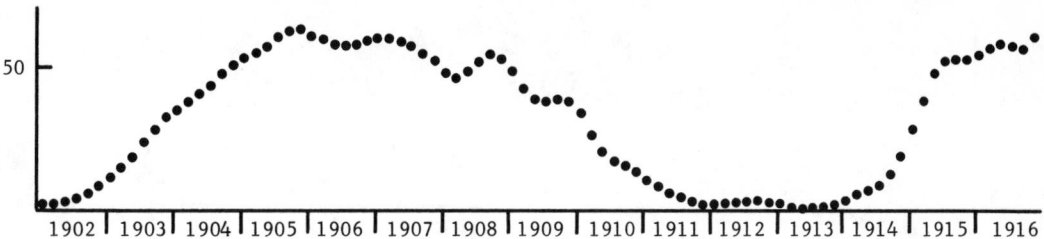

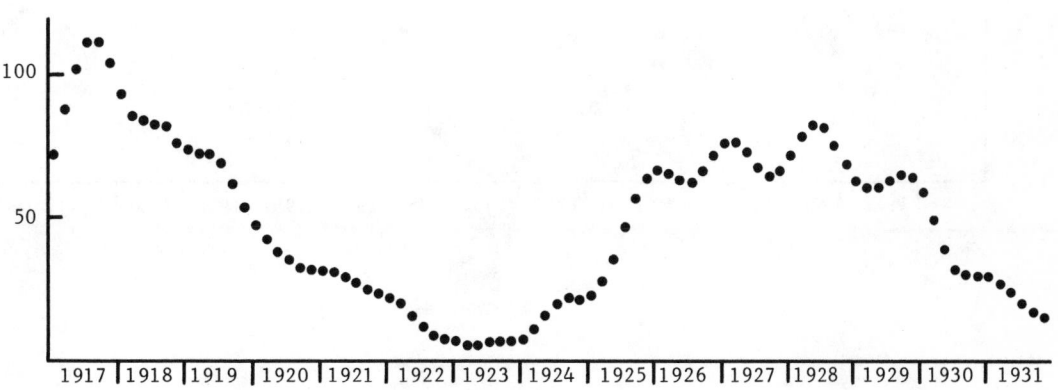

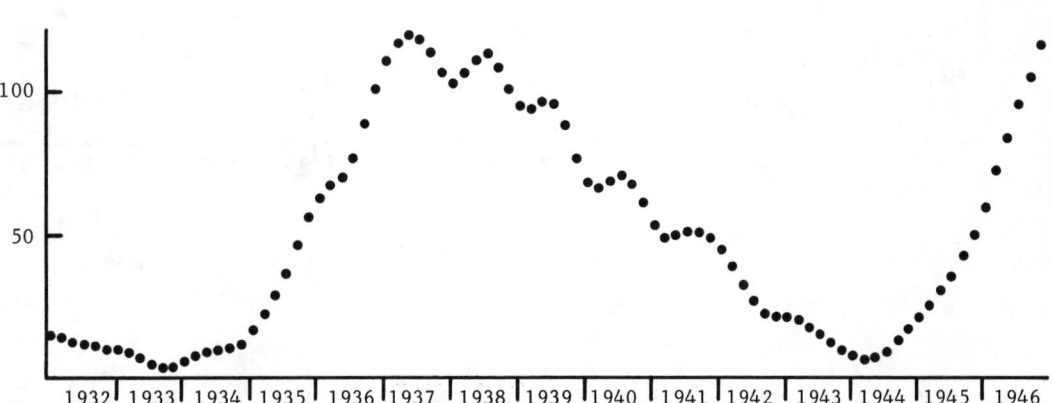

SUNSPOT ACTIVITY

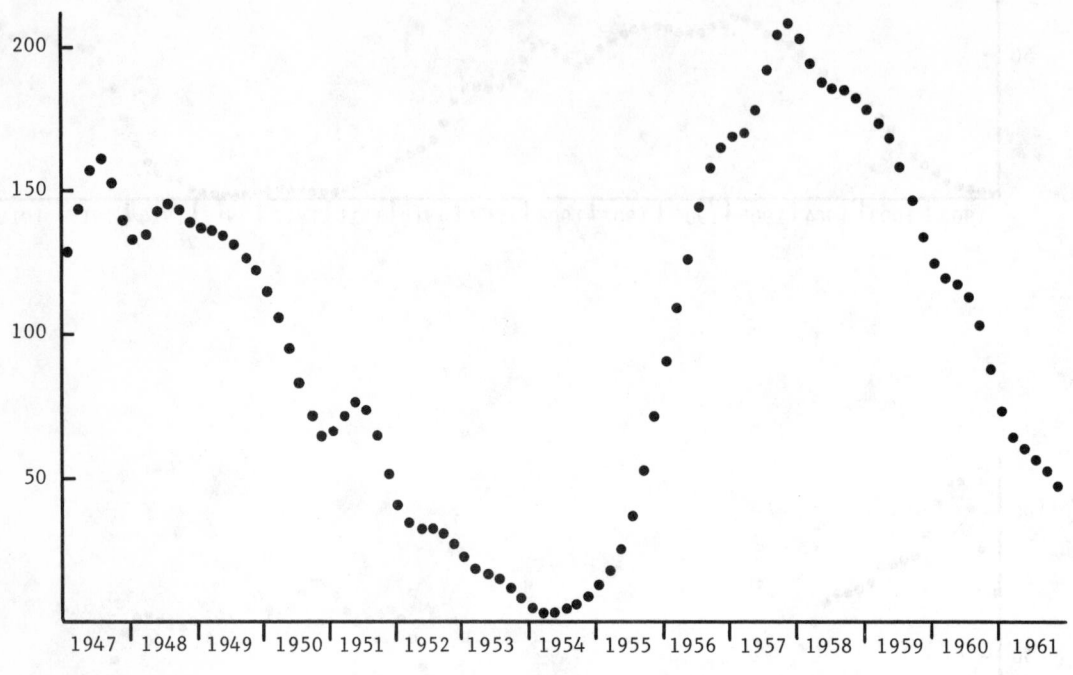

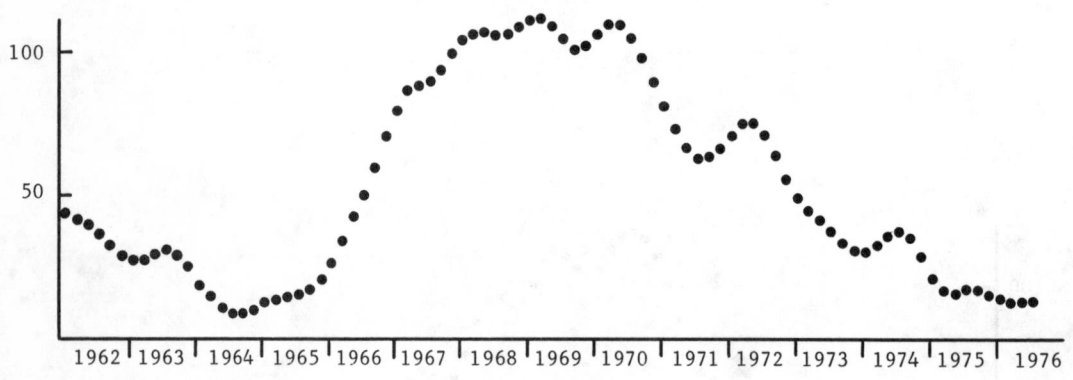

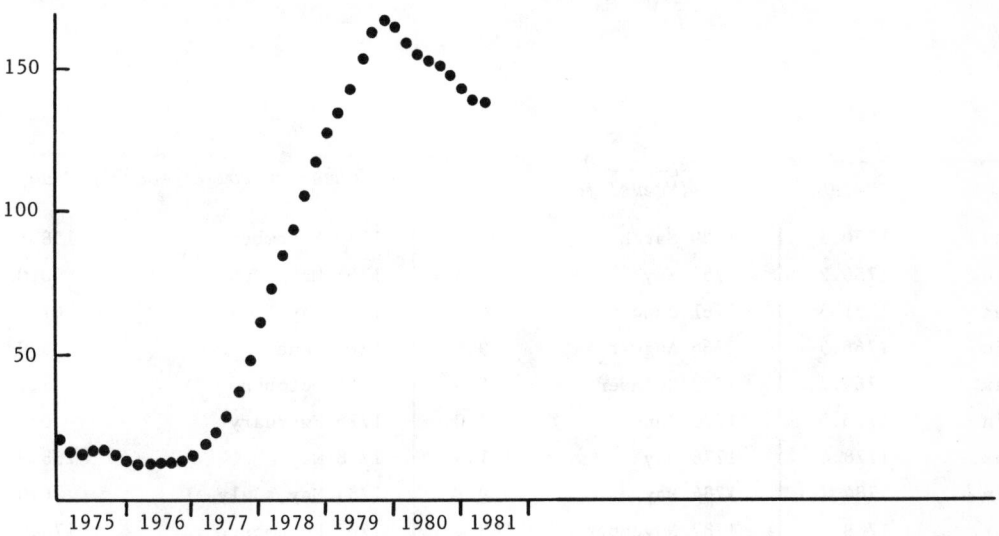

IV

EPOCHS OF MAXIMA AND MINIMA

	Epoch (Zürich)	Epoch and Value (Meeus' formula)		Highest or Lowest Monthly Mean	
Max.	1750.3	1750 March	90.3	1749 November	158.6
min.	1755.2	1755 May	6.8	1755 May - June	0.0
Max.	1761.5	1761 June	90.4	1761 May	107.2
min.	1766.5	1766 August	9.6	1766 June	3.0
Max.	1769.7	1769 October	125.3	1769 October	158.2
min.	1775.5	1775 June	7.0	1775 February	0.0
Max.	1778.4	1778 May	161.8	1778 May	238.9
min.	1784.7	1784 May	9.1	1784 May, July	6.0
Max.	1788.1	1787 November	143.4	1787 December	174.0
min.	1798.3	1798 June	2.8	1798 May - July	0.0
Max.	1805.2	1804 December (a)	52.5	1804 October	62.3
min.	1810.6	1810 July ?	0.0	1809 Oct. — 1811 June	0.0
Max.	1816.4	1816 March (b)	50.8	1817 March	96.2
min.	1823.3	1823 April	0.1	1822 Nov. — 1823 Feb. 1823 April — June 1823 August — November	0.0
Max.	1829.9	1829 June (c)	71.5	1830 April	106.3
min.	1833.9	1833 August	7.4	1833 June	1.0
Max.	1837.2	1837 February	152.8	1836 December	206.2
min.	1843.5	1843 July	10.7	1843 February	3.5
Max.	1848.1	1847 November (d)	131.3	1847 October	180.4
min.	1856.0	1856 January	3.3	1855 Sept., 1856 May	0.0
Max.	1860.1	1860 July (e)	98.5	1860 July	116.7
min.	1867.2	1867 April	4.3	1867 January	0.0
Max.	1870.6	1870 July	144.8	1870 May	176.0
min.	1878.9	1878 December	2.0	1878 Aug., 1879 March	0.0
Max.	1883.9	1884 January	78.1	1882 April	95.8
min.	1889.6	1890 February	4.0	1889 November	0.2
Max.	1894.1	1893 August	89.5	1893 August	129.2
min.	1901.7	1901 September	2.8	1901 April, December 1902 February, April	0.0
Max.	1907.0	1905 October (f)	63.9	1907 February	108.2
min.	1913.6	1913 June	1.1	1913 May — June	0.0

Epochs of Maxima and Minima

	Epoch (Zürich)	Epoch and Value (Meeus' formula)		Highest or Lowest Monthly Mean	
Max.	1917.6	1917 August	112.1	1917 August	154.5
min.	1923.6	1923 April	5.6	1923 Aug., 1924 Jan.	0.5
Max.	1928.4	1928 June (*g*)	82.0	1929 December	108.0
min.	1933.8	1933 September	2.9	1933 August	0.2
Max.	1937.4	1937 May (*h*)	119.8	1938 July	165.3
min.	1944.2	1944 April	6.5	1944 April	0.3
Max.	1947.5	1947 July	161.2	1947 May	201.3
min.	1954.3	1954 April	3.2	1954 January, June	0.2
Max.	1957.9	1957 November	208.4	1957 October	253.8
min.	1964.7	1964 August	8.5	1964 July	3.1
Max.	1968.9	1969 February (*i*)	111.6	1969 March	135.8
min.		1976 March	12.4	1976 July	1.9
Max.		1979 November	167.1	1979 September	188.4

REMARKS

(*a*) Secondary maxima : 1802 August (45.5) and 1803 December (47.0).

(*b*) Secondary maximum : 1817 March (46.1).

(*c*) Secondary maxima : 1828 June (68.8) and 1830 April (69.8).

(*d*) Secondary maximum : 1848 October (128.6).

(*e*) Secondary maximum : 1859 September (97.1).

(*f*) Secondary maximum : 1907 February (60.5).

(*g*) Secondary maximum : 1927 February (76.3).

(*h*) Secondary maximum : 1938 July (113.1).

(*i*) Secondary maximum : 1970 April (109.9).

V
NUMBER OF SPOTLESS DAYS

Year	Jan	Feb	Mar	Apr	May	Jun	Jul	Aug	Sep	Oct	Nov	Dec	Total
1850	0	0	0	1	0	0	5	0	0	0	1	0	7
1852	0	0	0	0	0	1	2	0	1	0	0	0	4
1853	0	0	2	1	0	1	1	0	1	0	0	0	6
1854	6	8	7	4	8	1	5	4	6	16	0	5	70
1855	17	9	6	21	19	21	30	25	30	9	21	26	234
1856	30	19	30	17	31	22	20	17	21	22	15	17	261
1857	4	14	18	13	0	6	6	9	0	0	0	0	70
1858	1	0	0	1	0	0	0	0	0	0	0	0	2
1861	0	0	0	0	0	0	0	0	0	2	0	0	2
1862	0	0	1	0	0	0	0	0	0	0	0	3	4
1863	0	0	0	0	0	0	0	0	2	0	0	0	2
1864	0	0	0	3	0	0	0	2	1	0	0	1	7
1865	0	0	2	2	2	2	2	1	7	10	4	12	44
1866	0	1	0	2	6	3	9	5	13	5	15	27	86
1867	31	26	12	20	24	26	18	20	16	13	9	7	222
1868	13	5	3	0	3	2	10	0	1	0	0	0	37
1869	0	0	0	1	0	0	1	0	0	0	0	0	2
1873	0	0	0	0	4	7	0	0	0	1	1	1	14
1874	0	0	0	3	0	0	0	0	2	2	0	5	12
1875	11	6	0	1	16	3	10	15	26	11	17	15	131
1876	17	9	8	26	21	26	11	17	14	7	15	19	190
1877	7	8	15	11	2	4	16	17	4	23	10	23	140
1878	22	20	15	29	22	18	30	31	18	28	18	29	280
1879	26	26	31	17	24	20	12	15	12	12	10	12	217

Number of spotless days

Year	Jan	Feb	Mar	Apr	May	Jun	Jul	Aug	Sep	Oct	Nov	Dec	Total
1880	7	6	5	2	4	0	5	0	0	0	3	1	33
1881	1	0	0	0	0	0	0	3	0	0	0	1	5
1882	0	0	0	0	0	0	0	0	0	0	0	2	2
1883	0	1	1	0	1	0	0	0	1	0	0	0	4
1885	1	0	1	0	0	0	0	0	0	3	3	5	13
1886	7	0	0	1	2	2	4	2	4	6	27	7	62
1887	8	10	15	11	5	0	4	11	12	11	13	4	104
1888	7	12	12	11	17	10	16	20	6	21	7	11	150
1889	28	15	15	17	24	17	11	8	13	20	27	17	212
1890	16	25	17	20	16	23	10	14	2	9	12	7	171
1891	11	3	5	2	0	0	0	1	0	0	0	2	24
1895	0	0	0	0	0	0	0	0	0	0	1	0	1
1896	0	0	0	3	1	0	0	2	0	1	0	0	7
1897	0	0	1	3	6	6	0	0	0	7	8	1	32
1898	1	3	5	9	0	1	11	2	0	0	0	7	39
1899	2	10	4	1	7	2	9	24	11	16	10	8	104
1900	10	13	14	3	8	9	14	21	13	8	16	29	158
1901	30	20	23	30	18	15	28	27	28	21	16	31	287
1902	20	28	17	30	20	25	28	22	18	4	17	28	257
1903	9	0	11	2	6	6	0	0	10	0	0	1	45
1904	1	0	0	0	0	0	0	0	0	0	0	0	1
1905	1	0	0	0	1	0	1	0	0	0	0	0	3
1906	0	0	0	0	0	0	0	0	0	4	0	0	4
1908	0	0	1	0	0	0	0	0	0	3	0	0	4
1909	0	0	0	0	0	1	2	1	0	0	0	2	6
1910	0	6	0	9	0	5	5	10	4	1	19	16	75
1911	19	13	14	5	11	22	19	17	20	22	15	23	200
1912	30	29	17	21	16	16	22	30	8	21	27	17	254
1913	22	19	29	27	31	30	25	30	27	21	27	23	311
1914	24	21	21	1	19	15	14	15	9	12	2	0	153
1915	0	1	0	2	9	0	0	0	0	0	0	0	12
1916	0	0	0	0	0	0	0	3	0	1	0	0	4
1920	0	0	0	2	0	0	0	1	4	0	0	0	7
1921	0	0	3	1	6	1	0	5	4	7	9	10	46
1922	10	4	3	13	14	14	13	17	12	14	9	11	134
1923	21	23	21	13	19	14	20	29	4	3	9	24	200
1924	29	22	25	17	5	0	1	0	0	0	8	9	116
1925	17	2	6	0	0	2	1	1	0	0	0	0	29

SUNSPOT ACTIVITY

Number of spotless days

Year	Jan	Feb	Mar	Apr	May	Jun	Jul	Aug	Sep	Oct	Nov	Dec	Total
1926	0	0	0	0	0	0	2	0	0	0	0	0	2
1930	0	0	0	0	0	0	0	0	1	0	1	1	3
1931	4	1	0	0	0	11	3	10	0	5	7	2	43
1932	7	11	8	12	5	1	10	14	16	6	12	6	108
1933	12	12	9	22	21	14	24	30	15	23	28	30	240
1934	21	8	18	10	4	17	12	13	18	11	11	11	154
1935	0	0	4	9	6	0	1	0	0	0	0	0	20
1941	0	0	0	0	0	0	0	0	0	0	3	2	5
1942	1	0	1	0	2	6	5	2	1	3	3	0	24
1943	4	2	1	0	0	9	7	3	13	10	10	6	65
1944	20	27	12	29	26	10	13	5	5	2	6	4	159
1945	3	3	1	2	1	1	4	1	0	0	0	0	16
1950	0	0	0	0	0	0	0	0	0	0	0	3	3
1952	0	6	9	0	0	0	0	0	2	2	2	2	23
1953	7	17	11	8	8	0	14	9	1	9	25	22	131
1954	30	26	14	24	28	29	14	10	24	12	15	15	241
1955	0	3	18	11	4	4	3	1	1	3	0	0	48
1961	0	0	0	0	0	0	0	0	0	0	3	3	6
1962	1	0	0	0	0	0	1	3	0	0	1	4	10
1963	0	1	2	6	0	0	0	1	3	2	1	5	21
1964	1	8	2	7	4	10	20	11	18	15	10	6	112
1965	2	4	4	13	8	4	10	9	3	3	7	3	70
1966	4	0	2	0	1	0	0	1	0	0	0	0	8
1973	0	0	0	0	0	0	1	6	2	5	6	7	27
1974	7	2	1	0	5	0	0	0	1	0	0	4	20
1975	2	9	9	18	13	9	0	0	7	9	6	13	95
1976	16	18	5	2	6	7	24	0	3	3	15	6	105
1977	6	2	7	4	2	0	4	0	0	0	0	0	25

VI

BEGINNING OF THE SUN'S SYNODIC ROTATIONS

Rotation No.	Date of beginning	Rotation No.	Date of beginning	Rotation No.	Date of beginning
1316	1952 Jan 22.96	1356	1955 Jan 17.96	1396	1958 Jan 12.96
1317	Feb 19.30	1357	Feb 14.30	1397	Feb 9.30
1318	Mar 17.63	1358	Mar 13.63	1398	Mar 8.63
1319	Apr 13.92	1359	Apr 9.93	1399	Apr 4.94
1320	May 11.16	1360	May 7.18	1400	May 2.20
1321	Jun 7.37	1361	Jun 3.39	1401	May 29.41
1322	Jul 4.57	1362	Jun 30.59	1402	Jun 25.61
1323	Jul 31.78	1363	Jul 27.80	1403	Jul 22.82
1324	Aug 28.01	1364	Aug 24.03	1404	Aug 19.04
1325	Sep 24.28	1365	Sep 20.28	1405	Sep 15.30
1326	Oct 21.56	1366	Oct 17.57	1406	Oct 12.57
1327	Nov 17.87	1367	Nov 13.87	1407	Nov 8.87
1328	Dec 15.18	1368	Dec 11.18	1408	Dec 6.19
1329	1953 Jan 11.52	1369	1956 Jan 7.51	1409	1959 Jan 2.51
1330	Feb 7.86	1370	Feb 3.85	1410	Jan 29.85
1331	Mar 7.19	1371	Mar 2.19	1411	Feb 26.19
1332	Apr 3.50	1372	Mar 29.50	1412	Mar 25.51
1333	Apr 30.76	1373	Apr 25.77	1413	Apr 21.79
1334	May 27.98	1374	May 23.00	1414	May 19.02
1335	Jun 24.18	1375	Jun 19.20	1415	Jun 15.22
1336	Jul 21.38	1376	Jul 16.40	1416	Jul 12.42
1337	Aug 17.60	1377	Aug 12.62	1417	Aug 8.64
1338	Sep 13.86	1378	Sep 8.87	1418	Sep 4.88
1339	Oct 11.13	1379	Oct 6.14	1419	Oct 2.15
1340	Nov 7.43	1380	Nov 2.44	1420	Oct 29.44
1341	Dec 4.74	1381	Nov 29.75	1421	Nov 25.75
		1382	Dec 27.07	1422	Dec 23.07
1342	1954 Jan 1.07				
1343	Jan 28.41	1383	1957 Jan 23.41	1423	1960 Jan 19.40
1344	Feb 24.75	1384	Feb 19.75	1424	Feb 15.75
1345	Mar 24.07	1385	Mar 19.07	1425	Mar 14.07
1346	Apr 20.35	1386	Apr 15.36	1426	Apr 10.37
1347	May 17.58	1387	May 12.60	1427	May 7.62
1348	Jun 13.79	1388	Jun 8.81	1428	Jun 3.83
1349	Jul 10.98	1389	Jul 6.01	1429	Jul 1.03
1350	Aug 7.20	1390	Aug 2.21	1430	Jul 28.23
1351	Sep 3.44	1391	Aug 29.45	1431	Aug 24.46
1352	Sep 30.71	1392	Sep 25.72	1432	Sep 20.72
1353	Oct 28.00	1393	Oct 23.00	1433	Oct 18.01
1354	Nov 24.31	1394	Nov 19.31	1434	Nov 14.31
1355	Dec 21.63	1395	Dec 16.63	1435	Dec 11.63

Rotation No.	Date of beginning	Rotation No.	Date of beginning	Rotation No.	Date of beginning
1436	1961 Jan 7.96	1490	1965 Jan 19.85	1543	1969 Jan 4.40
1437	Feb 4.30	1491	Feb 16.19	1544	Jan 31.74
1438	Mar 3.63	1492	Mar 15.52	1545	Feb 28.08
1439	Mar 30.94	1493	Apr 11.81	1546	Mar 27.39
1440	Apr 27.21	1494	May 9.06	1547	Apr 23.66
1441	May 24.44	1495	Jun 5.27	1548	May 20.89
1442	Jun 20.64	1496	Jul 2.46	1549	Jun 17.10
1443	Jul 17.84	1497	Jul 29.67	1550	Jul 14.29
1444	Aug 14.06	1498	Aug 25.90	1551	Aug 10.51
1445	Sep 10.31	1499	Sep 22.16	1552	Sep 6.76
1446	Oct 7.58	1500	Oct 19.45	1553	Oct 4.03
1447	Nov 3.88	1501	Nov 15.75	1554	Oct 31.32
1448	Dec 1.19	1502	Dec 13.07	1555	Nov 27.63
1449	Dec 28.51			1556	Dec 24.95
		1503	1966 Jan 9.40		
1450	1962 Jan 24.85	1504	Feb 5.74	1557	1970 Jan 21.29
1451	Feb 21.19	1505	Mar 5.08	1558	Feb 17.63
1452	Mar 20.51	1506	Apr 1.38	1559	Mar 16.96
1453	Apr 16.80	1507	Apr 28.65	1560	Apr 13.25
1454	May 14.04	1508	May 25.87	1561	May 10.49
1455	Jun 10.24	1509	Jun 22.07	1562	Jun 6.70
1456	Jul 7.44	1510	Jul 19.27	1563	Jul 3.90
1457	Aug 3.65	1511	Aug 15.50	1564	Jul 31.11
1458	Aug 30.89	1512	Sep 11.75	1565	Aug 27.34
1459	Sep 27.16	1513	Oct 9.02	1566	Sep 23.61
1460	Oct 24.44	1514	Nov 5.32	1567	Oct 20.89
1461	Nov 20.75	1515	Dec 2.63	1568	Nov 17.19
1462	Dec 18.07	1516	Dec 29.95	1569	Dec 14.51
1463	1963 Jan 14.40	1517	1967 Jan 26.29	1570	1971 Jan 10.84
1464	Feb 10.74	1518	Feb 22.63	1571	Feb 7.18
1465	Mar 10.08	1519	Mar 21.95	1572	Mar 6.52
1466	Apr 6.38	1520	Apr 18.24	1573	Apr 2.83
1467	May 3.63	1521	May 15.48	1574	Apr 30.09
1468	May 30.85	1522	Jun 11.68	1575	May 27.31
1469	Jun 27.05	1523	Jul 8.88	1576	Jun 23.51
1470	Jul 24.25	1524	Aug 5.09	1577	Jul 20.71
1471	Aug 20.48	1525	Sep 1.33	1578	Aug 16.93
1472	Sep 16.73	1526	Sep 28.60	1579	Sep 13.18
1473	Oct 14.02	1527	Oct 25.89	1580	Oct 10.46
1474	Nov 10.31	1528	Nov 22.19	1581	Nov 6.76
1475	Dec 7.63	1529	Dec 19.51	1582	Dec 4.07
				1583	Dec 31.40
1476	1964 Jan 3.95	1530	1968 Jan 15.84		
1477	Jan 31.29	1531	Feb 12.19	1584	1972 Jan 27.74
1478	Feb 27.63	1532	Mar 10.52	1585	Feb 24.08
1479	Mar 25.95	1533	Apr 6.82	1586	Mar 22.40
1480	Apr 22.22	1534	May 4.07	1587	Apr 18.68
1481	May 19.46	1535	May 31.29	1588	May 15.91
1482	Jun 15.66	1536	Jun 27.49	1589	Jun 12.12
1483	Jul 12.86	1537	Jul 24.69	1590	Jul 9.32
1484	Aug 9.07	1538	Aug 20.92	1591	Aug 5.53
1485	Sep 5.32	1539	Sep 17.17	1592	Sep 1.77
1486	Oct 2.59	1540	Oct 14.46	1593	Sep 29.04
1487	Oct 29.88	1541	Nov 10.76	1594	Oct 26.33
1488	Nov 26.19	1542	Dec 8.07	1595	Nov 22.63
1489	Dec 23.51			1596	Dec 19.95

Rotation No.	Date of beginning	Rotation No.	Date of beginning	Rotation No.	Date of beginning
1597	1973 Jan 16.29	1651	1977 Jan 28.18	1704	1981 Jan 12.73
1598	Feb 12.63	1652	Feb 24.52	1705	Feb 9.07
1599	Mar 11.96	1653	Mar 23.84	1706	Mar 8.40
1600	Apr 8.26	1654	Apr 20.12	1707	Apr 4.71
1601	May 5.51	1655	May 17.35	1708	May 1.97
1602	Jun 1.73	1656	Jun 13.56	1709	May 29.19
1603	Jun 28.92	1657	Jul 10.75	1710	Jun 25.39
1604	Jul 26.13	1658	Aug 6.97	1711	Jul 22.59
1605	Aug 22.36	1659	Sep 3.21	1712	Aug 18.81
1606	Sep 18.61	1660	Sep 30.48	1713	Sep 15.07
1607	Oct 15.90	1661	Oct 27.77	1714	Oct 12.35
1608	Nov 12.20	1662	Nov 24.08	1715	Nov 8.64
1609	Dec 9.51	1663	Dec 21.40	1716	Dec 5.96
1610	1974 Jan 5.84	1664	1978 Jan 17.73	1717	1982 Jan 2.28
1611	Feb 2.18	1665	Feb 14.07	1718	Jan 29.62
1612	Mar 1.52	1666	Mar 13.40	1719	Feb 25.96
1613	Mar 28.83	1667	Apr 9.70	1720	Mar 25.28
1614	Apr 25.10	1668	May 6.95	1721	Apr 21.56
1615	May 22.33	1669	Jun 3.16	1722	May 18.79
1616	Jun 18.53	1670	Jun 30.36	1723	Jun 14.99
1617	Jul 15.73	1671	Jul 27.57	1724	Jul 12.19
1618	Aug 11.95	1672	Aug 23.80	1725	Aug 8.41
1619	Sep 8.20	1673	Sep 20.06	1726	Sep 4.65
1620	Oct 5.47	1674	Oct 17.34	1727	Oct 1.92
1621	Nov 1.76	1675	Nov 13.64	1728	Oct 29.21
1622	Nov 29.07	1676	Dec 10.95	1729	Nov 25.52
1623	Dec 26.40			1730	Dec 22.84
		1677	1979 Jan 7.28		
1624	1975 Jan 22.73	1678	Feb 3.62	1731	1983 Jan 19.17
1625	Feb 19.07	1679	Mar 2.96	1732	Feb 15.52
1626	Mar 18.40	1680	Mar 30.27	1733	Mar 14.84
1627	Apr 14.69	1681	Apr 26.54	1734	Apr 11.14
1628	May 11.93	1682	May 23.77	1735	May 8.39
1629	Jun 8.14	1683	Jun 19.97	1736	Jun 4.60
1630	Jul 5.34	1684	Jul 17.17	1737	Jul 1.80
1631	Aug 1.55	1685	Aug 13.39	1738	Jul 29.01
1632	Aug 28.78	1686	Sep 9.64	1739	Aug 25.24
1633	Sep 25.05	1687	Oct 6.91	1740	Sep 21.50
1634	Oct 22.33	1688	Nov 3.21	1741	Oct 18.78
1635	Nov 18.64	1689	Nov 30.51	1742	Nov 15.08
1636	Dec 15.95	1690	Dec 27.84	1743	Dec 12.40
1637	1976 Jan 12.29	1691	1980 Jan 24.18	1744	1984 Jan 8.73
1638	Feb 8.63	1692	Feb 20.52	1745	Feb 5.07
1639	Mar 6.96	1693	Mar 18.84	1746	Mar 3.41
1640	Apr 3.27	1694	Apr 15.13	1747	Mar 30.71
1641	Apr 30.53	1695	May 12.37	1748	Apr 26.98
1642	May 27.75	1696	Jun 8.58	1749	May 24.21
1643	Jun 23.95	1697	Jul 5.78	1750	Jun 20.41
1644	Jul 21.15	1698	Aug 1.99	1751	Jul 17.61
1645	Aug 17.37	1699	Aug 29.22	1752	Aug 13.83
1646	Sep 13.63	1700	Sep 25.49	1753	Sep 10.08
1647	Oct 10.90	1701	Oct 22.77	1754	Oct 7.35
1648	Nov 7.20	1702	Nov 19.08	1755	Nov 3.65
1649	Dec 4.51	1703	Dec 16.40	1756	Nov 30.96
1650	Dec 31.84			1757	Dec 28.28

Rotation No.	Date of beginning	Rotation No.	Date of beginning	Rotation No.	Date of beginning
1758	1985 Jan 24.62	1811	1989 Jan 9.17	1865	1993 Jan 21.06
1759	Feb 20.96	1812	Feb 5.51	1866	Feb 17.40
1760	Mar 20.28	1813	Mar 4.85	1867	Mar 16.73
1761	Apr 16.57	1814	Apr 1.16	1868	Apr 13.02
1762	May 13.81	1815	Apr 28.42	1869	May 10.27
1763	Jun 10.02	1816	May 25.65	1870	Jun 6.48
1764	Jul 7.21	1817	Jun 21.85	1871	Jul 3.67
1765	Aug 3.43	1818	Jul 19.05	1872	Jul 30.88
1766	Aug 30.66	1819	Aug 15.27	1873	Aug 27.12
1767	Sep 26.93	1820	Sep 11.52	1874	Sep 23.38
1768	Oct 24.22	1821	Oct 8.79	1875	Oct 20.66
1769	Nov 20.52	1822	Nov 5.09	1876	Nov 16.97
1770	Dec 17.84	1823	Dec 2.40	1877	Dec 14.28
		1824	Dec 29.72		
1771	1986 Jan 14.17			1878	1994 Jan 10.61
1772	Feb 10.51	1825	1990 Jan 26.06	1879	Feb 6.96
1773	Mar 9.85	1826	Feb 22.40	1880	Mar 6.29
1774	Apr 6.15	1827	Mar 21.72	1881	Apr 2.60
1775	May 3.41	1828	Apr 18.01	1882	Apr 29.86
1776	May 30.62	1829	May 15.25	1883	May 27.08
1777	Jun 26.82	1830	Jun 11.45	1884	Jun 23.28
1778	Jul 24.03	1831	Jul 8.65	1885	Jul 20.48
1779	Aug 20.25	1832	Aug 4.86	1886	Aug 16.71
1780	Sep 16.51	1833	Sep 1.10	1887	Sep 12.96
1781	Oct 13.79	1834	Sep 28.37	1888	Oct 10.24
1782	Nov 10.09	1835	Oct 25.66	1889	Nov 6.53
1783	Dec 7.40	1836	Nov 21.96	1890	Dec 3.84
		1837	Dec 19.28	1891	Dec 31.17
1784	1987 Jan 3.73				
1785	Jan 31.07	1838	1991 Jan 15.62	1892	1995 Jan 27.51
1786	Feb 27.41	1839	Feb 11.96	1893	Feb 23.85
1787	Mar 26.72	1840	Mar 11.29	1894	Mar 23.17
1788	Apr 23.00	1841	Apr 7.59	1895	Apr 19.45
1789	May 20.23	1842	May 4.84	1896	May 16.69
1790	Jun 16.43	1843	Jun 1.06	1897	Jun 12.89
1791	Jul 13.63	1844	Jun 28.26	1898	Jul 10.09
1792	Aug 9.85	1845	Jul 25.46	1899	Aug 6.30
1793	Sep 6.09	1846	Aug 21.69	1900	Sep 2.54
1794	Oct 3.36	1847	Sep 17.95	1901	Sep 29.81
1795	Oct 30.65	1848	Oct 15.23	1902	Oct 27.10
1796	Nov 26.96	1849	Nov 11.53	1903	Nov 23.40
1797	Dec 24.28	1850	Dec 8.84	1904	Dec 20.72
1798	1988 Jan 20.62	1851	1992 Jan 5.17	1905	1996 Jan 17.06
1799	Feb 16.96	1852	Feb 1.51	1906	Feb 13.40
1800	Mar 15.29	1853	Feb 28.85	1907	Mar 11.73
1801	Apr 11.58	1854	Mar 27.16	1908	Apr 8.03
1802	May 8.83	1855	Apr 23.43	1909	May 5.28
1803	Jun 5.04	1856	May 20.67	1910	Jun 1.50
1804	Jul 2.24	1857	Jun 16.87	1911	Jun 28.70
1805	Jul 29.44	1858	Jul 14.07	1912	Jul 25.90
1806	Aug 25.68	1859	Aug 10.28	1913	Aug 22.13
1807	Sep 21.94	1860	Sep 6.53	1914	Sep 18.39
1808	Oct 19.22	1861	Oct 3.80	1915	Oct 15.67
1809	Nov 15.52	1862	Oct 31.09	1916	Nov 11.97
1810	Dec 12.84	1863	Nov 27.40	1917	Dec 9.28
		1864	Dec 24.72		

Rotation No.	Date of beginning	Rotation No.	Date of beginning	Rotation No.	Date of beginning
1918	1997 Jan 5.61	1972	2001 Jan 17.50	2025	2005 Jan 2.05
1919	Feb 1.95	1973	Feb 13.84	2026	Jan 29.39
1920	Mar 1.29	1974	Mar 13.17	2027	Feb 25.73
1921	Mar 28.60	1975	Apr 9.47	2028	Mar 25.05
1922	Apr 24.87	1976	May 6.72	2029	Apr 21.33
1923	May 22.10	1977	Jun 2.94	2030	May 18.56
1924	Jun 18.31	1978	Jun 30.13	2031	Jun 14.77
1925	Jul 15.50	1979	Jul 27.34	2032	Jul 11.97
1926	Aug 11.72	1980	Aug 23.57	2033	Aug 8.18
1927	Sep 7.97	1981	Sep 19.83	2034	Sep 4.42
1928	Oct 5.24	1982	Oct 17.11	2035	Oct 1.69
1929	Nov 1.54	1983	Nov 13.41	2036	Oct 28.98
1930	Nov 28.84	1984	Dec 10.73	2037	Nov 25.29
1931	Dec 26.17			2038	Dec 22.61
		1985	2002 Jan 7.05		
1932	1998 Jan 22.50	1986	Feb 3.40	2039	2006 Jan 18.95
1933	Feb 18.85	1987	Mar 2.73	2040	Feb 15.29
1934	Mar 18.17	1988	Mar 30.04	2041	Mar 14.62
1935	Apr 14.46	1989	Apr 26.31	2042	Apr 10.91
1936	May 11.70	1990	May 23.54	2043	May 8.16
1937	Jun 7.91	1991	Jun 19.74	2044	Jun 4.37
1938	Jul 5.11	1992	Jul 16.94	2045	Jul 1.57
1939	Aug 1.32	1993	Aug 13.16	2046	Jul 28.78
1940	Aug 28.55	1994	Sep 9.41	2047	Aug 25.01
1941	Sep 24.82	1995	Oct 6.68	2048	Sep 21.27
1942	Oct 22.10	1996	Nov 2.98	2049	Oct 18.55
1943	Nov 18.41	1997	Nov 30.29	2050	Nov 14.85
1944	Dec 15.72	1998	Dec 27.61	2051	Dec 12.17
1945	1999 Jan 12.06	1999	2003 Jan 23.95	2052	2007 Jan 8.50
1946	Feb 8.40	2000	Feb 20.29	2053	Feb 4.84
1947	Mar 7.73	2001	Mar 19.61	2054	Mar 4.18
1948	Apr 4.04	2002	Apr 15.90	2055	Mar 31.49
1949	May 1.30	2003	May 13.14	2056	Apr 27.75
1950	May 28.52	2004	Jun 9.35	2057	May 24.98
1951	Jun 24.72	2005	Jul 6.55	2058	Jun 21.18
1952	Jul 21.92	2006	Aug 2.76	2059	Jul 18.38
1953	Aug 18.14	2007	Aug 29.99	2060	Aug 14.60
1954	Sep 14.40	2008	Sep 26.26	2061	Sep 10.85
1955	Oct 11.68	2009	Oct 23.55	2062	Oct 8.12
1956	Nov 7.97	2010	Nov 19.85	2063	Nov 4.42
1957	Dec 5.28	2011	Dec 17.17	2064	Dec 1.73
				2065	Dec 29.05
1958	2000 Jan 1.61	2012	2004 Jan 13.50		
1959	Jan 28.95	2013	Feb 9.84	2066	2008 Jan 25.39
1960	Feb 25.29	2014	Mar 8.18	2067	Feb 21.73
1961	Mar 23.61	2015	Apr 4.48	2068	Mar 20.05
1962	Apr 19.89	2016	May 1.74	2069	Apr 16.34
1963	May 17.12	2017	May 28.96	2070	May 13.58
1964	Jun 13.33	2018	Jun 25.16	2071	Jun 9.79
1965	Jul 10.53	2019	Jul 22.36	2072	Jul 6.99
1966	Aug 6.74	2020	Aug 18.58	2073	Aug 3.20
1967	Sep 2.98	2021	Sep 14.84	2074	Aug 30.43
1968	Sep 30.25	2022	Oct 12.12	2075	Sep 26.70
1969	Oct 27.54	2023	Nov 8.42	2076	Oct 23.99
1970	Nov 23.85	2024	Dec 5.73	2077	Nov 20.29
1971	Dec 21.17			2078	Dec 17.61

SUNSPOT ACTIVITY

Rotation No.	Date of beginning	Rotation No.	Date of beginning	Rotation No.	Date of beginning
2079	2009 Jan 13.94	2106	2011 Jan 20.39	2133	2013 Jan 25.84
2080	Feb 10.29	2107	Feb 16.73	2134	Feb 22.18
2081	Mar 9.62	2108	Mar 16.06	2135	Mar 21.50
2082	Apr 5.92	2109	Apr 12.35	2136	Apr 17.78
2083	May 3.18	2110	May 9.60	2137	May 15.02
2084	May 30.40	2111	Jun 5.81	2138	Jun 11.23
2085	Jun 26.59	2112	Jul 3.01	2139	Jul 8.42
2086	Jul 23.80	2113	Jul 30.22	2140	Aug 4.64
2087	Aug 20.02	2114	Aug 26.45	2141	Aug 31.87
2088	Sep 16.28	2115	Sep 22.71	2142	Sep 28.14
2089	Oct 13.56	2116	Oct 19.99	2143	Oct 25.43
2090	Nov 9.86	2117	Nov 16.29	2144	Nov 21.73
2091	Dec 7.17	2118	Dec 13.61	2145	Dec 19.05
2092	2010 Jan 3.50	2119	2012 Jan 9.94	2146	2014 Jan 15.39
2093	Jan 30.84	2120	Feb 6.28	2147	Feb 11.73
2094	Feb 27.18	2121	Mar 4.62	2148	Mar 11.06
2095	Mar 26.49	2122	Mar 31.93	2149	Apr 7.36
2096	Apr 22.77	2123	Apr 28.19	2150	May 4.62
2097	May 20.00	2124	May 25.42	2151	May 31.83
2098	Jun 16.20	2125	Jun 21.62	2152	Jun 28.03
2099	Jul 13.40	2126	Jul 18.82	2153	Jul 25.24
2100	Aug 9.62	2127	Aug 15.04	2154	Aug 21.46
2101	Sep 5.86	2128	Sep 11.29	2155	Sep 17.72
2102	Oct 3.13	2129	Oct 8.56	2156	Oct 15.00
2103	Oct 30.42	2130	Nov 4.86	2157	Nov 11.30
2104	Nov 26.73	2131	Dec 2.17	2158	Dec 8.61
2105	Dec 24.05	2132	Dec 29.50		
				2159	2015 Jan 4.94

7

OTHER TABLES

OTHER TABLES

Contents

I
TABLE FOR CALCULATING THE JULIAN DATE

The table on the next page can be used for the calculation of the Julian Day Number for any date between the years −1900 and +2999. Simply add the three numbers corresponding to the century, year and month (from tables a to c), and then the day of the month. The result is the Julian Day Number at 0 h (Ephemeris Time or Universal Time) of the date.

Example . — Find the J.D. for 1981 February 15, at 0 h.

Table a	1900	2415 019.5
Table b	81	29 585
Table c	Feb	31
Day of month	....	15
	Sum :	2444 650.5

Note that the Julian Days begin at Greenwich mean $noon$ (12 h UT). If they are measured from 12 h ET, they are called Julian Ephemeris Days.

The number characterizing negative years should be split in such a way as to have the last two figures positive. For example, the year −328 should be split up as −400 and +72. Then table a should be consulted for −400, and table b for +72.

In table a, two values are given for the year 1500, the first for the Julian calendar (J), the second for the Gregorian calendar (G).

In table c, the months marked with (B) should be used in the case of a bissextile (leap) year.

Table a : Century Year Table b : Additional Year Table c: Month

Year	J.D.	Year	J.D.
−1900	1027 082.5	+600	1940 207.5
−1800	1063 607.5	700	1976 732.5
−1700	1100 132.5	800	2013 257.5
−1600	1136 657.5	900	2049 782.5
−1500	1173 182.5	1000	2086 307.5
−1400	1209 707.5	1100	2122 832.5
−1300	1246 232.5	1200	2159 357.5
−1200	1282 757.5	1300	2195 882.5
−1100	1319 282.5	1400	2232 407.5
−1000	1355 807.5	1500J	2268 932.5
− 900	1392 332.5	1500G	2268 922.5
− 800	1428 857.5	1600	2305 447.5
− 700	1465 382.5	1700	2341 971.5
− 600	1501 907.5	1800	2378 495.5
− 500	1538 432.5	1900	2415 019.5
− 400	1574 957.5	2000	2451 544.5
− 300	1611 482.5	2100	2488 068.5
− 200	1648 007.5	2200	2524 592.5
− 100	1684 532.5	2300	2561 116.5
0	1721 057.5	2400	2597 641.5
+ 100	1757 582.5	2500	2634 165.5
200	1794 107.5	2600	2670 689.5
300	1830 632.5	2700	2707 213.5
400	1867 157.5	2800	2743 738.5
500	1903 682.5	2900	2780 262.5

Year	J.D.	Year	J.D.	Year	J.D.
0	0	40	14 610	80	29 220
1	365	41	14 975	81	29 585
2	730	42	15 340	82	29 950
3	1 095	43	15 705	83	30 315
4	1 461	44	16 071	84	30 681
5	1 826	45	16 436	85	31 046
6	2 191	46	16 801	86	31 411
7	2 556	47	17 166	87	31 776
8	2 922	48	17 532	88	32 142
9	3 287	49	17 897	89	32 507
10	3 652	50	18 262	90	32 872
11	4 017	51	18 627	91	33 237
12	4 383	52	18 993	92	33 603
13	4 748	53	19 358	93	33 968
14	5 113	54	19 723	94	34 333
15	5 478	55	20 088	95	34 698
16	5 844	56	20 454	96	35 064
17	6 209	57	20 819	97	35 429
18	6 574	58	21 184	98	35 794
19	6 939	59	21 549	99	36 159
20	7 305	60	21 915		
21	7 670	61	22 280		
22	8 035	62	22 645		
23	8 400	63	23 010		
24	8 766	64	23 376		
25	9 131	65	23 741		
26	9 496	66	24 106		
27	9 861	67	24 471		
28	10 227	68	24 837		
29	10 592	69	25 202		
30	10 957	70	25 567		
31	11 322	71	25 932		
32	11 688	72	26 298		
33	12 053	73	26 663		
34	12 418	74	27 028		
35	12 783	75	27 393		
36	13 149	76	27 759		
37	13 514	77	28 124		
38	13 879	78	28 489		
39	14 244	79	28 854		

Month	J.D.
Jan.	0
Jan.(B)	−1
Feb.	31
Feb.(B)	30
March	59
April	90
May	120
June	151
July	181
Aug.	212
Sept.	243
Oct.	273
Nov.	304
Dec.	334

II

PERPETUAL CALENDAR

(TABLE FOR DETERMINING THE WEEKDAY OF A GIVEN DATE)

In the first table, take the number at the intersection of the secular part (left) and the last two figures (upper part) of the given year.

With that number, enter the second table, and take the number at the intersection with the given month. Note the special columns for January and February in the case of a bis-sextile year.

With that new number, enter the third table. The weekday is found at the intersection with the given day of the month.

00	01	02	03	–	04	05
06	07	–	08	09	10	11
–	12	13	14	15	–	16
17	18	19	–	20	21	22
23	–	24	25	26	27	–
28	29	30	31	–	32	33
34	35	–	36	37	38	39
–	40	41	42	43	–	44
45	46	47	–	48	49	50
51	–	52	53	54	55	–
56	57	58	59	–	60	61
62	63	–	64	65	66	67
–	68	69	70	71	–	72
73	74	75	–	76	77	78
79	–	80	81	82	83	–
84	85	86	87	–	88	89
90	91	–	92	93	94	95
–	96	97	98	99		

YEAR

0	7	14	17	21	6	0	1	2	3	4	5
1	8	15 J			5	6	0	1	2	3	4
2	9		18	22	4	5	6	0	1	2	3
3	10				3	4	5	6	0	1	2
4	11	15 G	19	23	2	3	4	5	6	0	1
5	12	16	20	24	1	2	3	4	5	6	0
6	13				0	1	2	3	4	5	6

J : until 1582 October 4 inclusively (Julian Calendar)

G : from 1582 October 15 onwards (Gregorian Calendar)

Month	May	Feb.(B) Aug.	Feb. March Nov.	June	Sept. Dec.	Jan.(B) April July	Jan. Oct.
1	2	3	4	5	6	0	1
2	3	4	5	6	0	1	2
3	4	5	6	0	1	2	3
4	5	6	0	1	2	3	4
5	6	0	1	2	3	4	5
6	0	1	2	3	4	5	6
0	1	2	3	4	5	6	0

(B) = Bissextile (leap) year

Day of Month	1 8 15 22 29	2 9 16 23 30	3 10 17 24 31	4 11 18 25	5 12 19 26	6 13 20 27	7 14 21 28
1	Sun.	Mon.	Tue.	Wed.	Thu.	Fri.	Sat.
2	Mon.	Tue.	Wed.	Thu.	Fri.	Sat.	Sun.
3	Tue.	Wed.	Thu.	Fri.	Sat.	Sun.	Mon.
4	Wed.	Thu.	Fri.	Sat.	Sun.	Mon.	Tue.
5	Thu.	Fri.	Sat.	Sun.	Mon.	Tue.	Wed.
6	Fri.	Sat.	Sun.	Mon.	Tue.	Wed.	Thu.
0	Sat.	Sun.	Mon.	Tue.	Wed.	Thu.	Fri.

Example : 1970 March 7.

In the first table, we find 5 at the intersection of 19 and 70.
In the second table, we find 1 at the intersection of 5 and March.
In the third table, we find *Saturday* at the intersection of 1 and 7.

III

DATE OF EASTER SUNDAY

In the Gregorian calendar, the earliest possible date for the Christian Easter Day is March 22, and the latest is April 25. During the period 1583 – 3000, these extreme Easter dates occur in the following years :

March 22 : in 1598, 1693, 1761, 1818, 2285, 2353, 2437, 2505, 2972 ;
April 25 : in 1666, 1734, 1886, 1943, 2038, 2190, 2258, 2326, 2410, 2573, 2630, 2782, 2877, 2945.

The following table gives the dates of Easter Sunday for the years 1583 to 2179 in the Gregorian calendar. The dates indicated by M correspond to March, and all other dates to April. For example in A.D. 1982 Easter Day was April 11, this date being found at the intersection of line 1980 and column 2.

Year	0	1	2	3	4	5	6	7	8	9
1580......				10	1	21	6	29M	17	2
1590......	22	14	29M	18	10	26M	14	6	22M	11
1600......	2	22	7	30M	18	10	26M	15	6	19
1610......	11	3	22	7	30M	19	3	26M	15	31M
1620......	19	11	27M	16	7	30M	12	4.	23	15
1630......	31M	20	11	27M	16	8	23M	12	4	24
1640......	8	31M	20	5	27M	16	1	21	12	4
1650......	17	9	31M	13	5	28M	16	1	21	13
1660......	28M	17	9	25M	13	5	25	10	i	21
1670......	6	29M	17	2	25M	14	5	18	10	2
1680......	21	6	29M	18	2	22	14	30M	18	10
1690......	26M	15	6	22M	11	3	22	7	30M	19
1700......	11	27M	16	8	23M	12	4	24	8	31M
1710......	20	5	27M	16	1	21	12	28M	17	9
1720......	31M	13	5	28M	16	1	21	13	28M	17
1730......	9	25M	13	5	25	10	1	21	6	29M
1740......	17	2	25M	14	5	18	10	2	14	6
1750......	29M	11	2	22	14	30M	18	10	26M	15
1760......	6	22M	11	3	22	7	30M	19	3	26M
1770......	15	31M	19	11	3	16	7	30M	19	4
1780......	26M	15	31M	20	11	27M	16	8	23M	12
1790......	4	24	8	31M	20	5	27M	16	8	24M
1800......	13	5	18	10	1	14	6	29M	17	2
1810......	22	14	29M	18	10	26M	14	6	22M	11

Year	0	1	2	3	4	5	6	7	8	9
1820......	2	22	7	30M	18	3	26M	15	6	19
1830......	11	3	22	7	30M	19	3	26M	15	31M
1840......	19	11	27M	16	7	23M	12	4	23	8
1850......	31M	20	11	27M	16	8	23M	12	4	24
1860......	8	31M	20	5	27M	16	1	21	12	28M
1870......	17	9	31M	13	5	28M	16	1	21	13
1880......	28M	17	9	25M	13	5	25	10	1	21
1890......	6	29M	17	2	25M	14	5	18	10	2
1900......	15	7	30M	12	3	23	15	31M	19	11
1910......	27M	16	7	23M	12	4	23	8	31M	20
1920......	4	27M	16	1	20	12	4	17	8	31M
1930......	20	5	27M	16	1	21	12	28M	17	9
1940......	24M	13	5	25	9	1	21	6	28M	17
1950......	9	25M	13	5	18	10	1	21	6	29M
1960......	17	2	22	14	29M	18	10	26M	14	6
1970......	29M	11	2	22	14	30M	18	10	26M	15
1980......	6	19	11	3	22	7	30M	19	3	26M
1990......	15	31M	19	11	3	16	7	30M	12	4
2000......	23	15	31M	20	11	27M	16	8	23M	12
2010......	4	24	8	31M	20	5	27M	16	1	21
2020......	12	4	17	9	31M	20	5	28M	16	1
2030......	21	13	28M	17	9	25M	13	5	25	10
2040......	1	21	6	29M	17	9	25M	14	5	18
2050......	10	2	21	6	29M	18	2	22	14	30M
2060......	18	10	26M	15	6	29M	11	3	22	14
2070......	30M	19	10	26M	15	7	19	11	3	23
2080......	7	30M	19	4	26M	15	31M	20	11	3
2090......	16	8	30M	12	4	24	15	31M	20	12
2100......	28M	17	9	25M	13	5	18	10	1	21
2110......	6	29M	17	2	22	14	29M	18	10	26M
2120......	14	6	29M	11	2	22	14	30M	18	10
2130......	26M	15	6	19	11	3	22	7	30M	19
2140......	3	26M	15	31M	19	11	3	16	7	30M
2150......	12	4	23	15	31M	20	11	27M	16	8
2160......	23M	12	4	24	8	31M	20	5	27M	16
2170......	1	21	12	4	17	9	31M	20	5	28M

IV

PERIGEE AND APOGEE OF THE MOON

1976

Perigee		km	Apogee		km
	h		Jan 8	17^h	404408
Jan 20	13	366874	Feb 5	13	405101
Feb 17	10	361310	Mar 4	4	406002
Mar 16	19	357635	Mar 31	10	406511
Apr 14	7	356935	Apr 27	12	406394
May 12	16	359174	May 24	24	405599
Jun 9	19	363553	Jun 21	17	404583
Jul 7	2	368377	Jul 19	11	404063
Aug 1	4	369106	Aug 16	6	404402
Aug 28	2	364544	Sep 12	23	405331
Sep 25	3	359839	Oct 10	12	406140
Oct 23	13	357155	Nov 6	15	406324
Nov 21	1	357493	Dec 3	18	405931
Dec 19	12	360955	Dec 31	9	405095

1977

Perigee		km	Apogee		km
Jan 16	10^h	366447	Jan 28	6^h	404371
Feb 11	4	370285	Feb 25	3	404358
Mar 8	23	366424	Mar 24	22	405090
Apr 5	21	361157	Apr 21	12	406022
May 4	5	357735	May 18	18	406539
Jun 1	15	357059	Jun 14	21	406388
Jun 29	24	359174	Jul 12	8	405602
Jul 28	2	363473	Aug 8	24	404654
Aug 24	9	368395	Sep 5	18	404229
Sep 18	9	369114	Oct 3	14	404655
Oct 15	9	364220	Oct 31	8	405615
Nov 12	12	359299	Nov 27	20	406369
Dec 10	23	356716	Dec 24	21	406508

1978

Perigee		km	Apogee		km
Jan 8	12^h	357462	Jan 21	2^h	406086
Feb 5	21	361346	Feb 17	18	405185
Mar 5	17	366862	Mar 17	14	404436
Mar 31	5	369933	Apr 14	10	404442
Apr 26	8	365947	May 12	4	405195
May 24	5	360938	Jun 8	18	406131
Jun 21	12	357668	Jul 5	24	406622
Jul 19	21	357035	Aug 2	3	406424
Aug 17	6	359195	Aug 29	13	405622
Sep 14	10	363645	Sep 26	6	404690
Oct 11	16	368778	Oct 24	1	404295
Nov 5	12	369023	Nov 20	22	404730
Dec 2	16	363684	Dec 18	16	405630
Dec 30	22	358856			

1979

Perigee		km	Apogee		km
	h		Jan 15	3^h	406288
Jan 28	10	356741	Feb 11	3	406397
Feb 25	22	357966	Mar 10	10	405898
Mar 26	6	361990	Apr 7	3	404926
Apr 22	22	367213	May 4	22	404189
May 18	9	369741	Jun 1	17	404251
Jun 13	16	365879	Jun 29	11	405059
Jul 11	12	361041	Jul 27	0	406019
Aug 8	19	357780	Aug 23	7	406488
Sep 6	5	357090	Sep 19	10	406265
Oct 4	15	359289	Oct 16	20	405482
Nov 1	20	363956	Nov 13	14	404595
Nov 28	24	369281	Dec 11	11	404254
Dec 23	16	368802			

1980

Perigee		km	Apogee		km
	h		Jan 8	8^h	404720
Jan 20	2	363253	Feb 5	2	405603
Feb 17	9	358700	Mar 3	11	406242
Mar 16	20	356927	Mar 30	12	406334
Apr 14	7	358288	Apr 26	20	405784
May 12	13	362197	May 24	11	404831
Jun 9	4	367186	Jun 21	6	404183
Jul 4	16	369614	Jul 19	0	404350
Jul 30	23	365834	Aug 15	18	405244
Aug 27	19	360873	Sep 12	9	406236
Sep 25	3	357438	Oct 9	15	406665
Oct 23	14	356757	Nov 5	17	406421
Nov 21	1	359257	Dec 3	4	405639
Dec 19	5	364339	Dec 30	23	404748

1981

Perigee		km	Apogee		km
Jan 15	4^h	369640	Jan 27	20^h	404404
Feb 8	23	368196	Feb 24	17	404856
Mar 8	12	362695	Mar 24	9	405713
Apr 5	19	358496	Apr 20	16	406337
May 4	5	357016	May 17	18	406387
Jun 1	14	358475	Jun 14	3	405779
Jun 29	19	362333	Jul 11	18	404830
Jul 27	9	367269	Aug 8	11	404227
Aug 21	21	369658	Sep 5	7	404447
Sep 17	4	365637	Oct 3	1	405359
Oct 15	2	360480	Oct 30	16	406293
Nov 12	11	357099	Nov 26	21	406636
Dec 11	0	356784	Dec 23	23	406362

Perigee (1982–1984)

Year	Date	h	km
1982	Jan 8	12^h	359759
	Feb 5	14	365058
	Mar 4	5	369884
	Mar 29	6	367760
	Apr 25	21	362575
	May 24	3	358706
	Jun 21	12	357328
	Jul 19	21	358703
	Aug 17	2	362481
	Sep 13	18	367511
	Oct 9	1	369906
	Nov 4	10	365466
	Dec 2	11	360166
	Dec 30	22	356958
1983	Jan 28	11	356983
	Feb 25	22	360199
	Mar 25	22	365417
	Apr 21	8	369747
	May 16	16	367518
	Jun 13	6	362550
	Jul 11	10	358686
	Aug 8	19	357175
	Sep 6	5	358487
	Oct 4	11	362429
	Nov 1	3	367775
	Nov 26	2	369863
	Dec 22	18	364853
1984	Jan 19	22	359558
	Feb 17	9	356711
	Mar 16	21	357150
	Apr 14	6	360541
	May 12	3	365604
	Jun 7	11	369568
	Jul 2	23	367298
	Jul 30	12	362358
	Aug 27	17	358442
	Sep 25	3	356962
	Oct 23	14	358506
	Nov 20	21	362825
	Dec 18	10	368390

Apogee (1982–1984)

Year	Date	h	km
1982	Jan 20	12^h	405512
	Feb 17	8	404557
	Mar 17	5	404202
	Apr 14	0	404667
	May 11	15	405537
	Jun 7	23	406158
	Jul 5	1	406173
	Aug 1	10	405555
	Aug 29	0	404651
	Sep 25	19	404124
	Oct 23	15	404422
	Nov 20	11	405367
	Dec 18	2	406258
1983	Jan 14	5^h	406560
	Feb 10	8	406263
	Mar 9	23	405370
	Apr 6	18	404438
	May 4	13	404164
	Jun 1	8	404717
	Jun 28	23	405650
	Jul 26	7	406286
	Aug 22	9	406276
	Sep 18	17	405670
	Oct 16	8	404814
	Nov 13	3	404350
	Dec 11	1	404691
1984	Jan 7	20^h	405613
	Feb 4	9	406435
	Mar 2	11	406711
	Mar 29	16	406345
	Apr 26	7	405382
	May 24	1	404457
	Jun 20	20	404232
	Jul 18	14	404831
	Aug 15	5	405778
	Sep 11	13	406373
	Oct 8	15	406314
	Nov 4	23	405691
	Dec 2	15	404816
	Dec 30	12	404331

Perigee (1985–1987)

Year	Date	h	km
1985	Jan 12	3^h	369590
	Feb 8	4	364272
	Mar 8	8	359354
	Apr 5	18	356972
	May 4	5	357622
	Jun 1	13	360917
	Jun 29	9	365768
	Jul 25	18	369644
	Aug 20	4	367359
	Sep 16	19	362298
	Oct 15	1	358282
	Nov 12	13	356870
	Dec 11	1	358682
1986	Jan 8	7^h	363304
	Feb 4	16	368820
	Mar 1	10	369173
	Mar 28	14	363957
	Apr 25	18	359352
	May 24	3	357096
	Jun 21	13	357668
	Jul 19	20	360847
	Aug 16	17	365720
	Sep 12	0	369753
	Oct 7	10	367199
	Nov 4	2	361815
	Dec 2	11	357736
	Dec 30	23	356615
1987	Jan 28	11	358894
	Feb 25	16	363778
	Mar 24	19	368977
	Apr 18	17	368642
	May 15	23	363624
	Jun 13	1	359221
	Jul 11	10	357042
	Aug 8	19	357643
	Sep 6	3	360926
	Oct 4	1	366025
	Oct 30	3	370090
	Nov 24	15	366810
	Dec 22	11	361259

Apogee (1985–1987)

Year	Date	h	km
1985	Jan 27	10^h	404640
	Feb 24	4	405508
	Mar 23	15	406291
	Apr 19	17	406539
	May 16	24	406103
	Jun 13	14	405125
	Jul 11	8	404257
	Aug 8	2	404116
	Sep 4	21	404792
	Oct 2	13	405763
	Oct 29	22	406312
	Nov 25	22	406227
	Dec 23	7	405601
1986	Jan 20	1^h	404718
	Feb 16	22	404256
	Mar 16	19	404610
	Apr 13	12	405519
	May 10	23	406328
	Jun 7	2	406562
	Jul 4	8	406103
	Jul 31	21	405165
	Aug 28	15	404380
	Sep 25	10	404333
	Oct 23	6	405073
	Nov 19	22	406028
	Dec 17	5	406507
1987	Jan 13	5^h	406400
	Feb 9	16	405715
	Mar 9	10	404778
	Apr 6	7	404318
	May 4	2	404699
	May 31	18	405626
	Jun 28	4	406428
	Jul 25	8	406620
	Aug 21	14	406126
	Sep 18	3	405188
	Oct 15	21	404425
	Nov 12	18	404400
	Dec 10	14	405117

Pages 7-8 to 7-12 contain, for the years 1976 to 2005, the times of perigee and apogee of the Moon, rounded to the nearest integer hour of *Universal Time*, together with the distance d (in kilometers) between the centers of Earth and Moon at these instants.

The corresponding value of the Moon's equatorial horizontal parallax π can be found from the formula $\sin \pi = 6378.14/d$, while the Moon's geocentric semidiameter, expressed in seconds of arc, is equal to $358\,482\,800/d$.

Perigee / Apogee 1988

Perigee 1988	h	km	Apogee 1988	h	km
Jan 19	21	357509	Jan 7	6	405981
Feb 17	10	356914	Feb 3	10	406395
Mar 16	20	359517	Mar 1	12	406250
Apr 13	23	364311	Mar 29	0	405472
May 10	22	369065	Apr 25	19	404509
Jun 4	24	368483	May 23	14	404096
Jul 2	6	363667	Jun 20	8	404541
Jul 30	8	359331	Jul 18	0	405514
Aug 27	17	357101	Aug 14	12	406319
Sep 25	4	357683	Sep 10	15	406475
Oct 23	12	361111	Oct 7	20	405977
Nov 20	10	366482	Nov 4	11	405071
Dec 16	4	370354	Dec 2	6	404355
			Dec 30	4	404375

Perigee / Apogee 1989

Perigee 1989	h	km	Apogee 1989	h	km
Jan 10	23	366382	Jan 27	0	405100
Feb 7	22	360935	Feb 23	14	405944
Mar 8	8	357540	Mar 22	18	406353
Apr 5	20	357191	Apr 18	21	406166
May 4	5	359785	May 16	9	405365
Jun 1	5	364388	Jun 13	2	404461
Jun 28	4	368959	Jul 10	21	404150
Jul 23	7	368431	Aug 7	15	404696
Aug 19	12	363570	Sep 4	8	405733
Sep 16	15	359045	Oct 1	20	406528
Oct 15	1	356713	Oct 28	22	406638
Nov 12	13	357468	Nov 25	4	406135
Dec 10	23	361309	Dec 22	19	405219

Perigee / Apogee 1990

Perigee 1990	h	km	Apogee 1990	h	km
Jan 7	19	366976	Jan 19	16	404493
Feb 2	3	370178	Feb 16	13	404507
Feb 28	8	365734	Mar 16	8	405214
Mar 28	8	360557	Apr 12	20	406039
Apr 25	17	357502	May 10	0	406430
May 24	3	357339	Jun 6	4	406184
Jun 21	11	359942	Jul 3	16	405354
Jul 19	11	364492	Jul 31	8	404479
Aug 15	10	369075	Aug 28	3	404218
Sep 9	11	368381	Sep 24	22	404804
Oct 6	18	363245	Oct 22	16	405823
Nov 3	23	358625	Nov 19	3	406529
Dec 2	11	356525	Dec 16	4	406584
Dec 30	24	357751			

Perigee / Apogee 1991

Perigee 1991	h	km	Apogee 1991	h	km
Jan 28	9	361980	Jan 12	11	406037
Feb 25	1	367580	Feb 9	4	405041
Mar 22	5	369899	Mar 9	1	404280
Apr 17	17	365440	Apr 5	21	404305
May 15	17	360627	May 3	15	405034
Jun 13	0	357779	May 31	3	405871
Jul 11	10	357599	Jun 27	7	406243
Aug 8	18	360099	Jul 24	11	405965
Sep 5	19	364661	Aug 20	23	405155
Oct 2	18	369418	Sep 17	15	404341
Oct 27	16	368397	Oct 15	11	404161
Nov 24	2	362941	Nov 12	8	404809
Dec 22	9	358353	Dec 10	2	405822

Perigee / Apogee 1992

Perigee 1992	h	km	Apogee 1992	h	km
Jan 19	22	356549	Jan 6	12	406470
Feb 17	11	358096	Feb 2	12	406508
Mar 16	18	362410	Feb 29	21	405916
Apr 13	7	367731	Mar 28	14	404904
May 8	12	369586	Apr 25	10	404202
Jun 4	2	365329	May 23	5	404320
Jul 2	1	360619	Jun 19	22	405130
Jul 30	8	357673	Jul 17	11	406012
Aug 27	18	357371	Aug 13	16	406372
Sep 25	3	359928	Sep 9	19	406079
Oct 23	5	364778	Oct 7	6	405298
Nov 19	0	369741	Nov 3	23	404538
Dec 13	21	367971	Dec 1	20	404411
			Dec 29	17	405070

Perigee / Apogee 1993

Perigee 1993	h	km	Apogee 1993	h	km
Jan 10	12	362264	Jan 26	10	406028
Feb 7	20	357905	Feb 22	18	406627
Mar 8	9	356528	Mar 21	19	406631
Apr 5	19	358381	Apr 18	5	405951
May 4	0	362697	May 15	22	404908
May 31	11	367759	Jun 12	16	404241
Jun 25	17	369357	Jul 10	11	404412
Jul 22	8	365146	Aug 7	4	405257
Aug 19	7	360388	Sep 3	17	406124
Sep 16	15	357407	Sep 30	21	406425
Oct 15	2	357243	Oct 27	24	406102
Nov 12	12	360142	Nov 24	13	405302
Dec 10	14	365357	Dec 22	8	404515

During the period 1976-2005 considered here, the smallest and the largest distances between the centers of Earth and Moon are : 356 525 km on 1990 December 2, and 406 711 km on 1984 March 2.

On the other hand, during the longer period of time 1750 to 2125, the extreme distances are 356 375 km on 1912 January 4, and 406 720 km on 2125 February 3 (see *Sky and Telescope*, Vol. 62, No. 2, pages 110-111 ; August 1981).

Perigee

1994			
Jan 6	1^h	370142	km
Jan 31	4	367408	
Feb 27	22	361845	
Mar 28	6	357958	
Apr 25	17	356928	
May 24	3	358816	
Jun 21	7	362954	
Jul 18	18	367865	
Aug 12	23	369464	
Sep 8	14	365148	
Oct 6	14	360244	
Nov 3	24	357240	
Dec 2	12	357264	
Dec 30	23	360488	

1995	h	km
Jan 27	23	365886
Feb 23	2	370181
Mar 20	13	366972
Apr 17	8	361701
May 15	15	358040
Jun 13	1	357009
Jul 11	10	358774
Aug 8	14	362863
Sep 5	1	367912
Sep 30	4	369505
Oct 26	21	364790
Nov 23	23	359669
Dec 22	10	356798

1996	h	km
Jan 19	23	357251
Feb 17	9	360883
Mar 16	6	366289
Apr 11	3	369909
May 6	22	366531
Jun 3	16	361491
Jul 1	22	357949
Jul 30	8	356948
Aug 27	17	358777
Sep 24	22	363053
Oct 22	9	368349
Nov 16	5	369459
Dec 13	4	364237

Apogee

1994			
Jan 19	5^h	404361	km
Feb 16	2	404980	
Mar 15	17	405891	
Apr 11	24	406468	
May 9	2	406423	
Jun 5	13	405693	
Jul 3	5	404676	
Jul 30	23	404086	
Aug 27	18	404342	
Sep 24	12	405245	
Oct 22	2	406098	
Nov 18	5	406346	
Dec 15	8	406018	

1995			
Jan 11	22^h	405204	km
Feb 8	18	404418	
Mar 8	15	404305	
Apr 5	10	404974	
May 3	1	405925	
May 30	8	406515	
Jun 26	11	406441	
Jul 23	20	405717	
Aug 20	12	404763	
Sep 17	6	404262	
Oct 15	2	404602	
Nov 11	21	405531	
Dec 9	10	406324	

1996			
Jan 5	12^h	406524	km
Feb 1	16	406163	
Feb 29	7	405274	
Mar 28	3	404463	
Apr 24	22	404374	
May 22	16	405071	
Jun 19	6	406032	
Jul 16	14	406597	
Aug 12	16	406477	
Sep 9	2	405735	
Oct 6	18	404791	
Nov 3	14	404311	
Dec 1	11	404654	
Dec 29	5	405522	

Perigee

1997			
Jan 10	9^h	359233	km
Feb 7	21	356847	
Mar 8	9	357758	
Apr 5	17	361498	
May 3	11	366625	
May 29	7	369788	
Jun 24	5	366493	
Jul 21	23	361581	
Aug 19	5	358019	
Sep 16	15	356966	
Oct 15	2	358860	
Nov 12	8	363380	
Dec 9	17	368873	

1998			
Jan 3	9^h	369239	km
Jan 30	14	363800	
Feb 27	20	359087	
Mar 28	7	357025	
Apr 25	18	358041	
May 24	0	361659	
Jun 20	17	366593	
Jul 16	14	369700	
Aug 11	12	366446	
Sep 8	6	361374	
Oct 6	13	357633	
Nov 4	1	356614	
Dec 2	12	358842	
Dec 30	18	363783	

1999	h	km
Jan 26	21	369257
Feb 20	15	368653
Mar 20	0	363267
Apr 17	5	358894
May 15	15	357093
Jun 13	1	358191
Jul 11	6	361776
Aug 7	24	366705
Sep 2	18	369819
Sep 28	17	366251
Oct 26	13	360949
Nov 23	22	357277
Dec 22	11	356653

Apogee

1997			
Jan 25	17^h	406224	km
Feb 21	17	406395	
Mar 20	24	405958	
Apr 17	15	405003	
May 15	10	404210	
Jun 12	5	404185	
Jul 9	23	404944	
Aug 6	14	405936	
Sep 2	21	406479	
Sep 29	23	406330	
Oct 27	9	405603	
Nov 24	2	404695	
Dec 21	23	404261	

1998			
Jan 18	21^h	404635	km
Feb 15	15	405496	
Mar 15	1	406188	
Apr 11	2	406347	
May 8	9	405861	
Jun 4	24	404926	
Jul 2	17	404220	
Jul 30	12	404300	
Aug 27	6	405148	
Sep 23	22	406169	
Oct 21	5	406669	
Nov 17	6	406493	
Dec 14	17	405757	

1999			
Jan 11	12^h	404829	km
Feb 8	9	404387	
Mar 8	5	404751	
Apr 4	22	405595	
May 2	6	406277	
May 29	8	406398	
Jun 25	16	405859	
Jul 23	6	404925	
Aug 19	23	404262	
Sep 16	19	404391	
Oct 14	14	405255	
Nov 11	6	406216	
Dec 8	11	406624	

PERIGEE AND APOGEE OF THE MOON

Perigee			Apogee			Perigee			Apogee		
2000	h	km	**2000**			**2003**	h	km	**2003**		
			Jan 4	12^h	406418 km				Jan 11	1^h	404343 km
Jan 19	23	359361	Feb 1	1	405608	Jan 23	22	369898	Feb 7	22	404551
Feb 17	3	364494	Feb 28	21	404615	Feb 19	16	364845	Mar 7	17	405382
Mar 14	24	369533	Mar 27	17	404167	Mar 19	19	359816	Apr 4	4	406209
Apr 8	22	368257	Apr 24	12	404558	Apr 17	5	357156	May 1	8	406529
May 6	9	363168	May 22	4	405428	May 15	16	357449	May 28	13	406168
Jun 3	13	359089	Jun 18	13	406112	Jun 12	23	360425	Jun 25	2	405232
Jul 1	22	357362	Jul 15	16	406200	Jul 10	22	365144	Jul 22	20	404328
Jul 30	8	358375	Aug 11	22	405652	Aug 6	14	369433	Aug 19	14	404102
Aug 27	14	361906	Sep 8	13	404761	Aug 31	19	367925	Sep 16	9	404713
Sep 24	8	366961	Oct 6	7	404170	Sep 28	6	362835	Oct 14	2	405692
Oct 19	22	370115	Nov 3	3	404377	Oct 26	12	358547	Nov 10	12	406300
Nov 14	23	366047	Nov 30	24	405273	Nov 23	23	356811	Dec 7	12	406278
Dec 12	22	360602	Dec 28	15	406193	Dec 22	12	358338			
2001	h	km	**2001**		h	**2004**	h	km	**2004**		
									Jan 3	20^h	405707 km
Jan 10	9	357130	Jan 24	19	406562	Jan 19	19	362770	Jan 31	14	404807
Feb 7	22	356852	Feb 20	22	406331	Feb 16	8	368322	Feb 28	11	404258
Mar 8	9	359775	Mar 20	11	405474	Mar 12	4	369506	Mar 27	7	404521
Apr 5	10	364809	Apr 17	6	404505	Apr 8	2	364547	Apr 24	0	405403
May 2	4	369419	May 15	1	404144	May 6	5	359811	May 21	12	406264
May 27	7	368032	Jun 11	20	404629	Jun 3	13	357246	Jun 17	16	406574
Jun 23	17	363132	Jul 9	11	405566	Jul 1	23	357448	Jul 14	21	406192
Jul 21	21	359027	Aug 5	21	406268	Jul 30	6	360323	Aug 11	10	405291
Aug 19	6	357157	Sep 1	23	406330	Aug 27	6	365105	Sep 8	3	404464
Sep 16	16	358128	Sep 29	6	405787	Sep 22	21	369589	Oct 5	22	404326
Oct 14	23	361860	Oct 26	20	404935	Oct 17	24	367758	Nov 2	18	404997
Nov 11	17	367256	Nov 23	16	404394	Nov 14	14	362311	Nov 30	11	405953
Dec 6	23	370117	Dec 21	13	404633	Dec 12	21	357983	Dec 27	19	406488
2002	7^h	365406 km	**2002**	9^h	405505 km	**2005**	10^h	356570 km	**2005**	19^h	406444 km
Jan 2			Jan 18			Jan 10			Jan 23		
Jan 30	9	359995	Feb 14	22	406363	Feb 7	22	358564	Feb 20	5	405805
Feb 27	20	356897	Mar 14	1	406707	Mar 8	4	363232	Mar 19	23	404846
Mar 28	8	357010	Apr 10	5	406407	Apr 4	11	368492	Apr 16	19	404304
Apr 25	16	360084	May 7	19	405482	Apr 29	10	369029	May 14	14	404600
May 23	16	364983	Jun 4	13	404521	May 26	11	364241	Jun 11	6	405506
Jun 19	7	369308	Jul 2	8	404210	Jun 23	12	359672	Jul 8	18	406363
Jul 14	13	367847	Jul 30	2	404743	Jul 21	20	357158	Aug 4	22	406631
Aug 10	23	362927	Aug 26	18	405694	Aug 19	6	357393	Sep 1	3	406212
Sep 8	3	358746	Sep 23	3	406352	Sep 16	14	360405	Sep 28	15	405307
Oct 6	13	356918	Oct 20	5	406360	Oct 14	14	365449	Oct 26	10	404494
Nov 4	1	358153	Nov 16	11	405797	Nov 10	0	370009	Nov 23	6	404370
Dec 2	9	362289	Dec 14	4	404914	Dec 5	5	367364	Dec 21	3	405013
Dec 30	1	367902									

V

TABLE FOR CALCULATING THE ILLUMINATED FRACTION
OF THE MOON'S DISK

The illuminated fraction of the Moon's disk, as seen from the Earth, for any instant between the years −900 and +2999, can be found by means of the tables on pages 7-13 to 7-17.

Firstly, express the instant in Ephemeris Time. (Between the years +1000 and +3000, the difference between UT and ET can be ignored for our purpose).

From the tables a to e, take the values of the quantities A, B, C and D corresponding to the century year, the additional year, the month, the day of the month, and the hour.

In table a, two values are given for the year 1500, the first one for the Julian calendar (J), and the second one for the Gregorian calendar (G).

In the case of a negative year, the year number should be split up in such a way as to have the last two figures positive. For example, the year −328 should be split up as −400 and +72. Then enter table a with −400, and table b with +72.

In table c the months marked with (B) should be used in the case of a bissextile (leap) year.

Add up the numbers in order to obtain the values of the arguments A, B, C, D. If an argument is larger than 1000, subtract 1000 or a convenient multiple.

With the arguments A, B, C, D, take from tables f to i the corrections I, II, III and IV, respectively. Then calculate

$$d = D + \text{corrections I to IV}$$

If d is larger than 1000, subtract 1000 or a convenient multiple. If d is negative, then add 1000.

Then the critical table k gives the illuminated fraction k of the Moon's disk as a function of d. For instance, if $d = 134$ then $k = 0.17$, because this value is situated between the values $d = 132$ and $d = 137$.

When d equals one of the tabular or "critical" values, the appropriate value of k is the upper of the two values. For example, if $d = 171$ then k is to be taken as 0.26. This rule is given as "In critical cases ascend", mentioned under the table.

The resulting value of k will never be more than 0.01 in error.

If d is between 0 and 500, then the Moon is waxing (phase increasing). If d is between 500 and 1000, the Moon is waning (phase decreasing).

Table a : Century Year

Year	A	B	C	D
−900	110	829	187	508
−800	107	380	343	361
−700	105	931	498	214
−600	102	482	653	68
−500	99	33	809	921
−400	97	584	964	774
−300	94	136	119	628
−200	91	687	275	481
−100	89	238	430	334
0	86	790	585	187
+100	83	341	740	41
200	81	893	895	894
300	78	444	50	747
400	76	996	205	600
500	73	548	359	453
600	70	99	514	307
700	68	651	669	160
800	65	203	823	13
900	62	755	978	866
1000	60	307	132	719
1100	57	859	286	573
1200	54	411	441	426
1300	52	963	595	279
1400	49	515	749	132
1500J	47	67	903	985
1500G	19	704	589	646
1600	17	256	743	500
1700	11	772	865	319
1800	6	288	988	138
1900	0	804	110	957
2000	998	357	264	810
2100	992	873	386	630
2200	987	389	509	449
2300	982	905	631	268
2400	979	458	784	121
2500	974	974	906	940
2600	968	491	28	759
2700	963	7	150	579
2800	960	560	303	432
2900	955	77	425	251

Table *b* :   Additional Year

Year	A	B	C	D	Year	A	B	C	D	Year	A	B	C	D
0	5	0	0	0	35	2	916	830	873	70	2	869	692	780
1	4	246	474	360	36	4	199	335	267	71	1	115	166	140
2	4	493	947	720	37	3	445	809	627	72	3	398	671	534
3	3	739	421	80	38	3	692	283	987	73	2	644	144	894
4	5	22	926	474	39	2	938	756	347	74	2	891	618	254
5	4	269	400	834	40	4	221	262	741	75	1	137	92	614
6	3	515	874	194	41	3	467	735	101	76	3	420	597	8
7	3	761	347	554	42	3	714	209	461	77	2	666	71	368
8	5	44	852	948	43	2	960	683	821	78	2	913	544	728
9	4	291	326	308	44	4	243	188	215	79	1	159	18	89
10	3	537	800	668	45	3	489	661	575	80	3	442	523	482
11	3	784	273	28	46	2	736	135	935	81	2	688	997	843
12	5	66	778	422	47	2	982	609	296	82	1	935	470	203
13	4	313	252	782	48	4	265	114	689	83	1	181	944	563
14	3	559	726	143	49	3	512	588	50	84	3	464	449	957
15	3	806	199	503	50	2	758	61	410	85	2	710	923	317
16	5	88	705	896	51	2	4	535	770	86	1	957	397	677
17	4	335	178	257	52	4	287	40	164	87	1	203	870	37
18	3	581	652	617	53	3	534	514	524	88	3	486	375	431
19	2	828	126	977	54	2	780	987	884	89	2	733	849	791
20	4	110	631	371	55	1	27	461	244	90	1	979	323	151
21	4	357	104	731	56	4	309	966	638	91	1	225	796	511
22	3	603	578	91	57	3	556	440	998	92	3	508	302	905
23	2	850	52	451	58	2	802	914	358	93	2	755	775	265
24	4	133	557	845	59	1	49	387	718	94	1	1	249	625
25	4	379	31	205	60	3	331	892	112	95	0	248	723	985
26	3	625	504	565	61	3	578	366	472	96	2	530	228	379
27	2	872	978	925	62	2	824	840	832	97	2	777	701	739
28	4	155	483	319	63	1	71	313	192	98	1	23	175	99
29	4	401	957	679	64	3	354	818	586	99	0	270	649	459
30	3	648	430	39	65	3	600	292	946					
31	2	894	904	399	66	2	846	766	306					
32	4	177	409	793	67	1	93	239	666					
33	3	423	883	153	68	3	376	745	60					
34	3	670	357	513	69	2	622	218	420					

Table *c* : Month

Month	A	B	C	D
Jan.	989	0	0	0
Jan. *(B)*	986	964	969	966
Feb.	74	125	974	50
Feb. *(B)*	71	89	943	16
March	151	141	855	998
April	235	266	829	48
May	318	355	772	64
June	402	480	747	113
July	485	569	690	129
August	569	694	664	179
Sept.	654	819	639	229
Oct.	736	908	582	245
Nov.	821	33	556	294
Dec.	903	121	499	310

Table *d* : Day of month

Day	A	B	C	D
1	3	36	31	34
2	5	73	63	68
3	8	109	94	102
4	11	145	126	135
5	14	181	157	169
6	16	218	189	203
7	19	254	220	237
8	22	290	251	271
9	25	327	283	305
10	27	363	314	339
11	30	399	346	372
12	33	435	377	406
13	36	472	409	440
14	38	508	440	474
15	41	544	472	508
16	44	581	503	542
17	47	617	534	576
18	49	653	566	610
19	52	690	597	643
20	55	726	629	677
21	57	762	660	711
22	60	798	692	745
23	63	835	723	779
24	66	871	754	813
25	68	907	786	847
26	71	944	817	880
27	74	980	849	914
28	77	16	880	948
29	79	52	912	982
30	82	89	943	16
31	85	125	974	50

Table *e* : Hour

Hour	A	B	C	D
0	0	0	0	0
1	0	2	1	1
2	0	3	3	3
3	0	5	4	4
4	0	6	5	6
5	1	8	7	7
6	1	9	8	8
7	1	11	9	10
8	1	12	10	11
9	1	14	12	13
10	1	15	13	14
11	1	17	14	16
12	1	18	16	17
13	1	20	17	18
14	2	21	18	20
15	2	23	20	21
16	2	24	21	23
17	2	26	22	24
18	2	27	24	25
19	2	29	25	27
20	2	30	26	28
21	2	32	28	30
22	3	33	29	31
23	3	35	30	32
24	3	36	31	34

EXAMPLE. — Find the illuminated fraction of the Moon's disk on 1982 November 7 at 20 h Universal Time (= 20 h ET).

		A	B	C	D
Table *a*	1900	0	804	110	957
Table *b*	82	1	935	470	203
Table *c*	Nov.	821	33	556	294
Table *d*	7	19	254	220	237
Table *e*	20 h	2	30	26	28
Sum		843	2056	1382	1719
			= 56	=382	=719

			D	719
Table *f*	A = 834		I	= +5
Table *g*	B = 56		II	= +7
Table *h*	C = 382		III	= +2
Table *i*	D = 719		IV	= +1
Sum			*d* = 734	

whence *k* = 0.55, waning Moon

Table f			Table g			Table h			Table i	
A	I		B	II		C	III		D	IV
0	0		0	0		0	0		0	0
20	−1		20	+ 2		20	+0		20	+0
40	−1		40	+ 5		40	+1		40	+1
60	−2		60	+ 7		60	+1		60	+1
80	−3		80	+ 9		80	+2		80	+2
100	−3		100	+11		100	+2		100	+2
120	−4		120	+13		120	+2		120	+2
140	−5		140	+14		140	+3		140	+2
160	−5		160	+15		160	+3		160	+2
180	−5		180	+16		180	+3		180	+1
200	−6		200	+17		200	+3		200	+1
220	−6		220	+17		220	+3		220	+1
240	−6		240	+17		240	+4		240	+0
260	−6		260	+17		260	+4		260	−0
280	−6		280	+17		280	+3		280	−1
300	−6		300	+16		300	+3		300	−1
320	−5		320	+15		320	+3		320	−1
340	−5		340	+14		340	+3		340	−2
360	−4		360	+13		360	+3		360	−2
380	−4		380	+11		380	+2		380	−2
400	−3		400	+10		400	+2		400	−2
420	−3		420	+ 8		420	+2		420	−2
440	−2		440	+ 6		440	+1		440	−1
460	−1		460	+ 4		460	+1		460	−1
480	−1		480	+ 2		480	+0		480	−0
500	0		500	0		500	0		500	0
520	+1		520	− 2		520	−0		520	+0
540	+1		540	− 4		540	−1		540	+1
560	+2		560	− 6		560	−1		560	+1
580	+3		580	− 8		580	−2		580	+2
600	+3		600	−10		600	−2		600	+2
620	+4		620	−11		620	−2		620	+2
640	+4		640	−13		640	−3		640	+2
660	+5		660	−14		660	−3		660	+2
680	+5		680	−15		680	−3		680	+1
700	+6		700	−16		700	−3		700	+1
720	+6		720	−17		720	−3		720	+1
740	+6		740	−17		740	−4		740	+0
760	+6		760	−17		760	−4		760	−0
780	+6		780	−17		780	−3		780	−1
800	+6		800	−17		800	−3		800	−1
820	+5		820	−16		820	−3		820	−1
840	+5		840	−15		840	−3		840	−2
860	+5		860	−14		860	−3		860	−2
880	+4		880	−13		880	−2		880	−2
900	+3		900	−11		900	−2		900	−2
920	+3		920	− 9		920	−2		920	−2
940	+2		940	− 7		940	−1		940	−1
960	+1		960	− 5		960	−1		960	−1
980	+1		980	− 2		980	−0		980	−0
1000	0		1000	0		1000	0		1000	0

Table k

d	k	d
0		1000
	0.00	
22		978
	0.01	
38		962
	0.02	
50		950
	0.03	
59		941
	0.04	
67		933
	0.05	
75		925
	0.06	
81		919
	0.07	
88		912
	0.08	
93		907
	0.09	
99		901
	0.10	
104		896
	0.11	
109		891
	0.12	
114		886
	0.13	
119		881
	0.14	
124		876
	0.15	
128		872
	0.16	
132		868
	0.17	
137		863
	0.18	
141		859
	0.19	
145		855
	0.20	
149		851
	0.21	
153		847
	0.22	
156		844
	0.23	
160		840
	0.24	
164		836

d	k	d
164		836
	0.25	
168		832
	0.26	
171		829
	0.27	
175		825
	0.28	
178		822
	0.29	
182		818
	0.30	
185		815
	0.31	
189		811
	0.32	
192		808
	0.33	
196		804
	0.34	
199		801
	0.35	
202		798
	0.36	
206		794
	0.37	
209		791
	0.38	
212		788
	0.39	
215		785
	0.40	
219		781
	0.41	
222		778
	0.42	
225		775
	0.43	
228		772
	0.44	
232		768
	0.45	
235		765
	0.46	
238		762
	0.47	
241		759
	0.48	
244		756
	0.49	
248		752

d	k	d
248		752
	0.50	
251		749
	0.51	
254		746
	0.52	
257		743
	0.53	
260		740
	0.54	
263		737
	0.55	
267		733
	0.56	
270		730
	0.57	
273		727
	0.58	
276		724
	0.59	
280		720
	0.60	
283		717
	0.61	
286		714
	0.62	
289		711
	0.63	
293		707
	0.64	
296		704
	0.65	
299		701
	0.66	
303		697
	0.67	
306		694
	0.68	
309		691
	0.69	
313		687
	0.70	
316		684
	0.71	
320		680
	0.72	
323		677
	0.73	
327		673
	0.74	
331		669

d	k	d
331		669
	0.75	
334		666
	0.76	
338		662
	0.77	
342		658
	0.78	
346		654
	0.79	
350		650
	0.80	
354		646
	0.81	
358		642
	0.82	
362		638
	0.83	
366		634
	0.84	
370		630
	0.85	
375		625
	0.86	
379		621
	0.87	
384		616
	0.88	
389		611
	0.89	
394		606
	0.90	
400		600
	0.91	
405		595
	0.92	
411		589
	0.93	
417		583
	0.94	
424		576
	0.95	
431		569
	0.96	
439		561
	0.97	
449		551
	0.98	
460		540
	0.99	
477		523
	1.00	
500		500

In critical cases ascend

VI

TABLE FOR COMPUTING
THE SELENOGRAPHIC COLONGITUDE OF THE SUN

Table a : Century Year

Year	Colong.	A
1600	89.86	17
1700	24.80	11
1800	319.72	6
1900	254.64	0
2000	201.76	998
2100	136.67	992
2200	71.58	987
2300	6.50	982

Table b : Additional Year

Year	Colong.	A	Year	Colong.	A	Year	Colong.	A
0	0.00	5	35	314.35	2	70	280.88	2
1	129.62	4	36	96.16	4	71	50.51	1
2	259.25	4	37	225.78	3	72	192.32	3
3	28.87	3	38	355.41	3	73	321.95	2
4	170.68	5	39	125.03	2	74	91.57	2
5	300.31	4	40	266.85	4	75	221.19	1
6	69.93	3	41	36.47	3	76	3.01	3
7	199.55	3	42	166.09	3	77	132.63	2
8	341.37	5	43	295.72	2	78	262.25	2
9	110.99	4	44	77.53	4	79	31.88	1
10	240.62	3	45	207.15	3	80	173.69	3
11	10.24	3	46	336.78	2	81	303.31	2
12	152.05	5	47	106.40	2	82	72.94	1
13	281.68	4	48	248.21	4	83	202.56	1
14	51.30	3	49	17.84	3	84	344.38	3
15	180.92	3	50	147.46	2	85	114.00	2
16	322.74	5	51	277.09	2	86	243.62	1
17	92.36	4	52	58.90	4	87	13.25	1
18	221.99	3	53	188.52	3	88	155.06	3
19	351.61	2	54	318.15	2	89	284.68	2
20	133.42	4	55	87.77	1	90	54.31	1
21	263.05	4	56	229.58	4	91	183.93	1
22	32.67	3	57	359.21	3	92	325.75	3
23	162.29	2	58	128.83	2	93	95.37	2
24	304.11	4	59	258.45	1	94	224.99	1
25	73.73	4	60	40.27	3	95	354.62	0
26	203.35	3	61	169.89	3	96	136.43	2
27	332.98	2	62	299.52	2	97	266.05	2
28	114.79	4	63	69.14	1	98	35.68	1
29	244.42	4	64	210.95	3	99	165.30	0
30	14.04	3	65	340.58	3			
31	143.66	2	66	110.20	2			
32	285.48	4	67	239.82	1			
33	55.10	3	68	21.64	3			
34	184.72	3	69	151.26	2			

The selenographic colongitude of the Sun is equal to 90° minus the selenographic longitude of the Sun. Its value and that of the selenographic latitude of the Sun are given each year in the great astronomical almanacs.

The selenographic latitude of the Sun never exceeds 1°.6 north or south, and may be neglected in a first approximation when the position of the terminator on the Moon is required.

The selenographic colongitude of the Sun may be calculated with a good approximation with the aid of the tables on pages 7-18 to 7-20. The result will rarely be more than 0.05 degree in error, and never more than 0.20 degree.

From the tables a to e, take the values of the Colong. and of the quantity A for the century year, the additional year, the month, the day of the month, and the hour (*Universal Time !*). Add the values thus found, in order to obtain the Colong. and the argument A for the given instant. If A exceeds 1000, subtract 1000 or a convenient multiple. If Colong. exceeds 360°, subtract 360°.

Table c : Month

Month	Colong.	A
Jan.	0.00°	989
Jan.(B)	347.81	986
Feb.	17.91	74
Feb.(B)	5.72	71
March	359.25	151
April	17.17	235
May	22.89	318
June	40.80	402
July	46.53	485
Aug.	64.44	569
Sept.	82.35	654
Oct.	88.07	736
Nov.	105.99	821
Dec.	111.71	903

Use the months marked by *(B)* for the bissextile (leap) years. All years whose millesime is divisible by 4 are bissextile years, except 1700, 1800; 1900, 2100, 2200, and 2300.

Table d : Day of month

Day	Colong.	A
1	12.19°	3
2	24.38	5
3	36.57	8
4	48.76	11
5	60.95	14
6	73.14	16
7	85.34	19
8	97.53	22
9	109.72	25
10	121.91	27
11	134.10	30
12	146.29	33
13	158.48	36
14	170.67	38
15	182.86	41
16	195.05	44
17	207.24	47
18	219.43	49
19	231.62	52
20	243.81	55
21	256.01	57
22	268.20	60
23	280.39	63
24	292.58	66
25	304.77	68
26	316.96	71
27	329.15	74
28	341.34	77
29	353.53	79
30	5.72	82
31	17.91	85

Table e : Hour

Hour	Colong.	A
0	0.00°	0
1	0.51	0
2	1.02	0
3	1.52	0
4	2.03	0
5	2.54	1
6	3.05	1
7	3.56	1
8	4.06	1
9	4.57	1
10	5.08	1
11	5.59	1
12	6.10	1
13	6.60	1
14	7.11	2
15	7.62	2
16	8.13	2
17	8.64	2
18	9.14	2
19	9.65	2
20	10.16	2
21	10.67	2
22	11.17	3
23	11.68	3
24	12.19	3

Table f

A	Corr.
0	0.00°
20	-0.25
40	-0.49
60	-0.72
80	-0.94
100	-1.15
120	-1.33
140	-1.50
160	-1.64
180	-1.75
200	-1.83
220	-1.89
240	-1.91
260	-1.91
280	-1.87
300	-1.81
320	-1.72
340	-1.60
360	-1.46
380	-1.29
400	-1.11
420	-0.91
440	-0.69
460	-0.47
480	-0.24
500	0.00
520	+0.24
540	+0.47
560	+0.69
580	+0.91
600	+1.11
620	+1.29
640	+1.46
660	+1.60
680	+1.72
700	+1.81
720	+1.87
740	+1.91
760	+1.91
780	+1.89
800	+1.83
820	+1.75
840	+1.64
860	+1.50
880	+1.33
900	+1.15
920	+0.94
940	+0.72
960	+0.49
980	+0.25
1000	0.00

With A as argument, take the correction by interpolation from table f, and add it to the Colong.

With the new value of the colongitude as argument, take the second correction from table g, and add it to the colongitude.

No interpolation is required for tables a to d. Table g is a so-called "critical table" which gives the value of the correction without interpolation.

Table g : Second Correction

Col.	Corr.	Col.	Col.	Corr.	Col.
°		°	°		°
0.0		360.0	90.0		270.0
	+0.15			0.00	
9.5		350.5	92.0		268.0
	+0.14			-0.01	
23.3		336.7	95.9		264.1
	+0.13			-0.02	
31.8		328.2	99.8		260.2
	+0.12			-0.03	
38.5		321.5	103.8		256.2
	+0.11			-0.04	
44.4		315.6	107.8		252.2
	+0.10			-0.05	
49.7		310.3	112.0		248.0
	+0.09			-0.06	
54.7		305.3	116.2		243.8
	+0.08			-0.07	
59.3		300.7	120.7		239.3
	+0.07			-0.08	
63.8		296.2	125.3		234.7
	+0.06			-0.09	
68.0		292.0	130.3		229.7
	+0.05			-0.10	
72.2		287.8	135.6		224.4
	+0.04			-0.11	
76.2		283.8	141.5		218.5
	+0.03			-0.12	
80.2		279.8	148.2		211.8
	+0.02			-0.13	
84.1		275.9	156.7		203.3
	+0.01			-0.14	
88.0		272.0	170.5		189.5
	0.00			-0.15	
90.0		270.0	180.0		180.0

As an example, let us compute the Sun's selenographic colongitude on 1963 November 22, at 17 h Universal Time.

We have :

		Colong.	A
		°	
Table a	1900	254.64	0
Table b	63	69.14	1
Table c	Nov.	105.99	821
Table d	22	268.20	60
Table e	17 h	8.64	2
Sum		706.°61	884
		= 346.°61	

With $A = 884$ as argument, we find in table f that the first correction is

$$+1.°29$$

Hence, the new value of the colongitude is

$$346.°61 + 1.°29 = 347.°90$$

With 347.°90 as argument, we find in table g that the second correction is

$$+0.°14$$

Our final result is then

colongitude = $347.°90 + 0.°14 = 348.°04$

The Sun's selenographic colongitude is *approximately* 270° at New Moon, 0° at First Quarter, 90° at Full Moon, and 180° at Last Quarter.

If on the Moon the selenographic longitudes λ are counted positively from the mean center of the lunar disk towards Mare Crisium, that is westward on the sky, then we have the following rules, c being the Sun's selenographic colongitude, and its latitude being neglected :

- for a place on the Moon whose selenographic longitude is λ, we have

sunrise when $c = 360° - \lambda$
noon when $c = 90° - \lambda$
sunset when $c = 180° - \lambda$

- if, at a given instant, the Sun's selenographic colongitude is c, then the morning (sunrise) terminator lies in longitude $\lambda = 360° - c$, while the evening (sunset) terminator is at longitude $\lambda = 180° - c$.

Tables VII to X list all inferior conjunctions of Venus, superior conjunctions of Venus, oppositions of Jupiter, and oppositions of Saturn, respectively, taking place during the period 0-2500.

The times are rounded to the nearest tenth of a day, Universal Time.

For instance, 1976 Jan 20.5 (in Table X) means that the opposition of Saturn with the Sun took place on 1976 January 20 at approximately 12 h UT.

Note that a date such as March 16.9 should n o t be rounded to March 17. Indeed, March 16.9 denotes an instant belonging to March 16, namely 0.9 day after March 16, 0 h UT ; consequently, the event takes place on March 16, not March 17.

VII

INFERIOR CONJUNCTIONS OF VENUS, 0 - 2500

1 Mar 30.8	60 May 23.1	119 Jul 16.8	178 Sep 9.7	237 Nov 5.5	297 Jan 1.7
2 Nov 3.0	61 Dec 30.2	121 Feb 23.7	180 Apr 18.5	239 Jun 11.8	298 Aug 4.8
4 Jun 8.3	63 Aug 2.3	122 Sep 26.8	181 Nov 23.0	241 Jan 18.8	300 Mar 14.6
6 Jan 16.3	65 Mar 12.2	124 May 4.7	183 Jun 28.1	242 Aug 21.4	301 Oct 16.6
7 Aug 19.0	66 Oct 14.1	125 Dec 10.4	185 Feb 4.7	244 Mar 31.0	303 May 24.3
9 Mar 28.5	68 May 20.8	127 Jul 14.4	186 Sep 7.3	245 Nov 3.0	304 Dec 30.2
10 Oct 31.5	69 Dec 27.8	129 Feb 21.3	188 Apr 16.2	247 Jun 9.5	306 Aug 2.4
12 Jun 6.0	71 Jul 30.9	130 Sep 24.4	189 Nov 20.5	249 Jan 16.3	308 Mar 12.3
14 Jan 13.9	73 Mar 9.8	132 May 2.4	191 Jun 25.7	250 Aug 19.0	309 Oct 14.1
15 Aug 16.6	74 Oct 11.6	133 Dec 7.9	193 Feb 2.3	252 Mar 28.6	311 May 22.0
17 Mar 26.2	76 May 18.5	135 Jul 12.1	194 Sep 4.9	253 Oct 31.5	312 Dec 27.7
18 Oct 29.0	77 Dec 25.3	137 Feb 19.0	196 Apr 13.9	255 Jun 7.2	314 Jul 31.0
20 Jun 3.7	79 Jul 28.5	138 Sep 21.9	197 Nov 18.0	257 Jan 13.9	316 Mar 9.9
22 Jan 11.5	81 Mar 7.5	140 Apr 30.0	199 Jun 23.4	258 Aug 16.7	317 Oct 11.6
23 Aug 14.2	82 Oct 9.2	141 Dec 5.5	201 Jan 30.9	260 Mar 26.3	319 May 19.7
25 Mar 23.8	84 May 16.2	143 Jul 9.7	202 Sep 2.5	261 Oct 29.0	320 Dec 25.3
26 Oct 26.5	85 Dec 22.8	145 Feb 16.6	204 Apr 11.5	263 Jun 4.9	322 Jul 28.7
28 Jun 1.4	87 Jul 26.2	146 Sep 19.5	205 Nov 15.5	265 Jan 11.5	324 Mar 7.6
30 Jan 9.0	89 Mar 5.1	148 Apr 27.7	207 Jun 21.1	266 Aug 14.3	325 Oct 9.1
31 Aug 11.8	90 Oct 6.7	149 Dec 3.0	209 Jan 28.4	268 Mar 24.0	327 May 17.4
33 Mar 21.5	92 May 13.9	151 Jul 5.1	210 Aug 31.0	269 Oct 26.5	328 Dec 22.8
34 Oct 24.0	93 Dec 20.4	153 Feb 14.2	212 Apr 9.2	271 Jun 2.6	330 Jul 26.3
36 May 30.1	95 Jul 23.8	154 Sep 17.0	213 Nov 13.0	273 Jan 9.0	332 Mar 5.2
38 Jan 6.6	97 Mar 2.8	156 Apr 25.4	215 Jun 18.8	274 Aug 11.9	333 Oct 6.7
39 Aug 9.4	98 Oct 4.2	157 Nov 30.5	217 Jan 26.0	276 Mar 21.6	335 May 15.1
41 Mar 19.2	100 May 11.6	159 Jul 5.1	218 Aug 28.6	277 Oct 24.0	336 Dec 20.3
42 Oct 21.6	101 Dec 17.9	161 Feb 11.8	220 Apr 6.9	279 May 31.2	338 Jul 23.9
44 May 27.8	103 Jul 21.5	162 Sep 14.6	221 Nov 10.5	281 Jan 6.6	340 Mar 2.9
46 Jan 4.1	105 Feb 28.4	164 Apr 23.1	223 Jun 16.4	282 Aug 9.5	341 Oct 4.2
47 Aug 7.0	106 Oct 1.7	165 Nov 28.0	225 Jan 23.6	284 Mar 19.3	343 May 12.8
49 Mar 16.8	108 May 9.3	167 Jul 2.7	226 Aug 26.2	285 Oct 21.5	344 Dec 17.9
50 Oct 19.1	109 Dec 15.4	169 Feb 9.4	228 Apr 4.6	287 May 28.9	346 Jul 21.6
52 May 25.5	111 Jul 19.1	170 Sep 12.2	229 Nov 8.0	289 Jan 4.1	348 Feb 29.5
54 Jan 1.7	113 Feb 26.1	172 Apr 20.8	231 Jun 14.1	290 Aug 7.1	349 Oct 1.8
55 Aug 4.7	114 Sep 29.3	173 Nov 25.5	233 Jan 21.2	292 Mar 17.0	351 May 10.4
57 Mar 14.5	116 May 7.0	175 Jun 30.4	234 Aug 23.8	293 Oct 19.1	352 Dec 15.4
58 Oct 16.6	117 Dec 12.9	177 Feb 7.0	236 Apr 2.3	295 May 26.6	354 Jul 19.3

INFERIOR CONJUNCTIONS OF VENUS

356 Feb 27.1	445 Sep 2.5	535 Mar 18.1	624 Sep 22.0	714 Apr 5.9	803 Oct 12.6
357 Sep 29.3	447 Apr 12.7	536 Oct 19.0	626 May 1.4	715 Nov 8.9	805 May 20.0
359 May 8.1	448 Nov 15.5	538 May 26.8	627 Dec 6.4	717 Jun 14.5	806 Dec 26.2
360 Dec 12.9	450 Jun 21.2	540 Jan 2.7	629 Jul 10.0	719 Jan 22.2	808 Jul 28.9
362 Jul 16.9	452 Jan 29.5	541 Aug 4.9	631 Feb 17.7	720 Aug 24.0	810 Mar 8.8
364 Feb 24.8	453 Aug 31.1	543 Mar 15.7	632 Sep 19.5	722 Apr 3.5	811 Oct 10.1
365 Sep 26.8	455 Apr 10.4	544 Oct 16.6	634 Apr 29.1	723 Nov 6.4	813 May 17.7
367 May 5.8	456 Nov 13.0	546 May 24.5	635 Dec 3.9	725 Jun 12.1	814 Dec 23.8
368 Dec 10.4	458 Jun 18.9	547 Dec 31.2	637 Jul 7.7	727 Jan 19.8	816 Jul 26.5
370 Jul 14.6	460 Jan 27.1	549 Aug 2.5	639 Feb 15.3	728 Aug 21.6	818 Mar 6.4
372 Feb 22.4	461 Aug 28.7	551 Mar 13.4	640 Sep 17.1	730 Apr 1.2	819 Oct 7.7
373 Sep 24.4	463 Apr 8.1	552 Oct 14.1	642 Apr 26.7	731 Nov 3.9	821 May 15.4
375 May 3.5	464 Nov 10.5	554 May 22.2	643 Dec 1.4	733 Jun 9.8	822 Dec 21.3
376 Dec 7.9	466 Jun 16.6	555 Dec 28.7	645 Jul 5.4	735 Jan 17.4	824 Jul 24.2
378 Jul 12.2	468 Jan 24.6	557 Jul 31.1	647 Feb 12.9	736 Aug 19.2	826 Mar 4.0
380 Feb 20.0	469 Aug 26.3	559 Mar 11.0	648 Sep 14.7	738 Mar 29.9	827 Oct 5.2
381 Sep 21.9	471 Apr 5.7	560 Oct 11.6	650 Apr 24.4	739 Nov 1.5	829 May 13.1
383 May 1.2	472 Nov 8.0	562 May 19.9	651 Nov 28.9	741 Jun 7.5	830 Dec 18.8
384 Dec 5.4	474 Jun 14.3	563 Dec 26.3	653 Jul 3.0	743 Jan 14.9	832 Jul 21.9
386 Jul 9.9	476 Jan 22.2	565 Jul 28.8	655 Feb 10.5	744 Aug 16.8	834 Mar 1.7
388 Feb 17.6	477 Aug 23.9	567 Mar 8.7	656 Sep 12.2	746 Mar 27.5	835 Oct 2.8
389 Sep 19.5	479 Apr 3.4	568 Oct 9.2	658 Apr 22.1	747 Oct 30.0	837 May 10.8
391 Apr 28.9	480 Nov 5.5	570 May 17.5	659 Nov 26.4	749 Jun 5.2	838 Dec 16.3
392 Dec 2.9	482 Jun 12.0	571 Dec 23.8	661 Jun 30.7	751 Jan 12.5	840 Jul 19.5
394 Jul 7.5	484 Jan 19.8	573 Jul 26.4	663 Feb 8.1	752 Aug 14.5	842 Feb 27.3
396 Feb 15.3	485 Aug 21.5	575 Mar 6.3	664 Sep 9.8	754 Mar 25.2	843 Sep 30.3
397 Sep 17.1	487 Apr 1.1	576 Oct 6.7	666 Apr 19.8	755 Oct 27.5	845 May 8.5
399 Apr 26.6	488 Nov 3.0	578 May 15.2	667 Nov 23.9	757 Jun 2.9	846 Dec 13.8
400 Nov 30.4	490 Jun 9.7	579 Dec 21.3	669 Jun 28.4	759 Jan 10.0	848 Jul 17.2
402 Jul 5.2	492 Jan 17.4	581 Jul 24.1	671 Feb 5.7	760 Aug 12.1	850 Feb 24.9
404 Feb 12.9	493 Aug 19.1	583 Mar 4.0	672 Sep 7.4	762 Mar 22.9	851 Sep 27.9
405 Sep 14.6	495 Mar 29.7	584 Oct 4.2	674 Apr 17.5	763 Oct 25.0	853 May 6.1
407 Apr 24.3	496 Oct 31.5	586 May 12.9	675 Nov 21.4	765 May 31.6	854 Dec 11.3
408 Nov 27.9	498 Jun 7.3	587 Dec 18.8	677 Jun 26.0	767 Jan 7.6	856 Jul 14.8
410 Jul 2.9	500 Jan 14.9	589 Jul 21.7	679 Feb 3.3	768 Aug 9.7	858 Feb 22.5
412 Feb 10.5	501 Aug 16.7	591 Mar 1.6	680 Sep 5.0	770 Mar 20.5	859 Sep 25.4
413 Sep 12.2	503 Mar 27.4	592 Oct 1.8	682 Apr 15.2	771 Oct 22.5	861 May 3.8
415 Apr 22.0	504 Oct 29.0	594 May 10.6	683 Nov 18.9	773 May 29.3	862 Dec 8.9
416 Nov 25.4	506 Jun 5.0	595 Dec 16.4	685 Jun 23.7	775 Jan 5.1	864 Jul 12.5
418 Jun 30.6	508 Jan 12.5	597 Jul 19.4	687 Jan 31.9	776 Aug 7.3	866 Feb 20.2
420 Feb 8.1	509 Aug 14.4	599 Feb 27.2	688 Sep 2.6	778 Mar 18.2	867 Sep 23.0
421 Sep 9.8	511 Mar 25.1	600 Sep 29.3	690 Apr 12.8	779 Oct 20.0	869 May 1.5
423 Apr 19.6	512 Oct 26.5	602 May 8.3	691 Nov 16.4	781 May 26.9	870 Dec 6.4
424 Nov 23.0	514 Jun 2.7	603 Dec 13.9	693 Jun 21.4	783 Jan 2.6	872 Jul 10.2
426 Jun 28.2	516 Jan 10.0	605 Jul 17.0	695 Jan 29.5	784 Aug 5.0	874 Feb 17.8
428 Feb 5.7	517 Aug 12.0	607 Feb 24.8	696 Aug 31.2	786 Mar 15.8	875 Sep 20.6
429 Sep 7.3	519 Mar 22.7	608 Sep 26.9	698 Apr 10.5	787 Oct 17.6	877 Apr 29.2
431 Apr 17.3	520 Oct 24.0	610 May 6.0	699 Nov 13.9	789 May 24.6	878 Dec 3.9
432 Nov 20.5	522 May 31.4	611 Dec 11.4	701 Jun 19.1	790 Dec 31.2	880 Jul 7.8
434 Jun 25.9	524 Jan 7.6	613 Jul 14.7	703 Jan 27.1	792 Aug 2.6	882 Feb 15.4
436 Feb 3.3	525 Aug 9.6	615 Feb 22.5	704 Aug 28.8	794 Mar 13.5	883 Sep 18.1
437 Sep 4.9	527 Mar 20.4	616 Sep 24.4	706 Apr 8.2	795 Oct 15.1	885 Apr 26.9
439 Apr 15.0	528 Oct 21.5	618 May 3.7	707 Nov 11.4	797 May 22.3	886 Dec 1.4
440 Nov 18.0	530 May 29.1	619 Dec 8.9	709 Jun 16.8	798 Dec 28.7	888 Jul 5.5
442 Jun 23.6	532 Jan 5.1	621 Jul 12.4	711 Jan 24.7	800 Jul 31.3	890 Feb 13.0
444 Jan 31.9	533 Aug 7.2	623 Feb 20.1	712 Aug 26.4	802 Mar 11.1	891 Sep 15.7

893 Apr 24.6	982 Nov 1.4	1072 May 13.2	1161 Nov 21.3	1251 Jun 1.9	1340 Dec 11.3
894 Nov 28.9	984 Jun 7.7	1073 Dec 18.8	1163 Jun 27.3	1253 Jan 7.5	1342 Jul 16.1
896 Jul 3.2	986 Jan 14.9	1075 Jul 23.0	1165 Feb 3.4	1254 Aug 10.9	1344 Feb 23.6
898 Feb 10.6	987 Aug 17.9	1077 Mar 1.7	1166 Sep 6.1	1256 Mar 20.7	1345 Sep 25.5
899 Sep 13.3	989 Mar 27.6	1078 Oct 2.8	1168 Apr 15.4	1257 Oct 22.5	1347 May 5.1
901 Apr 22.3	990 Oct 29.9	1080 May 10.9	1169 Nov 18.8	1259 May 30.6	1348 Dec 8.8
902 Nov 26.4	992 Jun 5.4	1081 Dec 16.3	1171 Jun 25.0	1261 Jan 5.1	1350 Jul 13.8
904 Jun 30.9	994 Jan 12.5	1083 Jul 20.6	1173 Feb 1.0	1262 Aug 8.5	1352 Feb 21.2
906 Feb 8.2	995 Aug 15.5	1085 Feb 27.4	1174 Sep 3.7	1264 Mar 18.3	1353 Sep 23.1
907 Sep 10.9	997 Mar 25.3	1086 Sep 30.3	1176 Apr 13.1	1265 Oct 20.0	1355 May 2.8
909 Apr 19.9	998 Oct 27.5	1088 May 8.6	1177 Nov 16.3	1267 May 28.3	1356 Dec 6.3
910 Nov 23.9	1000 Jun 3.0	1089 Dec 13.8	1179 Jun 22.7	1269 Jan 2.6	1358 Jul 11.4
912 Jun 28.5	1002 Jan 10.0	1091 Jul 18.3	1181 Jan 29.5	1270 Aug 6.2	1360 Feb 18.8
914 Feb 5.8	1003 Aug 13.2	1093 Feb 25.0	1182 Sep 1.3	1272 Mar 16.0	1361 Sep 20.6
915 Sep 8.5	1005 Mar 23.0	1094 Sep 27.9	1184 Apr 10.8	1273 Oct 17.5	1363 Apr 30.5
917 Apr 17.6	1006 Oct 25.0	1096 May 6.3	1185 Nov 13.9	1275 May 25.9	1364 Dec 3.8
918 Nov 21.4	1008 May 31.7	1097 Dec 11.3	1187 Jun 20.4	1276 Dec 31.1	1366 Jul 9.1
920 Jun 26.2	1010 Jan 7.6	1099 Jul 16.0	1189 Jan 27.1	1278 Aug 3.8	1368 Feb 16.4
922 Feb 3.4	1011 Aug 10.8	1101 Feb 22.6	1190 Aug 29.9	1280 Mar 13.6	1369 Sep 18.2
923 Sep 6.0	1013 Mar 20.6	1102 Sep 25.5	1192 Apr 8.4	1281 Oct 15.1	1371 Apr 28.2
925 Apr 15.3	1014 Oct 22.5	1104 May 4.0	1193 Nov 11.4	1283 May 23.6	1372 Dec 1.3
926 Nov 18.9	1016 May 29.4	1105 Dec 8.8	1195 Jun 18.1	1284 Dec 28.7	1374 Jul 6.8
928 Jun 23.9	1018 Jan 5.1	1107 Jul 13.6	1197 Jan 24.7	1286 Aug 1.5	1376 Feb 14.0
930 Jan 31.9	1019 Aug 8.4	1109 Feb 20.2	1198 Aug 27.5	1288 Mar 11.3	1377 Sep 15.8
931 Sep 3.6	1021 Mar 18.3	1110 Sep 23.0	1200 Apr 6.1	1289 Oct 12.6	1379 Apr 25.8
933 Apr 13.0	1022 Oct 20.0	1112 May 1.7	1201 Nov 8.9	1291 May 21.3	1380 Nov 28.8
934 Nov 16.4	1024 May 27.1	1113 Dec 6.3	1203 Jun 15.8	1292 Dec 26.2	1382 Jul 4.5
936 Jun 21.6	1026 Jan 2.6	1115 Jul 11.3	1205 Jan 22.2	1294 Jul 30.1	1384 Feb 11.6
938 Jan 29.5	1027 Aug 6.1	1117 Feb 17.8	1206 Aug 25.1	1296 Mar 8.9	1385 Sep 13.4
939 Sep 1.2	1029 Mar 15.9	1118 Sep 20.6	1208 Apr 3.8	1297 Oct 10.1	1387 Apr 23.5
941 Apr 10.6	1030 Oct 17.6	1120 Apr 29.3	1209 Nov 6.4	1299 May 19.0	1388 Nov 26.3
942 Nov 13.9	1032 May 24.8	1121 Dec 3.8	1211 Jun 13.5	1300 Dec 23.7	1390 Jul 2.1
944 Jun 19.2	1033 Dec 31.2	1123 Jul 9.0	1213 Jan 19.8	1302 Jul 27.8	1392 Feb 9.2
946 Jan 27.1	1035 Aug 3.7	1125 Feb 15.4	1214 Aug 22.8	1304 Mar 6.5	1393 Sep 11.0
947 Aug 29.8	1037 Mar 13.6	1126 Sep 18.2	1216 Apr 1.4	1305 Oct 7.7	1395 Apr 21.2
949 Apr 8.3	1038 Oct 15.1	1128 Apr 27.0	1217 Nov 3.9	1307 May 16.7	1396 Nov 23.8
950 Nov 11.4	1040 May 22.5	1129 Dec 1.3	1219 Jun 11.1	1308 Dec 21.2	1398 Jun 29.8
952 Jun 16.9	1041 Dec 28.7	1131 Jul 6.7	1221 Jan 17.3	1310 Jul 25.4	1400 Feb 6.8
954 Jan 24.7	1043 Aug 1.4	1133 Feb 13.0	1222 Aug 20.4	1312 Mar 4.2	1401 Sep 8.6
955 Aug 27.5	1045 Mar 11.2	1134 Sep 15.7	1224 Mar 30.1	1313 Oct 5.2	1403 Apr 18.9
957 Apr 6.0	1046 Oct 12.6	1136 Apr 24.7	1225 Nov 1.4	1315 May 14.4	1404 Nov 21.3
958 Nov 8.9	1048 May 20.2	1137 Nov 28.8	1227 Jun 8.8	1316 Dec 18.7	1406 Jun 27.5
960 Jun 14.6	1049 Dec 26.2	1139 Jul 4.3	1229 Jan 14.9	1318 Jul 23.1	1408 Feb 4.4
962 Jan 22.2	1051 Jul 30.0	1141 Feb 10.6	1230 Aug 18.0	1320 Mar 1.8	1409 Sep 6.2
963 Aug 25.1	1053 Mar 8.8	1142 Sep 13.3	1232 Mar 27.7	1321 Oct 2.8	1411 Apr 16.5
965 Apr 3.7	1054 Oct 10.2	1144 Apr 22.4	1233 Oct 29.9	1323 May 12.1	1412 Nov 18.8
966 Nov 6.4	1056 May 17.9	1145 Nov 26.3	1235 Jun 6.5	1324 Dec 16.2	1414 Jun 25.2
968 Jun 12.3	1057 Dec 23.7	1147 Jul 2.0	1237 Jan 12.5	1326 Jul 20.8	1416 Feb 2.0
970 Jan 19.8	1059 Jul 27.7	1149 Feb 8.2	1238 Aug 15.6	1328 Feb 28.4	1417 Sep 3.8
971 Aug 22.7	1061 Mar 6.5	1150 Sep 10.9	1240 Mar 25.4	1329 Sep 30.4	1419 Apr 14.2
973 Apr 1.3	1062 Oct 7.7	1152 Apr 20.1	1241 Oct 27.4	1331 May 9.8	1420 Nov 16.3
974 Nov 3.9	1064 May 15.5	1153 Nov 23.8	1243 Jun 4.2	1332 Dec 13.7	1422 Jun 22.9
976 Jun 10.0	1065 Dec 21.2	1155 Jun 29.7	1245 Jan 10.0	1334 Jul 18.4	1424 Jan 30.5
978 Jan 17.4	1067 Jul 25.3	1157 Feb 5.8	1246 Aug 13.3	1336 Feb 26.0	1425 Sep 1.4
979 Aug 20.3	1069 Mar 4.1	1158 Sep 8.5	1248 Mar 23.1	1337 Sep 27.9	1427 Apr 11.9
981 Mar 30.0	1070 Oct 5.2	1160 Apr 17.7	1249 Oct 25.0	1339 May 7.4	1428 Nov 13.8

1430 Jun 20.5	1520 Jan 1.1	1609 Jul 19.2	1699 Jan 30.8	1788 Aug 8.0	1878 Feb 21.2
1432 Jan 28.1	1521 Aug 3.9	1611 Feb 26.5	1700 Sep 2.9	1790 Mar 18.6	1879 Sep 24.1
1433 Aug 30.0	1523 Mar 14.7	1612 Sep 28.2	1702 Apr 13.6	1791 Oct 19.7	1881 May 3.4
1435 Apr 9.5	1524 Oct 15.1	1614 May 8.3	1703 Nov 15.8	1793 May 28.0	1882 Dec 6.7
1436 Nov 11.3	1526 May 23.8	1615 Dec 12.2	1705 Jun 22.4	1795 Jan 2.1	1884 Jul 12.1
1438 Jun 18.2	1527 Dec 29.6	1617 Jul 16.9	1707 Jan 29.3	1796 Aug 5.7	1886 Feb 18.8
1440 Jan 25.7	1529 Aug 1.6	1619 Feb 24.1	1708 Aug 31.5	1798 Mar 16.2	1887 Sep 21.7
1441 Aug 27.6	1531 Mar 12.3	1620 Sep 25.8	1710 Apr 11.3	1799 Oct 17.3	1889 May 1.1
1443 Apr 7.2	1532 Oct 12.6	1622 May 6.0	1711 Nov 13.3	1801 May 26.6	1890 Dec 4.2
1444 Nov 8.8	1534 May 21.5	1623 Dec 9.7	1713 Jun 20.1	1802 Dec 31.6	1892 Jul 9.8
1446 Jun 15.9	1535 Dec 27.1	1625 Jul 14.6	1715 Jan 26.8	1804 Aug 4.3	1894 Feb 16.4
1448 Jan 23.2	1537 Jul 30.2	1627 Feb 21.7	1716 Aug 29.2	1806 Mar 14.9	1895 Sep 19.3
1449 Aug 25.2	1539 Mar 10.0	1628 Sep 23.4	1718 Apr 8.9	1807 Oct 15.8	1897 Apr 28.7
1451 Apr 4.9	1540 Oct 10.2	1630 May 3.6	1719 Nov 10.9	1809 May 24.3	1898 Dec 1.7
1452 Nov 6.3	1542 May 19.1	1631 Dec 7.2	1721 Jun 17.8	1810 Dec 29.1	1900 Jul 8.5
1454 Jun 13.6	1543 Dec 24.7	1633 Jul 12.3	1723 Jan 24.4	1812 Aug 2.0	1902 Feb 15.0
1456 Jan 20.8	1545 Jul 27.9	1635 Feb 19.2	1724 Aug 26.8	1814 Mar 12.5	1903 Sep 17.9
1457 Aug 22.8	1547 Mar 7.6	1636 Sep 21.0	1726 Apr 6.6	1815 Oct 13.4	1905 Apr 27.4
1459 Apr 2.5	1548 Oct 7.7	1638 May 1.3	1727 Nov 8.4	1817 May 22.0	1906 Nov 30.2
1460 Nov 3.9	1550 May 16.8	1639 Dec 4.8	1729 Jun 15.5	1818 Dec 26.6	1908 Jul 6.1
1462 Jun 11.3	1551 Dec 22.2	1641 Jul 10.0	1731 Jan 21.9	1820 Jul 30.7	1910 Feb 12.5
1464 Jan 18.3	1553 Jul 25.6	1643 Feb 16.8	1732 Aug 24.5	1822 Mar 10.1	1911 Sep 15.5
1465 Aug 20.5	1555 Mar 5.2	1644 Sep 18.6	1734 Apr 4.2	1823 Oct 11.0	1913 Apr 25.1
1467 Mar 31.2	1556 Oct 5.3	1646 Apr 29.0	1735 Nov 5.9	1825 May 19.7	1914 Nov 27.7
1468 Nov 1.4	1558 May 14.5	1647 Dec 2.3	1737 Jun 13.2	1826 Dec 24.2	1916 Jul 3.8
1470 Jun 9.0	1559 Dec 19.7	1649 Jul 7.6	1739 Jan 19.5	1828 Jul 28.3	1918 Feb 10.1
1472 Jan 15.9	1561 Jul 23.2	1651 Feb 14.4	1740 Aug 22.1	1830 Mar 7.7	1919 Sep 13.1
1473 Aug 18.1	1563 Mar 2.8	1652 Sep 16.2	1742 Apr 1.8	1831 Oct 8.5	1921 Apr 22.7
1475 Mar 28.8	1564 Oct 2.8	1654 Apr 26.6	1743 Nov 3.4	1833 May 17.4	1922 Nov 25.2
1476 Oct 29.9	1566 May 12.2	1655 Nov 29.8	1745 Jun 10.9	1834 Dec 21.7	1924 Jul 1.5
1478 Jun 6.7	1567 Dec 17.2	1657 Jul 5.3	1747 Jan 17.0	1836 Jul 26.0	1926 Feb 7.6
1480 Jan 13.4	1569 Jul 20.9	1659 Feb 12.0	1748 Aug 19.7	1838 Mar 5.3	1927 Sep 10.7
1481 Aug 15.7	1571 Feb 28.4	1660 Sep 13.8	1750 Mar 30.5	1839 Oct 6.1	1929 Apr 20.4
1483 Mar 26.5	1572 Sep 30.4	1662 Apr 24.3	1751 Nov 1.0	1841 May 15.0	1930 Nov 23.2
1484 Oct 27.4	1574 May 9.9	1663 Nov 27.3	1753 Jun 8.5	1842 Dec 19.2	1932 Jun 29.2
1486 Jun 4.3	1575 Dec 14.7	1665 Jul 3.0	1755 Jan 14.5	1844 Jul 23.7	1934 Feb 5.2
1488 Jan 11.0	1577 Jul 18.6	1667 Feb 9.5	1756 Aug 17.4	1846 Mar 2.9	1935 Sep 8.4
1489 Aug 13.4	1579 Feb 26.1	1668 Sep 11.4	1758 Mar 28.1	1847 Oct 3.7	1937 Apr 18.0
1491 Mar 24.1	1580 Sep 27.9	1670 Apr 22.0	1759 Oct 29.5	1849 May 12.7	1938 Nov 20.3
1492 Oct 24.9	1582 May 7.6	1671 Nov 24.8	1761 Jun 6.2	1850 Dec 16.7	1940 Jun 26.9
1494 Jun 2.0	1583 Dec 22.2	1673 Jun 30.7	1763 Jan 12.0	1852 Jul 21.4	1942 Feb 2.7
1496 Jan 8.5	1585 Jul 26.2	1675 Feb 7.1	1764 Aug 15.0	1854 Feb 28.5	1943 Sep 6.0
1497 Aug 11.0	1587 Mar 5.7	1676 Sep 9.0	1766 Mar 25.8	1855 Oct 1.3	1945 Apr 15.7
1499 Mar 21.8	1588 Oct 5.5	1678 Apr 19.6	1767 Oct 27.1	1857 May 10.4	1946 Nov 17.8
1500 Oct 22.5	1590 May 15.2	1679 Nov 22.3	1769 Jun 3.9	1858 Dec 14.2	1948 Jun 24.6
1502 May 30.7	1591 Dec 19.7	1681 Jun 28.4	1771 Jan 9.6	1860 Jul 19.0	1950 Jan 31.3
1504 Jan 6.0	1593 Jul 23.9	1683 Feb 4.7	1772 Aug 12.7	1862 Feb 26.1	1951 Sep 3.6
1505 Aug 8.6	1595 Mar 3.3	1684 Sep 6.7	1774 Mar 23.4	1863 Sep 28.9	1953 Apr 13.3
1507 Mar 19.4	1596 Oct 3.1	1686 Apr 17.3	1775 Oct 24.6	1865 May 8.1	1954 Nov 15.3
1508 Oct 20.0	1598 May 12.9	1687 Nov 19.8	1777 Jun 1.6	1866 Dec 11.7	1956 Jun 22.2
1510 May 28.4	1599 Dec 17.2	1689 Jun 26.1	1779 Jan 7.1	1868 Jul 16.7	1958 Jan 28.8
1512 Jan 3.6	1601 Jul 21.6	1691 Feb 2.2	1780 Aug 10.4	1870 Feb 23.6	1959 Sep 1.3
1513 Aug 6.3	1603 Feb 28.9	1692 Sep 4.3	1782 Mar 21.0	1871 Sep 26.5	1961 Apr 11.0
1515 Mar 17.1	1604 Sep 30.7	1694 Apr 15.0	1783 Oct 22.1	1873 May 5.7	1962 Nov 12.8
1516 Oct 17.5	1606 May 10.6	1695 Nov 17.3	1785 May 30.3	1874 Dec 9.2	1964 Jun 19.9
1518 May 26.1	1607 Dec 14.7	1697 Jun 23.7	1787 Jan 4.6	1876 Jul 14.4	1966 Jan 26.4

1967 Aug 29.9	2057 Mar 13.5	2146 Sep 18.9	2236 Apr 3.6	2325 Oct 10.2	2415 Apr 24.5
1969 Apr 8.6	2058 Oct 14.4	2148 Apr 28.5	2237 Nov 4.9	2327 May 20.3	2416 Nov 25.7
1970 Nov 10.4	2060 May 23.1	2149 Dec 1.2	2239 Jun 13.8	2328 Dec 23.0	2418 Jul 3.5
1972 Jun 17.6	2061 Dec 27.6	2151 Jul 8.3	2241 Jan 18.4	2330 Jul 28.9	2420 Feb 9.1
1974 Jan 23.9	2063 Aug 1.8	2153 Feb 13.5	2242 Aug 22.6	2332 Mar 6.9	2421 Sep 11.5
1975 Aug 27.5	2065 Mar 11.1	2154 Sep 16.6	2244 Apr 1.2	2333 Oct 7.8	2423 Apr 22.2
1977 Apr 6.3	2066 Oct 12.0	2156 Apr 26.2	2245 Nov 2.5	2335 May 17.9	2424 Nov 23.2
1978 Nov 7.9	2068 May 20.8	2157 Nov 28.7	2247 Jun 11.5	2336 Dec 20.6	2426 Jul 1.1
1980 Jun 15.3	2069 Dec 25.1	2159 Jul 6.0	2249 Jan 15.9	2338 Jul 26.6	2428 Feb 6.6
1982 Jan 21.4	2071 Jul 30.5	2161 Feb 11.0	2250 Aug 20.3	2340 Mar 4.5	2429 Sep 9.2
1983 Aug 25.2	2073 Mar 8.7	2162 Sep 14.2	2252 Mar 29.8	2341 Oct 5.3	2431 Apr 19.8
1985 Apr 3.9	2074 Oct 9.6	2164 Apr 23.8	2253 Oct 31.0	2343 May 15.6	2432 Nov 20.7
1986 Nov 5.4	2076 May 18.5	2165 Nov 26.2	2255 Jun 9.2	2344 Dec 18.1	2434 Jun 28.8
1988 Jun 13.0	2077 Dec 22.6	2167 Jul 3.6	2257 Jan 13.5	2346 Jul 24.3	2436 Feb 4.2
1990 Jan 18.9	2079 Jul 28.1	2169 Feb 8.6	2258 Aug 17.9	2348 Mar 2.1	2437 Sep 6.8
1991 Aug 22.8	2081 Mar 6.3	2170 Sep 11.8	2260 Mar 27.5	2349 Oct 2.9	2439 Apr 17.5
1993 Apr 1.5	2082 Oct 7.1	2172 Apr 21.5	2261 Oct 28.6	2351 May 13.3	2440 Nov 18.2
1994 Nov 3.0	2084 May 16.2	2173 Nov 23.7	2263 Jun 6.9	2352 Dec 15.6	2442 Jun 26.5
1996 Jun 10.7	2085 Dec 20.1	2175 Jul 1.3	2265 Jan 11.0	2354 Jul 22.0	2444 Feb 1.7
1998 Jan 16.5	2087 Jul 25.8	2177 Feb 6.1	2266 Aug 15.6	2356 Feb 28.6	2445 Sep 4.4
1999 Aug 20.5	2089 Mar 3.9	2178 Sep 9.4	2268 Mar 25.1	2357 Sep 30.5	2447 Apr 15.1
2001 Mar 30.2	2090 Oct 4.7	2180 Apr 19.1	2269 Oct 26.1	2359 May 10.9	2448 Nov 15.8
2002 Oct 31.5	2092 May 13.8	2181 Nov 21.2	2271 Jun 4.5	2360 Dec 13.1	2450 Jun 24.2
2004 Jun 8.4	2093 Dec 17.6	2183 Jun 29.0	2273 Jan 8.5	2362 Jul 19.7	2452 Jan 30.3
2006 Jan 14.0	2095 Jul 23.5	2185 Feb 3.7	2274 Aug 13.2	2364 Feb 26.2	2453 Sep 2.1
2007 Aug 18.1	2097 Mar 1.5	2186 Sep 7.1	2276 Mar 22.7	2365 Sep 28.2	2455 Apr 12.7
2009 Mar 27.8	2098 Oct 2.3	2188 Apr 16.8	2277 Oct 23.7	2367 May 8.6	2456 Nov 13.3
2010 Oct 29.0	2100 May 12.5	2189 Nov 18.8	2279 Jun 2.2	2368 Dec 10.6	2458 Jun 21.9
2012 Jun 6.0	2101 Dec 16.1	2191 Jun 26.7	2281 Jan 6.0	2370 Jul 17.3	2460 Jan 27.8
2014 Jan 11.5	2103 Jul 22.2	2193 Feb 1.2	2282 Aug 10.9	2372 Feb 23.8	2461 Aug 30.7
2015 Aug 15.8	2105 Feb 28.1	2194 Sep 4.7	2284 Mar 20.3	2373 Sep 25.8	2463 Apr 10.4
2017 Mar 25.4	2106 Sep 30.9	2196 Apr 14.4	2285 Oct 21.3	2375 May 6.3	2464 Nov 10.8
2018 Oct 26.6	2108 May 10.2	2197 Nov 16.3	2287 May 30.9	2376 Dec 8.1	2466 Jun 19.6
2020 Jun 3.7	2109 Dec 13.6	2199 Jun 24.4	2289 Jan 3.5	2378 Jul 15.0	2468 Jan 25.3
2022 Jan 9.0	2111 Jul 19.9	2201 Jan 30.8	2290 Aug 8.6	2380 Feb 21.3	2469 Aug 28.4
2023 Aug 13.5	2113 Feb 25.6	2202 Sep 3.3	2292 Mar 17.9	2381 Sep 23.4	2471 Apr 8.0
2025 Mar 23.0	2114 Sep 28.5	2204 Apr 13.1	2293 Oct 18.8	2383 May 3.9	2472 Nov 8.4
2026 Oct 24.1	2116 May 7.8	2205 Nov 14.8	2295 May 28.6	2384 Dec 5.6	2474 Jun 17.3
2028 Jun 1.4	2117 Dec 11.1	2207 Jun 23.1	2297 Jan 1.4	2386 Jul 12.7	2476 Jan 22.8
2030 Jan 6.5	2119 Jul 17.5	2209 Jan 28.3	2298 Aug 6.2	2388 Feb 18.9	2477 Aug 26.0
2031 Aug 11.1	2121 Feb 23.2	2210 Sep 1.0	2300 Mar 16.5	2389 Sep 21.0	2479 Apr 5.6
2033 Mar 20.7	2122 Sep 26.1	2212 Apr 10.7	2301 Oct 17.4	2391 May 1.6	2480 Nov 5.9
2034 Oct 21.7	2124 May 5.5	2213 Nov 12.3	2303 May 27.3	2392 Dec 3.1	2482 Jun 14.9
2036 May 30.1	2125 Dec 8.6	2215 Jun 20.7	2304 Dec 30.5	2394 Jul 10.4	2484 Jan 20.4
2038 Jan 4.1	2127 Jul 15.2	2217 Jan 25.8	2306 Aug 4.9	2396 Feb 16.5	2485 Aug 23.7
2039 Aug 8.8	2129 Feb 20.8	2218 Aug 29.6	2308 Mar 14.1	2397 Sep 18.6	2487 Apr 3.3
2041 Mar 18.3	2130 Sep 23.7	2220 Apr 8.3	2309 Oct 15.0	2399 Apr 29.2	2488 Nov 3.5
2042 Oct 19.3	2132 May 3.2	2221 Nov 9.9	2311 May 24.9	2400 Nov 30.6	2490 Jun 12.6
2044 May 27.8	2133 Dec 6.2	2223 Jun 18.4	2312 Dec 28.0	2402 Jul 8.1	2492 Jan 17.9
2046 Jan 1.6	2135 Jul 12.9	2225 Jan 23.4	2314 Aug 2.6	2404 Feb 14.0	2493 Aug 21.4
2047 Aug 6.5	2137 Feb 18.4	2226 Aug 27.3	2316 Mar 11.7	2405 Sep 16.2	2495 Mar 31.9
2049 Mar 15.9	2138 Sep 21.3	2228 Apr 6.0	2317 Oct 12.6	2407 Apr 26.9	2496 Nov 1.0
2050 Oct 16.8	2140 Apr 30.8	2229 Nov 7.4	2319 May 22.6	2408 Nov 28.2	2498 Jun 10.3
2052 May 25.4	2141 Dec 3.7	2231 Jun 16.1	2320 Dec 25.5	2410 Jul 5.8	2500 Jan 15.4
2053 Dec 30.1	2143 Jul 10.6	2233 Jan 20.9	2322 Jul 31.3	2412 Feb 11.6	
2055 Aug 4.1	2145 Feb 15.9	2234 Aug 24.9	2324 Mar 9.3	2413 Sep 13.9	

VIII
Superior Conjunctions of Venus, 0 - 2500

0 Jun 10.7	59 Aug 2.4	118 Sep 25.2	177 Nov 22.1	237 Jan 20.7	296 Mar 18.4
2 Jan 18.2	61 Mar 16.0	120 May 8.2	179 Jun 29.7	238 Aug 20.8	297 Oct 14.9
3 Aug 18.3	62 Oct 12.4	121 Dec 10.3	181 Feb 7.4	240 Apr 3.9	299 May 27.3
5 Apr 1.5	64 May 23.9	123 Jul 15.3	182 Sep 6.1	241 Nov 1.5	300 Dec 31.1
6 Oct 30.0	65 Dec 28.6	125 Feb 24.7	184 Apr 20.0	243 Jun 11.9	302 Aug 2.5
8 Jun 8.5	67 Jul 31.1	126 Sep 22.8	185 Nov 19.5	245 Jan 18.2	304 Mar 16.0
10 Jan 15.7	69 Mar 13.6	128 May 5.9	187 Jun 27.5	246 Aug 18.5	305 Oct 12.4
11 Aug 16.0	70 Oct 9.9	129 Dec 7.7	189 Feb 4.9	248 Apr 1.5	307 May 25.1
13 Mar 30.1	72 May 21.7	131 Jul 13.1	190 Sep 3.8	249 Oct 30.0	308 Dec 28.5
14 Oct 27.4	73 Dec 26.0	133 Feb 22.3	192 Apr 17.7	251 Jun 9.7	310 Jul 31.3
16 Jun 6.3	75 Jul 28.8	134 Sep 20.4	193 Nov 16.9	253 Jan 15.6	312 Mar 13.6
18 Jan 13.1	77 Mar 11.2	136 May 3.7	195 Jun 25.3	254 Aug 16.2	313 Oct 9.9
19 Aug 13.7	78 Oct 7.4	137 Dec 5.1	197 Feb 2.4	256 Mar 30.2	315 May 22.8
21 Mar 27.8	80 May 19.4	139 Jul 10.9	198 Sep 1.4	257 Oct 27.4	316 Dec 25.9
22 Oct 24.9	81 Dec 23.4	141 Feb 19.8	200 Apr 15.4	259 Jun 7.5	318 Jul 29.0
24 Jun 4.1	83 Jul 26.6	142 Sep 18.0	201 Nov 14.3	261 Jan 13.0	320 Mar 11.2
26 Jan 10.5	85 Mar 8.8	144 May 1.4	203 Jun 23.0	262 Aug 13.9	321 Oct 7.5
27 Aug 11.5	86 Oct 5.0	145 Dec 2.5	205 Jan 30.9	264 Mar 27.9	323 May 20.6
29 Mar 25.4	88 May 17.2	147 Jul 8.6	206 Aug 30.1	265 Oct 24.9	324 Dec 23.3
30 Oct 22.4	89 Dec 20.8	149 Feb 17.3	208 Apr 13.1	267 Jun 5.2	326 Jul 26.8
32 Jun 1.8	91 Jul 24.3	150 Sep 15.6	209 Nov 11.8	269 Jan 10.5	328 Mar 8.8
34 Jan 8.0	93 Mar 6.4	152 Apr 29.1	211 Jun 20.8	270 Aug 11.6	329 Oct 5.0
35 Aug 9.2	94 Oct 2.5	153 Nov 29.9	213 Jan 28.3	272 Mar 25.5	331 May 18.3
37 Mar 23.1	96 May 14.9	155 Jul 6.4	214 Aug 27.7	273 Oct 22.4	332 Dec 20.7
38 Oct 19.9	97 Dec 18.2	157 Feb 14.9	216 Apr 10.8	275 Jun 3.0	334 Jul 24.5
40 May 30.6	99 Jul 22.1	158 Sep 13.2	217 Nov 9.2	277 Jan 7.9	336 Mar 6.4
42 Jan 5.4	101 Mar 4.0	160 Apr 26.9	219 Jun 18.6	278 Aug 9.3	337 Oct 2.6
43 Aug 6.9	102 Sep 30.1	161 Nov 27.3	221 Jan 25.8	280 Mar 23.1	339 May 16.1
45 Mar 20.7	104 May 12.7	163 Jul 4.2	222 Aug 25.4	281 Oct 19.9	340 Dec 18.1
46 Oct 17.4	105 Dec 15.6	165 Feb 12.4	224 Apr 8.5	283 May 31.8	342 Jul 22.3
48 May 28.4	107 Jul 19.8	166 Sep 10.8	225 Nov 6.6	285 Jan 5.3	344 Mar 4.0
50 Jan 2.8	109 Mar 1.6	168 Apr 24.6	227 Jun 16.4	286 Aug 7.1	345 Sep 30.1
51 Aug 4.6	110 Sep 27.6	169 Nov 24.7	229 Jan 23.3	288 Mar 20.8	347 May 13.9
53 Mar 18.4	112 May 10.4	171 Jul 1.9	230 Aug 23.1	289 Oct 17.4	348 Dec 15.5
54 Oct 14.9	113 Dec 12.9	173 Feb 9.9	232 Apr 6.2	291 May 29.6	350 Jul 20.0
56 May 26.1	115 Jul 17.6	174 Sep 8.5	233 Nov 4.1	293 Jan 2.7	352 Mar 1.6
57 Dec 31.2	117 Feb 27.1	176 Apr 22.3	235 Jun 14.2	294 Aug 4.8	353 Sep 27.7

355 May 11.6	444 Nov 14.3	534 May 29.7	623 Dec 6.0	713 Jun 16.8	802 Dec 26.7
356 Dec 12.9	446 Jun 23.2	536 Jan 3.6	625 Jul 11.3	715 Jan 24.1	804 Jul 29.4
358 Jul 17.8	448 Jan 31.8	537 Aug 5.0	627 Feb 20.8	716 Aug 23.4	806 Mar 12.3
360 Feb 28.1	449 Aug 30.2	539 Mar 19.4	628 Sep 18.2	718 Apr 7.4	807 Oct 8.6
361 Sep 25.3	451 Apr 14.2	540 Oct 14.9	630 May 2.7	719 Nov 5.0	809 May 20.9
363 May 9.4	452 Nov 11.7	542 May 27.5	631 Dec 3.4	721 Jun 14.5	810 Dec 24.1
364 Dec 10.3	454 Jun 21.0	544 Jan 1.0	633 Jul 9.0	723 Jan 21.6	812 Jul 27.1
366 Jul 15.5	456 Jan 29.3	545 Aug 2.7	635 Feb 18.3	724 Aug 21.1	814 Mar 9.9
368 Feb 25.7	457 Aug 27.9	547 Mar 17.1	636 Sep 15.8	726 Apr 5.0	815 Oct 6.1
369 Sep 22.9	459 Apr 11.9	548 Oct 12.4	638 Apr 30.4	727 Nov 2.5	817 May 18.7
371 May 7.1	460 Nov 9.2	550 May 25.3	639 Nov 30.8	729 Jun 12.3	818 Dec 21.5
372 Dec 7.6	462 Jun 18.8	551 Dec 29.4	641 Jul 6.8	731 Jan 19.0	820 Jul 24.9
374 Jul 13.3	464 Jan 26.8	553 Jul 31.5	643 Feb 15.8	732 Aug 18.8	822 Mar 7.5
376 Feb 23.2	465 Aug 25.6	555 Mar 14.7	644 Sep 13.4	734 Apr 2.7	823 Oct 3.7
377 Sep 20.5	467 Apr 9.6	556 Oct 10.0	646 Apr 28.1	735 Oct 31.0	825 May 16.4
379 May 4.8	468 Nov 6.6	558 May 23.0	647 Nov 28.2	737 Jun 10.1	826 Dec 18.9
380 Dec 5.0	470 Jun 16.6	559 Dec 26.8	649 Jul 4.6	739 Jan 16.5	828 Jul 22.6
382 Jul 11.1	472 Jan 24.2	561 Jul 29.2	651 Feb 13.3	740 Aug 16.5	830 Mar 5.0
384 Feb 20.8	473 Aug 23.3	563 Mar 12.3	652 Sep 11.1	742 Mar 31.3	831 Oct 1.3
385 Sep 18.1	475 Apr 7.3	564 Oct 7.5	654 Apr 25.9	743 Oct 28.5	833 May 14.2
387 May 2.6	476 Nov 4.1	566 May 20.8	655 Nov 25.6	745 Jun 7.8	834 Dec 16.3
388 Dec 2.4	478 Jun 14.3	567 Dec 24.2	657 Jul 2.3	747 Jan 13.9	836 Jul 20.4
390 Jul 8.8	480 Jan 21.6	569 Jul 26.9	659 Feb 10.8	748 Aug 14.2	838 Mar 2.6
392 Feb 18.3	481 Aug 21.0	571 Mar 9.9	660 Sep 8.7	750 Mar 29.0	839 Sep 28.9
393 Sep 15.7	483 Apr 5.0	572 Oct 5.1	662 Apr 23.6	751 Oct 25.9	841 May 11.9
395 Apr 30.3	484 Nov 1.5	574 May 18.5	663 Nov 23.0	753 Jun 5.6	842 Dec 13.7
396 Nov 29.8	486 Jun 12.1	575 Dec 21.6	665 Jun 30.1	755 Jan 11.3	844 Jul 18.2
398 Jul 6.6	488 Jan 19.1	577 Jul 24.7	667 Feb 8.3	756 Aug 12.0	846 Feb 28.1
400 Feb 15.8	489 Aug 18.7	579 Mar 7.4	668 Sep 6.4	758 Mar 26.6	847 Sep 26.5
401 Sep 13.3	491 Apr 2.6	580 Oct 2.6	670 Apr 21.3	759 Oct 23.4	849 May 9.6
403 Apr 28.0	492 Oct 30.0	582 May 16.3	671 Nov 20.4	761 Jun 3.4	850 Dec 11.1
404 Nov 27.2	494 Jun 9.9	583 Dec 19.0	673 Jun 27.9	763 Jan 8.7	852 Jul 15.9
406 Jul 4.4	496 Jan 16.5	585 Jul 22.4	675 Feb 5.8	764 Aug 9.7	854 Feb 25.7
408 Feb 13.4	497 Aug 16.4	587 Mar 5.0	676 Sep 4.0	766 Mar 24.3	855 Sep 24.1
409 Sep 10.9	499 Mar 31.3	588 Sep 30.2	678 Apr 19.0	767 Oct 20.9	857 May 7.4
411 Apr 25.7	500 Oct 27.4	590 May 14.0	679 Nov 17.8	769 Jun 1.1	858 Dec 8.5
412 Nov 24.6	502 Jun 7.7	591 Dec 16.4	681 Jun 25.7	771 Jan 6.5	860 Jul 13.7
414 Jul 2.1	504 Jan 14.0	593 Jul 20.2	683 Feb 3.3	772 Aug 7.4	862 Feb 23.2
416 Feb 10.9	505 Aug 14.1	595 Mar 2.6	684 Sep 1.7	774 Mar 21.9	863 Sep 21.7
417 Sep 8.6	507 Mar 28.9	596 Sep 27.8	686 Apr 16.7	775 Oct 18.4	865 May 5.1
419 Apr 23.4	508 Oct 24.9	598 May 11.8	687 Nov 15.3	777 May 29.9	866 Dec 5.9
420 Nov 22.0	510 Jun 5.4	599 Dec 13.8	689 Jun 23.4	779 Jan 3.5	868 Jul 11.4
422 Jun 29.9	512 Jan 11.4	601 Jul 18.0	691 Jan 31.8	780 Aug 5.2	870 Feb 20.7
424 Feb 8.4	513 Aug 11.8	603 Feb 28.1	692 Aug 30.4	782 Mar 19.5	871 Sep 19.3
425 Sep 6.2	515 Mar 26.6	604 Sep 25.4	694 Apr 14.4	783 Oct 16.0	873 May 2.8
427 Apr 21.1	516 Oct 22.4	606 May 9.5	695 Nov 12.7	785 May 27.7	874 Dec 3.3
428 Nov 19.5	518 Jun 3.2	607 Dec 11.2	697 Jun 21.2	786 Dec 31.9	876 Jul 9.2
430 Jun 27.7	520 Jan 8.8	609 Jul 15.7	699 Jan 29.2	788 Aug 2.9	878 Feb 18.3
432 Feb 5.9	521 Aug 9.5	611 Feb 25.7	700 Aug 28.0	790 Mar 17.1	879 Sep 16.9
433 Sep 3.9	523 Mar 24.2	612 Sep 23.0	702 Apr 12.0	791 Oct 13.5	881 Apr 30.5
435 Apr 18.9	524 Oct 19.9	614 May 7.2	703 Nov 10.1	793 May 25.4	882 Nov 30.7
436 Nov 16.9	526 Jun 1.0	615 Dec 8.6	705 Jun 19.0	794 Dec 29.3	884 Jul 7.0
438 Jun 25.5	528 Jan 6.2	617 Jul 13.5	707 Jan 26.7	796 Jul 31.6	886 Feb 15.8
440 Feb 3.3	529 Aug 7.2	619 Feb 23.2	708 Aug 25.7	798 Mar 14.7	887 Sep 14.5
441 Sep 1.5	531 Mar 21.8	620 Sep 20.6	710 Apr 9.7	799 Oct 11.0	889 Apr 28.3
443 Apr 16.6	532 Oct 17.4	622 May 5.0	711 Nov 7.6	801 May 23.2	890 Nov 28.1

892 Jul 4.8	982 Jan 16.4	1071 Jul 23.8	1161 Feb 5.7	1250 Aug 11.0	1340 Feb 26.6
894 Feb 13.3	983 Aug 17.7	1073 Mar 5.0	1162 Sep 5.3	1252 Mar 24.3	1341 Sep 24.2
895 Sep 12.2	985 Mar 31.4	1074 Oct 1.4	1164 Apr 19.2	1253 Oct 21.0	1343 May 8.6
897 Apr 26.0	986 Oct 28.5	1076 May 14.3	1165 Nov 17.8	1255 Jun 2.5	1344 Dec 8.3
898 Nov 25.5	988 Jun 8.0	1077 Dec 16.2	1167 Jun 27.0	1257 Jan 5.9	1346 Jul 15.1
900 Jul 2.5	990 Jan 13.8	1079 Jul 21.6	1169 Feb 3.2	1258 Aug 8.8	1348 Feb 24.1
902 Feb 10.8	991 Aug 15.4	1081 Mar 2.6	1170 Sep 3.0	1260 Mar 21.9	1349 Sep 21.9
903 Sep 9.8	993 Mar 29.0	1082 Sep 28.9	1172 Apr 16.8	1261 Oct 18.5	1351 May 6.4
905 Apr 23.7	994 Oct 25.9	1084 May 12.1	1173 Nov 15.2	1263 May 31.2	1352 Dec 5.8
906 Nov 22.9	996 Jun 5.8	1085 Dec 13.6	1175 Jun 24.8	1265 Jan 3.3	1354 Jul 12.8
908 Jun 30.3	998 Jan 11.2	1087 Jul 19.3	1177 Jan 31.6	1266 Aug 6.5	1356 Feb 21.7
910 Feb 8.3	999 Aug 13.1	1089 Feb 28.1	1178 Aug 31.6	1268 Mar 19.5	1357 Sep 19.5
911 Sep 7.5	1001 Mar 26.7	1090 Sep 26.5	1180 Apr 14.5	1269 Oct 16.0	1359 May 4.1
913 Apr 21.4	1002 Oct 23.4	1092 May 9.8	1181 Nov 12.6	1271 May 29.0	1360 Dec 3.2
914 Nov 20.4	1004 Jun 3.6	1093 Dec 11.0	1183 Jun 22.6	1272 Dec 31.8	1362 Jul 10.6
916 Jun 28.1	1006 Jan 8.6	1095 Jul 17.1	1185 Jan 29.1	1274 Aug 4.2	1364 Feb 19.2
918 Feb 5.8	1007 Aug 10.9	1097 Feb 25.7	1186 Aug 29.3	1276 Mar 17.1	1365 Sep 17.1
919 Sep 5.2	1009 Mar 24.3	1098 Sep 24.1	1188 Apr 12.2	1277 Oct 13.6	1367 May 1.8
921 Apr 19.1	1010 Oct 21.0	1100 May 7.5	1189 Nov 10.1	1279 May 26.7	1368 Nov 30.6
922 Nov 17.8	1012 Jun 1.3	1101 Dec 8.4	1191 Jun 20.4	1280 Dec 29.2	1370 Jul 8.4
924 Jun 25.8	1014 Jan 6.0	1103 Jul 14.9	1193 Jan 26.5	1282 Aug 2.0	1372 Feb 16.7
926 Feb 3.2	1015 Aug 8.6	1105 Feb 23.2	1194 Aug 27.0	1284 Mar 14.7	1373 Sep 14.8
927 Sep 2.8	1017 Mar 21.9	1106 Sep 21.8	1196 Apr 9.9	1285 Oct 11.1	1375 Apr 29.5
929 Apr 16.8	1018 Oct 18.5	1108 May 5.2	1197 Nov 7.6	1287 May 24.5	1376 Nov 28.0
930 Nov 15.2	1020 May 30.1	1109 Dec 5.8	1199 Jun 18.1	1288 Dec 26.6	1378 Jul 6.1
932 Jun 23.6	1022 Jan 3.4	1111 Jul 12.6	1201 Jan 24.0	1290 Jul 30.7	1380 Feb 14.2
934 Jan 31.7	1023 Aug 6.3	1113 Feb 20.7	1202 Aug 24.7	1292 Mar 12.3	1381 Sep 12.4
935 Aug 31.5	1025 Mar 19.5	1114 Sep 19.4	1204 Apr 7.5	1293 Oct 8.7	1383 Apr 27.2
937 Apr 14.4	1026 Oct 16.0	1116 May 2.9	1205 Nov 5.0	1295 May 22.2	1384 Nov 25.4
938 Nov 12.7	1028 May 27.8	1117 Dec 3.2	1207 Jun 15.9	1296 Dec 24.0	1386 Jul 3.9
940 Jun 21.4	1029 Dec 31.8	1119 Jul 10.4	1209 Jan 21.4	1298 Jul 28.5	1388 Feb 11.7
942 Jan 29.2	1031 Aug 4.1	1121 Feb 18.2	1210 Aug 22.4	1300 Mar 9.9	1389 Sep 10.1
943 Aug 29.2	1033 Mar 17.1	1122 Sep 17.0	1212 Apr 5.2	1301 Oct 6.3	1391 Apr 24.9
945 Apr 12.1	1034 Oct 13.5	1124 Apr 30.7	1213 Nov 2.5	1303 May 20.0	1392 Nov 22.8
946 Nov 10.1	1036 May 25.6	1125 Nov 30.6	1215 Jun 13.7	1304 Dec 21.4	1394 Jul 1.7
948 Jun 19.2	1037 Dec 29.2	1127 Jul 8.2	1217 Jan 18.9	1306 Jul 26.2	1396 Feb 9.1
950 Jan 26.6	1039 Aug 1.8	1129 Feb 15.7	1218 Aug 20.1	1308 Mar 7.4	1397 Sep 7.7
951 Aug 26.9	1041 Mar 14.7	1130 Sep 14.7	1220 Apr 2.8	1309 Oct 3.8	1399 Apr 22.6
953 Apr 9.8	1042 Oct 11.1	1132 Apr 28.4	1221 Oct 31.0	1311 May 17.7	1400 Nov 20.3
954 Nov 7.6	1044 May 23.3	1133 Nov 28.1	1223 Jun 11.4	1312 Dec 18.7	1402 Jun 29.4
956 Jun 17.0	1045 Dec 26.7	1135 Jul 6.0	1225 Jan 16.3	1314 Jul 24.0	1404 Feb 6.6
958 Jan 24.1	1047 Jul 30.6	1137 Feb 13.2	1226 Aug 17.8	1316 Mar 5.0	1405 Sep 5.4
959 Aug 24.6	1049 Mar 12.3	1138 Sep 12.3	1228 Mar 31.5	1317 Oct 1.4	1407 Apr 20.2
961 Apr 7.4	1050 Oct 8.6	1140 Apr 26.1	1229 Oct 28.5	1319 May 15.4	1408 Nov 17.7
962 Nov 5.0	1052 May 21.1	1141 Nov 25.5	1231 Jun 9.2	1320 Dec 16.1	1410 Jun 27.2
964 Jun 14.7	1053 Dec 24.0	1143 Jul 3.7	1233 Jan 13.7	1322 Jul 21.8	1412 Feb 4.1
966 Jan 21.5	1055 Jul 28.3	1145 Feb 10.7	1234 Aug 15.6	1324 Mar 2.5	1413 Sep 3.1
967 Aug 22.3	1057 Mar 9.9	1146 Sep 10.0	1236 Mar 29.1	1325 Sep 29.0	1415 Apr 17.9
969 Apr 5.1	1058 Oct 6.2	1148 Apr 23.8	1237 Oct 26.0	1327 May 13.2	1416 Nov 15.1
970 Nov 2.5	1060 May 18.8	1149 Nov 22.9	1239 Jun 7.0	1328 Dec 13.6	1418 Jun 25.0
972 Jun 12.5	1061 Dec 21.4	1151 Jul 1.5	1241 Jan 11.1	1330 Jul 19.5	1420 Feb 1.6
974 Jan 18.9	1063 Jul 26.1	1153 Feb 8.2	1242 Aug 13.3	1332 Feb 29.1	1421 Aug 31.8
975 Aug 20.0	1065 Mar 7.4	1154 Sep 7.6	1244 Mar 26.7	1333 Sep 26.6	1423 Apr 15.6
977 Apr 2.8	1066 Oct 3.8	1156 Apr 21.5	1245 Oct 23.5	1335 May 10.9	1424 Nov 12.6
978 Oct 31.0	1068 May 16.6	1157 Nov 20.3	1247 Jun 4.7	1336 Dec 11.0	1426 Jun 22.8
980 Jun 10.3	1069 Dec 18.8	1159 Jun 29.3	1249 Jan 8.5	1338 Jul 17.3	1428 Jan 30.0

1429 Aug 29.5	1519 Mar 18.1	1608 Sep 27.2	1698 Apr 16.3	1787 Oct 18.4	1877 May 7.0
1431 Apr 13.3	1520 Oct 13.6	1610 May 11.9	1699 Nov 13.5	1789 May 31.2	1878 Dec 5.7
1432 Nov 10.1	1522 May 26.9	1611 Dec 11.5	1701 Jun 25.0	1791 Jan 2.2	1880 Jul 14.0
1434 Jun 20.5	1523 Dec 30.1	1613 Jul 18.5	1703 Jan 30.7	1792 Aug 6.6	1882 Feb 21.0
1436 Jan 27.5	1525 Aug 2.2	1615 Feb 26.6	1704 Aug 31.4	1794 Mar 19.4	1883 Sep 21.0
1437 Aug 27.2	1527 Mar 15.7	1616 Sep 24.9	1706 Apr 14.9	1795 Oct 16.0	1885 May 4.7
1439 Apr 10.9	1528 Oct 11.2	1618 May 9.6	1707 Nov 12.0	1797 May 29.0	1886 Dec 3.2
1440 Nov 7.5	1530 May 24.6	1619 Dec 8.9	1709 Jun 22.7	1798 Dec 30.6	1888 Jul 11.8
1442 Jun 18.3	1531 Dec 27.5	1621 Jul 16.3	1711 Jan 28.1	1800 Aug 5.3	1890 Feb 18.5
1444 Jan 24.9	1533 Jul 30.9	1623 Feb 24.1	1712 Aug 29.2	1802 Mar 17.9	1891 Sep 18.7
1445 Aug 24.9	1535 Mar 13.3	1624 Sep 22.5	1714 Apr 12.5	1803 Oct 14.6	1893 May 2.4
1447 Apr 8.6	1536 Oct 8.7	1626 May 7.3	1715 Nov 9.5	1805 May 27.7	1894 Nov 30.6
1448 Nov 5.0	1538 May 22.4	1627 Dec 6.4	1717 Jun 20.5	1806 Dec 29.0	1896 Jul 9.5
1450 Jun 16.1	1539 Dec 24.9	1629 Jul 14.1	1719 Jan 25.5	1808 Aug 3.1	1898 Feb 15.9
1452 Jan 22.3	1541 Jul 28.7	1631 Feb 21.6	1720 Aug 26.9	1810 Mar 15.5	1899 Sep 16.3
1453 Aug 22.6	1543 Mar 10.9	1632 Sep 20.2	1722 Apr 10.1	1811 Oct 12.2	1901 May 1.1
1455 Apr 6.2	1544 Oct 6.3	1634 May 5.0	1723 Nov 7.0	1813 May 25.4	1902 Nov 29.1
1456 Nov 2.5	1546 May 20.1	1635 Dec 3.8	1725 Jun 18.3	1814 Dec 26.4	1904 Jul 8.3
1458 Jun 13.8	1547 Dec 22.3	1637 Jul 11.8	1727 Jan 22.9	1816 Jul 31.9	1906 Feb 14.4
1460 Jan 19.8	1549 Jul 26.4	1639 Feb 19.1	1728 Aug 24.6	1818 Mar 13.0	1907 Sep 15.0
1461 Aug 20.3	1551 Mar 8.4	1640 Sep 17.9	1730 Apr 7.8	1819 Oct 9.8	1909 Apr 28.7
1463 Apr 3.9	1552 Oct 3.9	1642 May 2.7	1731 Nov 4.5	1821 May 23.1	1910 Nov 26.6
1464 Oct 31.0	1554 May 17.8	1643 Dec 1.2	1733 Jun 16.0	1822 Dec 23.8	1912 Jul 6.1
1466 Jun 11.6	1555 Dec 19.7	1645 Jul 9.6	1735 Jan 20.3	1824 Jul 29.6	1914 Feb 11.8
1468 Jan 17.2	1557 Jul 24.2	1647 Feb 16.6	1736 Aug 22.3	1826 Mar 10.5	1915 Sep 12.7
1469 Aug 18.0	1559 Mar 6.0	1648 Sep 15.5	1738 Apr 5.4	1827 Oct 7.4	1917 Apr 26.4
1471 Apr 1.5	1560 Oct 1.5	1650 Apr 30.3	1739 Nov 2.0	1829 May 20.9	1918 Nov 24.0
1472 Oct 28.5	1562 May 15.6	1651 Nov 28.7	1741 Jun 13.8	1830 Dec 21.2	1920 Jul 3.8
1474 Jun 9.4	1563 Dec 14.5	1653 Jul 7.4	1743 Jan 17.7	1832 Jul 27.4	1922 Feb 9.3
1476 Jan 14.6	1565 Jul 21.9	1655 Feb 14.0	1744 Aug 20.1	1834 Mar 8.0	1923 Sep 10.4
1477 Aug 15.7	1567 Mar 3.5	1656 Sep 13.2	1746 Apr 3.0	1835 Oct 5.0	1925 Apr 24.0
1479 Mar 30.1	1568 Sep 29.1	1658 Apr 28.0	1747 Oct 30.6	1837 May 18.6	1926 Nov 21.5
1480 Oct 26.0	1570 May 13.3	1659 Nov 26.1	1749 Jun 11.5	1838 Dec 18.6	1928 Jul 1.6
1482 Jun 7.1	1571 Dec 14.5	1661 Jul 1.5	1751 Jan 15.1	1840 Jul 25.2	1930 Feb 6.7
1484 Jan 12.0	1573 Jul 19.7	1663 Feb 11.5	1752 Aug 17.8	1842 Mar 5.6	1931 Sep 8.1
1485 Aug 13.5	1575 Mar 1.0	1664 Sep 10.9	1754 Mar 31.5	1843 Oct 2.7	1933 Apr 21.7
1487 Mar 27.7	1576 Sep 26.7	1666 Apr 25.7	1755 Oct 28.1	1845 May 16.3	1934 Nov 19.0
1488 Oct 23.5	1578 May 11.0	1667 Nov 23.6	1757 Jun 9.3	1846 Dec 16.0	1936 Jun 29.4
1490 Jun 4.9	1579 Dec 11.9	1669 Jul 2.9	1759 Jan 12.6	1848 Jul 22.9	1938 Feb 4.1
1492 Jan 9.4	1581 Jul 17.5	1671 Feb 8.9	1760 Aug 15.6	1850 Mar 3.1	1939 Sep 5.9
1493 Aug 11.2	1583 Mar 8.6	1672 Sep 8.6	1762 Mar 29.1	1851 Sep 30.3	1941 Apr 19.3
1495 Mar 25.3	1584 Oct 4.3	1674 Apr 23.3	1763 Oct 25.7	1853 May 14.0	1942 Nov 16.5
1496 Oct 21.0	1586 May 18.7	1675 Nov 21.0	1765 Jun 7.0	1854 Dec 13.4	1944 Jun 27.1
1498 Jun 2.6	1587 Dec 19.3	1677 Jun 30.7	1767 Jan 10.0	1856 Jul 20.7	1946 Feb 1.6
1500 Jan 6.8	1589 Jul 25.2	1679 Feb 6.4	1768 Aug 13.3	1858 Feb 28.6	1947 Sep 3.6
1501 Aug 8.9	1591 Mar 6.1	1680 Sep 6.3	1770 Mar 26.7	1859 Sep 28.0	1949 Apr 16.9
1503 Mar 22.9	1592 Oct 1.9	1682 Apr 21.0	1771 Oct 23.2	1861 May 11.7	1950 Nov 14.0
1504 Oct 18.5	1594 May 16.5	1683 Nov 18.5	1773 Jun 4.8	1862 Dec 10.9	1952 Jun 24.9
1506 May 31.4	1595 Dec 16.7	1685 Jun 28.5	1775 Jan 7.4	1864 Jul 18.5	1954 Jan 30.0
1508 Jan 4.3	1597 Jul 23.0	1687 Feb 3.8	1776 Aug 11.1	1866 Feb 26.1	1955 Sep 1.3
1509 Aug 6.7	1599 Mar 3.6	1688 Sep 4.0	1778 Mar 24.3	1867 Sep 25.6	1957 Apr 14.5
1511 Mar 20.6	1600 Sep 29.6	1690 Apr 18.6	1779 Oct 20.8	1869 May 9.4	1958 Nov 11.5
1512 Oct 16.1	1602 May 14.2	1691 Nov 16.0	1781 Jun 2.5	1870 Dec 8.3	1960 Jun 22.7
1514 May 29.1	1603 Dec 14.1	1693 Jun 26.2	1783 Jan 4.8	1872 Jul 16.2	1962 Jan 27.4
1516 Jan 1.7	1605 Jul 20.8	1695 Feb 1.2	1784 Aug 8.8	1874 Feb 23.5	1963 Aug 30.0
1517 Aug 4.4	1607 Mar 1.1	1696 Sep 1.7	1786 Mar 21.8	1875 Sep 23.3	1965 Apr 12.2

SUPERIOR CONJUNCTIONS OF VENUS

1966 Nov 9.0	2056 May 26.5	2145 Nov 30.1	2235 Jun 16.8	2324 Dec 22.5	2414 Jul 5.9
1968 Jun 20.4	2057 Dec 27.3	2147 Jul 10.5	2237 Jan 18.9	2326 Jul 30.5	2416 Feb 10.5
1970 Jan 24.8	2059 Aug 3.0	2149 Feb 15.3	2238 Aug 23.1	2328 Mar 9.4	2417 Sep 11.4
1971 Aug 27.8	2061 Mar 14.0	2150 Sep 16.2	2240 Apr 4.5	2329 Oct 6.9	2419 Apr 25.7
1973 Apr 9.8	2062 Oct 10.9	2152 Apr 29.8	2241 Nov 1.2	2331 May 21.4	2420 Nov 22.0
1974 Nov 6.5	2064 May 24.2	2153 Nov 27.5	2243 Jun 14.5	2332 Dec 19.9	2422 Jul 3.7
1976 Jun 18.2	2065 Dec 24.7	2155 Jul 8.2	2245 Jan 16.4	2334 Jul 28.3	2424 Feb 7.9
1978 Jan 22.2	2067 Jul 31.8	2157 Feb 12.7	2246 Aug 20.9	2336 Mar 6.9	2425 Sep 9.1
1979 Aug 25.5	2069 Mar 11.5	2158 Sep 13.9	2248 Apr 2.1	2337 Oct 4.5	2427 Apr 23.3
1981 Apr 7.4	2070 Oct 8.5	2160 Apr 27.4	2249 Oct 29.8	2339 May 19.1	2428 Nov 19.5
1982 Nov 4.1	2072 May 21.9	2161 Nov 25.0	2251 Jun 12.3	2340 Dec 17.3	2430 Jul 1.4
1984 Jun 15.9	2073 Dec 22.1	2163 Jul 6.0	2253 Jan 13.8	2342 Jul 26.0	2432 Feb 5.4
1986 Jan 19.6	2075 Jul 29.6	2165 Feb 10.2	2254 Aug 18.6	2344 Mar 4.4	2433 Sep 6.9
1987 Aug 23.2	2077 Mar 9.0	2166 Sep 11.6	2256 Mar 30.7	2345 Oct 2.2	2435 Apr 21.0
1989 Apr 5.0	2078 Oct 6.1	2168 Apr 25.1	2257 Oct 27.3	2347 May 16.8	2436 Nov 17.0
1990 Nov 1.6	2080 May 19.7	2169 Nov 22.5	2259 Jun 10.0	2348 Dec 14.8	2438 Jun 29.2
1992 Jun 13.7	2081 Dec 19.5	2171 Jul 3.8	2261 Jan 11.2	2350 Jul 23.8	2440 Feb 2.8
1994 Jan 17.1	2083 Jul 27.3	2173 Feb 7.6	2262 Aug 16.4	2352 Mar 1.9	2441 Sep 4.6
1995 Aug 21.0	2085 Mar 6.5	2174 Sep 9.3	2264 Mar 28.2	2353 Sep 29.9	2443 Apr 18.6
1997 Apr 2.5	2086 Oct 3.8	2176 Apr 22.7	2265 Oct 24.9	2355 May 14.5	2444 Nov 14.5
1998 Oct 30.2	2088 May 17.4	2177 Nov 20.0	2267 Jun 7.7	2356 Dec 12.2	2446 Jun 26.9
2000 Jun 11.4	2089 Dec 17.0	2179 Jul 1.5	2269 Jan 8.6	2358 Jul 21.6	2448 Jan 31.2
2002 Jan 14.5	2091 Jul 25.1	2181 Feb 5.0	2270 Aug 14.1	2360 Feb 28.4	2449 Sep 2.3
2003 Aug 18.7	2093 Mar 4.0	2182 Sep 7.0	2272 Mar 25.8	2361 Sep 27.5	2451 Apr 16.2
2005 Mar 31.1	2094 Oct 1.4	2184 Apr 20.3	2273 Oct 22.5	2363 May 12.2	2452 Nov 12.0
2006 Oct 27.7	2096 May 15.0	2185 Nov 17.5	2275 Jun 5.5	2364 Dec 9.6	2454 Jun 24.7
2008 Jun 9.2	2097 Dec 14.4	2187 Jun 29.3	2277 Jan 6.0	2366 Jul 19.3	2456 Jan 28.6
2010 Jan 11.9	2099 Jul 22.9	2189 Feb 2.5	2278 Aug 11.9	2368 Feb 25.8	2457 Aug 31.1
2011 Aug 16.5	2101 Mar 2.5	2190 Sep 4.7	2280 Mar 23.3	2369 Sep 25.2	2459 Apr 13.8
2013 Mar 28.7	2102 Sep 30.1	2192 Apr 17.9	2281 Oct 20.1	2371 May 9.8	2460 Nov 9.6
2014 Oct 25.3	2104 May 13.7	2193 Nov 15.0	2283 Jun 3.2	2372 Dec 7.1	2462 Jun 22.4
2016 Jun 6.9	2105 Dec 12.8	2195 Jun 27.0	2285 Jan 3.4	2374 Jul 17.1	2464 Jan 26.0
2018 Jan 9.3	2107 Jul 21.6	2197 Jan 30.9	2286 Aug 9.7	2376 Feb 23.3	2465 Aug 28.8
2019 Aug 14.2	2109 Feb 28.0	2198 Sep 2.5	2288 Mar 20.9	2377 Sep 22.9	2467 Apr 11.4
2021 Mar 26.3	2110 Sep 27.7	2200 Apr 16.6	2289 Oct 17.7	2379 May 7.5	2468 Nov 7.1
2022 Oct 22.9	2112 May 11.4	2201 Nov 13.5	2291 May 31.9	2380 Dec 4.6	2470 Jun 20.2
2024 Jun 4.6	2113 Dec 10.3	2203 Jun 25.8	2292 Dec 31.8	2382 Jul 14.9	2472 Jan 23.4
2026 Jan 6.7	2115 Jul 19.4	2205 Jan 29.3	2294 Aug 7.4	2384 Feb 20.7	2473 Aug 26.5
2027 Aug 12.0	2117 Feb 25.4	2206 Sep 1.2	2296 Mar 18.4	2385 Sep 20.6	2475 Apr 8.9
2029 Mar 23.8	2118 Sep 25.4	2208 Apr 14.2	2297 Oct 15.3	2387 May 5.2	2476 Nov 4.7
2030 Oct 20.4	2120 May 9.1	2209 Nov 11.0	2299 May 29.6	2388 Dec 2.0	2478 Jun 17.9
2032 Jun 2.4	2121 Dec 7.7	2211 Jun 23.6	2300 Dec 30.2	2390 Jul 12.6	2480 Jan 20.8
2034 Jan 4.1	2123 Jul 17.2	2213 Jan 26.7	2302 Aug 6.2	2392 Feb 18.2	2481 Aug 24.3
2035 Aug 9.8	2125 Feb 22.9	2214 Aug 29.9	2304 Mar 16.9	2393 Sep 18.3	2483 Apr 6.5
2037 Mar 21.4	2126 Sep 23.1	2216 Apr 11.8	2305 Oct 14.0	2395 May 2.8	2484 Nov 2.2
2038 Oct 18.0	2128 May 6.8	2217 Nov 8.6	2307 May 28.3	2396 Nov 29.5	2486 Jun 15.7
2040 May 31.1	2129 Dec 5.1	2219 Jun 21.3	2308 Dec 27.6	2398 Jul 10.4	2488 Jan 18.3
2042 Jan 1.5	2131 Jul 14.9	2221 Jan 24.1	2310 Aug 4.0	2400 Feb 15.6	2489 Aug 22.0
2043 Aug 7.5	2133 Feb 20.4	2222 Aug 27.7	2312 Mar 14.4	2401 Sep 16.0	2491 Apr 4.1
2045 Mar 18.9	2134 Sep 20.8	2224 Apr 9.4	2313 Oct 11.6	2403 Apr 30.4	2492 Oct 30.8
2046 Oct 15.6	2136 May 4.4	2225 Nov 6.1	2315 May 26.0	2404 Nov 27.0	2494 Jun 13.4
2048 May 28.8	2137 Dec 2.6	2227 Jun 19.1	2316 Dec 25.0	2406 Jul 8.1	2496 Jan 15.7
2049 Dec 29.9	2139 Jul 12.7	2229 Jan 21.5	2318 Aug 1.7	2408 Feb 13.1	2497 Aug 19.8
2051 Aug 5.3	2141 Feb 17.8	2230 Aug 25.4	2320 Mar 11.9	2409 Sep 13.7	2499 Apr 1.6
2053 Mar 16.4	2142 Sep 18.5	2232 Apr 7.0	2321 Oct 9.2	2411 Apr 28.1	2500 Oct 29.4
2054 Oct 13.3	2144 May 2.1	2233 Nov 3.6	2323 May 23.7	2412 Nov 24.5	

IX

OPPOSITIONS OF JUPITER, 0 - 2500

0 Feb 26.4	50 May 18.3	100 Aug 17.9	150 Nov 18.4	201 Feb 6.8	251 Apr 27.7
1 Mar 28.4	51 Jun 22.4	101 Sep 24.2	151 Dec 21.3	202 Mar 9.5	252 May 30.5
2 Apr 29.7	52 Jul 28.2	102 Oct 30.3	153 Jan 20.7	203 Apr 9.8	253 Jul 5.3
3 Jun 2.7	53 Sep 3.7	103 Dec 3.3	154 Feb 20.5	204 May 11.6	254 Aug 11.5
4 Jul 7.7	54 Oct 10.6	105 Jan 3.5	155 Mar 23.4	205 Jun 15.3	255 Sep 18.0
5 Aug 14.1	55 Nov 14.8	106 Feb 3.5	156 Apr 23.3	206 Jul 21.8	256 Oct 23.4
6 Sep 20.5	56 Dec 16.9	107 Mar 6.2	157 May 26.8	207 Aug 28.3	257 Nov 26.8
7 Oct 26.7	58 Jan 17.4	108 Apr 5.4	158 Jul 1.3	208 Oct 3.4	258 Dec 29.2
8 Nov 28.8	59 Feb 17.1	109 May 8.0	159 Aug 7.3	209 Nov 7.9	260 Jan 29.4
9 Dec 31.0	60 Mar 18.9	110 Jun 11.4	160 Sep 12.8	210 Dec 11.4	261 Feb 28.0
11 Jan 31.0	61 Apr 19.6	111 Jul 17.7	161 Oct 19.3	212 Jan 12.2	262 Mar 31.1
12 Mar 1.7	62 May 23.1	112 Aug 23.1	162 Nov 23.0	213 Feb 11.0	263 May 2.4
13 Apr 1.8	63 Jun 27.5	113 Sep 29.4	163 Dec 25.7	214 Mar 13.8	264 Jun 4.4
14 May 4.3	64 Aug 2.5	114 Nov 4.2	165 Jan 25.0	215 Apr 14.3	265 Jul 10.4
15 Jun 7.7	65 Sep 9.1	115 Dec 7.9	166 Feb 24.7	216 May 16.3	266 Aug 16.8
16 Jul 12.9	66 Oct 15.7	117 Jan 7.8	167 Mar 27.7	217 Jun 20.2	267 Sep 23.1
17 Aug 19.3	67 Nov 19.5	118 Feb 7.7	168 Apr 27.8	218 Jul 26.8	268 Oct 28.2
18 Sep 25.6	68 Dec 21.3	119 Mar 10.4	169 May 31.6	219 Sep 2.3	269 Dec 1.3
19 Oct 31.5	70 Jan 21.7	120 Apr 9.7	170 Jul 6.3	220 Oct 8.3	271 Jan 2.6
20 Dec 3.3	71 Feb 21.3	121 May 12.6	171 Aug 12.5	221 Nov 12.6	272 Feb 2.7
22 Jan 4.3	72 Mar 23.2	122 Jun 16.3	172 Sep 18.0	222 Dec 15.9	273 Mar 4.3
23 Feb 4.3	73 Apr 24.2	123 Jul 22.9	173 Oct 24.3	224 Jan 16.5	274 Apr 4.6
24 Mar 6.0	74 May 27.9	124 Aug 28.5	174 Nov 27.7	225 Feb 15.3	275 May 7.0
25 Apr 6.3	75 Jul 2.6	125 Oct 4.6	175 Dec 30.1	226 Mar 18.1	276 Jun 9.3
26 May 9.0	76 Aug 7.7	126 Nov 9.0	177 Jan 29.3	227 Apr 18.8	277 Jul 15.5
27 Jun 12.6	77 Sep 14.2	127 Dec 12.4	178 Feb 28.9	228 May 21.0	278 Aug 21.8
28 Jul 18.0	78 Oct 20.6	129 Jan 12.2	179 Apr 1.0	229 Jun 25.2	279 Sep 28.1
29 Aug 24.5	79 Nov 24.1	130 Feb 11.9	180 May 2.3	230 Aug 1.0	280 Nov 2.0
30 Sep 30.6	80 Dec 25.7	131 Mar 14.6	181 Jun 5.4	231 Sep 7.5	281 Dec 5.9
31 Nov 5.3	82 Jan 25.9	132 Apr 14.2	182 Jul 11.4	232 Oct 13.3	283 Jan 6.9
32 Dec 7.9	83 Feb 25.6	133 May 17.3	183 Aug 17.8	233 Nov 17.3	284 Feb 6.9
34 Jan 8.7	84 Mar 27.6	134 Jun 21.4	184 Sep 23.2	234 Dec 20.3	285 Mar 8.7
35 Feb 8.6	85 Apr 28.8	135 Jul 28.2	185 Oct 29.3	236 Jan 20.8	286 Apr 9.0
36 Mar 10.3	86 Jun 1.8	136 Sep 2.7	186 Dec 2.4	237 Feb 19.5	287 May 11.6
37 Apr 10.8	87 Jul 7.7	137 Oct 9.6	188 Jan 3.6	238 Mar 22.3	288 Jun 14.2
38 May 13.6	88 Aug 12.9	138 Nov 13.7	189 Feb 2.5	239 Apr 23.2	289 Jul 20.6
39 Jun 17.4	89 Sep 19.3	139 Dec 16.9	190 Mar 5.2	240 May 25.7	290 Aug 27.0
40 Jul 23.0	90 Oct 25.5	141 Jan 16.4	191 Apr 5.4	241 Jun 30.1	291 Oct 3.2
41 Aug 29.5	91 Nov 28.7	142 Feb 16.1	192 May 6.9	242 Aug 6.2	292 Nov 6.8
42 Oct 5.5	92 Dec 30.1	143 Mar 19.0	193 Jun 10.3	243 Sep 12.8	293 Dec 10.4
43 Nov 10.0	94 Jan 30.2	144 Apr 18.7	194 Jul 16.6	244 Oct 18.4	295 Jan 11.3
44 Dec 12.4	95 Mar 1.9	145 May 22.1	195 Aug 23.1	245 Nov 22.1	296 Feb 11.1
46 Jan 13.1	96 Apr 1.0	146 Jun 26.4	196 Sep 28.4	246 Dec 24.9	297 Mar 12.8
47 Feb 12.9	97 May 3.4	147 Aug 2.3	197 Nov 3.1	248 Jan 25.1	298 Apr 13.3
48 Mar 14.6	98 Jun 6.5	148 Sep 7.9	198 Dec 6.9	249 Feb 23.8	299 May 16.2
49 Apr 15.2	99 Jul 12.6	149 Oct 14.5	200 Jan 7.8	250 Mar 26.7	300 Jun 19.0

OPPOSITIONS OF JUPITER

301 Jul 25.7	362 Sep 27.2	423 Nov 26.1	485 Jan 18.7	546 Mar 11.9	607 May 4.9
302 Sep 1.3	363 Nov 2.2	424 Dec 27.8	486 Feb 18.4	547 Apr 12.2	608 Jun 6.9
303 Oct 8.4	364 Dec 5.2	426 Jan 27.9	487 Mar 21.1	548 May 13.9	609 Jul 13.0
304 Nov 11.7	366 Jan 6.3	427 Feb 27.5	488 Apr 20.8	549 Jun 17.5	610 Aug 19.4
305 Dec 15.1	367 Feb 6.2	428 Mar 29.5	489 May 24.0	550 Jul 24.1	611 Sep 26.0
307 Jan 15.7	368 Mar 7.8	429 Apr 30.6	490 Jun 28.3	551 Aug 30.7	612 Oct 31.2
308 Feb 15.4	369 Apr 8.0	430 Jun 3.5	491 Aug 4.4	552 Oct 6.0	613 Dec 4.4
309 Mar 17.2	370 May 10.6	431 Jul 9.4	492 Sep 10.0	553 Nov 10.6	615 Jan 5.6
310 Apr 17.8	371 Jun 14.2	432 Aug 14.7	493 Oct 16.9	554 Dec 14.1	616 Feb 5.5
311 May 21.0	372 Jul 19.6	433 Sep 21.3	494 Nov 20.8	556 Jan 14.8	617 Mar 7.1
312 Jun 24.1	373 Aug 26.2	434 Oct 27.5	495 Dec 23.7	557 Feb 13.6	618 Apr 7.2
313 Jul 31.0	374 Oct 2.4	435 Nov 30.8	497 Jan 23.0	558 Mar 16.2	619 May 9.6
314 Sep 6.6	375 Nov 7.1	437 Jan 1.2	498 Feb 22.7	559 Apr 16.7	620 Jun 12.0
315 Oct 13.4	376 Dec 9.7	438 Feb 1.2	499 Mar 25.5	560 May 18.7	621 Jul 18.3
316 Nov 16.5	378 Jan 10.6	439 Mar 3.8	500 Apr 25.4	561 Jun 22.7	622 Aug 24.8
317 Dec 19.5	379 Feb 10.4	440 Apr 2.9	501 May 29.0	562 Jul 29.5	623 Oct 1.2
319 Jan 20.0	380 Mar 12.1	441 May 5.3	502 Jul 3.5	563 Sep 5.1	624 Nov 5.1
320 Feb 19.7	381 Apr 12.5	442 Jun 8.5	503 Aug 9.7	564 Oct 11.1	625 Dec 9.0
321 Mar 21.5	382 May 15.4	443 Jul 14.6	504 Sep 15.3	565 Nov 15.4	627 Jan 10.0
322 Apr 22.3	383 Jun 19.2	444 Aug 20.1	505 Oct 21.9	566 Dec 18.6	628 Feb 9.8
323 May 25.8	384 Jul 24.9	445 Sep 26.4	506 Nov 25.5	568 Jan 19.2	629 Mar 11.4
324 Jun 29.2	385 Aug 31.4	446 Nov 1.5	507 Dec 28.1	569 Feb 17.9	630 Apr 11.7
325 Aug 5.3	386 Oct 7.5	447 Dec 5.4	509 Jan 27.3	570 Mar 20.6	631 May 14.4
326 Sep 11.8	387 Nov 11.9	449 Jan 5.5	510 Feb 27.0	571 Apr 21.3	632 Jun 17.0
327 Oct 18.4	388 Dec 14.2	450 Feb 5.5	511 Mar 29.9	572 May 23.5	633 Jul 23.5
328 Nov 21.2	390 Jan 14.9	451 Mar 8.2	512 Apr 30.0	573 Jun 27.8	634 Aug 30.1
329 Dec 24.0	391 Feb 14.7	452 Apr 7.4	513 Jun 2.9	574 Aug 3.7	635 Oct 6.3
331 Jan 24.3	392 Mar 16.5	453 May 10.0	514 Jul 8.7	575 Sep 10.3	636 Nov 9.9
332 Feb 24.0	393 Apr 17.1	454 Jun 13.5	515 Aug 14.9	576 Oct 16.1	637 Dec 13.5
333 Mar 25.9	394 May 20.2	455 Jul 19.8	516 Sep 20.4	577 Nov 20.1	639 Jan 14.3
334 Apr 26.9	395 Jun 24.3	456 Aug 25.2	517 Oct 26.7	578 Dec 23.0	640 Feb 14.1
335 May 30.6	396 Jul 30.0	457 Oct 1.5	518 Nov 30.1	580 Jan 23.5	641 Mar 15.8
336 Jul 4.3	397 Sep 5.5	458 Nov 6.2	520 Jan 1.5	581 Feb 22.2	642 Apr 16.2
337 Aug 10.4	398 Oct 12.4	459 Dec 9.9	521 Jan 31.6	582 Mar 25.0	643 May 19.1
338 Sep 16.8	399 Nov 16.5	461 Jan 9.9	522 Mar 3.3	583 Apr 25.8	644 Jun 22.0
339 Oct 23.2	400 Dec 18.7	462 Feb 9.8	523 Apr 3.4	584 May 28.3	645 Jul 28.7
340 Nov 25.7	402 Jan 19.2	463 Mar 12.5	524 May 4.7	585 Jul 2.8	646 Sep 4.3
341 Dec 28.3	403 Feb 19.0	464 Apr 11.9	525 Jun 7.8	586 Aug 8.9	647 Oct 11.4
343 Jan 28.6	404 Mar 20.8	465 May 14.7	526 Jul 13.8	587 Sep 15.5	648 Nov 14.8
344 Feb 28.3	405 Apr 21.6	466 Jun 18.4	527 Aug 20.2	588 Oct 21.1	649 Dec 18.1
345 Mar 30.3	406 May 24.9	467 Jul 24.9	528 Sep 25.6	589 Nov 24.9	651 Jan 18.7
346 May 1.5	407 Jun 29.2	468 Aug 30.4	529 Oct 31.7	590 Dec 27.6	652 Feb 18.4
347 Jun 4.4	408 Aug 4.2	469 Oct 6.6	530 Dec 4.8	592 Jan 27.9	653 Mar 20.1
348 Jul 9.3	409 Sep 10.7	470 Nov 11.1	532 Jan 6.0	593 Feb 26.5	654 Apr 20.6
349 Aug 15.5	410 Oct 17.4	471 Dec 14.5	533 Feb 5.0	594 Mar 29.3	655 May 23.7
350 Sep 22.0	411 Nov 21.3	473 Jan 14.3	534 Mar 7.5	595 Apr 30.3	656 Jun 26.8
351 Oct 28.2	412 Dec 23.2	474 Feb 14.1	535 Apr 7.7	596 Jun 2.0	657 Aug 2.7
352 Nov 30.4	414 Jan 23.6	475 Mar 16.7	536 May 9.1	597 Jul 7.7	658 Sep 9.4
354 Jan 1.8	415 Feb 23.2	476 Apr 16.2	537 Jun 12.5	598 Aug 14.0	659 Oct 16.5
355 Feb 1.8	416 Mar 25.1	477 May 19.2	538 Jul 18.8	599 Sep 20.7	660 Nov 19.6
356 Mar 3.5	417 Apr 26.0	478 Jun 23.2	539 Aug 25.4	600 Oct 26.2	661 Dec 22.7
357 Apr 3.6	418 May 29.6	479 Jul 30.0	540 Sep 30.8	601 Nov 29.7	663 Jan 23.1
358 May 5.9	419 Jul 4.2	480 Sep 4.7	541 Nov 5.7	603 Jan 1.2	664 Feb 22.8
359 Jun 9.2	420 Aug 9.4	481 Oct 11.7	542 Dec 9.5	604 Feb 1.3	665 Mar 24.5
360 Jul 14.4	421 Sep 16.0	482 Nov 16.0	544 Jan 10.5	605 Mar 2.8	666 Apr 25.1
361 Aug 20.8	422 Oct 22.5	483 Dec 19.2	545 Feb 9.3	606 Apr 2.7	667 May 28.5

668 Jul 1.9	729 Sep 3.5	790 Nov 4.8	851 Dec 31.5	913 Feb 21.0	974 Apr 14.4
669 Aug 8.1	730 Oct 10.7	791 Dec 8.9	853 Jan 30.6	914 Mar 23.6	975 May 16.9
670 Sep 14.8	731 Nov 15.3	793 Jan 9.0	854 Mar 2.2	915 Apr 24.1	976 Jun 19.5
671 Oct 21.6	732 Dec 17.6	794 Feb 8.8	855 Apr 2.0	916 May 26.3	977 Jul 26.0
672 Nov 24.4	734 Jan 18.3	795 Mar 11.4	856 May 3.0	917 Jun 30.5	978 Sep 1.6
673 Dec 27.2	735 Feb 17.9	796 Apr 10.5	857 Jun 5.8	918 Aug 6.5	979 Oct 9.0
675 Jan 27.4	736 Mar 19.6	797 May 13.0	858 Jul 11.6	919 Sep 13.2	980 Nov 12.8
676 Feb 27.0	737 Apr 20.0	798 Jun 16.4	859 Aug 18.0	920 Oct 19.2	981 Dec 16.4
677 Mar 28.8	738 May 23.1	799 Jul 22.9	860 Sep 23.6	921 Nov 23.3	983 Jan 17.2
678 Apr 29.8	739 Jun 27.2	800 Aug 28.5	861 Oct 30.1	922 Dec 26.3	984 Feb 16.9
679 Jun 2.5	740 Aug 2.2	801 Oct 4.9	862 Dec 3.5	924 Jan 26.6	985 Mar 18.5
680 Jul 7.2	741 Sep 8.9	802 Nov 9.7	864 Jan 4.9	925 Feb 25.2	986 Apr 18.8
681 Aug 13.5	742 Oct 15.9	803 Dec 13.5	865 Feb 3.9	926 Mar 27.9	987 May 21.7
682 Sep 20.1	743 Nov 20.1	805 Jan 13.3	866 Mar 6.5	927 Apr 28.7	988 Jun 24.5
683 Oct 26.6	744 Dec 22.1	806 Feb 13.1	867 Apr 6.4	928 May 31.1	989 Jul 31.2
684 Nov 29.1	746 Jan 22.6	807 Mar 15.7	868 May 7.6	929 Jul 5.6	990 Sep 6.9
685 Dec 31.6	747 Feb 22.2	808 Apr 15.0	869 Jun 10.7	930 Aug 11.7	991 Oct 14.0
687 Jan 31.7	748 Mar 23.9	809 May 17.7	870 Jul 16.8	931 Sep 18.4	992 Nov 17.5
688 Mar 2.3	749 Apr 24.6	810 Jun 21.5	871 Aug 23.3	932 Oct 24.1	993 Dec 20.8
689 Apr 2.2	750 May 28.0	811 Jul 28.1	872 Sep 28.8	933 Nov 27.9	995 Jan 21.5
690 May 4.4	751 Jul 2.4	812 Sep 2.8	873 Nov 4.0	934 Dec 30.6	996 Feb 21.2
691 Jun 7.4	752 Aug 7.5	813 Oct 10.0	874 Dec 8.0	936 Jan 30.9	997 Mar 22.8
692 Jul 12.4	753 Sep 14.1	814 Nov 14.5	876 Jan 9.2	937 Mar 1.5	998 Apr 23.3
693 Aug 18.8	754 Oct 20.9	815 Dec 17.9	877 Feb 8.2	938 Apr 1.3	999 May 26.4
694 Sep 25.3	755 Nov 24.4	817 Jan 17.6	878 Mar 10.7	939 May 3.2	1000 Jun 29.5
695 Oct 31.5	756 Dec 26.6	818 Feb 17.3	879 Apr 10.8	940 Jun 4.9	1001 Aug 5.4
696 Dec 3.7	758 Jan 26.9	819 Mar 20.0	880 May 12.3	941 Jul 10.7	1002 Sep 12.1
698 Jan 5.0	759 Feb 26.5	820 Apr 19.5	881 Jun 15.6	942 Aug 17.0	1003 Oct 19.1
699 Feb 5.0	760 Mar 28.3	821 May 22.5	882 Jul 22.0	943 Sep 23.6	1004 Nov 22.2
700 Mar 6.6	761 Apr 29.2	822 Jun 26.5	883 Aug 28.5	944 Oct 29.1	1005 Dec 25.3
701 Apr 6.7	762 Jun 1.8	823 Aug 2.3	884 Oct 3.9	945 Dec 2.6	1007 Jan 25.8
702 May 9.1	763 Jul 7.5	824 Sep 8.0	885 Nov 8.8	947 Jan 4.1	1008 Feb 25.4
703 Jun 12.3	764 Aug 12.7	825 Oct 15.0	886 Dec 12.6	948 Feb 4.2	1009 Mar 27.2
704 Jul 17.5	765 Sep 19.3	826 Nov 19.3	888 Jan 13.6	949 Mar 5.8	1010 Apr 27.8
705 Aug 24.0	766 Oct 25.9	827 Dec 22.5	889 Feb 12.5	950 Apr 5.7	1011 May 31.1
706 Sep 30.4	767 Nov 29.5	829 Jan 22.0	890 Mar 15.1	951 May 7.8	1012 Jul 4.4
707 Nov 5.4	768 Dec 31.1	830 Feb 21.7	891 Apr 15.3	952 Jun 9.7	1013 Aug 10.4
708 Dec 8.4	770 Jan 31.2	831 Mar 24.4	892 May 16.9	953 Jul 15.6	1014 Sep 17.1
710 Jan 9.5	771 Mar 2.8	832 Apr 23.9	893 Jun 20.4	954 Aug 22.0	1015 Oct 23.9
711 Feb 9.4	772 Apr 1.7	833 May 27.1	894 Jul 26.6	955 Sep 28.5	1016 Nov 26.9
712 Mar 10.9	773 May 3.7	834 Jul 1.3	895 Sep 2.5	956 Nov 2.9	1017 Dec 29.8
713 Apr 11.1	774 Jun 6.5	835 Aug 7.3	896 Oct 8.9	957 Dec 7.2	1019 Jan 30.1
714 May 13.6	775 Jul 12.4	836 Sep 13.0	897 Nov 13.6	959 Jan 8.5	1020 Feb 29.7
715 Jun 17.1	776 Aug 17.7	837 Oct 20.0	898 Dec 17.2	960 Feb 8.5	1021 Mar 31.5
716 Jul 22.5	777 Sep 24.4	838 Nov 24.0	900 Jan 18.0	961 Mar 10.1	1022 May 2.3
717 Aug 29.1	778 Oct 30.8	839 Dec 27.0	901 Feb 16.8	962 Apr 10.1	1023 Jun 4.8
718 Oct 5.5	779 Dec 4.2	841 Jan 26.4	902 Mar 19.4	963 May 12.4	1024 Jul 9.4
719 Nov 10.3	781 Jan 4.6	842 Feb 26.0	903 Apr 19.7	964 Jun 14.5	1025 Aug 15.6
720 Dec 13.1	782 Feb 4.6	843 Mar 28.7	904 May 21.6	965 Jul 20.7	1026 Sep 22.3
722 Jan 13.9	783 Mar 7.1	844 Apr 28.5	905 Jun 25.4	966 Aug 27.2	1027 Oct 28.9
723 Feb 13.7	784 Apr 6.1	845 May 31.9	906 Aug 1.1	967 Oct 3.7	1028 Dec 1.6
724 Mar 15.3	785 May 8.3	846 Jul 6.4	907 Sep 7.8	968 Nov 7.8	1030 Jan 3.2
725 Apr 15.6	786 Jun 11.4	847 Aug 12.6	908 Oct 14.0	969 Dec 11.8	1031 Feb 3.3
726 May 18.3	787 Jul 17.6	848 Sep 18.3	909 Nov 18.5	971 Jan 12.8	1032 Mar 4.9
727 Jun 22.1	788 Aug 23.1	849 Oct 25.1	910 Dec 21.8	972 Feb 12.7	1033 Apr 4.7
728 Jul 27.8	789 Sep 29.6	850 Nov 28.8	912 Jan 22.3	973 Mar 14.2	1034 May 6.8

OPPOSITIONS OF JUPITER

1035 Jun 9.6	1096 Aug 9.3	1157 Oct 11.7	1218 Dec 10.0	1280 Feb 1.8	1341 Mar 24.7
1036 Jul 14.5	1097 Sep 16.0	1158 Nov 16.4	1220 Jan 11.2	1281 Mar 3.4	1342 Apr 25.1
1037 Aug 20.9	1098 Oct 23.0	1159 Dec 20.0	1221 Feb 10.1	1282 Apr 3.1	1343 May 28.1
1038 Sep 27.6	1099 Nov 27.0	1161 Jan 19.7	1222 Mar 12.6	1283 May 4.9	1344 Jul 1.2
1039 Nov 3.0	1100 Dec 28.9	1162 Feb 19.4	1223 Apr 12.6	1284 Jun 6.6	1345 Aug 7.1
1040 Dec 6.3	1102 Jan 29.2	1163 Mar 22.0	1224 May 14.0	1285 Jul 12.3	1346 Sep 13.9
1042 Jan 7.6	1103 Feb 28.8	1164 Apr 21.4	1225 Jun 17.3	1286 Aug 18.6	1347 Oct 21.0
1043 Feb 7.6	1104 Mar 30.5	1165 May 24.3	1226 Jul 23.6	1287 Sep 25.3	1348 Nov 24.2
1044 Mar 9.1	1105 May 1.3	1166 Jun 28.2	1227 Aug 30.1	1288 Oct 30.9	1349 Dec 27.4
1045 Apr 9.1	1106 Jun 3.8	1167 Aug 4.0	1228 Oct 5.6	1289 Dec 4.5	1351 Jan 27.8
1046 May 11.4	1107 Jul 9.4	1168 Sep 9.7	1229 Nov 10.6	1291 Jan 6.0	1352 Feb 27.4
1047 Jun 14.6	1108 Aug 14.6	1169 Oct 16.9	1230 Dec 14.5	1292 Feb 6.1	1353 Mar 29.1
1048 Jul 19.7	1109 Sep 21.3	1170 Nov 21.2	1232 Jan 15.5	1293 Mar 7.6	1354 Apr 29.7
1049 Aug 26.3	1110 Oct 27.9	1171 Dec 24.4	1233 Feb 14.3	1294 Apr 7.5	1355 Jun 2.0
1050 Oct 2.8	1111 Dec 1.6	1173 Jan 24.0	1234 Mar 16.9	1295 May 9.6	1356 Jul 6.3
1051 Nov 7.8	1113 Jan 2.2	1174 Feb 23.7	1235 Apr 17.1	1296 Jun 11.5	1357 Aug 12.4
1052 Dec 10.8	1114 Feb 2.4	1175 Mar 26.3	1236 May 18.7	1297 Jul 17.5	1358 Sep 19.2
1054 Jan 11.9	1115 Mar 5.0	1176 Apr 25.9	1237 Jun 22.3	1298 Aug 23.9	1359 Oct 26.0
1055 Feb 11.8	1116 Apr 3.9	1177 May 29.1	1238 Jul 28.8	1299 Sep 30.5	1360 Nov 29.0
1056 Mar 13.4	1117 May 5.9	1178 Jul 3.3	1239 Sep 4.4	1300 Nov 4.8	1361 Dec 31.8
1057 Apr 13.5	1118 Jun 8.7	1179 Aug 9.3	1240 Oct 10.7	1301 Dec 9.1	1363 Feb 1.1
1058 May 16.0	1119 Jul 14.5	1180 Sep 14.9	1241 Nov 15.4	1303 Jan 10.4	1364 Mar 2.7
1059 Jun 19.5	1120 Aug 19.8	1181 Oct 21.8	1242 Dec 19.0	1304 Feb 10.3	1365 Apr 2.4
1060 Jul 24.9	1121 Sep 26.4	1182 Nov 25.9	1244 Jan 19.9	1305 Mar 11.9	1366 May 4.3
1061 Aug 31.5	1122 Nov 1.8	1183 Dec 28.9	1245 Feb 18.6	1306 Apr 11.9	1367 Jun 6.9
1062 Oct 7.8	1123 Dec 6.2	1185 Jan 28.3	1246 Mar 21.2	1307 May 14.2	1368 Jul 11.4
1063 Nov 12.6	1125 Jan 6.6	1186 Feb 27.9	1247 Apr 21.6	1308 Jun 16.4	1369 Aug 17.6
1064 Dec 15.4	1126 Feb 6.7	1187 Mar 30.7	1248 May 23.4	1309 Jul 22.6	1370 Sep 24.3
1066 Jan 16.3	1127 Mar 9.3	1188 Apr 30.4	1249 Jun 27.2	1310 Aug 29.0	1371 Oct 30.9
1067 Feb 16.1	1128 Apr 8.3	1189 Jun 2.9	1250 Aug 2.9	1311 Oct 5.5	1372 Dec 3.6
1068 Mar 17.7	1129 May 10.5	1190 Jul 8.3	1251 Sep 9.5	1312 Nov 9.6	1374 Jan 5.2
1069 Apr 18.0	1130 Jun 13.5	1191 Aug 14.3	1252 Oct 15.6	1313 Dec 13.6	1375 Feb 5.4
1070 May 20.7	1131 Jul 19.5	1192 Sep 19.9	1253 Nov 20.1	1315 Jan 14.7	1376 Mar 7.0
1071 Jun 24.3	1132 Aug 24.9	1193 Oct 26.7	1254 Dec 23.5	1316 Feb 14.6	1377 Apr 6.8
1072 Jul 29.9	1133 Oct 1.4	1194 Nov 30.5	1256 Jan 24.2	1317 Mar 16.2	1378 May 8.9
1073 Sep 5.5	1134 Nov 6.6	1196 Jan 2.3	1257 Feb 22.9	1318 Apr 16.4	1379 Jun 11.7
1074 Oct 12.7	1135 Dec 10.7	1197 Feb 1.6	1258 Mar 25.6	1319 May 18.9	1380 Jul 16.5
1075 Nov 17.3	1137 Jan 11.0	1198 Mar 4.2	1259 Apr 26.0	1320 Jun 21.3	1381 Aug 22.9
1076 Dec 19.8	1138 Feb 11.0	1199 Apr 4.0	1260 May 28.1	1321 Jul 27.7	1382 Sep 29.5
1078 Jan 20.6	1139 Mar 13.6	1200 May 5.0	1261 Jul 2.2	1322 Sep 3.3	1383 Nov 4.9
1079 Feb 20.4	1140 Apr 12.7	1201 Jun 7.6	1262 Aug 8.0	1323 Oct 10.6	1384 Dec 8.3
1080 Mar 22.0	1141 May 15.0	1202 Jul 13.3	1263 Sep 14.7	1324 Nov 14.5	1386 Jan 9.7
1081 Apr 22.4	1142 Jun 18.3	1203 Aug 19.5	1264 Oct 20.7	1325 Dec 18.2	1387 Feb 9.7
1082 May 25.4	1143 Jul 24.6	1204 Sep 25.1	1265 Nov 24.9	1327 Jan 19.1	1388 Mar 11.2
1083 Jun 29.3	1144 Aug 30.1	1205 Oct 31.7	1266 Dec 28.0	1328 Feb 18.9	1389 Apr 11.1
1084 Aug 4.0	1145 Oct 6.5	1206 Dec 5.2	1268 Jan 28.5	1329 Mar 20.4	1390 May 13.3
1085 Sep 10.7	1146 Nov 11.5	1208 Jan 6.7	1269 Feb 27.1	1330 Apr 20.6	1391 Jun 16.4
1086 Oct 17.8	1147 Dec 15.4	1209 Feb 5.8	1270 Mar 29.8	1331 May 23.4	1392 Jul 21.5
1087 Nov 22.1	1149 Jan 15.4	1210 Mar 8.4	1271 Apr 30.4	1332 Jun 26.1	1393 Aug 28.1
1088 Dec 24.4	1150 Feb 15.2	1211 Apr 8.3	1272 Jun 1.7	1333 Aug 1.8	1394 Oct 4.7
1090 Jan 24.9	1151 Mar 17.7	1212 May 9.4	1273 Jul 7.1	1334 Sep 8.5	1395 Nov 9.9
1091 Feb 24.5	1152 Apr 16.9	1213 Jun 12.3	1274 Aug 13.2	1335 Oct 15.8	1396 Dec 13.0
1092 Mar 26.2	1153 May 19.6	1214 Jul 18.3	1275 Sep 20.0	1336 Nov 19.4	1398 Jan 14.2
1093 Apr 26.8	1154 Jun 23.2	1215 Aug 24.8	1276 Oct 25.9	1337 Dec 22.9	1399 Feb 14.0
1094 May 30.0	1155 Jul 29.7	1216 Sep 30.4	1277 Nov 29.8	1339 Jan 23.5	1400 Mar 15.5
1095 Jul 4.3	1156 Sep 4.4	1217 Nov 5.7	1279 Jan 1.6	1340 Feb 23.2	1401 Apr 15.5

1402 May 18.0	1463 Jul 16.6	1524 Sep 17.5	1585 Nov 28.5	1647 Jan 23.6	1708 Mar 16.9
1403 Jun 21.4	1464 Aug 22.1	1525 Oct 24.6	1587 Jan 1.1	1648 Feb 23.5	1709 Apr 16.5
1404 Jul 26.8	1465 Sep 28.8	1526 Nov 28.8	1588 Feb 1.9	1649 Mar 25.0	1710 May 18.2
1405 Sep 2.5	1466 Nov 4.3	1527 Dec 31.7	1589 Mar 3.6	1650 Apr 24.9	1711 Jun 20.7
1406 Oct 10.0	1467 Dec 8.8	1529 Jan 31.1	1590 Apr 3.1	1651 May 27.1	1712 Jul 25.4
1407 Nov 14.9	1469 Jan 9.2	1530 Mar 2.6	1591 May 4.3	1652 Jun 29.3	1713 Aug 31.7
1408 Dec 17.6	1470 Feb 9.2	1531 Apr 2.2	1592 Jun 5.1	1653 Aug 4.5	1714 Oct 8.6
1410 Jan 18.5	1471 Mar 11.7	1532 May 2.9	1593 Jul 10.0	1654 Sep 11.2	1715 Nov 14.3
1411 Feb 18.3	1472 Apr 10.5	1533 Jun 5.3	1594 Aug 15.8	1655 Oct 18.8	1716 Dec 18.0
1412 Mar 19.8	1473 May 12.7	1534 Jul 10.8	1595 Sep 22.6	1656 Nov 23.0	1718 Jan 19.6
1413 Apr 20.0	1474 Jun 15.8	1535 Aug 17.1	1596 Oct 28.9	1657 Dec 27.0	1719 Feb 19.7
1414 May 22.7	1475 Jul 21.9	1536 Sep 22.9	1597 Dec 3.4	1659 Jan 28.0	1720 Mar 21.1
1415 Jun 26.4	1476 Aug 27.5	1537 Oct 29.7	1599 Jan 5.7	1660 Feb 27.8	1721 Apr 20.9
1416 Aug 1.1	1477 Oct 4.1	1538 Dec 3.5	1600 Feb 6.2	1661 Mar 29.2	1722 May 22.8
1417 Sep 7.8	1478 Nov 9.3	1540 Jan 5.2	1601 Mar 7.8	1662 Apr 29.3	1723 Jun 25.7
1418 Oct 15.1	1479 Dec 13.4	1541 Feb 4.4	1602 Apr 7.4	1663 May 31.8	1724 Jul 30.6
1419 Nov 19.7	1481 Jan 13.6	1542 Mar 6.9	1603 May 8.9	1664 Jul 4.3	1725 Sep 6.1
1420 Dec 22.2	1482 Feb 13.5	1543 Apr 6.6	1604 Jun 9.9	1665 Aug 9.8	1726 Oct 13.8
1422 Jan 22.9	1483 Mar 16.0	1544 May 7.5	1605 Jul 15.1	1666 Sep 16.5	1727 Nov 19.2
1423 Feb 22.5	1484 Apr 15.0	1545 Jun 10.2	1606 Aug 21.1	1667 Oct 23.9	1728 Dec 22.6
1424 Mar 24.1	1485 May 17.4	1546 Jul 16.0	1607 Sep 27.8	1668 Nov 27.8	1730 Jan 24.0
1425 Apr 24.5	1486 Jun 20.8	1547 Aug 22.4	1608 Nov 2.9	1669 Dec 31.5	1731 Feb 23.9
1426 May 27.5	1487 Jul 27.1	1548 Sep 28.1	1609 Dec 8.1	1671 Feb 1.3	1732 Mar 25.4
1427 Jul 1.5	1488 Sep 1.7	1549 Nov 3.6	1611 Jan 10.1	1672 Mar 3.0	1733 Apr 25.3
1428 Aug 6.3	1489 Oct 9.1	1550 Dec 8.1	1612 Feb 10.5	1673 Apr 2.6	1734 May 27.5
1429 Sep 13.0	1490 Nov 14.1	1552 Jan 9.6	1613 Mar 12.1	1674 May 3.8	1735 Jun 30.6
1430 Oct 20.1	1491 Dec 17.9	1553 Feb 8.6	1614 Apr 11.8	1675 Jun 5.5	1736 Aug 4.7
1431 Nov 24.4	1493 Jan 17.9	1554 Mar 11.2	1615 May 13.4	1676 Jul 9.3	1737 Sep 11.3
1432 Dec 26.6	1494 Feb 17.7	1555 Apr 11.1	1616 Jun 14.7	1677 Aug 15.0	1738 Oct 18.9
1434 Jan 27.2	1495 Mar 20.3	1556 May 12.2	1617 Jul 20.1	1678 Sep 21.7	1739 Nov 24.1
1435 Feb 26.8	1496 Apr 19.5	1557 Jun 15.1	1618 Aug 26.3	1679 Oct 29.0	1740 Dec 27.3
1436 Mar 28.5	1497 May 22.1	1558 Jul 21.2	1619 Oct 3.1	1680 Dec 2.6	1742 Jan 28.4
1437 Apr 29.1	1498 Jun 25.7	1559 Aug 27.6	1620 Nov 7.9	1682 Jan 5.1	1743 Feb 28.3
1438 Jun 1.3	1499 Aug 1.3	1560 Oct 3.2	1621 Dec 12.8	1683 Feb 5.7	1744 Mar 29.7
1439 Jul 6.5	1500 Sep 6.9	1561 Nov 8.6	1623 Jan 14.7	1684 Mar 7.4	1745 Apr 29.7
1440 Aug 11.5	1501 Oct 14.3	1562 Dec 12.8	1624 Feb 14.9	1685 Apr 6.9	1746 Jun 1.0
1441 Sep 18.2	1502 Nov 19.0	1564 Jan 14.1	1625 Mar 16.4	1686 May 8.2	1747 Jul 5.3
1442 Oct 25.2	1503 Dec 22.6	1565 Feb 13.0	1626 Apr 16.1	1687 Jun 10.1	1748 Aug 9.7
1443 Nov 29.2	1505 Jan 22.3	1566 Mar 15.5	1627 May 17.8	1688 Jul 14.1	1749 Sep 16.4
1444 Dec 31.2	1506 Feb 22.0	1567 Apr 15.4	1628 Jun 19.4	1689 Aug 20.0	1750 Oct 24.0
1446 Jan 31.5	1507 Mar 24.6	1568 May 16.6	1629 Jul 25.0	1690 Sep 26.8	1751 Nov 29.0
1447 Mar 3.1	1508 Apr 23.8	1569 Jun 19.8	1630 Aug 31.4	1691 Nov 3.1	1752 Dec 31.9
1448 Apr 1.8	1509 May 26.6	1570 Jul 26.1	1631 Oct 8.2	1692 Dec 7.4	1754 Feb 1.8
1449 May 3.4	1510 Jun 30.5	1571 Sep 1.8	1632 Nov 12.9	1694 Jan 9.7	1755 Mar 4.6
1450 Jun 5.9	1511 Aug 6.3	1572 Oct 8.4	1633 Dec 17.6	1695 Feb 10.1	1756 Apr 3.0
1451 Jul 11.4	1512 Sep 12.1	1573 Nov 13.5	1635 Jan 19.2	1696 Mar 11.7	1757 May 4.1
1452 Aug 16.7	1513 Oct 19.4	1574 Dec 17.5	1636 Feb 19.3	1697 Apr 11.2	1758 Jun 5.7
1453 Sep 23.5	1514 Nov 23.9	1576 Jan 18.6	1637 Mar 20.7	1698 May 12.7	1759 Jul 10.3
1454 Oct 30.3	1515 Dec 27.2	1577 Feb 17.3	1638 Apr 20.5	1699 Jun 14.9	1760 Aug 14.9
1455 Dec 4.1	1517 Jan 26.8	1578 Mar 19.8	1639 May 22.4	1700 Jul 20.2	1761 Sep 21.7
1457 Jan 4.8	1518 Feb 26.4	1579 Apr 19.8	1640 Jun 24.3	1701 Aug 26.3	1762 Oct 29.2
1458 Feb 4.9	1519 Mar 28.9	1580 May 21.3	1641 Jul 30.2	1702 Oct 3.2	1763 Dec 3.9
1459 Mar 7.4	1520 Apr 28.3	1581 Jun 24.8	1642 Sep 5.8	1703 Nov 9.2	1765 Jan 5.5
1460 Apr 6.2	1521 May 31.4	1582 Jul 31.4	1643 Oct 13.5	1704 Dec 13.3	1766 Feb 6.2
1461 May 8.0	1522 Jul 5.6	1583 Sep 17.2	1644 Nov 18.0	1706 Jan 15.1	1767 Mar 8.8
1462 Jun 10.8	1523 Aug 11.7	1584 Oct 23.7	1645 Dec 22.3	1707 Feb 15.4	1768 Apr 7.3

OPPOSITIONS OF JUPITER

1769 May 8.5	1830 Jul 5.5	1891 Sep 5.9	1952 Nov 8.4	2014 Jan 5.9	2075 Feb 28.1
1770 Jun 10.4	1831 Aug 10.8	1892 Oct 12.8	1953 Dec 13.7	2015 Feb 6.8	2076 Mar 29.5
1771 Jul 15.4	1832 Sep 16.6	1893 Nov 18.4	1955 Jan 15.8	2016 Mar 8.4	2077 Apr 29.4
1772 Aug 20.3	1833 Oct 24.2	1894 Dec 23.1	1956 Feb 16.2	2017 Apr 7.9	2078 May 31.4
1773 Sep 27.1	1834 Nov 29.3	1896 Jan 24.6	1957 Mar 17.7	2018 May 9.0	2079 Jul 4.4
1774 Nov 3.4	1836 Jan 2.2	1897 Feb 23.6	1958 Apr 17.3	2019 Jun 10.6	2080 Aug 8.5
1775 Dec 8.8	1837 Feb 2.1	1898 Mar 26.0	1959 May 18.8	2020 Jul 14.3	2081 Sep 15.1
1777 Jan 10.0	1838 Mar 4.8	1899 Apr 25.8	1960 Jun 20.1	2021 Aug 20.0	2082 Oct 22.8
1778 Feb 10.5	1839 Apr 4.3	1900 May 27.8	1961 Jul 25.4	2022 Sep 26.8	2083 Nov 28.1
1779 Mar 13.0	1840 May 4.4	1901 Jun 30.7	1962 Aug 31.6	2023 Nov 3.2	2084 Dec 31.3
1780 Apr 11.6	1841 Jun 5.9	1902 Aug 5.7	1963 Oct 8.4	2024 Dec 7.9	2086 Feb 1.5
1781 May 13.0	1842 Jul 10.5	1903 Sep 12.3	1964 Nov 13.4	2026 Jan 10.4	2087 Mar 4.3
1782 Jun 15.2	1843 Aug 16.1	1904 Oct 19.0	1965 Dec 18.4	2027 Feb 11.0	2088 Apr 2.8
1783 Jul 20.5	1844 Sep 21.9	1905 Nov 24.4	1967 Jan 20.2	2028 Mar 12.6	2089 May 3.7
1784 Aug 25.6	1845 Oct 29.3	1906 Dec 28.6	1968 Feb 20.5	2029 Apr 12.2	2090 Jun 5.1
1785 Oct 2.4	1846 Dec 4.1	1908 Jan 29.9	1969 Mar 22.0	2030 May 13.5	2091 Jul 9.4
1786 Nov 8.4	1848 Jan 6.6	1909 Feb 28.8	1970 Apr 21.6	2031 Jun 15.4	2092 Aug 13.7
1787 Dec 13.5	1849 Feb 6.4	1910 Mar 31.3	1971 May 23.4	2032 Jul 19.3	2093 Sep 20.4
1789 Jan 14.4	1850 Mar 9.1	1911 May 1.2	1972 Jun 24.9	2033 Aug 25.2	2094 Oct 28.0
1790 Feb 14.7	1851 Apr 8.6	1912 Jun 1.4	1973 Jul 30.5	2034 Oct 2.0	2095 Dec 3.0
1791 Mar 17.2	1852 May 8.8	1913 Jul 5.6	1974 Sep 5.8	2035 Nov 8.2	2097 Jan 4.9
1792 Apr 15.9	1853 Jun 10.6	1914 Aug 10.9	1975 Oct 13.6	2036 Dec 12.6	2098 Feb 5.8
1793 May 17.6	1854 Jul 15.5	1915 Sep 17.5	1976 Nov 18.3	2038 Jan 14.8	2099 Mar 8.6
1794 Jun 20.0	1855 Aug 21.3	1916 Oct 24.1	1977 Dec 23.0	2039 Feb 15.3	2100 Apr 8.1
1795 Jul 25.6	1856 Sep 27.1	1917 Nov 29.2	1979 Jan 24.6	2040 Mar 16.9	2101 May 9.2
1796 Aug 30.8	1857 Nov 3.4	1919 Jan 2.2	1980 Feb 24.7	2041 Apr 16.5	2102 Jun 10.7
1797 Oct 7.6	1858 Dec 8.8	1920 Feb 3.3	1981 Mar 26.2	2042 May 18.0	2103 Jul 15.3
1798 Nov 13.4	1860 Jan 11.2	1921 Mar 5.1	1982 Apr 26.0	2043 Jun 20.1	2104 Aug 19.8
1799 Dec 18.2	1861 Feb 10.7	1922 Apr 4.6	1983 May 27.9	2044 Jul 24.3	2105 Sep 26.5
1801 Jan 19.9	1862 Mar 13.3	1923 May 5.6	1984 Jun 29.7	2045 Aug 30.3	2106 Nov 2.9
1802 Feb 20.1	1863 Apr 12.9	1924 Jun 6.0	1985 Aug 4.5	2046 Oct 7.1	2107 Dec 8.7
1803 Mar 22.6	1864 May 13.3	1925 Jul 10.4	1986 Sep 10.9	2047 Nov 13.1	2109 Jan 10.3
1804 Apr 21.3	1865 Jun 15.3	1926 Aug 15.8	1987 Oct 18.6	2048 Dec 17.2	2110 Feb 11.1
1805 May 23.1	1866 Jul 20.4	1927 Sep 22.5	1988 Nov 23.1	2050 Jan 19.2	2111 Mar 13.8
1806 Jun 25.7	1867 Aug 26.3	1928 Oct 29.0	1989 Dec 27.6	2051 Feb 19.6	2112 Apr 12.4
1807 Jul 31.5	1868 Oct 2.1	1929 Dec 4.0	1991 Jan 29.0	2052 Mar 21.2	2113 May 13.6
1808 Sep 5.8	1869 Nov 8.3	1931 Jan 6.7	1992 Feb 29.0	2053 Apr 20.8	2114 Jun 15.4
1809 Oct 13.6	1870 Dec 13.5	1932 Feb 7.6	1993 Mar 30.5	2054 May 22.5	2115 Jul 20.2
1810 Nov 19.3	1872 Jan 15.7	1933 Mar 9.3	1994 Apr 30.4	2055 Jun 24.8	2116 Aug 24.9
1811 Dec 23.9	1873 Feb 15.1	1934 Apr 8.8	1995 Jun 1.5	2056 Jul 29.3	2117 Oct 1.7
1813 Jan 24.4	1874 Mar 17.6	1935 May 10.0	1996 Jul 4.5	2057 Sep 4.5	2118 Nov 8.0
1814 Feb 24.4	1875 Apr 17.2	1936 Jun 10.7	1997 Aug 9.6	2058 Oct 12.3	2119 Dec 13.5
1815 Mar 26.8	1876 May 17.8	1937 Jul 15.3	1998 Sep 16.1	2059 Nov 18.1	2121 Jan 14.9
1816 Apr 25.6	1877 Jun 20.0	1938 Aug 21.0	1999 Oct 23.8	2060 Dec 22.0	2122 Feb 15.4
1817 May 27.7	1878 Jul 25.4	1939 Sep 27.8	2000 Nov 28.1	2062 Jan 23.7	2123 Mar 18.0
1818 Jun 30.6	1879 Aug 31.6	1940 Nov 3.2	2002 Jan 1.2	2063 Feb 23.9	2124 Apr 16.5
1819 Aug 5.6	1880 Oct 7.4	1941 Dec 8.8	2003 Feb 2.4	2064 Mar 25.3	2125 May 17.9
1820 Sep 11.2	1881 Nov 13.4	1943 Jan 11.3	2004 Mar 4.2	2065 Apr 25.0	2126 Jun 19.9
1821 Oct 18.9	1882 Dec 18.3	1944 Feb 11.9	2005 Apr 3.6	2066 May 26.8	2127 Jul 25.1
1822 Nov 24.3	1884 Jan 20.1	1945 Mar 13.5	2006 May 4.6	2067 Jun 29.5	2128 Aug 30.1
1823 Dec 28.5	1885 Feb 19.3	1946 Apr 13.0	2007 Jun 6.0	2068 Aug 3.3	2129 Oct 7.0
1825 Jan 28.7	1886 Mar 21.8	1947 May 14.3	2008 Jul 9.3	2069 Sep 9.8	2130 Nov 13.2
1826 Feb 28.6	1887 Apr 21.5	1948 Jun 15.3	2009 Aug 14.7	2070 Oct 17.6	2131 Dec 18.4
1827 Mar 31.0	1888 May 22.2	1949 Jul 20.3	2010 Sep 21.5	2071 Nov 23.2	2133 Jan 19.4
1828 Apr 29.9	1889 Jun 24.8	1950 Aug 26.3	2011 Oct 29.1	2072 Dec 26.7	2134 Feb 19.8
1829 Jun 1.2	1890 Jul 30.5	1951 Oct 3.2	2012 Dec 3.1	2074 Jan 28.2	2135 Mar 22.3

2136 Apr 20.8	2197 Jun 14.3	2258 Aug 14.5	2319 Oct 18.6	2380 Dec 18.0	2442 Feb 11.5
2137 May 22.4	2198 Jul 19.1	2259 Sep 21.2	2320 Nov 23.5	2382 Jan 20.3	2443 Mar 14.2
2138 Jun 24.7	2199 Aug 24.9	2260 Oct 27.9	2321 Dec 28.2	2383 Feb 20.7	2444 Apr 12.6
2139 Jul 30.2	2200 Oct 1.8	2261 Dec 3.2	2323 Jan 29.8	2384 Mar 22.2	2445 May 13.6
2140 Sep 4.5	2201 Nov 8.1	2263 Jan 6.2	2324 Feb 29.9	2385 Apr 21.7	2446 Jun 15.1
2141 Oct 12.3	2202 Dec 13.7	2264 Feb 7.3	2325 Mar 31.3	2386 May 23.1	2447 Jul 19.7
2142 Nov 18.2	2204 Jan 16.1	2265 Mar 9.0	2326 May 1.0	2387 Jun 25.3	2448 Aug 24.3
2143 Dec 23.1	2205 Feb 15.7	2266 Apr 8.5	2327 Jun 1.9	2388 Jul 29.5	2449 Oct 1.2
2145 Jan 23.8	2206 Mar 18.2	2267 May 9.5	2328 Jul 4.7	2389 Sep 4.7	2450 Nov 7.7
2146 Feb 24.0	2207 Apr 17.7	2268 Jun 9.9	2329 Aug 9.6	2390 Oct 12.6	2451 Dec 13.6
2147 Mar 26.5	2208 May 18.1	2269 Jul 14.3	2330 Sep 16.1	2391 Nov 18.7	2453 Jan 15.2
2148 Apr 25.2	2209 Jun 20.1	2270 Aug 19.8	2331 Oct 23.9	2392 Dec 22.8	2454 Feb 15.9
2149 May 27.0	2210 Jul 25.2	2271 Sep 26.6	2332 Nov 28.4	2394 Jan 24.8	2455 Mar 18.5
2150 Jun 29.6	2211 Aug 31.2	2272 Nov 2.1	2334 Jan 1.9	2395 Feb 25.0	2456 Apr 16.9
2151 Aug 4.4	2212 Oct 7.1	2273 Dec 8.0	2335 Feb 3.2	2396 Mar 26.5	2457 May 18.1
2152 Sep 9.8	2213 Nov 13.2	2275 Jan 10.8	2336 Mar 5.2	2397 Apr 26.1	2458 Jun 19.9
2153 Oct 17.5	2214 Dec 18.5	2276 Feb 11.6	2337 Apr 4.6	2398 May 27.7	2459 Jul 24.8
2154 Nov 23.2	2216 Jan 20.6	2277 Mar 13.3	2338 May 5.4	2399 Jun 30.1	2460 Aug 29.6
2155 Dec 27.7	2217 Feb 20.0	2278 Apr 12.8	2339 Jun 6.6	2400 Aug 3.7	2461 Oct 6.5
2157 Jan 28.2	2218 Mar 22.5	2279 May 13.9	2340 Jul 9.6	2401 Sep 9.9	2462 Nov 12.8
2158 Feb 28.3	2219 Apr 22.1	2280 Jun 14.6	2341 Aug 14.7	2402 Oct 17.8	2463 Dec 18.3
2159 Mar 30.7	2220 May 22.7	2281 Jul 19.3	2342 Sep 21.3	2403 Nov 23.6	2465 Jan 19.6
2160 Apr 29.6	2221 Jun 24.9	2282 Aug 25.0	2343 Oct 29.0	2404 Dec 27.4	2466 Feb 20.1
2161 May 31.6	2222 Jul 30.3	2283 Oct 1.8	2344 Dec 3.3	2406 Jan 29.1	2467 Mar 22.7
2162 Jul 4.5	2223 Sep 5.4	2284 Nov 7.1	2346 Jan 6.4	2407 Mar 1.3	2468 Apr 21.2
2163 Aug 9.4	2224 Oct 12.2	2285 Dec 12.7	2347 Feb 7.6	2408 Mar 30.8	2469 May 22.6
2164 Sep 14.9	2225 Nov 18.1	2287 Jan 15.2	2348 Mar 9.4	2409 Apr 30.5	2470 Jun 24.6
2165 Oct 22.6	2226 Dec 23.1	2288 Feb 15.9	2349 Apr 8.9	2410 Jun 1.3	2471 Jul 29.8
2166 Nov 27.9	2228 Jan 24.9	2289 Mar 17.5	2350 May 9.8	2411 Jul 5.0	2472 Sep 3.8
2168 Jan 1.3	2229 Feb 24.2	2290 Apr 17.1	2351 Jun 11.2	2412 Aug 8.8	2473 Oct 11.7
2169 Feb 1.6	2230 Mar 26.7	2291 May 18.4	2352 Jul 14.5	2413 Sep 15.2	2474 Nov 17.9
2170 Mar 4.5	2231 Apr 26.4	2292 Jun 19.3	2353 Aug 19.9	2414 Oct 23.0	2475 Dec 23.1
2171 Apr 4.0	2232 May 27.2	2293 Jul 24.3	2354 Sep 26.6	2415 Nov 28.6	2477 Jan 24.2
2172 May 3.9	2233 Jun 29.7	2294 Aug 30.2	2355 Nov 3.2	2417 Jan 1.2	2478 Feb 24.5
2173 Jun 5.2	2234 Aug 4.3	2295 Oct 7.0	2356 Dec 8.2	2418 Feb 2.6	2479 Mar 27.0
2174 Jul 9.3	2235 Sep 10.6	2296 Nov 12.2	2358 Jan 11.1	2419 Mar 5.6	2480 Apr 25.5
2175 Aug 14.5	2236 Oct 17.4	2297 Dec 17.6	2359 Feb 12.0	2420 Apr 4.0	2481 May 27.0
2176 Sep 20.1	2237 Nov 23.1	2299 Jan 19.8	2360 Mar 13.7	2421 May 4.7	2482 Jun 29.2
2177 Oct 27.7	2238 Dec 27.8	2300 Feb 20.3	2361 Apr 13.1	2422 Jun 5.7	2483 Aug 3.7
2178 Dec 2.9	2240 Jan 29.4	2301 Mar 22.8	2362 May 14.2	2423 Jul 8.9	2484 Sep 8.9
2180 Jan 5.9	2241 Feb 28.5	2302 Apr 22.3	2363 Jun 15.7	2424 Aug 13.7	2485 Oct 16.9
2181 Feb 6.0	2242 Mar 30.9	2303 May 23.7	2364 Jul 19.3	2425 Sep 20.4	2486 Nov 23.0
2182 Mar 8.7	2243 Apr 30.6	2304 Jun 24.9	2365 Aug 24.9	2426 Oct 28.2	2487 Dec 27.9
2183 Apr 8.2	2244 May 31.6	2305 Jul 30.2	2366 Oct 1.8	2427 Dec 3.7	2489 Jan 28.7
2184 May 8.2	2245 Jul 4.4	2306 Sep 5.3	2367 Nov 8.3	2429 Jan 5.9	2490 Feb 28.9
2185 Jun 9.6	2246 Aug 9.3	2307 Oct 13.3	2368 Dec 13.1	2430 Feb 7.1	2491 Mar 31.3
2186 Jul 14.1	2247 Sep 15.8	2308 Nov 18.3	2370 Jan 15.7	2431 Mar 10.0	2492 Apr 29.9
2187 Aug 19.6	2248 Oct 22.7	2309 Dec 23.5	2371 Feb 16.4	2432 Apr 8.3	2493 May 31.6
2188 Sep 25.4	2249 Nov 28.2	2311 Jan 25.4	2372 Mar 18.0	2433 May 9.2	2494 Jul 4.1
2189 Nov 2.0	2251 Jan 1.6	2312 Feb 25.6	2373 Apr 17.4	2434 Jun 10.4	2495 Aug 8.8
2190 Dec 7.9	2252 Feb 2.9	2313 Mar 27.1	2374 May 18.6	2435 Jul 14.6	2496 Sep 14.3
2192 Jan 10.6	2253 Mar 4.8	2314 Apr 26.7	2375 Jun 20.4	2436 Aug 19.0	2497 Oct 22.3
2193 Feb 10.4	2254 Apr 4.2	2315 May 28.3	2376 Jul 24.4	2437 Sep 25.8	2498 Nov 28.1
2194 Mar 13.0	2255 May 5.0	2316 Jun 29.7	2377 Aug 30.3	2438 Nov 2.5	2500 Jan 1.7
2195 Apr 12.5	2256 Jun 5.2	2317 Aug 4.3	2378 Oct 7.2	2439 Dec 8.7	
2196 May 12.6	2257 Jul 9.3	2318 Sep 10.7	2379 Nov 13.6	2441 Jan 10.6	

X
Oppositions of Saturn, 0 - 2500

0 Dec 6.3	48 Jul 13.2	96 Mar 5.1	143 Oct 9.8	191 May 25.2	239 Jan 11.3
1 Dec 20.5	49 Jul 25.9	97 Mar 17.7	144 Oct 23.0	192 Jun 5.2	240 Jan 24.9
3 Jan 3.5	50 Aug 7.8	98 Mar 30.1	145 Nov 6.3	193 Jun 17.4	241 Feb 6.3
4 Jan 17.2	51 Aug 20.9	99 Apr 11.4	146 Nov 20.7	194 Jun 29.6	242 Feb 19.4
5 Jan 29.7	52 Sep 2.3	100 Apr 22.6	147 Dec 5.0	195 Jul 12.0	243 Mar 4.2
6 Feb 12.0	53 Sep 15.9	101 May 4.6	148 Dec 18.2	196 Jul 23.6	244 Mar 15.8
7 Feb 24.9	54 Sep 29.8	102 May 16.7	150 Jan 1.2	197 Aug 5.5	245 Mar 28.3
8 Mar 8.7	55 Oct 13.9	103 May 28.7	151 Jan 15.0	198 Aug 18.6	246 Apr 9.5
9 Mar 21.2	56 Oct 27.1	104 Jun 8.8	152 Jan 28.5	199 Aug 31.9	247 Apr 21.6
10 Apr 2.6	57 Nov 10.4	105 Jun 20.9	153 Feb 9.8	200 Sep 13.6	248 May 2.7
11 Apr 14.8	58 Nov 24.8	106 Jul 3.2	154 Feb 22.9	201 Sep 27.4	249 May 14.6
12 Apr 25.9	59 Dec 9.0	107 Jul 15.7	155 Mar 7.6	202 Oct 11.5	250 May 26.6
13 May 7.9	60 Dec 22.1	108 Jul 27.4	156 Mar 19.2	203 Oct 25.7	251 Jun 7.7
14 May 19.9	62 Jan 5.1	109 Aug 9.4	157 Mar 31.6	204 Nov 8.1	252 Jun 18.8
15 Jun 1.0	63 Jan 18.8	110 Aug 22.5	158 Apr 12.9	205 Nov 22.4	253 Jul 1.0
16 Jun 12.1	64 Feb 1.3	111 Sep 5.0	159 Apr 25.0	206 Dec 6.7	254 Jul 13.5
17 Jun 24.2	65 Feb 13.5	112 Sep 17.7	160 May 6.1	207 Dec 20.9	255 Jul 26.1
18 Jul 6.6	66 Feb 26.4	113 Oct 1.6	161 May 18.1	209 Jan 2.9	256 Aug 7.0
19 Jul 19.1	67 Mar 11.1	114 Oct 15.7	162 May 30.1	210 Jan 16.7	257 Aug 20.1
20 Jul 30.8	68 Mar 22.6	115 Oct 29.9	163 Jun 11.2	211 Jan 30.2	258 Sep 2.5
21 Aug 12.8	69 Apr 4.0	116 Nov 12.3	164 Jun 22.4	212 Feb 12.5	259 Sep 16.1
22 Aug 26.1	70 Apr 16.2	117 Nov 26.6	165 Jul 4.7	213 Feb 24.4	260 Sep 29.0
23 Sep 8.6	71 Apr 28.3	118 Dec 10.8	166 Jul 17.2	214 Mar 9.2	261 Oct 13.1
24 Sep 21.3	72 May 9.3	119 Dec 24.9	167 Jul 30.0	215 Mar 21.7	262 Oct 27.4
25 Oct 5.3	73 May 21.3	121 Jan 6.8	168 Aug 10.9	216 Apr 2.1	263 Nov 10.7
26 Oct 19.5	74 Jun 2.3	122 Jan 20.5	169 Aug 24.1	217 Apr 14.3	264 Nov 24.1
27 Nov 2.8	75 Jun 14.4	123 Feb 2.9	170 Sep 6.6	218 Apr 26.5	265 Dec 8.4
28 Nov 16.2	76 Jun 25.7	124 Feb 16.1	171 Sep 20.4	219 May 8.5	266 Dec 22.5
29 Nov 30.5	77 Jul 8.0	125 Feb 28.0	172 Oct 3.3	220 May 19.5	268 Jan 5.5
30 Dec 14.7	78 Jul 20.6	126 Mar 12.7	173 Oct 17.5	221 May 31.5	269 Jan 18.3
31 Dec 28.8	79 Aug 2.4	127 Mar 25.1	174 Oct 31.7	222 Jun 12.6	270 Jan 31.8
33 Jan 10.7	80 Aug 14.4	128 Apr 5.4	175 Nov 15.1	223 Jun 24.8	271 Feb 14.0
34 Jan 24.3	81 Aug 27.7	129 Apr 17.6	176 Nov 28.4	224 Jul 6.2	272 Feb 27.0
35 Feb 6.6	82 Sep 10.2	130 Apr 29.7	177 Dec 12.6	225 Jul 18.7	273 Mar 10.7
36 Feb 19.8	83 Sep 24.0	131 May 11.7	178 Dec 26.7	226 Jul 31.4	274 Mar 23.2
37 Mar 3.6	84 Oct 7.0	132 May 22.7	180 Jan 9.6	227 Aug 13.4	275 Apr 4.6
38 Mar 16.3	85 Oct 21.2	133 Jun 3.8	181 Jan 22.2	228 Aug 25.7	276 Apr 15.8
39 Mar 28.7	86 Nov 4.5	134 Jun 15.9	182 Feb 4.6	229 Sep 8.2	277 Apr 27.9
40 Apr 9.0	87 Nov 18.9	135 Jun 28.1	183 Feb 17.7	230 Sep 22.0	278 May 9.9
41 Apr 21.2	88 Dec 2.2	136 Jul 9.5	184 Mar 1.6	231 Oct 6.0	279 May 21.9
42 May 3.3	89 Dec 16.4	137 Jul 22.1	185 Mar 14.3	232 Oct 19.2	280 Jun 1.9
43 May 15.3	90 Dec 30.5	138 Aug 3.9	186 Mar 26.7	233 Nov 2.5	281 Jun 14.0
44 May 26.3	92 Jan 13.3	139 Aug 17.0	187 Apr 8.0	234 Nov 16.8	282 Jun 26.2
45 Jun 7.3	93 Jan 25.9	140 Aug 29.3	188 Apr 19.1	235 Dec 1.1	283 Jul 8.5
46 Jun 19.5	94 Feb 8.2	141 Sep 11.9	189 May 1.2	236 Dec 14.3	284 Jul 20.1
47 Jul 1.8	95 Feb 21.3	142 Sep 25.7	190 May 13.2	237 Dec 28.4	285 Aug 1.9

286 Aug 14.9	343 Jul 22.4	400 Jun 28.7	457 Jun 6.5	514 May 15.5	571 Apr 23.4
287 Aug 28.2	344 Aug 3.1	401 Jul 11.0	458 Jun 18.6	515 May 27.5	572 May 4.5
288 Sep 9.7	345 Aug 16.2	402 Jul 23.5	459 Jun 30.7	516 Jun 7.5	573 May 16.5
289 Sep 23.6	346 Aug 29.5	403 Aug 5.3	460 Jul 12.1	517 Jun 19.6	574 May 28.5
290 Oct 7.6	347 Sep 12.1	404 Aug 17.4	461 Jul 24.6	518 Jul 1.8	575 Jun 9.5
291 Oct 21.8	348 Sep 24.9	405 Aug 30.7	462 Aug 6.4	519 Jul 14.1	576 Jun 20.5
292 Nov 4.1	349 Oct 9.0	406 Sep 13.3	463 Aug 19.4	520 Jul 25.6	577 Jul 2.7
293 Nov 18.4	350 Oct 23.2	407 Sep 27.1	464 Aug 31.7	521 Aug 7.4	578 Jul 15.0
294 Dec 2.8	351 Nov 6.5	408 Oct 10.2	465 Sep 14.3	522 Aug 20.4	579 Jul 27.6
295 Dec 17.0	352 Nov 19.9	409 Oct 24.4	466 Sep 28.2	523 Sep 2.7	580 Aug 8.3
296 Dec 30.0	353 Dec 4.2	410 Nov 7.7	467 Oct 12.2	524 Sep 15.3	581 Aug 21.3
298 Jan 12.9	354 Dec 18.4	411 Nov 22.1	468 Oct 25.5	525 Sep 29.1	582 Sep 3.6
299 Jan 26.5	356 Jan 1.5	412 Dec 5.4	469 Nov 8.8	526 Oct 13.2	583 Sep 17.2
300 Feb 8.9	357 Jan 14.3	413 Dec 19.7	470 Nov 23.2	527 Oct 27.4	584 Sep 30.0
301 Feb 20.9	358 Jan 27.9	415 Jan 2.7	471 Dec 7.5	528 Nov 9.8	585 Oct 14.1
302 Mar 5.8	359 Feb 10.3	416 Jan 16.6	472 Dec 20.7	529 Nov 24.1	586 Oct 28.3
303 Mar 18.3	360 Feb 23.4	417 Jan 29.2	474 Jan 3.8	530 Dec 8.5	587 Nov 11.6
304 Mar 29.7	361 Mar 7.2	418 Feb 11.5	475 Jan 17.6	531 Dec 22.7	588 Nov 25.0
305 Apr 11.0	362 Mar 19.7	419 Feb 24.6	476 Jan 31.3	533 Jan 4.7	589 Dec 9.3
306 Apr 23.1	363 Apr 1.1	420 Mar 8.4	477 Feb 12.6	534 Jan 18.6	590 Dec 23.5
307 May 5.1	364 Apr 12.3	421 Mar 21.0	478 Feb 25.7	535 Feb 1.2	592 Jan 6.6
308 May 16.0	365 Apr 24.4	422 Apr 2.3	479 Mar 10.5	536 Feb 14.5	593 Jan 19.4
309 May 28.0	366 May 6.4	423 Apr 14.5	480 Mar 22.1	537 Feb 26.6	594 Feb 2.0
310 Jun 9.0	367 May 18.3	424 Apr 25.6	481 Apr 3.4	538 Mar 11.5	595 Feb 15.4
311 Jun 21.2	368 May 29.3	425 May 7.6	482 Apr 15.6	539 Mar 24.0	596 Feb 28.5
312 Jul 2.4	369 Jun 10.3	426 May 19.5	483 Apr 27.7	540 Apr 4.4	597 Mar 12.3
313 Jul 14.8	370 Jun 22.4	427 May 31.5	484 May 8.7	541 Apr 16.6	598 Mar 24.9
314 Jul 27.5	371 Jul 4.7	428 Jun 11.5	485 May 20.6	542 Apr 28.7	599 Apr 6.3
315 Aug 9.3	372 Jul 16.1	429 Jun 23.6	486 Jun 1.6	543 May 10.7	600 Apr 17.5
316 Aug 21.5	373 Jul 28.7	430 Jul 5.8	487 Jun 13.6	544 May 21.6	601 Apr 29.6
317 Sep 3.9	374 Aug 10.6	431 Jul 18.3	488 Jun 24.7	545 Jun 2.6	602 May 11.6
318 Sep 17.6	375 Aug 23.7	432 Jul 29.9	489 Jul 6.9	546 Jun 14.6	603 May 23.6
319 Oct 1.5	376 Sep 5.2	433 Aug 11.8	490 Jul 19.3	547 Jun 26.7	604 Jun 3.5
320 Oct 14.6	377 Sep 18.9	434 Aug 24.9	491 Jul 31.9	548 Jul 7.9	605 Jun 15.6
321 Oct 28.9	378 Oct 2.8	435 Sep 7.3	492 Aug 12.8	549 Jul 20.3	606 Jun 27.7
322 Nov 12.2	379 Oct 16.9	436 Sep 20.0	493 Aug 25.9	550 Aug 1.9	607 Jul 9.9
323 Nov 26.6	380 Oct 30.2	437 Oct 3.9	494 Sep 8.4	551 Aug 14.8	608 Jul 21.3
324 Dec 9.9	381 Nov 13.6	438 Oct 18.1	495 Sep 22.0	552 Aug 26.9	609 Aug 2.9
325 Dec 24.0	382 Nov 27.9	439 Nov 1.3	496 Oct 5.0	553 Sep 9.3	610 Aug 15.8
327 Jan 7.0	383 Dec 12.2	440 Nov 14.7	497 Oct 19.1	554 Sep 23.0	611 Aug 28.9
328 Jan 20.8	384 Dec 25.4	441 Nov 29.1	498 Nov 2.4	555 Oct 6.9	612 Sep 10.3
329 Feb 2.2	386 Jan 8.4	442 Dec 13.4	499 Nov 16.7	556 Oct 20.0	613 Sep 23.9
330 Feb 15.5	387 Jan 22.1	443 Dec 27.5	500 Nov 30.1	557 Nov 3.3	614 Oct 7.8
331 Feb 28.4	388 Feb 4.6	445 Jan 9.5	501 Dec 14.4	558 Nov 17.7	615 Oct 21.9
332 Mar 12.1	389 Feb 16.8	446 Jan 23.2	502 Dec 28.5	559 Dec 2.0	616 Nov 4.2
333 Mar 24.6	390 Mar 1.7	447 Feb 5.7	504 Jan 11.5	560 Dec 15.3	617 Nov 18.5
334 Apr 5.9	391 Mar 14.4	448 Feb 18.9	505 Jan 24.2	561 Dec 29.5	618 Dec 2.9
335 Apr 18.1	392 Mar 25.9	449 Mar 2.9	506 Feb 6.7	563 Jan 12.4	619 Dec 17.2
336 Apr 29.2	393 Apr 7.2	450 Mar 15.6	507 Feb 19.9	564 Jan 26.1	620 Dec 30.3
337 May 11.2	394 Apr 19.4	451 Mar 28.0	508 Mar 3.9	565 Feb 7.6	622 Jan 13.2
338 May 23.2	395 May 1.4	452 Apr 8.3	509 Mar 16.6	566 Feb 20.8	623 Jan 27.0
339 Jun 4.2	396 May 12.4	453 Apr 20.5	510 Mar 29.1	567 Mar 5.8	624 Feb 9.5
340 Jun 15.3	397 May 24.4	454 May 2.5	511 Apr 10.4	568 Mar 17.5	625 Feb 21.7
341 Jun 27.5	398 Jun 5.4	455 May 14.5	512 Apr 21.5	569 Mar 30.0	626 Mar 6.6
342 Jul 9.8	399 Jun 17.5	456 May 25.5	513 May 3.6	570 Apr 11.3	627 Mar 19.3

628 Mar 30.8	685 Mar 7.5	742 Feb 11.2	799 Jan 17.1	855 Dec 22.3	912 Nov 25.3
629 Apr 12.2	686 Mar 20.2	743 Feb 24.4	800 Jan 30.7	857 Jan 4.4	913 Dec 9.6
630 Apr 24.3	687 Apr 1.7	744 Mar 8.4	801 Feb 12.2	858 Jan 18.2	914 Dec 23.7
631 May 6.4	688 Apr 13.0	745 Mar 21.1	802 Feb 25.4	859 Jan 31.9	916 Jan 6.7
632 May 17.4	689 Apr 25.2	746 Apr 2.6	803 Mar 10.3	860 Feb 14.3	917 Jan 19.6
633 May 29.4	690 May 7.3	747 Apr 15.0	804 Mar 22.1	861 Feb 26.5	918 Feb 2.2
634 Jun 10.4	691 May 19.3	748 Apr 26.2	805 Apr 3.6	862 Mar 11.4	919 Feb 15.6
635 Jun 22.5	692 May 30.3	749 May 8.2	806 Apr 15.9	863 Mar 24.2	920 Feb 28.8
636 Jul 3.6	693 Jun 11.3	750 May 20.3	807 Apr 28.1	864 Apr 4.7	921 Mar 12.7
637 Jul 16.0	694 Jun 23.4	751 Jun 1.3	808 May 9.3	865 Apr 17.0	922 Mar 25.4
638 Jul 28.5	695 Jul 5.6	752 Jun 12.3	809 May 21.3	866 Apr 29.2	923 Apr 6.9
639 Aug 10.2	696 Jul 16.9	753 Jun 24.4	810 Jun 2.3	867 May 11.4	924 Apr 18.2
640 Aug 22.2	697 Jul 29.5	754 Jul 6.6	811 Jun 14.4	868 May 22.4	925 Apr 30.4
641 Sep 4.5	698 Aug 11.2	755 Jul 19.0	812 Jun 25.5	869 Jun 3.5	926 May 12.6
642 Sep 18.1	699 Aug 24.2	756 Jul 30.5	813 Jul 7.7	870 Jun 15.6	927 May 24.6
643 Oct 1.9	700 Sep 5.5	757 Aug 12.3	814 Jul 20.1	871 Jun 27.7	928 Jun 4.7
644 Oct 14.9	701 Sep 19.1	758 Aug 25.3	815 Aug 1.7	872 Jul 9.0	929 Jun 16.8
645 Oct 29.2	702 Oct 2.9	759 Sep 7.6	816 Aug 13.5	873 Jul 21.4	930 Jun 29.0
646 Nov 12.5	703 Oct 16.9	760 Sep 20.1	817 Aug 26.5	874 Aug 3.0	931 Jul 11.3
647 Nov 26.9	704 Oct 30.1	761 Oct 3.9	818 Sep 8.8	875 Aug 15.8	932 Jul 22.7
648 Dec 10.2	705 Nov 13.4	762 Oct 17.9	819 Sep 22.3	876 Aug 27.8	933 Aug 4.3
649 Dec 24.4	706 Nov 27.8	763 Nov 1.1	820 Oct 5.1	877 Sep 10.1	934 Aug 17.2
651 Jan 7.4	707 Dec 12.1	764 Nov 14.4	821 Oct 19.1	878 Sep 23.7	935 Aug 30.3
652 Jan 21.3	708 Dec 25.3	765 Nov 28.7	822 Nov 2.3	879 Oct 7.5	936 Sep 11.6
653 Feb 2.9	710 Jan 8.3	766 Dec 13.0	823 Nov 16.6	880 Oct 20.5	937 Sep 25.2
654 Feb 16.2	711 Jan 22.1	767 Dec 27.2	824 Nov 29.9	881 Nov 3.7	938 Oct 9.0
655 Mar 1.3	712 Feb 4.7	769 Jan 9.2	825 Dec 14.2	882 Nov 17.9	939 Oct 23.0
656 Mar 13.2	713 Feb 17.0	770 Jan 23.0	826 Dec 28.3	883 Dec 2.2	940 Nov 5.2
657 Mar 25.8	714 Mar 2.1	771 Feb 5.6	828 Jan 11.3	884 Dec 15.5	941 Nov 19.4
658 Apr 7.2	715 Mar 15.0	772 Feb 19.0	829 Jan 24.1	885 Dec 29.6	942 Dec 3.7
659 Apr 19.4	716 Mar 26.6	773 Mar 3.1	830 Feb 6.7	887 Jan 12.6	943 Dec 17.9
660 Apr 30.5	717 Apr 8.1	774 Mar 15.9	831 Feb 20.0	888 Jan 26.3	944 Dec 31.0
661 May 12.5	718 Apr 20.3	775 Mar 28.6	832 Mar 4.1	889 Feb 7.9	946 Jan 14.0
662 May 24.5	719 May 2.4	776 Apr 9.0	833 Mar 16.9	890 Feb 21.2	947 Jan 27.7
663 Jun 5.5	720 May 13.5	777 Apr 21.3	834 Mar 29.6	891 Mar 6.2	948 Feb 10.2
664 Jun 16.5	721 May 25.5	778 May 3.4	835 Apr 11.0	892 Mar 18.0	949 Feb 22.5
665 Jun 28.6	722 Jun 6.5	779 May 15.5	836 Apr 22.3	893 Mar 30.7	950 Mar 7.5
666 Jul 10.9	723 Jun 18.5	780 May 26.5	837 May 4.5	894 Apr 12.1	951 Mar 20.3
667 Jul 23.3	724 Jun 29.6	781 Jun 7.5	838 May 16.5	895 Apr 24.4	952 Mar 31.9
668 Aug 3.9	725 Jul 11.9	782 Jun 19.6	839 May 28.6	896 May 5.6	953 Apr 13.4
669 Aug 16.8	726 Jul 24.3	783 Jul 1.7	840 Jun 8.6	897 May 17.7	954 Apr 25.7
670 Aug 29.9	727 Aug 6.0	784 Jul 13.0	841 Jun 20.7	898 May 29.7	955 May 7.8
671 Sep 12.3	728 Aug 17.8	785 Jul 25.4	842 Jul 2.9	899 Jun 10.8	956 May 18.9
672 Sep 24.9	729 Aug 30.9	786 Aug 7.1	843 Jul 15.2	900 Jun 21.9	957 May 31.0
673 Oct 8.8	730 Sep 13.3	787 Aug 20.0	844 Jul 26.7	901 Jul 4.1	958 Jun 12.1
674 Oct 22.9	731 Sep 27.0	788 Sep 1.1	845 Aug 8.3	902 Jul 16.5	959 Jun 24.2
675 Nov 6.1	732 Oct 9.8	789 Sep 14.5	846 Aug 21.3	903 Jul 29.0	960 Jul 5.5
676 Nov 19.5	733 Oct 23.9	790 Sep 28.2	847 Sep 3.4	904 Aug 9.7	961 Jul 17.9
677 Dec 3.8	734 Nov 7.1	791 Oct 12.0	848 Sep 15.8	905 Aug 22.7	962 Jul 30.4
678 Dec 18.0	735 Nov 21.4	792 Oct 25.1	849 Sep 29.5	906 Sep 4.9	963 Aug 12.2
680 Jan 1.2	736 Dec 4.7	793 Nov 8.3	850 Oct 13.3	907 Sep 18.3	964 Aug 24.2
681 Jan 14.1	737 Dec 19.0	794 Nov 22.6	851 Oct 27.4	908 Oct 1.0	965 Sep 6.4
682 Jan 27.8	739 Jan 2.1	795 Dec 6.9	852 Nov 9.6	909 Oct 14.8	966 Sep 19.9
683 Feb 10.3	740 Jan 16.0	796 Dec 20.1	853 Nov 23.9	910 Oct 28.9	967 Oct 3.6
684 Feb 23.5	741 Jan 28.7	798 Jan 3.2	854 Dec 8.1	911 Nov 12.1	968 Oct 16.5

969 Oct 30.5	1026 Oct 5.3	1083 Sep 10.6	1140 Aug 16.8	1197 Jul 24.6	1254 Jul 2.2
970 Nov 13.7	1027 Oct 19.2	1084 Sep 23.2	1141 Aug 29.9	1198 Aug 6.3	1255 Jul 14.5
971 Nov 28.0	1028 Nov 1.3	1085 Oct 7.0	1142 Sep 12.2	1199 Aug 19.2	1256 Jul 26.0
972 Dec 11.2	1029 Nov 15.5	1086 Oct 20.9	1143 Sep 25.8	1200 Aug 31.4	1257 Aug 7.7
973 Dec 25.3	1030 Nov 29.7	1087 Nov 4.1	1144 Oct 8.7	1201 Sep 13.7	1258 Aug 20.6
975 Jan 8.3	1031 Dec 13.9	1088 Nov 17.2	1145 Oct 22.6	1202 Sep 27.4	1259 Sep 2.7
976 Jan 22.1	1032 Dec 27.0	1089 Dec 1.5	1146 Nov 5.8	1203 Oct 11.2	1260 Sep 15.1
977 Feb 3.7	1034 Jan 10.0	1090 Dec 15.7	1147 Nov 20.0	1204 Oct 24.3	1261 Sep 28.8
978 Feb 17.1	1035 Jan 23.8	1091 Dec 29.8	1148 Dec 3.2	1205 Nov 7.4	1262 Oct 12.7
979 Mar 2.2	1036 Feb 6.3	1093 Jan 11.7	1149 Dec 17.4	1206 Nov 21.7	1263 Oct 26.7
980 Mar 14.1	1037 Feb 18.6	1094 Jan 25.5	1150 Dec 31.5	1207 Dec 5.9	1264 Nov 8.9
981 Mar 26.7	1038 Mar 3.7	1095 Feb 8.0	1152 Jan 14.5	1208 Dec 19.1	1265 Nov 23.2
982 Apr 8.2	1039 Mar 16.6	1096 Feb 21.3	1153 Jan 27.2	1210 Jan 2.2	1266 Dec 7.4
983 Apr 20.5	1040 Mar 28.2	1097 Mar 5.3	1154 Feb 9.7	1211 Jan 16.1	1267 Dec 21.6
984 May 1.8	1041 Apr 9.7	1098 Mar 18.1	1155 Feb 23.0	1212 Jan 29.8	1269 Jan 3.7
985 May 13.9	1042 Apr 22.0	1099 Mar 30.8	1156 Mar 7.0	1213 Feb 11.3	1270 Jan 17.6
986 May 26.0	1043 May 4.2	1100 Apr 11.2	1157 Mar 19.8	1214 Feb 24.6	1271 Jan 31.3
987 Jun 7.0	1044 May 15.3	1101 Apr 23.5	1158 Apr 1.3	1215 Mar 9.5	1272 Feb 13.8
988 Jun 18.2	1045 May 27.4	1102 May 5.6	1159 Apr 13.7	1216 Mar 21.3	1273 Feb 26.0
989 Jun 30.4	1046 Jun 8.4	1103 May 17.7	1160 Apr 25.0	1217 Apr 2.9	1274 Mar 11.0
990 Jul 12.7	1047 Jun 20.6	1104 May 28.8	1161 May 7.1	1218 Apr 15.2	1275 Mar 23.8
991 Jul 25.2	1048 Jul 1.8	1105 Jun 9.9	1162 May 19.2	1219 Apr 27.4	1276 Apr 4.3
992 Aug 5.8	1049 Jul 14.1	1106 Jun 22.0	1163 May 31.3	1220 May 8.6	1277 Apr 16.6
993 Aug 18.7	1050 Jul 26.6	1107 Jul 4.3	1164 Jun 11.3	1221 May 20.6	1278 Apr 28.8
994 Aug 31.8	1051 Aug 8.3	1108 Jul 15.6	1165 Jun 23.5	1222 Jun 1.7	1279 May 11.0
995 Sep 14.2	1052 Aug 20.3	1109 Jul 28.1	1166 Jul 5.7	1223 Jun 13.7	1280 May 22.0
996 Sep 26.8	1053 Sep 2.4	1110 Aug 9.9	1167 Jul 18.1	1224 Jun 24.9	1281 Jun 3.0
997 Oct 10.6	1054 Sep 15.8	1111 Aug 22.8	1168 Jul 29.6	1225 Jul 7.1	1282 Jun 15.1
998 Oct 24.6	1055 Sep 29.5	1112 Sep 4.0	1169 Aug 11.4	1226 Jul 19.5	1283 Jun 27.2
999 Nov 7.8	1056 Oct 12.3	1113 Sep 17.5	1170 Aug 24.3	1227 Aug 1.0	1284 Jul 8.4
1000 Nov 21.1	1057 Oct 26.4	1114 Oct 1.1	1171 Sep 6.6	1228 Aug 12.8	1285 Jul 20.8
1001 Dec 5.3	1058 Nov 9.5	1115 Oct 15.0	1172 Sep 19.0	1229 Aug 25.8	1286 Aug 2.3
1002 Dec 19.5	1059 Nov 23.8	1116 Oct 28.1	1173 Oct 2.7	1230 Sep 8.0	1287 Aug 15.1
1004 Jan 2.6	1060 Dec 7.0	1117 Nov 11.3	1174 Oct 16.7	1231 Sep 21.5	1288 Aug 27.1
1005 Jan 15.5	1061 Dec 21.2	1118 Nov 25.5	1175 Oct 30.8	1232 Oct 4.2	1289 Sep 9.4
1006 Jan 29.2	1063 Jan 4.3	1119 Dec 9.8	1176 Nov 13.0	1233 Oct 18.2	1290 Sep 22.9
1007 Feb 11.7	1064 Jan 18.2	1120 Dec 23.0	1177 Nov 27.2	1234 Nov 1.3	1291 Oct 6.6
1008 Feb 24.9	1065 Jan 30.9	1122 Jan 6.0	1178 Dec 11.5	1235 Nov 15.5	1292 Oct 19.6
1009 Mar 8.9	1066 Feb 13.3	1123 Jan 19.9	1179 Dec 25.6	1236 Nov 28.8	1293 Nov 2.7
1010 Mar 21.7	1067 Feb 26.5	1124 Feb 2.5	1181 Jan 7.7	1237 Dec 13.0	1294 Nov 16.9
1011 Apr 3.3	1068 Mar 10.4	1125 Feb 14.9	1182 Jan 21.5	1238 Dec 27.2	1295 Dec 1.2
1012 Apr 14.7	1069 Mar 23.2	1126 Feb 28.1	1183 Feb 4.1	1240 Jan 10.2	1296 Dec 14.5
1013 Apr 27.0	1070 Apr 4.8	1127 Mar 13.0	1184 Feb 17.5	1241 Jan 23.1	1297 Dec 28.6
1014 May 9.2	1071 Apr 17.1	1128 Mar 24.7	1185 Mar 1.7	1242 Feb 5.7	1299 Jan 11.6
1015 May 21.3	1072 Apr 28.4	1129 Apr 6.2	1186 Mar 14.6	1243 Feb 19.0	1300 Jan 25.5
1016 Jun 1.4	1073 May 10.6	1130 Apr 18.6	1187 Mar 27.2	1244 Mar 3.2	1301 Feb 7.1
1017 Jun 13.5	1074 May 22.7	1131 Apr 30.9	1188 Apr 7.7	1245 Mar 16.0	1302 Feb 20.4
1018 Jun 25.6	1075 Jun 3.8	1132 May 12.0	1189 Apr 20.1	1246 Mar 28.7	1303 Mar 5.5
1019 Jul 7.9	1076 Jun 14.9	1133 May 24.1	1190 May 2.3	1247 Apr 10.2	1304 Mar 17.4
1020 Jul 19.3	1077 Jun 27.0	1134 Jun 5.2	1191 May 14.4	1248 Apr 21.5	1305 Mar 30.0
1021 Jul 31.9	1078 Jul 9.3	1135 Jun 17.3	1192 May 25.5	1249 May 3.7	1306 Apr 11.5
1022 Aug 13.7	1079 Jul 21.8	1136 Jun 28.5	1193 Jun 6.6	1250 May 15.8	1307 Apr 23.8
1023 Aug 26.7	1080 Aug 2.4	1137 Jul 10.8	1194 Jun 18.7	1251 May 27.9	1308 May 5.0
1024 Sep 8.0	1081 Aug 15.2	1138 Jul 23.2	1195 Jun 30.9	1252 Jun 7.9	1309 May 17.1
1025 Sep 21.5	1082 Aug 28.3	1139 Aug 4.9	1196 Jul 12.2	1253 Jun 20.0	1310 May 29.1

1311 Jun 10.2	1368 May 18.2	1425 Apr 26.1	1482 Apr 3.3	1539 Mar 10.6	1596 Feb 24.0
1312 Jun 21.3	1369 May 30.3	1426 May 8.2	1483 Apr 15.8	1540 Mar 22.5	1597 Mar 8.4
1313 Jul 3.4	1370 Jun 11.3	1427 May 20.3	1484 Apr 27.1	1541 Apr 4.2	1598 Mar 21.5
1314 Jul 15.7	1371 Jun 23.4	1428 May 31.4	1485 May 9.2	1542 Apr 16.7	1599 Apr 3.4
1315 Jul 28.2	1372 Jul 4.6	1429 Jun 12.4	1486 May 21.3	1543 Apr 29.0	1600 Apr 15.0
1316 Aug 8.9	1373 Jul 16.9	1430 Jun 24.5	1487 Jun 2.4	1544 May 10.1	1601 Apr 27.5
1317 Aug 21.8	1374 Jul 29.3	1431 Jul 6.7	1488 Jun 13.4	1545 May 22.2	1602 May 9.8
1318 Sep 3.9	1375 Aug 11.0	1432 Jul 17.9	1489 Jun 25.5	1546 Jun 3.3	1603 May 22.0
1319 Sep 17.4	1376 Aug 22.9	1433 Jul 30.4	1490 Jul 7.6	1547 Jun 15.3	1604 Jun 2.1
1320 Sep 30.0	1377 Sep 5.1	1434 Aug 12.0	1491 Jul 19.9	1548 Jun 26.4	1605 Jun 14.2
1321 Oct 13.9	1378 Sep 18.5	1435 Aug 24.9	1492 Jul 31.4	1549 Jul 8.6	1606 Jun 26.3
1322 Oct 28.0	1379 Oct 2.2	1436 Sep 6.1	1493 Aug 13.0	1550 Jul 20.9	1607 Jul 8.4
1323 Nov 11.2	1380 Oct 15.1	1437 Sep 19.5	1494 Aug 25.9	1551 Aug 2.3	1608 Jul 19.5
1324 Nov 24.5	1381 Oct 29.2	1438 Oct 3.2	1495 Sep 8.1	1552 Aug 14.0	1609 Jul 31.8
1325 Dec 8.7	1382 Nov 12.4	1439 Oct 17.0	1496 Sep 20.4	1553 Aug 26.9	1610 Aug 13.3
1326 Dec 22.9	1383 Nov 26.6	1440 Oct 30.1	1497 Oct 4.1	1554 Sep 9.0	1611 Aug 26.0
1328 Jan 6.0	1384 Dec 9.9	1441 Nov 13.3	1498 Oct 18.0	1555 Sep 22.4	1612 Sep 6.8
1329 Jan 19.0	1385 Dec 24.1	1442 Nov 27.6	1499 Nov 1.0	1556 Oct 5.0	1613 Sep 19.9
1330 Feb 1.7	1387 Jan 7.2	1443 Dec 11.9	1500 Nov 14.2	1557 Oct 18.9	1614 Oct 3.3
1331 Feb 15.1	1388 Jan 21.1	1444 Dec 25.1	1501 Nov 28.5	1558 Nov 1.9	1615 Oct 16.9
1332 Feb 28.4	1389 Feb 2.8	1446 Jan 8.2	1502 Dec 12.8	1559 Nov 16.1	1616 Oct 29.8
1333 Mar 12.3	1390 Feb 16.3	1447 Jan 22.1	1503 Dec 27.0	1560 Nov 29.4	1617 Nov 12.8
1334 Mar 25.1	1391 Mar 1.5	1448 Feb 4.8	1505 Jan 9.1	1561 Dec 13.7	1618 Nov 27.0
1335 Apr 6.6	1392 Mar 13.5	1449 Feb 17.3	1506 Jan 23.0	1562 Dec 27.9	1619 Dec 11.3
1336 Apr 18.0	1393 Mar 26.3	1450 Mar 2.5	1507 Feb 5.7	1564 Jan 10.9	1620 Dec 24.5
1337 Apr 30.2	1394 Apr 7.8	1451 Mar 15.5	1508 Feb 19.2	1565 Jan 23.8	1622 Jan 7.7
1338 May 12.2	1395 Apr 20.1	1452 Mar 27.3	1509 Mar 3.4	1566 Feb 6.5	1623 Jan 21.8
1339 May 24.3	1396 May 1.3	1453 Apr 8.8	1510 Mar 16.4	1567 Feb 20.0	1624 Feb 4.7
1340 Jun 4.3	1397 May 13.4	1454 Apr 21.2	1511 Mar 29.2	1568 Mar 4.3	1625 Feb 17.4
1341 Jun 16.3	1398 May 25.4	1455 May 3.4	1512 Apr 9.8	1569 Mar 17.3	1626 Mar 2.8
1342 Jun 28.4	1399 Jun 6.4	1456 May 14.5	1513 Apr 22.1	1570 Mar 30.1	1627 Mar 16.1
1343 Jul 10.6	1400 Jun 17.5	1457 May 26.5	1514 May 4.3	1571 Apr 11.6	1628 Mar 28.1
1344 Jul 22.0	1401 Jun 29.6	1458 Jun 7.5	1515 May 16.4	1572 Apr 23.0	1629 Apr 9.9
1345 Aug 3.5	1402 Jul 11.8	1459 Jun 19.5	1516 May 27.5	1573 May 5.2	1630 Apr 22.5
1346 Aug 16.3	1403 Jul 24.1	1460 Jun 30.6	1517 Jun 8.5	1574 May 17.3	1631 May 4.9
1347 Aug 29.3	1404 Aug 4.7	1461 Jul 12.8	1518 Jun 20.5	1575 May 29.4	1632 May 16.1
1348 Sep 10.6	1405 Aug 17.4	1462 Jul 25.2	1519 Jul 2.6	1576 Jun 9.4	1633 May 28.3
1349 Sep 24.1	1406 Aug 30.4	1463 Aug 6.7	1520 Jul 13.8	1577 Jun 21.5	1634 Jun 9.3
1350 Oct 7.8	1407 Sep 12.7	1464 Aug 18.5	1521 Jul 26.2	1578 Jul 3.6	1635 Jun 21.4
1351 Oct 21.8	1408 Sep 25.2	1465 Aug 31.4	1522 Aug 7.7	1579 Jul 15.8	1636 Jul 2.4
1352 Nov 3.9	1409 Oct 8.9	1466 Sep 13.7	1523 Aug 20.5	1580 Jul 27.2	1637 Jul 14.6
1353 Nov 18.2	1410 Oct 22.5	1467 Sep 27.2	1524 Sep 1.4	1581 Aug 8.7	1638 Jul 26.8
1354 Dec 2.4	1411 Nov 6.0	1468 Oct 9.9	1525 Sep 14.7	1582 Aug 21.5	1639 Aug 8.2
1355 Dec 16.7	1412 Nov 19.2	1469 Oct 23.9	1526 Sep 28.1	1583 Sep 13.4	1640 Aug 19.7
1356 Dec 29.9	1413 Dec 3.5	1470 Nov 7.0	1527 Oct 11.9	1584 Sep 25.7	1641 Sep 1.5
1358 Jan 12.9	1414 Dec 17.8	1471 Nov 21.2	1528 Oct 24.8	1585 Oct 9.1	1642 Sep 14.5
1359 Jan 26.7	1415 Dec 31.9	1472 Dec 4.5	1529 Nov 7.9	1586 Oct 22.8	1643 Sep 27.7
1360 Feb 9.3	1417 Jan 14.0	1473 Dec 18.7	1530 Nov 22.1	1587 Nov 5.7	1644 Oct 10.1
1361 Feb 21.7	1418 Jan 27.8	1475 Jan 1.9	1531 Dec 6.4	1588 Nov 18.8	1645 Oct 23.8
1362 Mar 6.7	1419 Feb 10.4	1476 Jan 15.9	1532 Dec 19.6	1589 Dec 3.0	1646 Nov 6.7
1363 Mar 19.6	1420 Feb 23.7	1477 Jan 28.7	1534 Jan 2.8	1590 Dec 17.3	1647 Nov 20.8
1364 Mar 31.3	1421 Mar 7.8	1478 Feb 11.3	1535 Jan 16.8	1591 Dec 31.5	1648 Dec 4.0
1365 Apr 12.7	1422 Mar 20.7	1479 Feb 24.7	1536 Jan 30.6	1593 Jan 13.6	1649 Dec 18.2
1366 Apr 25.0	1423 Apr 2.3	1480 Mar 8.8	1537 Feb 12.2	1594 Jan 27.6	1651 Jan 1.4
1367 May 7.2	1424 Apr 13.8	1481 Mar 21.6	1538 Feb 25.5	1595 Feb 10.4	1652 Jan 15.5

1653 Jan 28.5	1710 Jan 3.5	1766 Dec 8.3	1823 Nov 13.5	1880 Oct 18.5	1937 Sep 25.2
1654 Feb 11.3	1711 Jan 17.6	1767 Dec 22.5	1824 Nov 26.6	1881 Nov 1.2	1938 Oct 8.6
1655 Feb 24.9	1712 Jan 31.5	1769 Jan 4.7	1825 Dec 10.7	1882 Nov 15.1	1939 Oct 22.1
1656 Mar 9.2	1713 Feb 13.3	1770 Jan 18.7	1826 Dec 24.9	1883 Nov 29.2	1940 Nov 3.9
1657 Mar 22.3	1714 Feb 26.9	1771 Feb 1.7	1828 Jan 8.0	1884 Dec 12.3	1941 Nov 17.8
1658 Apr 4.2	1715 Mar 12.2	1772 Feb 15.4	1829 Jan 21.1	1885 Dec 26.4	1942 Dec 1.8
1659 Apr 16.9	1716 Mar 24.3	1773 Feb 28.0	1830 Feb 4.0	1887 Jan 9.6	1943 Dec 16.0
1660 Apr 28.4	1717 Apr 6.2	1774 Mar 13.3	1831 Feb 17.7	1888 Jan 23.6	1944 Dec 29.1
1661 May 10.7	1718 Apr 18.8	1775 Mar 26.4	1832 Mar 2.2	1889 Feb 5.5	1946 Jan 12.2
1662 May 22.9	1719 May 1.3	1776 Apr 7.2	1833 Mar 15.5	1890 Feb 19.2	1947 Jan 26.2
1663 Jun 4.1	1720 May 12.7	1777 Apr 19.9	1834 Mar 28.6	1891 Mar 4.6	1948 Feb 9.1
1664 Jun 15.1	1721 May 24.9	1778 May 2.4	1835 Apr 10.4	1892 Mar 16.9	1949 Feb 21.7
1665 Jun 27.2	1722 Jun 6.0	1779 May 14.7	1836 Apr 22.1	1893 Mar 29.9	1950 Mar 7.2
1666 Jul 9.3	1723 Jun 18.1	1780 May 26.0	1837 May 4.6	1894 Apr 11.8	1951 Mar 20.4
1667 Jul 21.5	1724 Jun 29.2	1781 Jun 7.1	1838 May 16.9	1895 Apr 24.4	1952 Apr 1.4
1668 Aug 1.8	1725 Jul 11.4	1782 Jun 19.2	1839 May 29.1	1896 May 5.9	1953 Apr 14.2
1669 Aug 14.3	1726 Jul 23.6	1783 Jul 1.3	1840 Jun 9.3	1897 May 18.2	1954 Apr 26.8
1670 Aug 27.0	1727 Aug 4.9	1784 Jul 12.5	1841 Jun 21.4	1898 May 30.4	1955 May 9.3
1671 Sep 8.9	1728 Aug 16.4	1785 Jul 24.7	1842 Jul 3.6	1899 Jun 11.6	1956 May 20.6
1672 Sep 21.0	1729 Aug 29.1	1786 Aug 6.1	1843 Jul 15.7	1900 Jun 23.7	1957 Jun 1.8
1673 Oct 4.4	1730 Sep 11.0	1787 Aug 18.6	1844 Jul 27.0	1901 Jul 5.9	1958 Jun 14.0
1674 Oct 18.0	1731 Sep 24.1	1788 Aug 30.3	1845 Aug 8.4	1902 Jul 18.1	1959 Jun 26.1
1675 Oct 31.8	1732 Oct 6.5	1789 Sep 12.3	1846 Aug 20.9	1903 Jul 30.4	1960 Jul 7.3
1676 Nov 13.8	1733 Oct 20.1	1790 Sep 25.4	1847 Sep 2.7	1904 Aug 10.8	1961 Jul 19.5
1677 Nov 28.0	1734 Nov 2.9	1791 Oct 8.8	1848 Sep 14.6	1905 Aug 23.4	1962 Jul 31.8
1678 Dec 12.2	1735 Nov 17.0	1792 Oct 21.4	1849 Sep 27.8	1906 Sep 5.1	1963 Aug 13.2
1679 Dec 26.5	1736 Nov 30.1	1793 Nov 4.2	1850 Oct 11.2	1907 Sep 18.1	1964 Aug 24.8
1681 Jan 8.6	1737 Dec 14.3	1794 Nov 18.3	1851 Oct 24.8	1908 Sep 30.3	1965 Sep 6.7
1682 Jan 22.7	1738 Dec 28.5	1795 Dec 2.4	1852 Nov 6.7	1909 Oct 13.7	1966 Sep 19.7
1683 Feb 5.6	1740 Jan 11.7	1796 Dec 15.6	1853 Nov 20.7	1910 Oct 27.4	1967 Oct 2.9
1684 Feb 19.2	1741 Jan 24.7	1797 Dec 29.8	1854 Dec 4.8	1911 Nov 10.3	1968 Oct 15.4
1685 Mar 3.7	1742 Feb 7.6	1799 Jan 12.9	1855 Dec 19.0	1912 Nov 23.3	1969 Oct 29.1
1686 Mar 17.0	1743 Feb 21.3	1800 Jan 26.9	1857 Jan 1.2	1913 Dec 7.4	1970 Nov 11.9
1687 Mar 30.0	1744 Mar 5.7	1801 Feb 9.8	1858 Jan 15.3	1914 Dec 21.6	1971 Nov 26.0
1688 Apr 10.8	1745 Mar 18.9	1802 Feb 23.4	1859 Jan 29.3	1916 Jan 4.7	1972 Dec 9.1
1689 Apr 23.4	1746 Mar 31.9	1803 Mar 8.8	1860 Feb 12.1	1917 Jan 17.8	1973 Dec 23.3
1690 May 5.8	1747 Apr 13.7	1804 Mar 21.0	1861 Feb 24.7	1918 Jan 31.8	1975 Jan 6.4
1691 May 18.0	1748 Apr 25.3	1805 Apr 3.0	1862 Mar 10.1	1919 Feb 14.6	1976 Jan 20.5
1692 May 29.2	1749 May 7.8	1806 Apr 15.8	1863 Mar 23.3	1920 Feb 28.2	1977 Feb 2.4
1693 Jun 10.3	1750 May 20.0	1807 Apr 28.4	1864 Apr 4.3	1921 Mar 12.5	1978 Feb 16.2
1694 Jun 22.4	1751 Jun 1.2	1808 May 9.8	1865 Apr 17.0	1922 Mar 25.7	1979 Mar 1.7
1695 Jul 4.4	1752 Jun 12.3	1809 May 22.1	1866 Apr 29.6	1923 Apr 7.6	1980 Mar 14.1
1696 Jul 15.6	1753 Jun 24.4	1810 Jun 3.3	1867 May 12.0	1924 Apr 19.4	1981 Mar 27.2
1697 Jul 27.9	1754 Jul 6.5	1811 Jun 15.5	1868 May 23.3	1925 May 1.9	1982 Apr 9.1
1698 Aug 9.3	1755 Jul 18.7	1812 Jun 26.6	1869 Jun 4.5	1926 May 14.4	1983 Apr 21.8
1699 Aug 21.8	1756 Jul 30.0	1813 Jul 8.7	1870 Jun 16.7	1927 May 26.7	1984 May 3.3
1700 Sep 3.6	1757 Aug 11.4	1814 Jul 20.9	1871 Jun 28.8	1928 Jun 6.9	1985 May 15.8
1701 Sep 16.6	1758 Aug 24.0	1815 Aug 2.2	1872 Jul 10.0	1929 Jun 19.0	1986 May 28.0
1702 Sep 29.8	1759 Sep 5.8	1816 Aug 13.7	1873 Jul 22.2	1930 Jul 1.1	1987 Jun 9.2
1703 Oct 13.3	1760 Sep 17.8	1817 Aug 26.3	1874 Aug 3.5	1931 Jul 13.3	1988 Jun 20.4
1704 Oct 26.0	1761 Oct 1.1	1818 Sep 8.1	1875 Aug 16.0	1932 Jul 24.6	1989 Jul 2.5
1705 Nov 8.9	1762 Oct 14.6	1819 Sep 21.2	1876 Aug 27.7	1933 Aug 5.9	1990 Jul 14.7
1706 Nov 22.9	1763 Oct 28.3	1820 Oct 3.5	1877 Sep 9.6	1934 Aug 18.5	1991 Jul 27.0
1707 Dec 7.1	1764 Nov 10.1	1821 Oct 16.9	1878 Sep 22.7	1935 Aug 31.2	1992 Aug 7.4
1708 Dec 20.3	1765 Nov 24.2	1822 Oct 30.6	1879 Oct 6.0	1936 Sep 12.1	1993 Aug 19.9

1994 Sep 1.7	2051 Aug 9.8	2108 Jul 18.6	2165 Jun 26.6	2222 Jun 5.6	2279 May 14.3
1995 Sep 14.6	2052 Aug 21.4	2109 Jul 30.9	2166 Jul 8.7	2223 Jun 17.7	2280 May 25.6
1996 Sep 26.8	2053 Sep 3.2	2110 Aug 12.3	2167 Jul 20.9	2224 Jun 28.8	2281 Jun 6.8
1997 Oct 10.2	2054 Sep 16.2	2111 Aug 24.9	2168 Aug 1.2	2225 Jul 11.0	2282 Jun 18.9
1998 Oct 23.8	2055 Sep 29.4	2112 Sep 5.6	2169 Aug 13.6	2226 Jul 23.2	2283 Jul 1.0
1999 Nov 6.6	2056 Oct 11.8	2113 Sep 18.6	2170 Aug 26.2	2227 Aug 4.4	2284 Jul 12.1
2000 Nov 19.5	2057 Oct 25.4	2114 Oct 1.9	2171 Sep 8.0	2228 Aug 15.9	2285 Jul 24.3
2001 Dec 3.6	2058 Nov 8.3	2115 Oct 15.3	2172 Sep 20.0	2229 Aug 28.5	2286 Aug 5.6
2002 Dec 17.7	2059 Nov 22.3	2116 Oct 28.0	2173 Oct 3.3	2230 Sep 10.3	2287 Aug 18.0
2003 Dec 31.9	2060 Dec 5.3	2117 Nov 10.9	2174 Oct 16.8	2231 Sep 23.3	2288 Aug 29.6
2005 Jan 13.9	2061 Dec 19.5	2118 Nov 24.9	2175 Oct 30.5	2232 Oct 5.5	2289 Sep 11.4
2006 Jan 27.9	2063 Jan 2.6	2119 Dec 9.0	2176 Nov 12.4	2233 Oct 19.0	2290 Sep 24.4
2007 Feb 10.8	2064 Jan 16.7	2120 Dec 22.1	2177 Nov 26.4	2234 Nov 1.8	2291 Oct 7.7
2008 Feb 24.4	2065 Jan 29.7	2122 Jan 5.3	2178 Dec 10.5	2235 Nov 15.7	2292 Oct 20.2
2009 Mar 8.8	2066 Feb 12.5	2123 Jan 19.3	2179 Dec 24.7	2236 Nov 28.7	2293 Nov 2.9
2010 Mar 22.0	2067 Feb 26.1	2124 Feb 2.3	2181 Jan 6.8	2237 Dec 12.9	2294 Nov 16.8
2011 Apr 4.0	2068 Mar 10.5	2125 Feb 15.1	2182 Jan 20.9	2238 Dec 27.0	2295 Nov 30.9
2012 Apr 15.8	2069 Mar 23.7	2126 Feb 28.7	2183 Feb 3.8	2240 Jan 10.2	2296 Dec 14.1
2013 Apr 28.3	2070 Apr 5.6	2127 Mar 14.1	2184 Feb 17.6	2241 Jan 23.3	2297 Dec 28.2
2014 May 10.8	2071 Apr 18.3	2128 Mar 26.2	2185 Mar 2.2	2242 Feb 6.2	2299 Jan 11.4
2015 May 23.0	2072 Apr 29.9	2129 Apr 8.2	2186 Mar 15.6	2243 Feb 20.0	2300 Jan 25.5
2016 Jun 3.3	2073 May 12.3	2130 Apr 20.9	2187 Mar 28.7	2244 Mar 4.6	2301 Feb 8.4
2017 Jun 15.4	2074 May 24.5	2131 May 3.4	2188 Apr 9.6	2245 Mar 18.0	2302 Feb 22.2
2018 Jun 27.5	2075 Jun 5.7	2132 May 14.8	2189 Apr 22.3	2246 Mar 31.1	2303 Mar 7.8
2019 Jul 9.7	2076 Jun 16.9	2133 May 27.0	2190 May 4.8	2247 Apr 13.0	2304 Mar 20.2
2020 Jul 20.9	2077 Jun 29.0	2134 Jun 8.2	2191 May 17.2	2248 Apr 24.7	2305 Apr 2.3
2021 Aug 2.3	2078 Jul 11.1	2135 Jun 20.3	2192 May 28.4	2249 May 7.2	2306 Apr 15.2
2022 Aug 14.7	2079 Jul 23.4	2136 Jul 1.4	2193 Jun 9.5	2250 May 19.5	2307 Apr 27.9
2023 Aug 27.4	2080 Aug 3.7	2137 Jul 13.6	2194 Jun 21.6	2251 May 31.7	2308 May 9.4
2024 Sep 8.2	2081 Aug 16.2	2138 Jul 25.8	2195 Jul 3.8	2252 Jun 11.8	2309 May 21.7
2025 Sep 21.2	2082 Aug 28.8	2139 Aug 7.1	2196 Jul 14.9	2253 Jun 23.9	2310 Jun 2.9
2026 Oct 4.5	2083 Sep 10.7	2140 Aug 18.6	2197 Jul 27.1	2254 Jul 6.0	2311 Jun 15.0
2027 Oct 18.0	2084 Sep 22.8	2141 Aug 31.3	2198 Aug 8.5	2255 Jul 18.2	2312 Jun 26.1
2028 Oct 30.7	2085 Oct 6.1	2142 Sep 13.1	2199 Aug 21.0	2256 Jul 29.4	2313 Jul 8.2
2029 Nov 13.6	2086 Oct 19.6	2143 Sep 26.2	2200 Sep 2.6	2257 Aug 10.7	2314 Jul 20.3
2030 Nov 27.7	2087 Nov 8.6	2144 Oct 8.6	2201 Sep 15.5	2258 Aug 23.2	2315 Aug 1.5
2031 Dec 11.8	2088 Nov 15.3	2145 Oct 22.1	2202 Sep 28.6	2259 Sep 4.9	2316 Aug 12.9
2032 Dec 25.0	2089 Nov 29.3	2146 Nov 4.9	2203 Oct 11.9	2260 Sep 16.7	2317 Aug 25.3
2034 Jan 8.1	2090 Dec 13.5	2147 Nov 18.8	2204 Oct 24.5	2261 Sep 29.8	2318 Sep 7.0
2035 Jan 22.2	2091 Dec 27.6	2148 Dec 1.9	2205 Nov 7.3	2262 Oct 13.2	2319 Sep 19.9
2036 Feb 5.1	2093 Jan 9.8	2149 Dec 16.1	2206 Nov 21.2	2263 Oct 26.8	2320 Oct 2.0
2037 Feb 17.8	2094 Jan 23.8	2150 Dec 30.2	2207 Dec 5.3	2264 Nov 8.6	2321 Oct 15.3
2038 Mar 3.4	2095 Feb 6.7	2152 Jan 13.4	2208 Dec 18.5	2265 Nov 22.5	2322 Oct 28.9
2039 Mar 16.7	2096 Feb 20.4	2153 Jan 26.4	2210 Jan 1.7	2266 Dec 6.6	2323 Nov 11.7
2040 Mar 28.8	2097 Mar 4.9	2154 Feb 9.3	2211 Jan 15.8	2267 Dec 20.8	2324 Nov 24.6
2041 Apr 10.6	2098 Mar 18.2	2155 Feb 23.0	2212 Jan 29.8	2269 Jan 3.0	2325 Dec 8.7
2042 Apr 23.3	2099 Mar 31.3	2156 Mar 7.5	2213 Feb 11.7	2270 Jan 17.1	2326 Dec 22.9
2043 May 5.8	2100 Apr 13.1	2157 Mar 20.8	2214 Feb 25.4	2271 Jan 31.1	2328 Jan 6.1
2044 May 17.2	2101 Apr 25.8	2158 Apr 2.8	2215 Mar 10.9	2272 Feb 14.0	2329 Jan 19.2
2045 May 29.5	2102 May 8.3	2159 Apr 15.6	2216 Mar 23.2	2273 Feb 26.7	2330 Feb 2.3
2046 Jun 10.6	2103 May 20.6	2160 Apr 27.3	2217 Apr 5.2	2274 Mar 12.2	2331 Feb 16.1
2047 Jun 22.8	2104 May 31.9	2161 May 9.7	2218 Apr 18.0	2275 Mar 25.4	2332 Feb 29.8
2048 Jul 4.0	2105 Jun 13.1	2162 May 22.0	2219 Apr 30.6	2276 Apr 6.4	2333 Mar 14.3
2049 Jul 16.2	2106 Jun 25.2	2163 Jun 3.3	2220 May 12.0	2277 Apr 19.2	2334 Mar 27.6
2050 Jul 28.4	2107 Jul 7.4	2164 Jun 14.4	2221 May 24.4	2278 May 1.8	2335 Apr 9.6

2336 Apr 21.4	2364 Apr 3.4	2392 Mar 15.3	2420 Feb 25.2	2448 Feb 5.2	2476 Jan 16.3
2337 May 4.0	2365 Apr 16.3	2393 Mar 28.5	2421 Mar 9.8	2449 Feb 18.1	2477 Jan 29.3
2338 May 16.4	2366 Apr 29.0	2394 Apr 10.6	2422 Mar 23.2	2450 Mar 3.7	2478 Feb 12.3
2339 May 28.7	2367 May 11.5	2395 Apr 23.4	2423 Apr 5.3	2451 Mar 17.2	2479 Feb 26.1
2340 Jun 8.9	2368 May 22.8	2396 May 5.0	2424 Apr 17.2	2452 Mar 29.5	2480 Mar 10.7
2341 Jun 21.0	2369 Jun 4.0	2397 May 17.4	2425 Apr 29.9	2453 Apr 11.5	2481 Mar 24.0
2342 Jul 3.1	2370 Jun 16.1	2398 May 29.7	2426 May 12.5	2454 Apr 24.3	2482 Apr 6.2
2343 Jul 15.2	2371 Jun 28.2	2399 Jun 10.9	2427 May 24.8	2455 May 6.9	2483 Apr 19.1
2344 Jul 26.4	2372 Jul 9.3	2400 Jun 22.0	2428 Jun 5.0	2456 May 18.4	2484 Apr 30.8
2345 Aug 7.7	2373 Jul 21.4	2401 Jul 4.1	2429 Jun 17.1	2457 May 30.7	2485 May 13.4
2346 Aug 20.1	2374 Aug 2.6	2402 Jul 16.3	2430 Jun 29.2	2458 Jun 11.9	2486 May 25.7
2347 Sep 1.7	2375 Aug 14.9	2403 Jul 28.4	2431 Jul 11.3	2459 Jun 24.0	2487 Jun 7.0
2348 Sep 13.5	2376 Aug 26.4	2404 Aug 8.7	2432 Jul 22.4	2460 Jul 5.1	2488 Jun 18.1
2349 Sep 26.5	2377 Sep 8.1	2405 Aug 21.1	2433 Aug 3.6	2461 Jul 17.2	2489 Jun 30.2
2350 Oct 9.7	2378 Sep 20.9	2406 Sep 2.7	2434 Aug 16.0	2462 Jul 29.4	2490 Jul 12.3
2351 Oct 23.2	2379 Oct 4.0	2407 Sep 15.5	2435 Aug 28.4	2463 Aug 10.7	2491 Jul 24.4
2352 Nov 5.0	2380 Oct 16.4	2408 Sep 27.5	2436 Sep 9.1	2464 Aug 22.1	2492 Aug 4.6
2353 Nov 18.9	2381 Oct 29.9	2409 Oct 10.7	2437 Sep 22.0	2465 Sep 3.7	2493 Aug 17.0
2354 Dec 2.9	2382 Nov 12.7	2410 Oct 24.2	2438 Oct 5.0	2466 Sep 16.4	2494 Aug 29.4
2355 Dec 17.1	2383 Nov 26.7	2411 Nov 6.9	2439 Oct 18.4	2467 Sep 29.4	2495 Sep 11.1
2356 Dec 30.3	2384 Dec 9.8	2412 Nov 19.8	2440 Oct 30.9	2468 Oct 11.7	2496 Sep 23.0
2358 Jan 13.5	2385 Dec 23.9	2413 Dec 3.9	2441 Nov 13.7	2469 Oct 25.1	2497 Oct 6.0
2359 Jan 27.5	2387 Jan 7.1	2414 Dec 18.0	2442 Nov 27.6	2470 Nov 7.8	2498 Oct 19.3
2360 Feb 10.5	2388 Jan 21.2	2416 Jan 1.2	2443 Dec 11.7	2471 Nov 21.7	2499 Nov 1.9
2361 Feb 23.3	2389 Feb 3.3	2417 Jan 14.4	2444 Dec 24.9	2472 Dec 4.8	2500 Nov 15.6
2362 Mar 8.9	2390 Feb 17.1	2418 Jan 28.5	2446 Jan 8.0	2473 Dec 18.9	
2363 Mar 22.2	2391 Mar 2.8	2419 Feb 11.4	2447 Jan 22.1	2475 Jan 2.1	

During the period A.D. 0-2500, the three bright superior planets (Mars, Jupiter, Saturn) all reach opposition within an interval less than 30.0 days in the following cases :

Year	Interval	Year	Interval	Year	Interval
214	19.0 days	927	26.0 days	1503	4.4 days
451	25.9	929	18.8	1740-1741	28.5
453	27.7	1027	11.3	1743	12.0
551	27.1	1068	26.4	1980	18.3
690	16.2	1166	19.5	1982	25.6
788	16.6	1264	19.2	2219	20.8
829	28.3	1266	20.6	2458	18.6

During the same period 0-2500, in the following years neither Jupiter nor Saturn are in oppositions with the Sun :

591,   650,   709,   1445,   1504,   1563,   2239,   2298,   2357

In the years 591, 650, 1445, 1504, 2239, and 2357, there is no opposition of Mars, Jupiter, and Saturn.

XI

TRANSITS OF MERCURY, 1600 – 2300

The table on the next two pages lists all transits of Mercury over the Sun's disk taking place in the period A.D. 1600 to 2300. These data have been calculated in 1969 by Mr. Michel Walch, of Ville d'Avray, France, and had never been published before. The calculations are based on Newcomb's tables for the Sun and Mercury.

The first two columns give the date and the time t_m (Ephemeris Time) of the geocentric least distance between the centers of Mercury and the Sun. As short-period planetary perturbations were not taken into account, the times may be in error by 2 or 3 minutes. To obtain the Universal Time of the least distance, see the Note on Time Reckoning at the beginning of this book.

The third column provides the least geocentric distance d_m of the centers, in seconds or arc. This quantity is positive if Mercury passes to the north of the center of the Sun's disk, and negative if it passes to the south. The next columns contain the semidiameters of the Sun (s) and Mercury (s') in seconds of arc, and the right ascension α and declination δ of the center of the Sun in degrees and decimals.

Columns τ_e and τ_i give the geocentric semidurations for the exterior and interior contacts, respectively. The geocentric times of the exterior and interior contacts are found by subtracting and adding the values τ_e and τ_i to the instant t_m of the least distance. See the example given on page 7-49 for a transit of Venus.

The columns P_1 and P_2 give the geocentric position angles of the *center* of the disk of Mercury at ingress and egress, respectively, measured from the North Point of the solar disk in the usual way (eastward from North).

The last column indicates to what series (in the period of 46 years) the transit belongs. The transits indicated by b are the first ones of a 46-year series ; those marked with e are the last ones of a series.

There exists a longer and more accurate series with a period of 217 years. See, for instance, the transits of 1605, 1822, 2039, and 2256.

The transits taking place in November occur at the ascending node of the orbit of Mercury ; those of May take place at the descending node of this orbit.

Panorama of May transits

	1306	1523	1740	1957	2174	
	1352	1569	1786	2003	2220	2437
	1398	1615	1832	2049	2266	2483
:	1444	1661	1878	2095	2312	2529
:	1490	1707	1924	2141	2358	2575
1319	1536	1753	1970	2187	2404	:
1365	1582	1799	2016	2233	2450	:
1411	1628	1845	2062	2279	2496	
1457	1674	1891	2108	2325	2542	
	1937	2154	2371	2588		

The "panorama" given on this page groups all May transits of Mercury taking place in the period A.D. 1300-2610. In the vertical direction the transits are separated by intervals of 46 years ; horizontally the interval is 217 years.

A similar panorama can be constructed for the November transits.

For example, the transit of 1970 May 9 belongs to a 46-year series which began in 1740 and will end in 2154.

The transit of 1937 May 11 was the last of a 46-year series, and the first of a 217-year series which will live for many centuries. It was a partial transit for a part of the Earth's surface, while elsewhere no transit was visible.

The transit of 1999 November 15 will be the last one of a 46-year series which began in 1171. It will be a partial transit for a part of Earth.

Date			t_m	d_m	s	s'	α	δ	τ_e	τ_i	P_1	P_2	Ser.
			h m	$''$	$''$	$''$	°	°	h m	h m	°	°	
1605	Nov	1	20 04	−855	968.81	4.95	217.07	−14.68	1 19.2	1 15.7	179	235	D
1615	May	3	10 11	+467	950.34	5.99	40.00	+15.61	3 27.8	3 24.4	32	271	I
1618	Nov	4	13 45	−352	969.38	4.94	219.63	−15.49	2 33.9	2 32.1	138	275	C
1628	May	5	17 32	−571	949.70	6.01	43.03	+16.52	3 12.4	3 08.6	100	206	H
1631	Nov	7	7 22	+146	969.91	4.94	222.21	−16.27	2 42.6	2 41.0	107	304	B
1644	Nov	9	0 55	+640	970.41	4.94	224.80	−17.02	2 03.8	2 01.6	74	336	A
1651	Nov	3	0 52	−751	968.86	4.95	218.10	−15.01	1 45.4	1 42.7	167	246	D
1661	May	3	16 51	+266	950.28	5.99	41.10	+15.94	3 49.1	3 46.0	46	258	I
1664	Nov	4	18 33	−250	969.43	4.94	220.67	−15.81	2 39.4	2 37.7	131	281	C
1674	May	7	0 15	−774	949.64	6.01	44.14	+16.83	2 20.6	2 15.4	118	188	H e
1677	Nov	7	12 10	+248	969.96	4.94	223.25	−16.58	2 39.0	2 37.3	100	310	B
1690	Nov	10	5 43	+742	970.45	4.94	225.86	−17.31	1 46.5	1 43.9	65	344	A
1697	Nov	3	5 41	−648	968.92	4.95	219.14	−15.33	2 03.5	2 01.3	158	254	D
1707	May	5	23 32	+ 65	950.21	5.99	42.21	+16.27	3 58.1	3 55.1	58	246	I
1710	Nov	6	23 21	−146	969.48	4.94	221.71	−16.12	2 43.0	2 41.3	124	287	C
1723	Nov	9	16 57	+351	970.00	4.94	224.31	−16.88	2 33.4	2 31.6	94	316	B
1736	Nov	11	10 30	+844	970.49	4.94	226.92	−17.60	1 22.3	1 18.9	54	354	A
1740	Nov	2	23 03	+888	950.81	5.98	40.29	+15.68	1 28.4	1 20.1	353	311	J b
1743	Nov	5	10 30	−543	968.97	4.95	220.18	−15.65	2 17.2	2 15.2	150	262	D
1753	May	6	6 13	−139	950.15	6.00	43.32	+16.59	3 56.4	3 53.4	71	234	I
1756	Nov	7	4 10	− 43	969.53	4.94	222.76	−16.43	2 44.6	2 43.0	118	293	C
1769	Nov	9	21 45	+454	970.04	4.94	225.36	−17.17	2 25.4	2 23.6	87	322	B
1776	Nov	2	21 36	−943	968.45	4.95	218.65	−15.17	0 40.9	0 33.4	193	220	E b
1782	Nov	12	15 17	+945	970.53	4.93	227.98	−17.88	0 40.8	0 33.5	37	11	A e
1786	May	4	5 42	+688	950.74	5.98	41.39	+16.01	2 44.9	2 40.6	16	289	J
1789	Nov	5	15 19	−440	969.03	4.95	221.22	−15.96	2 27.4	2 25.6	143	269	D
1799	May	7	12 54	−342	950.08	6.00	44.43	+16.90	3 43.4	3 40.2	84	222	I
1802	Nov	9	8 58	+ 60	969.57	4.94	223.81	−16.73	2 44.4	2 42.7	111	299	C
1815	Nov	12	2 33	+555	970.08	4.94	226.42	−17.45	2 15.0	2 13.0	79	329	B
1822	Nov	5	2 26	−839	968.50	4.95	219.69	−15.49	1 24.0	1 20.6	176	236	E
1832	May	5	12 20	+489	950.67	5.98	42.49	+16.34	3 24.4	3 21.0	32	273	J
1835	Nov	7	20 08	−336	969.08	4.94	222.27	−16.27	2 35.0	2 33.2	136	275	D
1845	May	8	19 36	−546	950.02	6.00	45.55	+17.21	3 16.5	3 12.8	98	208	I
1848	Nov	9	13 46	+163	969.61	4.94	224.86	−17.02	2 42.3	2 40.6	105	304	C
1861	Nov	12	7 20	+658	970.12	4.94	227.48	−17.74	2 01.2	1 58.9	71	337	B
1868	Nov	5	7 15	−735	968.56	4.95	220.73	−15.81	1 48.6	1 46.0	165	247	E
1878	May	6	18 59	+289	950.60	5.99	43.60	+16.66	3 46.9	3 43.8	45	260	J
1881	Nov	8	0 56	−233	969.13	4.94	223.32	−16.58	2 40.3	2 38.6	129	281	D
1891	May	10	2 18	−750	949.96	6.01	46.67	+17.51	2 28.6	2 23.7	116	192	I
1894	Nov	10	18 34	+266	969.66	4.94	225.92	−17.31	2 38.3	2 36.6	99	310	C
1907	Nov	14	12 08	+759	970.16	4.94	228.55	−18.01	1 43.0	1 40.3	62	345	B
1914	Nov	7	12 04	−631	968.62	4.95	221.78	−16.12	2 06.1	2 03.9	156	255	E
1924	May	8	1 39	+ 87	950.53	5.99	44.72	+16.97	3 57.3	3 54.3	58	248	J
1927	Nov	10	5 45	−130	969.18	4.94	224.37	−16.87	2 43.6	2 41.9	123	287	D
1937	May	11	9 00	−955	949.90	6.01	47.80	+17.81	0 11	—	—	—	I e
1940	Nov	11	23 22	+368	969.71	4.94	226.98	−17.59	2 32.3	2 30.5	92	316	C
1953	Nov	14	16 55	+861	970.20	4.94	229.61	−18.28	1 16.9	1 13.3	51	356	B
1957	May	6	1 15	+907	951.14	5.97	42.78	+16.41	1 15.3	1 05.5	350	315	K b

Date	t_m	d_m	s	s'	α	δ	τ_e	τ_i	P_1	P_2	Ser.
	h m	"	"	"	°	°	h m	h m	°	°	
1960 Nov 7	16 53	−528	968.67	4.95	222.82	−16.42	2 19.1	2 17.1	148	262	E
1970 May 9	8 19	−115	950.47	5.99	45.84	+17.28	3 56.7	3 53.7	70	237	J
1973 Nov 10	10 33	− 26	969.23	4.94	225.42	−17.16	2 44.9	2 43.2	116	293	D
1986 Nov 13	4 10	+471	969.75	4.94	228.05	−17.87	2 24.0	2 22.1	85	323	C
1993 Nov 6	3 58	−928	968.15	4.95	221.29	−15.97	0 50.0	0 44.1	189	222	F b
1999 Nov 15	21 43	+963	970.24	4.93	230.69	−18.54	0 26.0	0 11.9	30	16	B e
2003 May 7	7 53	+709	951.06	5.97	43.90	+16.73	2 39.1	2 34.7	15	291	K
2006 Nov 8	21 42	−423	968.72	4.95	223.88	−16.72	2 29.0	2 27.1	141	269	E
2016 May 9	15 00	−319	950.41	6.00	46.96	+17.58	3 45.0	3 41.9	83	224	J
2019 Nov 11	15 22	+ 77	969.28	4.94	226.48	−17.45	2 44.4	2 42.7	110	299	D
2032 Nov 13	8 58	+573	969.79	4.94	229.11	−18.14	2 12.9	2 10.9	77	330	C
2039 Nov 7	8 48	−823	968.20	4.95	222.33	−16.27	1 28.6	1 25.4	174	237	F
2049 May 7	14 31	+509	951.00	5.98	45.01	+17.04	3 20.9	3 17.4	31	276	K
2052 Nov 9	2 31	−320	968.77	4.94	224.93	−17.02	2 36.2	2 34.4	134	275	E
2062 May 10	21 41	−521	950.34	6.00	48.09	+17.87	3 20.3	3 16.7	97	211	J
2065 Nov 11	20 10	+180	969.32	4.94	227.55	−17.73	2 41.9	2 40.2	103	305	D
2078 Nov 14	13 45	+675	969.83	4.94	230.18	−18.41	1 58.5	1 56.2	69	337	C
2085 Nov 7	13 37	−719	968.26	4.95	223.38	−16.58	1 51.8	1 49.3	163	247	F
2095 May 8	21 09	+310	950.93	5.98	46.13	+17.35	3 44.7	3 41.6	45	263	K
2098 Nov 10	7 20	−216	968.83	4.94	225.99	−17.31	2 41.2	2 39.5	127	282	E
2108 May 12	4 22	−726	950.28	6.00	49.22	+18.16	2 35.6	2 31.0	114	195	J
2111 Nov 14	0 58	+283	969.37	4.94	228.61	−18.01	2 37.6	2 35.9	97	311	D
2124 Nov 15	18 33	+778	969.95	4.94	231.26	−18.67	1 39.0	1 36.2	59	346	C
2131 Nov 9	18 26	−615	968.31	4.95	224.44	−16.87	2 08.5	2 06.4	154	255	F
2141 May 10	3 48	+109	950.86	5.98	47.25	+17.65	3 56.2	3 53.2	57	251	K
2144 Nov 11	12 08	−112	968.88	4.94	227.05	−17.59	2 44.1	2 42.4	121	287	E
2154 May 13	11 04	−930	950.21	6 01	50.35	+18.44	0 55.6	0 40.7	143	167	J e
2157 Nov 14	5 46	+386	969.41	4.94	229.68	−18.28	2 31.2	2 29.3	90	317	D
2170 Nov 16	23 20	+879	969.91	4.94	232.34	−18.92	1 11.0	1 07.0	47	357	C
2174 May 8	3 30	+927	951.47	5.96	45.31	+17.11	0 59.2	0 46.1	346	320	L b
2177 Nov 9	23 16	−510	968.37	4.95	225.49	−17.16	2 21.1	2 19.1	146	263	F
2187 May 11	10 27	− 93	950.80	5.99	48.38	+17.94	3 56.8	3 53.8	70	239	K
2190 Nov 12	16 57	− 9	968.93	4.94	228.12	−17.87	2 45.1	2 43.4	114	293	E
2203 Nov 16	10 34	+489	969.45	4.94	230.76	−18.54	2 22.4	2 20.5	83	323	D
2210 Nov 9	10 20	−911	967.83	4.95	223.95	−16.72	0 58.3	0 53.3	185	225	G b
2216 Nov 18	4 08	+981	969.96	4.93	233.42	−19.17	—	—	—	—	C e
2220 May 9	10 07	+728	951.40	5.97	46.43	+17.42	2 33.6	2 29.0	14	294	L
2223 Nov 12	4 05	−406	968.42	4.95	226.56	−17.45	2 30.5	2 28.7	139	269	F
2233 May 12	17 08	−295	950.73	5.99	49.51	+18.22	3 46.5	3 43.4	83	226	K
2236 Nov 13	21 46	+ 95	968.98	4.94	229.18	−18.14	2 44.2	2 42.6	108	299	E
2249 Nov 16	15 23	+592	969.50	4.94	231.83	−18.80	2 10.7	2 08.6	75	330	D
2256 Nov 9	15 10	−807	967.89	4.95	225.00	−17.02	1 32.9	1 29.8	171	238	G
2266 May 10	16 44	+530	951.33	5.97	47.55	+17.71	3 17.2	3 13.6	30	278	L
2269 Nov 12	8 54	−302	968.48	4.94	227.62	−17.73	2 37.4	2 35.6	132	276	F
2279 May 13	23 48	−499	950.66	5.99	50.65	+18.50	3 23.4	3 19.9	97	213	K
2282 Nov 15	2 34	+198	969.02	4.94	230.26	−18.41	2 41.5	2 39.8	101	305	E
2295 Nov 17	20 10	+694	969.54	4.94	232.91	−19.05	1 55.4	1 53.1	66	338	D

XII

Transits of Venus, 1300 - 4100

The table on the next page lists all transits of Venus over the Sun's disk taking place in the period A.D. 1300 to 4100. These data have been calculated in 1969 by Mr. Michel Walch, of Ville d'Avray, France ; they are based on Newcomb's tables for the Sun and Venus.

The arrangement is the same as for the transits of Mercury (see page 7-46). Note that the Julian Calendar is used before A.D. 1582.

As short-period planetary perturbations were not taken into account in the calculations, the times t_m may be in error by a few minutes.

For example, for the transit of 2004 June 8 we find the following times for a geocentric observer :

First contact (ingress, exterior contact) :	$8^h24^m - 3^h06^m = 5^h18^m$ ET
Second contact (ingress, interior contact) :	$8^h24^m - 2^h46^m = 5^h38^m$
Third contact (egress, interior contact) :	$8^h24^m + 2^h46^m = 11^h10^m$
Fourth contact (egress, exterior contact) :	$8^h24^m + 3^h06^m = 11^h30^m$

The last column of the table gives the numbering of the transit series in the period of 243 years. The series A and B occur at the ascending node of the orbit of Venus ; the series C and D take place at the descending node.

Remarks

1631 December 7. — The first transit of series A. For some places in the southern hemisphere the transit was a partial one.
   One period of 243 years earlier, in 1388, Venus grazed the Sun's northern limb. The geocentric least distance of the centers was +1042″, on 1388 November 26 at 6^h29^m ET. Geocentrically, the least distance between the center of Venus and the Sun's limb was 67″. For some places in the northern hemisphere, the least separation between the limbs was only 10″.

2611 December 13. — For some places in the northern hemisphere, the transit will be partial.

2854 December 14. — Last transit of series B. There will be no transit as seen from the center of the Earth. A very small partial transit at the Sun's southern limb will be visible from some places in Antarctica, according to Jean Meeus (*Journal of the British Astronomical Association*, Vol. 68, No. 3, page 107 ; April 1958).

3219 June 20. — For some places in the northern hemisphere, the transit will be a partial one.

3462 June 22. — For all places on the Earth's surface, the transit will be partial, at the Sun's southern limb.

3705 June 24. — Last transit of series C. There will be no transit as seen from the center of the Earth. There will be a partial transit for some places in the southern hemisphere.

5900 July 11. — The next smallest least distance of the centers, when this distance will be approximately zero at mid-transit (20^h46^m ET).

Details about earlier transits of Venus can be found in the article of Jean Meeus, published in the *Journal of the British Astronomical Association*, Vol. 68, No. 3, pages 98-108 (April 1958).

TRANSITS OF VENUS

Date	t_m	d_m	s	s'	α	δ	τ_e	τ_i	P_1	P_2	Ser.
	h m	"	"	"	°	°	h m	h m	°	°	
1396 Nov 23	19 29	−427	974.98	31.85	248.91	−22.10	3 41.6	3 24.4	133	261	B
1518 May 26	2 03	−506	944.80	29.08	71.87	+22.45	3 27.8	3 10.6	107	222	C
1526 May 23	19 18	+666	945.05	29.09	69.61	+22.18	2 57.4	2 36.9	29	298	D
1631 Dec 7	5 18	+936	974.97	31.84	253.68	−22.64	1 30.0	0 28.0	32	359	A b
1639 Dec 4	18 24	−526	974.66	31.83	251.06	−22.34	3 28.5	3 10.2	139	254	B
1761 Jun 6	5 19	−569	945.04	29.09	74.35	+22.69	3 17.7	2 59.5	112	218	C
1769 Jun 3	22 27	+610	945.30	29.10	72.07	+22.44	3 09.8	2 50.8	34	295	D
1874 Dec 9	4 02	+825	974.57	31.81	255.84	−22.82	2 20.1	1 50.9	47	343	A
1882 Dec 6	17 04	−639	974.27	31.80	253.21	−22.56	3 09.0	2 48.5	147	245	B
2004 Jun 8	8 24	−628	945.37	29.11	76.83	+22.89	3 05.9	2 46.5	118	215	C
2012 Jun 6	1 36	+553	945.67	29.13	74.54	+22.68	3 20.2	3 02.2	40	291	D
2117 Dec 11	2 51	+721	974.29	31.80	258.01	−22.97	2 50.4	2 27.4	56	332	A
2125 Dec 8	16 01	−740	973.95	31.78	255.38	−22.74	2 45.6	2 21.8	154	235	B
2247 Jun 11	11 43	−690	945.68	29.13	79.33	+23.04	2 51.8	2 30.5	124	210	C
2255 Jun 9	4 50	+492	945.96	29.15	77.03	+22.87	3 30.1	3 13.1	45	288	D
2360 Dec 13	1 52	+620	973.87	31.76	260.20	−23.09	3 12.8	2 52.7	63	323	A
2368 Dec 10	14 58	−839	973.51	31.74	257.57	−22.90	2 14.9	1 44.3	164	224	B
2490 Jun 12	14 46	−745	945.99	29.16	81.82	+23.16	2 37.1	2 13.4	130	206	C
2498 Jun 10	7 52	+443	946.33	29.18	79.51	+23.02	3 36.7	3 20.2	49	285	D
2603 Dec 16	0 47	+517	973.52	31.74	262.40	−23.18	3 27.7	3 09.6	70	314	A
2611 Dec 13	14 01	−939	973.16	31.72	259.77	−23.03	1 27.1	0 16.4	178	208	B
2733 Jun 15	18 00	−807	946.38	29.18	84.33	+23.24	2 16.7	1 48.8	138	201	C
2741 Jun 13	11 06	+384	946.69	29.20	82.02	+23.14	3 43.6	3 27.7	54	282	D
2846 Dec 16	23 59	+426	973.11	31.71	264.62	−23.24	3 41.6	3 24.4	75	307	A
2854 Dec 14	13 12	−1029	972.69	31.68	261.99	−23.12	—	—	—	—	B e
2976 Jun 16	21 00	−854	946.68	29.21	86.83	+23.28	1 57.6	1 23.4	145	196	C
2984 Jun 14	13 59	+339	947.05	29.23	84.51	+23.22	3 48.2	3 32.6	58	280	D
3089 Dec 18	22 55	+321	972.68	31.67	266.84	−23.27	3 52.0	3 35.6	81	300	A
3219 Jun 20	0 00	−907	947.11	29.24	89.34	+23.28	1 30.2	0 35.6	154	188	C
3227 Jun 17	17 01	+292	947.46	29.26	87.02	+23.25	3 52.0	3 36.7	62	278	D
3332 Dec 20	22 06	+230	972.29	31.64	269.07	−23.26	3 59.1	3 43.1	86	293	A
3462 Jun 22	2 54	−952	947.44	29.27	91.84	+23.24	0 54	—	—	—	C
3470 Jun 19	19 47	+251	947.82	29.29	89.51	+23.25	3 55.2	3 40.1	66	277	D
3575 Dec 23	21 11	+131	971.79	31.59	271.30	−23.23	4 02.8	3 47.2	91	286	A
3705 Jun 24	5 39	−990	947.88	29.30	94.33	+23.16	—	—	—	—	C e
3713 Jun 21	22 36	+213	948.28	29.33	92.00	+23.20	3 57.2	3 42.3	69	275	D
3818 Dec 25	20 27	+ 39	971.39	31.56	273.55	−23.16	4 04.3	3 48.9	95	280	A
3956 Jun 24	1 22	+176	948.65	29.36	94.50	+23.12	3 59.4	3 44.5	72	274	D
4061 Dec 26	19 50	− 45	970.86	31.51	275.81	−23.07	4 04.6	3 49.2	99	274	A

XIII
Solar Eclipses, 1951 - 2050

On the next pages the list is given of all solar eclipses taking place during the period A.D. 1951 - 2050, in two parts :

(A) The first part lists the total and annular eclipses. The successive columns give :

- the date ;

- the instant of the least distance of the axis of the lunar shadow cone to the center of the Earth, rounded to the nearest integer hour of Universal Time ;

- the type of the eclipse, for which the following symbols have been used :

  t = central total eclipse ;
  r = central annular eclipse (the first letter of the German word *ringförmig* = annular, a symbol used by Oppolzer in his famous *Canon der Finsternisse* in 1887) ;
  rt = annular-total eclipse. This is a central eclipse which is total for a part of the path, and annular for the rest ;
  (t) = non-central total eclipse ;
  (r) = non-central annular eclipse.

  At a (t)- or (r)-eclipse only a part of the umbral cone comes in contact with the surface of the Earth (within the polar regions), but the axis of the cone does not intersect it. Such eclipses therefore have no central line ;

- the maximum duration of the total or annular phase on the central line, in minutes and seconds. In the case of an eclipse of type rt, the maximum duration is that of the total phase ;

- the maximum width of the path of total or annular eclipse, in kilometers. Again, in the case of a rt-eclipse the maximum width refers to the part of the path where the eclipse is total ;

- the maximum altitude of the center of the solar disk at central eclipse, in degrees ;

- the maximum value of the ratio of the diameter of the Moon's disk to that of the Sun, at central eclipse. It should be noted that this ratio reaches a maximum near the middle of the central line and that it is sensibly less near the beginning and near the end of that line, where the eclipse occurs near the horizon. For instance, at the eclipse of 1991 July 11, the ratio Moon/Sun is 1.061 at the beginning of the central line, reaches a maximum value of 1.080 near the middle, and then decreases to 1.061 again towards the end of the line ;

- a brief description of the region where the total or annular phase is visible. The letters N, E, S, W signify "northern part of", etc.

(B) In the second part, all partial solar eclipses in the period 1951-2050 are given. These eclipses are nowhere total or annular. The successive column give :

- the date ;

- the instant of the least distance of the axis of the lunar shadow cone to the center of the Earth, rounded to the nearest integer hour of Universal Time ;

- the greatest magnitude on the Earth's surface (that is, the greatest possible fraction of the solar diameter obscured by the Moon) ;

- whether the eclipse is visible from the northern (N) or from the southern (S) hemisphere of the Earth.

(A) Total and Annular Eclipses

Date	UT	Type	Max. Dur.	Max. Width	Max. Alt.	Max. M/S	Area of Visibility
	h		m s	km	°		
1951 Mar 7	21	r	1 39	101	76	0.990	New Zealand, Pacific Ocean, Central America
1951 Sep 1	13	r	2 44	155	81	0.975	E of U.S.A., Atlantic Ocean, Africa, Madagascar
1952 Feb 25	9	t	3 10	142	62	1.037	Atlantic Ocean, Africa, Arabia, Asia
1952 Aug 20	15	r	6 41	309	52	0.942	South America
1954 Jan 5	3	r	1 49	401	21	0.972	Antarctica
1954 Jun 30	13	t	2 35	153	52	1.036	North America, S Greenland, Scandinavia, U.S.S.R., SW Asia
1954 Dec 25	8	r	7 40	346	75	0.932	Indian Ocean, Indonesia
1955 Jun 20	4	t	7 08	254	81	1.078	Indian Ocean, Ceylon, SE Asia, Philippines
1955 Dec 14	7	r	12 09	378	65	0.918	NE Africa, Indian Ocean, SE Asia
1956 Jun 8	21	t	4 44	431	26	1.058	S Pacific Ocean
1957 Apr 30	0	(r)	—	—	—	—	Arctic Ocean, Novaya Zemlya
1957 Oct 23	5	(t)	—	—	—	—	Very small portion of Antarctica
1958 Apr 19	3	r	7 07	297	74	0.941	Indian Ocean, SE Asia, Pacific Ocean
1958 Oct 12	21	t	5 11	210	73	1.061	Pacific Ocean, Chile
1959 Apr 8	3	r	7 26	301	63	0.940	Australia
1959 Oct 2	12	t	3 02	122	65	1.032	Atlantic Ocean, Canary Islands, Africa
1961 Feb 15	8	t	2 45	264	28	1.036	Europe, NW Asia
1961 Aug 11	11	r	6 35	522	27	0.937	S Atlantic Ocean
1962 Feb 5	0	t	4 08	147	78	1.043	Borneo, Celebes, New Guinea, Pacific Ocean
1962 Jul 31	12	r	3 33	162	84	0.972	N of South America, Atlantic Ocean, Africa, Madagascar
1963 Jan 25	14	r	1 07	89	60	0.995	S of South America, S Atlantic Ocean, S Africa, Madagascar
1963 Jul 20	21	t	1 40	101	49	1.022	N Pacific Ocean, North America
1965 May 30	21	t	5 16	199	65	1.054	New Zealand, Pacific Ocean
1965 Nov 23	4	r	4 05	187	67	0.966	Asia, Borneo, Celebes, New Guinea, Pacific Ocean
1966 May 20	10	r	0 59	69	70	0.999	Africa, Greece, Asia
1966 Nov 12	14	t	1 57	85	71	1.023	South America, S Atlantic Ocean
1967 Nov 2	6	(t)	—	—	—	—	S Atlantic Ocean
1968 Sep 22	11	t	0 40	110	19	1.010	Siberia, China

Date	UT	Type	Max. Dur.	Max. Width	Max. Alt.	Max. M/S	Area of Visibility
	h		m s	km	°		
1969 Mar 18	5	r	1 18	79	74	0.995	Indian Ocean, Indonesia, Pacific Ocean
1969 Sep 11	20	r	3 15	177	77	0.969	Pacific Ocean, South America
1970 Mar 7	18	t	3 28	158	63	1.041	Pacific Ocean, North America, N Atlantic Ocean
1970 Aug 31	22	r	6 48	310	57	0.940	Pacific Ocean
1972 Jan 16	11	r	1 58	450	20	0.969	Antarctica
1972 Jul 10	20	t	2 36	178	46	1.038	E Siberia, Alaska, Canada, Atlantic Oc.
1973 Jan 4	16	r	7 50	356	74	0.930	S Pacific Ocean, South America, Atlantic Ocean
1973 Jun 30	12	t	7 04	256	86	1.079	N South America, Atlantic Ocean, Africa, Indian Ocean
1973 Dec 24	15	r	12 03	379	65	0.917	Central America, N of South America, Atlantic Ocean, NW Africa
1974 Jun 20	5	t	5 08	346	34	1.059	Indian Ocean, extreme SW Australia
1976 Apr 29	10	r	6 41	302	70	0.942	Atlantic Ocean, N Africa, Turkey, Asia
1976 Oct 23	5	t	4 47	200	71	1.057	E Africa, Indian Ocean, S Australia
1977 Apr 18	11	r	7 05	272	66	0.945	S Atlantic Ocean, S Africa, Indian Oc.
1977 Oct 12	20	t	2 37	100	67	1.027	Pacific Ocean, N of South America
1979 Feb 26	17	t	2 49	308	26	1.039	U.S.A., Canada, Greenland
1979 Aug 22	17	r	6 02	1010	15	0.933	S Pacific Ocean, Antarctica
1980 Feb 16	9	t	4 08	149	77	1.043	Africa, Indian Ocean, S Asia
1980 Aug 10	19	r	3 23	158	79	0.973	Pacific Ocean, South America
1981 Feb 4	22	r	1 13	94	61	0.994	Tasmania, Pacific Ocean
1981 Jul 31	4	t	2 03	108	54	1.026	Siberia, Pacific Ocean
1983 Jun 11	5	t	5 11	200	60	1.052	Indian Ocean, Java, Celebes, New Guinea
1983 Dec 4	13	r	4 03	183	66	0.967	Atlantic Ocean, Africa
1984 May 30	17	r	1 05	72	74	0.998	E Pacific Ocean, Mexico, U.S.A., Atlantic Ocean, NW Africa
1984 Nov 22	23	t	2 00	85	72	1.024	New Guinea, S Pacific Ocean
1985 Nov 12	14	t	1 59	770	11	1.039	S Pacific Ocean
1986 Oct 3	19	rt	0 00	2	6	1.000	N Atlantic Ocean
1987 Mar 29	13	rt	0 08	5	72	1.001	S South America, Atlantic Ocean, Africa
1987 Sep 23	3	r	3 50	199	74	0.963	Asia, Pacific Ocean
1988 Mar 18	2	t	3 46	174	65	1.046	Sumatra, Borneo, Pacific Ocean
1988 Sep 11	5	r	6 57	316	62	0.938	Indian Ocean
1990 Jan 26	20	r	2 06	505	18	0.967	S Atlantic Ocean, Antarctica
1990 Jul 22	3	t	2 33	209	40	1.039	Finland, N and NE Siberia, N Pacific Oc.

Date	UT	Type	Max. Dur.	Max. Width	Max. Alt.	Max. M/S	Area of Visibility
	h		m s	km	°		
1991 Jan 15	24	r	7 55	361	74	0.929	SW Australia, Tasmania, New Zealand, Pacific Ocean
1991 Jul 11	19	t	6 54	258	90	1.080	Pacific Ocean, Hawaii, Mexico, Central and South America
1992 Jan 4	23	r	11 42	377	66	0.918	Pacific Ocean
1992 Jun 30	12	t	5 20	297	41	1.059	Atlantic Ocean
1994 May 10	17	r	6 14	311	66	0.943	E Pacific Ocean, North America, Atlantic Ocean, extreme NW Africa
1994 Nov 3	14	t	4 24	190	69	1.053	South America, S Atlantic Ocean
1995 Apr 29	18	r	6 38	247	70	0.950	Pacific Ocean, South America
1995 Oct 24	5	t	2 10	78	69	1.021	Asia, Borneo, Pacific Ocean
1997 Mar 9	1	t	2 50	371	23	1.042	Siberia
1998 Feb 26	17	t	4 08	152	76	1.044	Pacific Ocean, N of South America, Atlantic Ocean
1998 Aug 22	2	r	3 14	156	75	0.973	Sumatra, Borneo, Pacific Ocean
1999 Feb 16	7	r	1 19	96	62	0.993	Indian Ocean, Australia
1999 Aug 11	11	t	2 23	112	59	1.029	Atlantic Ocean, Europe, SE and S Asia
2001 Jun 21	12	t	4 56	201	55	1.050	Atlantic Ocean, S Africa, Madagascar
2001 Dec 14	21	r	3 54	177	66	0.968	Pacific Ocean, Central America
2002 Jun 10	24	r	1 13	78	78	0.996	Pacific Ocean
2002 Dec 4	8	t	2 04	87	72	1.024	S Africa, Indian Ocean, Australia
2003 May 31	4	r	3 37	—	3	0.938	Iceland
2003 Nov 23	23	t	1 57	545	15	1.038	Antarctica
2005 Apr 8	21	rt	0 42	27	70	1.007	Pacific Ocean, Central America, N of South America
2005 Oct 3	11	r	4 32	223	71	0.958	Atlantic Ocean, Spain, Africa, Indian Oc.
2006 Mar 29	10	t	4 07	189	67	1.051	Atlantic Ocean, Africa, Turkey, U.S.S.R.
2006 Sep 22	12	r	7 09	324	66	0.935	NE of South America, Atlantic Ocean, S Indian Ocean
2008 Feb 7	4	r	2 14	581	16	0.965	S Pacific Ocean, Antarctica
2008 Aug 1	10	t	2 28	251	34	1.039	N Canada, Arctic Ocean, Siberia, China
2009 Jan 26	8	r	7 56	363	73	0.928	S Atlantic Ocean, Indian Ocean, Sumatra, Borneo
2009 Jul 22	3	t	6 40	258	86	1.080	Asia, Pacific Ocean
2010 Jan 15	7	r	11 10	373	66	0.919	Africa, Indian Ocean, S and SE Asia
2010 Jul 11	20	t	5 20	262	47	1.058	Pacific Ocean, extreme S South America
2012 May 20	24	r	5 47	324	61	0.944	China, Japan, N Pacific Ocean, U.S.A.
2012 Nov 13	22	t	4 02	179	68	1.050	N Australia, Pacific Ocean

Date	UT	Type	Max. Dur.	Max. Width	Max. Alt.	Max. M/S	Area of Visibility
	h		m s	km	°		
2013 May 10	0	r	6 04	226	74	0.954	Australia, Pacific Ocean
2013 Nov 3	13	rt	1 40	58	71	1.016	Atlantic Ocean, Africa
2014 Apr 29	6	(r)	—	—	—	—	Antarctica
2015 Mar 20	10	t	2 47	487	19	1.045	N Atlantic Ocean, Svalbard, Arctic Oc.
2016 Mar 9	2	t	4 10	156	75	1.045	Sumatra, Borneo, Celebes, Pacific Oc.
2016 Sep 1	9	r	3 06	158	70	0.974	Atlantic Ocean, Africa, Madagascar, Indian Ocean
2017 Feb 26	15	r	1 22	96	63	0.992	S Pacific Ocean, South America, Atlantic Ocean, Africa
2017 Aug 21	18	t	2 40	115	64	1.031	Pacific Ocean, U.S.A., Atlantic Ocean
2019 Jul 2	19	t	4 32	201	50	1.046	Pacific Ocean, South America
2019 Dec 26	5	r	3 40	169	66	0.970	Arabia, India, Sumatra, Borneo, Pacific Ocean
2020 Jun 21	7	r	1 22	85	83	0.994	Africa, Arabia, Asia, Pacific Ocean
2020 Dec 14	16	t	2 10	90	73	1.025	Pacific Ocean, South America, S Atlantic Ocean
2021 Jun 10	11	r	3 51	692	23	0.944	Canada, Arctic Ocean, NE Siberia
2021 Dec 4	8	t	1 55	451	17	1.037	Antarctica, S Atlantic Ocean
2023 Apr 20	4	rt	1 16	49	67	1.013	Indian Ocean, New Guinea, Pacific Ocean
2023 Oct 14	18	r	5 18	245	68	0.952	U.S.A., Central and South America
2024 Apr 8	18	t	4 28	203	70	1.057	Pacific Ocean, North America, Atlantic Ocean
2024 Oct 2	19	r	7 25	333	69	0.933	Pacific Ocean, S of South America
2026 Feb 17	12	r	2 21	764	12	0.963	Antarctica
2026 Aug 12	18	t	2 18	318	26	1.039	Arctic Ocean, Greenland, N Atlantic Ocean, Spain
2027 Feb 6	16	r	7 54	361	73	0.928	S Pacific Ocean, South America, Atlantic Ocean
2027 Aug 2	10	t	6 23	259	82	1.079	Atlantic Ocean, N Africa, Arabia, Indian Ocean
2028 Jan 26	15	r	10 31	367	67	0.921	E Pacific Ocean, South America, Atlantic Ocean, Spain
2028 Jul 22	3	t	5 10	235	53	1.056	Indian Ocean, Australia, New Zealand
2030 Jun 1	6	r	5 21	345	56	0.944	N Africa, Greece, Asia, Japan
2030 Nov 25	7	t	3 44	169	67	1.047	S Africa, Indian Ocean, Australia
2031 May 21	7	r	5 26	208	79	0.959	Africa, Indian Ocean, S India, Borneo, Celebes
2031 Nov 14	21	rt	1 08	38	72	1.011	Pacific Ocean
2032 May 9	13	r	0 43	92	20	0.996	S Atlantic Ocean

Date	UT	Type	Max. Dur.	Max. Width	Max. Alt.	Max. M/S	Area of Visibility
	h		m s	km	°		
2033 Mar 30	18	t	2 37	830	11	1.046	Alaska, Arctic Ocean
2034 Mar 20	10	t	4 09	160	73	1.046	Atlantic Ocean, Africa, Arabia, Asia
2034 Sep 12	16	r	2 58	162	67	0.974	Pacific Ocean, South America, S Atlantic Ocean
2035 Mar 9	23	r	1 26	96	64	0.992	New Zealand, Pacific Ocean
2035 Sep 2	2	t	2 55	117	68	1.032	Asia, Japan, Pacific Ocean
2037 Jul 13	3	t	3 58	201	43	1.041	Indian Ocean, Australia, New Zealand
2038 Jan 5	14	r	3 19	160	65	0.973	Cuba, Haiti, Atlantic Ocean, Africa
2038 Jul 2	14	r	1 35	94	88	0.991	N of South America, Atlantic Ocean, Africa
2038 Dec 26	1	t	2 18	95	73	1.027	Australia, New Zealand, Pacific Ocean
2039 Jun 21	17	r	4 05	497	33	0.945	Alaska, Greenland, Scandinavia, U.S.S.R.
2039 Dec 15	16	t	1 52	402	18	1.036	Antarctica
2041 Apr 30	12	t	1 51	72	63	1.019	S Atlantic Ocean, Africa, Indian Ocean
2041 Oct 25	2	r	6 09	267	66	0.947	E Asia, Japan, Pacific Ocean
2042 Apr 20	2	t	4 51	215	73	1.061	Indian Ocean, Sumatra, Borneo, Philippines, Pacific Ocean
2042 Oct 14	2	r	7 45	343	72	0.930	SE Asia, Borneo, Celebes, Australia, New Zealand, S Pacific Ocean
2043 Apr 9	19	(t)	—	—	—	—	E Siberia
2043 Oct 3	3	(r)	—	—	—	—	Very small portion SW Indian Ocean
2044 Feb 28	20	r	2 27	—	4	0.960	S Atlantic Ocean
2044 Aug 23	1	t	2 04	500	16	1.036	Greenland, Canada, U.S.A.
2045 Feb 16	24	r	7 49	358	72	0.928	New Zealand, Pacific Ocean
2045 Aug 12	18	t	6 06	259	78	1.077	E Pacific Ocean, U.S.A., Haiti, N of South America
2046 Feb 5	23	r	9 46	358	68	0.923	New Guinea, Pacific Ocean, Hawaii, W of U.S.A.
2046 Aug 2	10	t	4 51	210	58	1.053	Atlantic Ocean, S Africa, S Indian Oc.
2048 Jun 11	13	r	4 59	376	49	0.944	North America, Greenland, Iceland, Scandinavia, U.S.S.R.
2048 Dec 5	16	t	3 28	160	66	1.044	Pacific Ocean, South America, S Atlantic Ocean, S Africa
2049 May 31	14	r	4 46	194	83	0.963	South America, Atlantic Ocean, Africa
2049 Nov 25	6	rt	0 38	21	73	1.006	Arabia, Indian Ocean, Sumatra, Borneo, Celebes, Pacific Ocean
2050 May 20	21	rt	0 21	26	29	1.004	S Pacific Ocean

(B) Partial Eclipses

Date	UT	Max. Magn.	
1953 Feb 14	1 h	0.760	N
1953 Jul 11	3	0.202	N
1953 Aug 9	16	0.373	S
1956 Dec 2	8	0.805	N
1960 Mar 27	7	0.705	S
1960 Sep 20	23	0.614	N
1964 Jan 14	20	0.559	S
1964 Jun 10	5	0.754	S
1964 Jul 9	11	0.322	N
1964 Dec 4	2	0.752	N
1967 May 9	15	0.720	N
1968 Mar 28	23	0.899	S
1971 Feb 25	10	0.787	N
1971 Jul 22	10	0.069	N
1971 Aug 20	23	0.508	S
1974 Dec 13	16	0.827	N
1975 May 11	7	0.864	N
1975 Nov 3	13	0.959	S
1978 Apr 7	15	0.788	S
1978 Oct 2	6	0.691	N
1982 Jan 25	5	0.566	S
1982 Jun 21	12	0.617	S
1982 Jul 20	19	0.464	N
1982 Dec 15	10	0.735	N
1985 May 19	21	0.841	N
1986 Apr 9	6	0.823	S
1989 Mar 7	18	0.827	N
1989 Aug 31	6	0.634	S
1992 Dec 24	1	0.842	N
1993 May 21	14	0.736	N
1993 Nov 13	22	0.928	S
1996 Apr 17	23	0.879	S
1996 Oct 12	14	0.758	N
1997 Sep 2	0	0.899	S
2000 Feb 5	13	0.579	S
2000 Jul 1	20	0.477	S
2000 Jul 31	2	0.603	N
2000 Dec 25	18	0.723	N

Date	UT	Max. Magn.	
2004 Apr 19	14 h	0.737	S
2004 Oct 14	3	0.929	N
2007 Mar 19	3	0.876	N
2007 Sep 11	13	0.750	S
2011 Jan 4	9	0.858	N
2011 Jun 1	21	0.602	N
2011 Jul 1	9	0.097	S
2011 Nov 25	6	0.904	S
2014 Oct 23	22	0.811	N
2015 Sep 13	7	0.787	S
2018 Feb 15	21	0.598	S
2018 Jul 13	3	0.336	S
2018 Aug 11	10	0.737	N
2019 Jan 6	2	0.715	N
2022 Apr 30	21	0.640	S
2022 Oct 25	11	0.862	N
2025 Mar 29	11	0.938	N
2025 Sep 21	20	0.855	S
2029 Jan 14	17	0.872	N
2029 Jun 12	4	0.458	N
2029 Jul 11	16	0.230	S
2029 Dec 5	15	0.891	S
2032 Nov 3	6	0.855	N
2033 Sep 23	14	0.689	S
2036 Feb 27	5	0.628	S
2036 Jul 23	11	0.199	S
2036 Aug 21	17	0.862	N
2037 Jan 16	10	0.706	N
2040 May 11	4	0.531	S
2040 Nov 4	19	0.808	N
2047 Jan 26	2	0.891	N
2047 Jun 23	11	0.313	N
2047 Jul 22	23	0.360	S
2047 Dec 16	24	0.881	S
2050 Nov 14	13	0.887	N

XIV

Lunar Eclipses, 1951 - 2050

This table mentions all lunar eclipses taking place from A.D. 1951 to 2050, including the penumbral eclipses. The data, except the last column, are taken from the *Canon of Lunar Eclipses -2002 to +2526* by Jean Meeus and Hermann Mucke (Astronomisches Büro, Wien, Austria ; 1979).

The first two columns give the date and the instant of the maximum of the eclipse, in *Ephemeris Time*. These times may be 1 minute in error.

The next two columns give the semidurations, in minutes, of the partial phase and of the total phase in the umbra, respectively. These values are not given, of course, in the case of a penumbral eclipse, that is when the Moon enters only the penumbra of the Earth. The semiduration of the partial phase (p) is half of the time elapsed between the first and the last exterior contacts of the Moon with the umbra. The semiduration of the total phase (t), given only for total eclipses in the umbra, is half of the time elapsed between the first and the last interior contacts of the Moon with the umbra.

To obtain the times t_1 and t_4 of the beginning and end of the partial phase, subtract or add the semiduration p from the time of maximum eclipse. The times t_2 and t_3 of the beginning and end of the total phase are obtained by subtracting or adding the semiduration t from the time of maximum eclipse.

The fifth column gives the (greatest) magnitude of the eclipse, the Moon's diameter being taken as unity. The eclipse is total if the magnitude is greater than 1. In the case of a penumbral eclipse, the magnitude in the penumbra is given, and this value is placed between parentheses. In all other cases, the magnitude in the umbra is given.

It should be noted that penumbral eclipses are not observable unless their magnitude is larger than approximately 0.7. Small partial penumbral eclipses, such as that of 1984 June 13, are undistinguishible and are given only for reason of completeness ; the statistics of eclipses would be incomplete without them.

Observation has shown that the atmosphere of the Earth has the effect of increasing the apparent radius of the Earth's shadow by about one fiftieth. Following an old tradition, most astronomical almanacs increase by 1/50 the geometric radii of both the umbra and the penumbra.

However, since 1951 the French almanac *Connaissance des Temps* uses another theory for the computation of the radii. A. Danjon correctly pointed out that the only reasonable way to take into account the presence of an opaque atmospheric layer around the Earth is to increase the Earth's radius, which can be performed by increasing proportionally the Moon's *parallax*. In that case, the radii of umbra and penumbra will undergo the same *absolute* correction, and not the same relative correction as the traditional rule states. Thus, even if the rule of 1/50 were true for the umbra, it cannot be correct for the penumbra. As a consequence, the magnitudes of lunar eclipses, as calculated using the traditional rule, are too large (as compared to the values given by the *Connaissance des Temps*) by about 0.005 for umbral eclipses, but by about 0.026 for penumbral ones.

In the present list, the magnitudes of the eclipses have been calculated in accordance with the rule used by the *Connaissance des Temps*. This should be taken into account when our results are compared with the data of other sources.

As a consequence, small penumbral "eclipses", found by using the classical rule of 1/50 for the enlargement of the penumbra, do in fact not exist. This was the case for the "eclipse" of 1951 February 21. It will occur again at the Full Moons of 2016 August 18 and 2042 October 28.

Finally, the last column indicates whether the center of the Moon passes to the North (N) or to the South (S) of the axis of the Earth's shadow.

Date	ET	p	t	Magn.		Date	ET	p	t	Magn.	
	h m	m	m				h m	m	m		
1951 Mar 23	10 38	–	–	(0.642)	S	1970 Aug 17	3 24	66	–	0.409	S
1951 Aug 17	3 14	–	–	(0.121)	S	1971 Feb 10	7 45	112	41	1.307	N
1951 Sep 15	12 27	–	–	(0.804)	N	1971 Aug 6	19 44	107	50	1.729	S
1952 Feb 11	0 40	35	–	0.083	N	1972 Jan 30	10 54	101	17	1.049	S
1952 Aug 5	19 48	73	–	0.532	S	1972 Jul 26	7 16	80	–	0.542	N
1953 Jan 29	23 48	113	42	1.331	N	1973 Jan 18	21 18	–	–	(0.865)	S
1953 Jul 26	12 21	108	50	1.864	S	1973 Jun 15	20 51	–	–	(0.469)	S
1954 Jan 19	2 33	101	14	1.030	S	1973 Jul 15	11 39	–	–	(0.104)	N
1954 Jul 16	0 21	70	–	0.404	N	1973 Dec 10	1 45	34	–	0.102	N
1955 Jan 8	12 33	–	–	(0.855)	S	1974 Jun 4	22 16	97	–	0.826	S
1955 Jun 5	14 23	–	–	(0.622)	S	1974 Nov 29	15 14	104	38	1.290	N
1955 Nov 29	17 00	37	–	0.120	N	1975 May 25	5 49	107	44	1.426	N
1956 May 24	15 32	102	–	0.965	S	1975 Nov 18	22 24	104	20	1.063	S
1956 Nov 18	6 48	104	39	1.317	N	1976 May 13	19 55	38	–	0.122	N
1957 May 13	22 32	105	39	1.297	N	1976 Nov 6	23 02	–	–	(0.837)	S
1957 Nov 7	14 27	103	14	1.030	S	1977 Apr 4	4 20	47	–	0.194	S
1958 Apr 4	4 01	–	–	(0.016)	S	1977 Sep 27	8 29	–	–	(0.901)	N
1958 May 3	12 13	10	–	0.009	N	1978 Mar 24	16 24	109	45	1.453	S
1958 Oct 27	15 28	–	–	(0.782)	S	1978 Sep 16	19 04	103	39	1.326	N
1959 Mar 24	20 12	55	–	0.265	S	1979 Mar 13	21 09	98	–	0.852	N
1959 Sep 17	1 03	–	–	(0.988)	N	1979 Sep 6	10 54	96	22	1.094	S
1960 Mar 13	8 28	109	47	1.513	S	1980 Mar 1	20 46	–	–	(0.653)	N
1960 Sep 5	11 22	105	43	1.424	N	1980 Jul 27	19 09	–	–	(0.255)	N
1961 Mar 2	13 28	96	–	0.799	N	1980 Aug 26	3 31	–	–	(0.709)	S
1961 Aug 26	3 08	93	–	0.987	S	1981 Jan 20	7 51	–	–	(1.014)	S
1962 Feb 19	13 04	–	–	(0.611)	N	1981 Jul 17	4 48	81	–	0.550	N
1962 Jul 17	11 55	–	–	(0.393)	N	1982 Jan 9	19 56	102	39	1.331	S
1962 Aug 15	19 57	–	–	(0.597)	S	1982 Jul 6	7 32	117	53	1.717	S
1963 Jan 9	23 20	–	–	(1.019)	S	1982 Dec 30	11 29	98	30	1.182	N
1963 Jul 6	22 02	90	–	0.705	N	1983 Jun 25	8 23	67	–	0.334	S
1963 Dec 30	11 08	102	39	1.336	S	1983 Dec 20	1 49	–	–	(0.890)	N
1964 Jun 25	1 07	116	50	1.556	S	1984 May 15	4 41	–	–	(0.807)	N
1964 Dec 19	2 37	97	29	1.175	N	1984 Jun 13	14 27	–	–	(0.065)	S
1965 Jun 14	1 50	50	–	0.175	S	1984 Nov 8	17 57	–	–	(0.900)	S
1965 Dec 8	17 10	–	–	(0.882)	N	1985 May 4	19 57	99	34	1.237	N
1966 May 4	21 13	–	–	(0.917)	N	1985 Oct 28	17 43	107	22	1.072	S
1966 Oct 29	10 13	–	–	(0.951)	S	1986 Apr 24	12 44	99	32	1.201	S
1967 Apr 24	12 08	101	39	1.336	N	1986 Oct 17	19 19	108	37	1.244	N
1967 Oct 18	10 16	109	30	1.141	S	1987 Apr 14	2 20	–	–	(0.778)	S
1968 Apr 13	4 48	97	24	1.111	S	1987 Oct 7	4 02	–	–	(0.986)	N
1968 Oct 6	11 42	107	31	1.168	N	1988 Mar 3	16 14	–	–	(1.092)	N
1969 Apr 2	18 34	–	–	(0.703)	S	1988 Aug 27	11 05	56	–	0.292	S
1969 Aug 27	10 48	–	–	(0.015)	S	1989 Feb 20	15 37	111	39	1.275	N
1969 Sep 25	20 10	–	–	(0.901)	N	1989 Aug 17	3 09	107	48	1.600	S
1970 Feb 21	8 31	26	–	0.046	N	1990 Feb 9	19 13	102	21	1.073	S

Date	ET	p	t	Magn.		Date	ET	p	t	Magn.	
	h m	m	m				h m	m	m		
1990 Aug 6	14 12	88	–	0.676	N	2009 Dec 31	19 24	30	–	0.078	N
1991 Jan 30	6 00	–	–	(0.880)	S	2010 Jun 26	11 40	81	–	0.537	S
1991 Jun 27	3 16	–	–	(0.313)	S	2010 Dec 21	8 18	104	36	1.257	N
1991 Jul 26	18 09	–	–	(0.254)	N	2011 Jun 15	20 13	109	50	1.700	N
1991 Dec 21	10 34	32	–	0.088	N	2011 Dec 10	14 33	106	25	1.104	S
1992 Jun 15	4 58	90	–	0.683	S	2012 Jun 4	11 04	63	–	0.370	N
1992 Dec 9	23 45	104	37	1.271	N	2012 Nov 28	14 34	–	–	(0.914)	S
1993 Jun 4	13 02	109	48	1.561	N	2013 Apr 25	20 10	14	–	0.017	S
1993 Nov 29	6 26	105	23	1.087	S	2013 May 25	4 12	–	–	(0.015)	N
1994 May 25	3 32	52	–	0.242	N	2013 Oct 18	23 51	–	–	(0.766)	N
1994 Nov 18	6 45	–	–	(0.881)	S	2014 Apr 15	7 48	107	39	1.291	S
1995 Apr 15	12 19	36	–	0.112	S	2014 Oct 8	10 55	99	29	1.165	N
1995 Oct 8	16 05	–	–	(0.827)	N	2015 Apr 4	12 02	104	–	0.998	N
1996 Apr 4	0 11	108	43	1.379	S	2015 Sep 28	2 48	100	36	1.276	S
1996 Sep 27	2 55	101	35	1.240	N	2016 Mar 23	11 49	–	–	(0.774)	N
1997 Mar 24	4 41	101	–	0.918	N	2016 Sep 16	18 56	–	–	(0.907)	S
1997 Sep 16	18 47	98	31	1.190	S	2017 Feb 11	0 45	–	–	(0.990)	S
1998 Mar 13	4 22	–	–	(0.707)	N	2017 Aug 7	18 22	57	–	0.247	N
1998 Aug 8	2 26	–	–	(0.122)	N	2018 Jan 31	13 31	101	38	1.316	S
1998 Sep 6	11 11	–	–	(0.813)	S	2018 Jul 27	20 23	117	51	1.609	N
1999 Jan 31	16 20	–	–	(1.005)	S	2019 Jan 21	5 13	98	31	1.196	N
1999 Jul 28	11 34	71	–	0.396	N	2019 Jul 16	21 32	89	–	0.652	S
2000 Jan 21	4 45	101	38	1.325	S	2020 Jan 10	19 11	–	–	(0.896)	N
2000 Jul 16	13 57	118	53	1.769	N	2020 Jun 5	19 26	–	–	(0.569)	S
2001 Jan 9	20 22	98	30	1.188	N	2020 Jul 5	4 31	–	–	(0.354)	S
2001 Jul 5	14 57	79	–	0.494	S	2020 Nov 30	9 44	–	–	(0.828)	S
2001 Dec 30	10 30	–	–	(0.893)	N	2021 May 26	11 20	93	7	1.010	N
2002 May 26	12 05	–	–	(0.691)	N	2021 Nov 19	9 05	104	–	0.974	S
2002 Jun 24	21 29	–	–	(0.209)	S	2022 May 16	4 12	103	42	1.414	S
2002 Nov 20	1 47	–	–	(0.860)	S	2022 Nov 8	11 00	110	42	1.357	N
2003 May 16	3 41	97	26	1.128	N	2023 May 5	17 25	–	–	(0.964)	S
2003 Nov 9	1 20	105	11	1.017	S	2023 Oct 28	20 15	38	–	0.121	N
2004 May 4	20 32	101	38	1.303	S	2024 Mar 25	7 15	–	–	(0.956)	N
2004 Oct 28	3 05	109	40	1.307	N	2024 Sep 18	2 45	31	–	0.085	S
2005 Apr 24	9 57	–	–	(0.865)	S	2025 Mar 14	7 01	109	33	1.179	N
2005 Oct 17	12 04	28	–	0.062	N	2025 Sep 7	18 12	104	41	1.362	S
2006 Mar 14	23 49	–	–	(1.031)	N	2026 Mar 3	11 36	103	29	1.148	S
2006 Sep 7	18 52	45	–	0.184	S	2026 Aug 28	4 14	99	–	0.929	N
2007 Mar 3	23 22	110	37	1.233	N	2027 Feb 20	23 14	–	–	(0.926)	S
2007 Aug 28	10 38	106	45	1.477	S	2027 Jul 18	16 04	–	–	(0.003)	S
2008 Feb 21	3 27	102	25	1.106	S	2027 Aug 17	7 15	–	–	(0.545)	N
2008 Aug 16	21 11	94	–	0.806	N	2028 Jan 12	4 14	28	–	0.067	N
2009 Feb 9	14 39	–	–	(0.899)	S	2028 Jul 6	18 22	71	–	0.391	S
2009 Jul 7	9 40	–	–	(0.157)	S	2028 Dec 31	16 53	104	36	1.246	N
2009 Aug 6	0 40	–	–	(0.402)	N	2029 Jun 26	3 24	109	51	1.842	N

Date	ET	p	t	Magn.		Date	ET	p	t	Magn.	
	h m	m	m				h m	m	m		
2029 Dec 20	22 43	106	27	1.116	S	2040 Nov 18	19 05	110	44	1.396	N
2030 Jun 15	18 34	72	–	0.502	N	2041 May 16	0 44	29	–	0.063	S
2030 Dec 9	22 29	–	–	(0.940)	S	2041 Nov 8	4 34	45	–	0.170	N
2031 May 7	3 52	–	–	(0.883)	S	2042 Apr 5	14 31	–	–	(0.869)	N
2031 Jun 5	11 45	–	–	(0.130)	N	2042 Sep 29	10 45	–	–	(0.954)	S
2031 Oct 30	7 47	–	–	(0.718)	N	2043 Mar 25	14 32	107	27	1.113	N
2032 Apr 25	15 15	105	33	1.191	S	2043 Sep 19	1 52	103	36	1.257	S
2032 Oct 18	19 04	98	24	1.104	N	2044 Mar 13	19 39	104	33	1.202	S
2033 Apr 14	19 14	107	24	1.093	N	2044 Sep 7	11 21	103	17	1.044	N
2033 Oct 8	10 56	101	39	1.350	S	2045 Mar 3	7 43	–	–	(0.962)	S
2034 Apr 3	19 08	–	–	(0.853)	N	2045 Aug 27	13 55	–	–	(0.681)	N
2034 Sep 28	2 47	13	–	0.014	S	2046 Jan 22	13 03	25	–	0.055	N
2035 Feb 22	9 07	–	–	(0.966)	S	2046 Jul 18	1 06	57	–	0.246	S
2035 Aug 19	1 12	38	–	0.103	N	2047 Jan 12	1 27	104	35	1.235	N
2036 Feb 11	22 14	101	37	1.300	S	2047 Jul 7	10 36	109	50	1.752	S
2036 Aug 7	2 52	115	48	1.454	N	2048 Jan 1	6 54	107	28	1.126	S
2037 Jan 31	14 02	98	32	1.207	N	2048 Jun 26	2 03	79	–	0.638	N
2037 Jul 27	4 10	96	–	0.808	S	2048 Dec 20	6 28	–	–	(0.960)	S
2038 Jan 21	3 50	–	–	(0.899)	N	2049 May 17	11 27	–	–	(0.766)	S
2038 Jun 17	2 45	–	–	(0.443)	N	2049 Jun 15	19 15	–	–	(0.250)	N
2038 Jul 16	11 36	–	–	(0.500)	S	2049 Nov 9	15 52	–	–	(0.682)	N
2038 Dec 11	17 45	–	–	(0.805)	S	2050 May 6	22 33	103	22	1.078	S
2039 Jun 6	18 55	89	–	0.885	N	2050 Oct 30	3 21	96	17	1.054	N
2039 Nov 30	16 56	103	–	0.941	S						
2040 May 26	11 47	105	46	1.534	S						

During the twentieth century, the lunar eclipse with the largest
magnitude in the umbra is that of 1953 July 26, with a value of
1.864. The next eclipse with a still larger magnitude, namely 1.869,
will be that of 2264 November 4.

XV

EQUINOXES AND SOLSTICES ON MARS

This table gives the times of the equinoxes and solstices on the planet Mars for the years 1646 to 2060. These times have been calculated by Christian Steyaert, Vereniging voor Sterrenkunde, Belgium.

On each line one finds, from the left to the right, respectively : the time of the spring equinox for the *northern* hemisphere of Mars (corresponding to the March equinox on the Earth), that of the summer solstice, that of the autumn equinox, and finally the time of the winter solstice. In each case, the Julian Date is given first, followed by the ordinary calendar date. The Julian Date is given to the nearest hundredth of a day. It should be remembered that the Julian Dates are counted from noon, Universal Time. Thus, for example,

J.D. 2443 452.64 corresponds to 1977 November 5.14,

J.D. 2443 651.28 corresponds to 1978 May 22.78.

The times so found are expressed in Ephemeris Time, but in practice they may be considered as being expressed in Universal Time as well, because the difference between ET and UT is not expected to exceed 4 minutes by the year 2060.

For the computation of the heliocentric positions of Mars, use was made of Newcomb's theory for this planet (*Astron. Papers American Ephemeris*, Vol. VI, Part 4 ; Washington, 1898) with corrections by Ross (*ibid.*, Vol. IX, Part 2 ; Washington, 1917). For the North Pole of Mars, Steyaert adopted the position obtained from radio tracking from the Viking 1 lander (*Science*, Vol. 193, page 803), namely $\alpha_0 = 317°35 \pm 0°06$ and $\delta_0 = +52°71 \pm 0°01$, these values being for the epoch 1976.55, but referred to the Earth's mean equator and equinox of 1950.0. Taking into account the precessional motions of both Mars and the Earth, Steyaert then obtained the following expressions for the positions of Mars' North Pole at the beginning of year t :

$$\alpha_0 = 317°378 + 0°006770 \ (t - 1950.0)$$
$$\delta_0 = +52°725 + 0°003530 \ (t - 1950.0)$$

The following equinoxes and solstices (for Mars' northern hemisphere) occur less than ten days from an opposition of Mars :

Spring Equinox :   1691 Dec 14, 1770 Dec 13, 1849 Dec 12, 1896 Dec 19, 1975 Dec 19,
2054 Dec 16

Summer Solstice :   1713 Mar 10, 1760 Mar 17, 1792 Mar 7, 1839 Mar 17, 1918 Mar 15,
1997 Mar 13, 2044 Mar 20

Autumn Equinox :   1717 Jun 13, 1796 Jun 11, 1875 Jun 11, 1922 Jun 19, 2001 Jun 17

Winter Solstice :   1672 Sep 15, 1751 Sep 15, 1830 Sep 14, 2035 Sep 20

Finally, the durations of the astronomical seasons, for the northern hemisphere of Mars, are as follows :

	Epoch 1650	Epoch 1850	Epoch 2050
Spring	199.06 days	198.80 days	198.52 days
Summer	182.35	182.99	183.65
Autumn	146.35	146.52	146.70
Winter	159.21	158.66	158.10

Spring Equinox		Summer Solstice		Autumn Equinox		Winter Solstice	
2322545.66	1646 Oct 24	2322744.73	1647 May 11	2322927.09	1647 Nov 9	2323073.41	1648 Apr 3
2323232.60	1648 Sep 10	2323431.68	1649 Mar 28	2323614.05	1649 Sep 26	2323760.41	1650 Feb 19
2323919.60	1650 Jul 29	2324118.64	1651 Feb 13	2324301.00	1651 Aug 14	2324447.36	1652 Jan 7
2324606.58	1652 Jun 15	2324805.62	1652 Dec 31	2324987.98	1653 Jul 1	2325134.32	1653 Nov 24
2325293.54	1654 May 3	2325492.60	1654 Nov 18	2325674.96	1655 May 19	2325821.32	1655 Oct 12
2325980.52	1656 Mar 20	2326179.58	1656 Oct 5	2326361.94	1657 Apr 5	2326508.29	1657 Aug 29
2326667.50	1658 Feb 4	2326866.55	1658 Aug 23	2327048.95	1659 Feb 21	2327195.27	1659 Jul 17
2327354.44	1659 Dec 23	2327553.50	1660 Jul 9	2327735.91	1661 Jan 8	2327882.28	1661 Jun 3
2328041.42	1661 Nov 9	2328240.46	1662 May 27	2328422.85	1662 Nov 26	2328569.24	1663 Apr 21
2328728.41	1663 Sep 27	2328927.43	1664 Apr 13	2329109.83	1664 Oct 13	2329256.19	1665 Mar 8
2329415.37	1665 Aug 14	2329614.39	1666 Mar 1	2329796.81	1666 Aug 31	2329943.18	1667 Jan 24
2330102.34	1667 Jul 2	2330301.37	1668 Jan 17	2330483.75	1668 Jul 18	2330630.13	1668 Dec 11
2330789.32	1669 May 19	2330988.37	1669 Dec 4	2331170.77	1670 Jun 5	2331317.09	1670 Oct 29
2331476.26	1671 Apr 6	2331675.33	1671 Oct 22	2331857.78	1672 Apr 22	2332004.13	1672 Sep 15
2332163.24	1673 Feb 21	2332362.28	1673 Sep 8	2332544.73	1674 Mar 10	2332691.12	1674 Aug 3
2332850.25	1675 Jan 9	2333049.25	1675 Jul 27	2333231.69	1676 Jan 26	2333378.07	1676 Jun 20
2333537.22	1676 Nov 26	2333736.22	1677 Jun 13	2333918.67	1677 Dec 13	2334065.04	1678 May 8
2334224.18	1678 Oct 14	2334423.20	1679 May 1	2334605.62	1679 Oct 31	2334752.02	1680 Mar 25
2334911.16	1680 Aug 31	2335110.19	1681 Mar 18	2335292.61	1681 Sep 17	2335438.96	1682 Feb 10
2335598.11	1682 Jul 19	2335797.14	1683 Feb 3	2335979.62	1683 Aug 5	2336125.97	1683 Dec 29
2336285.07	1684 Jun 5	2336484.10	1684 Dec 21	2336666.57	1685 Jun 22	2336812.96	1685 Nov 15
2336972.06	1686 Apr 23	2337171.07	1686 Nov 8	2337353.54	1687 May 10	2337499.93	1687 Oct 3
2337659.04	1688 Mar 10	2337858.04	1688 Sep 25	2338040.54	1689 Mar 27	2338186.91	1689 Aug 20
2338346.01	1690 Jan 26	2338545.01	1690 Aug 13	2338727.49	1691 Feb 11	2338873.90	1691 Jul 8
2339033.00	1691 Dec 14	2339232.00	1692 Jun 30	2339414.45	1692 Dec 29	2339560.84	1693 May 25
2339719.96	1693 Oct 31	2339918.96	1694 May 18	2340101.47	1694 Nov 16	2340247.82	1695 Apr 12
2340406.89	1695 Sep 18	2340605.91	1696 Apr 4	2340788.43	1696 Oct 3	2340934.83	1697 Feb 27
2341093.87	1697 Aug 5	2341292.87	1698 Feb 20	2341475.37	1698 Aug 21	2341621.79	1699 Jan 15
2341780.87	1699 Jun 23	2341979.84	1700 Jan 8	2342162.35	1700 Jul 9	2342308.74	1700 Dec 3
2342467.83	1701 May 11	2342666.82	1701 Nov 26	2342849.34	1702 May 27	2342995.74	1702 Oct 21
2343154.79	1703 Mar 29	2343353.81	1703 Oct 14	2343536.31	1704 Apr 13	2343682.71	1704 Sep 7
2343841.79	1705 Feb 13	2344040.79	1705 Aug 31	2344223.32	1706 Mar 1	2344369.68	1706 Jul 26
2344528.74	1707 Jan 1	2344727.74	1707 Jul 19	2344910.30	1708 Jan 17	2345056.69	1708 Jun 12
2345215.71	1708 Nov 18	2345414.70	1709 Jun 5	2345597.25	1709 Dec 4	2345743.67	1710 Apr 30
2345902.70	1710 Oct 6	2346101.67	1711 Apr 23	2346284.21	1711 Oct 22	2346430.61	1712 Mar 17
2346589.67	1712 Aug 23	2346788.64	1713 Mar 10	2346971.20	1713 Sep 8	2347117.60	1714 Feb 2
2347276.63	1714 Jul 11	2347475.61	1715 Jan 26	2347658.14	1715 Jul 27	2347804.57	1715 Dec 21
2347963.62	1716 May 28	2348162.61	1716 Dec 13	2348345.14	1717 Jun 13	2348491.52	1717 Nov 7
2348650.57	1718 Apr 15	2348849.57	1718 Oct 31	2349032.17	1719 May 1	2349178.54	1719 Sep 25
2349337.53	1720 Mar 2	2349536.52	1720 Sep 17	2349719.12	1721 Mar 18	2349865.55	1721 Aug 12
2350024.54	1722 Jan 18	2350223.48	1722 Aug 4	2350406.07	1723 Feb 3	2350552.49	1723 Jun 29
2350711.52	1723 Dec 6	2350910.45	1724 Jun 21	2351093.05	1724 Dec 21	2351239.46	1725 May 16
2351398.47	1725 Oct 22	2351597.42	1726 May 9	2351780.01	1726 Nov 8	2351926.44	1727 Apr 3
2352085.44	1727 Sep 9	2352284.41	1728 Mar 26	2352466.98	1728 Sep 25	2352613.38	1729 Feb 18
2352772.41	1729 Jul 27	2352971.37	1730 Feb 11	2353153.99	1730 Aug 13	2353300.37	1731 Jan 6

Spring Equinox		Summer Solstice		Autumn Equinox		Winter Solstice	
2353459.35	1731 Jun 14	2353658.32	1731 Dec 30	2353840.95	1732 Jun 30	2353987.38	1732 Nov 23
2354146.33	1733 May 1	2354345.29	1733 Nov 16	2354527.91	1734 May 18	2354674.34	1734 Oct 11
2354833.32	1735 Mar 19	2355032.26	1735 Oct 4	2355214.91	1736 Apr 4	2355361.32	1736 Aug 28
2355520.30	1737 Feb 3	2355719.24	1737 Aug 21	2355901.88	1738 Feb 20	2356048.32	1738 Jul 16
2356207.27	1738 Dec 22	2356406.23	1739 Jul 9	2356588.83	1740 Jan 8	2356735.26	1740 Jun 2
2356894.25	1740 Nov 8	2357093.20	1741 May 26	2357275.84	1741 Nov 25	2357422.23	1742 Apr 20
2357581.19	1742 Sep 26	2357780.15	1743 Apr 13	2357962.83	1743 Oct 13	2358109.25	1744 Mar 7
2358268.16	1744 Aug 13	2358467.10	1745 Feb 28	2358649.77	1745 Aug 30	2358796.22	1746 Jan 23
2358955.16	1746 Jul 1	2359154.07	1747 Jan 16	2359336.74	1747 Jul 18	2359483.16	1747 Dec 11
2359642.13	1748 May 18	2359841.05	1748 Dec 3	2360023.73	1749 Jun 4	2360170.16	1749 Oct 28
2360329.09	1750 Apr 5	2360528.04	1750 Oct 21	2360710.69	1751 Apr 22	2360857.14	1751 Sep 15
2361016.08	1752 Feb 21	2361215.02	1752 Sep 7	2361397.70	1753 Mar 9	2361544.10	1753 Aug 2
2361703.05	1754 Jan 8	2361901.97	1754 Jul 26	2362084.69	1755 Jan 25	2362231.10	1755 Jun 20
2362390.00	1755 Nov 26	2362588.93	1756 Jun 12	2362771.63	1756 Dec 12	2362918.09	1757 May 7
2363076.99	1757 Oct 13	2363275.89	1758 Apr 30	2363458.59	1758 Oct 30	2363605.04	1759 Mar 25
2363763.96	1759 Aug 31	2363962.86	1760 Mar 17	2364145.57	1760 Sep 16	2364292.01	1761 Feb 9
2364450.92	1761 Jul 18	2364649.82	1762 Feb 2	2364832.52	1762 Aug 4	2364978.99	1762 Dec 28
2365137.90	1763 Jun 5	2365336.82	1763 Dec 21	2365519.50	1764 Jun 21	2365665.93	1764 Nov 14
2365824.85	1765 Apr 22	2366023.80	1765 Nov 7	2366206.53	1766 May 9	2366352.94	1766 Oct 2
2366511.80	1767 Mar 10	2366710.75	1767 Sep 25	2366893.50	1768 Mar 26	2367039.96	1768 Aug 19
2367198.81	1769 Jan 25	2367397.71	1769 Aug 12	2367580.45	1770 Feb 10	2367726.92	1770 Jul 7
2367885.80	1770 Dec 13	2368084.68	1771 Jun 30	2368267.43	1771 Dec 29	2368413.87	1772 May 24
2368572.75	1772 Oct 30	2368771.65	1773 May 17	2368954.40	1773 Nov 15	2369100.87	1774 Apr 11
2369259.72	1774 Sep 17	2369458.63	1775 Apr 4	2369641.35	1775 Oct 3	2369787.81	1776 Feb 27
2369946.70	1776 Aug 4	2370145.61	1777 Feb 19	2370328.37	1777 Aug 20	2370474.79	1778 Jan 14
2370633.64	1778 Jun 22	2370832.56	1779 Jan 7	2371015.34	1779 Jul 8	2371161.80	1779 Dec 2
2371320.63	1780 May 9	2371519.53	1780 Nov 24	2371702.29	1781 May 25	2371848.77	1781 Oct 19
2372007.61	1782 Mar 27	2372206.50	1782 Oct 11	2372389.29	1783 Apr 12	2372535.74	1783 Sep 6
2372694.60	1784 Feb 12	2372893.47	1784 Aug 28	2373076.27	1785 Feb 27	2373222.74	1785 Jul 24
2373381.57	1785 Dec 30	2373580.45	1786 Jul 16	2373763.21	1787 Jan 15	2373909.69	1787 Jun 11
2374068.55	1787 Nov 17	2374267.43	1788 Jun 2	2374450.21	1788 Dec 2	2374596.64	1789 Apr 28
2374755.49	1789 Oct 3	2374954.38	1790 Apr 20	2375137.20	1790 Oct 20	2375283.65	1791 Mar 16
2375442.45	1791 Aug 21	2375641.33	1792 Mar 7	2375824.15	1792 Sep 6	2375970.64	1793 Jan 31
2376129.44	1793 Jul 8	2376328.29	1794 Jan 23	2376511.10	1794 Jul 25	2376657.58	1794 Dec 19
2376816.42	1795 May 26	2377015.27	1795 Dec 11	2377198.09	1796 Jun 11	2377344.56	1796 Nov 5
2377503.37	1797 Apr 12	2377702.25	1797 Oct 28	2377885.06	1798 Apr 29	2378031.56	1798 Sep 23
2378190.36	1799 Feb 28	2378389.24	1799 Sep 15	2378572.06	1800 Mar 17	2378718.51	1800 Aug 11
2378877.33	1801 Jan 16	2379076.20	1801 Aug 3	2379259.06	1802 Feb 2	2379405.51	1802 Jun 29
2379564.27	1802 Dec 4	2379763.15	1803 Jun 21	2379946.02	1803 Dec 21	2380092.51	1804 May 16
2380251.26	1804 Oct 21	2380450.11	1805 May 8	2380632.97	1805 Nov 7	2380779.46	1806 Apr 2
2380938.25	1806 Sep 8	2381137.08	1807 Mar 26	2381319.95	1807 Sep 25	2381466.42	1808 Feb 18
2381625.21	1808 Jul 26	2381824.05	1809 Feb 10	2382006.91	1809 Aug 12	2382153.42	1810 Jan 5
2382312.18	1810 Jun 13	2382511.04	1810 Dec 29	2382693.87	1811 Jun 30	2382840.35	1811 Nov 23
2382999.16	1812 Apr 30	2383198.03	1812 Nov 15	2383380.90	1813 May 17	2383527.35	1813 Oct 10
2383686.09	1814 Mar 18	2383884.98	1814 Oct 3	2384067.89	1815 Apr 4	2384214.38	1815 Aug 28

Spring Equinox		Summer Solstice		Autumn Equinox		Winter Solstice	
2384373.09	1816 Feb 3	2384571.93	1816 Aug 20	2384754.83	1817 Feb 19	2384901.35	1817 Jul 15
2385060.08	1817 Dec 21	2385258.90	1818 Jul 8	2385441.80	1819 Jan 7	2385588.29	1819 Jun 2
2385747.05	1819 Nov 8	2385945.87	1820 May 25	2386128.77	1820 Nov 24	2386275.28	1821 Apr 19
2386434.01	1821 Sep 25	2386632.85	1822 Apr 12	2386815.72	1822 Oct 12	2386962.24	1823 Mar 7
2387120.99	1823 Aug 13	2387319.83	1824 Feb 28	2387502.73	1824 Aug 29	2387649.19	1825 Jan 22
2387807.93	1825 Jun 30	2388006.79	1826 Jan 15	2388189.72	1826 Jul 17	2388336.20	1826 Dec 10
2388494.89	1827 May 18	2388693.74	1827 Dec 3	2388876.66	1828 Jun 3	2389023.18	1828 Oct 27
2389181.89	1829 Apr 4	2389380.72	1829 Oct 20	2389563.66	1830 Apr 21	2389710.15	1830 Sep 14
2389868.87	1831 Feb 20	2390067.68	1831 Sep 7	2390250.64	1832 Mar 8	2390397.14	1832 Aug 1
2390555.84	1833 Jan 7	2390754.66	1833 Jul 25	2390937.58	1834 Jan 24	2391084.12	1834 Jun 19
2391242.82	1834 Nov 25	2391441.65	1835 Jun 12	2391624.57	1835 Dec 12	2391771.06	1836 May 6
2391929.77	1836 Oct 12	2392128.61	1837 Apr 29	2392311.58	1837 Oct 29	2392458.06	1838 Mar 24
2392616.72	1838 Aug 30	2392815.56	1839 Mar 17	2392998.52	1839 Sep 16	2393145.06	1840 Feb 9
2393303.72	1840 Jul 17	2393502.52	1841 Feb 1	2393685.48	1841 Aug 2	2393832.01	1841 Dec 27
2393990.70	1842 Jun 4	2394189.49	1842 Dec 19	2394372.47	1843 Jun 20	2394518.97	1843 Nov 14
2394677.66	1844 Apr 21	2394876.47	1844 Nov 5	2395059.44	1845 May 7	2395205.98	1845 Oct 1
2395364.64	1846 Mar 9	2395563.46	1846 Sep 23	2395746.42	1847 Mar 25	2395892.94	1847 Aug 19
2396051.62	1848 Jan 25	2396250.43	1848 Aug 10	2396433.44	1849 Feb 9	2396579.92	1849 Jul 6
2396738.57	1849 Dec 12	2396937.38	1850 Jun 28	2397120.39	1850 Dec 28	2397266.93	1851 May 24
2397425.55	1851 Oct 30	2397624.34	1852 May 15	2397807.34	1852 Nov 14	2397953.88	1853 Apr 10
2398112.53	1853 Sep 16	2398311.31	1854 Apr 2	2398494.32	1854 Oct 2	2398640.83	1855 Feb 26
2398799.50	1855 Aug 3	2398998.27	1856 Feb 18	2399181.28	1856 Aug 19	2399327.83	1857 Jan 13
2399486.46	1857 Jun 20	2399685.25	1858 Jan 5	2399868.24	1858 Jul 7	2400014.77	1858 Dec 1
2400173.43	1859 May 8	2400372.24	1859 Nov 23	2400555.26	1860 May 24	2400701.74	1860 Oct 18
2400860.37	1861 Mar 25	2401059.20	1861 Oct 10	2401242.26	1862 Apr 11	2401388.78	1862 Sep 5
2401547.36	1863 Feb 10	2401746.15	1863 Aug 28	2401929.20	1864 Feb 27	2402075.76	1864 Jul 23
2402234.36	1864 Dec 28	2402433.12	1865 Jul 15	2402616.16	1866 Jan 14	2402762.71	1866 Jun 10
2402921.33	1866 Nov 15	2403120.10	1867 Jun 2	2403303.15	1867 Dec 2	2403449.69	1868 Apr 27
2403608.29	1868 Oct 2	2403807.06	1869 Apr 19	2403990.10	1869 Oct 19	2404136.66	1870 Mar 15
2404295.27	1870 Aug 20	2404494.05	1871 Mar 7	2404677.09	1871 Sep 6	2404823.61	1872 Jan 31
2404982.23	1872 Jul 7	2405181.01	1873 Jan 22	2405364.09	1873 Jul 24	2405510.61	1873 Dec 18
2405669.18	1874 May 25	2405867.97	1874 Dec 10	2406051.05	1875 Jun 11	2406197.61	1875 Nov 5
2406356.17	1876 Apr 11	2406554.94	1876 Oct 27	2406738.03	1877 Apr 28	2406884.57	1877 Sep 22
2407043.17	1878 Feb 27	2407241.91	1878 Sep 14	2407425.02	1879 Mar 16	2407571.56	1879 Aug 10
2407730.13	1880 Jan 15	2407928.88	1880 Aug 1	2408111.97	1881 Jan 31	2408258.54	1881 Jun 27
2408417.11	1881 Dec 2	2408615.88	1882 Jun 19	2408798.94	1882 Dec 19	2408945.48	1883 May 14
2409104.07	1883 Oct 20	2409302.84	1884 May 6	2409485.95	1884 Nov 5	2409632.47	1885 Mar 31
2409791.01	1885 Sep 6	2409989.78	1886 Mar 24	2410172.91	1886 Sep 23	2410319.47	1887 Feb 16
2410477.99	1887 Jul 25	2410676.74	1888 Feb 9	2410859.85	1888 Aug 10	2411006.43	1889 Jan 3
2411164.98	1889 Jun 11	2411363.71	1889 Dec 27	2411546.84	1890 Jun 28	2411693.38	1890 Nov 21
2411851.94	1891 Apr 29	2412050.69	1891 Nov 14	2412233.82	1892 May 15	2412380.38	1892 Oct 8
2412538.91	1893 Mar 16	2412737.67	1893 Oct 1	2412920.79	1894 Apr 2	2413067.35	1894 Aug 26
2413225.90	1895 Feb 1	2413424.65	1895 Aug 19	2413607.80	1896 Feb 18	2413754.32	1896 Jul 13
2413912.85	1896 Dec 19	2414111.60	1897 Jul 6	2414294.77	1898 Jan 5	2414441.33	1898 May 31
2414599.82	1898 Nov 6	2414798.56	1899 May 24	2414981.71	1899 Nov 23	2415128.30	1900 Apr 18

Spring Equinox		Summer Solstice		Autumn Equinox		Winter Solstice	
2415286.81	1900 Sep 24	2415485.53	1901 Apr 11	2415668.69	1901 Oct 11	2415815.25	1902 Mar 6
2415973.78	1902 Aug 12	2416172.50	1903 Feb 26	2416355.67	1903 Aug 29	2416502.24	1904 Jan 22
2416660.74	1904 Jun 29	2416859.47	1905 Jan 13	2417042.61	1905 Jul 16	2417189.20	1905 Dec 9
2417347.73	1906 May 17	2417546.47	1906 Dec 1	2417729.62	1907 Jun 3	2417876.15	1907 Oct 27
2418034.67	1908 Apr 3	2418233.43	1908 Oct 18	2418416.64	1909 Apr 20	2418563.19	1909 Sep 13
2418721.65	1910 Feb 19	2418920.38	1910 Sep 5	2419103.59	1911 Mar 8	2419250.19	1911 Aug 1
2419408.65	1912 Jan 7	2419607.35	1912 Jul 23	2419790.54	1913 Jan 23	2419937.13	1913 Jun 18
2420095.63	1913 Nov 24	2420294.32	1914 Jun 10	2420477.53	1914 Dec 11	2420624.10	1915 May 6
2420782.58	1915 Oct 12	2420981.29	1916 Apr 27	2421164.48	1916 Oct 27	2421311.08	1917 Mar 23
2421469.56	1917 Aug 29	2421668.28	1918 Mar 15	2421851.46	1918 Sep 14	2421998.02	1919 Feb 8
2422156.52	1919 Jul 17	2422355.24	1920 Jan 31	2422538.47	1920 Aug 1	2422685.01	1920 Dec 26
2422843.46	1921 Jun 2	2423042.19	1921 Dec 18	2423225.43	1922 Jun 19	2423372.02	1922 Nov 13
2423530.45	1923 Apr 20	2423729.16	1923 Nov 5	2423912.39	1924 May 6	2424058.99	1924 Sep 30
2424217.44	1925 Mar 7	2424416.12	1925 Sep 22	2424599.38	1926 Mar 24	2424745.96	1926 Aug 18
2424904.41	1927 Jan 23	2425103.09	1927 Aug 10	2425286.34	1928 Feb 9	2425432.95	1928 Jul 5
2425591.38	1928 Dec 10	2425790.08	1929 Jun 27	2425973.29	1929 Dec 27	2426119.89	1930 May 23
2426278.35	1930 Oct 28	2426477.05	1931 May 15	2426660.31	1931 Nov 14	2426806.86	1932 Apr 9
2426965.29	1932 Sep 14	2427164.00	1933 Apr 1	2427347.29	1933 Oct 1	2427493.88	1934 Feb 25
2427652.27	1934 Aug 2	2427850.96	1935 Feb 17	2428034.23	1935 Aug 19	2428180.85	1936 Jan 13
2428339.27	1936 Jun 19	2428537.93	1937 Jan 4	2428721.20	1937 Jul 6	2428867.79	1937 Nov 30
2429026.23	1938 May 7	2429224.91	1938 Nov 22	2429408.19	1939 May 24	2429554.79	1939 Oct 18
2429713.19	1940 Mar 24	2429911.89	1940 Oct 9	2430095.16	1941 Apr 10	2430241.77	1941 Sep 4
2430400.19	1942 Feb 9	2430598.88	1942 Aug 27	2430782.16	1943 Feb 26	2430928.73	1943 Jul 23
2431087.14	1943 Dec 28	2431285.83	1944 Jul 14	2431469.15	1945 Jan 13	2431615.74	1945 Jun 9
2431774.10	1945 Nov 14	2431972.78	1946 Jun 1	2432156.10	1946 Dec 1	2432302.72	1947 Apr 27
2432461.10	1947 Oct 2	2432659.75	1948 Apr 18	2432843.06	1948 Oct 18	2432989.67	1949 Mar 14
2433148.07	1949 Aug 19	2433346.72	1950 Mar 6	2433530.04	1950 Sep 5	2433676.65	1951 Jan 30
2433835.03	1951 Jul 7	2434033.69	1952 Jan 22	2434216.99	1952 Jul 23	2434363.63	1952 Dec 17
2434522.01	1953 May 24	2434720.68	1953 Dec 9	2434903.98	1954 Jun 10	2435050.57	1954 Nov 4
2435208.96	1955 Apr 11	2435407.65	1955 Oct 27	2435591.00	1956 Apr 27	2435737.58	1956 Sep 21
2435895.92	1957 Feb 26	2436094.59	1957 Sep 13	2436277.96	1958 Mar 15	2436424.59	1958 Aug 9
2436582.92	1959 Jan 14	2436781.55	1959 Aug 1	2436964.91	1960 Jan 31	2437111.54	1960 Jun 26
2437269.90	1960 Dec 1	2437468.53	1961 Jun 18	2437651.89	1961 Dec 18	2437798.50	1962 May 13
2437956.86	1962 Oct 19	2438155.49	1963 May 5	2438338.85	1963 Nov 5	2438485.49	1964 Mar 30
2438643.83	1964 Sep 5	2438842.48	1965 Mar 22	2439025.81	1965 Sep 22	2439172.43	1966 Feb 15
2439330.80	1966 Jul 24	2439529.45	1967 Feb 7	2439712.82	1967 Aug 10	2439859.41	1968 Jan 3
2440017.74	1968 Jun 10	2440216.40	1968 Dec 25	2440399.80	1969 Jun 27	2440546.42	1969 Nov 20
2440704.72	1970 Apr 28	2440903.37	1970 Nov 12	2441086.75	1971 May 15	2441233.40	1971 Oct 8
2441391.72	1972 Mar 15	2441590.34	1972 Sep 29	2441773.75	1973 Apr 1	2441920.37	1973 Aug 25
2442078.70	1974 Jan 31	2442277.31	1974 Aug 17	2442460.72	1975 Feb 17	2442607.36	1975 Jul 13
2442765.66	1975 Dec 19	2442964.29	1976 Jul 4	2443147.67	1977 Jan 4	2443294.32	1977 May 30
2443452.64	1977 Nov 5	2443651.28	1978 May 22	2443834.68	1978 Nov 22	2443981.27	1979 Apr 17
2444139.58	1979 Sep 23	2444338.22	1980 Apr 8	2444521.66	1980 Oct 9	2444668.29	1981 Mar 4
2444826.55	1981 Aug 10	2445025.18	1982 Feb 24	2445208.60	1982 Aug 27	2445355.27	1983 Jan 20
2445513.55	1983 Jun 28	2445712.14	1984 Jan 12	2445895.56	1984 Jul 14	2446042.21	1984 Dec 7

Spring Equinox			Summer Solstice			Autumn Equinox			Winter Solstice		
2446200.52	1985 May	15	2446399.12	1985 Nov	29	2446582.56	1986 Jun	1	2446729.19	1986 Oct	25
2446887.47	1987 Apr	1	2447086.10	1987 Oct	17	2447269.52	1988 Apr	18	2447416.19	1988 Sep	11
2447574.47	1989 Feb	16	2447773.08	1989 Sep	3	2447956.52	1990 Mar	6	2448103.14	1990 Jul	30
2448261.43	1991 Jan	4	2448460.04	1991 Jul	22	2448643.51	1992 Jan	22	2448790.14	1992 Jun	16
2448948.37	1992 Nov	21	2449146.99	1993 Jun	8	2449330.46	1993 Dec	8	2449477.13	1994 May	4
2449635.36	1994 Oct	9	2449833.95	1995 Apr	26	2450017.41	1995 Oct	26	2450164.08	1996 Mar	21
2450322.34	1996 Aug	26	2450520.92	1997 Mar	13	2450704.40	1997 Sep	12	2450851.05	1998 Feb	6
2451009.30	1998 Jul	14	2451207.89	1999 Jan	29	2451391.35	1999 Jul	31	2451538.04	1999 Dec	25
2451696.28	2000 May	31	2451894.88	2000 Dec	16	2452078.33	2001 Jun	17	2452224.97	2001 Nov	11
2452383.25	2002 Apr	18	2452581.87	2002 Nov	3	2452765.36	2003 May	5	2452911.98	2003 Sep	29
2453070.19	2004 Mar	5	2453268.81	2004 Sep	20	2453452.34	2005 Mar	22	2453599.00	2005 Aug	16
2453757.19	2006 Jan	21	2453955.77	2006 Aug	8	2454139.28	2007 Feb	7	2454285.96	2007 Jul	4
2454444.18	2007 Dec	9	2454642.74	2008 Jun	25	2454826.26	2008 Dec	25	2454972.91	2009 May	21
2455131.14	2009 Oct	26	2455329.71	2010 May	13	2455513.23	2010 Nov	12	2455659.91	2011 Apr	8
2455818.11	2011 Sep	13	2456016.69	2012 Mar	30	2456200.18	2012 Sep	29	2456346.86	2013 Feb	23
2456505.09	2013 Jul	31	2456703.67	2014 Feb	15	2456887.19	2014 Aug	17	2457033.82	2015 Jan	11
2457192.03	2015 Jun	18	2457390.62	2016 Jan	3	2457574.17	2016 Jul	4	2457720.83	2016 Nov	28
2457879.00	2017 May	5	2458077.58	2017 Nov	20	2458261.12	2018 May	22	2458407.81	2018 Oct	16
2458565.99	2019 Mar	23	2458764.55	2019 Oct	8	2458948.11	2020 Apr	8	2459094.78	2020 Sep	2
2459252.97	2021 Feb	7	2459451.52	2021 Aug	25	2459635.09	2022 Feb	24	2459781.77	2022 Jul	21
2459939.94	2022 Dec	26	2460138.49	2023 Jul	12	2460322.03	2024 Jan	12	2460468.73	2024 Jun	7
2460626.91	2024 Nov	12	2460825.48	2025 May	29	2461009.02	2025 Nov	29	2461155.67	2026 Apr	25
2461313.86	2026 Sep	30	2461512.43	2027 Apr	16	2461696.02	2027 Oct	17	2461842.68	2028 Mar	12
2462000.82	2028 Aug	17	2462199.38	2029 Mar	3	2462382.97	2029 Sep	3	2462529.67	2030 Jan	28
2462687.81	2030 Jul	5	2462886.35	2031 Jan	19	2463069.92	2031 Jul	22	2463216.62	2031 Dec	16
2463374.79	2032 May	22	2463573.33	2032 Dec	6	2463756.92	2033 Jun	8	2463903.59	2033 Nov	2
2464061.75	2034 Apr	9	2464260.31	2034 Oct	24	2464443.89	2035 Apr	26	2464590.60	2035 Sep	20
2464748.73	2036 Feb	25	2464947.29	2036 Sep	10	2465130.87	2037 Mar	13	2465277.55	2037 Aug	7
2465435.71	2038 Jan	12	2465634.26	2038 Jul	29	2465817.88	2039 Jan	29	2465964.54	2039 Jun	25
2466122.65	2039 Nov	30	2466321.21	2040 Jun	15	2466504.84	2040 Dec	16	2466651.54	2041 May	12
2466809.64	2041 Oct	17	2467008.17	2042 May	3	2467191.79	2042 Nov	3	2467338.50	2043 Mar	30
2467496.63	2043 Sep	4	2467695.14	2044 Mar	20	2467878.77	2044 Sep	20	2468025.46	2045 Feb	13
2468183.59	2045 Jul	22	2468382.10	2046 Feb	5	2468565.73	2046 Aug	8	2468712.45	2047 Jan	1
2468870.56	2047 Jun	9	2469069.09	2047 Dec	24	2469252.69	2048 Jun	25	2469399.39	2048 Nov	18
2469557.53	2049 Apr	26	2469756.08	2049 Nov	10	2469939.71	2050 May	13	2470086.37	2050 Oct	6
2470244.47	2051 Mar	13	2470443.02	2051 Sep	28	2470626.71	2052 Mar	30	2470773.40	2052 Aug	23
2470931.46	2053 Jan	28	2471129.97	2053 Aug	15	2471313.64	2054 Feb	15	2471460.38	2054 Jul	11
2471618.45	2054 Dec	16	2471816.94	2055 Jul	3	2472000.61	2056 Jan	3	2472147.32	2056 May	28
2472305.42	2056 Nov	2	2472503.91	2057 May	20	2472687.59	2057 Nov	20	2472834.30	2058 Apr	15
2472992.37	2058 Sep	20	2473190.88	2059 Apr	7	2473374.53	2059 Oct	8	2473521.27	2060 Mar	2

XVI

POSITIONS OF BRIGHT ZODIACAL STARS

This table, by J. Denoyelle (Royal Observatory, Uccle, Belgium) and J. Meeus, was first published in the Belgian journal *Ciel et Terre*, Vol. 88, No. 6, pages 467 – 473 (November-December 1972).

For 48 bright zodiacal stars, the ecliptical longitude and latitude are given from the year -1600 to +2800. These coordinates, referred to the ecliptic and mean equinox of the date, are expressed in degrees and decimals. In the calculation, the precession (including the rotation of the ecliptic) and the proper motions of the stars have been taken into account.

The tabular interval is 400 years. The positions are given for the beginning of the year mentioned in the first column ; they can be linearly interpolated, and then rounded to the nearest hundredth of a degree. The celestial longitude of each star increases by approximately five-and-a-half degrees in four centuries, while the latitude varies very slowly. These latitudes are not exactly constant due to the slow rotation of the ecliptical plane and due to the star's proper motions.

For the table we see, for instance, that δ Cancri was exactly on the ecliptic near the year +20 ; presently, this star is 5′ to the north of the ecliptic. The star ρ Leonis was on the ecliptic near the year -720. On the other hand, the latitude of η Virginis reached a maximum value (+1°25′) about the year -800, while that of β Virginis will be a maximum (+0°42′) in +2600.

YEAR	ETA PSC		ETA TAU		GAMMA TAU	
-1600	336.941	+5.209	10.110	+3.703	15.778	-6.112
-1200	342.447	+5.220	15.616	+3.737	21.302	-6.073
- 800	347.960	+5.234	21.131	+3.773	26.833	-6.033
- 400	353.483	+5.249	26.654	+3.810	32.373	-5.992
0	359.014	+5.266	32.186	+3.849	37.922	-5.951
+ 400	4.555	+5.286	37.728	+3.888	43.480	-5.908
800	10.106	+5.306	43.279	+3.928	49.047	-5.865
1200	15.666	+5.329	48.841	+3.968	54.624	-5.821
1600	21.236	+5.353	54.412	+4.010	60.210	-5.777
2000	26.816	+5.378	59.993	+4.051	65.806	-5.733
2400	32.406	+5.405	65.584	+4.094	71.412	-5.688
2800	38.006	+5.433	71.185	+4.137	77.027	-5.643

YEAR	EPSILON TAU		ALPHA TAU		BETA TAU	
-1600	18.463	-2.949	19.842	-5.706	32.678	+5.093
-1200	23.983	-2.909	25.357	-5.683	38.185	+5.124
- 800	29.511	-2.869	30.879	-5.658	43.701	+5.156
- 400	35.048	-2.828	36.410	-5.632	49.226	+5.189
0	40.594	-2.786	41.950	-5.606	54.760	+5.222
+ 400	46.149	-2.743	47.499	-5.579	60.303	+5.254
800	51.714	-2.700	53.057	-5.552	65.856	+5.287
1200	57.288	-2.656	58.625	-5.524	71.419	+5.320
1600	62.872	-2.612	64.202	-5.496	76.992	+5.353
2000	68.465	-2.568	69.789	-5.468	82.575	+5.386
2400	74.069	-2.523	75.386	-5.439	88.168	+5.418
2800	79.682	-2.478	80.993	-5.411	93.771	+5.450

YEAR	ZETA TAU		ETA GEM		MU GEM	
-1600	34.890	-2.648	43.615	-1.355	45.355	-1.187
-1200	40.398	-2.598	49.115	-1.302	50.868	-1.145
- 800	45.915	-2.548	54.623	-1.250	56.390	-1.104
- 400	51.440	-2.498	60.139	-1.197	61.921	-1.062
0	56.973	-2.448	65.665	-1.145	67.461	-1.021
+ 400	62.517	-2.397	71.200	-1.093	73.010	-0.980
800	68.069	-2.347	76.745	-1.041	78.568	-0.939
1200	73.631	-2.296	82.299	-0.990	84.137	-0.899
1600	79.203	-2.246	87.863	-0.939	89.714	-0.860
2000	84.785	-2.196	93.436	-0.889	95.302	-0.821
2400	90.376	-2.147	99.020	-0.839	100.900	-0.782
2800	95.978	-2.097	104.613	-0.790	106.507	-0.745

YEAR	GAMMA GEM		EPSILON GEM		DELTA GEM	
-1600	49.176	-7.177	50.053	+1.614	58.656	-0.626
-1200	54.689	-7.127	55.560	+1.666	64.160	-0.573
- 800	60.209	-7.078	61.075	+1.718	69.673	-0.521
- 400	65.738	-7.029	66.598	+1.770	75.194	-0.470
0	71.276	-6.980	72.131	+1.821	80.724	-0.419
+ 400	76.823	-6.931	77.673	+1.872	86.264	-0.370
800	82.379	-6.883	83.225	+1.923	91.813	-0.321
1200	87.945	-6.836	88.786	+1.973	97.372	-0.272
1600	93.520	-6.789	94.358	+2.022	102.941	-0.225
2000	99.105	-6.743	99.939	+2.070	108.519	-0.179
2400	104.699	-6.698	105.530	+2.118	114.108	-0.134
2800	110.304	-6.653	111.131	+2.165	119.706	-0.089

YEAR	LAMBDA GEM		BETA GEM		KAPPA GEM	
-1600	58.943	-6.055	63.919	+6.393	63.790	+2.687
-1200	64.445	-6.006	69.360	+6.429	69.295	+2.734
- 800	69.955	-5.957	74.809	+6.464	74.809	+2.780
- 400	75.474	-5.909	80.267	+6.498	80.331	+2.826
0	81.001	-5.861	85.734	+6.531	85.863	+2.870
+ 400	86.538	-5.814	91.211	+6.564	91.404	+2.914
800	92.084	-5.768	96.697	+6.595	96.955	+2.957
1200	97.639	-5.723	102.194	+6.626	102.515	+2.999
1600	103.204	-5.679	107.700	+6.656	108.086	+3.039
2000	108.779	-5.636	113.216	+6.685	113.666	+3.079
2400	114.364	-5.594	118.742	+6.712	119.256	+3.117
2800	119.958	-5.552	124.278	+6.739	124.857	+3.155

YEAR	DELTA CNC		OMIKRON LEO		ETA LEO	
-1600	78.789	-0.087	94.505	-3.980	97.984	+4.588
-1200	84.301	-0.064	99.996	-3.948	103.494	+4.626
- 800	89.822	-0.041	105.495	-3.918	109.013	+4.662
- 400	95.351	-0.020	111.003	-3.890	114.540	+4.697
0	100.889	-0.001	116.520	-3.864	120.077	+4.729
+ 400	106.436	+0.019	122.046	-3.839	125.623	+4.760
800	111.993	+0.036	127.582	-3.816	131.179	+4.789
1200	117.560	+0.051	133.127	-3.795	136.745	+4.817
1600	123.136	+0.065	138.683	-3.776	142.320	+4.842
2000	128.722	+0.077	144.247	-3.758	147.905	+4.866
2400	134.318	+0.088	149.822	-3.742	153.501	+4.888
2800	139.924	+0.098	155.407	-3.727	159.106	+4.908

YEAR	ALPHA LEO		RHO LEO		BETA VIR	
-1600	100.171	+0.278	106.505	-0.067	126.479	+0.586
-1200	105.652	+0.306	112.011	-0.036	132.075	+0.607
- 800	111.142	+0.333	117.526	-0.006	137.678	+0.626
- 400	116.641	+0.357	123.050	+0.023	143.291	+0.642
0	122.148	+0.379	128.582	+0.049	148.913	+0.656
+ 400	127.665	+0.400	134.124	+0.073	154.544	+0.668
800	133.192	+0.419	139.676	+0.095	160.184	+0.678
1200	138.728	+0.436	145.237	+0.115	165.835	+0.685
1600	144.274	+0.452	150.808	+0.134	171.495	+0.691
2000	149.829	+0.465	156.389	+0.150	177.164	+0.695
2400	155.395	+0.477	161.980	+0.165	182.844	+0.696
2800	160.971	+0.488	167.581	+0.178	188.534	+0.696

YEAR	ETA VIR		GAMMA VIR		ALPHA VIR	
-1600	134.978	+1.412	140.753	+3.062	153.989	-1.849
-1200	140.481	+1.415	146.204	+3.039	159.492	-1.864
- 800	145.993	+1.416	151.664	+3.015	165.004	-1.881
- 400	151.513	+1.415	157.133	+2.989	170.524	-1.901
0	157.042	+1.411	162.611	+2.960	176.053	-1.922
+ 400	162.581	+1.406	168.098	+2.930	181.591	-1.945
800	168.129	+1.399	173.594	+2.897	187.139	-1.970
1200	173.687	+1.390	179.100	+2.863	192.697	-1.997
1600	179.255	+1.379	184.616	+2.828	198.264	-2.025
2000	184.832	+1.366	190.141	+2.790	203.842	-2.055
2400	190.420	+1.351	195.677	+2.752	209.429	-2.086
2800	196.017	+1.335	201.222	+2.711	215.026	-2.119

YEAR	ALPHA LIB		DELTA SCO		PI SCO	
-1600	175.266	+0.748	192.691	-1.555	193.072	-5.033
-1200	180.765	+0.708	198.197	-1.599	198.576	-5.078
- 800	186.273	+0.666	203.711	-1.645	204.088	-5.125
- 400	191.789	+0.622	209.234	-1.691	209.610	-5.173
0	197.314	+0.577	214.766	-1.739	215.141	-5.222
+ 400	202.849	+0.531	220.308	-1.787	220.681	-5.271
800	208.393	+0.483	225.859	-1.836	226.231	-5.322
1200	213.946	+0.434	231.420	-1.886	231.791	-5.373
1600	219.510	+0.384	236.991	-1.936	237.361	-5.424
2000	225.083	+0.333	242.571	-1.987	242.940	-5.476
2400	230.666	+0.282	248.162	-2.038	248.530	-5.528
2800	236.259	+0.229	253.762	-2.089	254.130	-5.581

YEAR	BETA SCO		SIGMA SCO		ALPHA SCO	
-1600	193.288	+1.443	197.929	-3.581	199.888	-4.108
-1200	198.797	+1.399	203.434	-3.629	205.393	-4.157
- 800	204.314	+1.353	208.947	-3.678	210.906	-4.206
- 400	209.840	+1.306	214.469	-3.728	216.429	-4.257
0	215.375	+1.258	220.000	-3.778	221.960	-4.308
+ 400	220.919	+1.209	225.540	-3.829	227.501	-4.360
800	226.472	+1.160	231.090	-3.881	233.052	-4.412
1200	232.035	+1.110	236.650	-3.933	238.612	-4.464
1600	237.608	+1.059	242.220	-3.985	244.182	-4.517
2000	243.190	+1.008	247.800	-4.038	249.763	-4.571
2400	248.783	+0.957	253.389	-4.091	255.353	-4.624
2800	254.385	+0.905	258.989	-4.144	260.953	-4.677

YEAR	TAU SCO		ETA OPH		THETA OPH	
-1600	201.592	-5.654	208.030	+7.564	211.505	-1.352
-1200	207.095	-5.704	213.544	+7.525	217.012	-1.406
- 800	212.608	-5.754	219.066	+7.485	222.528	-1.460
- 400	218.129	-5.805	224.596	+7.445	228.052	-1.514
0	223.659	-5.856	230.135	+7.404	233.585	-1.569
+ 400	229.199	-5.909	235.683	+7.363	239.128	-1.624
800	234.749	-5.961	241.241	+7.322	244.680	-1.679
1200	240.308	-6.014	246.808	+7.281	250.242	-1.734
1600	245.878	-6.068	252.384	+7.240	255.813	-1.789
2000	251.457	-6.121	257.970	+7.198	261.395	-1.844
2400	257.046	-6.175	263.565	+7.157	266.987	-1.899
2800	262.646	-6.228	269.170	+7.116	272.588	-1.953

YEAR	GAMMA SGR		DELTA SGR		LAMBDA SGR	
-1600	221.429	-6.325	224.655	-5.969	226.474	-1.477
-1200	226.929	-6.399	230.165	-6.025	231.976	-1.551
- 800	232.437	-6.474	235.684	-6.082	237.486	-1.626
- 400	237.955	-6.548	241.212	-6.139	243.005	-1.700
0	243.481	-6.623	246.749	-6.196	248.533	-1.774
+ 400	249.018	-6.697	252.296	-6.252	254.071	-1.847
800	254.564	-6.771	257.852	-6.308	259.618	-1.920
1200	260.120	-6.845	263.418	-6.363	265.174	-1.993
1600	265.686	-6.919	268.995	-6.418	270.741	-2.065
2000	271.261	-6.992	274.581	-6.473	276.317	-2.136
2400	276.848	-7.064	280.177	-6.526	281.904	-2.207
2800	282.444	-7.136	285.784	-6.580	287.500	-2.277

YEAR	PHI SGR		SIGMA SGR		KSI SGR	
-1600	230.231	-3.475	232.478	-2.920	233.526	+2.153
-1200	235.744	-3.530	237.987	-2.981	239.037	+2.096
- 800	241.266	-3.585	243.504	-3.042	244.557	+2.040
- 400	246.797	-3.639	249.030	-3.102	250.085	+1.984
0	252.337	-3.693	254.565	-3.162	255.623	+1.929
+ 400	257.886	-3.747	260.110	-3.221	261.169	+1.874
800	263.445	-3.799	265.664	-3.279	266.725	+1.820
1200	269.014	-3.852	271.228	-3.337	272.291	+1.766
1600	274.593	-3.904	276.802	-3.394	277.866	+1.714
2000	280.182	-3.954	282.385	-3.450	283.451	+1.662
2400	285.780	-4.005	287.979	-3.505	289.046	+1.611
2800	291.389	-4.054	293.583	-3.560	294.651	+1.561

YEAR	ZETA SGR		TAU SGR		PI SGR	
-1600	233.754	-6.711	235.022	-4.379	236.368	+1.945
-1200	239.260	-6.765	240.520	-4.460	241.875	+1.886
- 800	244.774	-6.819	246.026	-4.541	247.390	+1.828
- 400	250.297	-6.872	251.541	-4.621	252.913	+1.770
0	255.829	-6.925	257.066	-4.701	258.446	+1.713
+ 400	261.371	-6.978	262.600	-4.780	263.988	+1.656
800	266.923	-7.029	268.144	-4.859	269.539	+1.600
1200	272.485	-7.080	273.697	-4.936	275.101	+1.545
1600	278.057	-7.130	279.261	-5.013	280.671	+1.491
2000	283.638	-7.179	284.834	-5.089	286.252	+1.437
2400	289.230	-7.228	290.418	-5.164	291.842	+1.385
2800	294.832	-7.275	296.011	-5.239	297.443	+1.334

YEAR	ALPHA 2 CAP		BETA CAP		THETA CAP	
-1600	253.934	+7.360	254.136	+5.012	263.888	-0.139
-1200	259.446	+7.307	259.646	+4.960	269.402	-0.195
- 800	264.966	+7.256	265.164	+4.910	274.925	-0.249
- 400	270.494	+7.205	270.691	+4.860	280.456	-0.302
0	276.031	+7.156	276.227	+4.812	285.997	-0.353
+ 400	281.578	+7.109	281.772	+4.764	291.547	-0.403
800	287.134	+7.062	287.326	+4.719	297.107	-0.451
1200	292.699	+7.017	292.890	+4.674	302.676	-0.498
1600	298.274	+6.973	298.464	+4.631	308.255	-0.543
2000	303.859	+6.931	304.048	+4.589	313.844	-0.586
2400	309.453	+6.890	309.641	+4.549	319.443	-0.628
2800	315.057	+6.850	315.244	+4.510	325.052	-0.669

YEAR	GAMMA CAP		DELTA CAP		LAMBDA AQR	
-1600	271.715	-2.151	273.490	-1.926	291.662	-0.233
-1200	277.242	-2.203	279.015	-2.008	297.172	-0.259
- 800	282.778	-2.253	284.548	-2.089	302.690	-0.282
- 400	288.323	-2.302	290.090	-2.167	308.217	-0.303
0	293.877	-2.349	295.641	-2.244	313.753	-0.322
+ 400	299.440	-2.394	301.202	-2.319	319.298	-0.339
800	305.013	-2.438	306.772	-2.393	324.853	-0.354
1200	310.596	-2.479	312.353	-2.464	330.418	-0.367
1600	316.188	-2.519	317.943	-2.534	335.992	-0.378
2000	321.791	-2.558	323.543	-2.602	341.576	-0.387
2400	327.403	-2.594	329.153	-2.668	347.170	-0.395
2800	333.026	-2.629	334.773	-2.733	352.774	-0.400

XVII

Conjunctions of the Sun with Bright Stars

This table provides the UT times of conjunction in *longitude* of the Sun with twelve bright zodiacal stars, to the nearest tenth of a day.

These times are given for century years only. For other years, add the correction given on the table at right ; this correction is always positive.

For example, calculate the time of the conjunction of the Sun with α Tauri in 1975. We find :

$$
\begin{array}{lr}
1900 \ldots\ldots\ldots & \text{May } 30.3 \\
75 \ldots\ldots\ldots & 1.2 \\
\hline
\text{Sum} & \text{May } 31.5
\end{array}
$$

Numbers characterizing negative years should be split up in such a way as to have the last two figures positive. For example, the year −328 should be split up as −400 and +72 ; then, take −400 from the main table and add the correction for +72 years.

For each star, two values are given for the year 1500, the first one for the Julian calendar (*J*), and the second one for the Gregorian calendar (*G*).

Correction for additional years					
Year	Corr.	Year	Corr.	Year	Corr.
0	$0^{d}.0$	35	$1^{d}.0$	70	$0^{d}.9$
1	0.3	36	0.2	71	1.2
2	0.5	37	0.5	72	0.5
3	0.8	38	0.7	73	0.7
4	0.0	39	1.0	74	1.0
5	0.3	40	0.3	75	1.2
6	0.5	41	0.5	76	0.5
7	0.8	42	0.8	77	0.7
8	0.1	43	1.0	78	1.0
9	0.3	44	0.3	79	1.3
10	0.6	45	0.5	80	0.5
11	0.8	46	0.8	81	0.8
12	0.1	47	1.0	82	1.0
13	0.3	48	0.3	83	1.3
14	0.6	49	0.6	84	0.5
15	0.8	50	0.8	85	0.8
16	0.1	51	1.1	86	1.0
17	0.4	52	0.3	87	1.3
18	0.6	53	0.6	88	0.6
19	0.9	54	0.8	89	0.8
20	0.1	55	1.1	90	1.1
21	0.4	56	0.4	91	1.3
22	0.6	57	0.6	92	0.6
23	0.9	58	0.9	93	0.8
24	0.2	59	1.1	94	1.1
25	0.4	60	0.4	95	1.4
26	0.7	61	0.6	96	0.6
27	0.9	62	0.9	97	0.9
28	0.2	63	1.2	98	1.1
29	0.4	64	0.4	99	1.4
30	0.7	65	0.7		
31	0.9	66	0.9		
32	0.2	67	1.2		
33	0.5	68	0.4		
34	0.7	69	0.7		

Year	α Ari *Hamal*	η Tau *Alcyone*	α Tau *Aldebaran*	β Tau *Elnath*	β Gem *Pollux*	α Leo *Regulus*
-1000	Mar 26$.^d$3	Apr 18$.^d$6	Apr 28$.^d$8	May 12$.^d$2	Jun 13$.^d$9	Jul 21$.^d$8
- 900	26.9	19.2	29.4	12.9	14.6	22.4
- 800	27.6	19.9	30.1	13.5	15.2	23.1
- 700	28.2	20.5	30.7	14.2	15.8	23.7
- 600	28.9	21.1	May 1.4	14.8	16.4	24.3
- 500	Mar 29.5	Apr 21.8	May 2.0	May 15.4	Jun 17.1	Jul 25.0
- 400	30.2	22.4	2.6	16.1	17.7	25.6
- 300	30.8	23.1	3.3	16.7	18.3	26.3
- 200	31.5	23.7	3.9	17.4	18.9	26.9
- 100	Apr 1.1	24.4	4.6	18.0	19.5	27.5
0	Apr 1.8	Apr 25.0	May 5.2	May 18.6	Jun 20.2	Jul 28.2
+ 100	2.4	25.7	5.9	19.3	20.8	28.8
200	3.1	26.3	6.5	19.9	21.4	29.4
300	3.7	26.9	7.1	20.6	22.0	30.0
400	4.4	27.6	7.8	21.2	22.6	30.7
500	Apr 5.0	Apr 28.2	May 8.4	May 21.8	Jun 23.2	Jul 31.3
600	5.7	28.8	9.1	22.5	23.9	31.9
700	6.3	29.5	9.7	23.1	24.5	Aug 1.6
800	7.0	30.1	10.3	23.7	25.1	2.2
900	7.6	30.8	11.0	24.4	25.7	2.8
1000	Apr 8.2	May 1.4	May 11.6	May 25.0	Jun 26.3	Aug 3.4
1100	8.9	2.0	12.2	25.6	26.9	4.1
1200	9.5	2.7	12.9	26.3	27.5	4.7
1300	10.2	3.3	13.5	26.9	28.1	5.3
1400	10.8	3.9	14.1	27.5	28.8	5.9
1500 *J*	Apr 11.4	May 4.6	May 14.8	May 28.2	Jun 29.4	Aug 6.5
1500 *G*	21.4	14.6	24.8	Jun 7.2	Jul 9.4	16.5
1600	22.1	15.2	25.4	7.8	10.0	17.2
1700	23.7	16.8	27.0	9.4	11.6	18.8
1800	25.4	18.5	28.7	11.0	13.2	20.4
1900	Apr 27.0	May 20.1	May 30.3	Jun 12.7	Jul 14.8	Aug 22.0
2000	27.6	20.7	30.9	13.3	15.4	22.6
2100	29.3	22.4	Jun 1.6	14.9	17.0	24.3
2200	30.9	24.0	3.2	16.5	18.6	25.9
2300	May 2.5	25.6	4.8	18.2	20.2	27.5

Year	γ Vir Porrima	α Vir Spica	α Lib Zubenelgenubi	β Sco Acrab	α Sco Antares	σ Sgr Nunki
-1000	Sep 1^d.2	Sep 14^d.6	Oct 5^d.8	Oct 23^d.6	Oct 30^d.1	Dec 1^d.0
- 900	1.9	15.3	6.5	24.3	30.7	1.7
- 800	2.5	15.9	7.1	24.9	31.4	2.3
- 700	3.1	16.6	7.8	25.6	Nov 1.1	3.0
- 600	3.8	17.2	8.4	26.3	1.7	3.7
- 500	Sep 4.4	Sep 17.9	Oct 9.1	Oct 26.9	Nov 2.4	Dec 4.3
- 400	5.1	18.5	9.7	27.6	3.1	5.0
- 300	5.7	19.2	10.4	28.2	3.7	5.7
- 200	6.3	19.8	11.0	28.9	4.4	6.3
- 100	7.0	20.5	11.7	29.5	5.0	7.0
0	Sep 7.6	Sep 21.1	Oct 12.4	Oct 30.2	Nov 5.7	Dec 7.7
+ 100	8.2	21.8	13.0	30.9	6.3	8.3
200	8.8	22.4	13.6	31.5	7.0	9.0
300	9.5	23.1	14.3	Nov 1.2	7.6	9.6
400	10.1	23.7	14.9	1.8	8.3	10.3
500	Sep 10.7	Sep 24.4	Oct 15.6	Nov 2.5	Nov 9.0	Dec 10.9
600	11.4	25.0	16.2	3.1	9.6	11.6
700	12.0	25.6	16.9	3.8	10.3	12.3
800	12.6	26.3	17.5	4.4	10.9	12.9
900	13.2	26.9	18.2	5.1	11.6	13.6
1000	Sep 13.9	Sep 27.6	Oct 18.8	Nov 5.7	Nov 12.2	Dec 14.2
1100	14.5	28.2	19.5	6.4	12.9	14.9
1200	15.1	28.8	20.1	7.0	13.5	15.5
1300	15.7	29.5	20.7	7.7	14.1	16.2
1400	16.3	30.1	21.4	8.3	14.8	16.8
1500 J	Sep 17.0	Sep 30.7	Oct 22.0	Nov 9.0	Nov 15.4	Dec 17.5
1500 G	27.0	Oct 10.7	Nov 1.0	19.0	25.4	27.5
1600	27.6	11.4	1.7	19.6	26.1	28.1
1700	29.2	13.0	3.3	21.2	27.7	29.8
1800	30.8	14.7	4.9	22.9	29.4	31.4
1900	Oct 2.4	Oct 16.3	Nov 6.6	Nov 24.5	Dec 1.0	Dec 33.1
2000	3.0	16.9	7.2	25.2	1.7	33.7
2100	4.7	18.5	8.8	26.8	3.3	35.4
2200	6.3	20.2	10.5	28.4	4.9	37.0
2300	7.9	21.8	12.1	30.1	6.6	38.7

BIBLIOGRAPHY

SUN, MOON

Herget, P., Solar Coordinates 1800-2000; Astronomical Papers, Vol. XIV (Washington, 1953).

Goldstine, H.H., New and Full Moons, 1001 B.C. to A.D. 1651; Mem. American Phil. Soc., Vol. 94 (Philadelphia, PA, 1973).

Meeus, J. Tables of the Moon and Sun, Kessel-lo (Belgium), Kesselberg Sterrenwaccht, 1962.

Tables to compute precise geocentric co-ordinates of the Moon and Sun for 1900 B.C. to 3000 A.D. Formulae for computation of occultations and eclipses.

ECLIPSES

Meeus, J., and Mucke, H., Canon of Lunar Eclipses, -2002 to +2526; Astronomisches Buro, Wien (1979).

Mucke, H., and Meeus, J., Canon of Solar Eclipses; Astronomisches Buro, Wien. (In preparation).

PLANETS

Tuckerman, B., Planetary, Lunar, and Solar Positions, 601 B.C. to A.D. 1, Mem. American Phil. Soc., Vol. 56 (Philadelphia, PA, 1962).

Tuckerman, B., Planetary, Lunar, and Solar Positions, A.D. 2 to A.D. 1649, Mem. American Phil. Soc., Vol. 59 (Philadelphia, PA, 1964).

Hunger, H. and Dvorak, R., Ephemeriden von Sonne, Mond und hellen Planeten von -1000 bis -601; Osterreichische Akademie der Wissenschaften, Philosophisch-Historische Klasse, Denkschriften, Band 155 (Wien, 1981).

These three books give the celestial longitudes and latitudes (to 0.01 degree) of the Sun and the five naked-eye planets.

Stahlman, W.D., and Gingerich, O., Solar and Planetary Longitudes for Years -2500 to +2000; University of Wisconsin Press, Madison (1963).

Gives the celestial longitudes (to the nearest degree) of the Sun and the five naked-eye planets at intervals of 10 days.

Duncombe, R.L., Tufekcioglu, Z., and Larson, G., Rectangular Coordinates of Mercury 1800-2000; U.S. Naval Obs., Circular No. 106 (1965).

Herget, P., Coordinates of Venus 1800-2000; Astronomical Papers, Vol. XV, Part III (Washington, 1955).

Duncombe, R.L., Provisional Ephemeris of Mars 1800-1950; U.S. Naval Obs., Circular No. 95 (1964).

Duncombe, R.L., and Clemence, G.M., Provisional Ephemeris of Mars 1950-2000; U.S. Naval Obs., Circular No. 90 (1960).

Eckert, W.J., Brouwer, D., and Clemence, G.M., Coordinates of the Five Outer Planets 1653-2060; Astronomical Papers, Vol. XII (Washington, 1951).

Contains the rectangular heliocentric equatorial coordinates of the planets Jupiter to Pluto at intervals of 40 days.

Kaplan, G.H., Seidelmann, P.K., and Smith, E., Astrometric Ephemeris of Pluto 1970-1990; U.S. Naval Obs., Circular No. 139 (1972).

MINOR PLANETS, COMETS

Pilcher, F., Tables of Minor Planets; Willmann-Bell, Inc. Richmond, Virginia (In preparation).

Ephemerides of Minor Planets, issued annually by the Institute of Theoretical Astronomy, Academy of Sciences, Leningrad.

Contains the orbital elements and ephemerides of all numbered minor planets.

Duncombe, R.L., Heliocentric Coordinates of Ceres, Pallas, Juno, Vesta, 1928-2000; Astronomical Papers, Vol. XX, Part II (Washington, 1969).

Marsden, B.G., Catalogue of Cometary Orbits; fourth edition (1982). Minor Planet Center, 60 Garden Street, Cambridge, MA.

STARS

Smithsonian Astrophysical Observatory: Star Catalog; Smithsonian Institution, Washington (1966).

This 4 volume set contains the positions and proper motions of 258,997 stars for the epoch and equinox of 1950.0.

Hirshfeld, A., and Sinnott, R.W., Sky Catalogue 2000.0; Vol. 1: Stars to Magnitude 8.0; Sky Publishing Corporation, Cambridge, Mass. (1982).

Hoffleit, D., Bright Star Catalogue; Yale University Observatory; fourth revised edition (New Haven, CT, 1982).

Robertson, J., Catalog of 3539 Zodiacal Stars; Astronomical Papers, Vol. X, Part II (Washington, 1940).
> Also known as the 'Zodiacal Catalog' (ZC), this is the reference work for occultations of stars by the Moon.

Wepner, W., 291 Doppelstern-Ephemeriden fur die Jahre 1975-2000; Treugesell Verlag, Dusseldorf (1976).

SERIAL OR ANNUAL PUBLICATIONS

Victor, R., and Pon, J., Sky Calendar, Abrams Planetarium, Michigan State University, East Lansing, MI 48824.
> Each month has its own page divided into daily boxes showning horizon at dawn or dusk. Positions of the moon, visible planets and bright stars are shown. Mailed 4 times a year, subscription of $5.00 starts delivery anytime.

Royal Astronomical Society of Canada; Observer's Handbook 19--. Toronto, Ontario, Canada.
> This annual volume is intended for the serious observer who wants the yearly observing opportunities readily at hand. While the emphasis is primarily on specific events occurring during the year much other information is included.

British Astronomical Association; Handbook 19--, London, England.
> Similar to Observer's Handbook above but more easily obtained in Europe.

Ottewell, G., Astronomical Calendar 19--; Department of Physics, Furman University, Greenville, SC.
> Large format, profusely illustrated with sketches. Appeals to both the beginner and experienced observer.

Nautical Almanac Office, U. S. Naval Observatory. The Astronomical Almanac 19--. Published jointly with the Royal Greenwich Observatory. Government Printing Office, Washington, D.C. and Her Majesty's Stationery Office, London.

Nautical Almanac Office, U.S. Naval Observatory, Astronomical Phenomena for the Year 19--, Government Printing Office, Washington, D.C.
> Pamphlet which is, in part, a reprint of selected pages from The Astronomical Almanac 19--.

Astronomisches Rechen-Institut, Apparent Places of Fundamental Stars; Heidelberg, West Germany, 19--.

CALENDARS AND CHRONOLOGY

Ginzel, F., Handbuch der mathematischen un technischen Chronologie; Leipzig, Hinrich, 1906-1914. 3 vols. Reprinted: Zwichau, Ullmann, 1958.

The calendar (In, Nautical Almanac Office, Royal Greenwich Observatory. Explanatory supplement to the Astronomical Ephemeris and the American Ephemeris and Nautical Almanac 1966. Her Majesty's Stationery Office, London.